D0208200

Organic Pollutants in Water

ADVANCES IN CHEMISTRY SERIES **214**

Organic Pollutants in Water

Sampling, Analysis, and Toxicity Testing

I. H. (Mel) Suffet, EDITOR
Drexel University

Murugan Malaiyandi, EDITOR
Health and Welfare Canada

Developed from a symposium sponsored by
the Divisions of Environmental and Analytical Chemistry
at the 188th Meeting
of the American Chemical Society,
Philadelphia, Pennsylvania,
August 29–31, 1984

American Chemical Society, Washington, DC 1987

Library of Congress Cataloging-in-Publication Data

Organic pollutants in water.
(Advances in chemistry series; 214)

"Developed from a symposium sponsored by the Divisions of Environmental and Analytical Chemistry at the 188th Meeting of the American Chemical Society, Philadelphia, Pennsylvania, August 29–31, 1984."

Includes bibliographies and index.

1. Organic water pollutants—Analysis.

I. Suffet, I. H. II. Malaiyandi, Murugan, 1923– III. American Chemical Society. Division of Environmental Chemistry. IV. American Chemical Society. Division of Analytical Chemistry. V. Series.

QD1.A355 no. 214 540 s [628.1'61] 86-22218
[TD427.07]
ISBN 0-8412-0951-0

PRINTED IN THE UNITED STATES OF AMERICA

Advances in Chemistry Series

M. Joan Comstock, *Series Editor*

FOREWORD

The ADVANCES IN CHEMISTRY SERIES was founded in 1949 by the American Chemical Society as an outlet for symposia and collections of data in special areas of topical interest that could not be accommodated in the Society's journals. It provides a medium for symposia that would otherwise be fragmented because their papers would be distributed among several journals or not published at all. Papers are reviewed critically according to ACS editorial standards and receive the careful attention and processing characteristic of ACS publications. Volumes in the ADVANCES IN CHEMISTRY SERIES maintain the integrity of the symposia on which they are based; however, verbatim reproductions of previously published papers are not accepted. Papers may include reports of research as well as reviews, because symposia may embrace both types of presentation.

ABOUT THE EDITORS

I. H. (MEL) SUFFET is Professor of Chemistry and Environmental Science at Drexel University. He received his Ph.D. from Rutgers University, his M.S. in chemistry from the University of Maryland, and his B.S. in chemistry from Brooklyn College. Among his numerous awards are the American Chemical Society Zimmerman Award in Environmental Science, Drexel University Research Achievement Award, Pennsylvania Water Pollution Control Association Service Award, and Instrument Society of America Service Award.

Suffet has coauthored more than 80 research papers and monograph chapters on environmental and analytical chemistry. His research expertise is the field of environmental chemistry and focuses on phase equilibria and transfer of hazardous chemicals. This expertise allows him to work on the analysis, fate, and treatment of hazardous chemicals and has led to his current studies on the isolation of chemicals for toxicity testing.

He has organized and chaired numerous technical society meetings. In addition, he has served on the Safe Drinking Water Committee of the National Academy of Sciences, for which he chaired the Subcommittee on Adsorption, and was a consultant to the Water Reuse Panel of the National Academy of Sciences.

Suffet coedited with M. J. McGuire ADVANCES IN CHEMISTRY SERIES No. 202, *Treatment of Water by Granular Activated Carbon,* as well as a two-volume set, *Activated Carbon Adsorption of Organics from the Aqueous Phase.* He also edited a two-volume treatise, *The Fate of Pollutants in the Air and Water Environments;* he was a journal editor for a special issue of the *Journal of Environmental Science and Health, Part A—Environmental Science and Engineering;* and he served on the editorial board of the companion journal, *Journal of Environmental Science and Health, Part B—Pesticides, Food Contaminants, and Agricultural Wastes.* He serves on the editorial boards of the journals *Chemosphere* and *CHEMTECH.* He is now completing a 4-year term as treasurer of the ACS Division of Environmental Chemistry.

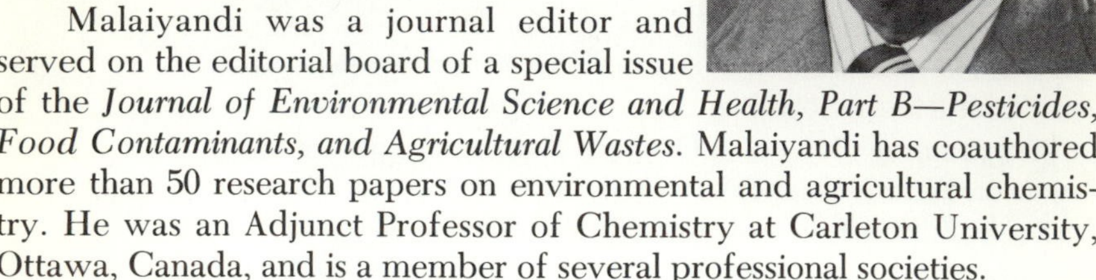

MURUGAN MALAIYANDI, Research Scientist, Environmental Health Directorate, Health and Welfare Canada, Ottawa, Canada, after graduating from Madras University, received his B.Sc. (Hons) and M.Sc. in chemistry from the University of Mysore, India, and his Ph.D. from the University of Toronto, Toronto, Canada. After 5 years of post-doctoral training, he joined the Canadian Department of Agriculture in 1965 and the Department of Health and Welfare Canada in 1976.

Malaiyandi was a journal editor and served on the editorial board of a special issue of the *Journal of Environmental Science and Health, Part B—Pesticides, Food Contaminants, and Agricultural Wastes*. Malaiyandi has coauthored more than 50 research papers on environmental and agricultural chemistry. He was an Adjunct Professor of Chemistry at Carleton University, Ottawa, Canada, and is a member of several professional societies.

His expertise in the field of reverse osmosis involved him in organizing two sessions on reverse osmosis in separation processes for the Gordon Research Conferences in 1982. He initiated, organized, and acted as chairman of the first session on reverse osmosis and ultrafiltration processes of the Gordon Research Conferences in 1983 and was an invited speaker on reverse osmosis at the Gordon Research Conferences in 1985. He is a member of the ACS Committee on Analytical Reagents and the American Society for Testing and Materials, D-19.06.

CONTENTS

INDEXES

PREFACE

THE PRESENCE OF ORGANIC CHEMICALS in drinking and natural waters and their associated health hazards have long been a concern to the scientific and engineering community. The public shares this concern as shown by the passage of safe drinking water legislation in the United States and other countries. Safe drinking water legislation in the United States has led to the development of maximum contaminant levels (MCLs) for volatile organic chemicals such as trihalomethanes and several pesticides. In 1985, MCLs for more than 80 other chemicals were proposed. Alternative or supplementary approaches being debated are tiered toxicological screening tests on aqueous extracts (e.g., Ames mutagenicity tests). Both approaches are dependent on quantitative methods to properly sample, isolate, concentrate, and sometimes fractionate trace organic chemicals from water.

This book will begin to answer some of the most important questions concerning the sampling, isolation, and analysis methods used to determine the chemicals of health concern that are in our drinking waters, waste waters, and natural waters. The book presents several schemes for isolating and concentrating trace contaminants. Viewpoints on which analytical scheme is best are presented. Is it a broad spectrum approach that attempts to determine everything that is present on the basis of many different isolation methods? Is it an approach that determines everything in a sample as the master analytical scheme proposes? Is it an approach that selects specific chemicals such as priority pollutants for quantitative isolation and analysis? Is it a combination of these approaches that is best? Other questions about sampling are presented. Should composite or grab sampling be the choice? Which approach will best describe the extent of the problem, temporally or spatially selected samples? What are the artifacts produced while sampling?

In general, this book deals with developing analytical protocols for concentrating organics for toxicity testing; isolating nonpolar and polar organics from water; and using reverse osmosis, synthetic polymers, and other methods for composite samples. The book is a modest effort to explore the expanding amount of data from research on sampling, isolating, concentrating, and fractionating organic chemicals from natural and treated water for mutagenic, carcinogenic, and toxicity testing. All of the analytical methods discussed are based on phase-transfer processes in which the compound is isolated by a second phase (e.g., solvent or resin) or separated by a membrane phase.

Regulatory aspects of using biological testing are also presented. Although the U.S. Environmental Protection Agency (USEPA) is proposing MCLs for specific chemicals in drinking water (*Federal Register*), the agency is still interested in seeing if a surrogate toxicity measure can be used to replace specific chemical analysis for chemicals of health concern. Also, the Denver Water Board is currently developing a reuse water treatment system and is planning to isolate organics from the drinking water for the purpose of biological testing. Denver projects that it will use 10% direct drinking water reuse of reclaimed waste water if the reuse water can be proved to be as healthy as its present water supply.

A panel discussion at the end of the book describes the potential biological hazards of drinking water and the needs and applications of the analytical methods presented in the book. This panel discussion is essential to the reader's understanding of the often complex chemistry–toxicology–water treatment–regulators interface. We hope that the reader will enjoy the panel discussion, not only for the technical content, but also for insight into the personal philosophies of the participants.

Acknowledgments

A project such as this one requires the expertise and help of many people. Unfortunately, there is not sufficient space to thank everyone. We know and appreciate how important their contributions were. The support of David B. Preston, Retired Executive Director, American Water Works Association (AWWA) and Michael J. McGuire, AWWA Liaison, is gratefully acknowledged. The USEPA provided financial support for this project (Cooperative Agreement No. CR 811168), making it possible to record the entire symposium, have court stenographers record the panel discussion, and plan the meeting. We are grateful for the guidance of Paul Ringhand, Project Officer, and Fred Kopfler, Chief, Chemical and Statistical Support Branch, Toxicology and Microbiology Division, USEPA, Office of Research and Development, Health Effects Research Laboratory, Cincinnati, Ohio. Although many of the chapters in this book describe research that was fully or partially funded by U.S. federal agencies, the individual chapters do not reflect the views of the agencies and no endorsement should be inferred.

The ACS Division of Environmental Chemistry also provided financial assistance for travel funds and guest registrations of scientists coming from other countries. We gratefully acknowledge the division support. Students from Drexel University, Environmental Chemistry

Group—Ronald Baker, Brian Brady, Gail Caron, Chuck Hertz, Eva Ibrahim, Kathy Hunchak, Lee Lippincott, and Areta Wowk—operated the tape recorders, microphones, and audio-visual system during the symposium from which this book was developed.

We are indebted to the reviewers who worked many hours to ensure that a high-quality book was produced. We are deeply indebted to the speakers and session chairpersons of the symposium for their contributions and efforts before, during, and after the symposium. We thank the discussion participants for their open and frank opinions; we feel that their contributions and the ideas exchanged will generate a better understanding of the subject for the readers.

The task of managing the review process, as well as all production phases of this book, was expertly handled by Janet S. Dodd of the ACS Books Department. We enjoyed working with Janet Dodd, Robin Giroux, Karen McCeney, and the entire ACS staff. We appreciate their patience and professionalism.

At both institutions where we are associated, several people must be acknowledged for their editing, typing, and clerical help. Suffet expresses his gratitude for the work of Beverly Henderson, Evelyn Soto, and Ronald Brown at Drexel University, who handled the correspondence and typing chores with professional dedication, and the work of Susan Sorace, who helped edit the panel discussion. Malaiyandi expresses his gratitude for the support and encouragement of P. Toft, Chief, Division of Monitoring and Criteria, Health and Welfare Canada, and Jean Ireland and Irmgard Schierfeld for handling correspondence.

We both gratefully acknowledge the understanding, support, and encouragement of our wives, Eileen Suffet and Pathali Malaiyandi. They remain a source of perspective as professionals in their own right, and without their concern and help, this book would not have been possible.

I. H. SUFFET
Drexel University
Philadelphia, PA 19104

MURUGAN MALAIYANDI
Health and Welfare Canada
Ottawa, K1A OL2, Canada

May 1986

To Our Wives and Families
Eileen Suffet, Alison, and Jeffrey
and
Pathali Malaiyandi

PROTOCOLS

1

Concentration Techniques for Isolating Organic Constituents in Environmental Water Samples

R. L. Jolley[1] and I. H. Suffet[2]

[1]Oak Ridge National Laboratory, Oak Ridge, TN 37831
[2]Drexel University, Philadelphia, PA 19104

Solutions must be concentrated or the constituents must be isolated before trace amounts of the various organics present as complex mixtures in environmental water samples can be chemically analyzed or tested for toxicity. A major objective is to concentrate or isolate the constituents with minimum chemical alteration to optimize the generation of useful information. Factors to be considered in selecting a concentration technique include the nature of the constituents (e.g., volatile, nonvolatile), volume of the sample, and analytical or test system to be used. The principal methods currently in use involve (1) concentration processes to remove water from the samples (e.g., lyophilization, vacuum distillation, and passage through a membrane) and (2) isolation processes to separate the chemicals from the water (e.g., solvent extraction and resin adsorption). Selected methods are reviewed and evaluated.

THE SAGE OF BALTIMORE, H. L. Mencken, once said (*1*), "All complicated problems have a simple solution. It is usually wrong."

Similarly, simple solutions may not be readily available to the real-life problems of analysis and toxicity testing of environmentally important waters. Many of these samples contain complex mixtures of many organic compounds, some of which may be present at very low concentrations. These aqueous solutions must be concentrated and/or the constituents must be isolated before most of the various organic constituents present can be chemically analyzed. This step is necessary so that a sufficient mass of chemicals can be obtained for separation and identification of individual components. An analogous situation

0065-2393/87/0214/0003$06.00/0

exists for determining the toxicity of such trace constituents (many unknown) in waters of environmental or human-health concern.

For chemical identification and toxicity testing efforts to be meaningful, a major objective of concentration methods must be the separation or concentration of the organic constituents with minimum chemical alteration. This objective is probably more important for toxicity testing than for chemical identification because chemical identification is often accomplished by fitting the pieces of information (e.g., functional group analyses) together like a puzzle. Other factors to be considered in the selection of a concentration method include the nature of the constituents of interest (e.g., volatile, semivolatile, nonvolatile, polar, nonpolar, acidic, basic), volume of the water sample, analytical or toxicity test methods to be used, desired concentrate media (e.g., "pure" compound, aqueous solution, solvent solution), necessary purity of concentrate (i.e., can solvent or salt concentration be tolerated?), and stability of concentrate or isolated constituents during storage.

Kopfler (*2*) indicated several areas of concern in preparing representative concentrates. For example, organic residues can change during storage of concentrates between preparation and biological testing or chemical analysis. In addition, humic material can bind lower molecular weight organic substances. Therefore, their recovery is necessary to ensure the integrity of the sample for such bound constituents.

Concentration Methods Used To Prepare for Chemical Analysis

Because of the large variety of chemical compounds present either naturally or as industrial contaminants in water samples, no single concentration method currently available is adequate for concentrating all organic constituents in a water sample. Consequently, in an attempt to concentrate or isolate as much of the organic matter as possible, most researchers use several methods. The combinations can become quite complex and are technically difficult to achieve (*3*).

Concentration methods used as preparation for chemical analysis may be divided into two major groups (*2*):

1. Concentration. This group consists of those processes in which water is removed and the dissolved substances are left behind. Examples are freeze concentration, lyophilization (freeze-drying), vacuum distillation, and membrane processes such as reverse osmosis and ultrafiltration. A common disadvantage of these methods is that inorganic species are concentrated along with the organic constituents.
2. Isolation. This group consists of those processes in which the chemicals are removed from water. Examples are solvent extraction and adsorption on resins.

Table I summarizes most of the practical methods used to achieve the concentration of organic compounds from water samples and indicates the utility of each method, the major advantages and disadvantages, and selected reference citations (*3–33*). These methods have been discussed previously in more detail (*3*). Although the methods focus primarily on trace organic chemicals, many of the principles and concerns apply to inorganic constituents as well.

Concentration Methods Used To Prepare for Toxicity Testing

Several of the methods listed in Table I have been used to prepare concentrates for biological testing. Table II presents selected examples of these methods, along with the principal biological test used and the reference citation (*3, 7, 10, 12, 34–43*). Although the ability of the method to concentrate all organic matter may be limited, significant positive results were achieved in most of the studies. Major questions arise as to whether a single concentration method or a combination of methods can be developed to achieve concentration of all organic matter in a water sample, or whether concentration of all organic matter is necessary to obtain meaningful biological tests; that is, it may be simpler and more economical to use several different concentration methods to concentrate and isolate organic matter for biological testing. This procedure, however, raises the possibility that a highly toxic substance may be omitted because the concentration methods used are not adequate to isolate a specific compound.

The selection of the concentration method must be based on the toxicity test to be conducted. For example, a long-term feeding study using mice could require organic material from many thousands of liters of water; a study with rats might require even more. In addition, the concentration method must be chosen on the basis of the chemical and physical properties of the organic constituents to be tested. Because of the technical difficulties involved, the toxicological testing of highly volatile constituents has generally been done on specific chemical compounds rather than on concentrates. Moderately volatile and less volatile constituents may be tested as concentrates. The concentration of volatile constituents by solvent extraction may be appropriate for the preparation of concentrates for biological testing. Apparently, reverse osmosis, in combination with other methods as used by Kopfler et al. (*12*), is currently the most useful method for preparing large quantities of nonvolatile organic concentrates. However, if thousands of liters per day must be processed, a simpler method should be developed. Thus, XAD resin or other sorption systems may be preferred. However, XAD resins must be meticulously cleaned and stored to avoid "bleeding" of unpolymerized constituents and other contaminants. For toxicological

Table I. Current Useful Methods for Concentrating and Isolating Organic Compounds from Water Solutions

Method	*Type of Compound Concentrated*	*Advantage*	*Disadvantage*	*Selected References*
Concentration Techniques				
Freeze concentration	polar and nonpolar	low temperature	liter volumes, limited concentration, salt concentration	4–6
Freeze-drying	nonvolatile	low temperature, high concentration, low contamination	1–100-L volumes, salt concentration	7, 8
Vacuum distillation	nonvolatile	ambient or near ambient temperature, high concentration, low contamination	1–100-L volumes, salt concentration	9–11
Reverse osmosis	molecular weight >200	ambient temperature, ≥100-L volumes	salt concentration, contamination	12–14
Ultrafiltration	molecular weight >1000	ambient temperature, ≥100-L volumes	limited throughput	15–17

Isolation Techniques				
Gas stripping	volatile, semivolatile	small samples	low volumes	21, 33
Solvent extraction	nonpolar, volatile, semivolatile	ambient temperature, large volumes, low salt concentration	solvent contamination, specificity, concentrate storage	18–21
Activated carbon	nonpolar, volatile, nonvolatile	ambient temperature, large volumes	limited recovery of adsorbed organics, concentrate storage, artifacts	22–23
Ion exchange	polar, nonpolar	large volumes, organic recovery 70–90%	resin preparation and elution	18, 24, 25
XAD resin	nonpolar to polar, volatile, nonvolatile	convenient, ambient temperature, large volumes	contamination, limited recovery of adsorbed organics, resin preparation	24, 27–30
Precipitation	humic materials, specific chemicals	suitable for large masses	specificity	31
Centrifugation	macromolecules	—	specificity (i.e., large molecules, low volumes)	32

SOURCE: Adapted from reference 3.

Table II. Concentration Methods for Biological Testing

Water Type	*Concentration Method*	*Biological Testing*	*References*
Drinking water	reverse osmosis	Ames test	12, 34–37
Drinking water	reverse osmosis	initiation–promotion study (SENCAR mice)	37
Drinking water	reverse osmosis	BALB-3T3 fibroblasts	38
Drinking water	adsorption on XAD	Ames test	39
Drinking water	adsorption on XAD	initiation–promotion study (SENCAR mice)	37
Drinking water	solvent extraction (combination)	cytotoxicity, promotion	40
Drinking water	solvent extraction	Ames test, cytotoxicity, promotion	41
Waste-water effluents	lyophilization and vacuum distillation	Ames test	7, 42
Waste-water effluents	ion exchange (combination)	Ames test	43
Waste-water effluents	ion exchange and distillation	Ames test	10
Paper-plant effluents	adsorption on XAD	Ames test	44

NOTE: SENCAR means sensitive to carcinogens; BALB-3T3 means a cell line from the bag albino mouse.
SOURCE: Adapted from reference 3.

testing, the concentration method can be tailored to isolate specific classes or individual compounds. For example, humic acids can be separated from water samples by alkaline extraction and acid precipitation.

The usefulness of toxicological tests of organic concentrates to estimate the hazards associated with water depends on the degree to which the concentrate represents the organic materials actually present in the water; that is, to estimate the total hazard, the organic concentrate should be representative of all the organic substances present in the water.

Practical Methods Recommended for Toxicity Testing

The procedures recommended in this section to prepare concentrates for toxicity testing represent practical choices within stated limitations (*45*). Two initial criteria are (1) not to concentrate volatile organics and (2) to minimize salt formation for physiological reasons (i.e., if possible, keep salt concentrations in the concentrates below 1%). Currently, it is not practical to efficiently isolate volatile organics below a boiling point of about 100 °C. However, these chemicals may be analyzed readily by liquid–liquid extraction and by purge and trap methods coupled with gas chromatography and gas chromatography–mass spectrometry (*46–48*). If desired, the identified volatile constituents can be reconstituted just before toxicity testing.

Before beginning a toxicity testing program, each of the procedures that are recommended must first be tested at the water-treatment site (sample-collection site) to ensure the adequacy of the concentration method (e.g., solubility of the components, minimization of artifacts, development of a quality assurance program). A mass balance based upon total organic carbon (TOC) is desired during this initial testing phase.

Sample Volume ≤100 L, High Concentration Factors. Lyophilization is a feasible process for concentrating limited numbers of 50–100-L samples to relatively high degrees of concentration (e.g., 3000-fold to dryness). Thus, it is one method of choice for the preparation of samples for in vivo tests (including teratology tests).

Bacterial mutagenesis tests have been conducted with distilled water solutions of the freeze-dried residues [concentrated up to 3000-fold (*7*)] and partially freeze-dried samples [concentrated 10-fold (*49*)]. High salt concentrations in such concentrates may cause toxicity problems in the bacterial tests. The use of dimethyl sulfoxide, methanol, or supercritical carbon dioxide to extract the organics from the freeze-dried residues for mutagenicity test purposes should be investigated.

Dialysis of concentrated solutions of the freeze-dried residues to remove the high salt concentration may result in unacceptable loss of low molecular weight organics. Ultrafiltration with 1000-MW cutoff membranes will remove essentially all inorganic salts (*50*) but will result in unacceptable loss of low molecular weight organics. The 200-MW cutoff ultrafiltration membranes reject only 5% of the inorganic salts; consequently, the concentrates would have to be diluted ≥20-fold with distilled water during the ultrafiltration process (diafiltration) to desalt the concentrate. This use of large volumes of distilled or deionized water to "rinse" the concentrate may be unacceptable because of the possible introduction of artifacts or toxic materials (*39*).

A second method of choice is adsorption or solid adsorbents. Adsorption on XAD resins adequately cleaned to remove artifacts (*28*) and extraction of the adsorbed organics with methanol or acetone (*51–54*) may be conveniently used for 50–100-L sample volumes. In addition to such solvents, some investigators use dilute acid and base washes to facilitate the removal of adsorbed organics from XAD resins followed by neutralization of the washes and ether extraction. However, the recovery of organics from water samples by XAD adsorption is limited (e.g., 5–20% of the TOC), although this recovery includes 60–80% of the neutral organic compounds (*49*). The use of XAD resin followed by activated carbon adsorption significantly increased the recovery of soluble organics from several water samples (*55*).

The resin bed that is used should have sufficient capacity for the sample to be concentrated. If this capacity is not carefully determined for a particular source water, the sample that is collected will only contain the well-adsorbed species in the sample because of competitive adsorption (*56*). The dechlorination of a sample by sulfite before the application of the water to a resin column has been shown to decrease the mutagenic activity of the sample (*57*). Therefore, the use of a dechlorination agent before a resin bed must be tested versus the artifact formation in a resin bed from the reaction of chlorine with the resin matrix.

The use of solvent extraction also represents a potentially feasible process. Solvent extraction is an engineering unit operation that is adapted effectively to continuous processing. It has been used with success for the isolation of nonpolar compounds of bp ≥100 °C (*58*). Solvent extraction (continuous liquid–liquid extraction) may represent a useful process for routinely concentrating 50–100 L of water. The major problem with solvent extraction is the evaporation and recovery for reuse of large volumes of the organic solvent. Other problem areas that must be considered are purification of sufficient solvent and minimization of artifact formation by heat.

Sample Volumes ≥100 L, Medium Concentration Factors. Long-term daily concentration of large-volume samples (e.g., for lifetime in vivo tests) requires that the concentration procedure be reliable, relatively easy to maintain, and capable of being operated continuously for long periods of time. Thus, because of proven use in industrial applications, ultrafiltration is one method of choice for concentrating large-volume samples. It is particularly effective if the toxicity tests can be conducted on concentrates of ≥1000-MW organic constituents. Ultrafiltration with 1000-MW cutoff membranes and essentially complete salt rejection can be used to routinely process large volumes of water for the preparation of essentially salt-free concentrates (*50*).

The use of ultrafiltration membranes with increasingly lower molecular weight cutoffs produces concentrates with increasingly higher salt concentrations. For example, 200-MW cutoff membranes reject only 5% salt. Consequently, the concentrates can be diluted with distilled water and ultrafiltered again. The diafiltration process is conducted continuously but may require large-scale dilution or rinsing of the concentrate. However, the use of very large volumes of dilution water may be unacceptable.

Reverse osmosis is essentially the same process as ultrafiltration with low salt rejection. Kopfler and co-workers (*12*) used reverse osmosis in combination with solvent extraction and XAD adsorption. Solvent extraction with pentane and methylene chloride was used to remove organics from the reverse-osmosis concentrate (i.e., for desalting), and XAD was used for adsorption of the intractables from the extracted concentrate (*12*). Although it is limited in the recovery of organics (30–40%), this combination procedure has been used successfully to concentrate large volumes of water.

The uses of both solid adsorbents (e.g., XAD) and solvent extraction represent feasible engineering processes for the concentration of large volumes of water. However, scale-up, cleaning, and preparing large quantities of solid adsorbent may be technically difficult to accomplish. As an engineering process, solvent extraction should be readily applicable to the continuous processing of large-volume samples. However, the caveats mentioned earlier are applicable particularly to large-volume processing. More work is needed in this area.

General Considerations

Sampling. Composite or continuous sampling furnishes water samples that are more representative than grab or batch samples. If possible, such samples should be used and processed immediately to avoid storage problems. However, the collection of small-volume samples may be more easily accomplished by batch sampling.

Control and renovated water samples should be processed quickly after sampling to avoid possible changes during storage. Should sample storage for short periods prior to concentration be required, the sample should be kept at 0 °C. For storage over long periods, the samples must be frozen, preferably at −40 °C or lower. Long-term storage of large samples should be avoided because of the considerable length of time required to thaw 50–100-L samples.

Samples of Concentrates. Concentrates or dried organic residues from extraction or lyophilization processes should be stored at −70 °C or lower. Little information is available regarding the stability of samples stored cryogenically. Thus, research should be conducted in this area to determine the best method for the storage of such samples to prevent the degradation and development of artifacts.

Recommendations

Efforts should be continued to develop simple but efficient methods for concentrating all the organic materials actually present in water samples. However, meeting this goal may not be technically possible at present.

Methods to evaluate and compare different concentration techniques should be developed and applied. This effort will require increased emphasis on chemical analysis to identify much of the organic material in the concentrates. Measurements of general parameters, such as TOC, cannot be used reliably to compare concentration techniques. All concentration techniques should be compared by determining the recovery of a series of compounds of various solubilities, selected to include a range of chemical classes, functional groups, and molecular weights (including aquatic humic substances). These same materials should be measured before and after their reaction with chlorine. Contaminants unique to the concentration process should be identified and analyzed to determine their effects on the toxicity test and to permit proper evaluation of the test on the concentrate itself. Possible artifact production by the concentration methods should be evaluated. Experiments should be performed to determine the source of these artifacts, for example, whether they may result from the use of membrane concentration techniques, from the reaction of constituents with peroxides in the ether used for eluting adsorption columns, or from other treatment steps. The stability of concentrates during storage should also be studied. An appropriate quality assurance program for sampling and concentration should be used to ensure the validity of analyses and resulting data.

On the basis of the state of the art for concentration methodologies, lyophilization is recommended for sample volumes <100 L when concentrates with high concentration factors are necessary. Ultrafiltration may also be used. For sample volumes >100 L when medium-range concentration factors are required, successive ultrafiltration and dialysis, or reverse osmosis, are currently feasible.

Acknowledgments

Oak Ridge National Laboratory is operated by Martin Marietta Energy Systems, Inc., for the U.S. Department of Energy under Contract No. DE-AC05-84OR21400.

Literature Cited

1. Schneiderman, M. In *Water Chlorination: Environmental Impact and Health Effects, Vol. 2;* Jolley, R. L.; Gorchev, H.; Hamilton, D. H., Jr., Eds.; Ann Arbor Science: Ann Arbor, MI, 1978; p 509.
2. Kopfler, F. C. *Environ. Sci. Res.* **1981,** *22,* 141.
3. Jolley, R. L. *Environ. Sci. Technol.* **1981,** *15(8),* 874.
4. Engdahl, G. E. U.S. Patent 4 314 455, 1982.
5. Mallevialle, J. *Techniques et Sciences Municipales* **1972,** *67(11),* 1.
6. Baker, R. A. *Water Res.* **1970,** *4,* 559.
7. Jolley, R. L. et al. *Nonvolatile Organics in Disinfected Wastewater Effluents: Chemical Characterization and Mutagenicity;* MERL. U.S. Environmental Protection Agency. U.S. Government Printing Office: Washington, DC, 1982; EPA-600/2-82-017.
8. Hayes, D. W. *Design and Evaluation of a Freeze-Dry Apparatus for Remov-*

ing Free Water for Tritium Analysis; DP-1634, Savannah River Laboratory: Aiken, SC, 1982.
9. Hall, K. J.; Lee, G. F. *Water Res.* **1974,** *8,* 239.
10. Johnston, J. B.; Ferdeyen, M. K. In *Chemistry and Analysis of Wastewater Intended for Reuse, Vol. 2;* Cooper, W. J., Ed.; Ann Arbor Science: Ann Arbor, MI, 1981; p 171.
11. Pitt, W. W.; Jolley, R. L.; Katz, S. *Automated Analysis of Individual Refractory Organics in Polluted Water;* SERL. U.S. Environmental Protection Agency. U.S. Government Printing Office: Washington, DC, 1974; EPA 660/2-74-076.
12. Kopfler, F. C. et al. *Ann. N.Y. Acad. Sci.* **1977,** *298,* 20.
13. Fang, H. H. P.; Chian, E. S. K. *Environ. Sci. Technol.* **1976,** *10,* 364.
14. Hinden, E. et al. *Water Sewage Works* **1969,** *116,* 466.
15. Hwang, S. T.; Kammermeyer, K. *Treatise Anal. Chem.* **1982,** *1(5),* 185.
16. Ohno, S. *Kobunshi Kako* **1983,** *32(1),* 34; *Chem. Abstr.* **1983,** *98,* 180550.
17. Scott, J. *Membrane and Ultrafiltration Technology;* Noyes Data Corp.: Park Ridge, NJ, 1980.
18. Ursted, J. F.; Borgen, G. *Ion Exchange and Solvent Extraction;* Society of Chemical Industry: London, 1982.
19. Sheldon, L. S.; Hites, R. A. *Environ. Sci. Technol.* **1976,** *12(10),* 1188.
20. Suffet, I. H. et al. *Environ. Sci. Technol.* **1976,** *10(13),* 1273.
21. Pellizzari, E. D. et al. *Master Analytical Scheme for Organic Compounds in Water;* U.S. Environmental Protection Agency. U.S. Government Printing Office: Washington, DC, 1985; EPA/600/54-85.
22. Suffet, I. H. *Activated Carbon Adsorption of Organics from the Aqueous Phase, Vol. 1;* Ann Arbor Science: Ann Arbor, MI, 1980.
23. McGuire, M. J. *Activated Carbon Adsorption of Organics from the Aqueous Phase, Vol. 2;* Ann Arbor Science: Ann Arbor, MI, 1980.
24. Nellor, M. H. et al. *Health Effects Study, Final Report;* County Sanitation Districts of Los Angeles County: Whittier, CA, 1984.
25. Junk, G. A.; Richard, J. J. *Prepr. Pap. Am. Chem. Soc. Div. Environ. Chem.* **1980,** *20(2),* 277.
26. Smith, C. C. In *Application of Short-Term Bioassays in the Fractionation and Analysis of Complex Environmental Mixtures;* Sandbu, S. S.; Claxton, L., Eds.; Plenum: New York, 1978.
27. Thurman, E. M. et al. *Anal. Chem.* **1978,** *50,* 775.
28. Junk, G. A. et al. In *Identification and Analysis of Organic Pollutants in Water;* Keith, L. H., Ed.; Ann Arbor Science: Ann Arbor, MI, 1976; p 135.
29. Lawrence, J.; Tosine, H. M. *Environ. Sci. Technol.* **1976,** *10,* 381.
30. Aiken, G. R. et al. *Anal. Chem.* **1979,** *151(11),* 1799.
31. Schnitzer, M.; Khan, S. U. *Soil Organic Matter;* Elsevier Scientific: New York, 1978.
32. Amburgey, J. W. *Humic Acid Investigations;* K-L-2549, Union Carbide Corp.: Oak Ridge, TN, 1967.
33. Pfaender, F. K. et al. *Environ. Sci. Technol.* **1978,** *12,* 438.
34. Tardiff, R. G. et al. In *Water Chlorination: Environmental Impact and Health Effects, Vol. 1;* Jolley, R. L., Ed.; Ann Arbor Science: Ann Arbor, MI, 1978; p 195.
35. Tabor, M. W. et al. In *Water Chlorination: Environmental Impact and Health Effects, Vol. 3;* Jolley, R. L. et al., Eds.; Ann Arbor Science: Ann Arbor, MI, 1980; p 899.
36. Loper, J. C. et al. *J. Toxicol. Environ. Health* **1978,** *5,* 919.
37. Neal, R. A. In *Water Chlorination: Environmental Impact and Health*

Effects, Vol. 3; Jolley, R. L. et al., Eds.; Ann Arbor Science: Ann Arbor, MI, 1980; p 1007.
38. Loper, J. C. In *Water Chlorination: Environmental Impact and Health Effects, Vol. 3;* Jolley, R. L. et al., Eds.; Ann Arbor Science: Ann Arbor, MI, 1980; p 937.
39. Cheh, A. H. et al. In *Water Chlorination: Environmental Impact and Health Effects, Vol. 4;* Jolley, R. L. et al., Eds.; Ann Arbor Science: Ann Arbor, MI, 1983; p 1221.
40. Hemon, D. et al. *Rev. Epid. Sante. Pub.* **1978,** *26,* 441.
41. Cabrindenc, R.; Sdika, A. Presented at the European Symposium on Analysis of Micropollutants of Water, Berlin, Dec. 11, 1979.
42. Cumming, R. B. et al. In *Water Chlorination: Environmental Impact and Health Effects, Vol. 3;* Jolley, R. L. et al., Eds.; Ann Arbor Science: Ann Arbor, MI, 1980; p 881.
43. Baird, R. et al. In *Water Chlorination: Environmental Impact and Health Effects, Vol. 3;* Jolley, R. L. et al., Eds.; Ann Arbor Science: Ann Arbor, MI, 1980; p 925.
44. Douglas, G. R. et al. In *Water Chlorination: Environmental Impact and Health Effects, Vol. 3;* Jolley, R. L. et al., Eds.; Ann Arbor Science: Ann Arbor, MI, 1980; p 865.
45. Panel on Quality Criteria for Water Reuse *Quality Criteria for Water Reuse;* National Research Council; National Academy of Sciences: Washington, DC, 1982.
46. Bellar, T. A.; Lichtenberg, J. J. *J. Am. Water Works Assoc.* **1974,** *66,* 739.
47. *Fed. Regist.* **1979,** *44(233),* 69463.
48. "Determination of Acetone and Methyl Ethyl Ketone in Water," *Water-Resources Investigation* 78-123, PB 291151, 1981.
49. Forster, R.; Wilson, I. *J. Inst. Water Eng. Sci.* **1981,** *35(3),* 259.
50. Lightly, F., Osmonics, Inc. personal communication, Jan. 15, 1982.
51. McCarty, P. L. et al. *Mutagenicity Activity and Chemical Characterization for the Palo Alto Reclamation and Groundwater Facility,* U.S. Environmental Protection Agency. U.S. Government Printing Office: Washington, DC, 1981; EPA 600/1-81-029.
52. Neal, M. W. et al. *Assessment of Mutagenic Potential of Mixtures of Organic Substances in Renovated Water;* U.S. Environmental Protection Agency. U.S. Government Printing Office: Washington, DC, 1980; EPA 600/1-81-016.
53. Flanagan, E. P.; Allen, H. E. *Bull. Environ. Contam. Toxicol.* **1981,** *27,* 764.
54. Kopperman, H. L. et al. In *Water Chlorination: Environmental Impact and Health Effects, Vol. 1;* Jolley, R. L., Ed.; Ann Arbor Science: Ann Arbor, MI, 1978; p 311.
55. Lee, N. E., Oak Ridge National Laboratory, personal communication, March 1982.
56. Najar, B. A.; Gibs, J.; Suffet, I. H. *Preprints of Papers, ACS Div. Environ. Chem.* **1982,** *22(1),* 90.
57. Cheh, A. H., Skochdopole, J.; Hellig, C.; Koski, P.M.; Cole, L. In *Water Chlorination: Environmental Impact and Health Effects, Vol. 3;* Jolley, R. L.; Brungs, W. A.; Cumming, R. B., Eds.; Ann Arbor Science: Ann Arbor, MI, 1980; p 803.
58. Yohe, T. L.; Suffet, I. H.; Grochowski, R. J. In *Measurement of Organic Pollutants in Water and Wastewater,* ASTM STP 686; Van Hall, C. E., Ed.; American Society for Testing and Materials: Philadelphia, PA, 1979; pp 47-67.

RECEIVED for review August 14, 1985. ACCEPTED December 26, 1985.

2

Interim Procedures for Preparing Environmental Samples for Mutagenicity (Ames) Testing

Paul J. Marsden[1], Donald F. Gurka[2], Llewellyn R. Williams[2], Jeffrey S. Heaton[3], and Jonathan P. Hellerstein[3]

[1]S-CUBED, Division of Maxwell Laboratories, San Diego, CA 92121
[2]Environmental Monitoring Systems Laboratory, U.S. Environmental Protection Agency, Las Vegas, NV 89114
[3]Life Systems Incorporated, Cleveland, OH 44122

Protocols for preparing six environmental sample types prior to the Ames Salmonella *assay were proposed at a recent panel discussion sponsored by the U.S. Environmental Protection Agency (USEPA) and the U.S. Army. Air particles, soil-sediment, and solid waste are extracted with dichloromethane, concentrated, and solvent exchanged into dimethyl sulfoxide (DMSO). Organics in water and waste water are absorbed onto XAD columns, then eluted with hexane-acetone, solvent reduced, and exchanged into DMSO. Nonaqueous liquids are assayed directly and as concentrates before they are solvent exchanged to DMSO. If bacterial toxicity or lack of dose response is observed in the Ames assay of extracts, the extracts are fractionated prior to solvent exchange. These are interim methods and have not been subjected to policy review of the USEPA or the U.S. Army.*

STANDARDIZED PROCEDURES for performing the Ames *Salmonella* assay, a widely used short-term procedure for assessing the mutagenic hazard posed by chemicals, chemical mixtures, and environmental samples, were published (*1*) to foster comparability among laboratories performing mutagenicity testing in support of, or in compliance with, U.S. Environmental Protection Agency (USEPA) regulations. However, those procedures do not include methods used in preparing environmental samples for assay. This chapter reports protocols proposed for sample preparation developed by a panel of experts (chemists and

0065-2393/87/0214/0015$6.75/0

genotoxicologists) during a meeting entitled, "Mutagenicity Sample Preparation Protocols Panel Meeting".[1]

The panel meeting was organized and conducted to facilitate development of sample preparation protocols for mutagenicity testing of six media: air, drinking water, nonaqueous liquid wastes, soils and sediments, waste solids, and waste water. The meeting objectives were established by the sponsors and were as follows:

1. Evaluate the adequacy and validity of the selected and reviewed sample preparation protocol for each medium.
2. Prepare revised sample preparation protocols in accordance with review comments and recommendations from the panel meeting participants.
3. Present a scientific basis for any disagreements or unresolvable issues.
4. Recommend additional research to support further development of medium-specific sample preparation protocols.

General Issues, Definitions, and Limitations

A number of issues were addressed in the general sessions that were common to all six of the proposed protocols. These included definitions of each of the sample media, limitations of the protocols, guidelines for the fractionation of samples found to be toxic in the *Salmonella* assay, and analytical quality control considerations.

Media Definitions. Sample preparation media were defined by the participants to assist laboratories in selecting the appropriate protocol for a specific sample. These definitions were intended to encompass the continuum of wastes and environmental materials, to reduce overlap between individual medium definitions, and to reflect the scope of the protocols. The following media definitions were developed:

1. Air particulate matter: Particulates collected from the air that are $\leq 20\ \mu m$ in diameter.
2. Drinking water: Water intended for human consumption.
3. Environmental waters and waste waters: Waters containing $<50\%$ suspended solids by weight, including water from lakes, ponds, lagoons, estuaries, rivers, streams, effluents, and ground water.

[1] The meeting was cosponsored by the USEPA Environmental Monitoring Systems Laboratory, Las Vegas, Nevada (EMSL-LV) and the U.S. Army Medical Bioengineering Research and Development Laboratory (MBRDL). The meeting was held July 23–25, 1984, in Palo Alto, CA.

4. Nonaqueous liquid wastes: Liquids whose major component is not water. These materials range from soluble organic liquids (e.g., acetone) to insoluble organic liquids (e.g., dichloromethane) and liquids such as light to heavy oils.
5. Soil and sediments: Soils are the unconsolidated material on the earth's surface capable of supporting plant growth, and sediments are soil material deposited and remaining in an aquatic environment.
6. Waste solids: Waste or complex mixtures consisting of <50% water and having relative firmness and coherence of particles or persistence of form as matter that is not liquid or gaseous at 25 °C. This category includes tarry material that is sticky or viscous, such as coal tar, adhesive waste, sludge, airborne particulates (e.g., re-entrained particles blown from a waste site), and solids partitioned from waste water.

Examples and further elaboration of these definitions are contained in each protocol.

Scope and Limitations of Protocols. Discussions to determine the scope and limitations provided the following key points, requirements, and resolutions for the protocols:

1. They must provide end products that meet minimal input requirements for Ames mutagenicity testing.
2. They should be quantitative to enable evaluation of any dose-response relationships.
3. They must be developed with consideration of the ultimate user (i.e., comprehensive, stepwise procedures designed for the inexperienced user).
4. They will be used to determine whether samples are mutagenically active or nonactive (i.e., positive or negative types of determinations).
5. They will be applied to screen large numbers of complex mixtures within a variety of media.
6. Development and standardization of the protocols will be dynamic processes that involve establishing interim procedures that are to be optimized and modified as a result of validation testing.

Bacterial Toxicity. A variety of criteria are used by testing laboratories to determine toxicity during the Ames test and when additional sample preparation (i.e., fractionation) is required to separate toxic components from potential mutagenic components of a complex mixture. However, no uniform criteria are available or accepted. The following criteria were proposed and accepted by panel meeting participants as the appropriate conceptual basis for determining lack of toxicity during the Ames test: (1) a dose level of ≥1 mg/plate of residue organic matter is achieved, (2) the background lawn is not affected, and (3) spontaneous revertants are countable and comparable to control plates.

When these criteria are met, the material is not toxic, and fractionation may not be required. Establishment of the 1-mg/plate level as the definition of toxicity and the trigger for fractionation was arbitrary but was considered the most reasonable choice by the panel meeting participants. It was recommended that toxicity determinations be performed with the entire set of bacterial strains (with and without enzymatic activation) being used. Concerns expressed regarding the above criteria included the following: (1) possible misinterpretation by the user that testing at the ≥1-mg/plate level is required to prove nonmutagenicity and (2) limited applicability to the Air Protocol because the resulting material may not be sufficient to allow dosing of ≥1 mg/plate with multiple strains and replicate plates.

Sample Fractionation. Although declining to recommend using a high-performance liquid chromatographic (HPLC) technique in lieu of the acid-base extraction procedure, the panel meeting participants identified it as a potential sample preparation technique if toxicity is detected. The available techniques to reduce or separate toxic activity from potentially mutagenic components of a mixture included the following: (1) liquid-liquid acid-base extraction fractionation, (2) thin-layer chromatography (TLC), (3) magnesium silicate (Florisil) chromatography, (4) low-pressure chromatography or cartridge column cleanup, (5) dilution of neat material with dimethyl sulfoxide (DMSO), and (6) HPLC fractionation.

The effect on reactive compounds and potential alteration of mutagenic activity of the acid-base extraction procedure were major concerns. The TLC method was not recommended because of possible mutagen loss due to reactions on the plate. The Florisil and low-pressure chromatographic methods were considered to be research methods.

The HPLC technique was viewed as the most recent technology that may be a viable alternative to acid-base extractions. However, the following issues were raised concerning HPLC: (1) time required to run a solvent gradient, (2) ability to screen large numbers of samples, (3) reliability and column life, (4) equipment downtime, (5) operating costs, (6) training level of operator, and (7) incompatibility of certain compounds to HPLC.

The panel meeting participants concluded that both the acid-base extraction and HPLC techniques should be identified in the protocols, and the choice of fractionation methods should be at the discretion of the user.The HPLC technique is described in greater detail later in this chapter.

Quality Control. Quality control procedures must be incorporated into each protocol to ensure comparability of interlaboratory testing data. Each laboratory using these protocols should have a quality assurance program based on guidelines set forth in "Good Laboratory Practice Regulations" (2) and *Handbook for Analytical Control in Water and Waste Water Laboratories* (3). As a minimum, the quality assurance program should include the following:

1. Chain-of-custody records on all samples including the source of the sample, date of collection, storage conditions, and records of any subsampling (e.g., how samples were homogenized, date, and technician's name).
2. Development of written standard operating procedures for any laboratory procedures used.
3. Maintenance of laboratory notebooks for recording all laboratory procedures and volumes and types of solvents used.
4. Determination of revertants per plate by using reagent control blanks.
5. Extraction and determination of samples spiked with positive controls performed as quality control checks for each bioassay run.
6. Maintenance of an independent quality assurance unit within the laboratory to be responsible for a master schedule, sample receipt, and periodic inspections of the conduct and reporting of each laboratory study.

The laboratory's quality assurance staff should include, or have immediate access to, a statistician experienced in the interpretation of biological testing and chemical analytical data.

Summaries of Protocols

Air Particulates. The air protocol (Figure 1) is restricted to sample preparation of particulate organic material (POM) and excludes vapor-phase and volatile organic compounds. Mutagenicity testing of volatiles and vapor-phase organics is an area of current research, and no routine screening or monitoring method was identified by the panel participants. The definition of air particulates, presented earlier, clarifies the scope of the protocol.

The protocol focuses on the preparation of collected particulate samples and provides guidance in the selection of a suitable collection technique. Particulate collection methods were not specified because the selection and implementation of those methods are source-dependent. Potential air-sampling techniques include the standard high-volume (Hi-Vol) samplers (*4*), massive-volume samplers (*5*), medium-volume samplers, low-volume samplers, and ultra-high-volume samplers (*6*).

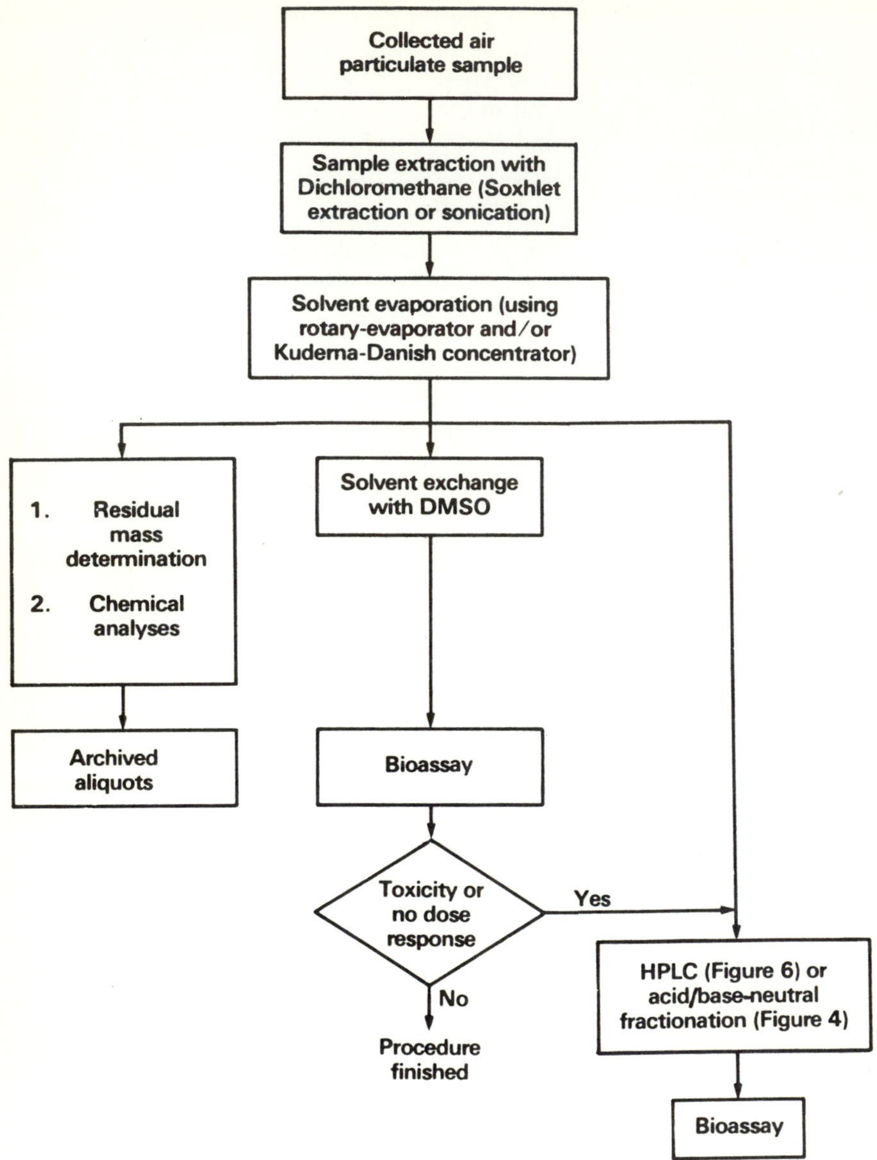

Figure 1. Flow scheme for the preparation of samples of ambient air particles.

Collected air particles (1 g) are extracted twice with dichloromethane (100 mL) by either Soxhlet extraction (5–10 cycles/h for 16 h at 40 °C) or sonication (at room temperature for 15 min each time) (7). The dichloromethane extract is filtered through a 0.5-μm filter (Millipore Corporation or equivalent) into a round-bottom evaporating flask and concentrated to 10 mL by using either a rotary evaporator or a Kuderna-Danish concentrator. One portion of this extract (5 mL) is saved for mass determination by gravimetric analysis and chemical analyses. The other portion of the extract (5 mL) is solvent exchanged with DMSO (5 mL) under a gentle stream of purified nitrogen, and the final volume is recorded. This DMSO solution may be further diluted as appropriate for bioassay measurements.

The air samples should be stored in the dark at −20 °C, and the solvent (dichloromethane or DMSO) extracts should be stored in amber-colored, Teflon-lined, screw-capped bottles at −20 °C. These air samples and solvent extracts should come to room temperature prior to use.

No sample fractionation procedure is given in the protocol, but several possible techniques were included in the literature review, including acid/base-neutral extraction, solvent fractionation (8), column chromatography, TLC, and HPLC (presented elsewhere in this chapter).

The panel meeting participants unanimously agreed that the air particulates protocol is a proven method with an adequate data base to demonstrate applicability to the preparation of air particulates for mutagenicity testing. However, the participants agreed that several areas require method validation.

Drinking Water. The drinking water protocol (Figure 2) is applicable for all types of finished drinking water and can measure the mutagenicity of residue organics that are adsorbed on XAD resins and recovered by solvent elution techniques. The protocol may not be suitable for highly polar, ionic, highly volatile, nor low molecular weight organic compounds that are not adsorbed or recovered from the collection apparatus described. The type of disinfection treatment (if any) used on the water is not considered to be a critical issue. A glass wool prefilter, a diatomaceous earth (Celite) column, and a bacterial filter are incorporated into the sampler when waters contain more than 5% solids, are not disinfected, or contain more than 20 ppm of total organic carbon.

A sample of drinking water is passed through specially designed accumulator columns containing polystyrene-divinylbenzene copolymers and polymethacrylate polymer stationary phases at 30–35 lb/in.^{2}. After passage of the sample through the columns, the residue organics are eluted from the sampling system components with hexane:acetone (85:15). The solvents are removed by rotary evaporation, and the remaining nonvolatile residue organics are stored in acetone under nitrogen at −20 °C until mutagenicity testing is conducted. If extract fractionation of the extract is required, the HPLC technique discussed later is recommended for these samples. Prior to Ames assay, the extracts and/or fractions are solvent exchanged to DMSO.

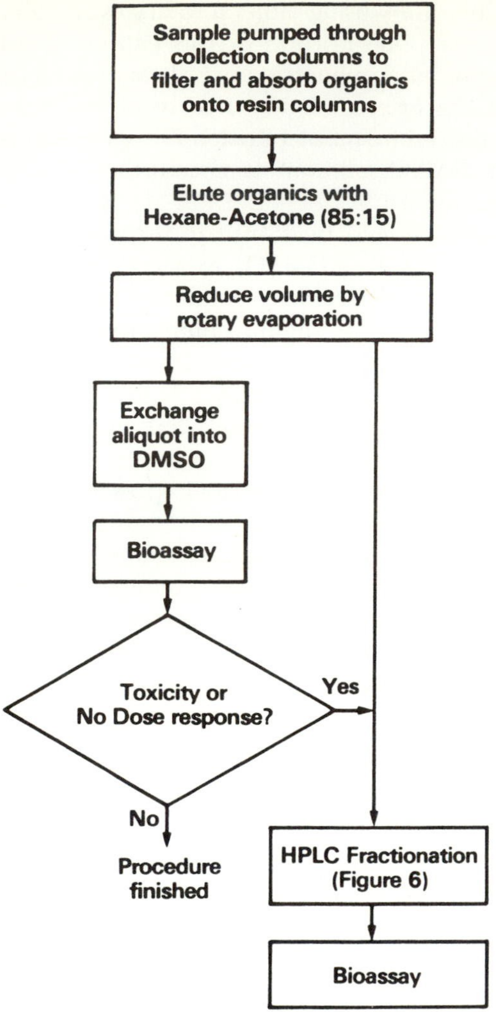

Figure 2. Flow scheme for the preparation of drinking water samples.

The method described in this protocol is based on existing methods used for the isolation of residue organics from drinking water (*9–15*). These studies established operating parameters such as flow rates and line pressures. The results of these studies show that the method provides reproducible qualitative recoveries of mutagenic residue organics from a wide variety of drinking waters prepared from ground and surface sources.

The resins used in this procedure, XAD-2 and XAD-7 (Rohm and Haas), are available from numerous distributors. To avoid method interferences, they must be cleaned prior to use. After removal of the fines, the moist resin is transferred to a glass extraction thimble fitted with a fritted disc, and then the thimble is inserted into the Soxhlet extraction apparatus. For each 500 g of resin, the following extraction sequence is conducted by using 1 L of solvent for each extraction. First, the resin is extracted with 1 L of methanol for 8 h, followed by a 14-h extraction with an additional 1 L of fresh methanol. The 22-h methanol extraction is followed by extraction with two 1-L portions of dichloromethane, 8 h and 14 h, respectively. The 22-h dichloromethane extraction is followed by extraction with two 1-L portions of hexane, 8 h and 14 h, respectively. Finally, the resin is extracted with two 1-L portions of acetone, 8 h and 14 h, respectively. The resin is rinsed from the thimble into an amber bottle with a fresh portion of acetone. At this point, 20 mL of resin is taken for resin blank analysis; the rest is stored under acetone.

The panel meeting participants unanimously agreed that the drinking water protocol is a well-established method with a sufficient data base to demonstrate the adequacy of its application to isolating residue organics from drinking water for mutagenicity testing. This consensus to recommend the drinking water protocol was made with cognizance of the inherent limitation that highly polar and ionic species or highly volatile, low molecular weight organics may not be recovered. The panel meeting participants recommended that the protocol be subjected to validation studies by various laboratories. Additional research on the method was not considered necessary.

Environmental Waters and Waste Waters. This medium is multiphasic and covers a wide range of constituents, including aqueous and nonaqueous liquids and dissolved and suspended solids. The protocol (Figure 3) is limited to solvent-extractable organic compounds; however, not all compounds will be recoverable and/or stable under the protocol's methods. An overall scheme was developed and incorporated in the protocol to link the variety of components of this medium to the other protocols.

Each discrete phase of this multiphasic medium is collected, separated, weighed, and concentrated or extracted during sample preparation. An initial gravity separation (24-h duration) of the sample is used to separate the three possible phases (aqueous liquids, nonaqueous liquids, and solids). The protocol for drinking water is used to prepare aqueous liquids containing <5% solids (Figure 2). Aqueous liquids containing >5% solids are prepared by either (1) liquid–liquid extraction of the entire sample followed by concentration and solvent exchange or (2) further separation of the aqueous fraction by using filtration or centrifugation (the liquids are processed according to the

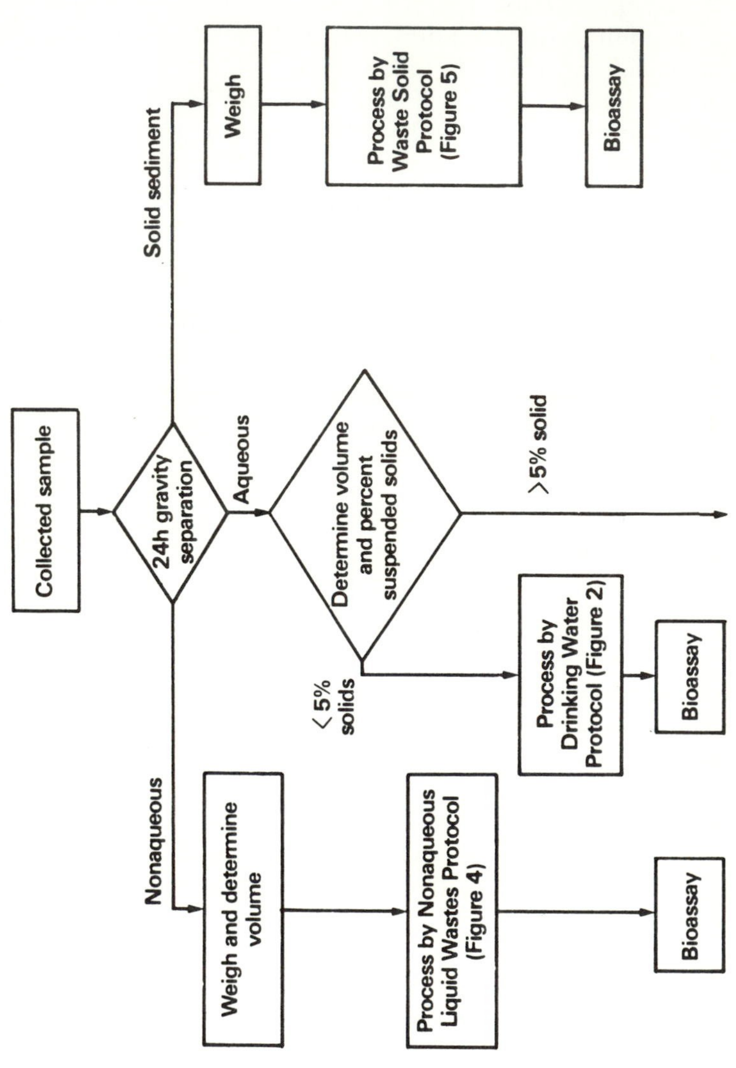
Collected sample
24h gravity separation
Solid sediment
Weigh
Process by Waste Solid Protocol (Figure 5)
Bioassay
Aqueous
Determine volume and percent suspended solids
>5% solid
<5% solids
Process by Drinking Water Protocol (Figure 2)
Bioassay
Nonaqueous
Weigh and determine volume
Process by Nonaqueous Liquid Wastes Protocol (Figure 4)
Bioassay

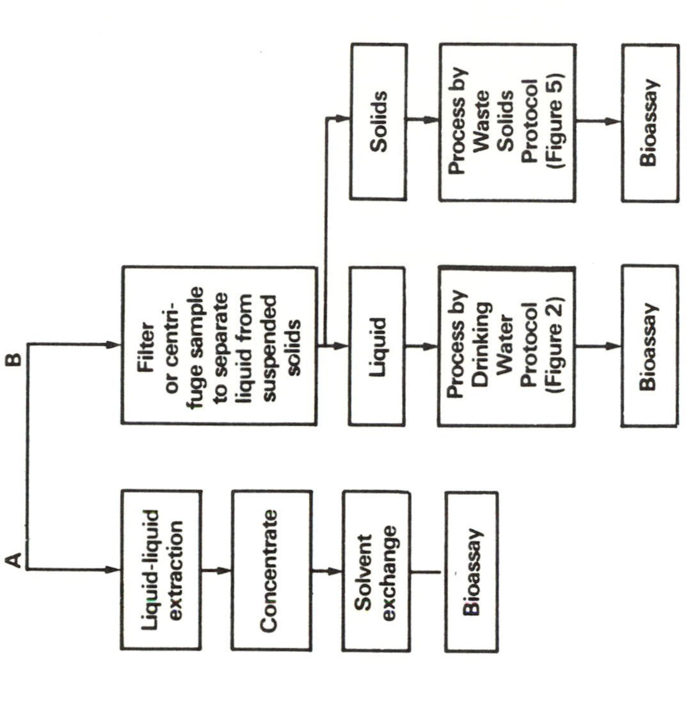

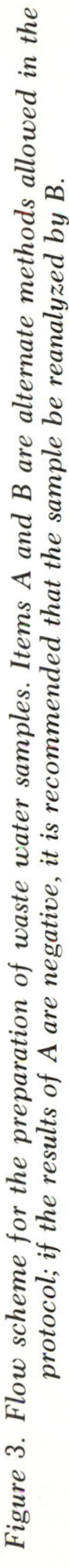

Figure 3. Flow scheme for the preparation of waste water samples. Items A and B are alternate methods allowed in the protocol; if the results of A are negative, it is recommended that the sample be reanalyzed by B.

drinking water protocol and the solids are processed according to the waste solids protocol). The solid sediments from the initial 24-h gravity separation are processed according to the protocol for waste solids. Nonaqueous liquid wastes partitioned from the samples are processed via that protocol.

The panel meeting participants unanimously agreed that the environmental waters and waste water protocol is a synthesized protocol that assembles portions of a number of established methods. There is a good indication that this combination of components will work together, but a comprehensive data base demonstrating the efficacy of the hybridized combination of methods needs to be generated and evaluated.

Nonaqueous Liquid Wastes Protocol. Nonaqueous liquid wastes were defined to include samples that range from water-soluble organic liquids to immiscible oils. Only a limited amount of data are available on the applicability of this protocol (Figure 4) to compounds other than oils or petroleum products. This medium differs from other environmental media because mutagenic materials are often concentrated in organic liquids. Therefore, this protocol incorporates dilution steps rather than the concentration techniques used in the other media protocols. This protocol is also unique because of the opportunity to test neat samples or samples diluted with DMSO rather than sample components isolated with an absorbent or extracted with a solvent. For this reason, samples treated with this protocol should contain polar compounds and/or volatile compounds that would be lost when the other protocols are used.

To determine the mutagenic potential of nonaqueous liquids as measured by the Ames *Salmonella*/mammalian-enzyme assay, the following protocol is recommended for the sample preparation. In step 1, the desiccator assay is performed on the neat material. The desiccator assay allows the detection of volatile mutagens (such as chlorinated solvents) that are often missed in the plate incorporation and pre-incubation assays (*16*, *17*). In addition, a suspension of the neat material (20 mg/mL) is prepared by ultrasonication (5 min at room temperature) in high-purity DMSO (*18*, *19*) and tested in the normal plate incorporation assay as well as in a pre-incubation Ames assay (*20*). The pre-incubation assay allows the detection of certain mutagens, such as dimethylnitrosamine, that require additional time for activation by mammalian or bacterial enzymes. A positive response in any of these three assays indicates the presence of mutagenic components, and the evaluation process is completed.

If a toxic response or a combination of toxic and negative responses is observed in all three assays, the neat material is diluted and the assay is repeated. Dilutions of 10, 100, 1000, and greater fold are made until a negative or positive response is observed.

If a negative response is observed in all three assays of step 1, then the liquid is concentrated by rotary evaporation at 40 °C in step 2. These conditions are gentle enough so that although most volatile solvents (acetone, methanol, etc.) are removed, the integrity of concentrated, thermally labile components is maintained. If the sample volume (or weight) can be reduced by a factor of at least 10 by the evaporation procedure, the concentrate is retested by using the plate incorporation and the pre-incubation assays. If a solid residue is produced, then it is dissolved in a minimum amount of DMSO prior to assay.

If the sample is not effectively concentrated by the evaporation procedure or produces a negative response, it is partitioned into acidic, basic, and neutral fractions in step 3 by using the fractionation scheme given in Figure 4. Initially, the liquid is dissolved in dichloromethane and extracted with 1 N HCl. The aqueous phase (AQ_1) is adjusted to pH 11 with 1 N NaOH and extracted with dichloromethane to isolate the basic compounds from the sample. This fraction is dried over anhydrous sodium sulfate, filtered, and rotary evaporated at 40 °C to yield an oil or residue, which is weighed and dissolved in a minimum of DMSO (DMSO-bases fraction). The organic layer (O_1) from the initial acid extraction is extracted with 1 N NaOH to produce aqueous (AQ_2) and organic (O_2) fractions. The AQ_2 fraction is adjusted to pH 2 with 1 N HCl and extracted with dichloromethane (O_3). The O_3 fraction is dried over anhydrous sodium sulfate, filtered, and evaporated to a residue. The residue is weighed and dissolved in a minimum of DMSO (DMSO-acids fraction). The O_2 fraction is handled in a fashion similar to that for the O_3 fraction and is worked up to become the DMSO-neutrals fraction. All extractions are performed three times with equal volumes of base or acid. The DMSO-acids, DMSO-bases, and DMSO-neutrals fractions are then evaluated for mutagenicity in the plate incorporation and the pre-incubation assays.

The liquid-liquid, acid-base fractionation method was preferred over HPLC by the liquid waste panel members because it is a better validated technique for the isolation of components that are toxic or that might otherwise interfere with the assay. Also, the distribution of activity in the acid, base, and neutral fractions provides a preliminary estimate of the types of chemicals responsible for the mutagenicity.

The panel meeting participants unanimously agreed that the nonaqueous liquid wastes protocol is a synthesized protocol that assembles portions of a number of proven methods. There is a good indication that the various hybridized components will be applicable to a range of nonaqueous liquids; however, this observation is based on data from the testing of discrete types of oily compounds. Further

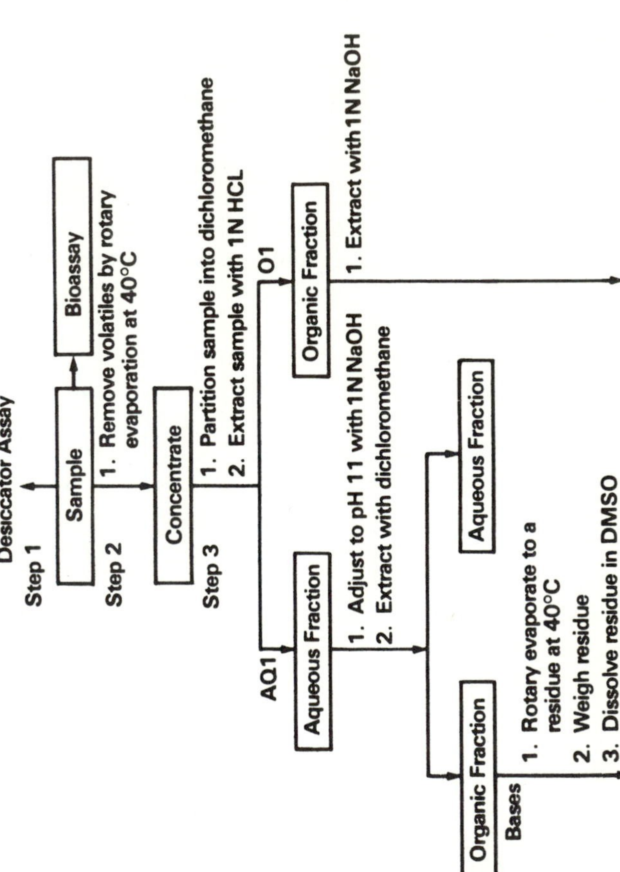
Step 1
Desiccator Assay
Sample
Bioassay
Step 2
1. Remove volatiles by rotary evaporation at 40°C
Concentrate
Step 3
1. Partition sample into dichloromethane
2. Extract sample with 1N HCL
AQ1
Aqueous Fraction
O1
Organic Fraction
1. Extract with 1N NaOH
1. Adjust to pH 11 with 1N NaOH
2. Extract with dichloromethane
Organic Fraction
Aqueous Fraction
Bases
1. Rotary evaporate to a residue at 40°C
2. Weigh residue
3. Dissolve residue in DMSO

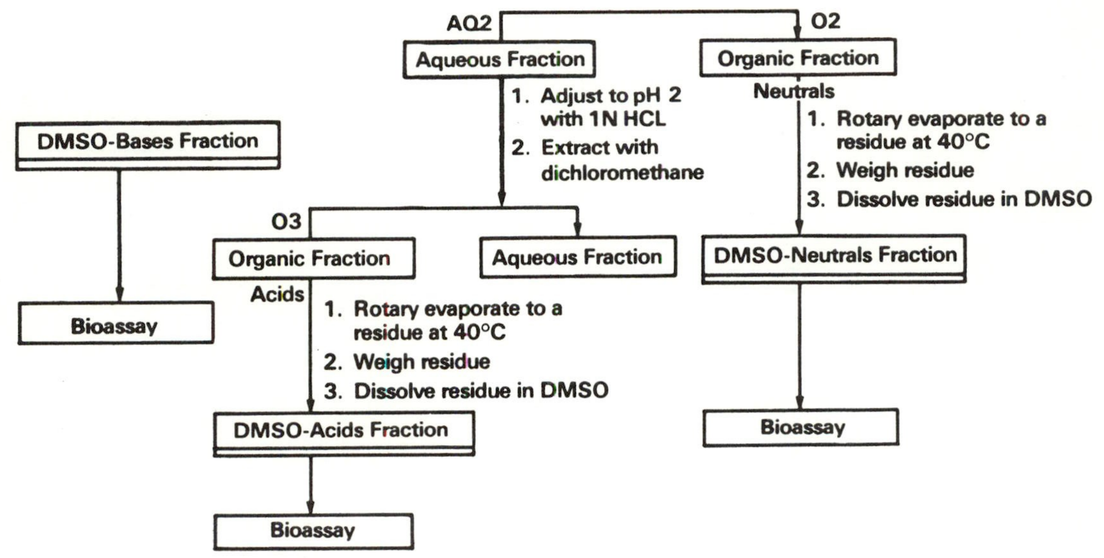

Figure 4. Flow scheme for the preparation of nonaqueous liquid wastes.

studies were recommended to evaluate applicability to a variety of nonaqueous liquids and to determine the optimum solvent for separation or fractionation of these compounds.

Soils and Sediments. This protocol (Figure 5) was developed for the preparation of soil and sediment samples for mutagenicity testing. It was designed to provide samples that accurately represent the mutagenic potential of the soil or sediment sample initially extracted and, if necessary, sufficiently fractionate the original material to isolate bioactive materials. Fractionation should be required only if the toxicity of the crude extract prevents determination of the mutagenic potential.

Prior to extraction, soil and sediment samples should first be sieved to remove primary particles >2 mm in diameter (it may be necessary to break aggregates by dicing or crushing). Excess water in samples should also be removed by filtration under a nitrogen blanket or by centrifugation prior to extraction. If this water is to be analyzed, it should be done by using the drinking water protocol (Figure 2). The weight of the samples should be determined before and after serving, filtration, or centrifugation.

The sample is extracted with dichloromethane by using one of three techniques: Soxhlet extraction (10 g of sample with 300 mL of solvent for 16 h), blender extraction (25 g of sample with sodium sulfate and 3 × 150 mL of solvent), or sonication (25 g of sample with sodium sulfate and 3 × 150 mL of solvent) (*21*). Once the sample has been extracted, the solvent volume is reduced by using either rotary evaporation or a Kuderna–Danish apparatus.

The concentrated extract is then split. To one portion, 1.0 mL of DMSO is added, and the dichloromethane is removed under a stream of nitrogen. The resulting DMSO solution is used for the direct assay of the extract. If fractionation of the remaining extract is required, the investigator is given the option of using the acid/base–neutral extraction scheme described in step 3 of the nonaqueous liquid protocol or the HPLC technique described in the sample fractionation methods section.

There was consensus among the panel meeting participants that the soils and sediments protocol is a synthesized protocol that assembles a combination of well-established techniques. Further, the data base demonstrating adequacy of components in this combination is not extensive. There is a good indication that the protocol is applicable for the preparation of both soils and sediments for mutagenicity testing. One dissenting participant was concerned with the use of water-insoluble solvents for the extraction of soils and sediments. The recovery efficiencies of methods using sodium sulfate or silica gel as drying agents and extraction with dichloromethane are not well-documented and were the basis of the disagreement. It was recommended that

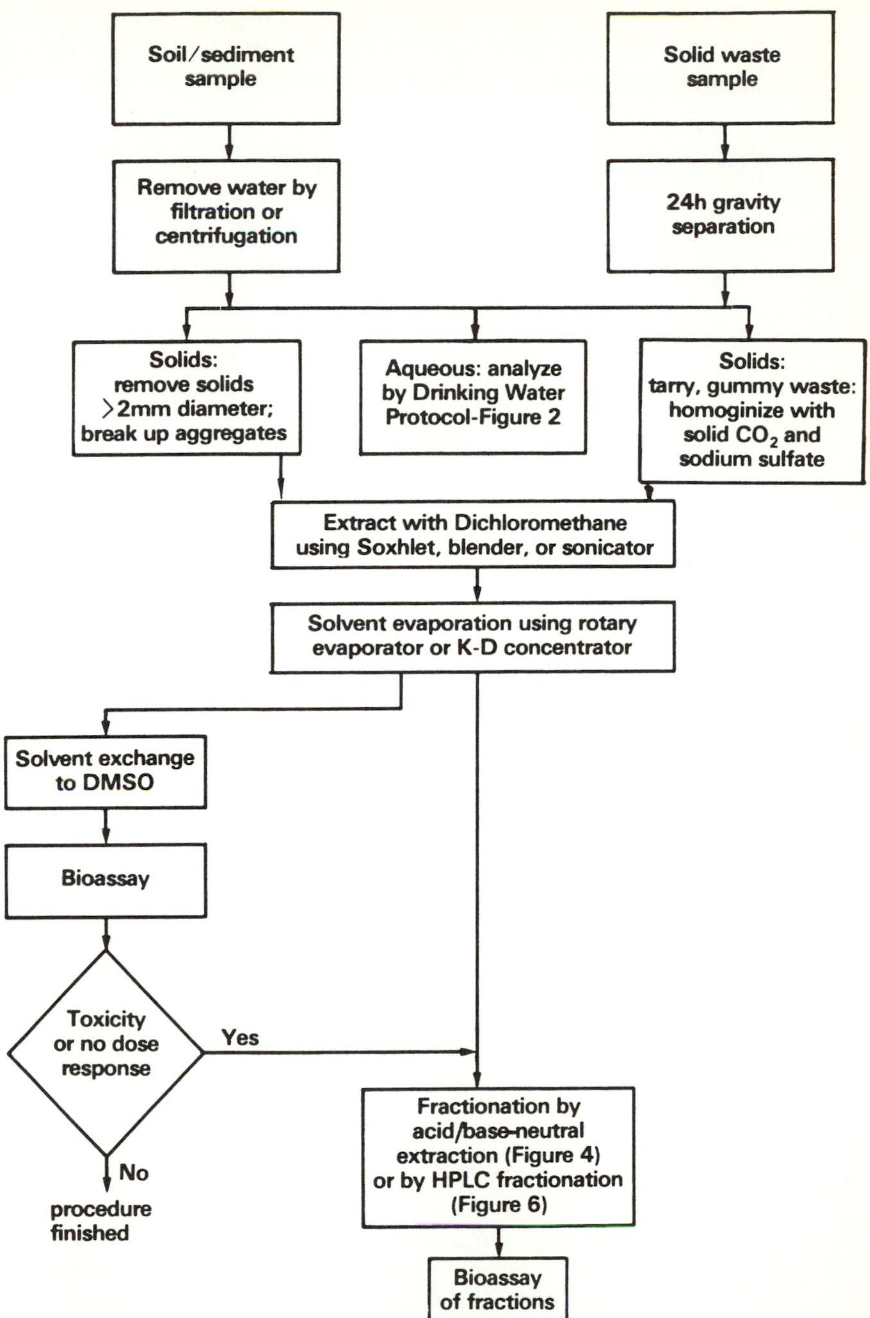

Figure 5. Flow scheme for the preparation of soil, sediment, or solid waste samples.

information be gathered on the effect of pH adjustment on the solvent extraction efficiency of mutagens from soils and sediments during the validation of this protocol.

Waste Solids. The scope of the waste solids protocol (Figure 5) is limited to measuring the mutagenicity of organic components in waste solids that are solvent extractable and that remain stable through the sample preparation procedure. The protocol is based on well-established analytical chemical methods that have been used extensively to determine compositions of complex organic mixtures and environmental samples. However, only limited data are available to demonstrate the efficacy of such methods to prepare waste solids samples for mutagenicity assays. Vapor-phase organics and inorganic components of waste solids are, by intent, not addressed in the protocol.

Waste solids are defined as heterogeneous materials that range from sticky, viscous, or tarry material to dry solid particulates. Special techniques for the treatment of oily, gummy, and adhesive materials (e.g., addition of anhydrous sodium sulfate or silica gel) are specified in the protocol. The gravity phase-separation procedure (24 h at 4 °C) developed by the Environmental Waters and Waste Water Work Group is incorporated by reference to address the removal of liquids from waste solids samples. The waste solids protocol can also be applied to solids partitioned from aqueous or nonaqueous liquids or from gaseous media.

The preparation steps for waste solids samples are very similar to those required for soil and sediment samples. The same flow scheme (Figure 5) is used to illustrate the sample preparation requirements for both soil–sediment and soil waste samples. However, some differences are found between the methods. Freezing samples or adding dry ice, prior to grinding waste solids to a maximum particle size of 2 mm or less, is recommended to enhance the extraction of mutagens. Soxhlet extraction is designated as the primary method for sample preparation, but blender and sonication (*21*) techniques are acceptable alternatives. Additional research is recommended to evaluate extraction efficiencies and method equivalency of the three techniques as well as the possible use of low-temperature (4 °C) extractions to reduce the degradation of unstable compounds during work up. The addition of surfactants or other compounds to improve extraction efficiency is not recommended except in cases where supporting scientific data are available. Dichloromethane is the recommended solvent, but the efficiency of different individual solvents or mixtures of solvents for the extraction of waste solids remains a research issue to be studied during method development.

The Kuderna–Danish apparatus is the primary method for solvent removal to facilitate processing multiple samples. However, the rotary evaporator may be preferable for some samples. The solvent DMSO is used to redissolve an aliquot of the concentrated crude extract for direct Ames assay. The rest of the extract will be redissolved by using ethanol, acetone, or other solvents appropriate for the particular sample.

An overall testing scheme is presented in the protocol to clarify decisions regarding the toxicity and implementation of fractionation techniques. Initial applications of crude extract material at dose levels ranging from 10 ng to 10 mg/plate are recommended for nontoxic materials. Fractionation is recommended only in cases in which toxicity is observed or characterization of specific mutagens is desired. The liquid-liquid, acid-base extraction procedure is recommended, after modification, to reduce the potential degradation of components in complex mixtures. Modifications include (1) dissolving the sample in dichloromethane prior to fractionation, (2) acid partitioning prior to base partitioning, and (3) performing separations at 4 °C. Adjustment of pH by hydrochloric acid is preferred to sulfuric acid to reduce chemical decomposition. Although the Work Group acknowledges that acid-base extraction can potentially alter sample components, implementation of the alternative HPLC fractionation method is not warranted because of a lack of suitable validation data for the intended objective of evaluating the overall sample mutagenicity.

Storage at 4 °C or less is recommended for all samples, and storage at −20 °C or less is recommended for crude and processed extracts. The recommended containers are amber glass bottles with Teflon tops, but Teflon containers may be used for particularly corrosive samples. Stainless steel containers are not considered acceptable because of the potential for sample contamination. Additional research to address optimum storage conditions (temperature, time, and storage containers) that preserve sample integrity is recommended. This study should include research on the storage of raw samples, crude extracts, and processed extracts.

The panel meeting participants unanimously agreed that the waste solids protocol is a synthesized protocol that integrates a number of well-established methods and offers a number of alternative techniques. There is a good indication that these methods are applicable for the preparation of soil and sediment samples for mutagenicity testing, but there is a need for additional research comparing extraction techniques (i.e., Soxhlet versus blender versus sonicator), choice of solvents (use of DMSO for crude extract, single solvent, solvent series, etc.), and drying techniques for extracts (silica gel versus sodium sulfate).

Sample Fractionation Methods. There was a consensus among panel participants that each of the protocols shared areas where further method evaluation was required. The most discussed area was the protocol of HPLC fractionation of sample extracts (Figure 6). The participants recognized that although the traditional acid/base-neutral extraction scheme (presented as part of the nonaqueous liquids protocol, Figure 4) could cause chemical degradation of some mutagens, the alternative of HPLC was more expensive and not as well-documented.

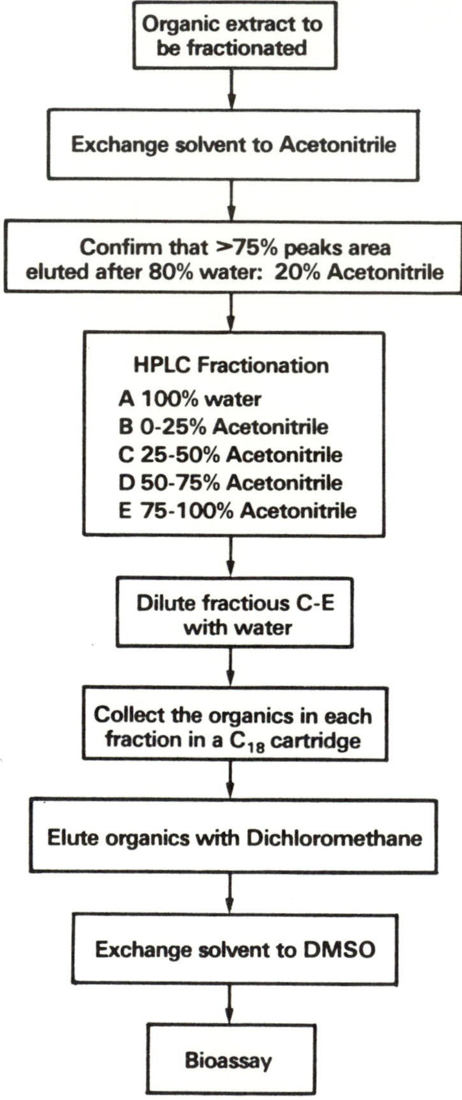

Figure 6. Flow scheme for the reversed-phase HPLC fractionation of sample extracts.

A protocol summary is offered for the HPLC fractionation, based on procedures currently used for the determination of mutagenic hazards associated with drinking and environmental waters (*10–14*, *22–27*). This procedure is developed more fully in Chapter 19 (*23*).

The sample extracts that show either toxicity or no dose response on initial testing should be fractionated. An aliquot of the extract is solvent exchanged to acetonitrile, and an initial analytical scale separation is made to assess the distribution of constituents in the sample. This separation is accomplished by using a C_{18} reversed-phase system eluted for 45 min with a linear gradient of 0–100% acetonitrile in water. If >75% of the sample elutes after the solvent composition of 80% and 20% acetonitrile, then the fractions are isolated by preparative reversed-phase HPLC. Fraction A is eluted with 100% water; fraction B is eluted with a linear mobile-phase gradient from 100% to 75% water and 25% acetonitrile; fractions C, D, and E are eluted with gradients with final compositions of 50%, 75%, and 100% acetonitrile.

The fractions are prepared for the Ames assay by dilution with water, adsorption of the organics on a C_{18} cleanup cartridge, elution of organics with 5 mL of dichloromethane, and exchange into DMSO.

For samples that are unsuitable for this reversed-phase separation (>75% of the sample elutes before the solvent composition of 80% water and 20% acetonitrile), a fractionation method involving normal-phase HPLC eluted with hexane–dichloromethane is used (*23*, *24*).

This HPLC fractionation technique should work for extracts of samples from all of the environmental media discussed here, but the method requires further validation. Because of the expense of HPLC equipment, an alternative procedure using low-pressure or flash chromatography should also be investigated.

Other Areas Requiring Method Development. Several other techniques common to all of the protocols require validation. These include the following:

1. The best solvents to use for the extraction of samples. Most of the protocols call for dichloromethane as an extraction solvent because it was deemed the most commonly used solvent by the panel participants. The comparative effectiveness of dichloromethane needs to be evaluated against other single- and mixed-solvent systems.
2. The best technique for reducing solvent volume, for example, by rotary evaporation or with a Kuderna–Danish apparatus.
3. The stability of samples and extracts under storage.
4. The identity of spikes and surrogates suitable for quality control checks of these determinations.

Conclusion

It was the consensus of the panel participants that the methods summarized here represent the best available technology to prepare environmental samples for the Ames assay. These methods were chosen, after group discussion, on the basis of the collective laboratory experience of the participants and a thorough review of the literature. However, these are interim protocols subject to laboratory validation. This validation effort has been initiated, and workers in the field of environmental mutagenesis are urged to test these interim protocols in their own laboratories.

The objective of this panel was to develop protocols that could be used today with a minimum of methods development. The panel members recognized that promising research is currently underway to develop new techniques for sample preparation. However, procedures such as supercritical fluid extractions or preparative ion chromatographic separation of polar fractions are still considered to be in a research phase and not yet ready for general laboratory application. The procedures presented will be periodically reviewed and updated as the state of available technology improves.

Acknowledgments

All participants in the panel meetings were, in a real sense, contributing authors to this chapter and are included in the list that follows. We acknowledge the efforts of Life Systems, Inc., Cleveland, Ohio, in organizing and facilitating the meetings, and the USEPA and U.S. Army for technical and funding support.

Although the research described in this chapter was wholly funded by the USEPA and the U.S. Army, it was not subjected to the policy review of either organization and therefore does not necessarily reflect the views of either or both organizations. Mention of trade names or commercial products is for identification only and does not constitute endorsement or recommendation for use.

Complete versions of the interim protocols summarized in this chapter are available from L. R. Williams, Environmental Monitoring Systems Laboratory, USEPA, Las Vegas, NV 89114.

List of Panel Participants

Air

V. M. S. Ramanujam
Peter Flessel
Martin S. Legator
Ray Merrill

Drinking Water

M. Wilson Tabor
Rodger B. Baird
Fred Kopfler
David T. Williams

Nonaqueous Liquid Wastes

Yi Wang
Stanton L. Gerson
Michael Guerin
Donald Gurka
Paul C. Howard
Ronald J. Spanggord

Soils and Sediments

Kirk Brown
Ken Loveday
Paul Marsden
Vincent I. Mastricola, Jr.

Waste Solids

Barry R. Scott
K. C. Donnelly
Elena C. McCoy
Bart Simmons

Environmental Waters and Waste Water

David J. Brusick
Dick Garnas
T. Kameswar Rao
Gary D. Stoner

Literature Cited

1. Williams, L. R.; Preston, J. E. *Interim Procedures for Conducting the* Salmonella-*Microsomal Mutagenicity Assay (Ames Test)*; Environmental Monitoring Systems Laboratory: Las Vegas, NV, 1983; EPA-600/4-82-068.
2. *Fed. Regist.* **1979,** *21.58*, 1-219.
3. *Handbook for Analytical Control in Water and Wastewater Laboratories*; U.S. Environmental Protection Agency. U.S. Government Printing Office: Washington, DC, 1979; EPA-600/4-79-019.
4. *Fed. Regist.* **1971,** *84.36*, 8191-8194.
5. Henry, W. M.; Mitchell, R. I. *Development of a Large Sampler Collector of Respirable Matter*; U.S. Environmental Protection Agency. U.S. Government Printing Office: Washington, DC, 1978; EPA-600/4-78-009.
6. Fitz, D. R.; Doyle, G. L.; Pitts, J. N., Jr. *J. Air Pollut. Control Assoc.* **1983,** *33*, 877-879.
7. Sawicki, E.; Belsky, T.; Friedel, R. R.; Hyde, D. L.; Monkman, J. L.; Rasmussen, A.; Ripperton, L. A.; White, L. O. *Health Lab. Sci.* **1975,** *12*, 407-414.
8. Pellizzari, E. D.; Little, L. W.; Sparacino, C.; Hughes, Y. I.; Claxton, L.; Waters, M. D. *Environ. Sci. Res.* **1979,** *15*, 333-351.
9. LeBel, G. L.; Williams, D. T.; Griffith, G.; Benoit, F. M. *J. Assoc. Off. Anal. Chem.* **1979,** *62(2)*, 241-249.
10. Loper, J. C.; Tabor, M. W.; Miles, S. K. In *Water Chlorination: Environmental Impact and Health Effects;* Jolley, R. L.; Brungs, W. A.; Cotruvo, J. A.; Cumming, R. B.; Mattice, J. S.; Jacobs, V. A.; Eds.; Ann Arbor Science: Ann Arbor, MI, 1983; Vol. 4, Book 2, pp 1199-1210.
11. Loper, J. C.; Tabor, M. W.; Rosenblum, L.; DeMarco, J. *Environ. Sci. Technol.* **1985,** *19(12)*, 333-339.
12. Baird, R. B.; Jacks, C. A.; Jenkins, R. L. In *Chemistry in Water Reuse*; Cooper, W. J., Ed.; Ann Arbor Science: Ann Arbor, MI, 1981; Vol. 2., pp 149-169.
13. Jenkins, R. L.; Jacks, C. A.; Baird, R. B. *Water Res.* **1983,** *17(11)*, 1569-1574.
14. Tabor, M. W.; Loper, J. C. *Int. J. Environ. Anal. Chem.* **1985,** *19*, 281-318.

15. Nellor, M. H.; Baird, R. B.; Smyth, J. R. *County Sanitation Districts of Los Angeles County Health Effects Study*; County Sanitation Districts of Los Angeles County: Whittier, CA, 1984.
16. Baden, J. M.; Kelley, M.; Whorton, R. S. *Anesthesiology* **1977,** *46*, 346-350.
17. Simmons, V. F.; Kauhanen, K.; Tardiff, R. G. In *Progress in Genetic Toxicology*; Elsevier/North Holland Biomedical: New York, 1978; pp 249-258.
18. Selby, C.; Calkins, J.; Enoch, H. *Mutat. Res.* **1983,** *124*, 53-60.
19. Wang, Y. Y.; Rapport, G. M.; Sawyer, R. F.; Talcott, R. E.; Wei, E. T. *Cancer Lett.* **1978,** *5*, 39-47.
20. Haworth, S.; Lawlor, T.; Mortelmans, T.; Speck, W.; Zeiger, E. *Environ. Mutagens Suppl.* **1983,** *1*, 3-142.
21. Warner, J. S.; Landes, M. C.; Slivon, L. E. In *Hazardous and Industrial Solid Waste Testing: Second Symposium, ASTM STP 805;* Conway, R. A.; Gulledge, W. P., Eds.; American Society for Testing and Materials: Philadelphia, PA, 1983; pp 203-213.
22. Tabor, M. W.; Loper, J. C.; Barone, K. In *Water Chlorination: Environmental Impact and Health Effects;* Jolley, R. L.; Brungs, W. A.; Cumming, R. B.; Jacobs, V. A., Eds.; Ann Arbor Science: Ann Arbor, MI, 1980; Vol. 3, pp 899-912.
23. Tabor, M. W.; Loper, J. C., Chapter 19 in this book.
24. Tabor, M. W.; Loper, J. C.; Myers, B. L.; Rosenblum, L.; Daniels, F. B. In *Short-Term Genetic Bioassays in the Evaluation of Complex Environmental Mixtures;* Waters, M. D.; Sandhu, S.S.; Hueisingh, J. L.; Claxton, L., Eds.; Plenum: New York, 1984; Vol. 4.
25. Loper, J. C.; Tabor, M. W. In *Short-Term Bioassays in the Analysis of Complex Environmental Mixtures III;* Waters, M.D.; Sandhu, S. S.; Lewtas, J.; Claxton, L.; Chernoff, N.; Nesnow, S., Eds.; Plenum: New York, 1983; pp 165-181.
26. Alfheim, I.; Becher, G.; Honglso, J. K.; Ramadhl, T. *Environ. Mutagens* **1984,** *6*, 91-102.
27. Baird, R.; Gute, J.; Jacks, C. A. In *Water Chlorination: Environmental Impact and Health Effects;* Jolley, R. L.; Brungs, W. A.; Cumming, R. B.; Jacobs, V. A., Eds.; Ann Arbor Science: Ann Arbor, MI, 1980; Vol. 3, pp 925-935.

RECEIVED for review August 14, 1985. ACCEPTED May 7, 1986.

3

Concentration Techniques Aimed at the Assignment of Organic Priority Pollutants

G. J. Piet, J. A. Luijten, and R. C. C. Wegman

National Institute of Public Health and Environmental Hygiene, P.O. Box 150, 2260 AD Leidschendam, The Netherlands

Environmental research should preferably be effect-oriented to assign and define priority pollutants and compound-directed to control pollution within existing and projected legislation. In an integral concept, concentration techniques and instrumental measurement of organic priority pollutants must be appropriate and available for various environmental compartments. Multicomponent concentrates derived from environmental samples have been submitted to biological test systems to measure the effect parameters followed by analytical-chemical characterization. This chapter describes methodologies and criteria in assigning priority pollutants, the interface and transit function of concentration techniques, and the criteria for concentration techniques in these functions. As illustrative applications, toxicity and mutagenicity of surface water, mutagenicity of fish bile extract, and isolation and instrumental measurement of polychlorinated dibenzofurans and polychlorinated dioxins are presented.

CONCENTRATION TECHNIQUES, applied to assign priority pollutants, function as an interface between the environment, chemical analysis, and bioassays. The transition of harmful pollutants in an environmental system to multicomponent concentrates derived from that system is the principal challenge of concentration techniques aimed at the assignment of organic priority pollutants. A compatible combination of chemical and biological methods must be used. These methods can vary from the development of analytical procedures based on the observed biological activity of fractions of an environmental concentrate to the toxicological

0065-2393/87/0214/0039$06.50/0

assessment of multicomponent mixtures derived from the available data on single compounds.

Prerequisites are (1) a biological test system that is capable of registering the observed effect in an environmental system and (2) an applied concentration technique that acts as an interface between the environment and the test system. If biomonitoring indicates an unwanted exposure to chemicals, it must be translated in chemical terms. This chemical information can be used for control purposes to eliminate the exposure, preferably to a real no-effect level, so that no risk evaluation has to be made. This method requires a bioassay that is specific for an effect in an environmental system and a concentration technique that is specific for the collection and transition of compounds causing the effect.

The urgent need of cooperation between analytical chemists and toxicologists is thwarted in this respect by the complexity of environmental systems, the limited potential of short-term tests to observe effects in these systems, and the restrictions of a concentration technique as a transmitting interface between the environment and applied bioassays. Because of the intricate nature of the problem, it can be approached by developing and applying a series of specific, short-term bioassays with compatible, interfacing, concentration techniques. However, when priority pollutants are assigned, the results must be expressed in terms suited for legislation and control purposes.

The selection and assignment of priority pollutants are the first substantial steps in environmental control. The definition of priority pollutants, however, is evolving from unwanted chemicals in a single environmental system to a more integral concept. Therefore, analytical procedures for the collection and measurement of priority pollutants must be appropriate and available for different environmental systems.

This chapter deals with the assignment of priority pollutants, the interface and transit function of concentration techniques in environmental studies, and the criteria for concentration techniques in these functions; recent illustrative results are presented.

Assignment of Priority Pollutants

In a description of environmental problems, the coherence between processes in all compartments and organisms belonging to the compartments should be considered. Environmental policy in the past was often focused on a single compartment; there was no international agreement on the selection of criteria for priority pollutants (*1*). This situation is illustrated by the number of priority pollutants selected by authorities in the European community and the U.S. Environmental Protection Agency (USEPA). The European community lists 126 priority pollutants, and the USEPA lists 114; 59 pollutants are common to both lists (*1*).

Table I. Organic Priority Pollutants (Integral Concept)

	Source				
Pollutant	*Air*	*Soil*	*Waste*	*Surface Water*	*Drinking Water*
Mineral oil, oil products	++	++	++	+	++
Chlorophenols	++	+	++	++	+
Tetrachloroethane	++	+	+	+	+
Polychlorobiphenyls and PCTs		+	++	+	++
Polyaromatic hydrocarbons	++	++	++	++	
Polychlorinated dibenzodioxins	++	+	++	+	
Chloroform	++		+	+	++
1,2-Dichloroethane	++		+	+	+
Hexachlorocyclohexane		++	+	++	
Vinyl chloride	++		+	+	
Chlorofluorohydrocarbons	++				

NOTE: ++ means of major importance; + means less important, not neglectable.

When priority pollutants are selected on the basis of an integral concept—a method that is currently proposed for strategic purposes in The Netherlands (2)—the attention and research are focused on other compounds and other compartments for which appropriate concentration techniques are required. Table I illustrates this tendency.

The choice and definition of selection criteria for priority pollutants play a major role in developing suitable concentration techniques that are problem-oriented. The research to this end must precede and support legislation and enable the control of existing legislation. Concentration techniques are aimed at the measurement of compounds because existing legislation is mainly based on standard settings for chemicals. In this respect, a risk evaluation covers the combination of adverse effects and the likelihood that these effects will actually take place along the lines from emission to intake. Assessment of the risk of a chemical then implies a comparison of the exposure levels of a receptor organism with the separately established no-observed effects or similar doses from toxicological studies, usually based on animal experiments.

However, to assess risk according to an integral concept, the routes of exposure have to be taken into account and the application of concentration techniques must be route-directed. The selection criteria should imply this concept as indicated in Table II.

Taking into account the trends in environmental concepts enables chemists and toxicologists to make problem-oriented efforts to control environmental pollution (*see* Table III).

Complexity increases with an integral concept because more aspects must be taken into account. For a risk evaluation of chemicals in the environment, an integral concept is a prerequisite (*see* Figure 1).

Table II. Selection Criteria for Priority Pollutants

Criteria	*European Community*	*USEPA*	*Integral*
Known occurrence in water	–	+	+
Known suspected properties	+	+	+
Likelihood of human exposure	–	+	+
Propensity for bioaccumulation	+	+	+
Occurrence in other compartments	–	–	+
Occupational health	–	–	+
Major routes of human exposure	–	–	+

NOTE: + means considered as criterion; – means not considered as criterion.

More than in the past, a systematic setting of priorities is advocated by the authorities. Development takes place by two approaches: (1) an effect-oriented approach and (2) a source-oriented approach. These approaches are based on two issues: (1) Which effects are caused by which products-processes? (2) Which sources emit-discharge which products? In this regard, the scheme from emission to toxicological effect must be traced and concentration techniques must be developed. The close connection between chemical and biological characterization must be strongly stressed in this respect.

Interface Function of Concentration Techniques in the Environmental Policy Life Cycle

In the environmental policy life cycle, four phases can be discerned: 1, calling attention to the problem; 2, definition phase; 3, formulation of the solution and taking measures; and 4, control phase. The development of concentration techniques with an interface and control function is indispensable in phases 1, 2, and 4. This situation is illustrated in Figure 2.

In the first two phases, the concentration technique must transfer observed effects to concentrates compatible for bioassays so that the

Table III. Trends in Environmental Concepts

Aspects	*Characteristics of Priorities*	
	1970	*1985*
Nature of the problem	Sectoral	Multisectoral
Nature of the effect	Human health	Human health and ecosystem
Reversibility of effects	Relatively easy	Relatively difficult
Scale of effects	Local-regional	International-global
Impact on economy	Direct losses	Substantial in course of time

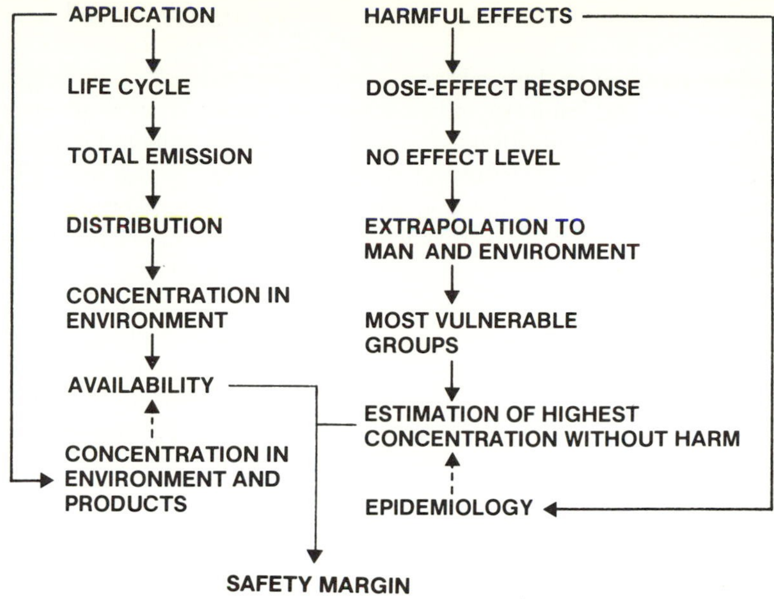

Figure 1. Risk evaluation of chemicals.

effect can be translated into chemical information to enable control of compounds.

In transport and transformation studies dealing with less known or unknown constituents, a severe demand on the ability of chemists to apply contemporary knowledge in the design of concentration techniques is made. Moreover, the effect-inducing compounds or precursors

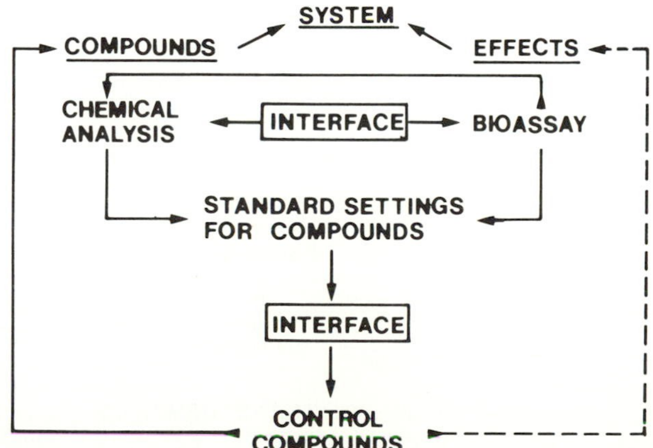

Figure 2. Interface function of a concentration technique.

of these compounds must be deduced from bioassays, and a close cooperation between several disciplines is needed for the definition of and a final solution to the problem.

In an integral concept, the control function of a concentration technique must be route-directed and applied to the selected compound and all of its possible transformation products and byproducts. Figure 3 demonstrates the need for a great versatility of concentration techniques applicable to diverse matrices. These techniques must be standardized and validated prior to use in the control phase.

The central dogma of control is "what you see depends on how and in which direction you look". This doctrine applies to any detection system, whether it is of a biological or a physicochemical (instrumental) nature. Both in toxicological research and in the chemical characterization of environmental samples, an enrichment step is required to see things. No active enrichment step is incorporated in epidemiological research.

The strength of epidemiology is that it provides direct information to show that a chemical or mixture of chemicals promotes a toxic effect. However, its weakness is reflected by its insensitivity, unless an effect rarely seen in a population system is revealed. Absence of hazards is not automatically implied by the failure to detect a toxic effect in a population. The same is valid for the absence of known toxic chemicals.

Two strategies can be followed to assess the health hazard of organic chemicals occurring as contaminants in an abiotic or a biotic matrix. In the chemical-oriented approach, compounds of known toxicity are monitored and a toxicological examination of identified organics of unknown toxicity is performed.

Merely listing compounds occurring in toxic fractions does not lead to an appropriate solution because toxicological (carcinogenic) examina-

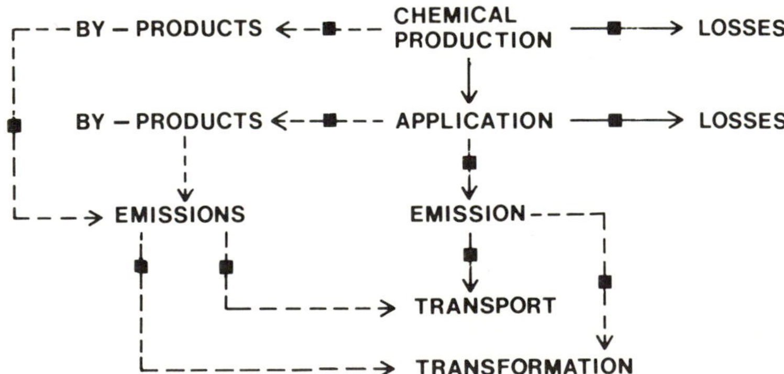

Figure 3. Control function of a concentration technique: ———, original compound; ----, derived product; and ■, accessible for control.

tion of all identified organics is extremely expensive. Alternatively, organic extracts or concentrate fractions of interest can be selected by biotest systems. Because of the complexity of concentrates, separation and identification of the effect-inducing compounds in such a system are not always possible, and the contribution of single compounds to the total effect cannot always be estimated.

In legislation, however, compounds or classes of compounds are selected for standard settings and imposed regulations. Suitable methodologies within budgets, costs, personnel expertise, labor, and degree of sophistication are developed as well as master schemes for a general analytical approach. Legislation based on compounds develops slowly because it is expensive. A study of the carcinogenicity of 10 major and chemically identified compounds present in coffee and tea (single or in combination) by the relatively cheap and rapid Ames test would take 1.5 billion plates and $10 billion. It would be a full-time job for 10,000 people for 4 years (*3*). Even if our sense of insecurity and our awareness of the risk of the carcinogenicity of coffee and tea were substantial, we could not afford to examine the extent of this risk. In practice we do not worry at all, however, because coffee and tea are natural. Very stringent standard settings and even prohibitive regulations for carcinogenic chemicals originate from our need to be protected against frightful diseases. At the same time, we have to realize that cancer mortality against the background of an integral concept will not be substantially reduced by standard settings for single chemicals but more by changing our habits and attitudes. Standard settings affect human and industrial activities and restrict environmental pollution. For this reason, in an integral concept, not only should chemists and toxicologists cooperate, but sociologists, dieticians, and other experts should cooperate as well. For specific carcinogenicity linked to apparent carcinogens such as asbestos and vinyl chloride monomer, standard settings can have a direct improving effect.

In selecting fields to be studied with priority, the complex pattern of human exposure to unwanted factors will not always mean that all responsible compounds must be known and identified. The knowledge of compounds that inhibit adverse reactions as well as health-promoting compounds is of substantial importance. Increased knowledge of the structure–effect relation of chemicals enables scientists to direct and limit their investigations to compounds of major interest.

Criteria for Concentration Techniques in Their Interface Function

By definition, concentration of a sample implies an alteration of its composition, often aimed at a specific removal of the major matrix constituent (water, soil material, air). A concentration procedure should not result in a qualitative and quantitative distortion of the organic constituents of interest in view of the chemical or biological constituents of interest and in view of the chemical or biological assay that follows. This statement means that recoveries, particularly for compounds of different physicochemical natures, must be known. Irreproducible recoveries must be avoided as well as transformation during collection. The quantitative determination of substances at ultratrace levels can be adversely affected by introducing unknown artifacts.

The medium submitted must be compatible with the detection system, and costs of materials and labor are decisive for the screening of large sample numbers. The capacity of a method can imperil the application because sufficient material of a uniform composition is needed for statistic reliability. Some basic criteria are mentioned in Table IV.

The chemical composition of a concentrate can be deliberately altered to improve the specific informative value by improving the selectivity of the concentration method. The intensity to which concentrates require fractionating is determined to a large extent by the information eventually required. Because interpretation of the results of operational systems is required, a coherence should exist between all phases prior to and after the concentration procedure, including sampling, sample treatment, conservation, and fractionation. Both for analytical and biological testing, the introduction of artifacts must be watched, and biodirected fractionating steps must be controlled by chemical analyses. In many cases, foreign materials introduced during a sample-treatment procedure interfere and lead to erroneous results, particularly when priority pollutants have to be measured at an ultratrace level.

Table IV. Criteria for Concentration Techniques

	Research		
Criteria	*Bioassay*	*Chemical Assay*	*Control*
Representative for effects	xx	x	x
Representative for compounds	x	x	xx
Sufficient material for statistic reliability	xx	x	x
Sufficient material for time series (trends)	xx	x	x
Compatible with test system	xx	x	x
Background and artifacts known	x	x	xx
Problem oriented	xx	x	xx
Maximum selectivity	xx	x	xx

NOTE: xx means of major importance.

Table V. Concentration Techniques for Organic Compounds

	Compartment						
Methodology	*Air*	*Soil Sediments*	*Waste*	*Surface Water*	*Tap Water*	*Biol. Sample*	*In Situ*
Bioconcentration		x	x	x			xxx
Adsorption–desorption	x			x	x	x	xxx
Cryogenic trapping	x						xxx
Electric fields	x						xxx
Membrane filtration	x			x	x	x	xx
Ion exchange				x		x	x
Centrifugation				x		x	x
Liquid extraction		x	x	x	x	x	
Gas stripping		x	x	x	x	x	
Freeze techniques	x			x			
Distillation	x	x	x	x			

NOTE: xxx means very suitable; x means applicable.

Concentration techniques are applied in the analyzing laboratory or in situ in the field. Field application enables progressive sampling and facilitates the handling of large amounts of sample. Fewer problems during transport over considerable distances are encountered. Concentration can take place by devices such as accumulator columns but also by living organisms (bioaccumulation or bioconcentration). Some major methodologies and fields of application are mentioned in Table V.

Concentration techniques also can be listed according to the isolating mechanism so that properties of compounds such as volatility, molecular size, and polarity are taken into account. Property-directed concentration techniques are recommended when the complexity of the matrices requires molecular sizing and polarity indexing of compounds. Versatile and manifold adsorption techniques enable the isolation of specific compounds. An example is shown in Figure 4.

The pentane extraction technique is comparable to a purging technique at 90 °C. Lipophilic compounds are extracted, however, up to higher molecular weights (up to 1500). The XAD-4 isolation (pH 7) is aimed at lipophilic and weak hydrophilic compounds having weak polar substituents. At pH 2, moderately hydrophilic compounds such as chloralcohols, hydroxyl ethers, and organic acids can be isolated. Figure 4 indicates that compounds of increasing molecular weight can be present in water when the hydrophilic character increases. To increase the selectivity of a technique, a sequential application, starting with gas stripping, pentane extraction, XAD adsorption, activated carbon adsorption, ion exchange, and freeze-drying can be used. Solvents with increasing polarity can be used, too. A substantial improvement is the

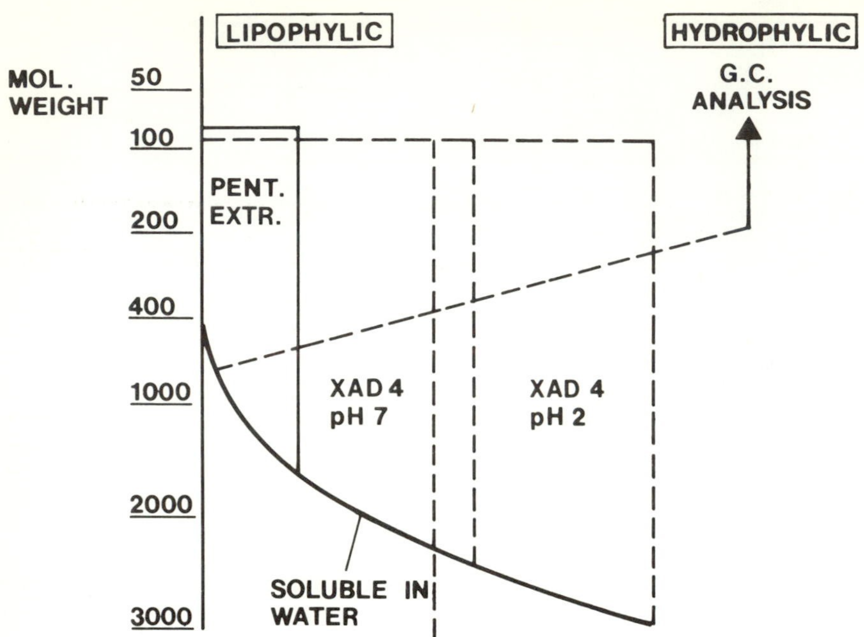

Figure 4. Selectivity of isolation techniques for water. (Reproduced with permission from reference 4.)

application of molecular sizing before further fractionation of subfractions takes place. This method enables the collection of classes of effect-inducing compounds in relatively narrow fractions. When isolation and fractionation systems are perpendicular to each other, selectivity is enhanced. XAD adsorption with solvent elution is most selective when it is followed by normal phase separation techniques (*5–6*). Several classes of specific compounds can be selectively isolated from a matrix when the isolation and subsequent fractionation technique are based on separation according to polarity, as illustrated in Figure 5.

A general rule for concentration techniques in their transit function is that selectivity of the transfer of a specific effect facilitates the identification of responsible compounds. A so-called "general concentration procedure" with an optimal recovery of all organic compounds turns out to be a utopian scheme in many cases. The interface is too broad and too extensive to transmit clearly distinctive signals. Such a broad interface, capable of transmitting simultaneously many different signals, is of use when toxicities of different environmental systems have to be compared with each other.

In the selection of a suitable isolation system, a deliberately balanced strategy should be followed. The physical properties of ef-

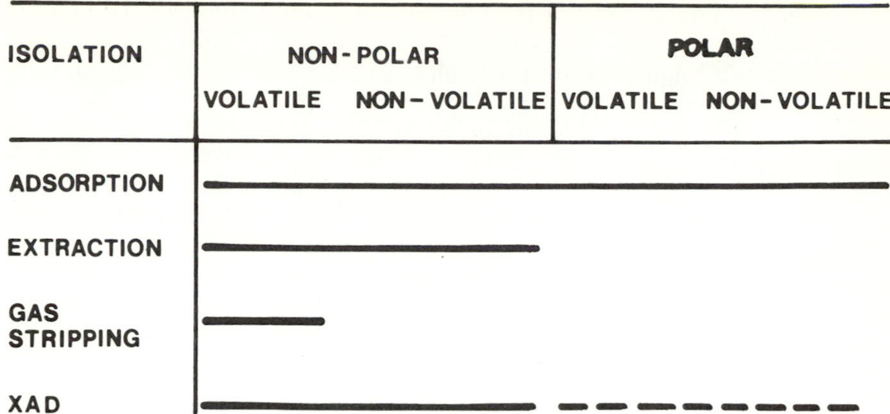

Figure 5. Relation between isolation techniques and physical properties.

fect-inducing compounds play a decisive role in the choice. If appropriate biodirected concentration techniques must be selected, then complementary methods, in which the enrichment step is irrespective of the physicochemical properties of the chemicals involved, are used (7). Complementary concentration procedures for the measurement of toxicity of water from the Rhine River are the following (the concentration factors at which 50% of the guppy fish [*Poecilla reticulata*] died within 48 h are given in parentheses): XAD-4:XAD-8, acetone elution (175); pentane extraction (500); dichloromethane extraction (370); ethyl ether extraction (117); and freeze-drying, elution with ether–acetone (218).

In the study just mentioned, the XAD concentration procedure was selected for monitoring on the basis of the comparison of the results of complementary methods.

One of the intricate problems of the interface function of concentration techniques is to arrive at completely standardized tests, both in technical and biological aspects. The mechanism of toxic action of compounds can be different for different organisms because susceptibility of many organisms is distinctly pollutant-specific. Consequently, not only should a set of tests be carried out with different test systems, but the concentration procedure must be a reproducible interface between the environment and the various test systems. This situation can be achieved by careful analysis of all coherent steps in the whole procedure from sampling to analysis.

Apart from the criteria for the concentration, equal attention should be paid to steps of manipulation and steps preceding the concentration. Important criteria for the biotests are: Where is the most pronounced effect observed? What is the source? Is the object of the investigation treatment or sanitation? For the chemical test, the nature, stability, and fate of the compounds are decisive for the method. These criteria stress the need for quality control for chemical as well as biological test systems.

Recent Studies

During the past 5 years, our institute has done extensive research on biological and toxicological surface and drinking water quality assessment and its relation to chemical pollution (*8*). Scientists discerned that water, along with other sources, has become the victim of the indifference of humans. One of the first multidisciplinary approaches of a water contamination problem was introduced in the thesis, "Sensory Assessment of Water Quality" (*9*).

Chemoreception indeed seems to function as a warning mechanism. Acquired data from 20 drinking water production plants suggested that a relationship may exist between taste assessment and the presence of suspected carcinogenic, mutagenic, or teratogenic compounds in drinking water when these compounds are accompanied by a wide range of organic substances. This situation will generally be the case when heavily contaminated surface water is used as a raw water source (*10*).

In situations where a single compound or a single source of contamination is responsible for drinking water pollution, it is highly unlikely that the increased level of toxic chemicals is accompanied by a bad taste and smell of water. The absence of an adverse taste of drinking water does not guarantee that the drinking water is without potential health hazard.

To determine the chemical composition of drinking water, concentration by headspace, extraction, and XAD-2 adsorption were applied to acquire concentrates. Sensory-directed fractionation of these concentrates has led to the identification of odor-intensive compounds of industrial origin.

In recent years, large consumer panels have been employed for sensory assessments of drinking water. As a result of these assessments, water treatment methods have been adjusted.

Along these lines, other test systems in combination with concentration methods were developed and applied. Although most chemicals have been introduced into the environment in ways that did not result in immediate effects on the environment, the total load of chemical pollutants has certainly contributed to observed changes in the structure and function of aquatic ecosystems. We now rely on aquatic toxicology to give reliable and proper information on the possible effects of man-made chemicals. This information enables the protection of aquatic ecosystems and, in particular, provides for the required scientific guidance in legislation and enforcement.

Organic concentrates of water samples from the Rhine River and Meuse River were tested for toxicity by using a 48-h mortality test on fish (*Poecilla reticulata*) at 3-month intervals for 1 year (*11*). The river samples were concentrated by adsorption on XAD followed by elution with acetone. Rhine water samples were more toxic than Meuse water samples in most cases (*7, 12, 13*) (*see* Figure 6).

On examining one Rhine location on a weekly basis, toxicity was found to vary with the seasons by more than a factor of 12 (*14*) (*see* Figure 7).

A rapid and relatively simple method was applied to determine the toxicity of water by using the concentration procedure prior to testing with the Ames test (*15*). The organic compounds were isolated, after filtering the water, on a mixture of XAD-4:XAD-8 (1:1). The acetone in the XAD eluate was removed by a stream of purified nitrogen.

The XAD procedure was selected on the basis of the comparison of results of complementary methods as mentioned earlier because it is effective in concentrating toxic as well as mutagenic compounds from Rhine water. The investigation demonstrates the application of short-cut biological methods needed for water quality control and complementary to chemical monitoring techniques.

As shown in Figure 7, the lowest toxicity levels were observed in winter (low water temperature and high river flow) and summer (high water temperature and low river flow), whereas the highest toxicity levels were noticed in early spring. The low toxicity in winter could be explained by the elevated degree of dilution of toxicants, whereas the

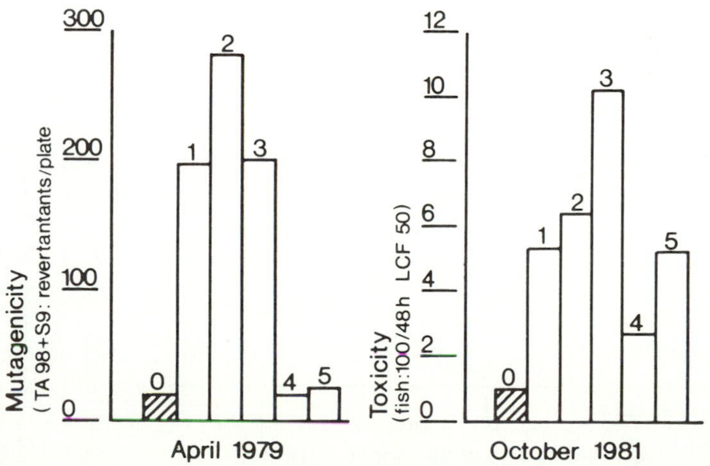

Figure 6. Toxicity and mutagenicity of the Rhine River and Meuse River in The Netherlands: 1–3, the Rhine River at different locations; 4–5, the Meuse River at different locations.

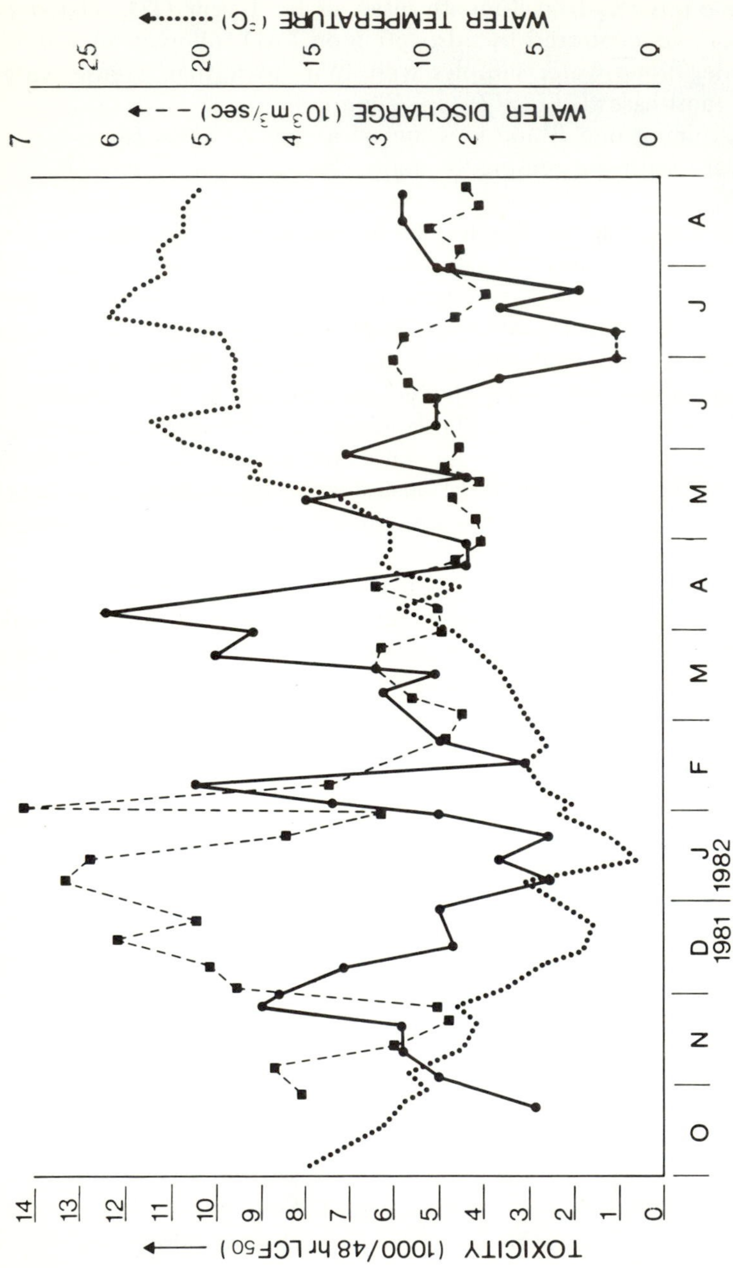

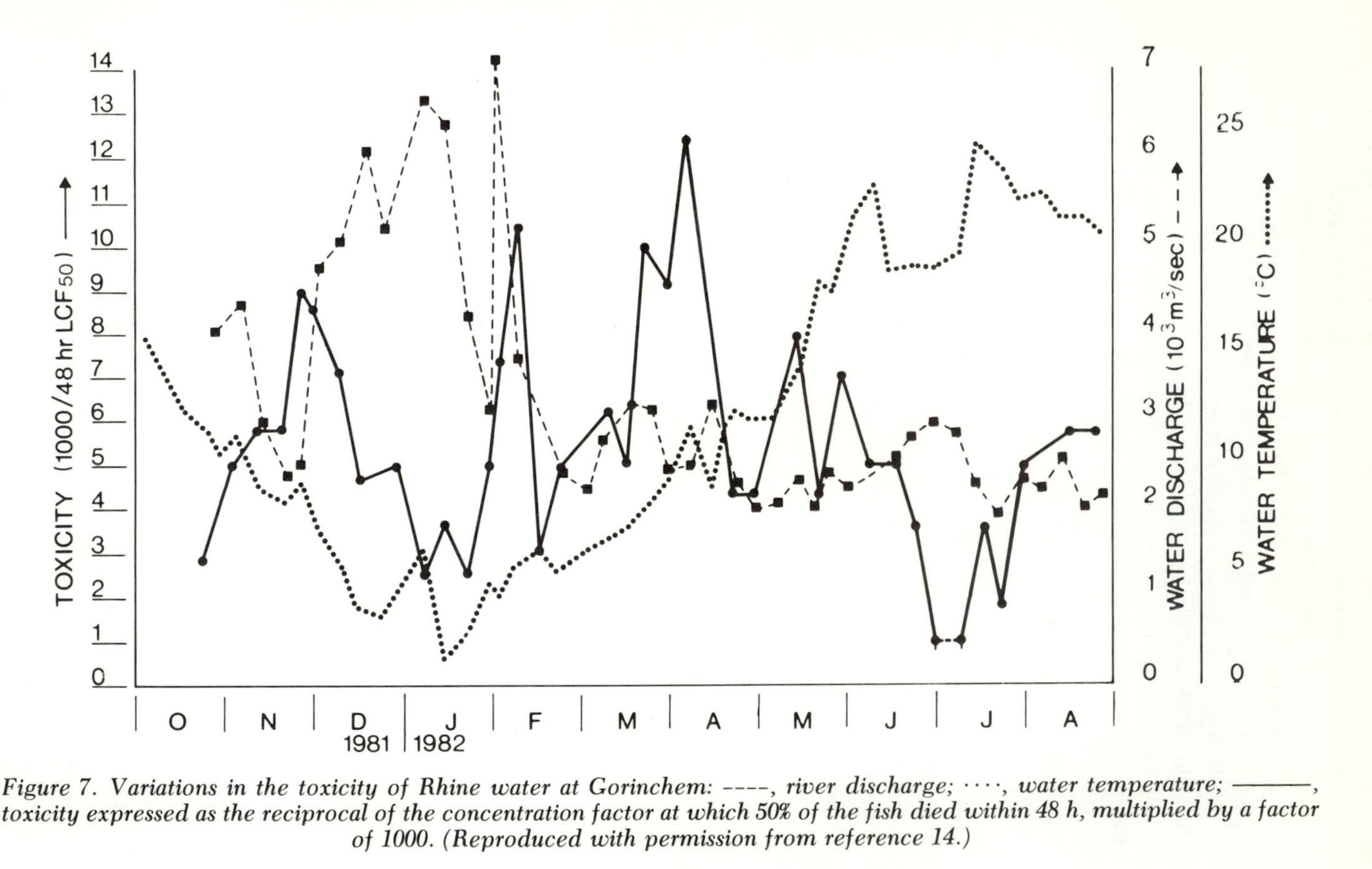

Figure 7. Variations in the toxicity of Rhine water at Gorinchem: ----, river discharge; ···, water temperature; ——, toxicity expressed as the reciprocal of the concentration factor at which 50% of the fish died within 48 h, multiplied by a factor of 1000. (Reproduced with permission from reference 14.)

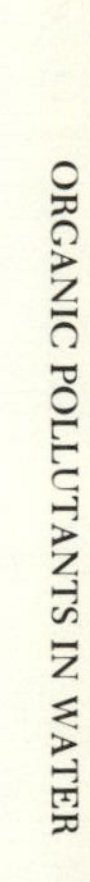

low toxic levels in summer could be the result of an increased inactivation of toxicants by elevated degradation processes. Both periods of low toxicity coincided with holiday periods during which the discharge of effluents may have been decreased (*14*).

Another investigation concerned the accumulation of mutagenic activity in bile fluid of Rhine River fish (*16*). Considerable mutagenic activity can be detected in the bile of natural fish (bream *Abramis brama*) from the Rhine River. Like the river water, organic mutagens can be extracted from aqueous bile by adsorption on XAD resins. The mutagenic activity of the fish bile in the Ames test resembles that of the river water with regard to strain specificity (TA98 and TA1538) and the effect of S-9 (enhancement). The activity in the bile is at least 10^4-fold higher than in Rhine River water. The study shows the biological significance of these yet unknown mutagens. Biliary excretion is known to be an important route for xenobiotics and their metabolites.

No mutagenic activity could be detected in the bile of bream from the Meuse River, which contains little activity in the water itself. Seasonal changes in the level of Rhine water mutagenicity roughly coincide with similar changes in the activity of bile from Rhine fish (*see* Figure 8).

Short-cut bioassays of appropriate organic concentrates of water can be used to show consistent differences between different surface waters. When these are ranked in order of decreasing mutagenic potency, the following sequence is obtained in The Netherlands: Rhine

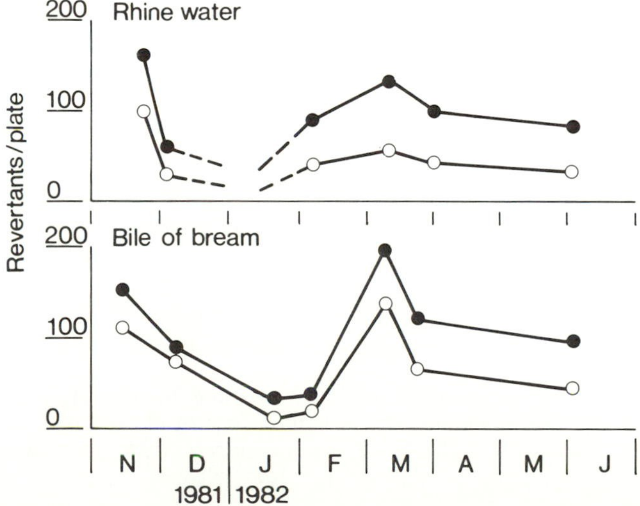

Figure 8. Variations in mutagenic activity of fish bile extracts of the Rhine River: ○——○, without S-9; ●——●, with rat liver S-9. (Reproduced with permission from reference 17.)

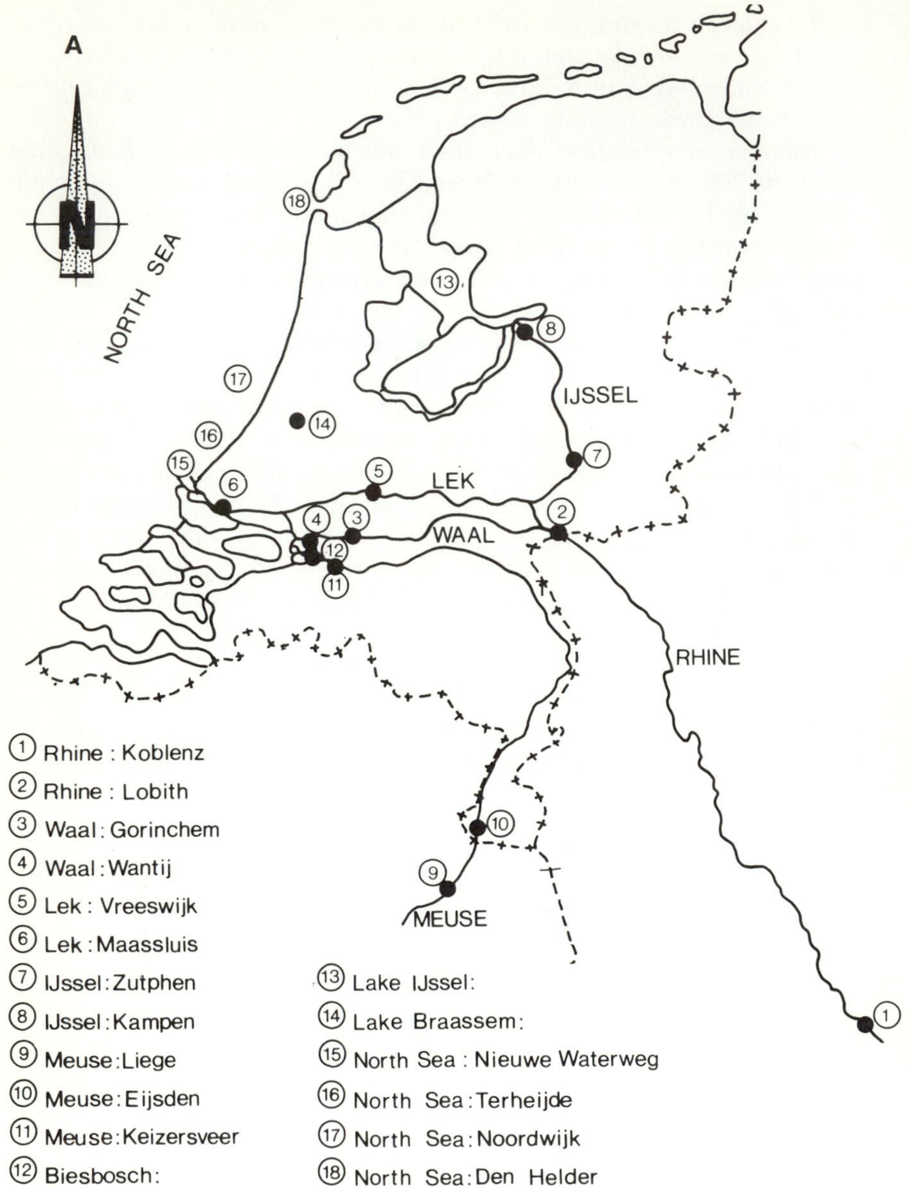

Figure 9. Mutagenic activity in Dutch surface waters: A, locations of sampling; B, mutagenic activity corresponding to concentrates of 1 L of water, as measured with TA98. (Reproduced with permission from reference 16.)

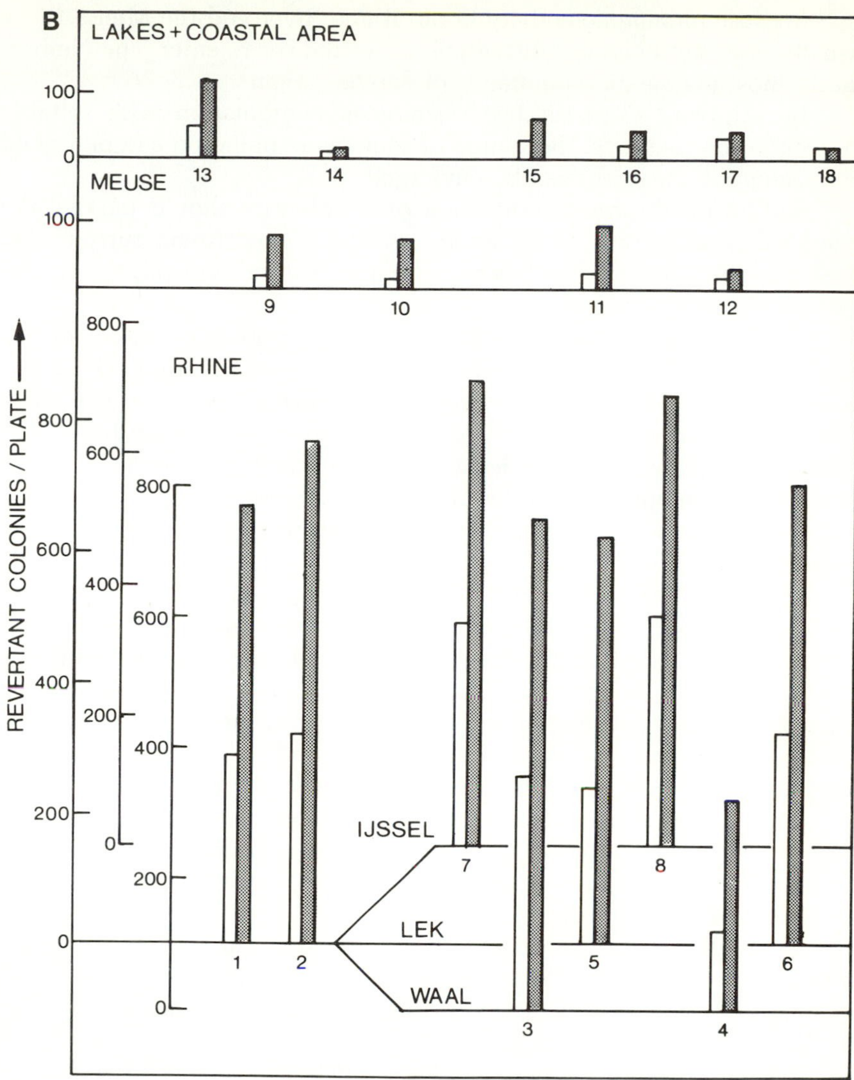

Figure 9B.

River and branches > Lake IJssel > Meuse River > North Sea > Lake Braassem (*16*) (*see* Figure 9).

Differences in mutagenic activity occur between the various locations. Local conditions may enhance self-purification processes of the river. Dilution processes play an important role, as is shown for Lake IJssel and the North Sea coast.

Because mutagenic activity in the Rhine River and the Meuse River usually does not change substantially after the rivers enter The Netherlands, most mutagenic pollution is of foreign origin.

By applying short-cut test systems in combination with suitable concentration methods, the source of mutagenic pollution can be traced and sanitation measures can be envisaged.

Studies on the bioaccumulation of xenobiotics should parallel the chemical qualification of the environmental compartment surrounding the organism. The concentration technique (natural and physicochemical) plays a crucial role.

For the quantitative determination of polychlorinated dibenzodioxins (PCDDs) and polychlorinated dibenzofurans (PCDFs), sample treatment and conservation play crucial roles, too. Only some of the 75 PCDD isomers and 135 PCDF isomers are highly toxic. The collection and analysis of the hazardous compounds present at ultratrace levels in environmental samples must preferably be isomer-specific. The exposure routes for these compounds originate from combustion processes (*18–19*).

In 1978, people of the Dow Company proposed the hypothesis that PCDDs are byproducts of common combustion and, therefore, have always been present in the environment. In one study (*20*), wood was selected as a combustion fuel for a survey study of PCDD emissions. This study suggested that additional research was necessary to confirm this hypothesis and to characterize the magnitude of this newly defined source. A recent review of the fire hypothesis is given in reference 21.

In another study (*22*), PCDDs were found in effluents from laboratory combustion of pine wood at 600 °C in the presence of hydrochloric acid vapor. In the presence of normal air, PCDDs and PCDFs could not be detected.

A sample of dried sludge of a municipal sewage treatment plant, which had been sealed in glass for exhibition purposes in 1933, was analyzed (*23*). It showed a broad range of PCDDs remarkably similar to that of more recent material.

All these investigations show that PCDDs and PCDFs can be expected in air, soil, sediments, organisms, human tissue, etc. Thus, a reliable concentration method for the quantitative recovery and determination of these compounds at ultratrace levels is necessary. The choice of the method depends on the matrix to be analyzed and the specificity, sensitivity, and degree of certainty required in the data. During the past several years, an increasing interest has been shown in the isomer-specific determination of dioxins and dibenzofurans. Specificity for the determination at parts-per-trillion levels makes it necessary to use sample preparation procedures and detection techniques having high resolving power, for example, a high-performance liquid chromatograph for cleanup and fractionation, and a high-resolution gas chromatograph coupled to a low- or high-resolution mass spectrometer. For a quantitative determination, isotopic dilution methods must be used, that is, labelled dioxins and dibenzofurans. The great number of different matrices makes it impossible to develop a universal method.

For example, in eel, high concentrations of polychlorinated biphenyls can be present and can interfere in the cleanup procedure. Fly ash is difficult to extract. Drastic concentration and cleanup procedures such as saponification can convert some isomers. Octachlorodibenzo-*p*-dioxin can easily be broken down during the cleanup procedure. For each matrix a specific isolation technique is necessary. The method has to be validated for all the isomers of the analytical program (*24*).

Rappe (*25*) summarized all the literature with regard to the analysis of polychlorinated dioxins and furans and concluded that, with the standards now available, isomer-specific analyses can be performed for all toxic PCDD and PCDF isomers. However, some attention still has to be focused on characteristics of analytical and concentration techniques that can be promising for the future (*26*).

If the combination of a high-resolution separation system with a physicochemical detection principle is specific to properties of many harmful (toxic) chemicals, then a straightforward approach may be possible (*27*). Electron capture detection (ECD) registers electronegative compounds. Since the early 1960s, Lovelock (*28*) has pointed at the possible link between the ability of a substance to capture electrons and its biological action.

Positive-negative-ion chemical ionization mass spectrometry enables the detection of harmful compounds because of their reaction with alkylating substances (*29–30*).

Experiments with negative-ion atmospheric pressure ionization mass spectrometry has suggested that a high electron affinity and the reaction with oxygen to form a neutral radical and a stable negative ion are characteristics for a compound to have toxic effects. A perusal of the USEPA list of priority pollutants reveals that of 114 organic compounds, only 7 substances are not significantly electron capturing. Apparently, many environmentally hazardous compounds have first been identified because of their selective response in ECD, in advance of epidemiological studies indicating the magnitude of some health risk.

The isolation and cleanup of biological macromolecules by means of affinity chromatography form another typical example of specific application of sample treatment. Affinity chromatography exploits specific functional properties of molecules as retardation of specific groups of solutes in the adsorption step takes place. Later on, adsorbed material can be removed by stepwise or gradient elution. The binding of a substrate to the active site of an enzyme, the complex formation between antigen and antibody, and the concentration of dilute solutions of a biological substance are some typical examples of this selective chromatographic technique.

Within the framework of the identification and analysis of priority pollutants, organic group parameters can be indicative for industrial pollution (*31–33*). Parameters such as extractable organic chlorine (EOCl) and particularly adsorbable organic chlorine (AOCl) are useful for screening and cleanup purposes (*34*). For some environmental systems, a relation between AOCl and mutagenicity exists (*35*). Many compounds contributing to AOCl response have a polar character, are of specific industrial origin, and behave persistently in the environment (*36*). There is a tendency to consider AOCl as a water quality parameter. In this respect, careful consideration and interpretation of group parameters and appropriate concentration procedures are of assistance for the selection of target compounds or groups of compounds.

Indicator parameters can be valid only for a certain time period, in a limited geographic area, and for a particular environmental system. On the other hand, a parameter such as drinking water taste will have more universal value as an indicative parameter (*10*).

For AOCl determination, the concentration method is of utmost importance and has to be standardized (*33*).

Conclusions

A combination of problem-oriented chemical and biological methods is strongly advocated as an approach to investigate adverse effects of man-made pollution on the environment. Because of the complexity of environmental systems, biodirected fractionation and analysis are applied to test relatively well-characterized fractions with appropriate biological systems.

A series of short-term biotests in combination with specific concentration methods, which transmit effects, are required to give an impression of the impact of pollution on an environmental system.

Although biotests do not always lead to unambiguous and easily interpretable results, standardization of chemical and biological procedures from sampling to final measurement will contribute to increased reliability in the assessment of pollution impact on different systems.

Standard settings and guidelines refer at this moment to compounds. In most cases, however, a complex load of organic compounds, among which are very hazardous constituents, is endangering the environment. Because of this complexity, legislation based on data of individual compounds requires expensive research, and major routes of exposure must be known and taken into account. This situation implies the application and development of concentration procedures.

Because of increasing knowledge of the fate of chemicals, structure–effect relations of chemicals, basic mechanisms leading to adverse effects, routes of exposure from emission to intake, etc., both the concentration method and the chemical tests with the short-term biotests can be optimized in selectivity.

The development of concentration techniques for multicomponent mixtures present in different environmental systems must be effect-oriented in the first approach and compound-directed in the control phase because of the existing legislation.

In the environmental policy life cycle (*see* Figure 10), the availability of concentration techniques plays a crucial role to define the nature, source, and extent of an observed problem before regulations and law enforcement can be formulated.

A close cooperation between chemists and ecotoxicologists is advocated to enable responsible authorities to formulate projects leading to the solution and future control of major issues.

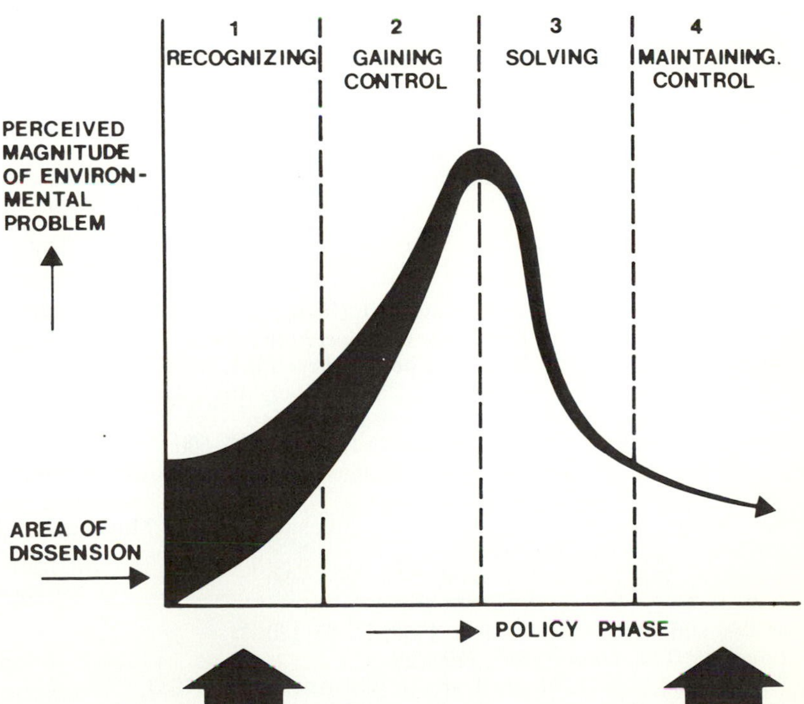

Figure 10. The environmental policy life cycle. (Reproduced with permission from reference 2.)

Literature Cited

1. Vrijhof, H. "Priority Black List Substances for the Rhine and Other Surface Waters of the European Community"; Report No. 840038002; National Institute of Public Health and Environmental Hygiene: Bilthoven, The Netherlands, 1984.
2. IMP-M (Indicative Multiannual Program Environmental Protection 1985–1989); Directorate General Environmental Protection, 1984; ISBN 90-12-04770-6 (in Dutch).
3. Rip, A. *Chem. Mag.* **1984,** *March*, 175–178 (in Dutch).
4. Noordsij, A.; Van Beveren, J.; Brandt, A. *H_2O* **1984,** *17*, 242–248.
5. Alfheim, I.; Becker, G.; Hongslo, J. K.; Romdahl, T. *Environ. Mutagens* **1984,** *6*, 91–102.
6. Tong, H. Y.; Shore, D. L.; Karasek, F. W.; Helland, P.; Jellum, E. *J. Chromatogr.* **1984,** *285*, 423–441.
7. Slooff, W. Ph.D. Thesis, University of Utrecht, The Netherlands, 1983.
8. Slooff, W.; Van Kreijl, C. F. *Aquat. Toxicol.* **1982,** *2*, 89–98.
9. Zoeteman, B. C. J. Ph.D. Thesis, University of Utrecht, The Netherlands, 1978.
10. Zoeteman, B. C. J.; Piet, G. J.; Postma, L. *J. Am. Water Works Assoc.* **1980,** *72*, 537–540.
11. Slooff, W.; De Zwart, D.; Van de Kerkhoff, J. F. J. *Aquat. Toxicol.* **1983,** *4*, 189–198.
12. Kool, H. J.; Van Kreijl, C. F.; Van Kranen, H. J.; De Greef, E. *Chemosphere* **1981,** *10*, 85–108.
13. Kool, H. J.; Van Kreijl, C. F.; Van Kranen, H. J.; De Greef, E. *Sci. Total Environ.* **1981,** *18*, 135.
14. Slooff, W.; De Zwart, D. *Environ. Monit. Assess.* **1983,** *3*, 237–245.
15. Van Kreijl, C. F.; Kool, H. J.; De Vries, M.; Van Kranen, H. J.; De Greef, E. *Sci. Total Environ.* **1980,** *15*, 137–147.
16. Van Kreijl, C. F.; Slooff, W. In *Mutagenicity Testing in Environmental Pollution Control;* Zimmermann, F. K.; Taylor-Mayer, R. E., Eds.; Ellis Horwood Limited: Chichester, 1985; pp 86–104.
17. Van Kreijl, C. F.; Van den Burg, A. C.; Slooff, W. In *Mutagens in Our Environment*; Sorsa, M.; Vainio, H., Eds.; A. R. Liss: New York, 1982; pp 287–296.
18. Olie, K.; Vemeulen, P. L.; Hutzinger, O. *Chemosphere* **1977,** *6*, 455–459.
19. Ahling, B.; Bjørseth, A.; Lunde, G. *Chemosphere* **1978,** *7*, 799–809.
20. Nestrick, T. J.; Lamparski, L. L. *Chemosphere* **1983,** *12*, 617–626.
21. Crummett, W. B.; Townsend, D. I. *Chemosphere* **1984,** *13*, 777–788.
22. Tierman, T. O.; Taylor, M. L.; Garrett, J. H.; Van Ness, G. F.; Solch, J. G.; Deis, D. A.; Wagel, D. J. *Chemosphere* **1983,** *12*, 595–606.
23. Lamparski, L. L.; Nestrick, T. J.; Stenger, V. A. *Chemosphere* **1984,** *13*, 361–365.
24. Korfmacher, W. A.; Moler, G. F.; Delongchamp, R. R.; Mitchum, R. K.; Harless, R. L. *Chemosphere* **1984,** *13*, 669–685.
25. Rappe, C. *Environ. Sci. Technol.* **1984,** *18*, 78–90.
26. Wilkins, C. L. *Science* **1983,** *222*, 291–296.
27. Macdonald, T. L. *CRC Toxicol.* **1983,** *11*, 85–120.
28. Lovelock, J. E. *Nature* **1961,** *189*, 729.
29. Smit, A. L. C. Ph.D. Thesis, University of Amsterdam, 1983.
30. Bruins, A. P. *Biomed. Mass Spectrom.* **1983,** *10*, 46.

31. Glaze, W. H.; Peyton, G. R.; Rowley, R. *Environ. Sci. Technol.* **1977,** *11*, 685-690.
32. Jekel, M. R.; Roberts, P. V. *Environ. Sci. Technol.* **1980,** *14*, 970-975.
33. Piet, G. J.; Wegman, R. C. C.; Quaghebeur, D. "Guidelines for the Application of Organic Group Parameters and Determination of Phenols in the Environment"; Environment Research Programme. Report OMP/37/83; Commission of the European Communities: Brussels, 1983.
34. Wegman, R. C. C. In *Analysis of Organic Micropollutants in Water*; Angeletti, G., Ed.; D. Reidel, Kluwer: Boston, 1981; pp 249-269.
35. Kool, H. J. Ph.D. Thesis, Agricultural University, Wageningen, The Netherlands, 1983.
36. Schnitzler, M.; Kühn, W. In *Analysis of Organic Micropollutants in Water*; Angeletti, G.; Bjorseth, A., Eds.; D. Reidel, Kluwer: Boston, 1983; pp 191-205.

RECEIVED for review August 14, 1985. ACCEPTED January 28, 1986.

4

Analytical Methods for the Determination of Volatile Nonpolar Organic Chemicals in Water and Water-Related Environments

James J. Lichtenberg, James E. Longbottom, and Thomas A. Bellar

Environmental Monitoring and Support Laboratory, U.S. Environmental Protection Agency, Cincinnati, OH 45268

This chapter reviews state-of-the-art methods for the analysis of volatile nonpolar organic chemicals that require monitoring as a result of the Safe Drinking Water and Clean Water Acts. Methods for the determination of purgeable volatile and semivolatile organic priority pollutants and other organic chemicals identified as hazardous or toxic are discussed. Recommended procedures for sample collection, preparation, identification, and quantification are presented. The emphasis is on compound-specific methods such as gas chromatography employing packed and capillary columns with conventional and mass spectrometric detectors. High-performance liquid chromatographic methods are also included. Accuracy, precision, and detection limit data are presented or discussed for many of the analytes of interest. Quality control practices recommended for proper application of the methods are presented. Mandatory quality control practices related to proposed rule making under the Clean Water Act are reviewed.

THE DEVELOPMENT OF ANALYTICAL METHODS for the specific organic pollutants listed in the Consent Decree of 1976 between the U.S. Environmental Protection Agency (USEPA) Administrator and several environmental groups is reviewed in this chapter (*1*). Also discussed is the current status of analytical methods for quantitative determination of these pollutants as well as methods for the analysis of drinking waters as required by the Safe Drinking Water Act (*2*).

A series of 15 test procedures was developed for 114 specific organic priority pollutants and proposed for use in the National Pollution Discharge Elimination System (NPDES) permits program, for state certifications, and for compliance monitoring under the Clean Water Act (*3*). These procedures employ 11 conventional gas chromatographic (GC) and two liquid chromatographic (LC) techniques for quantitative measurement of specific organic compounds in effluents. In addition to the relatively low-cost GC procedures employing conventional detectors, three GC–mass spectrometric (GC–MS) procedures were also provided for routine monitoring of a broad spectrum of pollutants. The high-performance liquid chromatographic (HPLC) methods were proposed for certain analytes not amenable to GC methods. Each method has been evaluated for applicability to a variety of industrial and municipal effluents, and interlaboratory validation studies have been carried out on each of the methods to develop a formal definition of the accuracy and precision of each and to define the method detection limit (*4*).

A summary of the initial priority pollutant methods development and ongoing and proposed future methods development research is presented in this chapter.

Initial Protocol

The initial analytical protocol (*5*) provided sampling procedures as well as GC–MS methods for the determination of the 114 organic priority pollutant compounds that were divided into two major classes of compounds: purgeable volatiles and solvent-extractable semivolatiles. Purgeable volatiles were further divided into halogenated and nonhalogenated purgeables; solvent-extractable semivolatiles were further divided into pesticides, base–neutrals, and acids. The protocol was written and used in the initial phases of the USEPA Effluent Guidelines Division's program to survey 22 industrial categories and was intended primarily as a qualitative screening tool. However, estimates of the pollutant concentrations found were reported.

Priority Pollutant Methods

The objective of the subsequent methods development by the Environmental Monitoring and Support Laboratory–Cincinnati (EMSL–Cincinnati) was to provide a series of methods that could reliably quantify the results of the analysis of relevant waste waters, that is, waters where known specific compounds were expected to be present. Because of the relatively high cost of GC–MS analysis and its general lack of availability in smaller laboratories in 1976 and 1977, efforts were directed toward the more conventional GC methods. At that time, the state of the art in capillary GC was still developing, and few analysts were equipped or able to use it on a routine basis. For these reasons, capillary columns were not included in the initial developmental efforts. However, in the interim between the initial work on these methods and the present time, the state of the art in capillary column technology has advanced greatly, and many laboratories now have the capability to use these columns. Because of these advances, the USEPA now allows the option to use capillary columns. To be acceptable, the analyst must have data on file to demonstrate that the results obtained with the capillary columns are at least as good as those reported for the approved method.

The list of 114 organic compounds was divided into 12 categories as follows (the numbers in parentheses indicate the number of compounds in the category: 1, phthalate esters (6); 2, haloethers (7); 3, chlorinated hydrocarbons (9); 4, nitrobenzenes (3) and isophorone; 5, nitrosamines (3); 6, 2,3,7,8-tetrachlorodibenzo-*p*-dioxin (2,3,7,8-TCDD); 7, benzidines (3); 8, phenols (11); 9, polynuclear aromatics (16); 10, pesticides and polychlorinated biphenyls (PCBs) (21); 11, halocarbon purgeables (26) and aromatic purgeables (3); and 12, acrolein and acrylonitrile. These categories are divided primarily along organic functional group lines, although categories 4, 11, and 12 deviate somewhat from this scheme. The methods fall into two major classifications: (1) GC and (2) HPLC. With the exception of the method for 2,3,7,8-TCDD, all of the GC methods employ conventional low-cost detectors. The method for 2,3,7,8-TCDD requires the use of capillary column GC–MS. Three of the methods use purge and trap GC and are modifications of the technique reported by Bellar and Lichtenberg (*6*) in 1974. Two methods require HPLC, and seven use various conventional, more or less selective, GC detectors. One of the methods, that for phenols, also includes a derivatization step. Tables I and II present summaries of the methods.

The purpose of this chapter is not to give complete details for each of the test methods but rather to give a summary of each to indicate the scope and application.

Brief descriptions of each of the test methods are presented in the following sections. The complete methods are available on request from EMSL–Cincinnati. Sample collection procedures for purgeable volatile analytes are unique and are described in the individual methods. Sample collection procedures for analytes other than the purgeable volatiles can be found in reference 7.

Purge and Trap Methods for Purgeable Volatiles

Three purge and trap methods are used to determine 29 halocarbons (Method 601), seven aromatics (Method 602, including four of the halocarbons), and acrolein and acrylonitrile (Method 603). The three methods are distinctly different in the sorbent trap materials, GC columns, and

Table I. Summary of Conditions for Purge and Trap Analyses

EPA Method No.	*Parameters*	*Purge Temp.*	*Sorbent Trap*	*Column Packing*	*GC Temp. (°C)*	*Detector*[a]
601	halocarbons	ambient	OV-1, Tenax, silica gel, charcoal	Carbopak-B + 1% SP-1000	45–220	OHD
602	aromatics	ambient	OV-1, Tenax	5% SP-1200 + 1.75% Bentone-34	50–90	PID
603	acrolein-acrylonitrile	85 °C	Tenax	Porapak-Q5	110–150	FID

[a] Detector abbreviations are as follows: OHD, organohalide detector; PID, photoionization detector; and FID, flame ionization detector.

Table II. Summary of Methods for Semivolatile Organic Priority Pollutants

EPA Method No.	*Parameters*	*Method*	*Cleanup*	*Column*	*Temp. (°C)*	*Detector*[a]
604	phenols (acids)	GC	silica gel	1% SP-1240 DA + 5% OV-17	80-150, 200	FID, EC
605	benzidines (bases)	LC	H_2SO_4	Lichrosorb RP-2	ambient	ED
606	phthalates (neutrals)	GC	Florisil, aluminum oxide	1.5% SP-2250 + 1.95% SP-2401	180 and 220	FID, EC
607	nitrosamines	GC	10% HCl, Florisil, aluminum oxide	10% Carbowax 20 M + 2% KOH	110 and 220	NPD, TEA
608	pesticides and PCBs	GC	Florisil	1.5% SP-2250 + 1.95% SP-2401	160 and 200	EC
609	nitroaromatics and isophorone	GC	Florisil	1.5% OV-17 + 1.95% QF-1	85 and 145	EC or FID
610	polynuclear aromatics	LC	silica gel	HC-ODS Sil-X + 3% OV-17	ambient	UV, F
		GC	silica gel	HC-ODS Sil-X + 3% OV-17	100-280	FID
611	haloethers	GC	Florisil	SP-1000	60-230	OHD
612	chlorinated hydrocarbons	GC	Florisil	1.5% OV-1 + 2.4% OV-225	75, 100, or 165	EC
613	2,3,7,8-TCDD	GC	NaOH, H_2SO_4, silica gel, aluminum oxide	SP-2330 capillary	200-250	MS

[a] Detector abbreviations are as follows: FID, flame ionization detector; EC, electron capture detector; ED, electrochemical detector; NPD, nitrogen-phosphorus detector; UV, ultraviolet detector; F, fluorescence detector; OHD, organohalide detector; MS, mass spectrometric detector; and TEA, thermal energy analyzer.

analytical conditions employed. In addition, the acrolein–acrylonitrile method uses an elevated purge temperature of 85 °C, whereas the other two methods use an ambient temperature. Table I summarizes the analytical conditions for this group of methods. Problems with stability of the trapping material, Porapak-N, prompted additional work on Method 603, which was subsequently modified for final rule making (8).

The difference in the trap sorbent materials is required because of the variable capacity of the sorbents to retain the compounds of interest. EMSL–Cincinnati studies showed that the purge and trap method, as originally published, was not capable of trapping the very volatile priority pollutant compounds such as vinyl chloride and dichlorodifluoromethane, which are gases at room temperature. It was experimentally determined that activated carbon and silica gel were required to successfully trap the freons. The trap composition and dimensions, beginning with the inlet end, are as follows: 1 cm of 3% OV-1 on Chomosorb-W, 7.7 cm of Tenax-GC, 7.7 cm of silica gel, and 7.7 cm of activated charcoal. A trap containing 1 cm of the 3% OV-1 material and 23 cm of Tenax-GC was found to be satisfactory for trapping the aromatic compounds.

The higher polarity and thus higher water solubility of acrolein and acrylonitrile required an elevated purge temperature (85 °C) to produce acceptable recoveries. The trap composition was initially 24 cm of the porous polymer, Poropak N. It was subsequently changed to 1 cm of 3% OV-1 on Chromosorb W and 23 cm of Tenax-GC.

Other important parameters in providing successful GC are the column packing, temperature conditions, and selection of a detector as specific to the analyte as possible. Maximum resolution of the halocarbons is achieved with an 8-ft × 0.1-in. i.d. column of Carbopack-B coated with 1% SP-1000. The initial temperature of 45 °C is held for 3 min and then programmed at 8 °C/min to 220 °C. An organohalogen detector (OHD) is used. The aromatics are best resolved with a 6-ft × 0.085-in. i.d. column of Supelcoport coated with 5% SP-1200 plus 1.75% Bentone-34. They are measured with a photoionization detector. The temperature conditions are as follows: 50 °C for 2 min then programmed at 6 °C/min to 90 °C. A 10-ft × 2-mm i.d. Porapak-QS (80–100 mesh) column at a temperature of 110 °C for 1.5 min and rapidly heated to 150 °C is now used for acrolein and acrylonitrile. This method employs a flame ionization detector (FID).

Methods for Semivolatile Organic Priority Pollutants

Table II summarizes the methods for the semivolatile organic compounds. A brief description of these methods follows.

Phenols (Method 604). Two approaches are given for the determination of phenols and, although we believe these to be state-of-the-art methods, some problems remain to be solved. The sample is extracted under basic conditions to remove potential interferences. The water phase is then acidified and extracted three times with methylene chloride. The extract is concentrated and analyzed by FID–GC by using a 1.8-m × 2-mm i.d. column packed with Supelcoport 80–100 mesh coated with 1% SP–1240 DA. The column temperature is programmed from 80 to 150 °C at 8 °C/min. The resolution provided by this column is not adequate to separate all 11 of the phenols; if all are present, the second approach, derivatization and electron capture–GC (EC–GC), must be used. After conversion to their pentafluorobenzyl bromide derivatives and silica gel column chromatographic cleanup, the phenols are analyzed on a 1.8-m × 2.0-mm i.d. glass column containing Chromosorb W–AW–DMCS coated with 5% OV–17 at 200 °C. Neither approach is 100% successful for resolving all of the phenols. Poor chromatographic resolution is achieved between 2-chlorophenol and 2-nitrophenol, 2,4-dimethylphenol and 2,4-dichlorophenol, and 2,4-dinitrophenol and 4,6-dinitro-3-methylphenol.

Benzidines (Method 605). The method chosen for the determination of benzidine and 3,3′-dichlorobenzidine uses HPLC. Lichrosorb RP–2 (5 μm) is used as the analytical column, and acetonitrile and an acetate buffer are used as the mobile phase. A relatively selective electrochemical detector is used to detect and measure the benzidines. The instability of 1,2-diphenylhydrazine, which decomposes to azobenzene, caused it to be eliminated from consideration.

Phthalate Esters (Method 606). These compounds are extracted with methylene chloride, concentrated, and solvent exchanged to hexane for Florisil or aluminum oxide column cleanup and EC–GC determination by using a mixed phase columnn of 1.5% SP–2250 and 1.95% SP–2401. This method is essentially the same as that for organochlorine pesticides.

Nitrosamines (Method 607). The nitrosamines are extracted with methylene chloride, treated with HCl, concentrated, and solvent exchanged to methanol for direct nitrogen–phosphorus or thermal energy analyzer (TEA) detection. Provision is made for Florisil or aluminum oxide column cleanup prior to GC analysis. The GC column liquid phase is 10% Carbowax 20 M plus 2% KOH. *N*-Nitrosodiphenylamine thermally degrades to diphenylamine in the GC and is measured as diphenylamine after prior removal of any diphenylamine occurring, as such, in the extract. Florisil or aluminum oxide column cleanup is used for this purpose.

Pesticides and Polychlorinated Biphenyls (Method 608). The methods for organochlorine pesticides and polychlorinated biphenyls (PCBs) differ from earlier 304(h) methods only by the extracting solvent employed. Methylene chloride is used instead of the mixed solvent (15%) methylene chloride in hexane. The CG columns, conditions, and detector are very close to the same.

Nitroaromatics and Isophorone (Method 609). The GC column used for determination of these compounds is essentially the same as that used for the pesticides and PCBs. Nitrobenzene and isophorone are determined at 85 °C by using a FID. 2,6-Dinitrotoluene and 2,4-dinitrotoluene are determined at 145 °C by using an EC detector.

Polynuclear Aromatic Hydrocarbons (Method 610). The polynuclear aromatic hydrocarbons (PAHs) are analyzed by HPLC by using a reverse-phase HC-ODS Sil-X column with UV and fluorescence detectors in series. The option to use FID-GC is given. The column is 1.8-ft × 2-mm i.d. containing Chromosorb W-AW-DCMS coated with 3% OV-17. It is operated isothermally at 100 °C for 4 min and then programmed at 8 °C/min to 280 °C.

Haloethers (Method 611). Bis(chloromethyl) ether is known to have a very short half-life in water (38 s). Therefore, it was eliminated from the priority pollutant list and from consideration in EMSL-Cincinnati method studies. Some of the other haloethers are very volatile, and care must be taken to prevent their loss during the concentration step. The haloethers are determined by OHD-GC by using a 1.8-m × 2-mm i.d. glass column containing Supelcoport coated with 3% SP-1000. The temperature is held at 60 °C for 2 min and then programmed at 8 °C/min to 230 °C.

Chlorinated Hydrocarbons (Method 612). These include chlorinated compounds other than those classified as pesticides, PCBs, or purgeables. One of these compounds, hexachlorocyclopentadiene, has been found to be unstable in the extracting solvent, methylene chloride. Thus, recoveries for this compound are variable and low. The chlorinated hydrocarbon compounds are determined by EC-GC by using a 1.8-m × 2-mm i.d. glass column containing Supelcoport (80-100 mesh) coated with 1.5% OV-1 plus 2.4% OV-225. The oven temperature for six of the earlier eluting compounds is 75 °C. For two later eluting compounds, the oven temperature is 160 °C. Hexachlorocyclopentadiene is chromatographed at 100 °C.

2,3,7,8-TCDD (Method 613). 2,3,7,8-TCDD is a very hazardous compound. Special handling facilities are required when working with the standard material or samples containing 2,3,7,8-TCDD. A section on safe handling practice for 2,3,7,8-TCDD, issued by the Dow Chemical Company (9), has been included with this method. The sample is extracted with methylene chloride and solvent exchanged to hexane. Cleanup is accomplished by washing the extract with sodium hydroxide followed by sulfuric acid and water. The extract is concentrated and further cleaned by using either of the optional column chromatographic procedures: silica gel or aluminum oxide. These may be used independently or in series as needed.

Determination of 2,3,7,8-TCDD is by capillary column GC–selective ion monitoring MS. The capillary column is 60-m × 0.25-mm i.d. glass-fused silica coated with SP-2330 to a thickness of 0.2 μm. This column gives a unique separation of 2,3,7,8-TCDD from the other 21 tetrachlorodibenzodioxin isomers.

Broad Spectrum Methods (Methods 624 and 625). Two broad-spectrum GC–MS methods, Method 624 for purgeable volatile organics and Method 625 for semivolatile organics, were included in the development and evaluation studies. They were the successors of the "Sampling and Analysis Procedures for Survey of Industrial Effluents for Priority Pollutants" (5).

METHOD 624 FOR PURGEABLE VOLATILES. This method, similar to Methods 601 and 602, simultaneously determines the chlorinated aliphatic and aromatic purgeable priority pollutants. The sorbent trap employed here consists of 1 cm of 3% OV-1 coated solid support, 15 cm of Tenax-GC, and 8 cm of silica gel. The GC column is 6-ft × 0.1-in. i.d. packed with 1% SP-1000 Carbopack B (60–80 mesh) operated under the following oven temperature conditions: initial temperature of 45 °C for 3 min, then programmed at 8 °C/min to 220 °C.

The analytes are determined by acquiring a full mass scan and obtaining the extracted ion current profiles (EICP) for the primary mass-to-charge ratio and at least two secondary masses of each analyte. Ions recommended for this purpose are listed in the EMSL methods.

METHOD 625 FOR SEMIVOLATILES. This method is a solvent extraction method intended to determine as many of the organic semivolatile priority pollutants as possible. To accomplish this, the sample is serially extracted, first at a pH greater than 11 and then at pH 2. Figure 1 shows a flow diagram of the procedure. The two fractions, base–neutrals and acids, are independently determined by using two separate GC columns. The base–neutrals are determined on a 1.8-m × 2-mm i.d. glass column packed with Supelcoport (100–120 mesh) coated with 3%

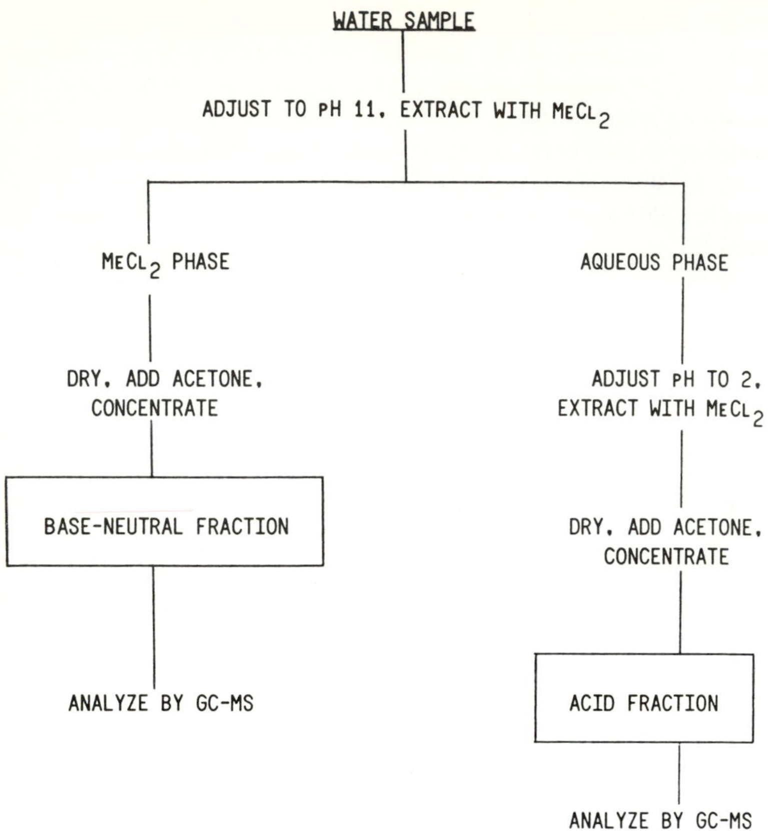

Figure 1. Extraction of a sample for base–neutral and acid compounds.

SP–2250 under programmed temperature conditions at an initial temperature of 50 °C for 4 min, then 8 °C/min to 270 °C. The acids are determined on a 1.8-m × 2-mm i.d. glass column packed with Supelcoport (100–120 mesh) coated with 1% SP–1240 DA under programmed temperature conditions. The initial temperature of 70 °C is held for 2 min and then is programmed at 8 °C/min to 200 °C.

The analytes are determined in a manner similar to Method 624 by using EICP of the primary characteristic mass-to-charge ratio and the internal standard approach for quantification. Characteristic ions for the analytes and internal standards are provided in the method.

Method Revisions

The foregoing methods, non-MS and MS, were first published in the *Federal Register* as proposed rule making in December 1979 (3). As a result of public review and comment following that proposal, many significant comments were received. USEPA considered these comments, and, as a result, a number of significant revisions were made to the methods. The revised methods were published for final rule making on October 26, 1984 (*10*). These methods supplement those approved earlier (1973), which have been updated and which are being reproposed in the current rule making. The initial series of eight USEPA methods [304(h) methods] and the parameters measured are the following (the numbers in parentheses indicate the number of compounds included in the method): 614, organophosphorus pesticides (8); 615, chlorinated herbicides (5); 616, benzidines (2); 617, organochlorine pesticides and PCBs (36); 618, chlorinated solvents (26); 619, triazine herbicides (11); 620, *O*-aryl carbamates (5); and 621, *N*-aryl carbamates and ureas (12). Six of the methods are for pesticides and herbicides, whereas the other two methods are for PCBs and organochlorine solvents. The methods for organochlorine and organophosphorus pesticides, PCBs, and chlorinated phenoxyacid and triazine herbicides are solvent-extraction GC methods. The methods for carbamates and ureas are thin-layer chromatography. A colorimetric method (Chloramine-T), formerly approved for benzidine and its salts, will be withdrawn.

Isotopic Dilution Methods (Methods 1624 and 1625)

In the interim between the initial development and application of Methods 624 and 625, the Effluent Guidelines Division, under contract, developed isotopic dilution methods for purgeable volatiles and semivolatile priority pollutants (*11*). The primary difference between Methods 624 and 625 and their 1624 and 1625 counterparts is that stable, isotopically labeled analogs of the compounds of interest are added to the sample prior to extraction, and quantitative determination is made by using the isotope ratio values for the compounds determined relative to their labeled analogs. In addition, Method 1625 is designed as a capillary column method and uses a 30-m × 0.25-mm silicone-bonded-phase fused silica column (J&W DB-5). Method 1624 uses the same column materials as Method 624. Methods 1624 and 1625 are presented as acceptable alternatives to Methods 624 and 625.

Additional methods considered for 304(h) rule making and the parameters measured are the following (the numbers in parentheses indicate the number of compounds included in the method): 622, organophosphorus pesticides (19); 623, 4,4′-methylene bis(2-chloroaniline); 626, acrolein and acrylonitrile (2); 627, dinitroaniline pesticides (5); 628, carbofuran; 629, cyanazine; 630, dithiocarbamates (15); 631, carbendazim and benomyl; 632, carbamate and urea pesticides (7); and 633, organonitrogen pesticides (7). Most of these are associated with point-source pesticide regulations and were cited in the *Federal Register* 40 CFR Part 455, November 1982 (*12*). They are a mixture of conventional GC and HPLC methods.

Interlaboratory Validation Studies

Each of the analytical methods directly associated with the Consent Decree, except Methods 603 and 1624, have been subjected to an interlaboratory study to define their accuracy and precision for each of the analytes of interest.

The interlaboratory method validation studies (IMVS) were designed according to the approach of Youden (*13*) in which pairs of samples having slightly different spiked concentrations of the compound of interest are analyzed. The IMVS were conducted by the Quality Assurance Branch of EMSL–Cincinnati. For each method, 15–20 laboratories were subcontracted through a prime contractor to perform the analyses. Six sample types were studied for each method: distilled water, drinking water, ambient surface water, and three relevant industrial waste effluents. The participating laboratories furnished the first three types from local sources. The prime contractor furnished the effluent samples. Each sample type was spiked at three concentration levels by using the Youden pair approach at each level. The individual spike of each Youden pair was similar to but measurably different from the other. The lower Youden pair was selected so that the concentration would be above the minimum detection limit of the method. The mean values of the three pairs were designed to spread over a usable and realistic range of concentrations for the method.

The results of these interlaboratory studies are reported in USEPA Method Validation Studies 14 through 24 (*14*). The data were reduced to four statistical relationships related to the overall study: 1, multilaboratory mean recovery for each sample; 2, accuracy expressed as relative error or bias; 3, multilaboratory standard deviation of the spike recovery for each sample; and 4, multilaboratory relative standard deviation. In addition, single-analyst standard deviation and relative standard deviation were calculated.

The interlaboratory studies supported the belief that, if a laboratory performs well with the methods using distilled water, it should be able to obtain good results with surface waters and industrial waste waters. On the basis of these studies, the multilaboratory regression equations for accuracy and single-analyst overall precision for distilled or reagent water have been incorporated into the quality assurance and quality control provisions of Methods 601, 602, 604–613, 624, and 625. These provisions will be discussed later.

Drinking Water Methods

A series of methods designed specifically for the analysis of drinking water has also been developed. Primary among these are the purge and trap and liquid-liquid extraction methods for trihalomethanes (THMs) (*15, 16*). These methods were prepared by EMSL-Cincinnati through the collaborative efforts of chemists from the Municipal Environmental Research Laboratory-Cincinnati (MERL-Cincinnati) and the Office of Drinking Water, Technical Support Division. They were published in the *Federal Register* as part of an amendment to the Interim Primary Drinking Water Regulations (*2*).

The analysis of THMs is reviewed below to point out general principles of analysis for specific pollutants. Precision and accuracy are specified where possible. The THMs of primary concern in drinking water consist of chloroform, dichlorobromomethane, dibromochloromethane, and bromoform. USEPA methods for the analysis of drinking water and the parameters measured are the following (the numbers in parentheses indicate the number of compounds included in the method): 501.1, THMs—purge and trap GC (4); 501.2, THMs—liquid-liquid extraction GC (4); 501.3, THMs—purge and trap GC-MS (4); 502, chlorinated hydrocarbons—purge and trap (47); 503, aromatic hydrocarbons—purge and trap (33); and 524, chlorinated and aromatic hydrocarbons—purge and trap GC-MS (28). Methods 502, 503, and 524 have been revised as methods 502.1, 503.1, and 524.1 and are cited in the proposed National Primary Drinking Water Regulations. Routine monitoring methods employ batch analyses and include the purge and trap procedure (*6, 17*), liquid-liquid extraction (*18–22*), and direct water injection. The limitations of direct water injection have been documented by Pfaender et al. (*23*). THMs are now routinely measured in many laboratories, primarily by the purge and trap and liquid-liquid extraction methods.

Keith et al. (*24*) reported that, by using the purge and trap method, most of the analyses of both 10-μg/L and 1-μg/L THM standards agree within 30% of the calculated values. The variances due to instrument background during routine monitoring were judged to be 0.6 and 1.7 μg/L for chloroform and bromodichloromethane, respectively. Reding et al. (*25*) obtained comparable results regarding the overall precision of the purge and trap and liquid-liquid extraction methods for routine monitoring of THMs in drinking water at levels less than 1 μg/L. Single laboratory precision and accuracy developed by Beller (*16*) are significantly better than this value.

Results of interlaboratory studies conducted by the American Society for Testing and Materials (ASTM) Committee D-19 for the purge and trap technique (*26*) and liquid-liquid extraction (*27*) have been summarized (*28*) for chloroform, 1,2,3-trichlorobenzene, and chlorobenzene (benzene and ethylbenzene were also included in the study). For the purge and trap technique for the range 2.5–450 μg/L, the recoveries were generally 100% ± 20%. Single operator precision varied from 20–40% at the lowest concentration levels to 6–15% at the highest concentration levels.

The liquid–liquid extraction procedure was also subjected to an interlaboratory study by the ASTM Committee D-19. Two water matrices were used: a purified water and a matrix water (a water simulating natural conditions). Bromoform, bromodichloromethane, chlorodibromomethane, chloroform, tetrachloroethylene, and 1,1,1-trichloroethane were studied. For the range 1.9–99 μg/L, recoveries were from 90% to 120%. The relative standard deviation ranged from 10–27% at the lowest concentration to 3.8–8.0% at the highest concentration.

In addition to the THM methods, EMSL-Cincinnati has developed purge and trap methods for selected halogenated (*29*) and aromatic (*30*) compounds that are considered to be chemical indicators of industrial contamination. The methods are applicable to 47 halogenated compounds (Method 502) and 33 compounds that have ionization potentials less than 10.2 eV and that are aromatic or contain a doubly bonded carbon (Method 503). Seven of these compounds are halogenated and are also included in the method for halogenated compounds. Another method, Method 524 (*31*), provides for GC–MS determination of 28 purgeable volatiles. Single laboratory precision and accuracy data for these compounds are provided in the EMSL methods.

The halogenated method employs a packed column of 1% SP-1000 on Carbopak-B (60–80 mesh) as its primary analytical column. The column is 8-ft × 0.1-in. i.d. It is operated at a helium flow rate of 40 mL/min under programmed temperature conditions of 45 °C isothermal for 3 min, then 8 °C/min to 220 °C, and then held at 220 °C for 15 min or until all compounds have eluted. An electrolytic conductivity detector operated in the halide-specific mode is used for measurement.

The aromatic method uses as its primary analytical column a packed column of 5% SP-1200 + 1.75% Bentone 34 on Supelcoport (100–120 mesh). The carrier gas is helium at a flow rate of 30 mL/min. The temperature is programmed as follows: (for lower boiling compounds) 50 °C isothermal for 2 min, then 6 °C/min, then 6 °C/min to 90 °C, and then held until all compounds have eluted; (for a higher boiling range of compounds) 50 °C isothermal for 2 min, then 3 °C/min to 110 °C, and then held until all compounds have eluted. A photoionization detector with a 10.2-eV lamp is used for measurement.

Additional methods for pesticides and herbicides in drinking water are available (*32*).

In addition to the specific methods just discussed, EMSL–Cincinnati has developed or evaluated additional methods for the determination of toxic and hazardous organic chemicals in water and waste waters. Together, these methods are applicable to approximately 390 compounds, including 169 pesticides.

Analytical Quality Control

Analytical quality control is an important aspect of any environmental analysis. It is particularly important when the data generated are to be used in determining the health effects of the compounds measured, effectiveness of a treatment process, or compliance with regulations limiting discharge in waste waters or allowable concentrations in drinking waters. The quality of the data produced, that is, the accuracy and precision with which the measurements are made, must be known and of the highest quality possible in order to reach sound decisions in these areas. This requirement is well-recognized in the environmental analytical community. Thus, it is no surprise that public comment on the December 1979 proposed rule making strongly agreed, in principle, with USEPA's proposal for mandatory quality control when analytical methods are used for regulatory purposes. The program should consist of both intralaboratory and interlaboratory quality control and should be cost effective.

Intralaboratory Quality Control. In its efforts to establish minimum intralaboratory quality control, the USEPA has incorporated certain mandatory quality control practices into each of the 600-series organic methods promulgated as part of final rule making on October 26, 1984 (*10*). These quality control practices are found in Section 8 of each EMSL method.

Laboratories using these methods for regulatory purposes are required to operate a formal quality control program. The minimum requirements of the program consist of an initial demonstration of laboratory capability and an ongoing analysis of spiked samples to evaluate and document data quality. The laboratory must maintain records to document the quality of data that is generated. Ongoing data quality checks are compared with established performance criteria to determine whether or not the results of analyses meet the demonstrated performance characteristics of the method. When results of spike sample analyses indicate atypical method performance, a quality control check standard must be analyzed to confirm that the measurements were performed in an in-control mode of operation.

INITIAL PHASE OF THE PROGRAM. As an initial or start-up test, the analyst must use the method to analyze four spiked distilled or reagent-water samples to demonstrate the ability to generate acceptable accuracy and precision.

Acceptability is determined for each parameter by comparison of the standard deviation (s) and the average recovery ($\bar{X}$) with the corresponding acceptance criteria for precision and accuracy as published in the method for the analytes of interest. If s and $\bar{X}$ for all parameters of interest meet the acceptance criteria, the system performance is acceptable and analysis of actual samples may begin. If any individual s exceeds the precision limit, or if any individual $\bar{X}$ falls outside the range for accuracy, the system performance is unacceptable for that parameter. The analyst must locate and correct the source of the problem and repeat the test for all parameters that failed.

Ongoing Quality Control. The laboratory is required to carry out an ongoing quality control program with a requirement to analyze a method blank each time a set of samples is extracted or when a change of reagents occurs. There is an additional requirement to assess accuracy by spiking and analyzing at least 10% of the samples from each site being monitored. This requirement means analyzing at least one spiked sample for every sample or set of samples of 10 or less. The spiked concentration should be at the regulatory concentration limit or 1–5 times the background concentration in the sample, whichever is higher.

To assess the accuracy of the measurement, the analyst should compare the recovery (P) for each parameter with the corresponding quality control acceptance criteria published in the method for each analyte of interest. The analyst must use either the published quality control acceptance criteria or the optional quality control acceptance criteria as defined in the *Federal Register* (*10*).

If any individual P falls outside the designated range for recovery, that parameter has failed the acceptance criteria. When this situation occurs, a quality control check standard containing each parameter that failed the criteria must be analyzed independent of the matrix, that is, spiked reagent water, to demonstrate that the laboratory is operating in control. If this second test is failed, the sample results for those parameters are judged to be out of control, and the problem must be immediately identified and corrected. The analytical results for those parameters in the unspiked sample are suspect and may not be reported for regulatory purposes.

As another phase of the ongoing quality control program, method accuracy for waste water samples must be assessed and records must be maintained. After the analysis of five spiked waste water samples as in the accuracy check just described, the average ($\bar{P}$) and the standard deviation of the percent recovery (s_p) are calculated. The accuracy is expressed as a percent interval or range from $\bar{P} - 2s_p$ to $\bar{P} + 2s_p$. The accuracy assessment of each parameter of interest is updated on a regular basis, for example, after every 5–10 additional accuracy measurements.

An additional quality control check that consists of the spiking of all samples with surrogate standards is required for Methods 624 and 625. Surrogate standards may be labeled or unlabeled compounds similar in structure and chemical and chromatographic behavior to the parameters of interest. Examples provided in the methods are deuterated and fluorinated compounds. A minimum of three surrogate compounds should be selected to cover the retention time range of the parameters being determined. The percent recovery for each surrogate is calculated and records are maintained. To date, no performance criteria or required action have been set for the surrogate results.

The foregoing mandatory quality control requirements represent a minimal formal program that should be applied only after the laboratory-analyst has performed all of the usual standard operating procedures for eliminating or, to the extent possible, minimizing sources of determinate errors including proper cleaning of glassware and calibration of analytical instruments or systems. It is recommended that the laboratory adopt additional quality assurance practices, such as the analysis of field duplicates, to assess the precision of the sampling process.

Interlaboratory Quality Control. In addition to the mandatory quality control practices just outlined, the laboratory is encouraged to participate in interlaboratory programs such as relevant performance evaluation (PE) studies, analysis of standard reference materials, and split sample analyses. Participation in interlaboratory analytical method validation studies is also encouraged.

In keeping with this recommendation, the EMSL–Cincinnati in cooperation with the USEPA Office of Water Enforcement conducts a quality assurance program to evaluate the performance of laboratories doing analyses required in major NPDES permits. The studies conducted under this program are referred to as Discharge Monitoring Report–Quality Assurance (DMR–QA) studies (33). Under the NPDES, major permit holders who generate self-monitoring data are required to participate in this program. Participants are provided with sample concentrates (reference standards in solution) that are diluted to a prescribed volume in reagent water for analyses. The results are used to identify laboratories having significant analytical problems that need to be corrected.

The EMSL–Cincinnati conducts other PE studies under the USEPA Mandatory Quality Control Program for drinking water analysis certification and general water analysis.

Literature Cited

1. Consent Decree Settlement of Litigation Relating to Toxic Pollutant Effluent Standards and Pretreatment Standards FWPCA, 307(a) and 307(b), February, 1976.
2. *Fed. Regist.* **1979,** *44(231)* 68624–68707.
3. *Fed. Regist.* **1979,** *44(233),* 69464–69575.
4. Glaser, J. A.; Foerst, D. L.; McKee, G. D.; Quave, S. A.; Budde, W. L. *Environ. Sci. Technol.* **1981,** *15,* 1426.
5. *Sampling and Analysis Procedures for Screening of Industrial Effluents for Priority Pollutants;* Environmental Monitoring and Support Laboratory. U.S. Environmental Protection Agency: Cincinnati, OH, 1977.
6. Bellar, T. A.; Lichtenberg, J. J. *J. Am. Water Works Assoc.* **1974,** *66,* 739.
7. *ASTM Book of Standards,* D3370; American Society for Testing and Materials: Philadelphia, 1985; Part 11.02, p 87.
8. *Evaluation of EPA Method 603 (Modified);* Environmental Monitoring and Support Laboratory. U.S. Environmental Protection Agency: Cincinnati, OH, 1984.
9. *Precautions for Safe Handling of 2,3,7,8-Tetrachlorodibenzo-para-Dioxin (TCDD) in the Laboratory;* Dow Chemical: 1978.
10. *Fed. Regist.* **1984,** *49(209),* 43234–43442.
11. Colby, B. N.; Beimer, R. G.; Rushneck, D. R.; Telliard, W. A. *Isotope Dilution Gas Chromatography-Mass Spectrometry for the Determination of Priority Pollutants in Industrial Effluents;* Effluent Guidelines Division. U.S. Environmental Protection Agency. U.S. Government Printing Office: Washington, DC, 1980.
12. *Fed. Regist.* **1983,** *48(29).*
13. Youden, W. J. *Statistical Techniques for Collaborative Tests;* The Association of Official Analytical Chemists: Arlington, VA, 1967.
14. Interlaboratory Methods Studies for EPA Methods 601, 602, 604–613 and 624 and 625. Studies 14–24; Environmental Monitoring and Support Laboratory. U.S. Environmental Protection Agency: Cincinnati, OH.
15. *The Analysis of Trihalomethanes in Finished Waters by the Purge and Trap Method;* Environmental Monitoring and Support Laboratory. U.S. Environmental Protection Agency: Cincinnati, OH, 1979; Method 501.1.
16. *The Analysis of Trihalomethanes in Drinking Water by Liquid-Liquid Extraction;* Environmental Monitoring and Support Laboratory. U.S. Environmental Protection Agency: Cincinnati, OH, 1979; Method 501.2.
17. Bellar, T. A.; Lichtenberg, J. J.; Kroner, R. C. *J. Am. Water Works Assoc.* **1974,** *66,* 703–706.
18. Henderson, J. E.; Peyton, G. R.; Glaze, W. H. In *Identification and Analysis of Organic Pollutants in Water;* Keith, L. H., Ed.; Ann Arbor Science: Ann Arbor, MI, 1976; pp 105–133.
19. Mieure, J. P. *J. Am. Water Works Assoc.* **1977,** *69,* 60.
20. Richard, J. J.; Junk, G. A. *J. Am. Water Works Assoc.* **1977,** *69,* 62.
21. Kaiser, K. L. E.; Oliver, B. G. *Anal. Chem.* **1976,** *48,* 2207.
22. Nicholson, A. A.; Meresz, C.; Lemyk, B. *Anal. Chem.* **1977,** *49,* 814.
23. Pfaender, F. K.; Stevens, A. A.; Moore, L.; Hass, J. R. *Environ. Sci. Technol.* **1978,** *12,* 438.
24. Keith, L. H.; Lee, K. W.; Provost, L. P.; Present, D. L. In *Measurement of Organic Pollutants in Water and Wastewater;* VanHall, C. E., Ed.; American Society for Testing and Materials: Philadelphia, 1980; ASTM STP 686, pp 85–107.

25. Reding, R.; Kullman, W. B.; Weisner, M. J.; Brass, H. J. In *Measurement of Organic Pollutants in Water and Wastewater;* VanHall, C. E., Ed.; American Society for Testing and Materials: Philadelphia, 1980; ASTM STP 686, pp. 34–46.
26. *ASTM Book of Standards,* D-3871-79; American Society for Testing and Materials: Philadelphia, 1985; Vol. 11.02, p 611.
27. *ASTM Book of Standards,* D-3973-85; American Society for Testing and Materials: Philadelphia, 1985; Vol. 11.02, p 179.
28. Mieure, J. P. In *Chemistry in Water Reuse;* Cooper, W. J., Ed.; Ann Arbor Science: Ann Arbor, MI, in press.
29. *The Determination of Halogenated Chemical Indicators of Industrial Contamination in Water by the Purge and Trap Method;* Environmental Monitoring and Support Laboratory. U.S. Environmental Protection Agency: Cincinnati, OH, 1979; Method 502.
30. *The Analysis of Aromatic Chemical Indicators of Industrial Contamination in Water by the Purge and Trap Method;* Environmental Monitoring and Support Laboratory. U.S. Environmental Protection Agency: Cincinnati, OH, 1980; Method 503.
31. Alford-Stevens, A.; Eichelberger, J. W.; Budde, W. L. *Measurement of Purgeable Organic Compounds in Drinking Water by Gas Chromatography-Mass Spectrometry;* Environmental Monitoring and Support Laboratory. U.S. Environmental Protection Agency: Cincinnati, OH, 1983; Method 524.
32. *Methods for Organochlorine Pesticides and Chlorophenoxy Acid Herbicides in Drinking Water and Raw Source Water;* Environmental Monitoring and Support Laboratory. U.S. Environmental Protection Agency: Cincinnati, OH; EPA 600/4-81-053.
33. Britton, P. W. In *Quality Assurance for Environmental Measurements,* ASTM, STP 867; Taylor, J.; Stanley, T., Eds.; American Society for Testing and Materials: Philadelphia, 1985; pp 206–215.

RECEIVED for review August 14, 1985. ACCEPTED January 27, 1986.

5

Application of the Master Analytical Scheme to Polar Organic Compounds in Drinking Water

A. W. Garrison[1] and E. D. Pellizzari[2]

[1]Environmental Research Laboratory, U.S. Environmental Protection Agency, Athens, GA 30613
[2]Research Triangle Institute, Research Triangle Park, NC 27709

The U.S. Environmental Protection Agency Master Analytical Scheme (MAS) for Organic Compounds in Water provides for comprehensive qualitative-quantitative analysis of gas chromatographable organic compounds in many types of water. This chapter emphasizes the analysis of polar and ionic organic compounds—the more water-soluble compounds in the MAS repertoire—in raw and treated drinking water. Mean recoveries from drinking water made by using the MAS protocols that handle polar and ionic compounds were as follows: 84% for neutral water-soluble organic compounds (25 compounds spiked at 1 μg/L), 89% for extractable semivolatile strong acids (24 compounds spiked at 50–100 μg/L), 82% for volatile strong acids (18 compounds spiked at 0.3 μg/L), and 81% for strong primary and secondary amines (11 compounds spiked at 35 μg/L). The protocol for nonvolatile acids has not yet been applied to spiked drinking water, but recoveries should be higher than the average of 85% (14 compounds spiked at 50 μg/L) obtained for these acids in industrial–municipal effluents.

ORGANIC CONTAMINANTS IN WATER can be divided into three groups: 1, volatile (gas chromatographable) nonpolar; 2, polar of intermediate volatility; and 3, nonvolatile polar. This grouping roughly parallels Neal's (*1*) division of contaminants into three classes: 1, lipid soluble with molecular weight (MW) < 500; 2, more water soluble with MW $<$ 500; and 3, lipid soluble and water soluble with MW > 500. Any differences in categorization are the result of current analytical operational limitations; for example, most contaminants in Neal's first group

0065-2393/87/0214/0083$06.00/0

are best separated by gas chromatography (GC), but as the MW increases from 300 to 500, volatility decreases so that high-performance liquid chromatography (HPLC) often becomes the preferred technique. The third groups in both classification schemes are limited to HPLC separation.

This chapter is limited to compounds in group 2 of both classifications, that is, polar, even ionic, water-soluble compounds of MW < 500 that are less lipid-soluble than those of group 1. These polar compounds, in most cases, are volatile enough for GC separation or can be made volatile enough by derivatization. However, some of MW < 500 are still not volatile enough for GC under any conditions; these must be separated by HPLC. These compounds are covered elsewhere in this book (2).

The U.S. Environmental Protection Agency (USEPA) Master Analytical Scheme (MAS) for Organic Compounds in Water (3) provides for comprehensive qualitative–quantitative analysis of most gas chromatographable organic compounds in many types of water. These compounds include the purgeable and solvent-extractable, relatively nonpolar, lipid-soluble compounds. They also include the more water-soluble polar and ionic compounds that usually cannot be isolated from water with conventional solvent extraction and are usually not gas chromatographable without derivatization. Again, this chapter is limited to polar and ionic compounds, in raw and treated drinking water, that are amenable to MAS analysis.

MAS Protocols

Figure 1 is a flow diagram of MAS procedures. The polar compounds discussed here are covered by six protocols:

1. NEWS: neutral, water-soluble, low-molecular-weight compounds concentrated by azeotropic distillation then isolated by a heated purge and trap procedure (alcohols, aldehydes, nitriles, etc.).
2. ESSA: extractable, semivolatile, strong acids extracted at pH 1 (less volatile carboxylic acids, strong phenols, etc.).
3. VOSA: volatile, strong acids isolated on anion-exchange resin (volatile carboxylic acids, generally <C-9).
4. NOVA: nonvolatile, strong acids isolated on anion-exchange resin (certain carboxylic acids, sulfonic acids, etc.).
5. SAM-PT and SAM-S: strong amines isolated on cation-exchange resin (primary, tertiary, and secondary amines, primarily aliphatic).
6. WABN-SC: weak acids, bases, and neutrals isolated at pH 8.0 on XAD-4 resin (weak phenols, anilines, and a wide variety of neutral compounds).

All protocol extracts or isolates are analyzed by capillary column GC–mass spectrometry (GC–MS); all except the NEWS and WABN must be derivatized before analysis.

Recovery, Precision, and Sensitivity

During development of the MAS, extensive recovery studies were conducted to verify protocols and to form a data bank for future use in quantitative analysis. Two hundred and seventeen different model compounds from a wide variety of chemical classes and physical property groups were dosed into drinking water; accuracy (recovery) and precision were determined for each compound and for each class of organic within each analytical protocol. [Recovery data for 327 organics in several types of water obtained by using all of the MAS protocols were presented at the 186th meeting of the American Chemical Society in Washington, DC, in September 1983 (*4*).]

Table I shows the summarized recovery data with spiking levels and precision. This table includes, for purposes of comparison with polar compounds, some data for nonpolar compounds [volatile organics (VO) and the neutrals in WABN]; also, the recoveries for NOVA are not for drinking water. The mean recovery for 217 compounds was 82%; the mean relative standard deviation (RSD) (for three or more measurements) was 12%. Several observations can be made in regard to these data: (1) The best recoveries and precision are obtained for VO and ESSA compounds. (2) NEWS organics, a new class of analytes, are

Table I. Summary of MAS Recovery Data for Organics in Drinking Water by Protocol Class

Protocol Class	*No. of Compounds*	*Spiking Range (ppb)*[a]	*Nominal Spiking Level (ppb)*[a]	*Mean Recovery (%)*	*Mean RSD (%)*
VO	52	0.2–1.8	1	90	10
NEWS	25	0.8–1.2	1	84	16
WABN-SC	87	0.5–5	1	74	12
ESSA	24	50–100	55	89	9
VOSA	18	0.3	0.3	82	10
SAM-PT and SAM-S	11	35	35	81	12
Total	217	—	—	82[b]	12[b]
NOVA	14	50	50	85	20

NOTE: For all protocol classes except NOVA, triplicate determinations were made from drinking water. For NOVA, triplicate determinations were made from industrial–municipal effluents.

[a] Values indicate level spiked into water sample.

[b] These mean recoveries and mean RSDs were calculated from the individual values for the 217 compounds.

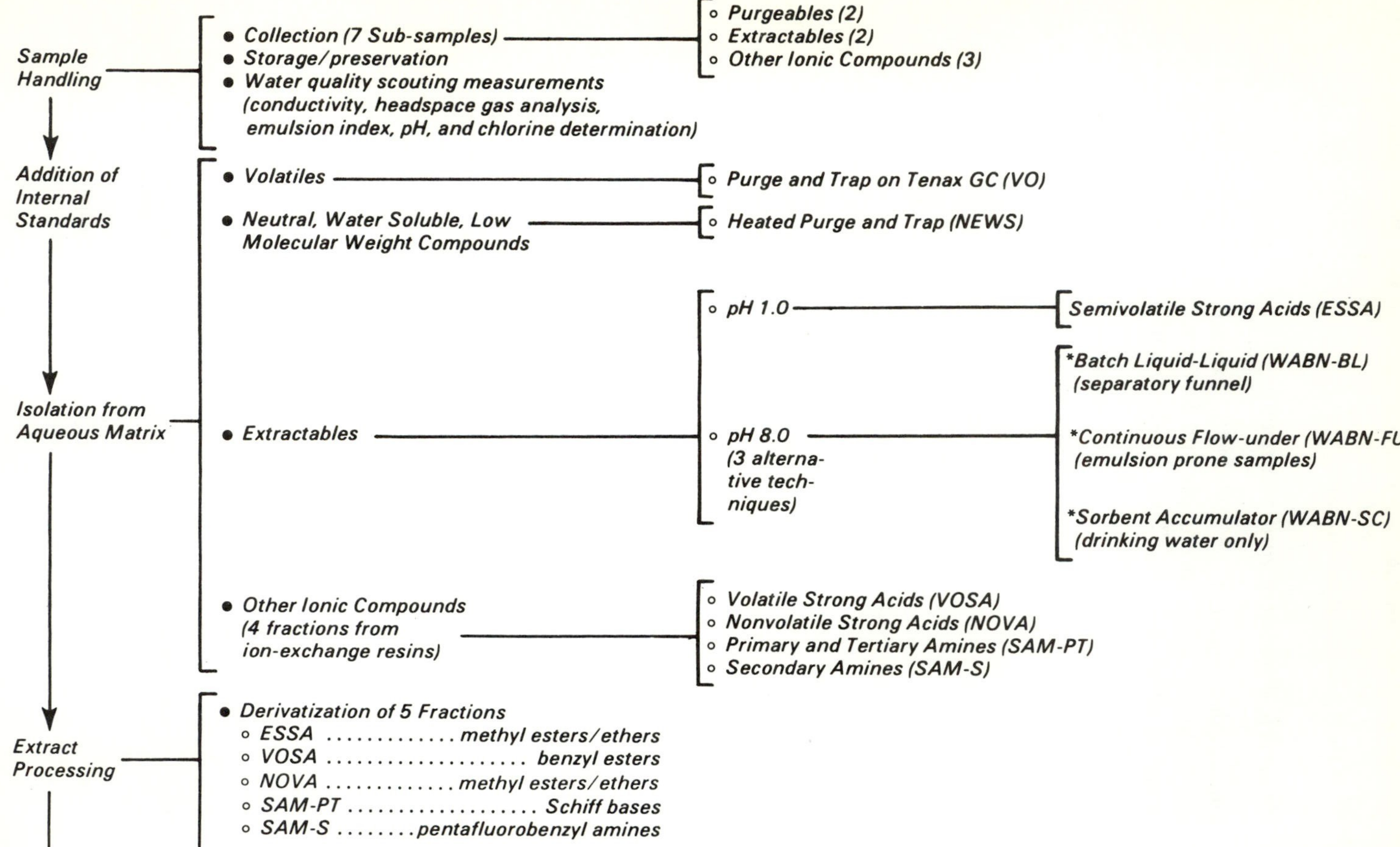
Sample Handling
• Collection (7 Sub-samples)
◦ Purgeables (2)
◦ Extractables (2)
◦ Other Ionic Compounds (3)
• Storage/preservation
• Water quality scouting measurements (conductivity, headspace gas analysis, emulsion index, pH, and chlorine determination)
Addition of Internal Standards
• Volatiles
◦ Purge and Trap on Tenax GC (VO)
• Neutral, Water Soluble, Low Molecular Weight Compounds
◦ Heated Purge and Trap (NEWS)
Isolation from Aqueous Matrix
• Extractables
◦ pH 1.0
Semivolatile Strong Acids (ESSA)
◦ pH 8.0 (3 alternative techniques)
*Batch Liquid-Liquid (WABN-BL) (separatory funnel)
*Continuous Flow-under (WABN-FU (emulsion prone samples)
*Sorbent Accumulator (WABN-SC) (drinking water only)
• Other Ionic Compounds (4 fractions from ion-exchange resins)
◦ Volatile Strong Acids (VOSA)
◦ Nonvolatile Strong Acids (NOVA)
◦ Primary and Tertiary Amines (SAM-PT)
◦ Secondary Amines (SAM-S)
Extract Processing
• Derivatization of 5 Fractions
◦ ESSA .... methyl esters/ethers
◦ VOSA .... benzyl esters
◦ NOVA .... methyl esters/ethers
◦ SAM-PT .... Schiff bases
◦ SAM-S .... pentafluorobenzyl amines

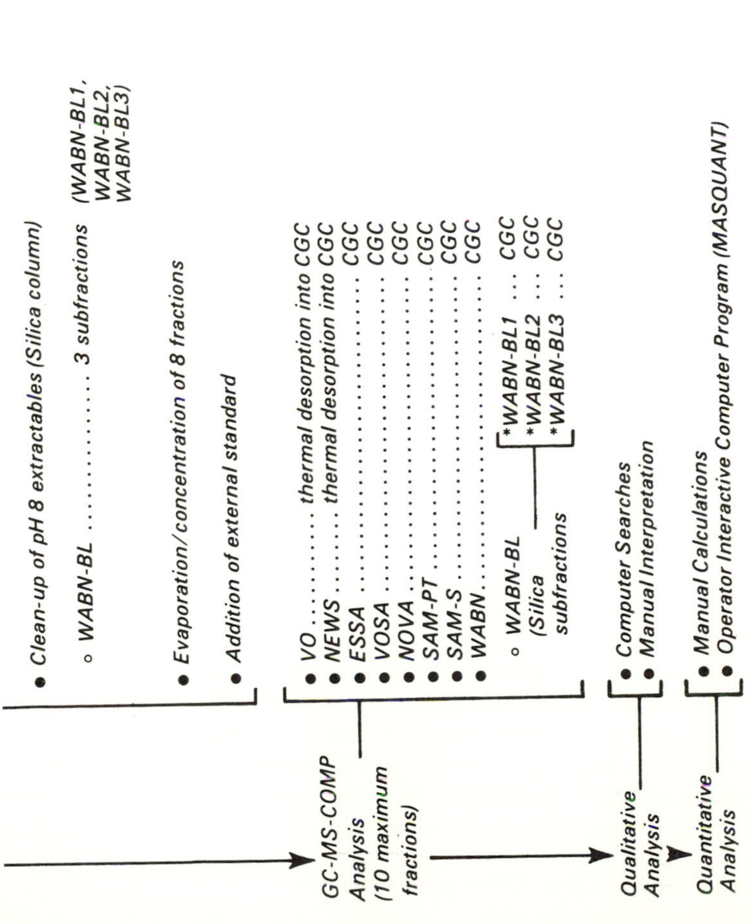

Figure 1. Master analytical scheme flow diagram.

recovered well with adequate precision. (3) The three classes of ionizable compounds that are isolated on ion-exchange resins (VOSA, SAM, and NOVA) are recovered well with adequate precision. (4) The lowest recoveries are for the MAS protocol fraction containing the largest number of compounds—WABN-SC.

Comparisons can be made with recoveries from other matrices, especially industrial and municipal effluents. For example, recoveries of VO and NEWS compounds are practically the same for drinking water and for effluents; this result indicates the lack of matrix effects. On the other hand, the two classes of ionizable organics, VOSA and SAM, that were studied in both matrices were recovered significantly better and with better precision from drinking water than from the effluents. Thus, we deduce that the recoveries of NOVA compounds will be even better in drinking water than the 85% average recovery from effluents. Finally, the relatively poor recovery and precision (74% with RSD of 12%) of WABN organics from drinking water, by using the prescribed accumulator-column isolation technique for added sensitivity, are still better than those from the effluents (69% with RSD of 20%), where batch liquid–liquid extraction is used.

Recovery data for the separate chemical classes in each MAS protocol are given in Table II. Data are included for all protocols, including those for nonpolar compounds. These data are for all types of water studied, not just drinking water; footnotes to Table II give information on sample matrices and spiking levels. Chapter 1 of the MAS protocols (*3*) gives recovery values for each individual analyte, including separate values for each analyte in drinking water for each protocol except nonvolatile acids. The MAS protocols (*3*) and experimental reports (*5*) provide some insight into experimental difficulties that may cause some of the poorer recoveries and precision values of Table II.

Nominal detection limits for the MAS as applied to drinking water range from 0.1 μg/L for VO and NEWS compounds to 5 μg/L for ESSA compounds.

Conclusions

Application of the MAS to drinking water should considerably broaden the scope of organic compounds detected and measured, relative to previously available analytical methods. This conclusion is especially true for the polar compounds of relatively low MW (<500); however, a few of these compounds are not recovered well by the MAS extraction and isolation techniques or are not gas chromatographable, even after derivatization. HPLC methods offer the most promise for separation and analysis of these compounds as well as those of high molecular weight.

Table II. Summary of MAS Recovery Data (Including Deuterated Internal Standards)

Protocol Class Chemical Class (Examples)	*Compounds Studied*	*Recovery Range (%)*	*Mean Recovery (%)*	*RSD Range (%)*	*Mean RSD (%)*	*Footnotes*
VO						
Aromatic hydrocarbons (benzene; naphthalene)	9	59–113	85	3–35	11	*a*
Halogenated aromatics (chlorobenzene; 1,2,4-trichlorobenzene)	7	91–106	100	2–30	15	*a*
Misc. aromatic compounds (anisole)	1	—	68	—	12	*a*
Aliphatic and alicyclic hydrocarbons (cyclohexane; *n*-tridecane)	11	44–120	82	1–40	16	*a*
Halogenated aliphatic hydrocarbons (chloroform; 1,4-dibromobutane)	7	77–118	90	4–16	9	*a*
Misc. oxygen and sulfur compounds (ethyl ether; hexyl ether)	5	70–115	93	3–26	11	*a*
Deuterated standards (d_5-bromoethane; 2,4,6-d_3-anisole; d_5-chlorobenzene; d_8-naphthalene)	4	57–120	90	10–25	18	*a*

Continued on next page.

Table II. Continued

Protocol Class Chemical Class (Examples)	*Compounds Studied*	*Recovery Range (%)*	*Mean Recovery (%)*	*RSD Range (%)*	*Mean RSD (%)*	*Footnotes*
NEWS						
Alcohols (1-propanol; 1-heptanol)	4	106–131	115	18–25	22	*b*
Aldehydes (*n*-butyraldehyde; crotonaldehyde)	3	72–83	79	25–32	29	*b*
Esters (methyl formate; ethyl butyrate)	7	45–93	74	4–17	7	*c*
Ethers (tetrahydrofuran; dioxane)	2	50–79	65	14–31	23	*b*
Ketones (methyl ethyl ketone; cyclohexanone)	2	65–69	67	2–32	17	*b*
Nitriles (acrylonitrile; benzonitrile)	4	57–95	83	6–14	9	*b*
Nitro compounds (nitromethane; nitrobenzene)	3	70–96	86	5–13	9	*b*
Deuterated standards (d_9-*t*-butyl alcohol; d_5-nitrobenzene)	2	93–98	96	7–10	9	*b*

WABN-SC and WABN-BL						
Weak acids						
(phenol; 2,4-dichlorophenol)						
accumulator column	6	55-95	77	5-23	11	*d*
batch liquid-liquid	12	49-118	71	5-27	13	*e*
Weak bases						
(aniline; carbazole)						
accumulator column	16	53-95	82	0-16	7	*d*
batch liquid-liquid	15	40-86	64	6-40	24	*e*
Alkanes						
(*n*-decane; *n*-tridecane)						
accumulator column	8	42-66	52	6-20	15	*d*
batch liquid-liquid	11	45-82	62	13-31	21	*f*
Aliphatic ketones, alcohols, and esters						
(fenchone; methyl stearate)						
accumulator column	7	49-94	75	1-22	10	*d*
batch liquid-liquid	9	49-111	71	10-43	25	*e*
Misc. aliphatic compounds						
(di-*t*-butyl disulfide; tributyl phosphate)						
accumulator column	6	40-92	72	2-36	18	*d*
batch liquid-liquid	4	57-104	75	9-36	21	*e*
Aromatic hydrocarbons						
(2-methylnaphthalene, pyrene)						
accumulator column	10	60-87	74	1-27	16	*d*
batch liquid-liquid	7	48-118	79	13-41	24	*e*
Halogenated aromatics						
(*o*-chloroanisole; hexachlorobenzene)						
accumulator column	10	56-100	73	1-42	16	*d*
batch liquid-liquid	9	43-107	68	13-33	20	*e*

Continued on next page.

Table II. Continued

Protocol Class Chemical Class (Examples)	*Compounds Studied*	*Recovery Range (%)*	*Mean Recovery (%)*	*RSD Range (%)*	*Mean RSD (%)*	*Footnotes*
Aromatic aldehydes and ketones (*o*-tolualdehyde; acetophenone)						
accumulator column	3	88–96	92	2–17	12	*d*
batch liquid–liquid	4	43–105	69	6–19	13	*e*
Aromatic esters and sulfonates (benzyl acetate, ethyl *p*-toluene-sulfonate)						
accumulator column	6	46–87	69	7–17	12	*d*
batch liquid–liquid	7	55–138	84	3–19	11	*e*
Misc. aromatic compounds (nitrobenzene; tetraphenyltin)						
accumulator column	6	47–89	74	6–24	13	*d*
batch liquid–liquid	10	48–105	68	8–33	17	*e*
Deuterated standards (d_{10}-xylene; d_5-phenol; d_5-acetophenone; d_5-phenylethanol; d_5-nitrobenzene; d_5-propiophenone; d_8-naphthalene; d_9-acridine; d_{12}-perylene)						
accumulator column	9	55–93	78	4–21	10	*d*
batch liquid–liquid	8	40–78	58	11–40	26	*e*

ESSA						
Carboxylic acids (benzoic acid; palmitic acid)	17	63–110	89	2–20	9	*g*
Phenols (2-nitro-*p*-cresol; pentachlorophenol)	5	88–100	94	4–19	8	*g*
Deuterated standards (d_{13}-heptanoic acid; d_5-benzoic acid)	2	65–92	79	6–12	9	*g*
VOSA						
Volatile carboxylic acids (acrylic acid; 1-octanoic acid)	16	46–90	65	4–34	19	*h*
Deuterated standards (d_7-butyric acid)	1	—	85	—	14	*h*
NOVA						
Carboxylic acids (succinic acid; 2,4,5-trichlorophenoxyacetic acid)	6	42–87	64	2–45	18	*i*
Sulfonic acids (benzenesulfonic acid; 2-naphthalenesulfonic acid)	4	84–110	96	7–45	23	*i*
Misc. nonvolatile acids (benzenephosphoric acid; pentachlorophenol)	3	62–140	102	11–50	25	*i*
Deuterated standards (2-naphthalenesulfonic acid-$d_7 \cdot H_2O$)	1	—	110	—	14	*i*

Continued on next page.

Table II. Continued

Protocol Class Chemical Class (Examples)	Compounds Studied	Recovery Range (%)	Mean Recovery (%)	RSD Range (%)	Mean RSD (%)	Footnotes
SAM–PT and SAM–S						
Primary and tertiary amines (*n*-butylamine; tri-*n*-butylamine)	11	58–86	72	12–41	24	*j*
Secondary amines (diallylamine; 2-methylpiperidine)	6	40–98	63	20–53	36	*j*
Deuterated standards (d_9-butylamine; d_4-phenylethylamine; *N*-ethyl-2-fluorobenzylamine)	1	—	75	—	27	*j, k*

[a] Mean recoveries are for triplicate determinations from drinking water, spiked at 0.2–1.8 ppb (nominally 1 ppb), plus triplicate determinations from a 60–40 industrial–municipal waste water, spiked at 30–87 ppb (nominally 50 ppb).
[b] Mean recoveries are for triplicate determinations from drinking water, spiked at 0.8–1.2 ppb (nominally 1 ppb), plus triplicate determinations from a 60–40 industrial–municipal waste water, spiked at 40–63 ppb (nominally 50 ppb).
[c] Mean recoveries are for triplicate determinations from 60–40 industrial–municipal waste water only, spiked at 40–63 ppb (nominally 50 ppb).
[d] Mean recoveries are for triplicate determinations from drinking water, spiked at 0.5–5 ppb (nominally 1 ppb), using XAD–4 resin sorbent columns.
[e] Mean recoveries are for triplicate determinations from a 60–40 industrial–municipal waste water or, for about 25% of the total compounds, from reagent water spiked at 15–50 ppb (nominally 25 ppb), using batch liquid–liquid extraction, with cleanup included.
[f] Mean recoveries are for triplicate determinations from reagent water only, with cleanup step included. (Interferences prevented recovery determinations from waste water.)
[g] Mean recoveries are from triplicate determinations from drinking water only, spiked at 50–100 ppb (nominally 55 ppb). Recoveries were not determined from more complex waters.
[h] Mean recoveries are for triplicate determinations from drinking water, spiked at 0.3 ppb, plus triplicate determinations from a 60–40 industrial–municipal waste water, spiked at 120 ppb.
[i] Mean recoveries are for triplicate determinations from several industrial and municipal effluents.
[j] Mean recoveries are for triplicate determinations from three industrial and two municipal effluents spiked at 110 ppb, and including, in some cases, triplicate determinations from drinking water spiked at 35 ppb.
[k] Recoveries determined for only one (d_9-butylamine) of the three internal standards.

Literature Cited

1. Neal, R. A. *Environ. Sci. Technol.* **1983,** *17(3),* 113A.
2. Graham, J., Chapter 6 in this book.
3. Pellizzari, E. D. et al. *Master Analytical Scheme for Organic Compounds in Water. Part I: Protocols;* U.S. Environmental Protection Agency: Athens, GA, 1985; EPA-600/4-84-010a.
4. Pellizzari, E. D.; Garrison, A. W. *Prepr. Pap. Am. Chem. Soc. Div. Environ. Chem.* **1983,** *23(2),* 427–430.
5. Pellizzari, E. D. et al. *Experimental Development of the Master Analytical Scheme for Organic Compounds in Water. Part I: Text;* U.S. Environmental Protection Agency: Athens, GA, 1985; EPA-600/4-85-007a.

RECEIVED for review August 14, 1985. ACCEPTED February 13, 1986.

6

High-Performance Liquid Chromatography for Determination of Trace Organic Compounds in Aqueous Environmental Samples

Assessment of Current and Future Capabilities

Jeffrey A. Graham

Monsanto Agricultural Company, Life Sciences Research Center, St. Louis, MO 63198

A review of high-performance liquid chromatographic (HPLC) instrumentation, techniques, and methodologies for the determination of trace organic compounds in water is presented. The review includes approaches to sample cleanup or analyte isolation for those compounds likely to be candidates for analysis by HPLC. Column technology, as it contributes to the use of HPLC for trace organic analyses, is discussed. Finally, various techniques for quantitative and qualitative detection of analytes are discussed.

"WHAT IS IN THIS WATER?" This seemingly simple, yet horrendously complex question is being posed by increasing numbers of well-informed and not so well-informed individuals. In the recent U.S. Environmental Protection Agency (USEPA) report, "National Water Quality Inventory Report to Congress", two issues of national concern were identified: (1) pollution resulting from toxic substances and (2) contamination and depletion of ground water. More than half the states reported ground water problems stemming from waste disposal, landfill seepage, and excessive depletion of ground water supplies. A diverse collection of highly skilled scientists and engineers will be needed to effectively address these ubiquitous problems and to provide meaningful answers to the question "What is in this water?"

As more people ask this question, a greater burden will fall on the

0065-2393/87/0214/0097$10.50/0

shoulders of the chemist. The quality and quantity of the information that the chemist will be able to provide not only depend upon integrity and ability, but also upon the tools available to perform the task at hand. High-performance liquid chromatography (HPLC) is readily available to solve many of the problems encountered in providing reliable information on the identity and quantity of organic compounds in aqueous environmental samples.

The use of modern HPLC technology for the analysis of trace organic compounds in aqueous environmental samples has been steadily increasing. The mode of use primarily has been to quantify a single organic compound or group of closely related compounds. Thus far, the technology has not been commonly used to determine organic compounds in a given sample in a broad nonspecific sense. However, gas chromatography–mass spectrometry (GC–MS) has been used in this way. This situation exists because currently no detection system can be readily employed on-line with HPLC that can provide adequate sensitivity and structurally interpretable data.

In this chapter, the current and future capabilities of HPLC for the determination of trace organic compounds in aqueous environmental samples will be assessed. This assessment will include approaches to sample cleanup or analyte isolation for those species likely to be candidates for analysis by HPLC. Column technology, as it contributes to the use of HPLC for trace organic analyses, will be surveyed. Finally, detection of the compounds eluting from the system will be examined. The ultimate detector will always adequately identify and measure the compounds of interest.

Throughout this chapter, continued reference will be made to the article, "Principles of Environmental Analysis", by Keith et al. (*1*), which should be read by those who request that a method be developed or an analysis be performed. Without elaborating upon the contents, the two most important points of the article are the following: (1) "It cannot be assumed that the person requesting an analysis will also be able to define the objectives of the analysis properly." (2) "Analytical chemists must always emphasize to the public that the single most important characteristic of any result obtained from one or more analytical measurements is an adequate statement of its uncertainty interval (2)."

Sample Cleanup and Analyte Isolation

Assuming a properly preserved aqueous sample has been received from a particular location, and assuming that the sampling program has been designed to provide an adequate number of representative samples (*1*), what steps are necessary to prepare the sample for analysis by HPLC and associated detection systems? The answer depends upon that nature of the compounds that are targeted for the analysis as well as the characteristics of the matrix. Concentration or isolation techniques that exploit volatility, for example, purge and trap, headspace sampling, or closed-loop gas stripping, will unlikely be used because compounds amenable to these isolation modes should be more than adequately determined by gas chromatography (GC). Some form of isolation from the aqueous matrix is generally performed, and proper design of the step can serve to selectively or nonselectively concentrate the compounds of interest (*3*, *4*). Rosenfeld et al. (*5*) recently reported an approach to the simultaneous extraction and derivatization of organic acids from water. By using XAD-2 macroreticular resin impregnated with pentafluorobenzyl bromide, carboxylic acids and phenols could be extracted from water and derivatized with the products retained upon the resin. Innovative designs of methods using integrated steps such as this one serve to reduce the cost per analysis as well as to eliminate sample handling steps that can only contribute additional imprecision.

Liquid-liquid extraction and sorbent accumulation are the most commonly employed isolation-concentration methodologies. In their ideal forms, these methods readily extract or accumulate relatively hydrophobic compounds such as polynuclear aromatic hydrocarbons (PAHs) or polychlorinated biphenyls. However, when HPLC is likely to be the method of choice, the compounds are likely to be highly polar or ionic. In these common cases, adaptation of the traditional methodology can readily serve to carry out the necessary isolation.

Liquid-liquid extraction is used extensively and successfully (*6*). If the analytes are acidic or basic, as is often the case when HPLC is the analytical method selected, appropriate ionization suppression can be employed to affect the desired extraction. Back extraction of the analytes into an appropriately buffered aqueous volume can then serve to isolate and concentrate. Anionic and cationic surfactants, or so-called ion-pairing reagents, can be added prior to extraction to increase the partition coefficients of the trace organic ionic compounds.

The highly manual nature of the liquid-liquid extraction process has been one impetus for the development of alternative isolation-extraction techniques such as sorbent accumulation (*7–10*). Junk and Richard (*11*) noted that methodology for the accumulation of highly hydrophilic compounds has not kept pace with that associated with hydrophobic compounds. They modified XAD-4 resin to produce an anion exchanger and applied this resin to the accumulation of anionic and neutral organic compounds. Among their conclusions was that the anion-exchange resin was more convenient and efficient than either solvent extraction or neutral resin sorption. However, they also noted that because of the finite number of exchange sites, extremely large volumes of water could not be accumulated as has been done with neutral resins such as XAD-2, -4, and -7.

Many different ion-exchange materials are readily available and can be efficiently applied to the accumulation, isolation, or cleanup of water samples prior to analysis. The box below lists ion exchangers available from Bio-Rad (Richmond, CA). The wide variety of exchange moieties allows great flexibility in choosing a material with the optimum selectivity necessary for isolating a given analyte from a particular matrix. Most of the resins listed are based upon a relatively hydrophobic polymer backbone. This structure can either help the situation, as was the case reported by Junk and Richard (*11*), or it can present serious difficulties. If difficulties are encountered, the use of cellulose- or silica-based exchangers can generally solve the problem. Recently, silica derivatized not only with ion-exchange moieties but also with a wide variety of other useful ligands has become available.

Derivatized-silica bonded phases are available in disposable 1-, 3-, and 5-mL columns as follows: nonpolar phases (ethyl, octyl, cyclohexyl,

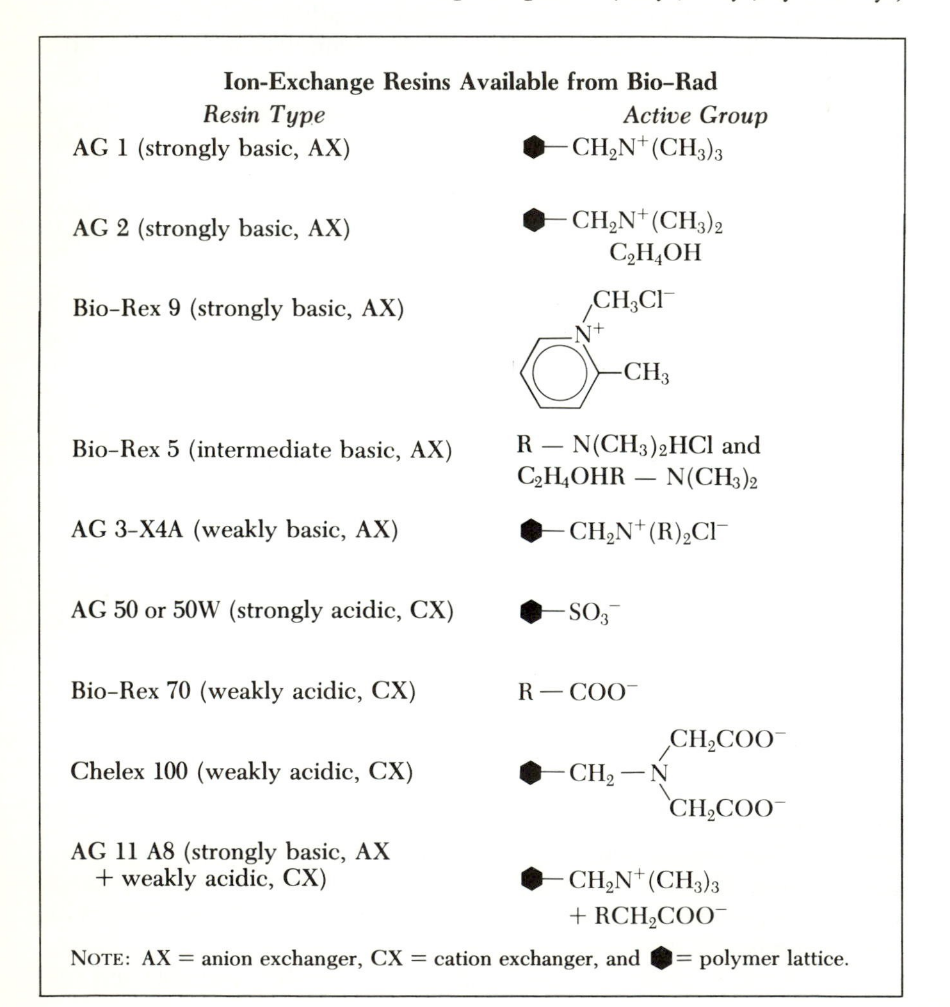

Ion-Exchange Resins Available from Bio-Rad

Resin Type	*Active Group*
AG 1 (strongly basic, AX)	⬣—$CH_2N^+(CH_3)_3$
AG 2 (strongly basic, AX)	⬣—$CH_2N^+(CH_3)_2$ C_2H_4OH
Bio-Rex 9 (strongly basic, AX)	pyridinium ring with N^+—CH_3Cl^- and —CH_3
Bio-Rex 5 (intermediate basic, AX)	R — $N(CH_3)_2HCl$ and C_2H_4OHR — $N(CH_3)_2$
AG 3-X4A (weakly basic, AX)	⬣—$CH_2N^+(R)_2Cl^-$
AG 50 or 50W (strongly acidic, CX)	⬣—SO_3^-
Bio-Rex 70 (weakly acidic, CX)	R — COO^-
Chelex 100 (weakly acidic, CX)	⬣—CH_2 — N(CH_2COO^-)(CH_2COO^-)
AG 11 A8 (strongly basic, AX + weakly acidic, CX)	⬣—$CH_2N^+(CH_3)_3$ + RCH_2COO^-

NOTE: AX = anion exchanger, CX = cation exchanger, and ⬣ = polymer lattice.

phenyl, octadecyl); polar and weak ion-exchange phases (cyanopropyl, aminopropyl, propylcarboxylic acid, diol, nonbonded silica); and strong ion-exchange phases (benzenesulfonic acid, quarternary amine). These columns are mounted in a vacuum manifold capable of processing 10 columns simultaneously. The manufacturer claims a 75% reduction in the time it takes to carry out liquid–liquid extraction. Uniform derivatization of the silica, which also has a narrow particle and pore size distribution, yields consistent and reproducible extractions and cleanups. West and Day (*12*) used octadecylsilica (ODS) for extraction of the herbicide fluridone, 1-methyl-3-phenyl-5-(3-trifluoromethylphenyl)-4-(1*H*)-pyridinone, in pond water from 1–100 ng/mL. Recoveries averaged better than 94% throughout the range. In similar work, Saner and Gilbert (*13*) compared liquid–liquid extraction to accumulation of the pesticide Dursban, *O*,*O*-diethyl-*O*-(3,5,6-trichloro-2-pyridyl)phosphorothioate, from contaminated environmental waters. Figure 1 shows the chromatograms of the water samples enriched on ODS and extracted with methylene chloride. The disparity in the chromatograms was explained by the presence of a large fraction of clay particulates contained within these samples. The drastically higher concentration of Dursban found with the ODS enrichment of sample number 3 in Figure 1 was explained by the fact that the enrichment cartridges acted as mechanical filters so that the clay particulates could effectively accumulate at the head of the cartridge. The subsequent methanol rinse, which was designed to elute the Dursban that was enriched on the ODS, also served to co-extract the Dursban adsorbed on the clay particulates.

Kirkland (*14*) described the reduction in detectable quantities for HPLC-based analyses by loading large samples on the analytical column. Later Little and Fallick (*15*) used this concept and concentrated nonpolar organics from river water directly on the analytical column. The term *trace enrichment* was coined to describe the injection–sampling technique. Since then, trace enrichment has been used to describe a wide variety of accumulation techniques, both directly on the analytical column and off-line on disposable columns such as those described earlier. Saner (*16*) extensively reviewed trace-enrichment techniques that were classified as direct injection, off-line enrichment, or on-line enrichment. Direct injection of a larger than typical volume on the analytical column has as its primary goal greater sensitivity. Off-line enrichment serves to concentrate the analytes as well as to provide a certain degree of cleanup. Off-line enrichment is readily carried out with the derivatized-silica disposable extraction columns described earlier. It is not limited to these by any means. The off-line enrichment could be carried out with a column inserted in line with the analytical column at the appropriate time. On-line enrichment has as its goal

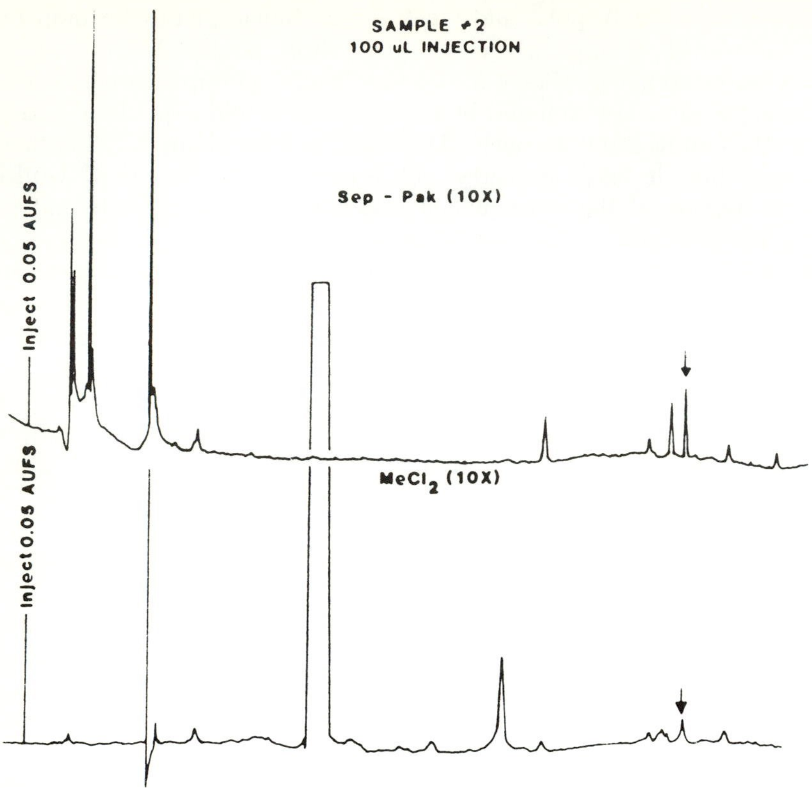

Figure 1. Sep–Pak adsorption and methylene chloride extraction of Dursban (see arrows) from water samples 2 and 3. (The 10× denotes the concentration factor for Dursban above the original concentration in the water samples.) (Reproduced with permission from reference 13. Copyright 1980 Marcel Dekker.)

automation of the loading of the sample into the HPLC instrumentation, sample concentration, analyte isolation, and removal of interferences. The extensive time–cost savings, as well as the reduction in systematic error due to substantially minimized operator sample handling, have many implications.

Graham and Garrison (*17*) evaluated on-line trace enrichment for the determination of trace organic compounds in aqueous environmental samples. These workers were primarily interested in nonvolatile and thermally labile compounds that were not readily analyzed by GC methodology. A 2-mm i.d. × 70-mm long stainless steel precolumn was packed with 30–75 μm diameter octadecyl-derivatized silica. This precolumn was substituted for the sample loop in a conventional, high-pressure, six-port valve. Water samples, 10–100 mL, were pumped directly on the precolumn. After loading, the valve was switched to

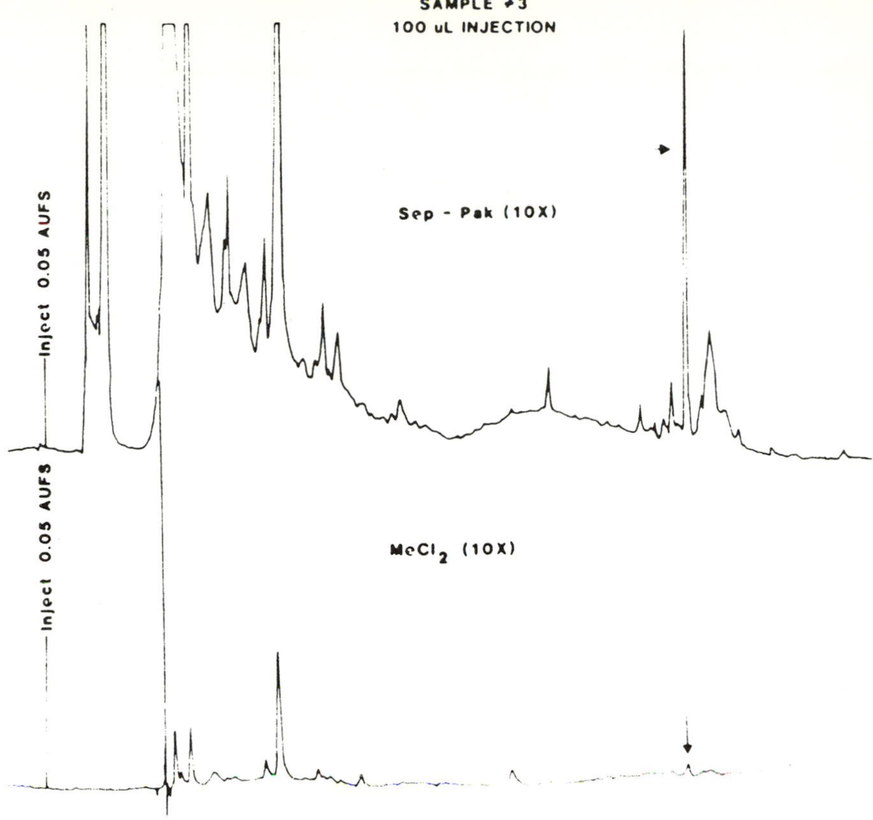

Figure 1. Continued.

place the precolumn in series with the solvent delivery system and the analytical column. Gradient elution was used for the subsequent analytical separation, and UV spectrometric absorption at two wavelengths was used for detection. Figure 2 shows results obtained from fortified and blank drinking water (Athens, GA). Recoveries from the drinking water for five of the seven fortified compounds were not statistically different from the recoveries obtained when the same compounds were fortified in distilled-deionized water. The other two compounds were completely lost. Graham and Garrison (*17*) also evaluated an approach to enrich ionic organic compounds. Acidic organic compounds could be efficiently recovered from water samples with the same ODS precolumn if tetrabutylammonium phosphate and an appropriate buffer were added to the water sample prior to loading of the precolumn. By using model compounds, such as *m*-chlorobenzoic acid and *o*-chlorophenylacetic acid, recoveries typically decreased with

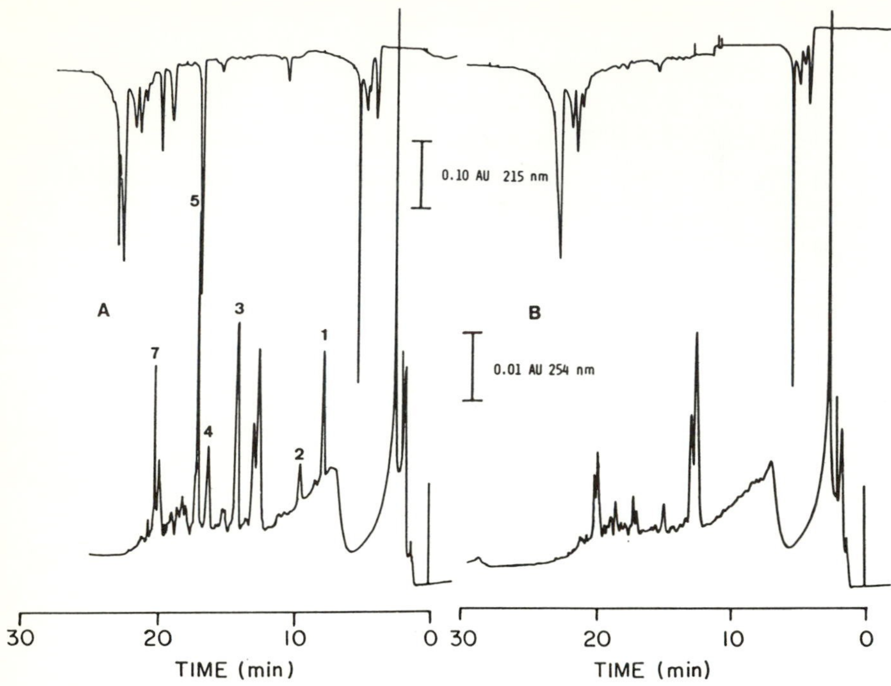

Figure 2. A, chromatogram from elution of 100 mL of enriched drinking water (Athens, GA) fortified with 1, 0.26 µg of caffeine; 2, 0.050 µg of m-*nitroaniline; 3, 0.44 µg of atrazine; 4, 0.75 µg of 2,6-dichloroaniline; 5, 0.43 µg of* N-*nitrosodiphenylamine; 6, 0.85 µg of decafluorobiphenyl (not detected); and 7, 0.41 µg of disperse red dye 13. B, chromatogram from elution of 100 mL of enriched drinking water (Athens, GA). Conditions for both enrichments: 100-mL samples enriched on an ODS-packed precolumn at 5 mL/min. Analytical separation was on Partisil-10, ODS-2, 250-mm × 4.6-mm i.d. column. Mobile-phase gradient was 10% to 90% (v/v) acetonitrile in distilled-deionized water at 5%/min, and flow rate was 1.0 mL/min. Detection was at 254 nm. (Reproduced with permission from reference 17. Copyright 1981 Ann Arbor Science.)*

real samples compared with fortifications in distilled and deionized water.

Miller et al. (*18*) evaluated an approach to simultaneous on-line enrichment of neutral, acidic, and basic organic compounds by serially connecting three precolumns containing ODS, a strong anion-exchange derivatized silica, and a strong cation-exchange derivatized silica, respectively. Table I lists the recoveries of representative compounds that were fortified into 10 mL of drinking water (Athens, GA) and river water (Oconee River) and subsequently on-line enriched on the three serial precolumns. Also listed in Table I are the deviations found with each matrix with respect to the recoveries obtained from fortified

Table I. Recoveries of Standards from Drinking and River Waters

Standard (Level)	*Recovery from Athens Drinking Water*[a]	*AD from DDIW Water*[b]	*Recovery from Oconee River Water*[a]	*AD from DDIW Water*[b]
ODS Group				
Pyrene (11 ppb)	81 (3)	+5	74 (4)	−2
Caffeine (24 ppb)	94 (0)	−7	88 (2)	−13
Atrazine (30 ppb)	89 (1)	+7	80 (3)	−2
N-Nitrosodiphenylamine (24 ppb)	90 (3)	−7	97 (6)	0
2,6-Dichloroaniline (52 ppb)	90 (7)	+2	89 (3)	+1
SCX Group				
m-Nitroaniline (6 ppb)	91 (2)	+2	86 (2)	−3
Aniline (42 ppb)	93 (4)	0	90 (2)	−3
SAX Group				
Pentachlorophenol (20 ppb)	74 (4)	−15	86 (2)	−3
Phenoxyacetic acid (87 ppb)	84 (9)	−3	91 (10)	+4
p-Chlorobenzoic acid (120 ppb)	87 (6)	0	90 (7)	+3
2-Naphthalenesulfonic acid (33 ppb)	95 (10)	+3	90 (3)	−2

NOTE: All values are percentages.

[a] Values in parentheses are standard deviations.

[b] AD denotes absolute deviation, and DDIW denotes distilled-deionized water.

SOURCE: Reproduced with permission from reference 18.

distilled-deionized water. Chromatograms of elution of each of the three precolumn types are shown in Figures 3, 4, and 5. Distilled-deionized water, to which fulvic acid had been added at a level of 5 mg of carbon per liter of water, was also used as a test sample matrix. Recoveries of the various compounds from fortified water containing fulvic acid are listed in Table II, and a representative chromatogram is shown in Figure 6. The interference of the fulvic acid was minimal. Corresponding limits of detection were calculated for each test compound on the basis of the recoveries obtained with both distilled-deionized water and fulvic acid fortified water. These are listed in Table III.

Goewie et al. (*19*) developed an organometallic-silica bonded phase for the selective retention of phenylurea herbicides and anilines from water. Seven-micrometer diameter silica was derivatized with 2-amino-1-cyclopentene-1-dithiocarboxylic acid (ACDA), resulting in Structure **I**. Capacity factors of phenylurea herbicides and corresponding anilines were measured on the ACDA-silica and ACDA-metal-loaded silica. The platinum-loaded material was found to selectively retain the anilines. Anilines could be eluted with acetonitrile, but not with methanol or tetrahydrofuran, because of the strength with which acetonitrile forms complexes with platinum and thus displaces the anilines. Application of the ACDA-Pt precolumn in series with an ODS-silica precolumn for

Table II. Recoveries of Standards from Water in the Presence of 5 mg of Carbon per Liter as Fulvic Acid

Standard (Level)	*Recovery*[a]	*AD from DDIW Water*[b]
ODS Group		
Pyrene (5.5 ppb)	79 (3)	+3
Caffeine (14 ppb)	97 (4)	−4
Atrazine (15 ppb)	82 (3)	0
N-Nitrosodiphenylamine (14 ppb)	93 (2)	−4
2,6-Dichloroaniline (26 ppb)	90 (3)	+2
SCX Group		
Quinoline (42 ppb)	92 (2)	−5
p,p′-Methylenedianiline (7 ppb)	82 (2)	+7
Aniline (38 ppb)	88 (5)	−5
SAX Group		
Pentachlorophenol (15 ppb)	79 (1)	−10
Phenoxyacetic acid (98 ppb)	90 (5)	+3
p-Chlorobenzoic acid (44 ppb)	95 (1)	+8
2-Naphthalenesulfonic acid (32 ppb)	91 (2)	−1

NOTE: All values are percentages.
[a] Values in parentheses are standard deviations.
[b] AD denotes absolute deviation, and DDIW denotes distilled-deionized water.
SOURCE: Reproduced with permission from reference 18.

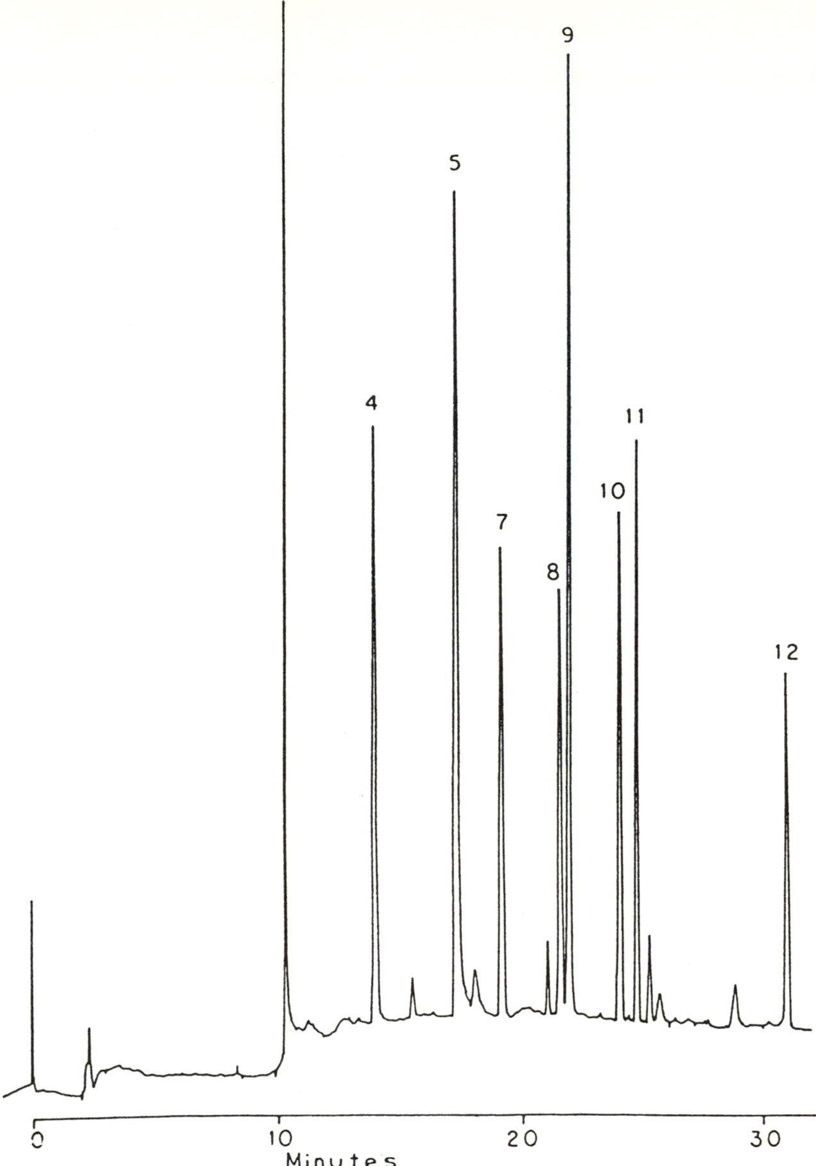

Figure 3. Standards recovered from 10 mL of distilled-deionized water on an ODS precolumn. Peak identities: 4, 0.14 μg of caffeine; 5, 0.20 μg of pentachlorophenol; 7, 0.061 μg of m-*nitroaniline; 8, 0.15 μg of atrazine; 9, 0.40 μg of quinoline; 10, 0.26 μg of 2,6-dichloroaniline; 11, 0.14 μg of* N-*nitrosodiphenylamine; and 12, 0.055 μg of pyrene. Conditions for concentration: 10-mL sample enriched on an ODS-packed precolumn. Analytical separation was on Zorbax ODS, 250-mm by 4.6-mm i.d. column. Mobile-phase gradient was 100% pH 7, 0.1 M acetate buffer for 2 min followed by ramp to 90% acetonitrile/10% pH 7, 0.1 M acetate buffer (v/v) in 20 min at 1.0-mL/min flow rate. Detection was at 254 nm. (Reproduced with permission from reference 18.)*

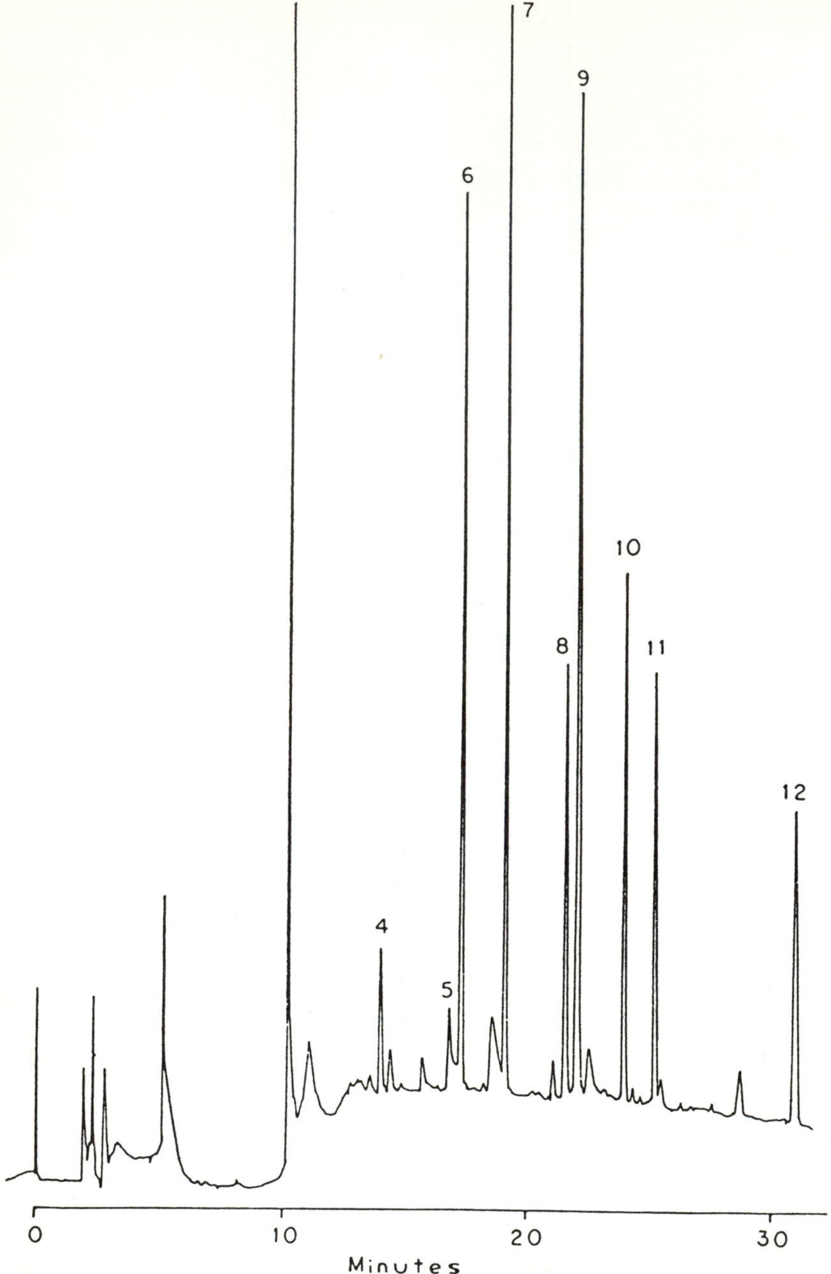

Figure 4. Standards recovered from 10 mL of distilled-deionized water on a strong-cation-exchanger packed precolumn. Peak identities: 4, 0.14 μg of caffeine; 5, 0.20 μg of pentachlorophenol; 6, 0.42 μg of aniline; 7, 0.061 μg of m-*nitroaniline; 8, 0.15 μg of atrazine; 9, 0.40 μg of quinoline; 10, 0.26 μg of 2,6-dichloroaniline; 11, 0.14 μg of* N-*nitrosodiphenylamine; and 12, 0.055 μg of pyrene. Conditions for concentration, analytical separation, mobile-phase gradient, and detection were the same as in Figure 3. Reproduced with permission from reference 18.)*

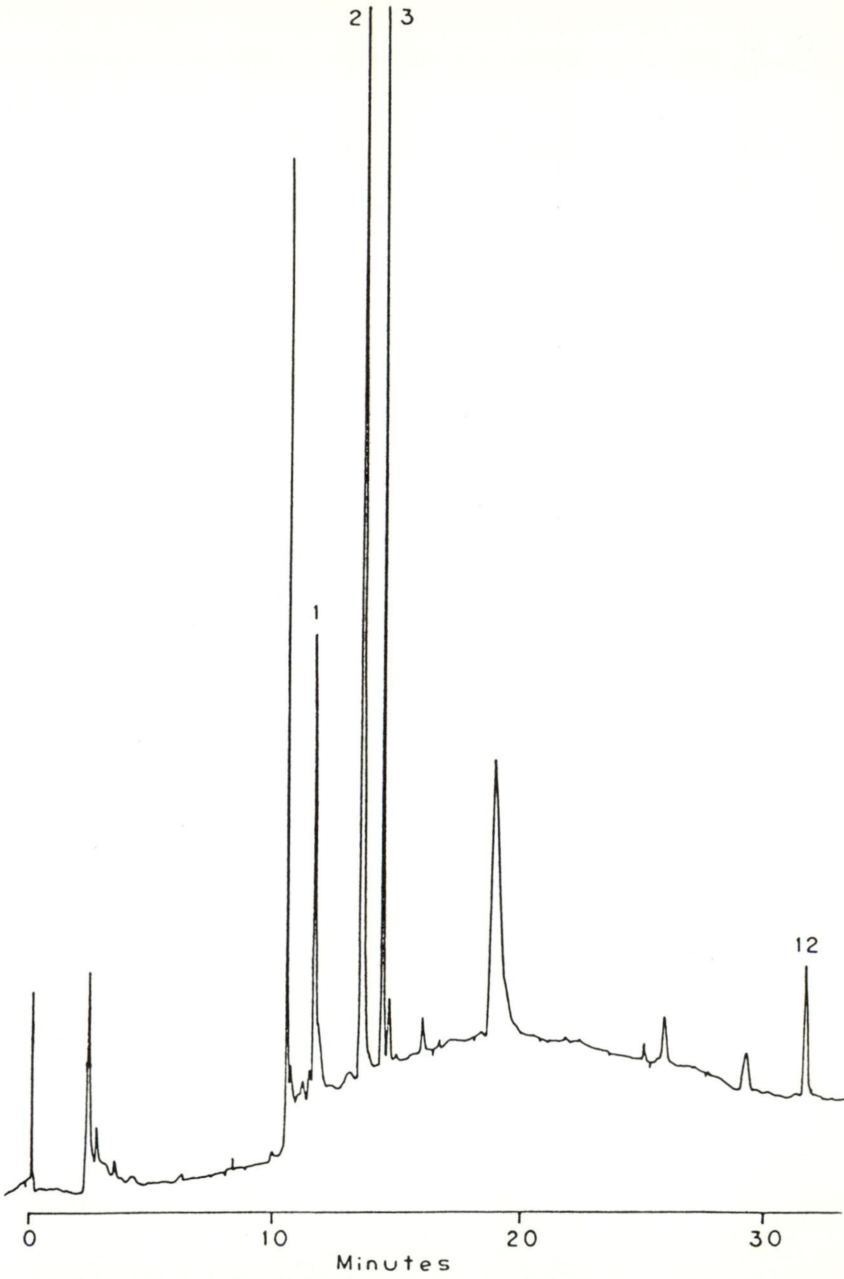

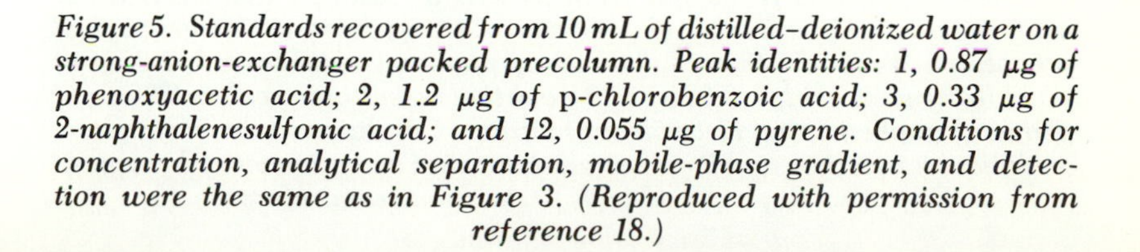

Figure 5. Standards recovered from 10 mL of distilled–deionized water on a strong-anion-exchanger packed precolumn. Peak identities: 1, 0.87 μg of phenoxyacetic acid; 2, 1.2 μg of p*-chlorobenzoic acid; 3, 0.33 μg of 2-naphthalenesulfonic acid; and 12, 0.055 μg of pyrene. Conditions for concentration, analytical separation, mobile-phase gradient, and detection were the same as in Figure 3. (Reproduced with permission from reference 18.)*

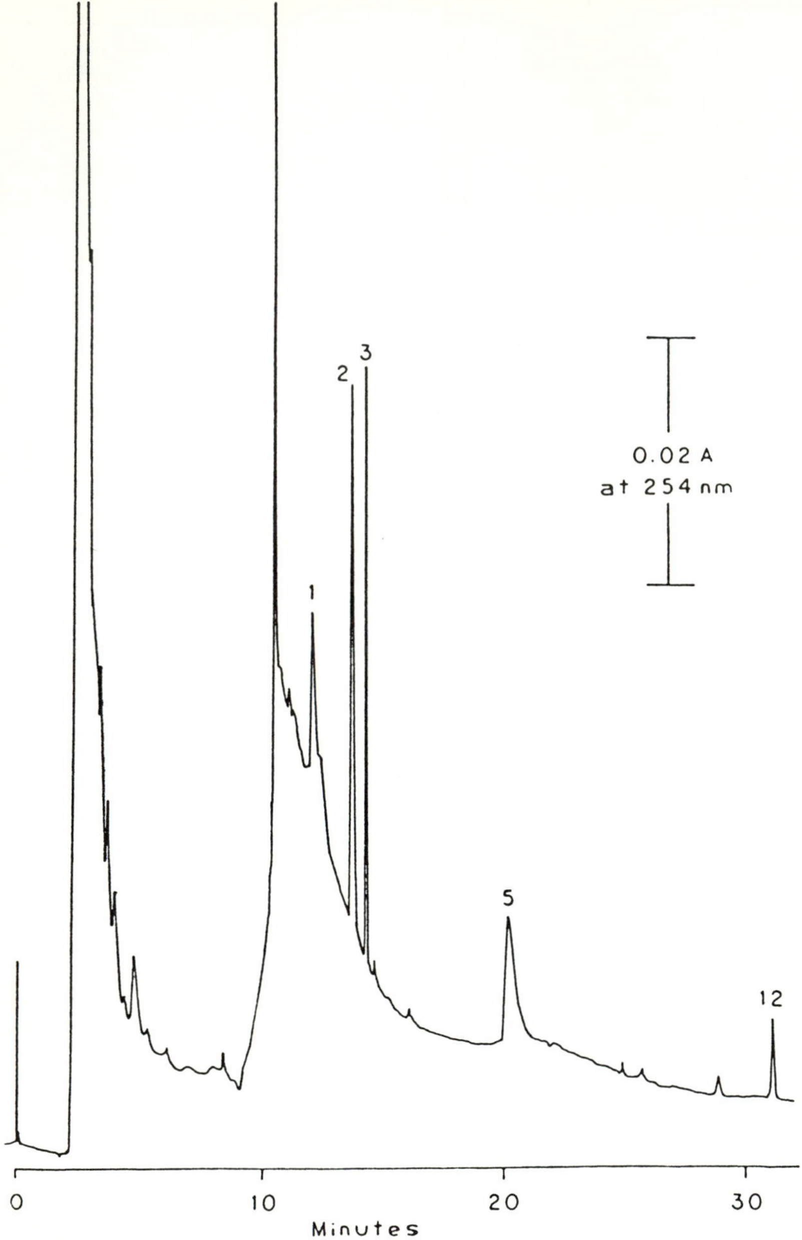

Figure 6. Standards recovered from 10 mL of distilled-deionized water containing 5 mg/L carbon as fulvic acid on a strong-anion-exchanger packed precolumn. Peak identities: 1, 0.87 μg of phenoxyacetic acid; 2, 1.2 μg of p-*chlorobenzoic acid; 3, 0.33 μg of 2-naphthalenesulfonic acid; 5, 0.20 μg of pentachlorophenol; and 12, 0.055 μg of pyrene. Conditions for concentration, analytical separation, mobile-phase gradient, and detection were the same as in Figure 3. (Reproduced with permission from reference 18.)*

Table III. Method Detection Limits

Standard	*DDIW Water (ppb)*[a]	*5 mg of Carbon/L of Fulvic Acid (ppb)*
ODS Group		
Pyrene	0.01	0.04
Caffeine	0.02	0.06
Atrazine	0.1	0.1
N-Nitrosodiphenylamine	0.03	0.05
2,6-Dichloroaniline	0.06	0.2
SCX Group		
Quinoline	0.07	0.2
p,p'-Methylenedianiline	0.02	0.1
Aniline	0.2	0.5
SAX Group		
Pentachlorophenol	0.1	0.1
Phenoxyacetic acid	1	1
p-Chlorobenzoic acid	0.3	0.7
2-Naphthalenesulfonic acid	0.2	0.2

[a] DDIW denotes distilled-deionized water.
SOURCE: Reproduced with permission from reference 18.

the enrichment of the phenylurea herbicides and anilines in river water was demonstrated.

Analytichem International has taken trace enrichment a step further with the development of the Analytichem Automated Sample Preparation LC Module, AASP LCM. This unit uses derivatized silica in specially configured cassettes on which the sample is off-line enriched by using an inexpensive vacuum manifold. After enrichment, the cassette is placed in the unit, which holds up to 10 cassettes containing 10 samples each, and the enriched samples are eluted directly on the analytical column. Provisions allow for early- and late-eluting bands to be switched to waste. The unit also provides for external control of

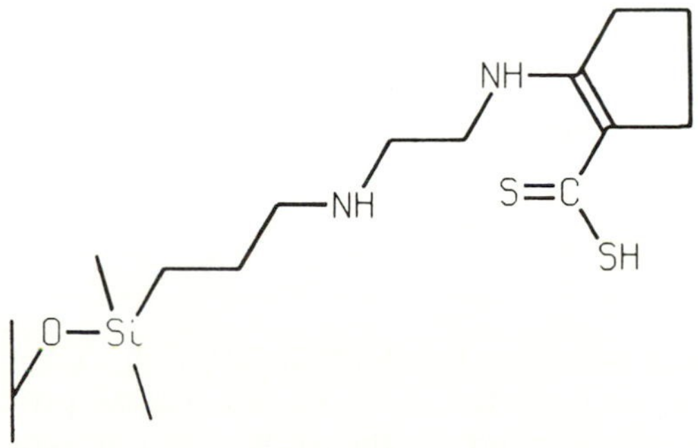

Structure **I.**

other ancillary equipment, for example, integrators, auto-zero controls on detectors, and data acquisition systems.

The sample cleanup or analyte isolation step is often the major contributor to the overall systematic error of the analysis. In most cases, the variation in the recovery of the analyte, or class of analytes, determines the magnitude of the precision of the analysis. Thus, these steps must be adequately validated (*1*). Because check material is usually unavailable, fortifications to determine the effect of the matrix on recovery have to be made. This procedure generally requires prescreening a sample to determine an approximate analyte level so that the fortification level is not drastically greater than or less than the analyte level. Validation of the spiking or fortification procedure is sometimes necessary.

Analytical Separation

The column is the heart of any chromatography-based analytical method. Developments in column technology during the past 15 years have contributed to the high performance of HPLC. Initial developments focused upon reduction of the mean particle diameter to achieve the column efficiencies predicted by Giddings (*20*). The trend toward smaller particles still exists. For example, in the recent review on liquid chromatography (LC) by Majors et al. (*21*), the reviewers examined the use of various particle diameters reported in 369 randomly selected articles in 10 popular chromatographically oriented journals published in 1982–83. This survey found that 2.6% of the papers reported using particle diameters greater than 10 μm, 54% reported using 10-μm packing, 7.5% reported using 7.5-μm packing, 35% reported using 5-μm packing, and 1.4% reported using 3-μm packing. The reviewers noted that the 10-μm diameter packings were the workhorses, but the 5-μm diameter packings were gaining popularity. This trend should continue, but many using the 5-μm diameter packings may change to the 3-μm diameter packings for various reasons. What impact does particle diameter, as well as other column dimensions and various operating parameters, have on using HPLC for trace analysis? Examination of eq 1 derived by Karger et al. (22) for the dependence of various HPLC operating parameters on the lower limit of detection provides some useful insight:

$$\mathrm{LDC} = 5(\sqrt{2\pi})(1/N)(1/\mathrm{SNR})(1/V_i)[V_o(1 + k')] \quad (1)$$

where LDC is the lowest detectable concentration, N is the number of theoretical plates, SNR is the ratio of the signal produced by the detection of the solute band to the root mean square noise of the detection system, V_i is the volume of sample injected, V_o is the void volume of the column and associated tubing, and k' is the capacity factor of the eluting solute band. Eq 1 assumes isocratic separation, a Gaussian profile solute band, no interfering or coeluting bands, and a minimum detectable signal of five times the root mean square noise of the detection system.

Eq 1 predicts the LDC to decrease inversely proportionally to the volume of sample injected. Trace enrichment, discussed earlier, has as one of its primary goals the effective increase of V_i by accumulation of large volumes and direct elution of the total accumulated sample onto the analytical column. This job must be achieved without substantially increasing the variance of the injection band contribution to the overall solute band variance. Solute bands with small capacity factors, k', result in greater sensitivity. For complex multicomponent separations, greater detectability for late eluting solute bands is achieved by using gradient elution for which the effective k' of all but the most strongly retained solutes is small when the solute band elutes (*23*, *24*). Also, a substantial reduction in the time for separation occurs with respect to isocratic elution. In most cases, the signal-to-noise ratio for a particular detection system is set, and little can be done to substantially affect greater sensitivity for a given analyte or class of analytes. However, use of post- and precolumn reactions to produce a solute with greater sensitivity for a given detection system appears to be increasing.

Before column efficiency, N, and V_o are discussed, eq 2 will be introduced:

$$\sigma(\text{total}) = \sigma(\text{inj}) + \sigma(\text{col}) + \sigma(\text{det}) + \sigma(\text{tub}) + \sigma(\text{elec}) + \sigma(\text{da}) \quad (2)$$

Eq 2 states that all components of the system—injector, column, detector flow cell, associated tubing and fittings, detector electronics, and associated signal recording and/or data acquisition systems—contribute to the total solute band variance or width. This equation also indicates that improvements in all areas must be made to fully realize an overall improvement. Thus, use of 3-μm packings is often pointless unless these highly efficient columns are connected to appropriate injection, detection, and data recording or acquisition systems. As a result, manufacturers of HPLC instrumentation are producing new systems designed to operate with the smaller diameter packings or microbore columns or are providing retrofits for previously designed instrumentation. This situation may be one reason a relatively low percentage of researchers reported using the 3-μm diameter packings (*21*).

What can be done with HPLC columns to optimize the parameters N and V_o to achieve the lowest detectable concentration limits? The answer is not simple or straightforward because of the interaction of other components in the system. This interaction is one reason why many have the mistaken conception that microbore columns, with very low V_o, offer operating conditions that provide lower limits of detection than the more conventional 4.6-mm i.d. columns. The detectability comparison must be made with careful control of many parameters, for example, sample injection volume, mobile-phase flow velocity, detector flow-cell geometry, band variance, detector-cell path length, and signal-to-noise ratio. Cooke et al. (25) recently made a critical comparison of microbore to standard HPLC columns with respect to sample detectability using UV-absorbance detectors. Cooke et al. (25) showed that for the case in which the sample size is exceedingly small, that is, less than approximately 15 μL, microbore columns can provide three to six times lower limits of detection. Figure 7 shows three series of three chromatograms from Cooke et al. (25) that illustrate when the sample injection volume is held constant at 0.5 μL, the 1-mm i.d. column provides six times the sensitivity of the 4.6-mm i.d. column. The volume of the detector flow cell used to record each chromatogram should be noted. The 1- and 2-mm i.d. columns need to use a smaller volume detector cell to realize the greater response. This situation relates back to eq 2 because in cases in which the flow cell was standard (first series of chromatograms in Figure 7), the variance contribution of the flow cell was a much greater amount relative to the overall solute band variance. As such, the small solute band variances produced by the 1- and 2-mm i.d. columns were negated by the larger solute band variance contribution of the standard flow cell. When the detector flow-cell variance is reduced to a contribution in line with those contributions from the 1- and 2-mm i.d. columns, substantial improvement in the total solute band variance or width is realized (second series of chromatograms

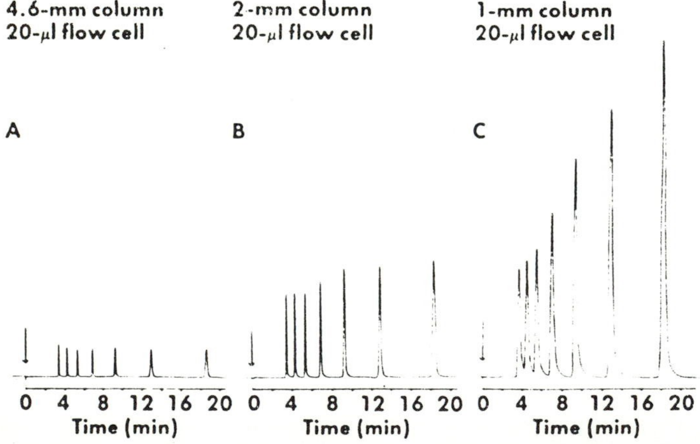

Figure 7. Detectability comparisons among 4.6-mm, 2.0-mm, and 1.0-mm i.d. columns using 20- and 5-μL flow cells at a full-scale detector range of 0.05 absorbance units. The same seven-component sample was used for all injections but was diluted by one-half for chromatograms A, B, and C. (Reproduced with permission from reference 25. Copyright 1984 LC Magazine.)

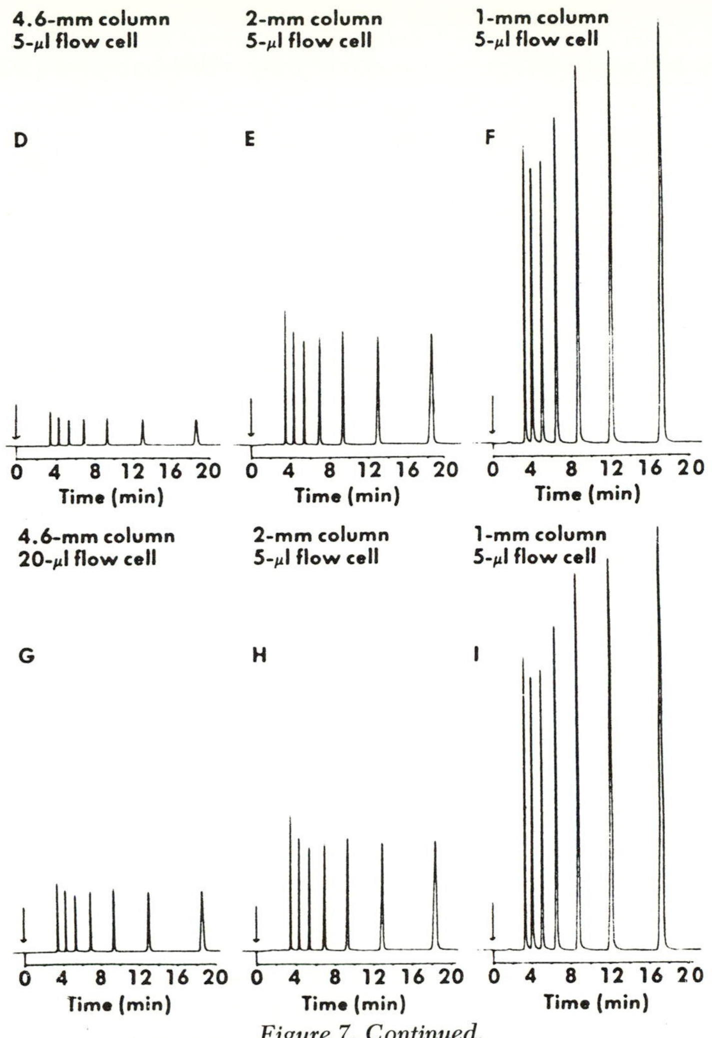

Figure 7. Continued.

in Figure 7). However, this improvement requires reduction of the detector flow-cell optical path length by one-half.

What happens when the sample injection volume constraint is removed? Calculations can be made from the data presented by Cooke et al. (25). For these calculations, the data required is that relating to the maximum sample volumes that can be injected on the various columns while a given number of theoretical plates is maintained. This data, shown in Table IV, was calculated by using eq 3:

$$V_{mi} = (Q/N_r)(1+k')V_oB \quad (3)$$

where V_{mi} is the maximum injection volume maintaining a constant given efficiency, Q is the fractional loss in plates resulting from the

Table IV. Maximal Injection Volumes

Column i.d. (mm)	N_r = *20,000*	N_r = *10,000*	N_r = *5000*
1	np	7	14
2	12	36	60
4.6	85	200	320

NOTE: N_r is the number of plates necessary for the separation; np denotes not possible. All values, except column i.d., are given in microliters.
SOURCE: Reproduced with permission from reference 25. Copyright 1984 *LC Magazine*.

maximized injection volume, N_r is the number of plates necessary for the separation, k' is the solute capacity factor, V_o is the column void volume, and B is a parameter relating the profile of the sample injection. Eq 3 is derivable from an equation that expresses extracolumn solute band broadening by using component variances (*26*). Table IV also shows the maximum injection volumes allowed on the various columns by assuming a solute capacity factor of 4, $k' = 4$, and $B = 2.45$ (*25, 27*).

By assuming that a proportional increase in the amount of sample injected results in a proportional increase in the detector response for the solute band of interest, the detector response for chromatogram I in Figure 7 will increase 14 times when the maximum sample volume of 7 μL is injected. However, for the 4.6-mm i.d. column, the detector response will increase 400 times when the maximum sample volume of 200 μL is injected. By taking into account the relative detector responses for the 0.5-μL injection, at the maximum sample injection volumes, the 4.6-mm i.d. column with the 20-μL detector flow cell will produce approximately five times the detector response of the 1-mm i.d. column with the 5-μL flow cell. In most cases, studies can be designed to provide excess sample because aqueous environmental samples are seldom limited with respect to volume.

When one is deciding what column geometry is optimal for trace analysis with unlimited sample volume, two additional points should be evaluated. First, to what extent does the analysis require accurate and reproducible injections? Strict performance specifications may eliminate microbore columns from consideration. The accuracy and reproducibility of injection systems that deliver 0.1-, 0.2-, and 0.5-μL samples have not been adequately characterized. Second, if the analyte of interest requires postcolumn derivatization, construction of a postcolumn reaction system that is compatible with the exceedingly small band volumes characteristic of microbore columns may be extremely difficult, but not impossible. Apffel et al. (*28*) developed and evaluated both packed-bed and open tubular postcolumn reactors for use with 1-mm i.d. analytical columns. Catecholamines were postcolumn derivatized with *o*-phthal-aldehyde and detected spectrofluorometrically. The 5-μm particle diameter packed beds were found to have variance contributions between 50- and 100-μm i.d. open tubular reactors. The reaction times for 50-μm i.d. open tubular reactors were reported to be prohibitively high, even for short reactions, whereas the 5-μm particle diameter glass beads packed in the bed reactor had to be replaced every 40–50 h because of their solubility in the high pH solution, the *o*-phthalaldehyde reaction solution. Because the other components of the HPLC system had nonoptimized variance contributions, no significant decrease in resolution was apparent when the open tubular reactor was compared with the packed-bed reactor, which had three times greater V_o.

However, by no means are microbore HPLC columns nonoptimal when sample volumes are limited. Because of the substantially reduced mobile-phase volumes necessary to carry out a given separation, microbore columns are more easily interfaced to many useful detection systems, for example, electron-capture detectors, nitrogen-specific thermionic detectors, and mass spectrometers.

Additionally, the combination of trace enrichment and microbore columns can effectively increase the maximum sample volume injectable without seriously degrading efficiency. Slais et al. (*29*) evaluated this combination for the determination of polynuclear hydrocarbons and chlorinated phenols in water. By using reversed-phase HPLC and amperometric detection, Slais et al. (*29*) reported lower limits of detection from 20 to 280 ng/L of water (parts per trillion) when 1-mL sample enrichments were carried out directly on the analytical microbore column.

The wide variety of bonded stationary phases that are available to carry out a given HPLC separation allows tremendous amounts of flexibility. In the recent review of column LC by Majors et al. (*21*), the reviewers examined the different modes of usage by randomly examining articles in nine scientific journals traditionally containing chromatographic methodology. Table V lists the results for the periods 1980–81 and 1982–83. The most drastic change observed was a 6% reduction in the usage of ion-exchange columns and an 11% increase in the use of reversed-phase columns. The reviewers noted that these two important points may indicate that reversed-phase ion-pairing techniques may be making a strong intrusion into the separation of ionic and ionizable compounds.

Most of the articles in the scientific literature dealing with analytical methods using HPLC technology for the analysis of aqueous environmental samples employ the reversed-phase mode. This finding is not surprising, as the data in Table V attest to. The flexibility of the reversed-phase mode suits environmental analyses. Many metabolites of environmentally important organic compounds are ionic or ionizable and ideally suited to separation by ion-pair techniques. A variety of

Table V. HPLC Column-Mode Use

Chromatographic Mode	*Column Usage (%) (1982–83)*	*Column Usage (%) (1980–81)*
Reversed phase	72	61
C-18	54	48
C-8	9.6	8.7
C-2	2.5	1.0
Phenyl	2.1	1.1
Other	3.2	1.7
Normal bonded phase (cyano, amino)	5.9	6.8
Liquid solid (adsorption)	10	14
Ion exchange	6.3	12
Cation	2.3	5.7
Anion	4.0	6.5
Size exclusion	5.5	6.0
Organic	1.9	2.2
Aqueous	3.6	3.8

SOURCE: Reproduced from reference 21. Copyright 1984 American Chemical Society.

organic modifiers are compatible with the reversed-phase mode. Methanol, acetonitrile, and tetrahydrofuran are the most popular because of low absorbance in the UV. These three organic compounds can be combined in different ratios to provide separation selectivity unattainable with a single solvent.

Identifying the optimal mobile-phase composition for a given separation problem can be a frustrating experience for the inexperienced chromatographer. If the separation problem involves many components, identifying the optimal composition necessary to obtain acceptable resolution of all components may be trying for even the most experienced chromatographer. As a result, many workers have focused efforts upon development of statistically based optimization procedures. These procedures ideally could be carried out with automated instrumentation. Berridge (*30*) has reviewed many of the techniques for automated optimization of HPLC separations. These procedures are not without problems. Many require that the identity of each peak be known. This requirement can be impossible to achieve when the separation problem involves a complex matrix with hundreds of unidentified components, as is so often the case with environmental samples. All optimization approaches require that the resolution of the components of interest be measured in some way so as to quantify the quality of the chromatogram. Debets et al. (*31*) reviewed quality criteria for two-component and multicomponent chromatographic separations used in optimization schemes during 1963–83 and concluded that all suffer from severe shortcomings in expressing the quality of multicomponent chromatographic separations.

Nonetheless, these optimization techniques can be quite useful. The approach developed by Glajch et al. (*32*) is based on the solvent-selectivity triangle developed by Snyder (*33*) and a statistical Simplex design. The initial solvent strength is estimated from the results of a linear water-to-methanol gradient. Then, seven combinations of mobile phases having different selectivities but the same solvent strength are run by using a Simplex design. On the basis of the results of the seven experiments, an overlapping resolution map is constructed, which allows selection of the optimum mobile-phase composition. This optimization approach can be carried out for both reversed- and normal-phase isocratic modes, as well as gradient-elution modes. Goldberg and Nowakowska (*34*) adapted the overlapping resolution map approach of Glajch et al. (*32*) to the separation of ionic organic compounds by reversed-phase ion-pairing HPLC. This approach is useful because of the complicated dependence of retention volumes on the mobile-phase variables in ion-pairing separations, that is, concentration of ion-pair reagent, pH, and ionic strength. This optimization scheme has been incorporated in the Du Pont Instruments Sentinel System, which carries out the seven mobile-phase selectivity experiments unattended. Subsequent calculations are performed, and the optimum mobile-phase composition is presented.

Detection

In most situations, sensitive and selective detection of eluting solute bands is a necessity for successful environmental analysis. Unfortunately, such detection is currently the weakest link in HPLC technology. As mentioned earlier, this situation has resulted in a somewhat limited application of HPLC to the analysis of aqueous environmental samples (*3*). The role of HPLC technology in recent years has been to provide separation and detection of a specific analyte or class of analytes. Thus, HPLC technology cannot readily answer the question "What is in this water?" Instead, HPLC technology has been delegated to answer questions such as "Is aldicarb in this sample of ground water at a concentration greater than 5 μg/L?"

In the recent review on column LC by Majors et al. (*21*), a survey on the use of detector types was carried out in the same manner as that for the use of the various separation modes already mentioned. The results, shown in Table VI, were tabulated for the periods 1982–83 and 1980–81. The increased use of electrochemical and refractive index detectors is significant in these data. The authors speculated that the increased use of refractive index detectors resulted from the increased number of publications on the separation of carbohydrates. The increased use of electrochemical detection is probably a function of many different factors: cell designs that are easier to use, expanding sales

Table VI. HPLC Detector Use

Detector Type	*Use (%) (1982–83)*	*Use (%) (1980–81)*
Absorbance	67	71
UV fixed	22	28
UV filter	7.4	8.7
Spectrophotometric	37	34
Diode array	0.2	—
Fluorescence	14	15
Filter	5.6	15
Spectrophotometric	8.4	—
Refractive index	7.2	5.5
Electrochemical	9.1	4.4
Other	2.6	4.4

SOURCE: Reproduced from reference 21. Copyright 1984 American Chemical Society.

forces, and the never-ending search for greater sensitivity and selectivity.

Not surprisingly, the UV-absorption spectrometric detector was reported as being cited the most. New developments by HPLC instrument manufacturers have led to improved signal processing electronics providing better signal-to-noise ratios, flow cells and electronic time constants compatible with microbore and 3-μm particle diameter packed columns, better detector and lamp power supply designs, and microprocessor control providing programmable wavelength changes as well as partial spectrum recording on-the-fly of eluting peaks. A good example of some of these developments is the diode array HPLC detectors produced by a variety of manufacturers. The nature of the diode array is such that its signal-to-noise ratio for a small band, that is, 6 nm, is not as great as the conventional single-wavelength detectors. The great use of the diode array is a result of its ability to acquire a UV or UV and visible spectrum. Thus, the suspected identity of a specific solute band in a multicomponent chromatogram can be readily confirmed. Fell and Scott (*35*) recently reviewed the applications of rapid-scanning multichannel detectors in HPLC and thin-layer chromatography (TLC). The potential uses of three-dimensional projections of absorbance, wavelength, and time were discussed in the context of applications to pharmaceutical, bioanalytical, and forensic fields. Figure 8 is an example of the type of data readily obtained from the commercially available Hewlett-Packard Model 1040A diode array HPLC detector.

If appropriate cleanup is accomplished and if the analyte or analytes have sufficient absorption, UV-absorption detection is a highly reliable method of quantification. However, the UV-absorption detector

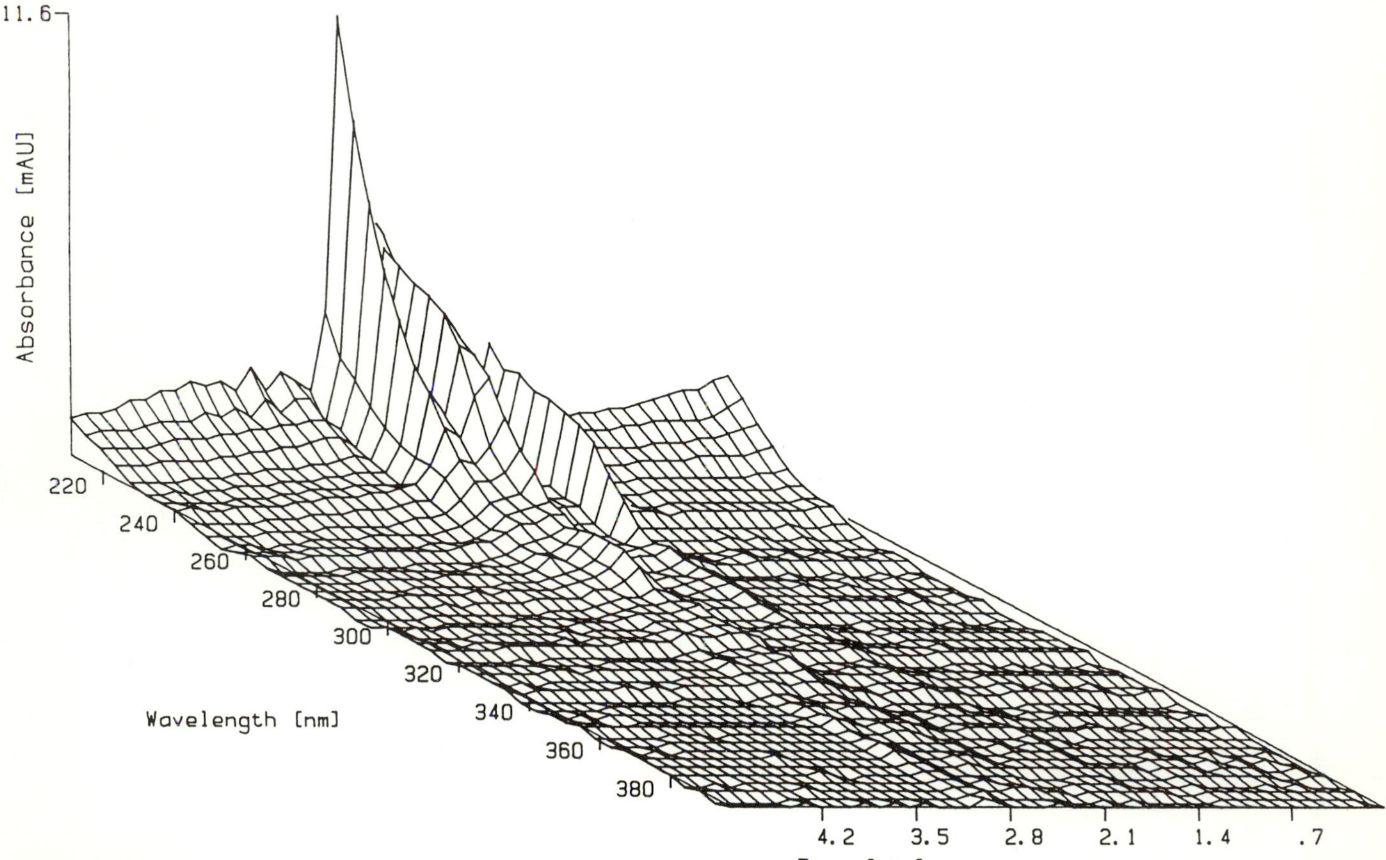

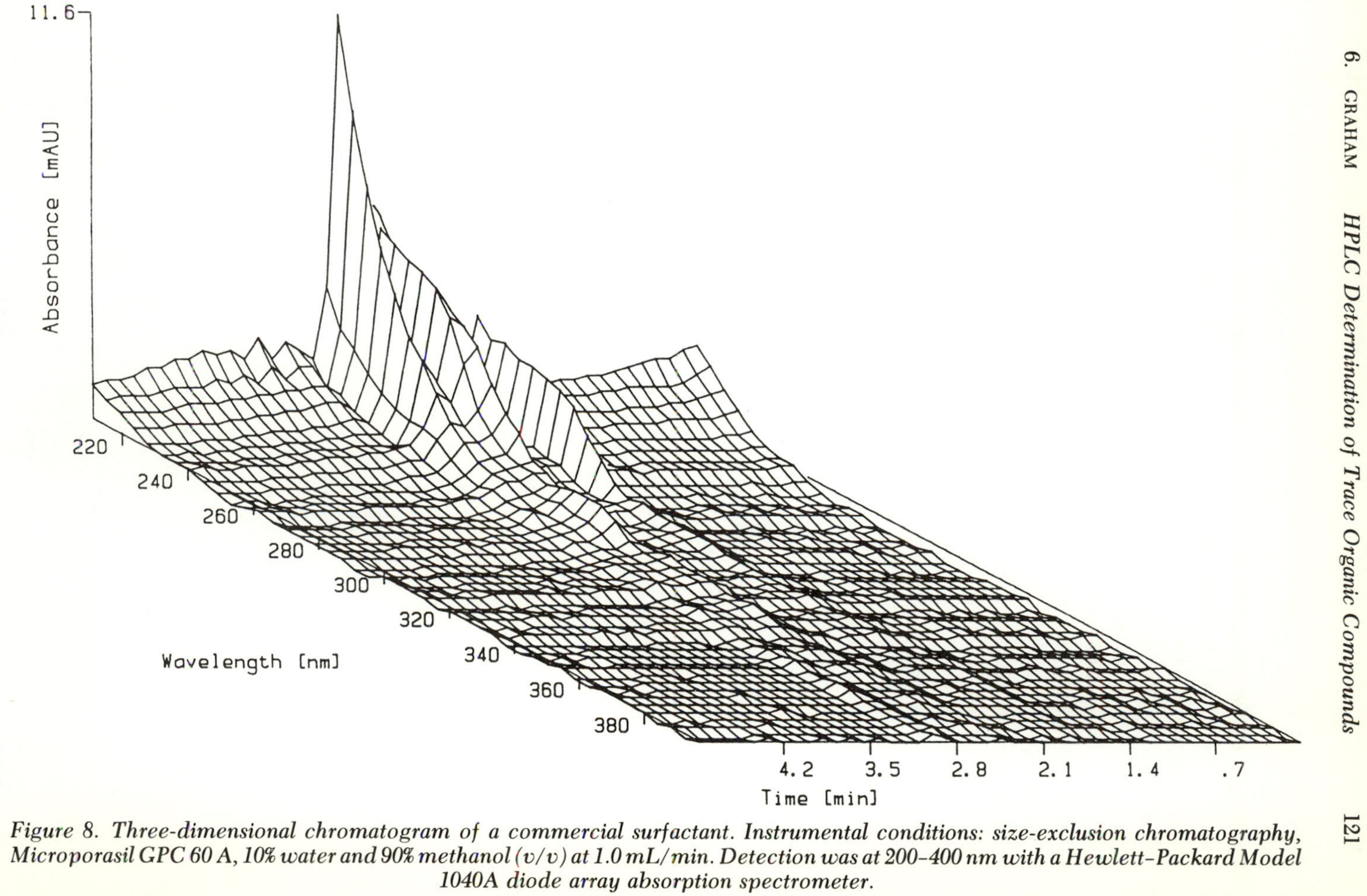

Figure 8. Three-dimensional chromatogram of a commercial surfactant. Instrumental conditions: size-exclusion chromatography, Microporasil GPC 60 A, 10% water and 90% methanol (v/v) at 1.0 mL/min. Detection was at 200–400 nm with a Hewlett–Packard Model 1040A diode array absorption spectrometer.

typically does not provide adequate selectivity unless the analyte or analytes have sufficient absorption at wavelengths greater than approximately 280 nm. Otherwise, chromatograms typically appear with large unresolved humps having protruding analyte peaks, especially when extremely low limits of detection are the goal, that is, less than 10 μg/L. A good example of application of UV-absorption detection for a wide variety of compounds is that reported by Pinkerton (*36*). Thirty-two priority pollutants, mostly common solvents, phthalates, and PAHs, were determined by direct injection of 2 mL of a filtered water sample, gradient elution from a reversed-phase column, and UV detection at 202 nm. Recoveries from a water sample fortified with 19 of the 32 compounds at 80–100 μg/L (ppb) ranged from a low of 71% for benz[*a*]anthracene to 103% for dimethyl phthalate. The overall average recovery was 92%; the relative standard deviation was 8%. Lower limits of detection were reported to be 10 μg/L for some compounds.

Fluorescence detection, because of the limited number of molecules that fluoresce under specific excitation and emission wavelengths, is a reasonable alternative if the analyte fluoresces. Likewise, amperometric detection can provide greater selectivity and very good sensitivity if the analyte is readily electrochemically oxidized or reduced. Brunt (*37*) recently reviewed a wide variety of electrochemical detectors for HPLC. Bulk-property detectors (i.e., conductometric and capacitance detectors) and solute-property detectors (i.e., amperometric, coulometric, polarographic, and potentiometric detectors) were discussed. Many flow-cell designs were diagrammed, and commercial systems were discussed.

Two analytical methods for priority pollutants specified by the USEPA (*38*) use HPLC separation and fluorescence or electrochemical detection. Method 605, 40 CFR Part 136, determines benzidine and 3,3-dichlorobenzidine by amperometric detection at +0.80 V, versus a silver/silver chloride reference electrode, at a glassy carbon electrode. Separation is achieved with a 1:1 (v/v) mixture of acetonitrile and a pH 4.7 acetate buffer (1 M) under isocratic conditions on an ethyl-bonded reversed-phase column. Lower limits of detection are reported to be 0.05 μg/L for benzidine and 0.1 μg/L for 3,3-dichlorobenzidine. Method 610, 40 CFR Part 136, determines 16 PAHs by either GC or HPLC. The HPLC method is required when all 16 PAHs need to be individually determined. The GC method, which uses a packed column, cannot adequately individually resolve all 16 PAHs. The method specifies gradient elution of the PAHs from a reversed-phase analytical column and fluorescence detection with an excitation wavelength of 280 nm and an emission wavelength of 389 nm for all but three PAHs: naphthalene, acenaphthylene, and acenaphthene. As a result of weak fluorescence, these three PAHs are detected with greater sensitivity by UV-absorption detection at 254 nm. Thus, the method requires that fluorescence and UV-absorption detectors be used in series. Lower limits of detection for the HPLC–fluorescence–UV method range from 5 μg/L for acenaphthylene to 0.04 μg/L for benzo[*a*]pyrene, benz[*a*]anthracene, benzo[*k*]fluoranthene, and benzo[*b*]fluoranthene.

Quite often, neither of the three aforementioned detection systems will provide adequate sensitivity and selectivity to complete the analysis with specified lower limits of detection. In these cases, an increasingly popular trend has been to either pre- or postcolumn derivatize the analyte or analytes to convert them to species that can be adequately detected with reliable and commercially available detectors. Tables VII and VIII list a variety of reagents for the derivatization of specific organic moieties that provide increased UV absorption and fluorescence, respectively (*39*).

Derivatization reactions for enhanced UV-absorption detection generally rely upon a shift of the absorption toward the visible region where greater specificity can generally be found. Lauren and Agnew (*40*) determined 2,4-dinitrophenyl ether derivatives of the phenolic metabolites of carbofuran, 2,3-dihydro-2,2-dimethyl-7-benzofuranyl methylcarbamate, by reversed-phase HPLC and UV-absorption detection at 280 nm. Derivatization to produce easily oxidized or reduced products that can be electrochemically detected either coulometrically or amperometrically is becoming increasingly popular. Tanaka et al. (*41*) reported a novel ferrocene reagent for the precolumn labeling of amines for electrochemical detection.

Precolumn derivatization is readily carried out in miniature glassware designed specifically for this purpose. Another approach to

Table VII. Derivatization Reagents for UV Detection

Functional Group	*Reagent*
Acids, carboxylic acids	*O*-*p*-Nitrobenzyl-*N*,*N*′-diisopropylisourea (PNBDI)
	p-Bromophenacyl bromide (PBPB)
Alcohols	3,5-Dinitrobenzoyl chloride (DNBC)
Aldehydes	*p*-Nitrobenzyloxyamine (PNBA)
Amines, primary and secondary	*N*-Succinimidyl-*p*-nitrophenylacetate (SNPA)
	3,5-Dinitrobenzoyl chloride (DNBC)
Amino acids	*N*-Succinimidyl-*p*-nitrophenylacetate (SNPA)
Isocyanates	*p*-Nitrobenzyl-*N*-*n*-propylamine hydrochloride (PNBPA)
Ketones	*p*-Nitrobenzoylamine hydrochloride (PNBA)
Phenols	3,5-Dinitrobenzoyl chloride (DNBC)
Thiols	3,5-Dinitrobenzoyl chloride (DNBC)

Table VIII. Derivatization Reagents for Fluorescence Detection

Functional Group	*Reagent*
Acids, carboxylic acids	4-Bromomethyl-7-methoxycoumarin (Br–MC)
Aldehydes, hydrazine	1-Dimethylaminonaphthalene-5-sulfonyl hydrazine (dansyl hydrazine)
Amines, primary and secondary	7-Chloro-4-nitrobenzo-2-oxa-1,3-diazole (NBD chloride)
	1-Dimethylaminonaphthalene-5-sulfonyl chloride (dansyl chloride)
	o-Phthalaldehyde (OPA)
Ketones	1-Dimethylaminonaphthalene-5-sulfonyl hydrazine (dansyl hydrazine)
Phenols	1-Dimethylaminonaphthalene-5-sulfonyl chloride (dansyl chloride)
Thiols	7-Chloro-4-nitrobenzo-2-oxa-1,3-diazole (NBD chloride)

precolumn derivatization is the application of some automated device to carry out the reaction without manual manipulation of the samples. Hodgin et al. (*42*) designed and evaluated a device that simultaneously mixes *o*-phthalaldehyde with a sample containing amino acids and immediately injects the fluorescent products into an HPLC. Separation of the amino acids is completed in 30 min with a peak-area precision of better than ±2.00% relative standard deviation. Calibration curve linearity extended from 10 to 500 pmol of injected sample. A lower limit of detection of 100 fmol of injected sample was reported.

Precolumn derivatization with a solid-phase derivatizing precolumn has also been reported. Xie et al. (*43*) applied polymeric permanganate oxidations of alcohols and aldehydes for the production of UV-absorbing species.

Precolumn derivatization is often inadequate for dirty samples. In these cases, application of a postcolumn reaction detection system will often suffice. Deelder et al. (*44*) and van der Wal (*45*) have examined different configurations for postcolumn reactors and defined optimal selections on the basis of reaction time and type and effect on resolution and sensitivity. Both studies preferred the packed-bed reactor to the open tubular reactors when conventional column geometries were employed for separation, that is, 4.6 mm i.d. × 15 or 25 cm.

Nondek et al. (*46*) reported an innovative approach to the analysis of *N*-methylcarbamates in river water using postcolumn reaction detection. Separation of the underivatized *N*-methylcarbamates was carried out on a reversed-phase column hooked directly to a bed reactor packed with Aminex A-28, a tetraalkylammonium anion-exchange resin. The packed bed catalytically base-hydrolyzed the carbamates and liberated methylamine, Figure 9. Liberated methylamine was subsequently derivatized with *o*-phthalaldehyde to produce a fluorophore, which was detected fluorometrically. Precision better than 2% relative standard deviation and subnanogram lower limits of detection were reported. Figure 10 shows chromatograms of river water and river water fortified with five different *N*-methylcarbamates.

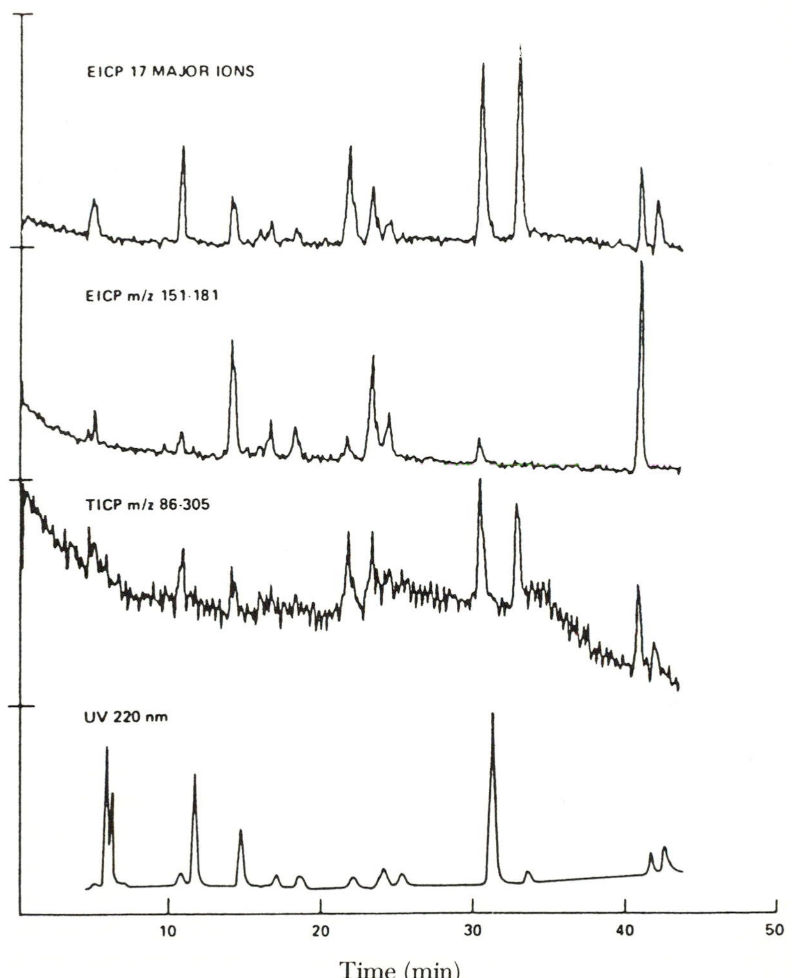

Figure 9. Illustration of the use of extracted ion-current profiles obtained with LC–MS, moving-belt interface, for the detection of carbamate and other pesticides. Top, extracted ion-current profile for 17 major ions; second from top, extracted ion-current profile for m/z = *151 to* m/z = *181; third from top, extracted ion-current profile for* m/z = *86 to* m/z = *305; bottom, UV absorption detection at 220 nm. (Reproduced with permission from reference 53. Copyright 1982 Preston Publications.)*

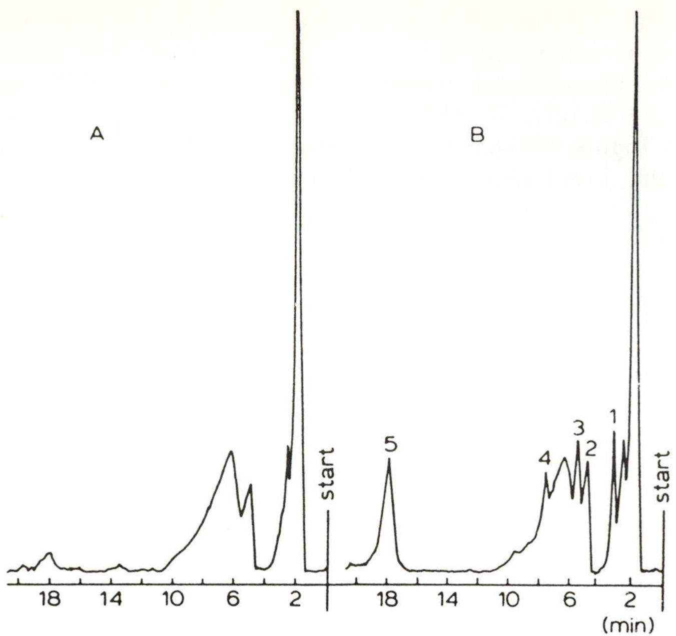

Figure 10. Reaction chromatograms for A, Amstel river water and B, Amstel river water fortified with 3 ng of aldicarb (peak 1); 3 ng of methomyl (peak 2); 5 ng of propoxur (peak 3); 5 ng of carbaryl (peak 4); and 10 ng of methiocarb (peak 5). Conditions: 150-mm × 4.6-mm i.d. column packed with Spherisorb ODS; mobile phase of 50% water and 50% methanol (v/v) at a flow rate of 1.0 mL/min; 60-mm × 4.6-mm i.d. reactor column packed with Aminex A-28; reaction temperature of 100 °C; OPA reagent flow rate of 30 μL/min; detection with Perkin-Elmer Model 204A fluorescence spectrometer; excitation wavelength of 340 nm; emission wavelength of 455 nm. (Reproduced with permission from reference 46. Copyright 1983 Elsevier Scientific Publishers.)

Postcolumn photochemical reactions are another approach to the detection problem. High-intensity UV light, generally provided by a Hg or Zn lamp, photolyzes the HPLC effluent, which passes through a Teflon (*47*) or quartz tube. The photolysis reaction determines the nature of the subsequent detection. If the compound has a UV chromophore, such as an aromatic ring, and an ionizable heteroatom, such as chlorine, then the products of the reaction can be detected conductometrically. Busch et al. (*48*) have examined more than 40 environmental pollutants for applicability to detection with photolysis and conductance detection. Haeberer and Scott (*49*) found the photoconductivity approach superior to precolumn derivatization for the determination of nitrosoamines in water and waste water. The primary limitation of this detection approach results from the inability to use mobile phases that contain ionic modifiers, that is, buffers and ion-pairing reagents. This approach also requires the use of an aqueous organic mobile phase; thus, normal-phase mode operation is eliminated. Therefore, the separation chemistry that can be employed is extremely limited.

Postcolumn photochemical reactions that detect photolysis products by modes other than conductivity have been reported. Werkhoven-Goewie et al. (*50*) determined chlorophenols by photolytic dehalogenation to fluorescent products. Chlorophenols in river water were on-line preconcentrated on PRP-1, a styrene-divinylbenzene copolymeric sorbent packed precolumn. The effluent of the analytical column was passed through Teflon tubing and photolyzed with a 200-W Xe-Hg lamp. The photolysis products flowed directly into the spectrofluorometer for detection. Figure 11 is a diagram of the system that was used by Werkhoven-Goewie et al. (*50*). Graham (*51*) developed a postcolumn photochemical reaction detector specific for photolabile organofluorine compounds. The system used air segmentation of the column effluent, which was pumped through a 1-m quartz coil that encircled a 450-W high-pressure Hg lamp. Photolyzed fluoride was detected colorimetrically with an alizarin-lanthanum complex. Ding and Krull (*52*) developed a system to selectively detect organic thiophosphate agricultural chemicals such as malathion and parathion. Detection of the photolysis products was accomplished amperometrically with dual,

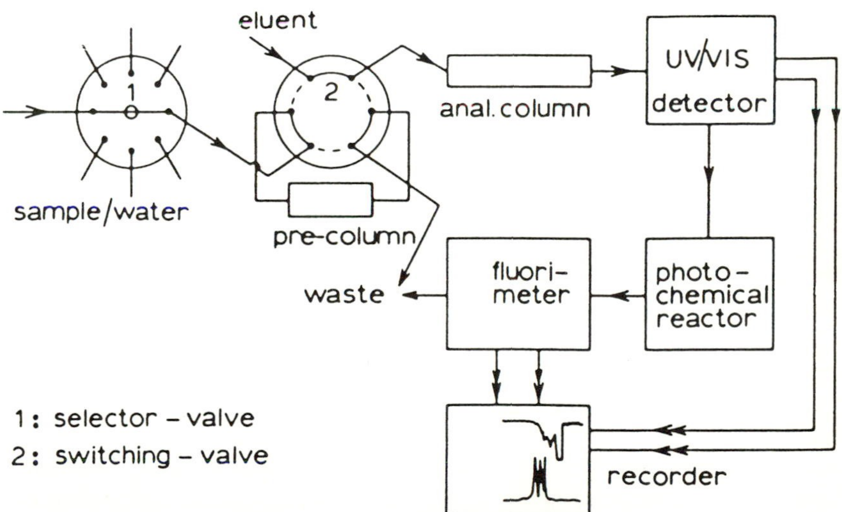

Figure 11. Diagram of system used to carry out automated preconcentration and analysis of water samples. Detection accomplished with UV absorption and photochemical dehalogenation with subsequent fluorescence detection of the photochemical reaction products. (Reproduced with permission from reference 50. Copyright 1982 Vieweg.)

parallel, glassy carbon electrodes. None of the photolysis products were identified. Lower limits of detection were reported for a number of agricultural chemicals and ranged from 20 to 80 μg/L. However, because of the excessive variance contribution of the Teflon tube used as the photoreactor, it was estimated that the photolysis–electrochemical detector produced peak heights from one to two orders of magnitude smaller than those produced by a UV detector.

The ultimate detector for any chromatographic technique provides high sensitivity and qualitative information regarding the nature of the eluting solute. GC–MS has been the mainstay of trace organic analysis for the past decade. Many researchers have sought to interface MS with HPLC in the past few years. However, the incompatibility of liquids with the operating conditions of MS has presented serious problems. The earliest successful attempts at interfacing LC and MS were obtained by depositing the LC effluent on a moving belt, evaporating the excess solvent, and ionizing the nonvolatile material that remained on the belt. Wright (*53*) evaluated this technique for the detection of carbamate pesticides separated by HPLC. However, even in the selective ion-monitoring mode, limits of detection were generally lower with UV-absorption detection. Figure 9 shows a comparison of chromatograms obtained for the carbamates displayed as extracted ion-current profiles (EICP), total ion-current profiles (TICP), and UV-absorption detection. The identities of the individual peaks are shown in the TICP chromatogram in Figure 12.

The solvent elimination problem became less of a problem with the commercialization of microbore columns. Hayes et al. (*54*) studied gradient HPLC–MS using microbore columns and a moving-belt interface. The heart of the system was the spray deposition device designed to be compatible with microbore-column flow rates. Nebulization of the eluent was found to be applicable to a variety of mobile-phase compositions and thus was readily compatible with gradient elution. Figure 13 shows a comparison of UV detection with that obtained with the HPLC–MS system. Applications of this system were demonstrated on water from coal gasification processes.

The next major advance in LC–MS interfacing was developed by Blakely and Vestal (*55, 56*). To circumvent the solvent elimination problem, Blakely et al. (*55*) developed the thermospray interface that operates with aqueous–organic mobile phase at typical 4.6-mm i.d. column flow rates, 1–2 mL/min. The thermospray technique works well with aqueous buffers. This feature is an advantage when the versatility of the reversed-phase mode is considered. In fact, with aqueous buffers, ions are produced when the filament is off. A recent improvement in the thermospray technique is the development of an electrically heated vaporizer that permits precise control of the vaporization (*56*). This

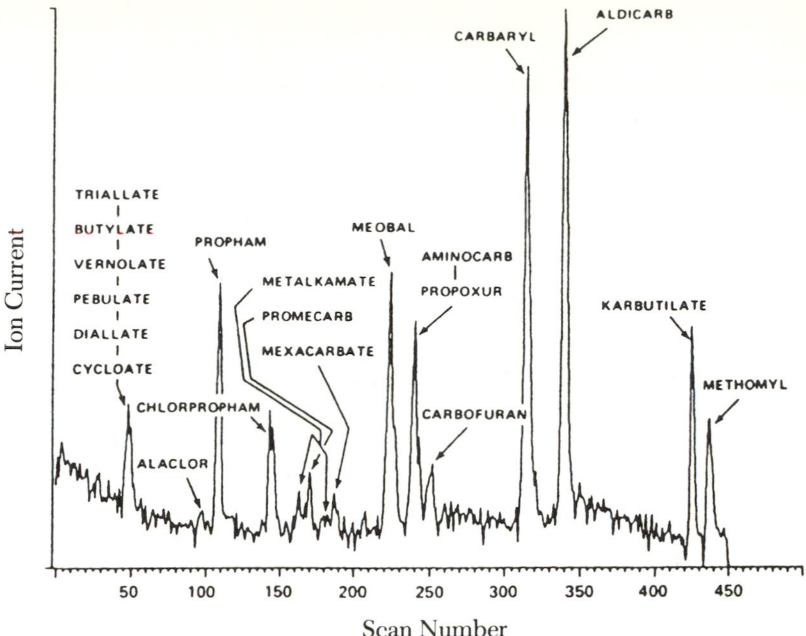

Figure 12. Total ion-current profiles obtained with LC–MS, moving-belt interface, for the detection of carbamates and other pesticides. (Reproduced with permission from reference 53. Copyright 1982 Preston Publications.)

vaporizer is also readily adaptable to any quadrupole MS that is equipped to operate under chemical ionization conditions.

Recently, Willoughby and Browner (57) reviewed the combination of LC with MS. They commented that the trend of future developments would be (1) simple inexpensive interfaces compatible with many different MS systems and (2) specialized LC–MS systems designed around new ionization sources intended for relatively nonvolatile compounds.

A number of other detection systems have been engineered for compatibility with microbore and open tubular columns. McGuffin and Novotny (58) developed a dual-flame thermionic detector for use with microbore columns, packed capillary columns, and open tubular columns. The total microcolumn effluent is aspirated into the primary flame. Nitrogen- and phosphorus-containing compounds are then selectively detected by the change in the conductivity of the secondary flame in the presence of a glass bead with a high rubidium content. Figure 14 shows a comparison of the detection of three organophosphorus pesticides by the thermionic detector and a UV-absorption detector. Sepaniak et al. (59) applied laser-excited fluorescence and thermal-lens

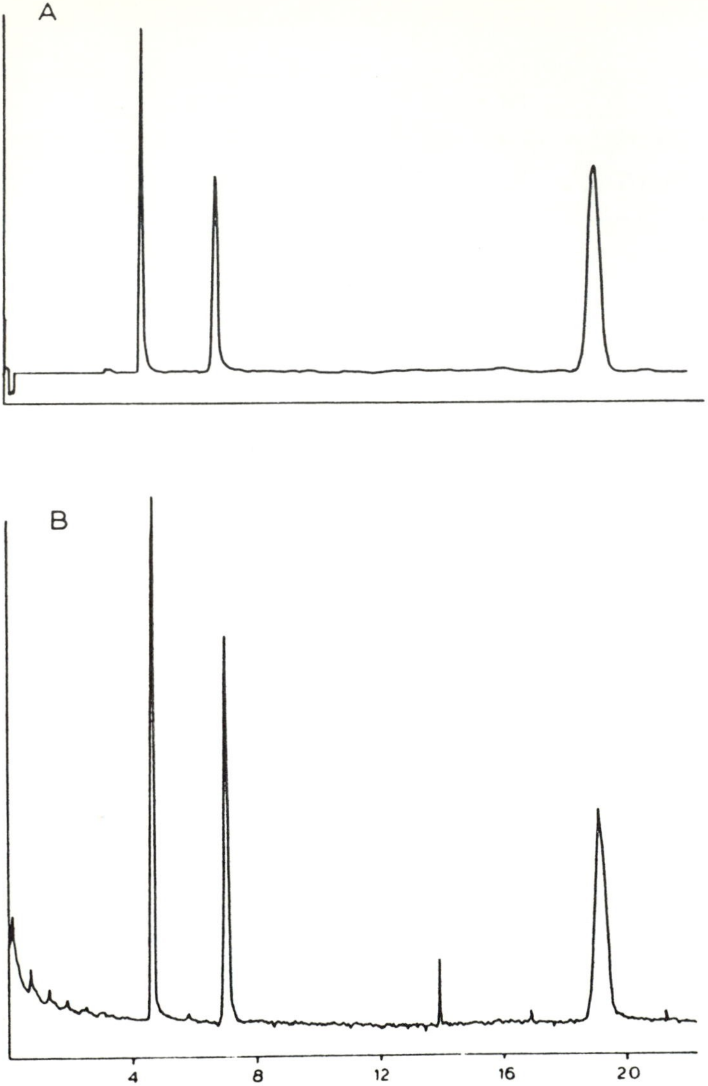

Figure 13. Comparison of UV and MS chromatograms using a 1.0-mm i.d. column packed with a 5-μm diameter Supelcosil C-18. LC-MS interface used aerosol spray deposition on a moving belt. Peaks correspond from left to right to 0.2 μg each of resorcinol, 1,5-dihydroxynaphthol, and 2-methylphenol. Conditions: 41% acetonitrile and 59% water (v/v) with 0.1% trifluoroacetic acid at a flow rate of 40 μL/min. A, UV trace at 280 nm, 0.015 AUFS; B, MS trace, selected ion chromatogram. (Reproduced from reference 54. Copyright 1984 American Chemical Society.)

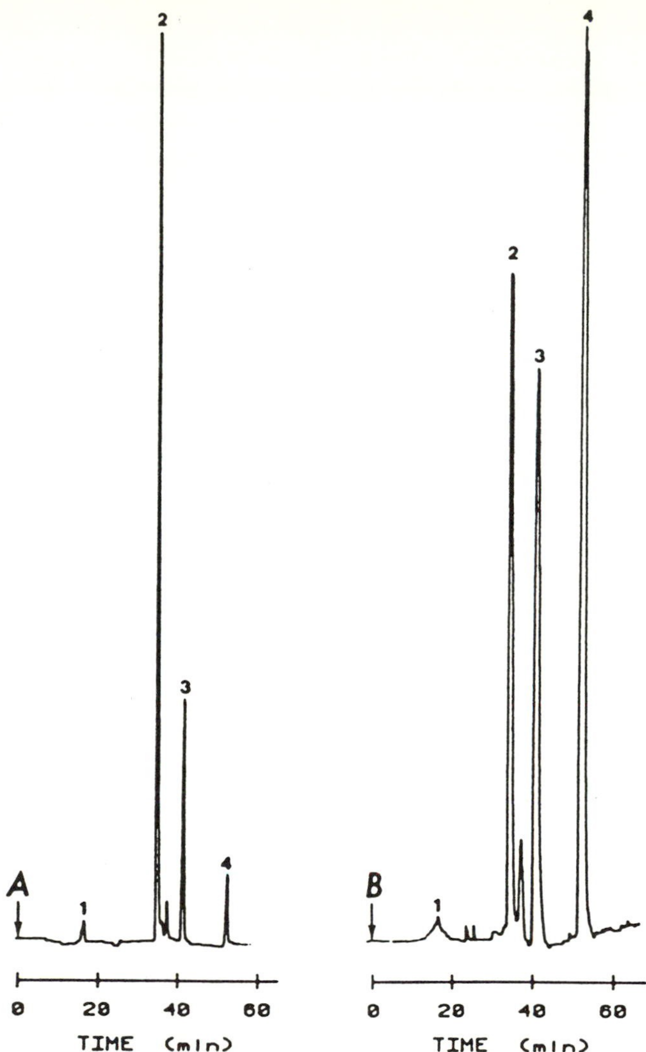

Figure 14. Chromatograms of organophosphorus pesticides obtained from a small-bore fused silica column (0.2 mm i.d., 1 m long) packed with 5-μm Spherisorb ODS. A, UV detection at 254 nm and B, dual-flame thermionic detector. Peaks correspond to 1, solvent containing phosphorus impurities; 2, guthion, 89 ng of phosphorus; 3, zolone, 71 ng of phosphorus; and 4, ethion, 144 ng of phosphorus. Mobile-phase conditions: 15% water and 85% methanol (v/v) at a flow rate of 1.6 μL/min. (Reproduced from reference 58. Copyright 1983 American Chemical Society.)

calorimetry (*60*) to the detection of solutes separated by open tubular column LC. Picogram sensitivities were obtained for fluorescent detection of 7-chloro-4-nitrobenzo-2-oxa-1,3-diazole (NBD) derivatives of alkylamines and thermal-lens detection of three nitroanilines. Figures 15 and 16 show the chromatograms obtained with each of these detection methodologies.

Conclusion

In the past 5 years the frequency of reports on the use of HPLC technology for the determination of trace organic compounds in aqueous environmental samples has been steadily increasing. Many innovative approaches to sample cleanup and analyte isolation have been reported. Reversed-phase separation, with its many mobile-phase adaptations, has been and continues to be the most popular HPLC separation mode. The development of fast columns and microbore columns should provide optimal configurations for particular applications. The operating characteristics of microbore columns also make

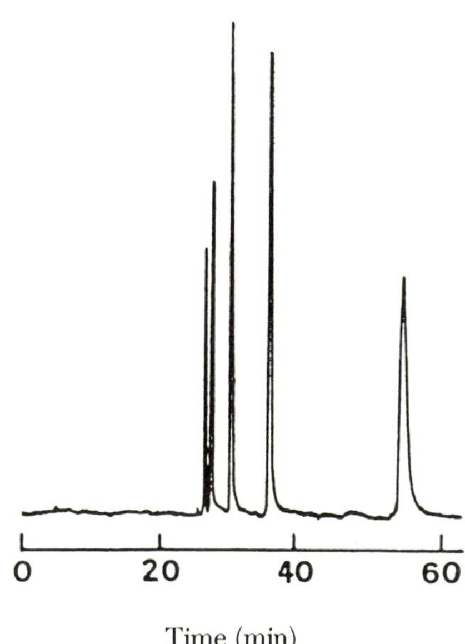

Figure 15. Open tubular liquid chromatography of amine–NBD derivatives using on-column fluorescence detection. Peaks correspond from left to right to ethylamine, n*-propylamine,* n*-butylamine, cyclohexylamine, and* n*-hexylamine. Conditions: 20-μm × 8.3-m column with C-18 bonded phase; 20% acetonitrile and 80% water (v/v) mobile phase at a linear velocity of 0.50 cm/s; on-column injection of 5 nL. (Reproduced from reference 59. Copyright 1984 American Chemical Society.)*

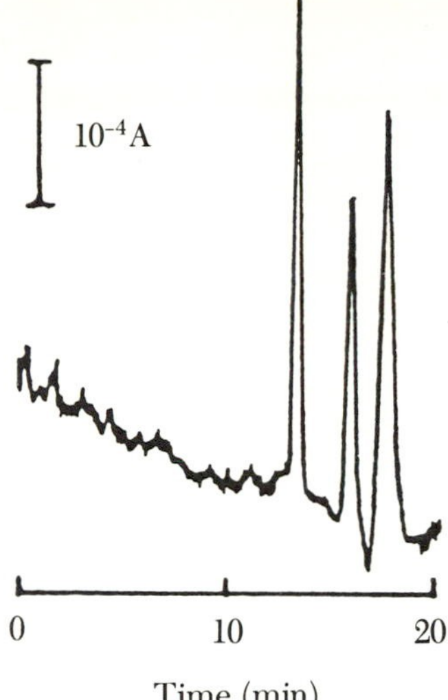

Figure 16. Open tubular liquid chromatography of nitroanilines using thermal lens detection. Peaks correspond from left to right to 410 pg of o-*nitroaniline, 310 pg of 4,5-dimethyl-2-nitroaniline, and 3600 pg of* N, N-*dimethyl-3-nitroaniline. Conditions: 20-μm × 7.5-m column with C-18 bonded phase; 40% methanol and 60% water (v/v) mobile phase at a linear velocity of 0.50 cm/s; on-column injection of 5 nL. (Reproduced from reference 59. Copyright 1984 American Chemical Society.)*

possible the application of many traditional detection methodologies not readily compatible with the flow rates typical of conventional HPLC columns. Commercially available detection systems, such as diode array detectors, continue to improve as well as provide new operating capabilities. Pre- and postcolumn reaction detection systems tailored to the physiochemical properties of the analyte can be readily developed to address the selectivity problems commonly encountered with complex matrices. MS and highly sensitive laser-based detection methodologies continue to evolve. Many developments are necessary before these detection methodologies can be routinely used as HPLC detection systems. However, the extraordinary detection capabilities will continue to drive the development of cost-effective and reliable systems for routine use.

An attempt has been made to survey the current status of technology in HPLC as it applies to the analysis of trace organic compounds in aqueous environmental samples. No doubt, some developments relative to this topic have been overlooked, but the overall assessment should provide a glimpse of what has been done and also of what is possible.

Literature Cited

1. Keith, L. H.; Crummett, W.; Deegan, J., Jr.; Libby, R. A.; Wentler, G. *Anal. Chem.* **1983,** *55,* 2210–2218.
2. Rogers, L. B., et al., Eds. *Chem. Eng. News* **1982,** *60(23),* 44.
3. Graham, J. A. In *Trace Analysis;* Lawrence, J. F., Ed.; Academic: New York, 1981; Vol. 1, pp 1–46.
4. Ballschmiter, K. *Pure Appl. Chem.* **1983,** *55,* 1943–1956.
5. Rosenfeld, J. M.; Mureika-Russell, M.; Phatak, A. *J. Chromatogr.* **1984,** *283,* 127–135.
6. *Advances in the Identification and Analysis of Organic Pollutants in Water;* Keith, L. H., Ed.; Ann Arbor Science: Ann Arbor, MI, 1981; Chapters 3, 15–18.
7. Burham, A. K.; Calder, G. U.; Junk, G. A.; Svec, H. J.; Willis, R. *Anal. Chem.* **1972,** *44,* 139–149.
8. Dressler, M. *J. Chromatogr.* **1979,** *165,* 167–206.
9. Van Rossum, P.; Webb, R. G. *J. Chromatogr.* **1978,** *152,* 329–340.
10. Tateda, A.; Fritz, J. S. *J. Chromatogr.* **1978,** *152,* 329–340.
11. Junk, G. A.; Richard, J. J. In *Advances in the Identification and Analysis of Organic Pollutants in Water;* Keith, L. H., Ed.; Ann Arbor Science: Ann Arbor, MI, 1981; pp 295–315.
12. West, S. D.; Day, E. W., Jr. *J. Assoc. Off. Anal. Chem.* **1981,** *64,* 1205–1207.
13. Saner, W. A.; Gilbert, J. *J. Liq. Chromatogr.* **1980,** *3,* 1753–1765.
14. Kirkland, J. J. *Analyst* **1974,** *99,* 859–885.
15. Little, J. N.; Fallick, G. J. *J. Chromatogr.* **1975,** *112,* 389–397.
16. Saner, W. A. In *Trace Analysis;* Lawrence, J. F., Ed.; Academic: New York, 1982; pp 152–222.
17. Graham, J. A.; Garrison, A. W. In *Advances in the Identification and Analysis of Organic Pollutants in Water;* Keith, L. H., Ed.; Ann Arbor Science: Ann Arbor, MI, 1981; pp 399–432.
18. Miller, J. W.; Garrison, A. W.; Rogers, L. B. *A General Approach to the Analysis of Water for Nonvolatile Organics by Precolumn Trace Enrichment High Performance Liquid Chromatography;* to be published.
19. Goewie, C. E.; Kwakman, P.; Frei, R. W.; Brinkman, U. A. Th.; Massfeld, W.; Seshadri, T.; Kettrup, A. *J. Chromatogr.* **1984,** *284,* 73–86.
20. Giddings, J. C. *Dynamics of Chromatography, Part 1;* Marcel Dekker: New York, 1965.
21. Majors, R. E.; Barth, H. G.; Lochmuller, C. H. *Anal. Chem.* **1984,** *56,* 300R–349R.
22. Karger, B. L.; Martin, M.; Guiochon, G. *Anal. Chem.* **1974,** *46,* 1640–1647.
23. Snyder, L. R.; Dolan, J. W.; Gant, J. R. *J. Chromatogr.* **1979,** *165,* 3–30.
24. Dolan, J. W.; Gant, J. R.; Snyder, L. R. *J. Chromatogr.* **1979,** *165,* 31–58.
25. Cooke, N. H. C.; Olsen, K.; Archer, B. G. *LC Magazine* **1984,** *2,* 514–524.
26. Cooke, N. H. C.; Archer, B. G.; Olsen, K.; Berick, A. *Anal. Chem.* **1982,** *54,* 2277–2283.
27. Stout, R. W.; DeStefano, J. J.; Snyder, L. R. *J. Chromatogr.* **1983,** *261,* 189–212.
28. Apffel, J. A.; Brinkman, U. A. Th.; Frei, R. W. *Chromatographia* **1983,** *17,* 125–131.
29. Slais, K.; Kourilova', D.; Krejci', M. *J. Chromatogr.* **1983,** *282,* 363–370.
30. Berridge, J. C. *Trends in Anal. Chem.* **1984,** *3,* 5–10.
31. Debets, H. J. G.; Bajema, B. L.; Doornbos, D. A. *Anal. Chim. Acta* **1983,** *151,* 131–141.

32. Glajch, J. L.; Kirkland, J. J.; Squire, K. M.; Minor, J. M. *J. Chromatogr.* **1980,** *199*, 57-79.
33. Snyder, L. R. *J. Chromatogr. Sci.* **1978,** *16*, 223.
34. Goldberg, A. P.; Nowakowska, E. L. *LC Magazine* **1984,** *2*, 458-462.
35. Fell, A. F.; Scott, H. P. *J. Chromatogr.* **1983,** *273*, 3-17.
36. Pinkerton, K. A. *J. High Resolut. Chromatogr. Chromatogr. Commun.* **1981,** *4*, 33-34.
37. Brunt, K. In *Trace Analysis;* Academic: New York, 1981.
38. *Fed. Regist.* **1979,** *44(233)*, 69464-69575.
39. Regis Chemical Company Catalog, pp 74-78.
40. Lauren, D. R.; Agnew, M. P. *J. Chromatogr.* **1984,** *292*, 439-443.
41. Tanaka, M.; Shimada, K.; Nambara, T. *J. Chromatogr.* **1984,** *292*, 410-411.
42. Hodgin, J. C.; Howard, P. Y.; Ball, D. M.; Cloete, C.; De Jager, L. *J. Chromatogr. Sci.* **1983,** *21*, 503-507.
43. Xie, K.-H.; Santasania, C. T.; Krull, I. S.; Neue, U.; Bidlingmeyer, B.; Newhart, A. *J. Liq. Chromatogr.* **1983,** *6*, 2109-2127.
44. Deelder, R. S.; Kuijpers, A. T. J. M.; Van Den Berg, J. H. M. *J. Chromatogr.* **1983,** *255*, 545-561.
45. van der Wal, Sj. *J. Liq. Chromatogr.* **1983,** *6*, 37-59.
46. Nondek, L.; Frei, R. W.; Brinkman, U. A. Th. *J. Chromatogr.* **1983,** *282*, 141-150.
47. Scholten, A. H. M. T.; Welling, P. L. M.; Brinkman, U. A. Th.; Frei, R. W. *J. Chromatogr.* **1980,** *199*, 239-248.
48. Bush, B.; Smith, R. M.; Narang, A. S.; Hong, C.-S. *Anal. Lett.* **1984,** *17*, 467-474.
49. Haeberer, A. F.; Scott, T. A. In *Advances in the Identification and Analysis of Organic Pollutants in Water;* Keith, L. H., Ed.; Ann Arbor Science: Ann Arbor, MI, 1981; Vol. 1.
50. Werkhoven-Goewie, C. E.; Boon, W. M.; Praat, A. J. J.; Frei, R. W.; Brinkman, U. A. Th.; Little, C. J.; *Chromatographia* **1982,** *16*, 53-59.
51. Graham, J. A., to be published.
52. Dong, X.-D.; Krull, I. S. *J. Agric. Food Chem.* **1984,** *32*, 622-628.
53. Wright, L. H. *J. Chromatogr. Sci.* **1982,** *20*, 1-6.
54. Hayes, M. J.; Schwartz, H. E.; Vouros, P.; Karger, B. L.; Thruston, A. D., Jr.; McGuire, J. M. *Anal. Chem.* **1984,** *56*, 1229-1236.
55. Blakley, C. R.; Carmody, J. J.; Vestal, M. L. *J. Am. Chem. Soc.* **1980,** *26*, 1467-1473.
56. Blakley, C. R.; Vestal, M. L. *Anal. Chem.* **1983,** *55*, 750.
57. Willoughby, R. C.; Browner, R. F. In *Trace Analysis;* Lawrence, J. F., Ed.; Academic: New York, 1982; Vol. 2.
58. McGuffin, V. L.; Novotny, M. *Anal. Chem.* **1983,** *55*, 2296-2302.
59. Sepaniak, M. J.; Vargo, J. D.; Kettler, C. N.; Maskarinec, M. P. *Anal. Chem.* **1984,** *56*, 1252-1257.
60. Dovichi, N. J.; Harris, J. M. *Anal. Chem.* **1979,** *51*, 728-732.

RECEIVED for review August 14, 1985. ACCEPTED February 2, 1986.

REVERSE OSMOSIS TO ISOLATE ORGANIC POLLUTANTS FROM WATER

7

A Fundamental Approach to Reverse-Osmosis Concentration and Fractionation of Organic Chemicals in Aqueous Solutions for Environmental Analysis

T. D. Nguyen, Takeshi Matsuura, and S. Sourirajan

Division of Chemistry, National Research Council of Canada, Ottawa, Ontario, K1A 0R9 Canada

The preferential sorption–capillary flow mechanism and its quantitative expression given by the surface force–pore flow model together offer a fundamental approach to the reverse-osmosis concentration of organic chemicals in aqueous solutions for environmental analysis. On the basis of this approach, the attractive and repulsive forces prevailing at the membrane material–aqueous solution interfaces, together with the average pore size and pore size distribution on the membrane surface, govern reverse-osmosis separations. Data on interfacial interaction force parameters and also data on average pore size and pore size distribution on the membrane surface have been experimentally determined with particular reference to a selected number of organic solutes, a cellulose acetate membrane, and an aromatic polyamidohydrazide membrane. The use of such data for the calculation of parameters relevant to the concentration process is illustrated.

THE SEPARATION, CONCENTRATION, AND FRACTIONATION of organic solutes in aqueous solutions by reverse osmosis are of practical interest from the points of view of water purification and collection of samples for environmental analysis. Although many experimental data on the separation of organic solutes are available in the literature (*1–2*), very few fundamental works have been accomplished so far. We have been studying this subject in the framework of the preferential sorp-

0065–2393/87/0214/0139$06.75/0

tion–capillary flow mechanism of reverse osmosis and its quantitative expression represented by the surface force–pore flow model. According to this mechanism, the transport of organic solute and solvent water through the membrane is governed by the interaction forces working in the solute–solvent membrane material system and the average pore size and the pore size distribution on the membrane surface. Therefore, the determinations of both the interfacial interaction forces and the pore size distribution on the membrane surface are of vital importance in our approach. Methods have been developed in our laboratory for these determinations as illustrated in this chapter. Furthermore, the interfacial interaction forces and the pore size distribution are incorporated into the transport equations on the basis of the surface force–pore flow model, which enables the prediction of the solute separation for a wide variety of organic solutes involved in the environmental analysis. These transport equations are applicable for the cases in which either solvent water or the organic solute molecule is preferentially sorbed at the membrane material–solution interface. For many organic solute compounds in sample waters, the solute may be weakly or strongly adsorbed to the membrane material and thereby give rise to the preferential sorption of solutes at the membrane material–solution interface. Therefore, the applicability of the transport equation to the preferential sorption of solutes is illustrated and the special features in the separation and concentration of such solute molecules by reverse-osmosis processes are discussed in this chapter.

Among the many organic solutes treated by reverse-osmosis processes, some may be highly separated, whereas others may permeate through the membrane to different extents depending on experimental conditions. Thus, the fractionation of organic solutes sometimes becomes very important to accomplish the concentration of selected organic solutes of interest. Because the solute separation is governed by the interfacial interaction force and the pore size distribution, and the interfacial interaction force is subject to change with a change in the chemical nature of the polymer membrane material, the optimization of membrane material and the pore size distribution may naturally be expected to solve a given fractionation problem effectively; this aspect of the subject will also be discussed in this chapter. Such optimization provides guidelines for selecting the polymer membrane material and for adjusting the membrane casting procedure to produce the membrane with the required pore size distribution. Thus, this approach contributes to rational membrane design for applications involving reverse-osmosis separation, concentration, and fractionation.

Experimental

Chromatography Experiments. The liquid chromatography (LC) model ALC202 of Waters Associates fitted with a differential refractometer was used in this work. The method of column preparation and the general experimental technique were the same as those reported previously (*3*). Briefly, solutes were injected into the solvent water stream that flows through a column packed with membrane polymer powder. The particle size was kept in the range of 38–54 μm by sieving, and the column length was 60 cm. Ten microliters of sample solution (1 wt% solute) was injected into the column, and the retention volume was determined. In the case of heavy water, 10 wt% solution was used as the sample. From the experimental retention volume of a solute A (V'_{R_A}) and of heavy water ($V'_{R_{\text{water}}}$), the specific surface excess ($\Gamma_A/c_{A,b}$) of the solute A was calculated:

$$\Gamma_A/c_{A,b} = (V'_{R_A}) - (V'_{R_{\text{water}}})/A_p \tag{1}$$

where A_p denotes the total surface area of the polymer membrane material in the chromatography column. The specific surface excess determined shows the nature and magnitude of the interfacial interaction force between solute and membrane material.

Reverse-Osmosis Membranes. Laboratory-made cellulose acetate (CA) membranes were used in this study. The method of membrane fabrication was the same as that described by Kutowy et al. (*4*). Briefly, CA (E-400-25 supplied from Eastman Chemicals) was dissolved in an acetone-water-magnesium perchlorate mixture and gelled in a gelation medium after a very short solvent evaporation period. Some of the membranes so produced were further shrunk at the shrinkage temperature ranging from 80 to 85 °C. The casting solution composition, in weight percent, was as follows: CA (E-400-25), 14.8; acetone, 63.0; magnesium perchlorate, 2.3; and water, 19.9. The casting solution and casting atmosphere were at ambient temperature, the humidity of the casting atmosphere was 35%, and the solvent evaporation time was <5 s. The gelation medium was an ethanol-water mixture at 25 °C, and the gelation period was 15 min in the gelation medium followed by a more than 1-h immersion in ice-cold water. All the membranes thus produced are characterized in Tables I and II together with CA-398-316-82 and aromatic polyamidohydrazide (PAH-09) films that were produced and characterized as reported in our previous work (*5*-*6*).

Reverse-Osmosis Experiments. All reverse-osmosis experiments were performed with continuous-flow cells. Each membrane was subjected to an initial pure water pressure of 2068 kPag (300 psig) for 2 h; pure water was used as feed to minimize the compaction effect. The specifications of all the membranes in terms of the solute transport parameter [$(D_{AM}/K\delta)_{\text{NaCl}}$], the pure water permeability constant (A), the separation, and the product rate (PR) are given in Table I. These were determined by Kimura-Sourirajan analysis (*7*) of experimental reverse-osmosis data with sodium chloride solution at a feed concentration of 0.06 *m* unless otherwise stated. All other reverse-osmosis experiments were carried out at laboratory temperature (23–25 °C), an operating pressure of 1724 kPag (250 psig), a feed concentration of 100 ppm, and a feed flow rate >400 cm^3/min. The fraction solute separation (f) is defined as follows:

Table I. Reverse-Osmosis Membrane Specifications

Film No.	Film Code	$A \times 10^7$ (kg-mol/ $m^2 \cdot s \cdot kPa$)	$(D_{AM}/K\delta)_{NaCl}$ $\times 10^7$ (m/s)	NaCl Data[a] Sepn (%)	PR $\times 10^3$ (kg/h)
—	CA-398-316-82[b]	2.120	1.684	96.1	26.8
—	PAH-09[c]	1.104	1.502	94.0	13.34
1	CA-400-15-85[d]	0.971	1.534	92.9	11.22
2	CA-400-15-0[d]	3.737	31.90	72.0	47.8
3	CA-400-5-85[d]	2.089	4.879	90.8	26.0
4	CA-400-10-85[d]	1.321	2.543	91.9	16.27
5	CA-400-25-0[d]	3.737	43.64	62.8	48.0
6	CA-400-20-0[d]	1.112	10.67	71.4	14.4
7	CA-400-25-0[d]	1.817	22.14	63.2	23.5
8	CA-400-20-0[d]	2.353	28.41	61.3	30.2
9	CA-400-20-80[d]	0.990	11.93	68.1	13.0
10	CA-400-10-0[d]	3.899	131.8	43.2	53.4

[a] Effective film area = 13.2 cm^2; data at 1724 kPag and 0.06 *m*.
[b] The first, second, and third numbers indicate acetyl content, membrane batch, and shrinkage temperature, respectively. Film is specified in reference 5.
[c] The number indicates the evaporation period (min). Film is specified in reference 6.
[d] The first, second, and third numbers indicate acetyl content, volume percent of ethanol in gelation bath, and shrinkage temperature, respectively. A zero as the third number means no shrinkage.

f = (ppm in feed solution − ppm in product solution)/ppm in feed solution

The value of f, PR, and the pure water permeation rate (PWP) for a given area of film surface (13.2 cm^2 in this work) were determined under the specified experimental conditions. The data on PR and PWP are corrected for 25 °C. Concentrations of NaCl were determined by conductance measurement, whereas concentrations of organic solutes were determined by the Beckman total carbon analyzer model 915-A.

Theoretical

Determination of Interfacial Interaction Force Constants. The transport equations based on the surface force-pore flow model; the procedure for the calculation of f, PWP, and PR, which are obtainable by reverse-osmosis experiments; the procedure for the determination of the interaction force constants from the LC data; and the procedure for calculating the average pore size and the pore size distribution on the surface layer of asymmetric reverse-osmosis membranes have all been described elsewhere (*8–10*). For the clarification of symbols used in this chapter, however, the framework of the theory is outlined in the following discussion.

The quantities obtainable from reverse osmosis such as f and PR can be calculated when data on PWP, pore size distribution, and

Table II. Average Pore Size and Pore Size Distribution on Membrane Surfaces

Film No.	Film Code	$\overline{R}_{b,1} \times 10^{10}$ (m)	$\sigma_1/\overline{R}_{b,1}$	$\overline{R}_{b,2} \times 10^{10}$ (m)	$\sigma_2/\overline{R}_{b,2}$	h_2
—	CA-398-316-82	7.5	0.002	36	0.195	0.001
—	PAH-09	4.6	0.001	—	—	0
1	CA-400-15-85	8.6	0.001	—	—	0
2	CA-400-15-0	9.6	0.001	—	—	0
3	CA-400-5-85	8.4	0.040	51	0.050	0.001
4	CA-400-10-85	8.6	0.003	51	0.050	0.001
5	CA-400-25-0	9.4	0.001	42	0.050	0.002
6	CA-400-20-0	9.0	0.050	55	0.100	0.003
7	CA-400-25-0	9.6	0.001	40	0.050	0.004
8	CA-400-20-0	9.6	0.001	40	0.050	0.005
9	CA-400-20-80	8.6	0.050	59	0.050	0.005
10	CA-400-10-0	9.1	0.001	50	0.050	0.006

interfacial interaction force constants are given under the operating conditions of the experiment such as feed concentrations, operating pressure, and feed flow rate (*5*). The pore size distribution is expressed in terms of one or more Gaussian normal distributions. For describing such pore size distributions, the distribution function of the *i*th distribution given as

$$Y_i(R_b) = (1/\sigma_i\sqrt{2\pi}) \exp\{-(R_b - \overline{R}_{b,i})^2/2\sigma_i^2\} \quad (2)$$

and a quantity defined as

$$h_i = n_i/n_1 \quad (3)$$

are necessary, where R_b is the pore radius, and $\overline{R}_{b,i}$, σ_i, and n_i denote the average pore size, the standard deviation, and the number of pores that belong to the *i*th distribution, respectively (*5*). Also, by definition, $\overline{R}_{b,i}$ becomes progressively larger as the number *i* increases.

With respect to nonionized organic solutes in aqueous solutions, the interfacial interaction force constants were expressed as constants that define the interfacial potential function as follows:

$$\phi(d) = \begin{cases} \text{very large} & \text{when } d \leq \underset{\sim}{D} \\ -(\underset{\sim}{B}/d^3)\,\underset{\sim}{R}T & \text{when } d > \underset{\sim}{D} \end{cases} \quad (4)$$

where d is the distance between the polymer surface and the center of the solute molecule, $\underset{\sim}{D}$ is a constant associated with the steric hindrance (distance of steric repulsion), $\underset{\sim}{B}$ expresses the nature and the magnitude of the van der Waals force, $\underset{\sim}{R}$ is the gas constant, and T is the absolute temperature. The quantity $\underset{\sim}{D}$ is always positive, and when the solute shape is assumed spherical, it can be approximated by the molecular radius such as the Stokes' law radius. The quantity $\underset{\sim}{B}$ may be either positive (corresponding to an attractive force) or negative (corresponding to a repulsive force). The parameters associated with the pore size distribution ($\overline{R}_{b,i}$, σ_i, and h_i) and the interfacial interaction force parameters ($\underset{\sim}{B}$ and $\underset{\sim}{D}$) are related to $\Gamma_A/c_{A,b}$, obtainable from the chromatographic experiment, and f, obtainable from reverse-osmosis experiments, by the following equations:

$$(\Gamma_A/c_{A,b})_j = \{g(\underset{\sim}{B},\underset{\sim}{D})\}_j \tag{5}$$

$$f_j = \{h(\overline{R}_{b,i}, \sigma_i, h_i, \underset{\sim}{B}, \underset{\sim}{D}) \text{ under given operating conditions}\}_j \tag{6}$$

where the subscript j indicates the jth solute, and $g(...)$ and $h(...)$ are some functional forms established in the surface force–pore flow model (*9, 11*). By using these equations, the numerical values for the pore size distribution parameters and the interfacial interaction force parameters can be determined as follows.

Cases in Which a Uniform Pore Size Distribution Is Involved. *Step 1 Calculation.* Determination of $\underset{\sim}{B}$ and $\underset{\sim}{D}$ values for a reference solute: Only one reference solute is considered; therefore, $j = 1$. For this solute, $\underset{\sim}{D}$ is approximated by the Stokes' law radius of the solute. The numerical value for $\underset{\sim}{B}$ can then be obtained from eq 5 by using experimental values of $\Gamma_A/c_{A,b}$ for the reference solute.

Step 2 Calculation. Determination of the average pore radius $\overline{R}_{b,i}$: Only a uniform pore of $\overline{R}_{b,i}$ is considered. $\overline{R}_{b,i>1}$, σ_i, and h_i can be considered to be equal to zero. Because $\underset{\sim}{B}$ and $\underset{\sim}{D}$ values are known from the first step, $\overline{R}_{b,1}$ can be immediately calculated from eq 6 by using available data of f_1.

Step 3 Calculation. Determination of $\underset{\sim}{B}$ and $\underset{\sim}{D}$ values for solutes other than the reference solute: Once $\overline{R}_{b,1}$ is known, $\underset{\sim}{B}$ and $\underset{\sim}{D}$ values for the solutes other than the reference solute can be obtained by simultaneous solution of eq 5 and eq 6 for each solute such that experimental $\Gamma_A/c_{A,b}$ and f data can be satisfied.

Cases in Which Two Normal Pore Size Distributions Are Considered. In two normal distributions, five parameters, namely $\overline{R}_{b,1}$, σ_1, $\overline{R}_{b,2}$, σ_2, and h_2, are necessary to characterize the pore size distribution on the membrane surface. Five or more reference solutes are necessary for obtaining such parameters. Let us use eight reference solutes. By setting eq 5 and eq 6 for each reference solute, we have eight eq 5's and eight eq 6's, corresponding to $j = 1,2,3...8$.

Step 4 Calculation. Determination of $\underset{\sim}{B}$ and $\underset{\sim}{D}$ values for reference solutes: Again, by setting $\underset{\sim}{D}_j$ equal to the Stokes' law radius of the jth reference solute, $\underset{\sim}{B}_j$ can be calculated from eq 5 established for the jth solute so that experimental $(\Gamma_A/c_{A,b})_j$ can be satisfied.

Step 5 Calculation. The five pore size distribution parameters mentioned earlier can be calculated by nonlinear regression analysis of eight eq 6's by using the $\underset{\sim}{B}$ and $\underset{\sim}{D}$ values obtained for each reference solute.

Calculation of Solute Separation and Product Rate. Once the pore size distribution parameters $\overline{R}_{b,1}$, σ_1, $\overline{R}_{b,2}$, σ_2, and h_2 are known for a membrane and the interfacial interaction force parameters $\underset{\sim}{B}$ and $\underset{\sim}{D}$ are known for a given system of membrane material–solute, solute separation f can be calculated by eq 6 for any combination of these parameters. Furthermore, because the PR-to-PWP ratio (PR/PWP) can also be calculated by the surface force–pore flow model (*9*), PR is obtained by multiplying experimental PWP data by this ratio.

Calculation of Processing Capacities. The processing capacity of a membrane (V_i/St) is defined as the volume of charge (feed solutions) that 1 m^2 of film surface can handle per day in a batch concentration process to increase the solute weight fraction from z_i to z_f at a given operating pressure. The following relationship has been derived by Sourirajan and Kimura (*12*):

$$\frac{V_i}{St} = \left[(\rho_1)_i \int_{z_i}^{z_f} \frac{\exp\left\{ \int_{z_i}^{z_f} \frac{dz}{z-a} \right\}}{G(z-a)} dz \right]^{-1} \tag{7}$$

where z and a denote weight fractions of solute in the solution at the high pressure side of the membrane and product, respectively, and $(\rho_1)_i$ and G are the density of the solution and product permeation flux, respectively. Subscripts i and f represent the initial and final states, respectively, of the given feed solutions on the high pressure side of the membrane. Eq 7 allows the calculation of the processing capacity by performing the numerical integration of the right side of the equation from $z = z_i$ to $z = z_f$ because G versus z data as well as a versus z data become available from the foregoing calculations. As can be seen from eq 7, the processing capacity can be given as a function of z_f/z_i.

Results and Discussion

Determination of $\underset{\sim}{B}$ and $\underset{\sim}{D}$ Values for Solutes Relevant to Membrane Concentrations of Organic Pollutants. A compilation of the data on $\underset{\sim}{B}$ and $\underset{\sim}{D}$ values for a large number of organic solutes was made (*11*) in which glycerol was chosen as a reference solute for both CA and aromatic PAH or PPPH8273 materials in the step 1 calculation. The average pore size on the membrane surface was further determined for each membrane by the step 2 calculation. Then the values for $\underset{\sim}{B}$ and $\underset{\sim}{D}$ were determined by step 3 for solutes other than the reference solute. Some of the results with respect to solutes relevant to the membrane concentration of organic pollutants are listed in Table III.

Determination of Pore Size Distributions on the Surface of CA Membranes and Aromatic PAH Membranes. The organic solutes listed in Table IV were chosen as reference solutes, and then $\underset{\sim}{D}$ and $\underset{\sim}{B}$ values with respect to CA-398 and PAH materials were obtained by step 4. The results are listed in Table IV. Then, by using these $\underset{\sim}{B}$ and $\underset{\sim}{D}$ values, the average pore size and the pore size distribution on surfaces of membranes under study were calculated by following step 5. In these calculations, $\underset{\sim}{B}$ and $\underset{\sim}{D}$ values for CA-400 material were assumed to be equal to those of CA-398 material because of the closeness of acetyl content. The results are listed in Table II.

Process Calculation of the Reverse-Osmosis Concentration of Organic Solutes. For this study we have chosen several interfacial interaction force parameters corresponding to different combinations of several chosen organic solutes with either CA-398 material or aromatic PAH material. Such parameters are summarized in Table V. For both membrane materials, solutes are numbered so that the attractive interaction force given as LC retention volume increases as the solute number increases. By using the pore size distributions of CA-398-316-82 and PAH-09, both listed in Table II, and the interfacial interaction force parameters given in Table V, the separations of each solute by the membranes were calculated at pressures ranging from 1724 kPag (250 psig) to 10342 kPag (1500 psig) and at concentrations ranging from 100 to 1000 ppm. The results are illustrated in Figures 1 and 2 for CA and PAH membranes, respectively. The mass transfer coefficient $k = \infty$ was used for the calculation of the separation by the CA membrane, whereas $k = 22 \times 10^{-6}$ m/s was used for the calculation with respect to the PAH membrane. PWP values of 32.4×10^{-3} and 10.4×10^{-3} kg/h were used for CA and PAH membranes, respectively, at the operating pressure of 1724 kPag (250 psig) with the assumption of a linear relationship between operating pressure and PWP.

Several features of reverse-osmosis separations are revealed by these calculations; some of them are the following:

1. Solute concentration does not affect solute separation significantly in the concentration range studied.
2. Positive separations are obtained in all cases except in cases V and VI.

Table III. Data on Interfacial Interaction Force Parameters for Some Potential Water Pollutants

Solute	Stokes' Law Radius × 10^{10} (m)	CA-398 Polymer $\underset{\sim}{D}$ × 10^{10} (m)	CA-398 Polymer $\underset{\sim}{B}$ × 10^{30} (m^3)	PAH (PPPH8273) Polymer $\underset{\sim}{D}$ × 10^{10} (m)	PAH (PPPH8273) Polymer $\underset{\sim}{B}$ × 10^{30} (m^3)
Methanol	1.45	1.85	8.31	1.41	5.71
Ethanol	2.05	2.03	21.54	1.94	21.99
1-Propanol	2.12	2.15	37.77	2.13	36.40
2-Propanol	2.26	2.45	38.40	3.06	89.74
1-Butanol	2.33	2.10	47.27	2.57	95.50
2-Butanol	2.33	2.45	62.15	3.14	134.6
2-Methyl-1-propanol	3.05	2.75	95.57	3.78	243.6
2-Methyl-2-propanol	3.35	3.67	134.3	3.79	132.2
1-Pentanol	2.63	—	—	3.75	361.3
1-Hexanol	1.97	1.97	56.76	—	—
Acetone	1.91	1.91	26.90	2.22	47.16
Methyl ethyl ketone	1.92	1.92	35.87	2.39	71.38
Methyl isopropyl ketone	2.59	2.45	83.67	4.02	360.9
Methyl isobutyl ketone	2.86	2.45	97.88	—	—
Cyclohexanone	2.77	2.72	120.8	—	—
Diisopropyl ketone	3.11	2.62	131.7	—	—
Methyl acetate	2.05	1.73	27.88	2.58	85.88
Ethyl acetate	2.39	1.83	36.41	2.90	132.7
Propyl acetate	2.68	—	—	3.46	265.3
Ethyl propionate	2.68	1.88	45.63	—	—
Ethyl butyl ether	2.93	2.53	86.45	—	—
Ethyl *tert*-butyl ether	2.93	5.30	901.2	—	—
Isopropyl *tert*-butyl ether	3.18	5.60	918.2	—	—
Propionamide	2.24	1.98	19.13	—	—
Acetonitrile	1.47	1.78	28.49	—	—
Propionitrile	1.85	1.78	34.20	—	—
Nitromethane	1.65	1.80	36.00	—	—
1-Nitropropane	2.00	2.05	65.76	—	—
Phenol	2.10	1.71	45.39	—	—
2-Chlorophenol[a]	2.58	2.58	126.4	—	—
3-Chlorophenol	2.58	1.50	75.0	2.35	130
4-Chlorophenol[a]	2.58	2.58	124.7	—	—
4-Nitrophenol[a]	2.69	2.69	135.1	—	—
2,4-Dimethylphenol[a]	2.90	2.90	113.8	—	—
3,4-Dimethylphenol[a]	2.90	2.90	107.3	—	—
Dichlorophenols	2.80	1.66	300	—	—
Resorcinol	2.74	1.73	45.16	—	—
Aniline	2.42	1.80	48.98	—	—
Dimethylaniline	2.99	2.50	125.8	—	—
1,2-Ethanediol	2.11	2.20	−16.82	—	—
Glycerol	2.30	2.30	−52.30	2.30	−2.99
2,3-Butanediol	2.46	2.75	−11.97	—	—

Continued on next page

Table III. Continued

Solute	Stokes' Law Radius × 10^{10} (m)	CA-398 Polymer		PAH (PPPH8273) Polymer	
		$D \times 10^{10}$ (m)	$B \times 10^{30}$ (m^3)	$D \times 10^{10}$ (m)	$B \times 10^{30}$ (m^3)
Xylitol	3.00	3.30	−135.0	—	—
1,2,6-Hexanetriol	3.07	2.82	−16.95	—	—
Dulcitol	3.31	—	—	2.16	−104.4
D-Sorbitol	3.30	4.64	−180.2	—	—
D-Glucose	3.66	3.36	−203.1	2.29	−67.9
D-Fructose	3.22	4.51	−181.8	—	—
Sucrose	4.67	5.11	−343.2	—	—
Maltose	4.98	4.98	−346.0	—	—

[a] D was assumed equal to Stokes' law radius.

3. Solute separation increases with an increase in operating pressure, except for cases IV, V, VI, and XI (*see* Table V); all these cases are characterized by high B values among the solutes considered.
4. When the feed concentration is as small as those considered in this study, the osmotic pressure of solutions is negligible. Therefore, PR/PWP may be expected to be nearly equal to unity. However, for cases V and VI, in which strong attractive forces work between membrane and solute, PR/PWP

Table IV. Some Surface Parameters Pertinent to Reference Solutes and CA and Aromatic PAH Materials

Solute	CA-398 Material[a]		Aromatic PAH (PPPH8273) Material[b]	
	$D \times 10^{10}$ (m)	$B \times 10^{30}$ (m^3)	$D \times 10^{10}$ (m)	$B \times 10^{30}$ (m^3)
Methanol	—	—	1.41	5.71
Ethanol	—	—	1.94	21.99
Trimethylene oxide	1.83	32.89	2.30	50.30
1,3-Dioxolane	1.96	42.43	2.41	55.60
p-Dioxane	2.23	53.87	2.80	5.76
Oxepane	2.53	46.70	—	—
12-Crown-4	3.19	28.68	4.00	−1.16
15-Crown-5	3.77	−321.8	—	—
18-Crown-6	4.29	−202.6	—	—
Glucose	3.36	−203	—	—

[a] Data are taken from reference 17.
[b] Data are taken from reference 6.

Table V. Interfacial Interaction Force Parameters Used

Case	$\underset{\sim}{D} \times 10^{10}$ *(m)*	$\underset{\sim}{B} \times 10^{30}$ *(m³)*	*Solute*
		CA-398 Material	
I	3.36	−203	glucose
II	2.45	38.4	2-propanol
III	2.15	37.8	1-propanol
IV	5.60	918	isopropyl *tert*-butyl ether
V	1.50	75.0	3-chlorophenol
VI	1.66	300	dichlorophenols
		Aromatic PAH Material	
VII	2.29	−67.9	glucose
VIII	1.94	22.0	ethanol
IX	3.06	89.4	2-propanol
X	2.13	36.4	1-propanol
XI	2.35	130	3-chlorophenol

was calculated to be ~0.7 at the feed concentration of 100 ppm. Thus, a strong attraction between the solute and the membrane surface reduces the ratio significantly.

All these features are essentially the same as those obtained by assuming a uniform pore size distribution (*13*); therefore, the major trend of the performance data with the change in the operating variables does not change by the pore size distribution. However, the absolute values of solute separation and PR/PWP do change with a change in average pore size and the pore size distribution (*13–14*). Later in this chapter this aspect will be investigated in more detail with respect to membranes with two normal distributions.

Processing Capacity. On the basis of the data on solute separation and product rate calculated in the foregoing section, the processing capacity was obtained with respect to all cases involved in Table V. The results are shown in Figures 3 and 4 for the CA-398-316-82 and the aromatic PAH-09 membranes, respectively. Because both the solute separations and the product rate are practically unchanged in the concentration range under study, the processing capacity given as function of parts per million in the concentrate solution per parts per million in the initial feed solution, which corresponds to z_f/z_i in eq 7, is applicable to the entire range of initial concentrations indicated in the figures. Although the generally higher processing capacities of the CA membrane reflect the higher permeation flux through the CA membrane than through the aromatic PAH membrane considered in this study, the concentration ratio of the CA membrane is lower than that of the aromatic PAH membrane. This result is obtained because the organic

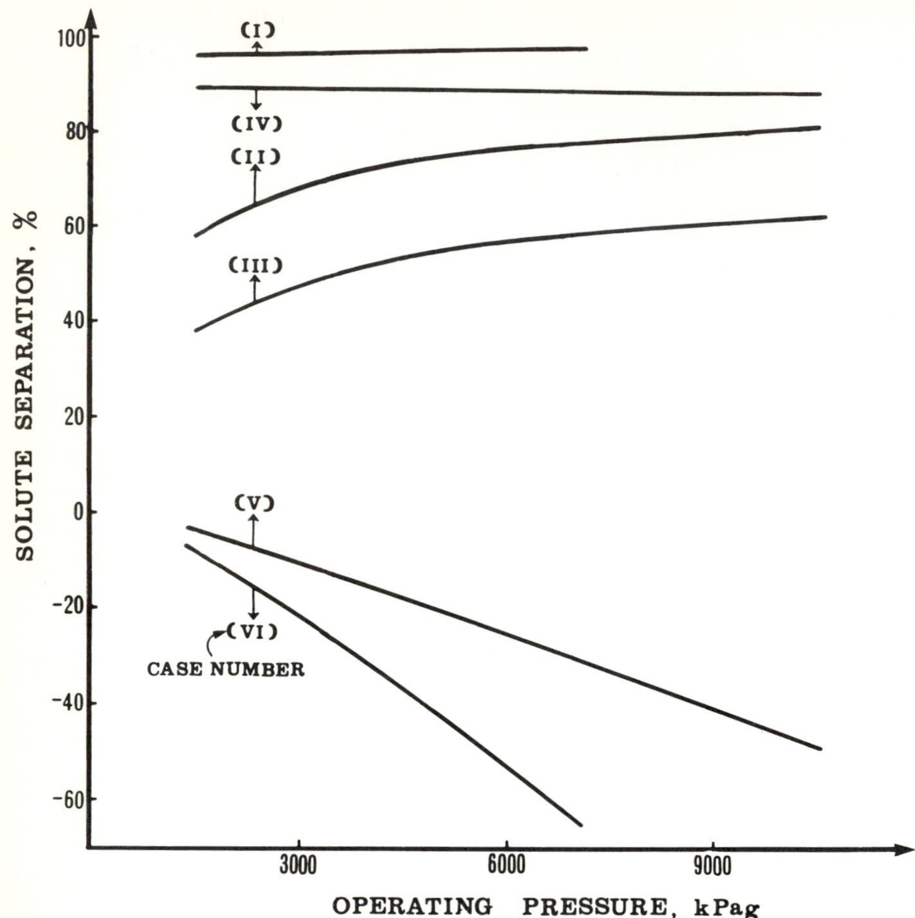

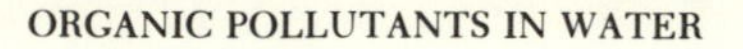

Figure 1. Effect of the operating pressure on the solute separation calculation based on k $= \infty$ *and PWP* $= 32.4 \times 10^{-3}$ kg/h *at 1724 kPag (250 psig). The membrane codes are the same as those given in Tables I and II; the case numbers are the same as those given in Table V. Membrane: CA-398-316-82; feed concentration = 0.1–100 ppm.*

solute separation by the CA membrane is smaller than that by the aromatic PAH membrane. Interestingly, concentration ratios of 3-chlorophenol and dichlorophenols can only be less than unity, regardless of the processing capacity, with the CA membrane used in this study. This finding reflects the negative separations of these solutes under the operating pressure of 1724 kPag (250 psig).

Fractionation of Glucose and *tert*-Butyl Isopropyl Ether Solute by Cellulose Acetate Membranes. The fractionation of glucose and *tert*-

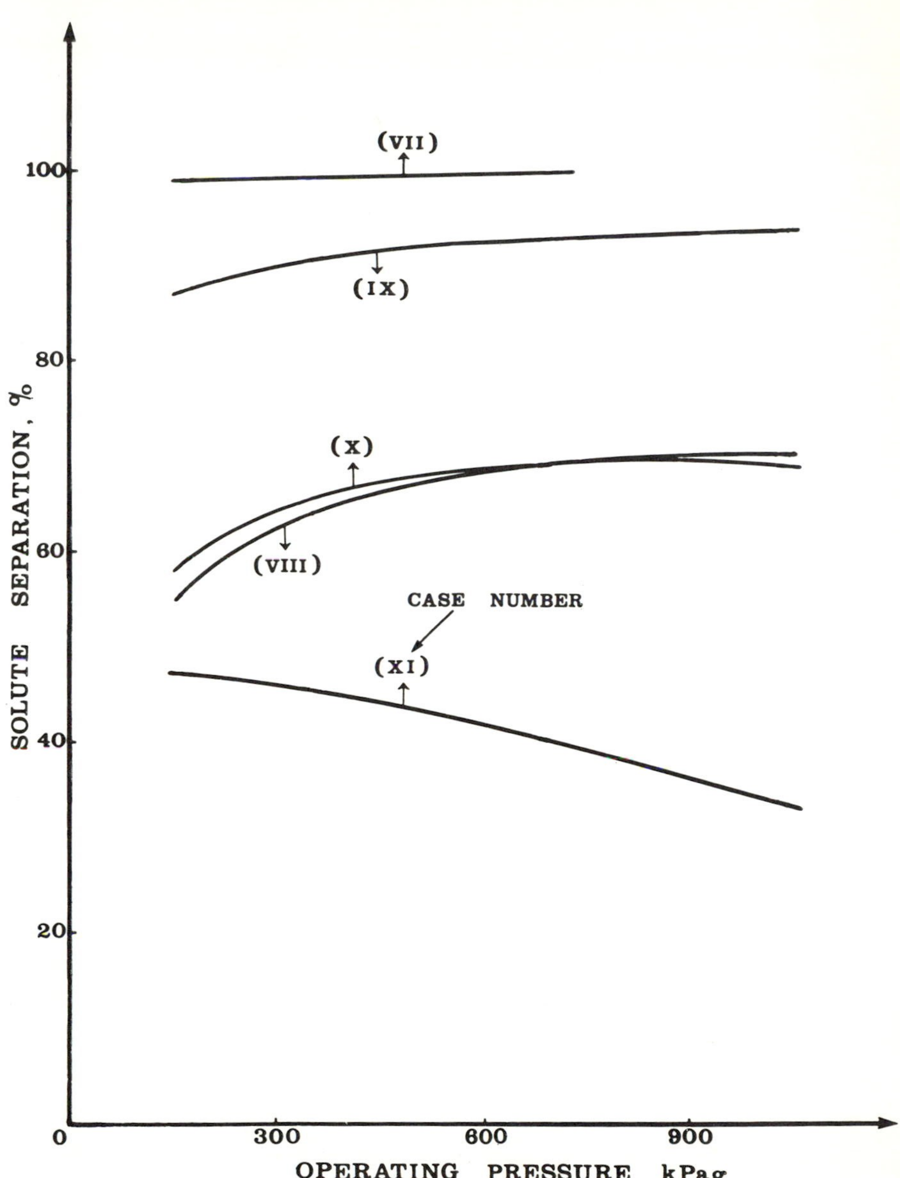

Figure 2. Effect of the operating pressure on the solute separation calculation based on k $= 22 \times 10^{-6}$ *m/s and PWP* $= 10.4 \times 10^{-3}$ *kg/h at 1724 kPag (250 psig). The membrane codes are the same as those given in Tables I and II; the case numbers are the same as those given in Table V. Membrane: PAH-09; feed concentration = 0.1–100 ppm.*

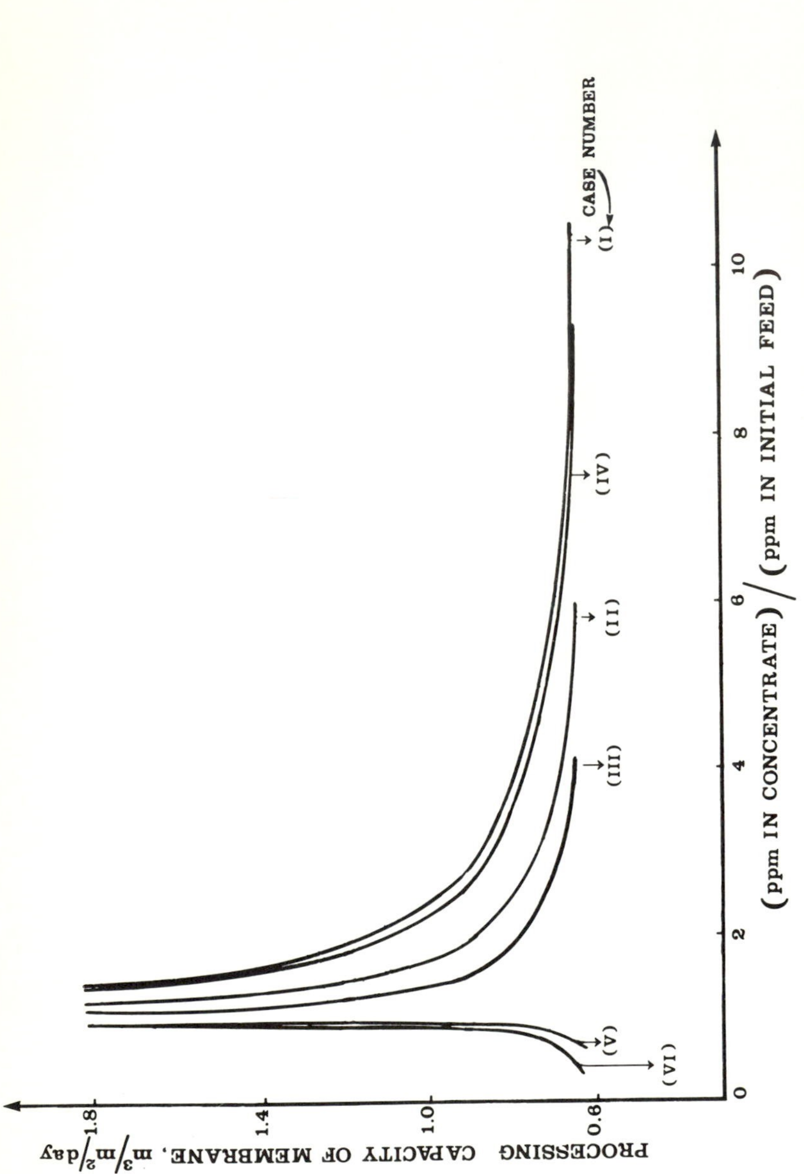

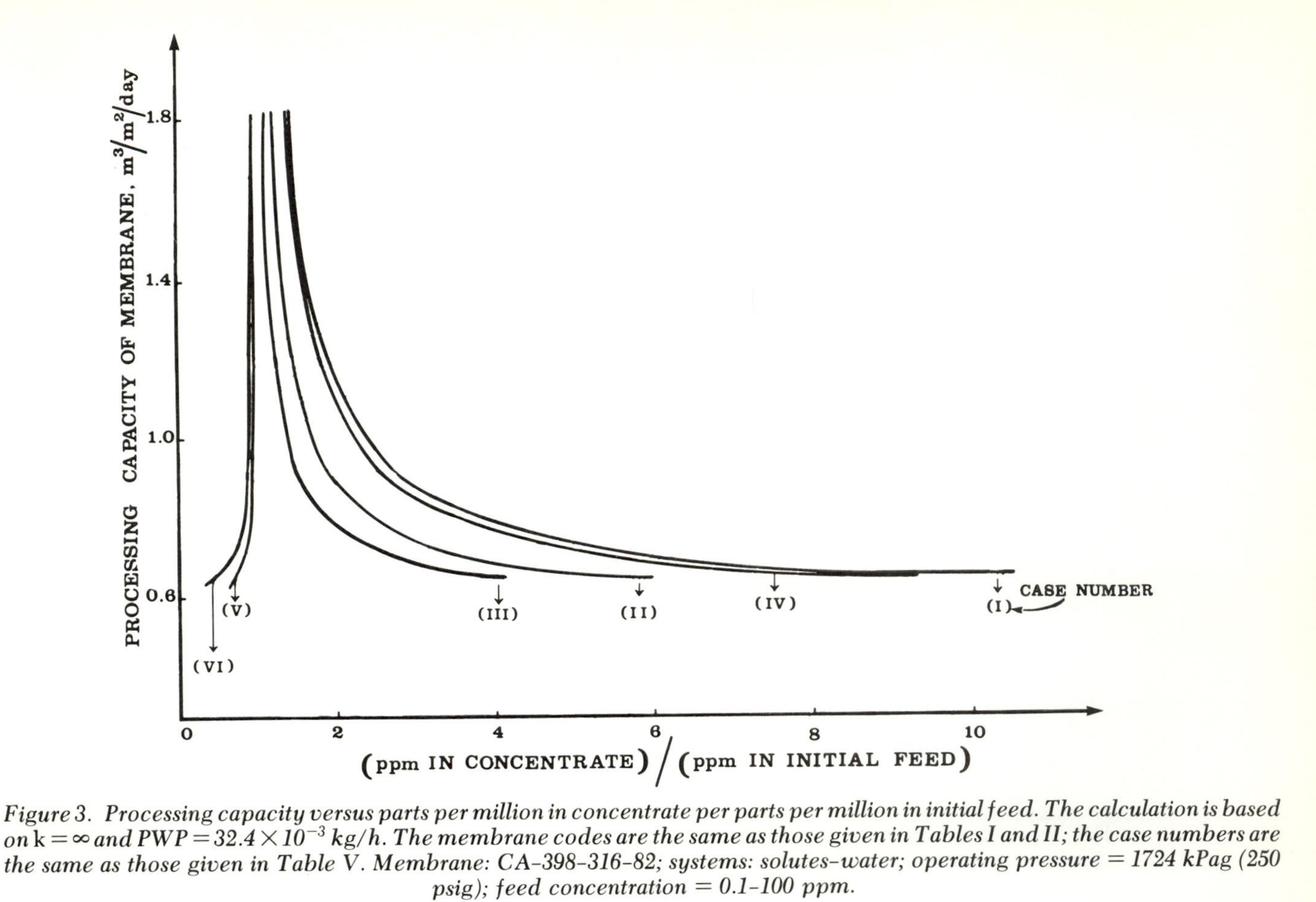

Figure 3. Processing capacity versus parts per million in concentrate per parts per million in initial feed. The calculation is based on $k = \infty$ *and* $PWP = 32.4 \times 10^{-3}$ *kg/h. The membrane codes are the same as those given in Tables I and II; the case numbers are the same as those given in Table V. Membrane: CA-398-316-82; systems: solutes–water; operating pressure = 1724 kPag (250 psig); feed concentration = 0.1–100 ppm.*

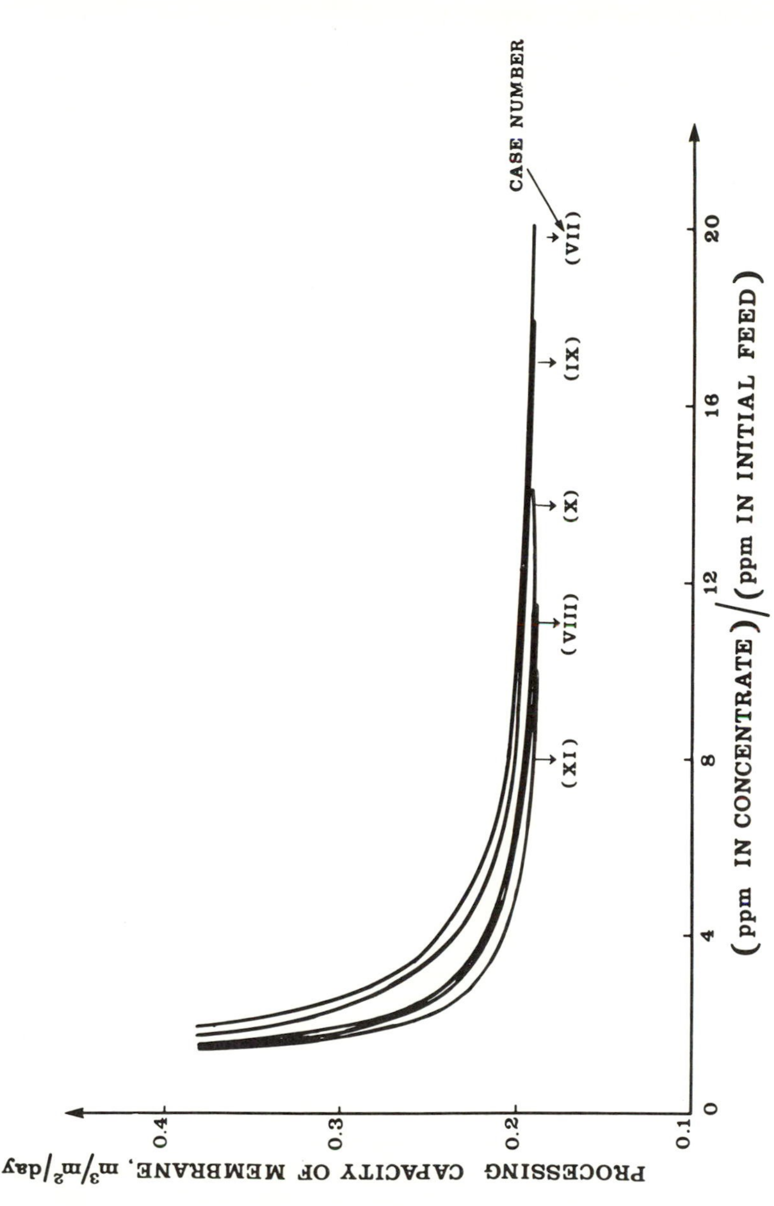

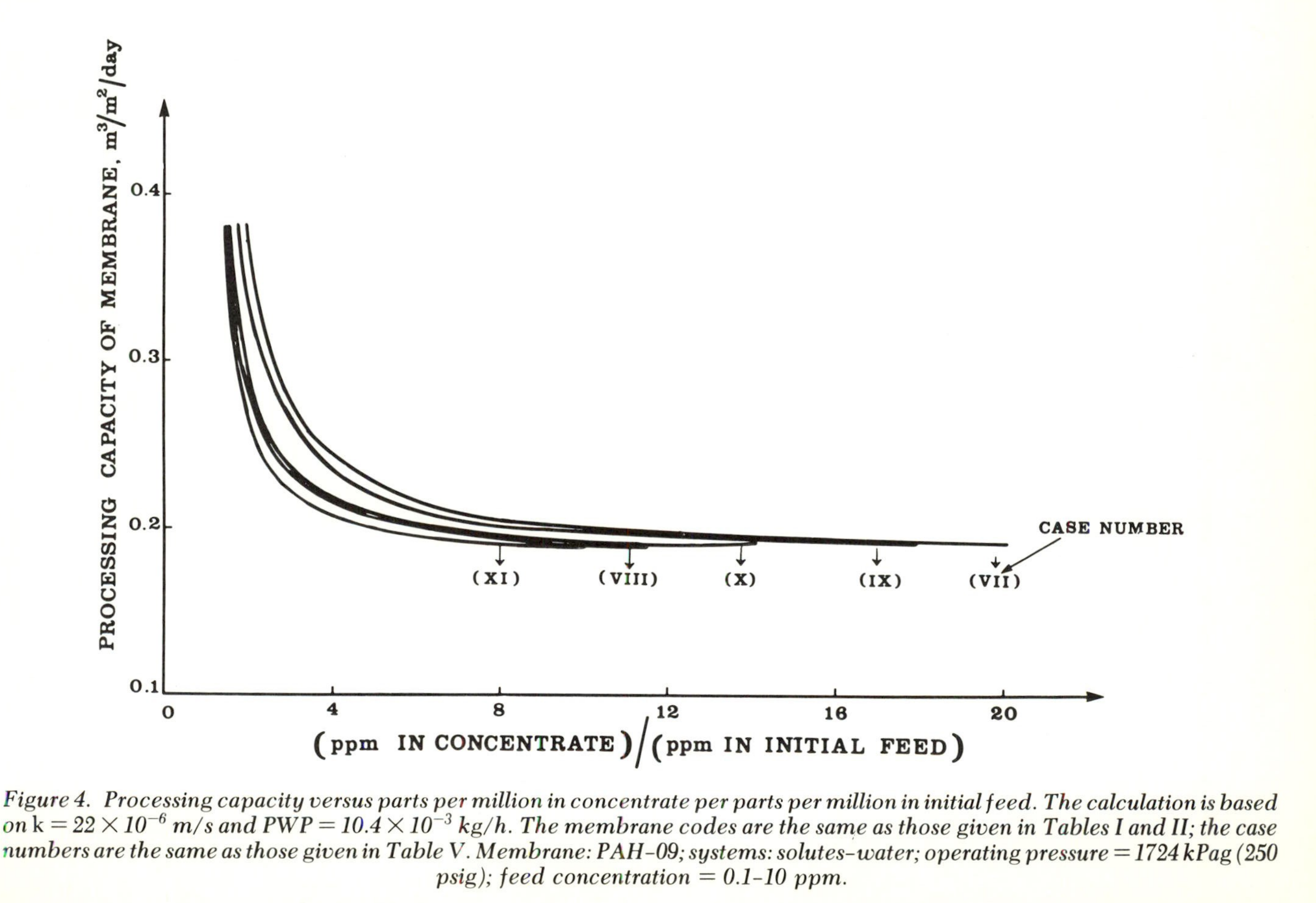

Figure 4. Processing capacity versus parts per million in concentrate per parts per million in initial feed. The calculation is based on k = 22×10^{-6} *m/s and PWP* = 10.4×10^{-3} *kg/h. The membrane codes are the same as those given in Tables I and II; the case numbers are the same as those given in Table V. Membrane: PAH-09; systems: solutes–water; operating pressure* = *1724 kPag (250 psig); feed concentration* = *0.1–10 ppm.*

butyl isopropyl ether by CA membranes was studied in detail both theoretically and experimentally. The reason for this detailed study is twofold: both glucose and *tert*-butyl isopropyl ether possess relatively large sizes among the organic solutes that constitute potential water pollutants, as indicated by large $\underset{\sim}{D}$ values of case I and case IV solutes in Table V. On the other hand, water is preferentially sorbed in the case of glucose solute, whereas the solute is preferentially sorbed in the case of *tert*-butyl isopropyl ether solute, as indicated by the large negative $\underset{\sim}{B}$ value of case I solute and by the large positive $\underset{\sim}{B}$ value of case IV solute in Table V. Thus, these solutes are good examples to demonstrate the contrast between solute separations when water is preferentially sorbed and when solute is preferentially sorbed at the membrane–solution interface. Both the theoretical and experimental solute separations as well as product rate data were obtained in this study with respect to the single solute systems. This method was used because separation and product rate data are unaffected by mixing solutes when they are undissociated polar organic solutes and the solution is maintained dilute (*15*). Because this study is on the fractionation of glucose and *tert*-butyl isopropyl ether by CA membranes, glucose will be called case I solute and *tert*-butyl isopropyl ether will be called case IV solute hereafter according to the listing of these solutes in Table V.

Figure 5 shows the results of some parametric studies on the separation of both case I and case IV solutes. In this calculation $\overline{R}_{b,2}$, $\sigma_1/\overline{R}_{b,1}$, and $\sigma_2/\overline{R}_{b,2}$ values were fixed at 50×10^{-10} m, 0.001, and 0.05, respectively, whereas $\overline{R}_{b,1}$ and h_2 values were changed. Furthermore, PR/PWP was forced to be equal to unity to facilitate the computer calculation. Thus, the calculation corresponds to the dilute solution. This assumption is not always valid particularly for case IV solute which is preferentially sorbed to the membrane polymer material. However, the general conclusion obtained from the calculated separation data is unchanged (*13*) on the basis of this assumption.

As shown in Figure 5a, the separations of case I solute are positive regardless of the $\overline{R}_{b,1}$ and h_2 values. This result is obtained because the rejective force is working between the membrane material and the solute (*13*). For an h_2 value of 0.2, solute separation increases as $\overline{R}_{b,1}$ increases, passes through a maximum at $\overline{R}_{b,1} = 13 \times 10^{-10}$ m, and starts decreasing thereafter. The initial increase of solute separation with the increase in $\overline{R}_{b,1}$ is due to the increased contribution to the overall membrane separation from the solution that permeates through the pores belonging to the first (smaller) pore size distribution. Because solute separation by the pores of the first distribution is significantly higher than that by the pores of the second distribution, the increase in the contribution of the flow through the pores of the first distribution

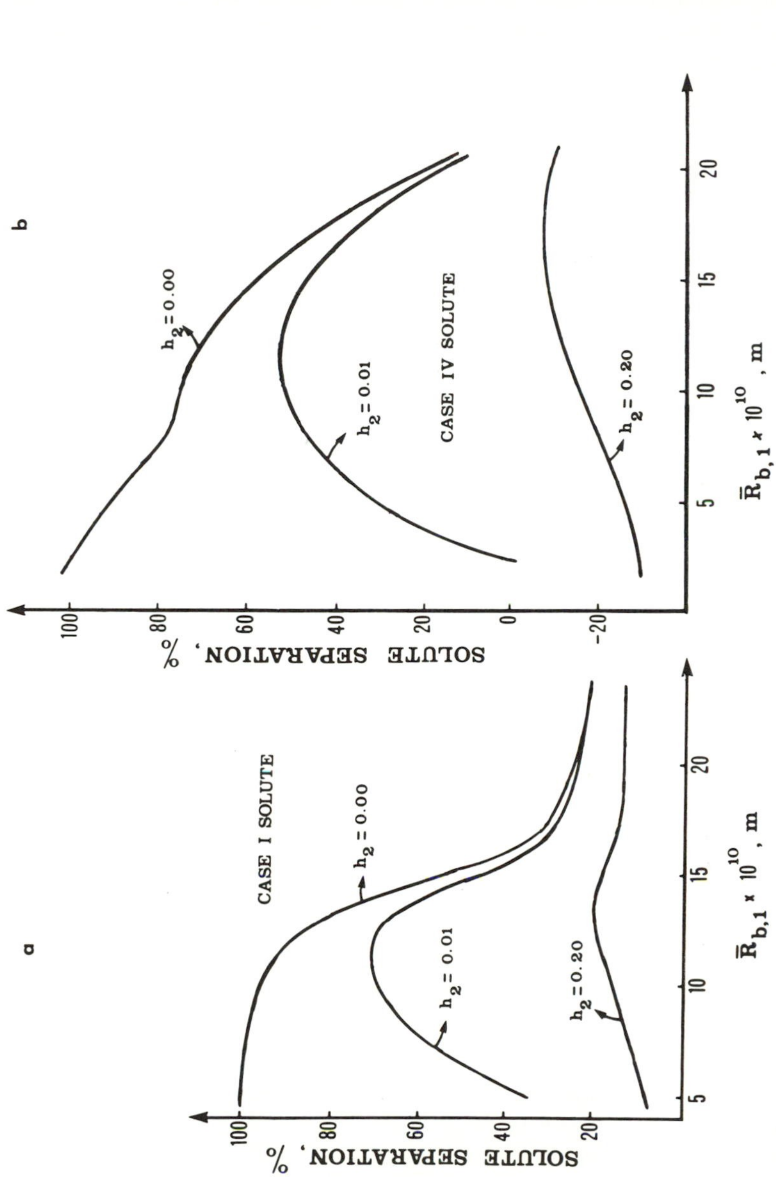

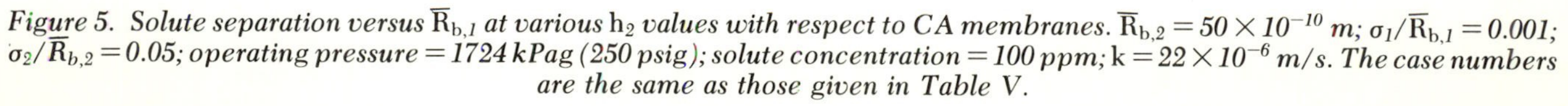

Figure 5. Solute separation versus $\overline{R}_{b,1}$ *at various* h_2 *values with respect to CA membranes.* $\overline{R}_{b,2} = 50 \times 10^{-10}$ *m;* $\sigma_1/\overline{R}_{b,1} = 0.001$*;* $\sigma_2/\overline{R}_{b,2} = 0.05$*; operating pressure = 1724 kPag (250 psig); solute concentration = 100 ppm;* $k = 22 \times 10^{-6}$ *m/s. The case numbers are the same as those given in Table V.*

increases the overall solute separation. When the size of the pores of the first distribution increases even further, however, the solute separation by the first pore tends to decrease. Therefore, the maximum found in the solute separation versus $\overline{R}_{b,1}$ curve is the result of the superimposition of the opposing tendencies.

Figure 5a also indicates that the $\overline{R}_{b,1}$ value corresponding to the maximum solute separation of case I solute shifts gradually to the smaller value as the h_2 value decreases and ultimately disappears when $h_2 = 0$.

The separation for case IV solute shown in Figure 5b is different from that for case I solute in several aspects. Comparing the separation data of both solutes at $h_2 = 0.2$, the separation of case IV solute is negative. This result reflects the strong attractive force working between the membrane material and solute as indicated in Table V by the large positive $\underset{\sim}{B}$ value of case IV solute. On the other hand, the maximum in the separation of case IV solute shifts toward the larger $\overline{R}_{b,1}$ value in comparison to case I solute at each level of h_2. This result reflects the larger size of case IV solute, which is again indicated in Table V by a larger $\underset{\sim}{D}$ value of case IV solute than that of case I solute. As a result, in some range of the large $\overline{R}_{b,1}$ values, the separation of case IV solute may surpass that of case I solute. Furthermore, as h_2 decreases, the $\overline{R}_{b,1}$ value corresponding to the maximum separation of case IV solute also decreases. Interestingly, the tendency to exhibit a maximum separation is preserved in the form of an inflection point at $\overline{R}_{b,1} = 9 \times 10^{-10}$ m, even when $h_2 = 0$.

The separation data of both case I and case IV solutes shown in Figure 5a and Figure 5b were then amalgamated into Figure 6. In Figure 6, the separation data approaches the left-bottom corner as h_2 increases. This behavior is not surprising because the increase in the average pore size (of the total pores on the membrane surface) caused by an increase in h_2 decreases the separation of both solutes. When $h_2 = 0.2$, separations of both case I solute and case IV solute increase as $\overline{R}_{b,1}$ increases. At $\overline{R}_{b,1} = 13 \times 10^{-10}$ m, the separation of case I solute starts decreasing, whereas that of case IV solute continues to increase. Furthermore, at $\overline{R}_{b,1} \geq 21 \times 10^{-10}$ m, separations of both solutes tend to decrease. As h_2 decreases, the general pattern of the curve is preserved, although the $\overline{R}_{b,1}$ value at which the separation of case I solute starts decreasing decreases from 13×10^{-10} m at $h_2 = 0.2$ to 9×10^{-10} m at $h_2 = 0.003$. When $h_2 = 0$, separations of both case I and case IV solutes tend to decrease monotonically as $\overline{R}_{b,1}$ increases. Figure 6 indicates that the fractionation of case I solute and case IV solute is possible by CA membranes under the given operating conditions at any point in the region surrounded by the triangular envelope generated by assembling all data points corresponding to different h_2 and $\overline{R}_{b,1}$ values. Such an

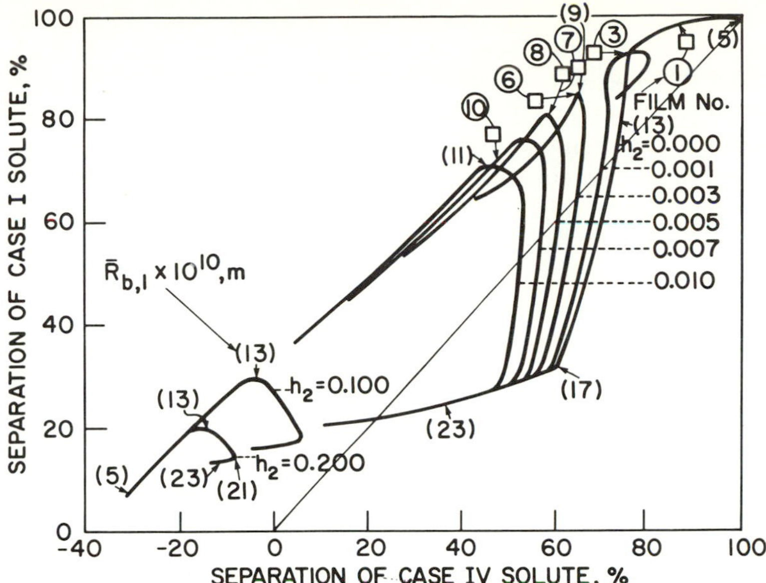

Figure 6. Separations of case I and case IV solutes at various $\bar{R}_{b,1}$ *and* h_2 *values with respect to CA membranes.* $\bar{R}_{b,2} = 50 \times 10^{-10}$ *m;* $\sigma_1/\bar{R}_{b,1} = 0.001$*;* $\sigma_2/\bar{R}_{b,2} = 0.05$*; operating pressure = 1724 kPag (250 psig); solute concentration = 100 ppm; k = 22 × 10^{-6} m/s. The case numbers are the same as those in Table V. Key: □, experimental points; ——, calculated values.*

assembly of data points will be called a data bank hereafter. The separation of case I solute may range from 30% to 82% at a given separation of case IV solute of 60%. Thus, the separation of case I solute may be more or less than that of case IV solute depending on the nature of the pore size distribution on the membrane surface. The upper limit of the data bank is constituted by the points corresponding to low $\bar{R}_{b,1}$ values, whereas the lower limit corresponds to high $\bar{R}_{b,1}$ values. Because the $\bar{R}_{b,2}$ value is fixed at 50×10^{-10} m in this calculation, a higher value of $\bar{R}_{b,1}$ means a more uniform pore size distribution. Therefore, the more uniform the pore size distribution, the less the separation of case I solute expected for a given separation of case IV solute. As the furthest extreme of this tendency, the correlation corresponding to the most uniform pore size distribution ($h_2 = 0$) constitutes the lower limit of the entire data bank.

Some of the calculated values shown in Figure 6 were experimentally verified by using films 1, 3, 6, 7, 8, and 10. As shown in Table II, $\bar{R}_{b,2}$ values of these membranes were in the range from 40×10^{-10} m to approximately 55×10^{-10} m, $\sigma_1/\bar{R}_{b,1}$ was 0.001 except for membranes 3 and 6, and $\sigma_2/\bar{R}_{b,2}$ was 0.05 except for membrane 6; thus, films 1, 7, 8, and 10 particularly satisfy the restriction under which Figure 6 was produced. On the other hand, $\bar{R}_{b,1}$ was in the range from 8.4×10^{-10} to 9.6×10^{-10}, whereas h_2 was in the range from 0 to 0.006. As expected from the relatively low $\bar{R}_{b,1}$ values, the experimental points should fall close to the upper envelope of the data bank; this situation was the case as indicated by experimental points in Figure 6. Furthermore, in Figure 6, the experimental points are connected by arrows to the corresponding values calculated on the basis of the pore size distribution given in Table II. The agreement between calculated and experimental fractionation data is indeed good considering the many variables involved in the calculation. Unfortunately, our capability of adjusting the pore size distribution in the real membrane is rather limited at present, and hence the experimental data points corresponding to the lower envelope of the data bank were not achievable. As mentioned earlier, a more uniform pore size distribution with larger $\bar{R}_{b,1}$ values is necessary for this purpose. The experimental data, however, indicate that the existence of a dual pore size distribution is necessary to explain the experimental data on the separations of case I and case IV solutes by the transport model employed in this study and the associated numerical parameters.

Our next study is concerned with the fractionation of glucose and 3-chlorophenol by CA membranes. 3-Chlorophenol was chosen for the study particularly because of its low $\underset{\sim}{D}$ value and high $\underset{\sim}{B}$ value, the combination of which often leads to the negative separation of the solute. Furthermore, 3-chlorophenol may represent the entire chlorophenols that are regarded as typical water pollutants (*16*). According to Table V, the system involving 3-chlorophenol and CA–398 material is called case V solute system. A parametric study was conducted again by fixing $\bar{R}_{b,2}$, $\sigma_1/\bar{R}_{b,1}$, and $\sigma_2/\bar{R}_{b,2}$ values at 50×10^{-10} m, 0.001, and 0.05, respectively, and by changing $\bar{R}_{b,1}$ and h_2 values. Again, the PR/PWP ratio was assumed to be unity to facilitate the computation. The calculated results are shown in Figure 7. At the h_2 value of zero, Figure 7 shows that the separation of case I solute decreased from ~100% to nearly equal to 20% when $\bar{R}_{b,1}$ was changed from 5 to 23×10^{-10} m, whereas the separation of case V solute decreased from +10% to −10% under the given conditions. When h_2 was 0.001, the separation of case I solute first increased with an increase in $\bar{R}_{b,1}$, passed maximum when $\bar{R}_{b,1} = 9 \times 10^{-10}$ m, and further decreased to approximately 20% at the $\bar{R}_{b,1}$ value of 23×10^{-10} m. Case V solute, on the other hand, decreased from +8% to −10% as $\bar{R}_{b,1}$ increased from 5×10^{-10} m to 23×10^{-10} m. The main feature of the curve corresponding to $h_2 = 0.001$ is preserved for the higher h_2 values. As a result, all the calculated points are bounded by the curve corresponding to $h_2 = 0$. Besides, at any h_2 value, the data for $\bar{R}_{1,b}$ values larger than 11×10^{-10} m are expected to fall on the boundary of the data bank, whereas the data are expected to move away from the

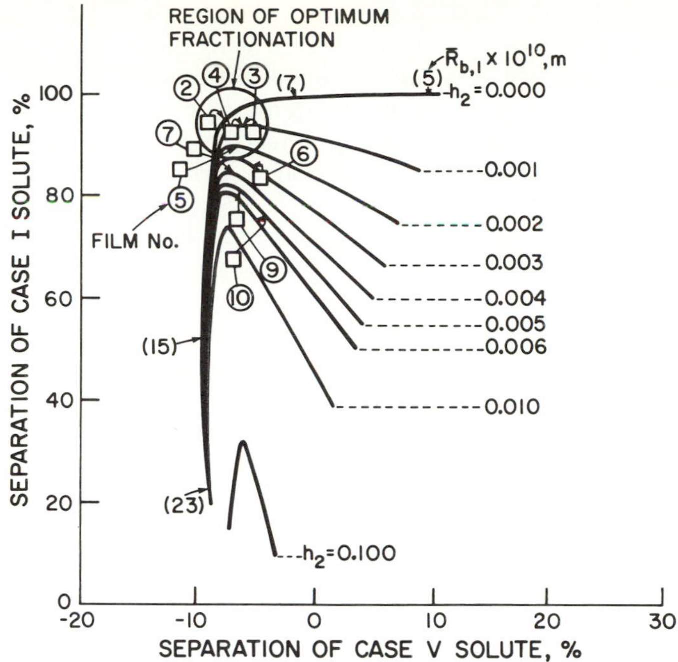

Figure 7. Separations of case I and case V solutes at various $\bar{R}_{b,1}$ and h_2 values with respect to CA membranes. $\bar{R}_{b,2} = 50 \times 10^{-10}$ m; $\sigma_1/\bar{R}_{b,1} = 0.001$; $\sigma_2/\bar{R}_{b,2} = 0.05$; operating pressure = 1724 kPag (250 psig); solute concentration = 100 ppm; $k = 22 \times 10^{-6}$ m/s. The case numbers are the same as those in Table V. Key: □, experimental points; ——, calculated values.

boundary to the right (further toward the inside of the data bank) as $\bar{R}_{b,1}$ decreases. No data can be on the left side of the boundary as far as CA membranes are used under the specified experimental conditions.

In Figure 7 the experimental data are also plotted. The plot of such data is justified because membranes 2–7, 9, and 10, which were considered in the study, possess $\bar{R}_{b,2}$ values in the range of $40 \sim 55 \times 10^{-10}$ m, $\sigma_1/\bar{R}_{b,1}$ value of <0.01 (except membranes 3 and 6), and $\sigma_2/\bar{R}_{b,2}$ value of 0.05 (except membrane 6). Thus, membranes 2, 4, 5, 7, 9, and 10 particularly satisfy the conditions that were assumed when the calculated data in Figure 7 were produced. Attention has to be paid to the two main features of these experimental points. First, experimental data were connected by arrows to the values calculated on the basis of the pore size distribution data given in Table II. The agreement of experimental and calculated data is again generally good. The data corresponding to membranes 2, 5, and 7 fall on the left side of the boundary of the data bank, whereas data from membranes 3, 4, 6, 9, and 10 fall on the right side of the boundary. $\bar{R}_{b,1}$ values of membranes 2, 5, and 7 are more than or equal to to 9.4×10^{-10} m. On the other hand, $\bar{R}_{b,1}$ values of membranes 3, 4, 6, 9, and 10 are less than or equal to 9.1×10^{-10} m. Therefore, membranes 2, 5, and 7 possess distinctively higher $\bar{R}_{b,1}$ values than membranes 3, 4, 6, 9, and 10. As predicted by numerical calculation, the data should move toward the right into the data bank such as shown in Figure 7 as $\bar{R}_{b,1}$ decreases. The actual experimental data indicate the predicted trend.

By defining the optimum fractionation of case I solute and case V solute as the maximum of the value $(f_I - f_V)^2$, the optimum membrane pore size distribution that corresponds to the maximum fractionation can be searched for. For example, the optimum fractionations are represented in Figure 7 by a circular region located at the left-top corner of the fractionation data bank. The necessary pore size distributions for achieving such optimum fractionations were produced because membranes 2, 3, and 4 and experimental data from the membranes fell precisely into the circular region. Thus, the fractionation of case I and case V solutes can be optimized by a proper design of the pore size and pore size distribution by computer analysis and the formation of membranes that possess the calculated pore size distributions.

Conclusion

The interfacial force constants available in the literature for many organic solutes that constitute potential pollutants in water enable one to calculate the separation of such solutes at various operating conditions by a membrane of a given average pore size and pore size distribution on the basis of the surface force–pore flow model. The product rate of the permeate solution can also be calculated. Such data further allow us to calculate the processing capacity of a membrane to achieve a preset ratio of concentration in the concentrate to concentration of the initial feed solution.

This approach also enables the design of the pore size distribution that optimizes the fractionation of different solutes. Furthermore, the correlation of the process of the membrane formation to the average pore size and the pore size distribution produced on the membrane surface enables one to produce membranes that are most appropriate for the fractionation of a given set of solutes. Therefore, this approach contributes to the rational design of membranes for the concentration of water pollutants of particular interest.

Acknowledgments

One of the authors (T.D.N.) gratefully acknowledges the award of research associateship from the National Research Council of Canada.

List of Symbols

A	pure water permeability constant (kg-mol/$m^2 \cdot s \cdot$ kPa)
A_p	the surface area of polymer powder in the chromatography column (m^2)
a	weight fraction of the solute in the product solution (ppm)
$\underset{\sim}{B}$	constant characterizing the van der Waals attraction force (m^3)
$c_{A,b}$	bulk concentration (mol/m^3)
$\underset{\sim}{D}$	constant characterizing the steric repulsion at the interface (m)
$(D_{AM}/K\delta)_{NaCl}$	transport parameter of reference sodium chloride in water (treated as single quantity) (m/s)
d	distance between polymer material surface and the center of solute molecule (m)
f	fraction solute separation based on the feed concentration
G	product permeation flux (kg/$m^2 \cdot$ day)
h_i	ratio defined by eq 3
n_i	number of pores belonging to the ith normal distribution
PR	product rate through given area of membrane surface (kg/h)
PWP	pure water permeation rate through given area of membrane surface (kg/h)
$\underset{\sim}{R}$	gas constant
R_b	pore radius (m)
$\overline{R}_{b,i}$	average pore radius of the ith distribution (m)
S	area of membrane surface (m^2)
T	absolute temperature (K)
t	operation period (days)
V_i	volume of charge on the high-pressure side of the membrane at $t = 0$ in the batch process (m^3)
V'_{RA}	chromatography retention volume of solute A (m^3)
z	weight fraction of solute in the selection on the high-pressure side of the membrane (ppm)
Γ_A	surface excess of solute A (mol/m^2)
ρ_1	density of the solution (kg/m^3)
σ_i	standard deviation of the ith normal pore size distribution (m)
ϕ	potential function of interaction force exerted on the solute from the pore wall (J/mol)

Literature Cited

1. Kurihara, M.; Harumiya, N.; Kanamura, N.; Tonomura, T.; Nakasatomi, M. *Desalination* **1981,** *38,* 449.
2. Riley, R. L.; Fox, R. L.; Lyons, C. R.; Milstead, C. E.; Soroy, M. W.; Tagani, M. *Desalination* **1976,** *19,* 113.
3. Matsuura, T.; Blais, P.; Pageau, L.; Sourirajan, S. *Ind. Eng. Chem. Process Des. Dev.* **1977,** *16,* 510.
4. Kutowy, O.; Thayer, W. L.; Sourirajan, S. *Desalination* **1978,** *26,* 195.
5. Chan, K.; Liu, T.; Matsuura, T.; Sourirajan, S. *Ind. Eng. Chem. Prod. Res. Dev.* **1984,** *23,* 124.
6. Chan, K., Matsuura, T.; Sourirajan, S. *Ind. Eng. Chem. Prod. Res. Dev.* **1984,** *23,* 492.
7. Sourirajan, S. *Reverse Osmosis;* Academic: New York, 1970; Chapter 3.
8. Matsuura, T.; Sourirajan, S. *Ind. Eng. Chem. Process Des. Dev.* **1981,** *20,* 273.
9. Matsuura, T.; Taketani, Y.; Sourirajan, S. In *Synthetic Membranes;* Turbak, A. F., Ed.; ACS Symposium Series 154; American Chemical Society: Washington, DC, 1981; Vol. 2, p 315.
10. Chan, K.; Matsuura, T.; Sourirajan, S. *Ind. Eng. Chem. Prod. Res. Dev.* **1982,** *21,* 605.
11. Sourirajan, S. *Lectures on Reverse Osmosis;* National Research Council of Canada: Ottawa, 1983.
12. Sourirajan, S. *Reverse Osmosis;* Academic: New York, 1970; Chapter 6.
13. Matsuura T.; Tweddle, T. A.; Sourirajan, S. *Ind. Eng. Chem. Process Des. Dev.* **1984,** *23,* 674.
14. Matsuura, T.; Sourirajan, S. Presented at the AIChE Annual Meeting (Diamond Jubilee), Washington, DC, Oct. 30–Nov. 4, 1983.
15. Matsuura, T.; Bednas, M. E.; Sourirajan, S. *J. Appl. Polym. Sci.* **1974,** *18,* 567.
16. Robertson, J. H.; Cowen, W. F.; Longfield, J. Y. *Chem. Eng.* **1980,** *June 30,* 102.
17. Nguyen, T. D.; Chan, K.; Matsuura, T.; Sourirajan, S. *Ind. Eng. Chem. Prod. Res. Dev.* **1984,** *23,* 501.

RECEIVED for review August 14, 1985. ACCEPTED December 18, 1985.

8

Concentration of Selected Organic Pollutants: Comparison of Adsorption and Reverse-Osmosis Techniques

Murugan Malaiyandi[1], R. H. Wightman[2], and C. LaFerriere[2]

[1]**Environmental Health Directorate, Health and Welfare Canada, Ottawa, Ontario, Canada, K1A 0L2**
[2]**Department of Chemistry, Carleton University, Ottawa, Ontario, Canada, K1A 5B6**

Polar organic pollutants such as 2,4-dichlorophenol, 2,4,5-trichlorophenol, 4-chloroaniline, and 3,3′-dichlorobenzidine and nonpolar organics such as α-hexachlorocyclohexane and bis(2-ethylhexyl) phthalate in aqueous solutions were concentrated by an adsorption–desorption technique using XAD-2 and XAD-4 resins and carbon-impregnated polyurethane foam. By using concentrations ranging from parts-per-million to parts-per-trillion levels, both resins behaved similarly in their concentration efficiency; however, the modified polyurethane foam was inadequate for 4-chloroaniline. Also compared was the reverse-osmosis technique as a potential method for concentrating the same organic pollutants from aqueous solutions. This study reemphasizes the general ineffectiveness of cellulose acetate membranes for rejecting small organic molecules in low concentrations, whereas polyamide hydrazide and polybenzimidazolone membranes seem to show promise for rejecting such compounds.

THE CONCENTRATION AND ANALYSIS of organic pollutants in environmental aqueous samples have been the focus of many studies (*1–6*). Included in these studies are (1) the large number of organic compounds and their functional diversity; (2) the variation in their levels; (3) the variety of methods to concentrate or separate these compounds from an aqueous matrix; and (4) techniques for detection, identification, or analysis. However, it is rather difficult to find many comparative studies involving all these factors.

We have attempted to provide a comparison of concentration and

0065-2393/87/0214/0163$06.00/0
Published 1987 American Chemical Society

analytical methods that include six common priority organic pollutants representing acidic functionalities [2,4-dichlorophenol (DCP) and 2,4,5-trichlorophenol (TCP)]; basic functionalities [4-chloroaniline (CA) and 3,3′-dichlorobenzidine (DCB)]; and neutral functionalities [α-1,2,3,4,5,6-hexachlorocyclohexane (BHC) and bis(2-ethylhexyl) phthalate (DEHP)]. Concentration procedures involving three of the more commonly used solid adsorbents (*7–9*) were compared to the reverse-osmosis (RO) technique. The solid adsorbents in this study were XAD-2 (*10*) and XAD-4 (*11*) macroreticular resins and polyurethane foam impregnated with 1% activated carbon (*12, 13*). For the RO studies, the membranes employed were cellulose triacetate (CAc), polyamide hydrazide (PA), and polybenzimidazolone (PBI) (*14–18*). This study also includes results from two types of analytical methodology, namely, capillary gas chromatography (GC) (*2, 19–21*) and reverse-phase high-performance liquid chromatography (HPLC) (*3, 22*). Furthermore, GC analysis of two field samples of drinking water and GC–mass spectrometric (GC–MS) identification of some of the compounds found in these water samples are also reported in this chapter.

Experimental

Materials and Reagents. All glassware was washed with chromic acid and thoroughly rinsed successively with water, glass-distilled acetone, and purified hexane (*23*). Selected organic compounds for this study were commercially available: DCP, TCP, CA, and DEHP (Aldrich Chemicals); DCB (Supelco); and BHC (Analabs). These compounds were purified, if necessary; checked for purity by IR, UV, GC, HPLC, and ^{1}H and ^{13}C NMR; and shown to be ≥97% pure. Anhydrous Na_2SO_4 (pesticide grade, Canlab) was prerinsed with purified solvents before use.

ADSORBENTS. Macroreticular Amberlite resins XAD-2 and XAD-4 (20–50 mesh; Lots 90721 and 89829, respectively) were obtained from BDH Chemicals Ltd. Polyurethane foam (upholstery grade, Woodbridge Foam Co.) and vegetable charcoal (Darco G-60; Lot 363-53; Matheson Canada Ltd.) were purchased.

MEMBRANES. Flat sheets of CAc, PA, and PBI membranes were cast at the National Research Council of Canada by using published procedures (*24*) and were selected to obtain two different porosities as determined by percent NaCl rejection. The membrane sheets (ca. 400 μm thickness) were cut into circles approximately 7.5 cm in diameter.

SOLVENTS. Hexanes (distilled in glass, Caledon Laboratories Ltd.) were purified by treating them with H_2SO_4 and $KMnO_4$ as previously described (*23*). Acetonitrile and acetone (HPLC grade, Caledon Laboratories Ltd.) were used as received.

GLASSWARE. All special borosilicate glassware was fabricated in-house. Kuderna-Danish (K-D) evaporators and the modified Snyder columns were constructed as per the design described previously (*25*); adsorption glass columns (60-cm × 1.1-cm i.d.) were fitted with coarse glass frits, polytetrafluoroethylene (PTFE) polymer stopcocks (2 mm), and ground glass standard-tapered 24/40 joints; reservoirs for the columns were 1-L separatory funnels with ground glass 24/40 joints and PTFE polymer stopcocks (4 mm); effluent receivers were 1-L Buchner filter flasks fitted with ground glass 24/40 joints. A 4-L reservoir for the RO system was provided with a ground glass joint 40/50 and an outlet on the side near the bottom of the reservoir.

Equipment. RO cells (Figure 1) were constructed of 316 stainless steel at the Science Technology Centre, Carleton University, Ottawa, Canada, as per design described by Matsuura et al. (*26*). The RO system consisted of six cells (assembled as shown in Figure 2), a variable-flow circulating pump and motor with 316 stainless steel valves, Viton diaphragm (BIF no. 1731-12-9820, rated at 13 gal/h at 950 psig), a surge tank with Viton diaphragm (Greer, 1 pt; Dynesco Equipment Sales), miscellaneous valves and gauges, and the 4-L borosilicate

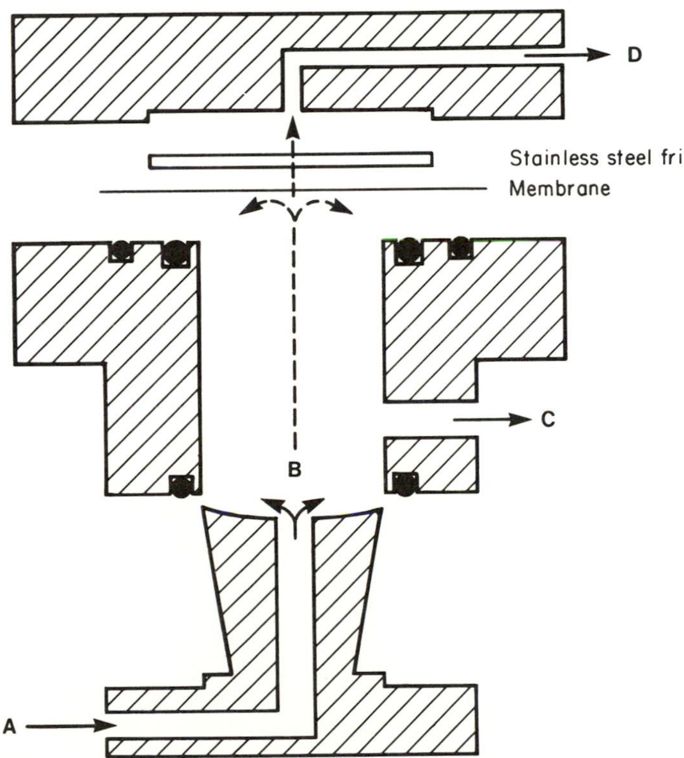

Figure 1. Exploded view of RO cell (to scale). The various components of the cell fit together, are compressed by machine bolts, and are sealed with Viton O-rings. The membrane (effective diameter = 3.8 cm) is compressed against a porous steel plate (1/16 in., porosity = 25 μm) and flushed with feed solution. A certain amount of water penetrates the membrane and is collected as the permeate water (D). The feed solution enters the cell (A) and washes across the membrane (B) before being forced out of the cell (C).

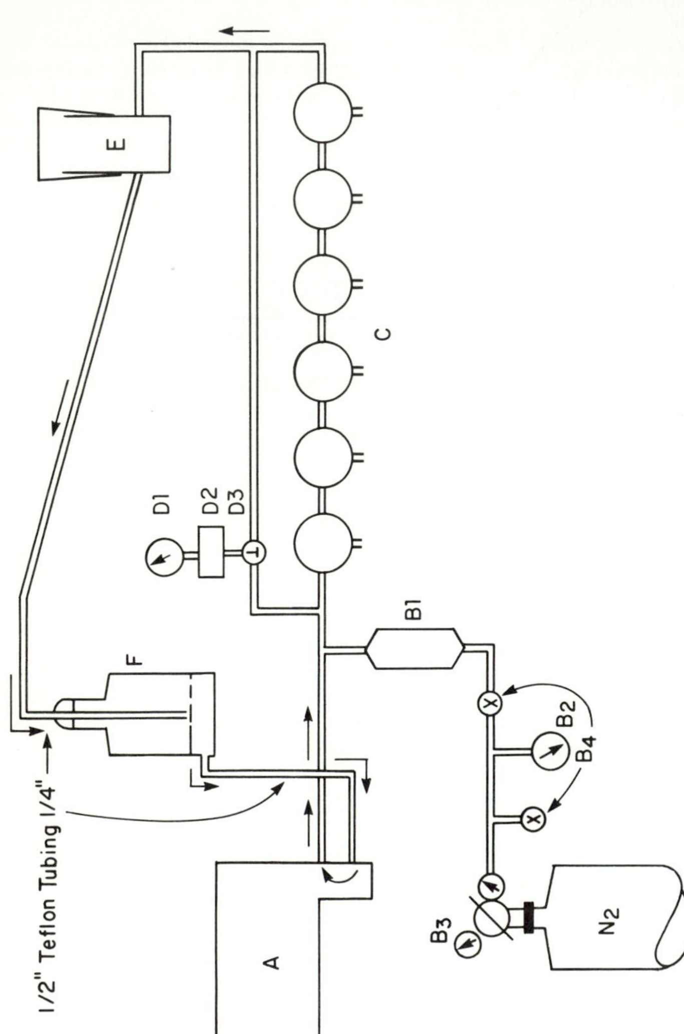

Figure 2. Schematic representation of RO system. Feed solution travels from reservoir (F) via pump (A) through cells (C) where a portion of the water permeates the cells. Pressure in the system is adjusted by the regulator (E) and monitored by gauge and valve system (D). Damping of pressure fluctuations is achieved by nitrogen pressure and the system (B).

reservoir. The connecting tubing was made of either 316 stainless steel or PTFE polymer.

Instrumentation and Analytical Procedures. GAS CHROMATOGRAPHY. A Vista 6000 (Varian) instrument equipped with a fused silica capillary column (DB-5, J & W Scientific, 14 m × 0.25 mm), a flame ionization detector (FID, 1×10^{-11} amperes full scale [AFS]), and an electron capture detector (ECD, Range 10) were used. The detectors were used as required. A splitless injection configuration with a purge flow of 40 mL/min was used at all times. Typical parameters were as follows: carrier gas, ultra-high-purity (UHP) helium (flow adjusted to give a linear velocity of 25 cm/sec for butane); makeup gas to the detector, nitrogen (zero gas, <0.5% hydrocarbons), 26 mL/min; injector, 280 °C; detector, 300 °C. A typical column oven temperature program was 50 °C, (hold for 1 min) to 75 °C at 20 °C/min; hold at 75 °C for 5 min; 75–250 °C at 10 °C/min; hold at 250 °C for 10 min. Under these conditions, the following retention times were observed: 9.3 min for DCP, 10.4 min for CA, 13.6 min for TCP, 18.3 min for BHC, 26.1 min for DCB, and 27.0 min for DEHP. One-microliter samples of concentrates or the working standard were injected with appropriate attenuations.

GAS CHROMATOGRAPHY–MASS SPECTROMETRY. A Finnigan 4500/Incos instrument with a 30-m × 0.32-mm i.d. capillary column coated with SP-B-5 was used. The GC parameters were as follows: injector, 270 °C; column oven temperature programmed, 50 °C (0.1 min, hold); 15 °C/min to 100 °C, 5 °C/min to 270 °C; internal standard, anthracene-d_{10}; helium flow, 3.0 mL/min; sample size, 3.0 μL. MS conditions were as follows: EI, 70 eV; scan (m/z), 35–650 daltons; source temperature, 250 °C; filament current, 0.5 A; sensitivity, 10^{-8} A/V. (NOTE: When the name of a compound is followed by "(confirmed)", it means that the standard material was analyzed for confirmation under conditions identical to those of the sample; when the name is followed by "(tentative)", it means that the mass fragmentography showed the best fit (>80%) based on the National Bureau of Standards [NBS] library computer search.)

LIQUID CHROMATOGRAPHY. A model M 6000A (Waters) instrument was used with a manually variable UV–vis detector (Schoeffel Instruments Co.) and a U6K injector, both supplied by Technical Marketing Associates. A Hamilton PRP-1, reverse-phase resin, 150-mm × 4.1-mm (10 μm) mesh column packing was used under the following conditions: mobile phase, acetonitrile/water (4:6 v/v); flow rate, 1 mL/min. The retention times were as follows: 5.35 min for CA (λ = 243 nm), 7.3 min for DCP (λ = 243 nm), and 13.4 min for TCP (λ = 257 nm). The composition of mobile phase was altered to 20% water in acetonitrile to give a retention time of 3.1 min for DCB (λ = 213 nm) at the flow rate of 2.0 mL/min. Injection volumes of actual samples and working standards and attenuations were varied as necessary. An electronic integrator (Spectra-Physics Minigrator) and a 1-m V strip chart recorder (Fisher Recordall 5000) were routinely employed in this study.

PREPARATIONS AND PURIFICATION. Purified water was prepared by passing distilled water through two borosilicate glass columns, each separately containing purified XAD-2 and XAD-4 resin (75 cm × 3 cm) connected in series. The resin-treated water was redistilled in glass from an alkaline $KMnO_4$ solution by using a Vigreux column (2 m) and was collected in precleaned amber-colored bottles with PTFE polymer-lined caps.

Stock Solutions. Aqueous stock solutions (1–10 ppm) of the polar organic compounds were prepared by adding known weights of pure compound(s) to an appropriate volume of purified water in a thoroughly rinsed 12-L round-bottomed flask. The mixture was heated to 50 °C and stirred with a PTFE polymer-coated magnetic stirrer for 24–48 h; it was then cooled, while being stirred, to room temperature for 24–48 h before it was filtered through a precleaned Millipore filter (HA, 0.4 μm) (*25*). The exact concentrations of organic compounds in these solutions were determined on the basis of peak heights of the compounds in samples and in working standards. The stock solutions of the nonpolar compounds (BHC and DEHP, 50–100 ppb) were prepared by adding aliquots of concentrated solutions of known concentrations of these organic compounds in acetone to the required volume of water.

Adsorbents. The macroreticular resins XAD-2 and XAD-4 were separately suspended in distilled water, and the suspensions were stirred to leave the fine particulates floating. These fines were removed by decantation of the supernatant layer. This operation was repeated until no opalescence was noticeable in the supernatant layer. After filtration through Whatman no. 1 filter paper and washing with methanol, the resins were dried at 70 °C in a convection oven prior to further purification. The average weight per milliliter of the resin was found to be 0.40 ± 0.02 g.

Polyurethane foam impregnated with 1% carbon was prepared as previously described (*13*) with slight modification. A slurry of a known weight of polyurethane foam in dichloromethane was prepared by blending for 2 min in a commercial Waring blender and filtering through Whatman no. 1 filter paper. The solids were resuspended in methanol. After the required amount of carbon was added, the mixture was slurried for another minute and then was filtered through Whatman no. 1 filter paper.

The resins and the carbon-impregnated polyurethane foam were separately placed in cellulose thimbles and Soxhlet extracted successively for 24 h with HPLC-grade hexane, dichloromethane, acetone, and methanol. The final methanol extract from each adsorbent was concentrated and analyzed by GC by using the FID.

Each adsorbent was slurry-packed in borosilicate columns (11-mm i.d.) fitted with glass frits and PTFE polymer stopcocks. Two columns of each adsorbent were packed to heights of 70 and 100 mm and were thoroughly washed with purified water.

Adsorption Studies. The general procedure for sorbing and desorbing organic compounds in fortified water samples was as follows: Approximately 1000 mL ± 20 mL aqueous solutions of each organic compounds or their mixtures were measured into six 1-L separatory funnel reservoirs. The reservoirs were then placed on top of six individual columns packed with the three solid adsorbents of two differing heights. The solutions were allowed to percolate through the column at flow rates of 1/3 bed volume/min. The adsorbates were stripped from the columns with 100 mL of acetone/hexanes (3:7 v/v) per column. The eluates were dried over prewashed, anhydrous Na_2SO_4 and concentrated to 3–5 mL with the aid of K-D evaporators with the modified Snyder columns. The concentrates were then analyzed by GC or HPLC. The adsorbents were then rinsed with acetone (100 mL), followed by seven rinses with 100 mL of water, and equilibrated with purified water in preparation for the subsequent run.

To measure the possible "breakthrough", the aqueous effluents from the columns were analyzed directly by HPLC for the presence of polar compounds at a sensitivity of 0.005 AUFS. The absorbance of earlier fractions was below 0.02 AUFS. If any nonpolar compounds were present in the column effluents, they were extracted with purified hexanes (8 × 25 mL). The combined extracts were then dried over anhydrous Na_2SO_4 and concentrated as described earlier for analysis by GC by using ECD to determine the presence or absence of any nonpolar compounds.

LARGE-SCALE DRINKING WATER STUDY. A borosilicate glass column (75-mm × 23-mm i.d.) containing purified XAD-2 resin was connected by means of a ground glass Ⓢ 24/40 joint to the top of a similar column containing XAD-4 resin. The 1-L separatory funnel reservoir was attached to the top of the XAD-2 column. Municipal tap water samples from Kingston (site 1) and Trenton (site 2), Ontario, Canada, were collected in precleaned 4-L amber-colored glass bottles, adjusted to pH 2.0 with phosphoric acid at the time of sampling, and stored in the dark at ambient temperatures prior to concentration and analysis.

During the concentration step, 1-L aliquots of water samples were periodically measured into the reservoir and allowed to percolate through the two resin beds at a rate of 1/3 bed volume/min. Thus, 30 L of the Kingston sample and 35 L of the Trenton sample were separately processed. Each column was individually stripped with acetone/hexanes (3:7 v/v, 150 mL), and the organic eluates were dried over prewashed, anhydrous Na_2SO_4 and concentrated to 3–5 mL before GC analysis. The resin packings were regenerated by rinsing with purified acetone (400 mL), washing thoroughly, and equilibrating with purified water before reuse. In this manner, two separate fractions (from XAD-2 and XAD-4 columns) of the concentrates for each sample were obtained. Appropriate solvent and system blanks were also obtained. All extracts were analyzed by GC by using both FID and ECD and were compared to a standard mixture of the six compounds of interest. In addition, the extracts were analyzed by GC–MS for further identification of the peaks of interest.

REVERSE-OSMOSIS STUDIES. The various components of the RO equipment were thoroughly cleaned with acetone and hexane before assembly. After assembly the system was further cleaned by successive circulation of hexane, ethanol, 10% aqueous ethanol (1 × 4 h each), and finally purified water (3 × 10 h).

Three pairs of membranes, each with two different porosities, were installed in the RO cells. The membranes used were PA (92% and 97%), CAc (85% and 91%), and PBI (89% and 99%). Each cell had an effective membrane diameter of 4.1 cm (area of 13.4 cm^2). The operating pressure for all runs was 260 ± 10 psig, and the flow rate was adjusted to 410 ± 10 mL/min. The system and membranes were washed by operating with an ethanol/water mixture (1:9 v/v; twice) for a 6–8-h period to get rid of any trace organic impurities in the system. The system was then cleaned twice with purified water and equilibrated with purified water (3 × 10 h). During the run, the temperature of the feed solution increased from 20–22 °C to 26–29 °C.

The reservoir was filled with 3–4 L of the stock solution (original), and the entire system was rinsed thoroughly with approximately 1 L of solution, which was discarded. The volume of the cells, tubing, etc. was estimated to be approximately 200 ± 10 mL, whereas the surge tank retained approximately 100 mL of fluid. Initially the concentration of the stock solution was determined to verify whether any loss had occurred during storage. After the run was started, the system was equilibrated by collecting at least 10 mL of the permeate from the slowest membrane, and the solution left in the reservoir was then called the "feed solution". The actual volume of the feed solution was determined at time "zero." Clean flasks were connected to the permeate end of each cell, and the run was continued until the volume of the feed solution was reduced to approximately one-third of its volume at "zero time". At the end of the run, the exact volumes of the feed solution and each permeate (100–500 mL) were measured, and aliquots of samples of feed and permeates were analyzed to determine the levels of organic compounds. The concentrations of the feed solution and permeates were denoted by C_e and C_p, respectively.

The RO system was flushed thoroughly with 1 L of pure water between runs. A new batch of purified water was added to the reservoir, and the system was run to cleanse the membranes by permeation as described earlier. The cleanliness of the system was checked by analyzing the feed solution and the permeates of each wash for the organics under investigation.

From the results, permeation rates for each membrane–solute combination were obtained, and the rejection characteristic of the membrane was calculated according to the following equation: percent rejection = $[(C_e - C_p)/C_e] \times 100$ where C_e is the concentration of the chemical in the feed solution after 90 min of equilibration and C_p is the concentration of the chemical in the permeate. (NOTE: A small or a negative value indicates that the membrane is not effective at preventing the solute from passing through the membrane, whereas a large value indicates that the membrane is effective at preventing the solute from permeating through the membrane.)

Results and Discussion

Adsorption Studies. Extreme care was taken in purifying the adsorbents. To verify the purity of the adsorbents, the final methanol concentrates of the purified adsorbent washings were analyzed to show only a few small peaks. One of the peaks was identified as DEHP by its retention time from the GC trace by using FID. This finding was confirmed by GC–MS. The flow rates and volumes of water samples percolated through the adsorbents were not allowed to exceed the loading limits of the adsorbents (*21*). After preliminary studies, an acetone/hexane (3:7 v/v) (*3*) mixture was chosen for stripping the adsorbents. After concentration, the samples were analyzed by GC with fused silica DB-5 column, which gave excellent resolution and peak shapes of the analytes.

At the concentrations of the six organic compounds studied, BHC was analyzed by GC by using ECD. CA and DEHP were preferentially detected by using FID. Analyses were performed in triplicate, and the data obtained were averaged for calculations.

The percent recoveries of the six organic compounds from the fortified water samples and for the three adsorbents are shown in Table I. The average recoveries for these organic compounds are considered acceptable (individually and in their mixtures). In separate experiments, the average percent recoveries from a set of two runs for individual columns of each adsorbent are reproducible within ±15% (*21*). Comparison of the data for 70- and 100-mm columns shows that the average percent recoveries are not consistent and bear no relationship

Table I. Percent Retention–Recovery of Organic Compounds from Various Adsorbents

Compound		*XAD-2*[a]		*XAD-4*[a]		*Polyurethane–Carbon*[a]	
		70	*100*	*70*	*100*	*70*	*100*
DP	ind	107	95	150	112	170	135
	mix	122	85	115	105	96	107
TCP	ind	102	118	82	93	103	106
	mix	111	79	106	102	90	100
CA	ind 1	88	109	101	—	27 (70)	44 (51)
	ind 2	92	99	88	103	17 (70)	36 (61)
	mix	111	79	105	93	23 (74)	46 (58)
DCB	ind	96	95	91	80	96	88
	mix	110	91	113	105	109	106
BHC	ind 1	93	74	105	70	67	78
	ind 2	96	95	85	71	81	103
	ind 3	103	70	84	97	93	92
	mix	89	89	69	55	92	57
DEHP	ind 1	21	18	24	18	18	21
	ind 2	16	13	15	15	19	15
	mix	34	30	42	36	36	45

NOTE: ind denotes individual, and mix denotes mixture; values in parentheses are the percent recovery measured in the effluent.
[a]The column packing lengths are 70 and 100 mm.

in regard to the height of the column packings. In the case of XAD-2 and XAD-4 resins, the recovery values for DCP, TCP, CA, and DCB vary markedly, depending on whether the compounds were present singly or in a mixture. The reason for this variation is unclear at this time. However, with respect to BHC, both XAD-2 and XAD-4 resins behaved similarly within experimental error under the conditions described earlier.

Two anomalies are distinctly observable in the recovery data. The first feature involves CA, which was not well-retained by the polyurethane–carbon adsorbent either from its individual solution or when mixed with the other five compounds. The effluent from the column contained more CA than was found sorbed onto the adsorbent. Although it is tempting to attribute this lack of sorption to the amino functionality, basicity cannot be the entire reason because DCB with two amino groups behaved normally. Perhaps water solubility could also be a contributing factor. In any event, this result indicates some ineffectiveness of the polyurethane–carbon mixed adsorbent system and shows the need for further investigations of various parameters affecting the recovery of CA or other similar compounds.

Secondly, the recoveries for DEHP are not satisfactory. This result is even more puzzling because extraction with ethyl acetate of a water sample similar to that used for the adsorption studies yielded the same magnitude of recovery. Recent studies (27) have shown that about 10–15% of DEHP was sorbed on the walls of the silylated containers, and this sorbed DEHP could not be recovered by rinsing with several solvents. Moreover, about 25–35% of DEHP in aqueous solutions breaks through from the macroreticular resin columns even at concentrations of a few micrograms per liter for a 500-mL sample and 5-mL adsorbents. Further, DEHP might be hydrolyzed in aqueous media, and the resulting acids were not easily desorbed by the solvents used for elution. These reasons explain the fact that a reasonably constant low recovery of DEHP was obtained from all adsorbents in any given run.

Drinking Water Samples. Two tap water samples from municipal sources in the Lake Ontario region, namely Kingston (site 1) and Trenton (site 2), Ontario, Canada, were extracted, concentrated, and analyzed by using a column of XAD-2 followed by a XAD-4 macroreticular resin column. The combination of the two resin columns was employed to ensure that any unretained organic compound by the XAD-2 column due to channeling, etc. would be trapped by the XAD-4 resin. Stripping the resins with acetone/hexane (3:7 v/v), drying, and concentrating in the usual manner produced the concentrates of the field water samples. Capillary GC analyses employing both FID and ECD indicated the probable presence of three of the organic substances under investigation, namely, TCP, BHC, and DEHP, in concentrations greater than the estimated detection threshold of 20–30 pg (Table II). These three compounds were also identified by GC–MS. The concentrates from the XAD-4 columns contained detectable amounts of organic substances having the same retention time as those from the

Table II. Analysis of Selected Compounds in Water Samples from Sites 1 and 2

Compound	*Site 1*	*Site 2*
DCP	nd	nd
TCP	40	990
CA	nd	nd
DCB	nd	nd
BHC	12	13
DEHP	100	30

NOTE: Values are expressed in parts per trillion; nd means none detected.

XAD–2 macroreticular resin columns. This situation implied that the XAD–2 column was not effective in retaining completely all the solutes present in the water samples. [NOTE: The XAD–2 resin column contained about 25% more packing, and the rate of percolation was about the same as that normally used for processing 200 L of tap water (*21*).] In addition, a variety of volatiles that appeared immediately following the solvent peak were also present. Subsequent analysis of these concentrates by GC–MS indicated the presence of 6-chloro-2,4-diamino-1,3,5-triazine (tentative), 2,5-diphenylisoxazole (tentative), tributoxyethyl phosphate (confirmed), bis(2-ethylhexyl) phthalate (confirmed), and dimethylbenzoic acid (confirmed) from site 1. The concentrate from site 2, however, showed the presence of 2,4,5-trichlorophenol (confirmed), BHC (confirmed), 2,5-diphenylisoxazole (tentative), bis(2-ethylhexyl) phthalate (confirmed), trimethylbenzene (confirmed), ethylbenzaldehyde (confirmed), ethylacetophenone (confirmed), hexanoic acid (confirmed), and 4-cyano-3,7,11-tridecatriene (tentative).

Reverse-Osmosis Study. The RO system consisted of six radial flow cells with flat sheet membranes (as discs), appropriate pumping, pressure regulators, and surge tank components along with stainless steel or PTFE polymer tubing and borosilicate glass reservoir. Each type of membrane, namely CAc, PA, and PBI, was represented by two different porosities as determined by standard NaCl rejection. Pure water permeation rates were determined at various times during the study. Accordingly, solutions of the six pollutants, singly and as mixtures, were subjected to a standardized RO run.

Permeation was allowed to proceed for 90 min to equilibrate the membranes. This method permitted at least 10 mL of solution to permeate through even the low-flux membrane. Instead of taking the original concentration for the calculation of percent rejection of the solutes, the concentration after equilibration was used.

Analysis of aqueous solutions of the polar compounds (DCP, TCP, CA, and DCB) at concentrations of 1–10 ppm was easily accomplished by direct aqueous injection liquid chromatography. The Hamilton PRP–1 reverse-phase column gave a better resolution of these compounds than the conventional reverse-phase columns. Acetonitrile/water mixtures have been found to be as effective as the buffered mobile phases recommended by the manufacturer (*28*). Analyses of the nonpolar compounds (BHC and DEHP) at concentrations of 25–100 ppb were achieved by XAD resin adsorption–desorption, concentration, and GC techniques.

Table III presents the percent rejection of the selected six organic compounds, both when they are present individually and when mixed with others in aqueous solutions, and for the three pairs of membranes

Table III. Percent Rejections of Various Pollutants (Individually and Mixed) by Membrane Types

Membrane Type		*PA (92)*[a]	*PA (97)*	*CAc (91)*	*CAc (85)*	*PBI (99)*	*PBI (89)*
DCP	ind	51	39	−32	−41	11	9
	mix 1	55	51	−7	−2	49	40
	mix 2	69	62	−35	−31	78	62
TCP	ind	65	57	−13	−11	60	34
	mix 1	70	65	0	4	81	67
	mix 2	80	75	−9	−21	88	82
CA	ind 1	44	35	−26	−16	0	−7
	ind 2	50	37	−20	−27	8	−4
	mix 1	60	41	−23	−26	35	25
	mix 2	61	54	−32	−46	47	30
DCB	ind	78	74	15	21	100	100
	mix 1	74	83	24	10	86	79
	mix 2	87	92	42	28	97	95
BHC	ind 1	97	95	55	31	100	98
	ind 2	92	95	37	33	94	96
	mix 1	90	90	34	24	84	87
	mix 2	95	98	31	23	98	95
DEHP	ind 1	73	34	50	80	50	82
	ind 2	50	(−)	−41	74	29	34
	mix	73	91	53	80	85	91

NOTE: ind denotes individual, and mix denotes mixture; (−) indicates sample lost.
[a] Values in parentheses indicate the percent rejection of NaCl.

with two different porosities (shown in parentheses as percent rejection of NaCl). The data show that the CAc membranes are quite ineffective for concentrating or rejecting organic compounds in general, and more so with respect to polar organic compounds which have low molecular weights and high solubilities. Among the polar organic compounds studied, DCB was the only compound rejected between 10% and 42% by the CAc membrane.

The PA membranes have distinctly better rejection properties for the individual organic compounds (when present in their respective aqueous solutions) in comparison to the two other membrane types. However, PBI membranes have equally good rejection properties except in the cases of DCB and CA. With respect to CA, the rejection behavior of PBI is analogous to the retention properties of polyurethane–carbon adsorbent (*see* Table I). The striking resemblance in percent rejection of PA and PBI membranes is their behavior toward DCB and BHC, and in these cases, their rejection is more than 75%. Comparatively, the percent rejection of these organic compounds in their mixtures by PA and PBI is better than when these compounds were present individually. The only possible synergistic effects might be noticeable between DCP and CA in the case of PBI membranes. Such mutually improved rejections might be ascribed to some ionic species separation after salt formation; however, this analysis is purely speculative and requires further investigation. The CAc membrane has shown some slight improvement in its rejection behavior of the compounds in their mixtures. Also, the α-isomer of BHC used in this study has been rejected by the CAc membrane to the extent of an average of 46%; this value is similar to the 60% reported previously for the γ-isomer (*17*).

Table IV presents the comparative data on the permeation rates of the three types of membranes with two different porosities for various aqueous organic solutions and for pure water as measured over the duration of the study. The data shown here represent the relative chronological order in which the samples were tested. In the beginning, even though the percent rejection of NaCl is high for PA and CAc (indicating small-size pores), the rates of permeation of pure water are higher for denser membranes than for membranes having lower percent rejection of NaCl. In the case of the PBI membrane, the reverse of this phenomenon is observed.

As usually observed, the pure water permeation rates for the membranes plateau at 65–80% of the original pure water permeation rate after several months of use probably because of compaction or other fouling mechanisms. Also to be expected is the general trend for individual solution runs to be somewhat slower than pure water runs. Apparently, there is little relation between the permeation rates of aqueous solutions containing trace levels of organic compounds and the rejection behavior of membranes except for the denser of the two PBI membranes. This finding indicates some unique properties of the PBI polymer. What is somewhat puzzling is the extremely slow rates of permeation for the mixture of compounds in the aqueous solution. This finding could be indicative of some aggregation phenomenon among the various components of the mixture due to an increase in viscosity.

Table IV. Water Permeation Rates of Pure Water and Pollutant Solutions through Membranes

Membrane	*PA(92)*[a]	*PA(97)*	*CAc(85)*	*CAc(91)*	*PBI(99)*	*PBI(89)*
Pure water (beginning)	15.2	18.6	6.5	13.8	5.7	11.4
Pure water (60 days)	13.0	16.1	5.9	12.4	5.1	10.1
Pure water (~150 days)	12.9	16.1	6.2	12.8	4.4	8.8
DCP (~16 ppm)	13.1	15.9	5.7	11.0	4.7	9.0
TCP (~19 ppm)	12.5	15.2	5.0	10.8	4.0	7.3
CA (~18 ppm)	13.2	16.3	5.8	12.3	5.1	10.0
CA repeat (~21 ppm)	13.2	16.2	5.8	12.4	4.6	9.7
DCB (~1 ppm)	13.6	17.6	6.3	13.1	5.1	10.0
BHC (~155 ppb)	13.3	17.4	6.0	12.6	5.6	10.6
DEHP (~30 ppb)	13.4	16.6	6.0	12.4	5.0	10.0
First mixture	11.9	14.7	4.4	9.6	2.8	5.3
Second mixture	11.2	13.2	4.2	9.0	3.0	5.5

NOTE: Values are expressed in grams per hour.
[a] Values in parentheses indicate the percent rejection of NaCl.

Also, the dense PBI membrane is slow (99%). Although this membrane exhibited, in general, the best rejection characteristics, the very low flux observed is a definite drawback of such dense membranes.

A comparison of the approximate mass balance calculations for the various compounds and for all the membranes combined is presented in Table V. These numbers indicate only an attempt to account for all the organic material in the various runs (concentration volume values at the beginning of the run, namely, the feed solution only; and at the end of the run, namely, the feed solution plus permeate). Again, for the small, water-soluble molecules (DCP, TCP, and CA), total accountability is quite good, whereas for the large, less soluble molecules (DCB, BHC, and DEHP), significant amounts of the compounds have disappeared during the course of the run. Such discrepancies for CAc membranes have been previously noted (*17, 18, 29*), and it is tempting to speculate that the membranes themselves are somehow retaining the more hydrophobic compounds. However, one can only begin to study this problem by ensuring that all components used in the feed system are composed of inert materials such as stainless steel, glass, or PTFE polymers.

Another interesting comparison is outlined in Table VI where the

Table V. Mass Balances

Compound (mass units)	*Initial Amount*	*Reservoir After*	*Permeate Total*	*Recovery*	*Percent Recovery*
		Individual Runs			
DCP (mg)	15.89	7.66	8.37	16.03	100
TCP (mg)	18.42	8.57	6.62	15.19	82
CA (mg)	18.43	8.59	9.54	18.13	98
CA (repeat, mg)	20.56	8.56	10.64	19.20	93
DCB (mg)	1.22	0.52	0.28	0.79	65
BHC (10^{-3} mg)	633	149	77	226	36
BHC (repeat, 10^{-3} mg)	135	52	20	72	53
DEHP (10^{-3} mg)	33.0	3.9	7.9	11.8	36
		Mixture			
DCP (mg)	16.04	8.54	6.63	15.17	91
TCP (mg)	16.93	8.98	5.23	18.70	110
CA (mg)	10.14	4.48	4.67	9.15	90
DCB (mg)	1.53	0.70	0.37	1.07	70
BHC (10^{-3} mg)	221	94	39	133	60
		Repeat Mixture			
DCP (mg)	9.84	4.86	4.12	8.98	91
TCP (mg)	10.43	5.08	3.32	8.40	81
CA (mg)	8.02	3.73	3.93	7.66	96
DCB (mg)	2.65	1.07	0.46	1.53	58
BHC (10^{-3} mg)	130	70	27	97	75
DEHP (10^{-3} mg)	123	7	16	23	19

Table VI. Comparison of Feed Concentrations

Compound		*Original*	*After Equilibration*	*Final*
DCP	ind	7.7	6.7	7.8
	mix	7.0	5.1	6.9
TCP	ind	9.2	8.7	9.5
	mix	7.3	5.1	7.2
CA	ind	8.2	7.6	8.6
	mix	2.5	2.4	3.1
DCB	ind	0.68	0.45	0.47
	mix	0.76	0.46	0.56
BHC	ind	0.05	0.05	0.05
	mix	0.06	0.07	0.08
DEHP	ind	0.10	0.11	0.00
	mix	0.10	0.04	0.01

NOTE: Values are expressed in parts per million; ind denotes individual, and mix denotes mixture.

concentrations of the selected organic compounds in the feed solution at various times throughout the run are compared. The word "original" denotes the concentration of the feed solution at the beginning of the run; "equilibration" indicates the concentration of the feed solution after 90 min of the RO run with permeation; and "final" denotes the concentration of the feed solution at the termination of the RO run. Although we are dealing with sometimes offsetting effects of various membranes, one should note the disappearance of most of the DEHP. A similar observation was noted during the adsorption studies.

Another indication of the somewhat complex initial few hours of interaction between membrane and the solutes in the feed solution, that is, the earlier part of the RO run, can be found in Table VII. Here, the rejection characteristics of the membranes after 90 min of permeation are compared with a final value obtained for the feed solution at the end of the RO run. In most cases, the membranes have not reached a saturation or complete equilibration value even after 10–20 mL of solution has been allowed to permeate the membranes. Although the mechanics of interaction between the membrane and the solute requires sufficient time, and such equilibrations may not be significant for extended runs for several weeks, these details should be considered in exploratory work on the RO process.

Conclusions

XAD-2 and XAD-4 macroreticular resin adsorbents were found to be adequate to accumulate the organic pollutants considered in the present study except for bis(2-ethylhexyl) phthalate. Polyurethane-carbon ad-

Table VII. Comparison of Membrane Rejections at 90 min and at End of Run

Compound	*PA(92)*	*PA(97)*	*CAc(91)*	*CAc(85)*	*PBI(99)*	*PBI(89)*
DCP	68/51	64/39	59/32	24/−41	100/11	100/9
TCP	76/65	72/57	57/−13	−32/−11	98/60	99/34
CA	48/44	48/35	−6/−26	−16/−16	77/0	8/−7
DCB	100/78	100/74	100/15	100/21	100/100	100/100

NOTE: The first value is the rejection at 90 min; the second value is the rejection at the end of the run.
[a] Values in parentheses indicate the percent rejection of NaCl.

sorbent was observed to be ineffective for concentrating 4-chloroaniline and bis(2-ethylhexyl) phthalate. Among the membranes currently investigated in the RO technique, PA membrane was found to be superior to PBI and CAc material. For the concentration of bis(2-ethylhexyl) phthalate, the RO technique proved to be superior to XAD resin adsorption. CAc membrane was noted to be ineffective for rejecting the investigated compounds.

Acknowledgments

The authors extend their sincere gratitude to F. M. Benoit and R. O'Grady for their GC–MS analysis, and we wish to thank J. Godin, M. Abedini, and K. Diedrich for technical assistance. We are indebted to S. Sourirajan, T. Matsuura, and A. Baxter for their generous advice on many phases of RO studies; G. L. LeBel and R. Otson for critically reviewing the manuscript; and Jean Ireland for word processing.

This chapter is abstracted from a report presented to Health and Welfare Canada as part of contract number 887–1982/83.

Literature Cited

1. Kool, H. J.; van Kreijl, C. F.; van Kranen, H. J.; De Greef, E. *Sci. Total Environ.* **1981,** *18,* 135.
2. Williams, D. T.; Nestmann, E. R.; LeBel, G. L.; Benoit, F. M.; Otson, R. *Chemosphere,* **1982,** *11,* 263.
3. Fishbein, L. *Toxicol. Environ. Chem. Rev.* **1980,** *3,* 145.
4. Volkoff, A. W.; Creed, C. *J. Liq. Chromatogr.* **1981,** *4,* 1459.
5. Drevenkar, V.; Frose, Z.; Stengl, B.; Tkalcevic, B. *Mikrochim. Acta* **1985,** *1,* 143.
6. Laane, R. W. P. M.; Manuels, M. W.; Staal, W. *Water Res.* **1984,** *18,* 163.
7. Suffet, I. H.; Brenner, L.; Coyle, J. T.; Cairo, P. R. *Environ. Sci. Technol.* **1978,** *12,* 1315.
8. Dressler, M. *J. Chromatogr.* **1979,** *165,* 167.
9. Harris, J. C.; Cohen, M. J.; Grosser, Z. A.; Hayes, M. J. *EPA Project No. PB-81-106585;* U.S. Environmental Protection Agency. U.S. Government Printing Office: Washington, DC, 1981.

10. Narang, A. S.; Eadon, G. *Int. J. Environ. Anal. Chem.* **1982,** *11,* 167.
11. Stuber, H. A.; Leenheer, J. A. *Anal. Chem.* **1983,** *55,* 111.
12. Suffet, I. H.; McGuire, M. J. *Activated Carbon Adsorption of Organics from the Aqueous Phase;* Ann Arbor Science: Ann Arbor, MI, 1981; Vols. 1 and 2.
13. Babjack, L. J.; Chau, A. S. Y. *J. Assoc. Off. Anal. Chem.* **1979,** *62,* 1174.
14. Sourirajan, S. In *Sythetic Membranes;* Turbak, A. Ed.; ACS Symposium Series 153; American Chemical Society: Washington, DC, 1981; Vol. 1, pp 11–62.
15. Kopfler, F. C.; Coleman, W. E.; Melton, R. G.; Tardiff, R. C.; Lynch, S. C.; Smith, J. K. *Ann. N.Y. Acad. Sci.* **1977,** *298,* 203.
16. Deinzer, M.; Melton, R.; Mitchell, D. *Water Res.* **1975,** *9,* 799.
17. Malaiyandi, M.; Blais, P.; Sastri, V. S. *Sep. Sci. Technol.* **1980,** *15,* 1483.
18. Fang, H. H. P.; Chian, E. S. K. *Environ. Sci. Technol.* **1976,** *10,* 364.
19. LeBel, G. L.; Williams, D. T. *Bull. Environ. Contam. Toxicol.* **1980,** *24,* 397.
20. Hurst, R. E.; Settine, R. L.; Fish, F.; Roberts, E. C. *Anal. Chem.* **1981,** *53,* 2175.
21. LeBel, G. L.; Williams, D. T.; Benoit, F. M. *J. Assoc. Off. Anal. Chem.* **1981,** *64,* 991.
22. Riggin, R. M.; Howard, C. C. *Anal. Chem.* **1979,** *44,* 139.
23. Malaiyandi, M.; Benoit, F. M. *J. Environ. Sci. Health* **1981,** *A16,* 215.
24. Nguyen, T. D.; Chan, K.; Matsuura, T.; Sourirajan, S. *I&EC Prod. Res. Dev.* **1984,** *23,* 501.
25. Malaiyandi, M. *J. Assoc. Off. Anal. Chem.* **1978,** *61,* 1459.
26. Matsuura, T.; Taketani, Y.; Sourirajan, S. *Proceedings of the 4th Bioenergy Research and Development Seminar, National Research Council of Canada* **1982,** *529.*
27. Malaiyandi, M.; Zhow, S., unpublished results.
28. Lee, D. P. *J. Chromatogr. Sci.* **1982,** *20,* 203.
29. Chian, E. S. K.; Bruce, W. N.; Fang, H. H. P. *Environ. Sci. Technol.* **1975,** *9,* 52.

RECEIVED for review August 14, 1985. ACCEPTED January 7, 1986.

9

Comparison of High Molecular Weight Organic Compounds Isolated from Drinking Water in Five Cities

E. S. K. Chian[1], M. F. Giabbai[2], J. S. Kim[1], J. H. Reuter[3], and F. C. Kopfler[4]

[1]School of Civil Engineering, Georgia Institute of Technology, Atlanta, GA 30332
[2]EnvironScience Laboratories, Inc., Atlanta, GA 30316
[3]School of Geophysical Sciences, Georgia Institute of Technology, Atlanta, GA 30332
[4]Health Effects Research Laboratory, U.S. Environmental Protection Agency, Cincinnati, OH 45268

Because the previously unidentifiable nonvolatile fraction of the chlorinated organic compounds in drinking water is of significant health concern, the physical and chemical characteristics of the high molecular weight organic compounds isolated from drinking water in five cities (New Orleans, Philadelphia, Miami, Seattle, and Ottumwa) were compared. From the results of this study, these organic compounds may be considered as part of a single group of natural organic compounds. Although the respective molecular structures of this group of compounds are still unknown, they showed many similarities in terms of molecular weight, solubility, and functional groups among the samples collected from the five cities. This result indicates that it is probably acceptable to base results of toxicological testing on chlorinated organic compounds from a limited number of drinking water sources.

SUBSTANCES INTRODUCED BY SOURCES exogenous to the drinking water treatment process were the focus of most studies, prior to 1975, of organic contaminants in drinking water. When it was reported that trihalomethanes were produced during the chlorination of drinking water (*1*, *2*), research began to focus more on the byproducts of disinfection because this step cannot be eliminated. Disinfection is necessary because (1) pathogenic bacteria must be controlled and (2)

0065-2393/87/0214/0181$06.00/0

the identifiable products occurring most often in chlorinated drinking water were trihalomethanes. Thus, it was necessary to determine if these and other currently unidentifiable reaction byproducts were of significant health importance. Of the greatest concern to this study was the nonvolatile fraction of the chlorinated organic compounds, which is known to compose the bulk of the chlorinated compounds in drinking water and which was previously unidentifiable with gas chromatography (GC) (*3*).

Because the cost of isolating sufficient humic matter for all of the desired biological tests was prohibitive (*4*), samples of a commercially available humic acid were chlorinated and used in these tests. The work presented here was conducted to compare samples of this humic acid chlorinated under two different pH conditions with samples of humiclike material isolated from five chlorinated drinking waters. This comparison was done to help decide if the results obtained from the biological testing of the former substance could be extrapolated to the latter ones. The drinking water-derived material was obtained during an earlier study (*5*) from the drinking water of five cities in the United States, which at the time were thought to be representative of water derived from sources containing various types and amounts of organic matter. Information regarding these water supplies is presented in Table I. The organic matter isolated from some of these waters (by liquid–liquid extraction and XAD-2 absorption from reverse-osmosis concentrate) was found to be mutagenic (*4*). Most of the organic matter in these samples was polar acidic compounds and could not be chromatographed on silica gel even with methanol or in a GC system (*3*). Because these samples were highly colored and acidic, they were thought to consist of humic or fulvic acid modified by the action of chlorine disinfectant. Thus, efforts were made to isolate the humic and

Table I. Drinking Water Samples from Five Cities

City	*Suspected Types of Contaminants*
Miami, FL	uncontaminated[a]
New Orleans, LA	industrial wastes
Ottumwa, IA	agricultural runoff
Philadelphia, PA	municipal wastes
Seattle, WA	uncontaminated[a]

NOTE: For all cities except Miami, the raw water source was surface water. For Miami, the raw water source was ground water.
[a] Uncontaminated means there was no known contamination from municipal, agricultural, and industrial wastes. However, there presumably was contamination from decomposition products of natural origin.

fulvic acid fractions and byproducts by using a series of solvent extractions; removal of the nonpolar constituents with hexane was followed by ether extraction. The residual solids after ether extraction were then dissolved in acetone and were found to contain mainly the polar acidic humic substances, according to the methods of a previous study (*6*). This acetone-soluble fraction of organic matter was subjected to a variety of physical and chemical tests to obtain as much information as possible on the similarities of the material isolated from the drinking water of the five cities, which covered widely geographical sources. These tests made it possible to better estimate the similarities in biological activity that may be expected with this material. If the isolated materials are similar in both physical-chemical characteristics and biological activities, then a large amount of high molecular weight drinking water organic compounds can be isolated from one conveniently located water treatment plant for more extensive biological tests. This method would greatly reduce the logistic problems of having to transfer qualified personnel and necessary equipment all over the country to collect representative samples for more in-depth studies on these materials.

In addition to characterizing the high molecular weight organic matter isolated from drinking water, samples of a naturally occurring aquatic humic acid and a commercially obtained humic acid chlorinated under both basic and neutral conditions in the laboratory were also characterized. This approach was taken as a best effort to determine if the presence of these chlorinated model humic acids in drinking water would constitute the same hazard to the health of the consumers as those isolated from actual drinking water samples (*4*). Therefore, comparison of the similarity of the two types of material is important. Because of the complex nature of these substances, the tests reported here provide only the best means of comparing chlorinated organic compounds obtained from widely different sources.

Materials and Methods

Concentration of Organic Matter from Drinking Water. Water samples were concentrated from the drinking water of five cities (Table I) by reverse osmosis by using cellulose acetate membranes in conjunction with the Donnan softening process (7). The volume processed depended on the organic carbon content of the water and varied from 1500 L in Miami, Florida, to 15,000 L in Seattle, Washington. The permeate from this system was reprocessed by a nylon, hollow-fiber, reverse-osmosis loop (7). The two concentrates were sequentially extracted at neutral pH with pentane and methylene chloride and again at pH 2 with methylene chloride. After removal of residual solvent by purging with purified nitrogen, the aqueous concentrates were passed through Amberlite XAD-2 columns.

The cross-sectional flow velocity through the resin columns was maintained at 10 cm/min. The columns were each rinsed with 2.5 bed volumes of 1 M HCl followed by 2.5 bed volumes of distilled water to remove metallic oxides and other inorganic constituents. Finally, the organic components were eluted with 2.5 bed volumes of 95% ethanol. Ethanol was selected as the eluent because it seemed to be the best choice as a solvent tolerated by animals used in proposed biological tests.

The ethanol was removed by vacuum distillation at 40 °C and pressures of 203–208 mm of mercury. The eluates from both columns were combined. In all cases, the majority of the organic matter was contained in the cellulose acetate membrane concentrate. These procedures have been more completely reported elsewhere (7).

Isolation of River Humic Substances. River water humic substances were isolated by this laboratory from the Satilla River by using a modification of a previously described adsorption technique involving XAD-7 resin followed by elution with triethylamine (8). Excess triethylamine was removed in a rotary evaporator, and the residual solution was continuously extracted with chloroform to remove neutral molecules. The resultant aqueous solution was reacted with BioRad AG 50WX8 cation-exchange resin (H^+-form) and then freeze-dried. About 70% of Satilla River humic substances as measured by total organic carbon (TOC) was recovered. The elemental composition of the aquatic humus is as follows: 50.77% carbon, 3.81% hydrogen, 0.79% nitrogen, and 2.28% ash. The characteristics of the aquatic humus based on oxidative degradation studies were reported by Reuter et al. (9).

Commercially Available Humic Acids. A large quantity of the commercially available humic acids was purchased from Fluka AG (Buch, Switzerland) by the Health Effects Research Laboratory (HERL) of the U.S. Environmental Protection Agency (USEPA), Cincinnati, Ohio. This coal-based humic substance is from West Germany.

Solvent Fractionation of Humic Substances. The organic matter from drinking water samples, Fluka humic acid, and chlorinated Fluka humic was fractionated by using a series of solvents, for example, hexane, ether, and acetone (distilled-in-glass grade, Burdick and Jackson) according to procedures developed by Kopfler (3). In general, these procedures involved the use of an ultrasonic bath to repeatedly extract the humic substances in a 10-mL serum bottle with appropriate solvents. The insoluble fractions were separated in centrifuge tubes, and the supernatant layer was dewatered over anhydrous sodium sulfate (previously muffled at 550 °C). The dehydrated solvent extracts were dried in an all stainless steel-glass-Teflon evaporator with ultrapure nitrogen. The nitrogen gas flow rate was adjusted to 300 mL/min to minimize the loss of the volatile sample components. The evaporator was immersed in a water bath set at 40 °C.

Procedures for Chlorination of Humic Acids. Four grams of humic acid (Fluka AG, Lot No. 215315) was placed in a 500-mL beaker to which 250 mL of distilled water was added. This step was followed by the slow addition, while stirring, of 0.1 N NaOH until the pH was brought to 7.0. The solution was stirred for 1 h after which the pH was checked and readjusted to 7.0 with 0.1 N NaOH if necessary. The sample was placed in an ultrasonic bath for 2 h to aid dissolution, and the pH was readjusted to 7.0 with 0.1 N NaOH if necessary. The sample beaker was covered and stirred overnight. The following morning the pH was checked and readjusted to 7.0 if necessary. The sample was then centrifuged in a Sorvall RC-2B centrifuge (Du Pont) at 16,300 $\times$ *g*. The supernate was decanted and diluted to 800 mL with distilled water. The solid residual was saved for gravimetric analysis.

The concentration of humic acid was determined by TOC analysis. TOCs for the humic stock solutions were normally 2.3 $\pm$ 0.1 g/L. Elemental analysis showed that the powdered humic substance was 52% carbon weight. This result indicates that the theoretical maximum of TOC for an 800-mL solution prepared from 4.0 g of humic substances is 2.6 g/L.

For chlorination of the humics at neutral pH, a solution of 11-13 g/L of chlorine at pH 7.5 was used. This solution was prepared by bubbling chlorine gas into a solution of 5.6 g of NaOH in 500 mL of water. Bubbling was continued until the solution was within the pH range 7.5 $\pm$ 0.1. For chlorination of the humics at high pH, a solution of 17-20 g/L of chlorine at pH 11.8 was used. This solution was prepared by bubbling chlorine gas into a solution of 11.3 g of NaOH in 500 mL of water. Bubbling was continued until the solution was within the pH range 11.8 $\pm$ 0.2. The concentration of chlorine was determined by the starch-iodine method.

The chlorination of humics was carried out immediately after preparation of the chlorine water. Enough of the concentrated humic solution was added to a 250-mL amber bottle so that when it was diluted to 200 mL, the TOC concentration would be 1.0 g/L. A ratio of one equivalent of chlorine (as Cl) per mole of carbon atoms (Cl:C = 1:1) was most commonly used because at this ratio more than 99% of the chlorine equivalents are consumed in a 90-h period. The solutions were added to the bottle as follows: humus, water, mix well, chlorine water, and final mix by shaking. The bottles were capped with Teflon-backed septa.

Initially, at room temperature, the chlorine reacted rapidly with the humus. This reaction was evident by the change in color of the solution from a thick black to a transparent brown immediately after the addition of the chlorine. A drastic drop of pH was observed after 17 h (e.g., a drop from 7.0 to 2.6 while being chlorinated at neutral pH). Any storage after 90 h was at 4 °C for no more than 48 additional hours. The chlorinated samples were lyophilized to dryness and stored in a refrigerator. The acetone-soluble fraction of each sample was subjected to the following analyses.

Elemental Analysis. The acetone fractions were analyzed for C, H, and N by an automatic analyzer. Sulfur and the halogens (i.e., Cl and Br) were analyzed by combustion and subsequent titration. The following standard compounds were used for quality assurance (QA) purposes: acetanilide (C, H, N), sulfanilamide (S), *p*-chlorobenzoic acid (Cl), and *p*-bromobenzoic acid (Br).

Spectral Studies. UV spectra were recorded in methanolic solution (40–60 ppm) on a Beckman Model 26 spectrophotometer; matched quartz cells having a 1-cm path length were used. A holmium oxide spectrum was recorded prior to the analysis of the samples for QA purposes.

IR spectra were recorded with a Beckman Acculab 6 spectrophotometer. Pellets were prepared by torque pressing a finely ground mixture composed of 1–2 mg of sample and 100–500 mg of spectroanalytical KBr. The pellets were translucent brown and showed good dispersion of the sample. Considerable attention was given to prevent moisture interference. The spectrum of a polystyrene film was recorded prior to the sample for QA purposes.

Fluorescence spectra were recorded in methanolic solution (40–50 ppm) on a Perkin–Elmer Model 204A spectrofluorometer. Mutually maximized excitation and emission wavelengths were recorded in nanometers. Water was used for QA purposes.

X-ray Scattering. Each sample was analyzed by a low-angle X-ray scattering technique in the laboratory of Wershaw and Pinkney (*10*).

Ultrafiltration and Size-Exclusion Chromatography. The apparent molecular weight distribution of each sample was determined by sequential use of ultrafiltration (UF) and size-exclusion chromatography (SEC). A modification of a procedure proposed by Chian and DeWalle (*11*) was used. Approximately 10 mg of a sample was dissolved in 200 mL of an alkaline buffer solution [pH 10.4; 12.1 g of tris(hydroxymethyl)aminomethane per liter of organic-free water]. The solution was then filtered through a UM05 UF membrane (Amicon) having a molecular weight cutoff at 500 daltons. Membrane ultrafiltrations were performed in a static test cell at a nitrogen pressure of 2.465 kg/cm^2. Fifty milliliters of retentate of each sample was collected. The UF retentate was further fractionated by SEC on Sephadex G-75 (100- × 1-cm i.d. glass column, 41-cm bed height) (Pharmacia Fine Chemicals). The flow rate was maintained at 18 mL/h by a peristaltic pump (Milton Roy), and the column effluent was monitored by a variable UV detector (Perkin-Elmer). The exclusion volume of the gel permeation column was found by chromatographing a mixture of blue dextran (MW of 2,000,000 daltons) and phenol and monitoring the column effluent with a variable UV detector (Model LC-65T, Perkin-Elmer) at 230 mn.

Thermal Gravimetric Analysis and Differential Thermal Analysis. Approximately 10 mg of accurately weighed sample was subjected to thermal gravimetric analysis (TGA) and differential thermal analysis (DTA). The percentage weight loss under a temperature program of 25 °C/min was automatically recorded on a Mettler Thermoanalyzer 2 and on a Perkin-Elmer TGS-2 thermogravimetric system. The mean weight loss of a $CaCO_3$ standard (three runs) was 42.9% (theoretical 44%); the precision of this measurement was ±1.0%. The minimum weight of 10 mg required for the analysis was determined by a test with aquatic humus on TGA.

Analysis of Acidic Functional Groups. CARBOXYL GROUP ANALYSIS. Carboxyl groups of the samples were determined by modifications of methods developed by Perdue et al. (*12*). Accurately weighed samples of approximately 10 mg were used for this determination. Eleven analyses of an aquatic humus sample gave a mean value of 4.76 meq/g; a standard deviation (s) of 0.14 and a coefficient of variation (cv) of 2.9% were calculated.

TOTAL ACIDITY. The procedure proposed by Schnitzer and Gupta (*13*) and Schnitzer and Kahn (*14*) was adopted with some modification for the determination of total acidity (TA). All operations were carried out under N_2 atmosphere. Accurately weighed samples of approximately 10 mg were used for this determination. A blank experiment and six repetitive analyses from the Satilla River were performed. The mean value was 11.1 meq/g; s = 0.35 and cv = 3.2% were calculated.

Table II lists samples investigated in this study.

Table II. Samples Investigated in This Study

Sample Number	*Sample Name*	*Detailed Descriptions*
1	Miami 1A	*see* Table I
2	Miami 1B	*see* Table I
3	Miami 2	*see* Table I
4	New Orleans 1A	*see* Table I
5	New Orleans 2	*see* Table I
6	Ottumwa 2	*see* Table I
7	Ottumwa 1	*see* Table I
8	Philadelphia 2	*see* Table I
9	Philadelphia 1	*see* Table I
10	Seattle	*see* Table I
11	CFH-1	low-pH chlorination of Fluka humic acid
12	CFH-2	low-pH chlorination of Fluka humic acid
13	CFH-3	high-pH chlorination of Fluka humic acid
14	Fluka humic acid	Fluka lot no. 215315 coal-based humus
15	aquatic humus (Satilla River)	naturally occurring surface water humus (*see* reference 9)
16	chlorinated aquatic humus (Satilla River)	

Results and Discussion

The results of the fractionation of the samples according to their solubility in hexane, ether, and acetone are presented in Table III. The recoveries of acetone fractions of the five city samples (samples 1–10) ranged between 22.5% and 59.6% of the total recovered weight; the ether fractions ranged between 20.3% and 56.2%; and the hexane fractions ranged between 2.5% and 4.8%. The two neutral-pH, chlorinated Fluka humic acid (CFH) samples gave a consistently lower percentage for three solvent fractions. As a result, the insoluble fraction of CFH became much higher (up to 83% compared with a range of 1–17% for the drinking water sample). Although the original Fluka humic acid gave a similar weight distribution in each of the solvent fractions as that of neutral-pH CFH (CFH-1 and CFH-2), the high-pH CFH (CFH-3) showed a much greater difference in solubility with extraction solvents. This result indicates that chlorination at neutral pH does not alter significantly the nature of the model Fluka humic acid. However, high-pH chlorination of Fluka humic acid rendered all organically soluble fractions insoluble. More than 99% of the resulting organic compounds upon high-pH chlorination of Fluka humic acid were insoluble. This re-

Table III. Fractionation of Samples Under Investigation

Sample Number[a]	*Initial Weight Fractionated (g)*	*Hexane-Soluble Fraction (g)*	*Ether-Soluble Fraction (g)*	*Acetone-Soluble Fraction (g)*	*Insoluble Fraction (g)*	*Percent Recovered from Total Amount Fractionated*
1	2.0595	0.0740 (4.4)	0.5943 (35.1)	1.0115 (59.6)	0.0148 (0.9)	82.2
2	1.0424	0.0447 (4.8)	0.3637 (39.4)	0.5111 (55.3)	0.0044 (0.5)	88.6
3	4.8950	0.0889 (3.8)	1.1041 (47.8)	1.0903 (47.2)	0.0279 (1.2)	47.2
4	1.8828	0.0324 (2.5)	0.6434 (50.0)	0.5221 (40.5)	0.0897 (7.0)	68.4
5	2.6387	0.0418 (2.8)	0.6711 (44.3)	0.6298 (41.5)	0.1734 (11.4)	57.5
6	2.3787	0.0523 (4.8)	0.4138 (38.2)	0.4425 (40.9)	0.1734 (16.1)	45.5
7	2.0784	0.0305 (2.4)	0.2544 (20.3)	0.8483 (67.7)	0.1202 (9.6)	60.3
8	1.5244	0.0280 (2.7)	0.4808 (47.4)	0.4575 (45.1)	0.0469 (4.6)	66.4
9	4.5167	0.0509 (2.5)	0.7937 (39.0)	1.1308 (55.5)	0.0595 (2.9)	45.1
10	0.5007	0.0065 (4.3)	0.0859 (56.2)	0.0344 (22.5)	0.0259 (17.0)	30.5
11	1.0268	0.0042 (0.4)	0.0387 (3.9)	0.1212 (12.3)	0.8244 (83.4)	96.3
12	1.0524	0.0043 (0.4)	0.0415 (4.3)	0.1225 (12.8)	0.7922 (82.5)	91.3
13	1.03326	0.000015 (0.01)	0.00188 (0.18)	0.00417 (0.41)	1.0047 (99.39)	97.8
14	1.01048	0.0001 (0.01)	0.0058 (0.63)	0.13713 (14.90)	0.77733 (84.46)	91.1

NOTE: Values in parentheses are the percentage of each fraction with respect to the total amount recovered.
[a] *See* Table II for sample names.

sult indicates a severe alteration of the original Fluka humic acid. Because our characterization studies used the acetone-soluble fraction, further characterization of the high-pH chlorinated CFH-3 became impossible because of a lack of samples. Characterization of the acetone-soluble fractions of other samples is described in the following sections.

Elemental Analysis. The elemental analyses are presented in Table IV. The atomic ratios H/C for all drinking water samples (nos. 1–10) were between 1.28 and 1.39. These values were comparable to humic acid derived from lake sediments. However, H/C ratios were much lower when compared to the chlorinated model humic substances (e.g., 1.04–1.08 for CFH-1 and CFH-2). Bromine was present in almost negligible quantities, whereas Cl varied between 0.3% and 2.4%, and S varied between 0.9% and 2.7% in the drinking water organic matter. All fractions from drinking water showed similar elemental composition. However, they differed from the elemental composition of the CFH samples in all respects, especially in chlorine content.

Spectral Studies. **UV ABSORPTION.** The acetone fractions derived from drinking water exhibited intense UV absorption between 208 and 214 nm; this behavior may be attributed to unsaturated acids or esters. The CFH samples presented maximum absorption at 254–256 nm, whereas an aquatic humus sample derived from the Satilla River showed a maximum absorption at 234 nm. Therefore, apparent differences in UV spectra among all sample groups were found.

Table IV. Elemental Analysis of Acetone-Soluble Fractions

Sample Number[a]	*C*	*H*	*N*	*S*	*Cl*	*Br*	*Ash*	*H/C Atomic Ratio*
1	54.41	6.04	1.74	2.74	2.19	0.39	0.54	1.33
2	57.51	6.11	2.27	2.07	1.45	0.24	0.91	1.28
3	57.20	6.19	1.85	1.13	1.42	0.36	1.06	1.30
4	56.11	6.50	1.47	0.87	2.24	0.14	1.91	1.39
5	55.58	6.41	1.74	0.92	1.35	0.23	0.92	1.38
6	58.31	6.73	2.95	0.94	0.30	<0.10	0.46	1.38
7	55.77	6.13	2.56	0.93	0.57	0.11	0.34	1.32
8	53.67	6.13	1.99	1.88	1.56	0.20	0.45	1.37
9	55.80	6.19	1.77	1.50	1.26	0.22	0.43	1.33
10	52.83	6.03	2.11	1.04	2.02	<0.10	0.97	1.37
11	41.27	3.70	0.53	2.43	14.61	<0.5	2.51	1.08
12	39.34	3.41	0.53	2.40	16.08	<0.5	2.80	1.04

NOTE: All values are percentages except for the H/C atomic ratios.
[a] *See* Table II for sample names.

FLUORESCENCE SPECTRA. Fluorescence data are presented in Table V. All samples derived from drinking water gave an emission maximum between 417 and 430 nm, whereas the excitation maximum ranged between 346 and 365 nm. Miami 1B showed an excitation maximum similar to the CFH samples, but the maximum was quite different from the one exhibited by the aquatic humic substances from the Satilla River. The emission maximum of these samples (Table V), however, was very similar.

IR SPECTRA. The IR spectra of all 10 samples derived from drinking water are almost identical. Figure 1 shows a representative spectrum. For comparison, the IR spectra of Satilla River aquatic humus and CFH are added. All samples exhibited broad bands in the 3400-cm^{-1} (—OH), 1720-cm^{-1} (C=O), and 1640-cm^{-1} (C=C) regions. However, the spectra from the 10 drinking water samples and CFH samples showed distinct differences from that of the Satilla River aquatic humus, which absorbs sharply near 1600 cm^{-1}. This result suggests that the samples derived from drinking water are strongly altered by chlorination and/or oxidation. Further confirmation of the spectral dissimilarities should be pursued by analysis of chlorinated alteration products of humus.

X-ray Scattering. The results of the low-angle X-ray scattering analyses are reported in Table VI. By comparing the results obtained

Table V. Fluorescence Spectra of Acetone-Soluble Fractions

Sample Number[a]	*Ex*	*Em*
0	335	450
1	355	425
2	370	463
3	352	426
4	355	425
5	348	426
6	348	417
7	355	420
8	350	430
9	346	426
10	350	430
11	370	450
12	370	460

NOTE: The solvent was methanol. Ex denotes excitation; Em denotes emission. Values given are λ_{max}(nm).
[a] *See* Table II for sample names.

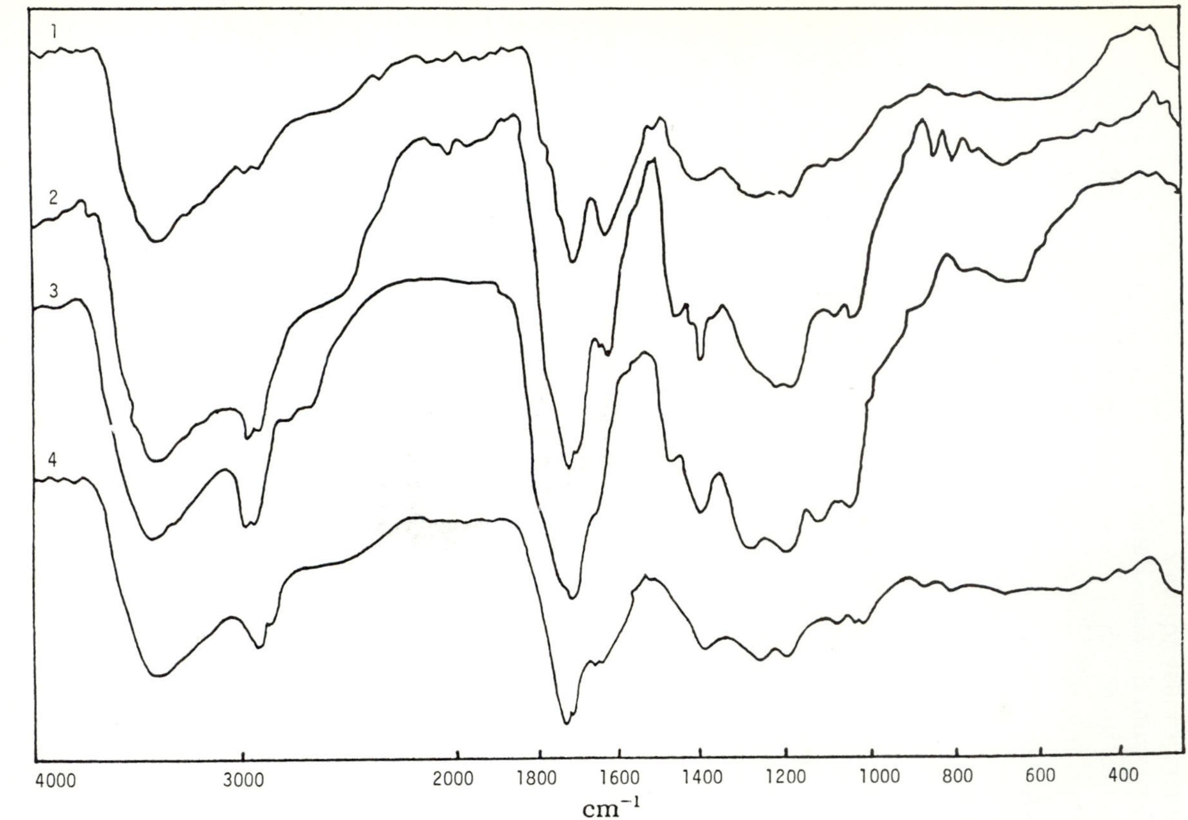

Figure 1. IR spectra of 1, aquatic river humus; 2, chlorinated aquatic river humus; 3, organic compounds from drinking water (Philadelphia 1, no. 9); and 4, CFH.

Table VI. X-ray Scattering of Acetone-Soluble Fractions

Sample Number[a]	R_g[b]	*pH*
1	5.6	11
2	7.6	9
3	7.4	11
4	6.4	11
5	8.4	11
6	5.4	10
7	5.3	11
8	7.2	11
9	10.9	11
10[c]	—	—
11	8.7	11
12	6.8	11

[a] *See* Table II for sample names.
[b] R_g is the radius of gyration expressed in angstroms.
[c] R_g and pH were not determined because of insufficient quantity of sample.

in this study with those obtained by Thurman et al. (*15*), based on the use of standards of polyglycols and polyacrylic acids, our aquatic samples along with the two neutral-pH CFHs all fell within the range reported by these authors (i.e., radius of gyration values ranged from 5.6 to 10.9 Å, which corresponds to molecular weights ranging from 800 to 3000 daltons). A brief discussion on the molecular weights of humics determined by small-angle X-ray scattering and those obtained by membrane UF and Sephadex gel permeation methods will be presented in the following section.

Ultrafiltration and Size-Exclusion Chromatography. The UF separation (i.e., drinking water organic matter, CFH) gave similar values for all samples. Eighty-nine percent or more of the initial weight was recovered in the retentate fraction (*see* Table VII). The subsequent analysis of the UF retentate by Sephadex G-75 showed differences in molecular weight distributions of the two sample categories. The typical elution patterns are shown in Figure 2, along with the calibration standard. The apparent molecular weights of drinking-water-derived samples ranged between 1000 and 5000 daltons, whereas the CFH samples had molecular weights >5000 daltons. According to the X-ray scattering results, the sizes of CFH samples fall within those of the drinking water samples (Table VI). Therefore, no agreement between the SEC and the X-ray scattering results in terms of molecular sizes can be reached. However, Thurman et al. (*15*) showed that Sephadex and UF methods gave molecular weights of humic substances <10,000,

Table VII. Ultrafiltration of Acetone-Soluble Fractions

Sample Number[a]	Original Wt (mg)	Retentate Wt (mg)	Retentate Recovery (%)	Permeate Wt (mg)	Permeate Recovery (%)	Total Recovery (%)
1	10.1	9.6	95	0	0	95
2	10.3	12.1	117	0.3	3.6	120
3	10.1	9.9	98	1.8	18.1	116
4	9.2	9.2	89	0	0	89
5	9.2	10.9	119	2.0	22.0	140
6	9.9	9.1	92	1.0	9.7	102
7	9.9	10.6	107	0	0	107
8	9.7	10.4	107	1.6	16.0	124
9	9.6	9.0	94	1.0	10.9	104
10[b]	—	—	—	—	—	—
11	11.0	11.1	101	0.6	5.5	107
12	11.0	11.1	101	0.6	5.5	107

NOTE: Volume of original solution was 200 mL; volume of retentate was 50 mL; and volume of permeate was 150 mL.
[a] *See* Table II for sample names.
[b] Weights and percent recoveries were not determined because of insufficient quantity of sample.

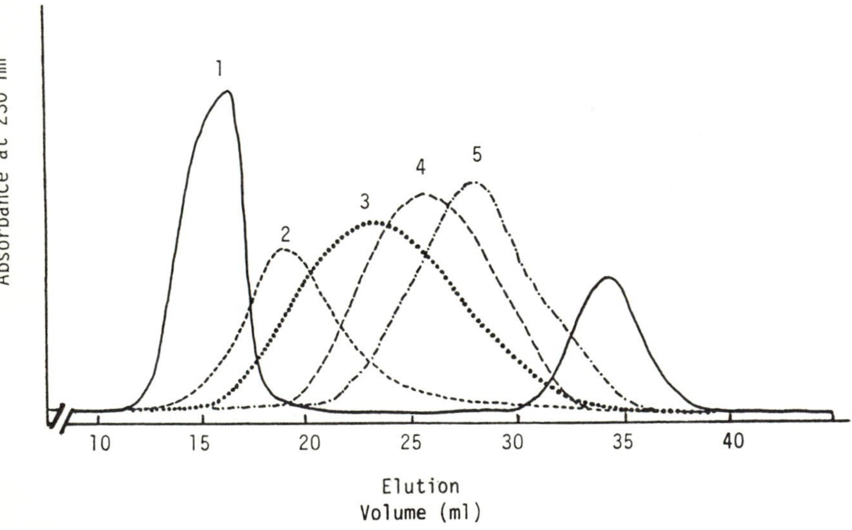

Figure 2. Gel permeation chromatography on Sephadex G-75 of 1, blue dextran and phenol standard; 2, aquatic river humus; 3, CFH; 4, chlorinated aquatic river humus; and 5, organic compounds from drinking water (Philadelphia 1, no. 9).

which are somewhat higher than those determined by the colligated properties methods (i.e., <3000). These authors (*15*) also concluded that the data from low-angle X-ray scattering are similar to the measurements of colligated properties. This finding appears to be in good agreement with our study here (i.e., the UF and Sephadex methods tend to give higher molecular weight determinations than the small-angle X-ray method).

Thermal Gravimetric Analysis and Differential Thermal Gravimetric Analysis. TGA and DTA of the samples derived from drinking water gave similar thermal maximum temperatures (thermal maximum temperatures were within 20 °C) and thermal maximum curves. On the other hand, the CFH samples, although similar among themselves, presented widely different thermal maximum temperatures and curve shapes when compared with the 10 samples derived from drinking water. Representative DTA curves for these two sample categories are shown in Figure 3.

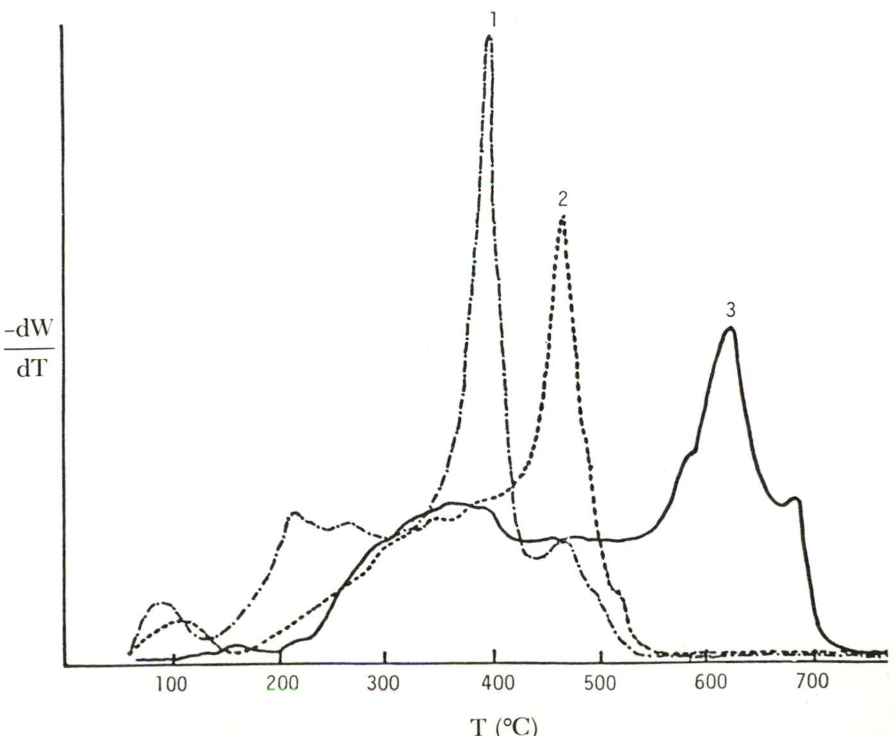

Figure 3. Differential thermogravimetric curves of 1, CFH; 2, aquatic river humus; and 3, organic compounds from drinking water (Ottumwa 1).

Acidic Group Analyses. [COOH] and total acidity values are reported in Table VIII. Two distinct groupings can be recognized according to the two sample categories, although one CFH sample presents a [COOH] value (i.e., CFH-2) similar to at least three values found for the drinking water samples.

Conclusions

The physical and chemical characteristics of all these samples indicate that they may be considered as part of a single group of natural organic compounds. However, they are still unidentifiable insofar as their molecular structures are concerned.

The results of the samples from the five cities showed great similarities. This finding indicates that it is probably acceptable to base results of toxicological testing on chlorinated humics from a limited number of drinking water sources.

Conversely, the coal-based Fluka humic acid may be too different from the aquatic humic matter in terms of molecular weight, solubility, and functional group to serve as the best model for toxicological evaluation of drinking water-derived substances.

When comparing the physical and chemical characteristics of natural organics, care should be taken to ascertain that identical methods are used to isolate and concentrate these materials.

Table VIII. Acidic Functions of Acetone-Soluble Fractions

Sample Number[a]	*COOH (meq/g)*	*Total Acidity (meq/g)*
1	2.5	5.9
2	2.4	6.1
3	2.7	5.8
4	1.9	7.0
5	2.0	5.6
6	2.3	5.3
7	2.7	7.3
8	2.3	5.2
9	1.9	7.4
10[b]	—	—
11	3.1	3.8
12	2.6	3.4

[a] *See* Table II for sample names.
[a] Acidic functions were not determined because of insufficient quantity of sample.

Literature Cited

1. Bellar, T. A.; Lichtenberg, J. J.; Kroner, R. C. *J. Am. Water Works Assoc.* **1974,** *66,* 703.
2. Rook, J. J. *Water Treat. Exam.* **1974,** *23,* 234.
3. Lin, D. C. K.; Melton, R. G.; Kopfler, F. C.; Lucas, S. V. In *Advances in the Identification and Analysis of Organic Pollutants in Water;* Keith, L. K., Ed.; Ann Arbor Science: Ann Arbor, MI, 1981; Vol. 2, p 861.
4. Loper, J. C.; Lang, D. R.; Smith, C. C.; Schoeny, R. S.; Kopfler, F. C.; Tardiff, R. C. In *Aquatic Pollutants: Transformation and Biological Effects;* Hutzinger, O. et al., Eds.; Pergamon: 1978; p 405.
5. Bull, R. J.; Kopfler, F. C.; McCabe, L. J. Presented at the Annual Am. Water Works Assoc. Conference, Atlantic City, NJ, 1978.
6. Radian Corp. Report to EPA, 1980; personal communication.
7. Smith, J. K.; Lunch, S. C.; Kopfler, F. C. In *Water Reuse;* Cooper, W. J., Ed.; Ann Arbor Science: Ann Arbor, MI, 1981; Vol. 2, p 191.
8. Perdue, E. M. *Chemical Modeling in Aqueous Systems;* Jeune, E. A., Ed.; ACS Symposium Series 93; American Chemical Society: Washington, DC, 1979; p 94.
9. Reuter, J. H.; Ghosal, M.; Chian, E. S. K.; Giabbai, M. F. In *Aquatic and Terrestial Humic Materials;* Christman, R. F.; Gjessing, E. T., Eds.; Ann Arbor Science: Ann Arbor, MI, 1983; p 107.
10. Wersham, R. C.; Pinkney, D. J. *J. Res. U.S. Geol. Surv.* **1973,** *1,* 707.
11. Chian, E. S. K.; DeWalle, F. B. *Environ. Sci. Technol.* **1977,** *11,* 158.
12. Perdue, E. M.; Reuter, J. H.; Ghosal, M. *Geochim. Cosmochim. Acta* **1980,** *44,* 1841.
13. Schnitzer, M.; Gupta, U. C. *Soil Sci. Soc. Am. Proc.* **1965,** *29,* 274.
14. Schnitzer, M.; Kahn, S. V. *Humic Substances in the Environment;* Marcel Dekker: New York, 1972.
15. Thurman, E. M.; Warshaw, R. L.; Malcolm, R. L.; Pinckney, D. L. *Org. Geochem.* **1982,** *4,* 27.

RECEIVED for review August 14, 1985. ACCEPTED April 14, 1986.

SYNTHETIC POLYMERS FOR CONCENTRATING ORGANIC CHEMICALS FROM WATER

10

Synthetic Polymers for Accumulating Organic Compounds from Water

G. A. Junk

Ames Laboratory, Iowa State University, Ames, IA 50011

*The increased use of solid synthetic polymers and the advantages of this procedure compared to other approaches for determining organic components present in water are described in this review. The procedure involves sorption of the organic components present in water onto a polymer by using simple column chromatography and subsequent removal of the sorbed components by elution with an appropriate solvent. Various polymers used for the sorption and different solvents used for elution are reviewed. The main polymers are styrene-divinylbenzenes (XAD-1, XAD-2, XAD-4, Porapaks, Chromosorbs), methacrylates (XAD-7, XAD-8), diphenyl-*p*-phenylene oxide (Tenax), polytetrafluoroethylenes (Teflon), polyurethanes, polypropylene, bonded phases, and ion-exchange resins; the primary eluents are methanol, ethyl ether, and acetone. These solvents and polymers are discussed relative to their characteristics and applications and the theoretical aspects of the methodology.*

THE USE OF SYNTHETIC POLYMERS to accumulate organic components from water for analytical and bioassay purposes is reviewed in this chapter. This review is given perspective by including a brief history of adsorption chromatography, the use of activated carbons in water research, and the recent introduction of bonded phases for aqueous sample preparations.

Adsorption chromatography, first introduced by Tswett at the turn of the 20th century, was used almost five decades later by Middleton and Rosen (*1–4*) in the first successful attempts to characterize the organic components present in water. Columns of carbon were used to isolate the organic constituents that were subsequently recovered by extracting the carbon with an organic solvent. These pioneering efforts led to the extensive use of carbon for both analytical and water

0065-2393/87/0214/0201$12.25/0

purification purposes as documented by some representative citations included in this review (*1-75*). These reports give prospective evidence of the past and current value of carbon as a solid adsorbent in the analytical chemistry and treatment of water. The heterogeneous nature of the activated carbons used in these early studies caused analytical problems that might have been resolved by using a more homogeneous solid adsorbent.

Synthetic polymers are homogeneous, and in 1969, Riley and Taylor (*76*) were the first to report the use of a styrene–divinylbenzene polymer, distributed by Rohm and Haas as XAD-1 of their Amberlite XAD series, to accumulate a variety of organic compounds from different test waters. This same polymer had been used earlier, in 1968, by Bradlow (*77*) for the determination of drugs in urine.

The research results from Riley and Taylor (*76*) and the technical reports from Rohm and Haas should have sparked an active interest in the use of polymers to accumulate organic compounds from water. However, this interest did not develop until after the publication, by Junk et al. (*78*) in 1974, of extensive test results and a detailed methodology for XAD-2, another styrene–divinylbenzene polymer. As examples of the rapid acceptance of this XAD-2 procedure, it was used in 9 of 36 symposium reports presented in 1975 and published in Keith's book (*79*) on organic compounds in water; it was used in 17 of 57 reports at a follow-up symposium held 5 years later in 1980 (*80*). A list of references to the use of the XAD series of styrene–divinylbenzene polymers—XAD-1, XAD-2, and XAD-4—is part of the bibliography (*6, 7, 9, 12, 17, 18, 22-25, 28, 30, 33, 39-41, 45, 48, 50, 52, 54-58, 60-66, 68, 74, 76-78, 81-318, 318a*). This list is comprehensive through 1980; it is not complete for 1981-84 because many studies were published in difficult-to-locate reports in which organic accumulation using XAD polymers was only a small, but vital, part of a total research effort. During the past decade, the use of other XAD polymers having the acrylate functionality have been reported (*6, 24, 25, 30, 48, 56, 57, 64, 68, 74, 89, 91, 104, 108, 111, 112, 114, 133, 143, 146, 164, 165, 167, 168, 181, 191, 204, 207, 211-213, 218, 224, 227, 228, 239, 240, 244, 252, 266, 283, 286, 291-293, 296, 303, 305, 315, 319-329*). Polymers from other suppliers have also been used: the Porapaks (*21, 26, 68, 119, 189, 240, 330-342*); Tenax (*21, 26, 36, 48, 53, 68, 82, 88, 101, 162, 178, 187, 191, 240, 289, 317, 331, 332, 338, 343-373*); the Chromosorbs (*21, 26, 53, 65, 68, 317, 336, 338, 340, 353, 357*); polytetrafluoroethylenes (PTFE) as sheets (*138a, 374, 375*), powders (*148, 220, 376, 377*), and porous aggregates (*378*); polyurethanes (*7, 19, 37, 38, 46, 48, 61, 83, 87, 104, 146, 379-411*); polyethylene (*412-414*); polypropylene (*415*); carbonaceous resins (*15, 16, 24, 27, 30, 57, 66, 68, 69, 71, 97, 181, 286*); coated solid supports (*48, 83, 102, 185, 245, 367, 381, 385, 386, 393, 394,*

396, 398, 402, 404, 406, 407, 410, 411, 416); miscellaneous other adsorbents (*11, 46, 70, 124, 126, 149, 218, 240, 281, 301, 308, 338, 404, 404a, 404b, 404c, 417–432*); and ion-exchange resins (*22, 24, 71, 80, 160, 167, 178, 181, 182, 208, 214, 239, 242, 250, 252, 257, 260, 280, 296, 305, 315, 321, 324, 325, 433–468*). None of these have the popularity of the XAD polymers primarily because of precedent. Next to the XAD polymers, Tenax and the polyurethanes are fairly popular because of the interest generated by the early reports by Leoni et al. (*346, 354*) and Gesser et al. (*383, 392*), respectively.

The recent availability of reverse-phase liquid chromatographic supports, in which organic units are chemically bonded to porous silica, has resulted in many investigations (*11, 21, 48, 67, 68, 82, 161, 220, 241, 256, 277, 334, 341, 390, 398, 409, 418, 438, 444, 463, 469–535*) of the use of these solid adsorbents. In 1975, Little and Fallick (*516*) and May et al. (*82*) first reported the applicability of these bonded phases to the accumulation of aromatic compounds from water. Although the bonded phases are not polymeric, they are included in the review because organic compounds are accumulated from water with an adsorption mechanism identical to that of the porous polymers. The number of published articles attests to the popularity of bonded phases especially for clinical applications to drugs, pharmaceuticals, and metabolites in urine and other biofluids. The bibliography of bonded phases is not complete for the same reasons explained earlier for the XAD literature of the past 3 years; many references are difficult to locate because the organic accumulation was only a small, but vital, element of more extensive research efforts.

Other published reviews discuss the use of solid adsorbents (*318, 423, 512, 536*) and other methodologies (*146, 318*) for the determination of trace levels of organic constituents in water. These reviews provide added introductory insight into the value of using synthetic polymers.

Theoretical Considerations

Any detailed discussion of the theory governing the accumulation of organic components from water by using solid adsorbents is beyond the scope of this review. However, a qualitative theoretical overview of the essence of the methodology is helpful in understanding the operational features and in selecting the proper adsorbent for the accumulation of the desired organic components.

Comparison to Solvent Extraction. The methodology for using solid adsorbents is described in detail in a later section. Briefly, it involves adsorption and desorption processes. In the adsorption process, the liquid matrix is water containing very small amounts of organic constituents. As a first approximation, the adsorption process can be considered analogous to liquid–liquid partitioning, popularly called solvent extraction. In solvent extraction, a favorable partitioning of the organic solutes to the immiscible organic solvent occurs. The "likes dissolve likes" concept is operative. The same holds true for extraction (adsorption) of organic solutes by using a synthetic polymer; the solid organic polymer is analogous to the immiscible organic liquid. Partitioning by adsorption on the surface of the polymer is analogous to dissolution in the organic liquid. This analogy brings to focus two of the advantages of solid adsorbents: first, the guarantee of phase separation when solid polymers are used; and second, the more favorable partitioning of the organic solutes to the solid phase in comparison to the partitioning to an immiscible liquid phase (*423*). The tendency to stretch the analogy beyond this point should be avoided.

Frontal Chromatography. The mechanisms involved in the adsorption of organic solutes onto solid organic polymers are complex and not yet well defined (*537*). Adsorption chromatography is really frontal chromatography but not the kind operable in gas chromatography in which an exact quantitative description of the important gas–liquid interactions can be achieved. When the mobile phase is liquid, frontal chromatography is more complex and defies an accurate quantitative treatment. Qualitatively, the adsorption of organic solutes from water solutions onto the polymer is determined by solute–water, solute–polymer, and water–polymer interactions. For effective adsorption, the solute–water and water–polymer interactions should be weak and the solute–polymer interaction should be strong. For water solutions of slightly soluble organic compounds, the solute–water interaction is certainly weak. If the polymer is hydrophobic, the water–polymer interaction is also weak. Provided a chemical similarity is found between the organic solute and the functional units of the polymer, the solute–polymer interaction is expected to be strong.

Once the organic solute is adsorbed onto the polymer, it can be desorbed by using an organic eluting solvent that has strong solvent–polymer and solute–solvent interactions. A better focus of this qualitative description can be obtained by considering the example of the adsorption and desorption of polychlorinated biphenyls (PCBs).

PCB ADSORPTION. PCBs are practically insoluble in water because of a very weak solute–water interaction; PCBs will have a strong solute–polymer interaction if a polymer such as styrene–divinylbenzene is used; water will have a weak interaction with styrene–divinylbenzene; thus, conditions for effective adsorption are present. Therefore, large volumes of water can be passed through a column packed with a styrene–divinylbenzene polymer, and the PCBs will be adsorbed (partitioned) efficiently.

PCB DESORPTION. PCBs are very soluble in a number of organic solvents. Because acetone is very effective in displacing the water from the pores of the polymer, it will be used in this example of desorption. A fairly strong interaction of acetone with the styrene–divinylbenzene surface can be predicted because acetone and benzene are miscible solvents. Consequently, a small amount of acetone will desorb the PCBs because strong solvent–solute and solvent–polymer interactions override the strong solute–polymer interaction. This desorption, commonly called elution, does not occur during the adsorption process because the matrix water is a poor eluent dictated by its weak interaction with hydrophobic polymers.

Thermodynamic View. Superficial thermodynamic considerations can also be used to explain the adsorption process. Two mechanisms that are not well understood are used in the thermodynamic view of the adsorption process. The first mechanism is surface adsorption. In the absence of a quantitative description, the reliable "likes adsorb likes" principle is invoked, and hydrophobic organic solutes will associate with the hydrophobic polymer in preference to dissolution in the water. The second mechanism is solubilization of hydrophobic solutes by water. In 1945, Frank and Evans (*538*) used the iceberg effect to explain the solubility of hydrophobic solutes in water. Briefly, this effect suggests that a nonpolar organic molecule is solubilized in water because of the orientation of many layers of water molecules around the organic solute. The necessity for orientation to accomplish dissolution forces the water to become more crystalline, hence the iceberg effect (*538a, 538b*). The important physical concept is the decrease in the entropy of the water system as it becomes more ordered in the partial crystalline form. When the dissolved organic molecules, now having water molecules ordered in crystalline fashion, are adsorbed on the polymer, the icebergs are partially dispersed and the entropy of the water system increases. Positive entropy changes are usually accompanied by negative free energy changes. The adsorption process is therefore favorable and spontaneous. This qualitative thermodynamic view is an aid in understanding the adsorption mechanism in a fundamentally acceptable manner consistent with the observed effectiveness of hydrophobic polymers for removing trace levels of hydrophobic organic solutes from water solutions.

Additional Background. Two other theoretical considerations provide background for a better understanding of the use of solid adsorbents for analytical and bioassay purposes. These considerations are irreversible adsorption and concentration plus solvent transfer.

IRREVERSIBLE ADSORPTION. A difficulty in the use of many activated carbons for analytical and bioassay purposes is irreversible ad sorption. This phenomenon occurs when very strong solute–adsorbent interactions cannot be overridden by the solvent–adsorbent and solvent–solute interactions during elution. When this situation occurs, the adsorbent has little value for analytical or bioassay value. However, the adsorbent still may be used very effectively for water treatment processes.

CONCENTRATION PLUS SOLVENT TRANSFER. Concentration of the organic solutes is essential to the determination of many organic contaminants present in water at very trace levels. The solvent transfer is needed for implementation of the separation and detection schemes that do not tolerate the water matrix. For bioassay work, concentration and solvent transfer are also needed because the amounts are too low for direct testing of the water solutions, and dimethyl sulfoxide (DMSO) is the preferred solvent. In bioassay studies that involve animal exposure, the concentration scheme must accommodate very large volumes of water. Theoretically and practically, these elements of the analytical and bioassay methodologies can be achieved by using solid adsorbents, especially synthetic polymers.

Generalized Methodology

Many different names have been used to describe the general methodology of using solid adsorbents for accumulating trace levels of organic components from water. Some of these descriptive names are the following: resin extraction, macroreticular resin extraction, synthetic polymer extraction, solid phase extraction, macroreticular resin sorption, resin sorption, XAD method, and adsorption–desorption. Independent of one's nomenclature preference, the methodology is a two-step procedure involving adsorption of the organic solutes onto a solid adsorbent and desorption either with an organic solvent or on occasion by thermal means.

Adsorption. The adsorption is efficiently accomplished by gravity flow column chromatography. A column packed with XAD-2 is shown in Figure 1. In oversimplified language, water containing one or more organic components enters at the top of the column, and water free of all traces of organic components is exhausted at the bottom of the column. For a single dissolved component, this favorable condition is achieved only if (1) the solute–adsorbent interaction is strong, (2) the solute–water interaction is weak, and (3) the water–adsorbent interaction is negligible. Then, a very low level of an organic solute (e.g., <1 ng/L) can be adsorbed from large volumes of water (>100 L). As suggested in the Theoretical Considerations section, the necessary inter-

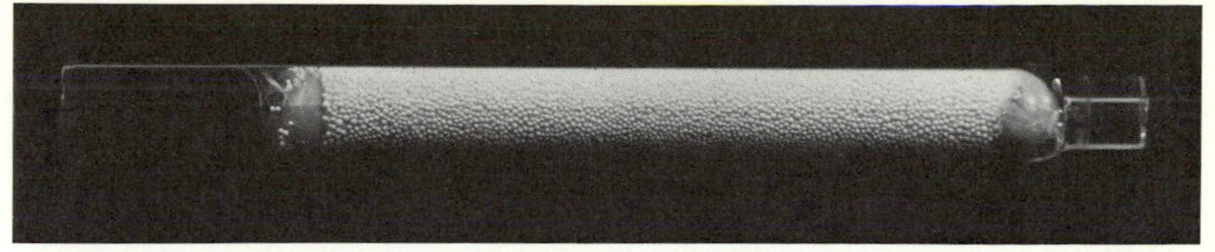

Figure 1. Photograph of XAD-2 accumulator column (~2 g of 40–60-mesh XAD-2 and glass wool retainers in a 0.13-cm o.d. × 9-cm long glass tube).

actions occur if (1) the organic solute is chemically and physically similar to the polymeric units, (2) the organic solute is very insoluble in water, and (3) the solid adsorbent is very hydrophobic (apolar).

In all cases except ion exchange, the van der Waals forces that bind the organic solute to the adsorbent are not amenable to an accurate quantitative description. Without this quantitative treatment, the qualitative description presented in the Theoretical Considerations section can be used to conclude that no single polymer will be efficient for all kinds of organic compounds at all concentration levels and with large volumes of water. If the organic compounds are very insoluble, then apolar adsorbents such as XAD-2 and XAD-4 can be used effectively for large volumes of water. These adsorbents will not be efficient for soluble hydrophilic compounds. The accumulation efficiency can be improved for soluble compounds by switching to more polar adsorbents, such as the acrylates, but high efficiency will never be attained even when polar polymers having very high surface areas are used. The shift in polymer polarity increases the water–polymer interaction such that the matrix water becomes an effective eluent and breakthrough occurs after passage of only a small amount of water through the column. Small volumes of water can be processed effectively, and relatively large amounts of hydrophilic solutes can be adsorbed only from quite concentrated solutions.

Polymeric adsorbents can be tailored in functionality and physical form to accumulate most organic solutes from small volumes of water with high efficiency. The scheme developed at the U.S. Geological Survey (USGS) laboratories (*167, 252*) is probably the best example of the use of tailored polymers for accumulating the widest possible variety of organic components. In some cases, ion-exchange polymers can be used to adsorb organic ions and neutral components (*80, 208, 433*) simultaneously, but this procedure, like the USGS scheme, is applicable only to small volumes.

Desorption. Desorption with an organic eluent is the most common and useful procedure for most applications. The solvents that have been employed are listed on page 208 along with the corresponding

Solvents Used for Desorbing (Eluting) Organic Compounds from Solid Adsorbents and Corresponding References

Methanol: 34, 43, 64, 77, 84, 88, 93, 96, 101, 103, 107, 127, 130, 134, 137, 148, 149, 153, 154, 156, 157, 163, 167, 178, 186, 191, 193, 194, 196, 199, 200, 207, 208, 210, 214, 222, 225, 227, 234, 237, 256, 267, 273, 275, 289, 309, 319, 322, 323, 332, 404a, 418, 419, 425, 427, 433, 436, 442, 444, 453, 454, 460, 463, 485, 499, 503, 508, 514, 517, 518, 520, 523, 529, 530.

Ethyl ether: 7a, 14, 23, 28, 30, 34, 36, 39, 40, 49, 52, 54, 60, 78, 79, 83, 89, 93, 103, 107, 112, 114, 117, 120, 121, 140, 141, 149, 163, 167, 171–173, 176–178, 184, 185, 191, 196, 199, 200, 203, 208, 213, 219, 222, 229, 230, 232–234, 243, 271–273, 275, 276, 280, 307, 309, 322, 339, 346, 354, 371, 372, 427, 433, 442.

Acetone: 6, 7, 9, 12, 28, 33, 41, 50, 52, 68, 75, 100–102, 105, 108, 110, 111, 113, 147, 149, 154, 159, 163–165, 171, 179, 180, 200, 201, 213, 217, 220–222, 254, 266, 269, 270, 278, 287, 289, 303, 311, 312, 337, 349, 380–385, 392–394, 396, 403, 404a, 404c, 405–407, 439, 449, 514.

Chloroform: 2, 3, 10, 20, 27, 39–41, 44, 47, 51, 52, 73, 74, 134, 135, 149, 154, 166, 169, 189, 205, 222, 226, 243, 274, 303, 314, 322, 493, 517.

Methylene chloride: 6, 25, 43, 56, 62, 86, 107, 110, 138, 144, 159, 161, 164, 165, 169, 200, 222, 228, 267, 269, 278, 287, 290, 312, 375, 378, 418, 470, 497, 536.

Hexane: 36, 83, 85, 100, 112, 138a, 179, 180, 201, 203, 221, 227, 236, 266, 270, 285, 311, 333, 365, 383, 385, 386, 392–394, 396, 403, 407.

Ethanol: 7, 10, 20, 32, 41, 47, 51, 62, 77, 102, 134, 150, 177, 182, 207, 225, 232, 255, 257, 263, 269, 283, 295, 297, 304, 312, 404b.

Acetonitrile: 5, 101, 116, 161, 178, 207, 236, 283, 296, 318a, 334, 432, 479, 485, 492, 505, 516, 527.

Sodium hydroxide: 22, 64, 70, 89, 90, 92, 98, 115, 143, 174, 178, 191, 324, 326, 327, 434.

Ethyl acetate: 95, 121, 122, 156, 175, 194, 222, 224, 226, 228, 244, 258, 310, 404a, 480, 518.

Isopropyl alcohol: 9, 129, 135, 144, 166, 175, 189, 195, 205, 226, 243, 258, 314, 404a, 470.

Miscellaneous: This category includes benzene (*5, 68, 81, 248, 384, 395, 405*); DMSO (*211–213, 216, 235, 328*); carbon disulfide (*12, 13, 28, 33, 60, 72*); tetrahydrofuran (*107, 195, 334, 415, 535*); pentane (*65, 101, 385, 478*); cyclohexane (*88, 380, 382*); hydrogen chloride (*64, 218, 254, 438*); toluene (*5, 310*); dichloroethane (*175, 258*); petroleum ether (*395, 416*); isopropylacetone (*332*); carbon tetrachloride (*401*); dioxane (*497*); pyridine (*118, 197*); sodium carbonate (*276*).

references. On the basis of the number of references, methanol, ethyl ether, and acetone are used most frequently. As with adsorption, for which the polymer can be chosen to enhance the accumulation of selected solutes, the organic solvent can be chosen to achieve the most efficient elution. Given a specified column dimension, the efficiency of the elution process will depend on the solvent–adsorbent and the solvent–solute interactions. With the proper choice of solvent, only a small amount is needed to achieve quantitative elution of the desired solutes.

Thermal desorption requires stability of both the polymer and the adsorbed solutes; this requirement has severely restricted its use except for the most volatile and chemically stable components.

Polymer Wetting. A critical feature of the successful use of synthetic polymers, particularly those having macroporosity, is the wetting of the entire polymeric surface with the test water. This wetting action has been called preconditioning by some and incorrectly referred to as hydration by others. The vital point is that adsorption of the organic solute can occur efficiently only if the water solution makes contact with the entire surface of the macroporous polymer. As will be shown later, 99+% of this surface is internal even for polymers ground to submicrometer particle size. Thus, the macroporous polymerization process dictates the surface area to the extent that particle size has a negligible influence.

An understanding of this macroporosity and the hydrophobicity of many polymers is useful for the selection of the solvent sequence used in wetting. This sequence must promote complete permeation of the water into all of the pores of the polymer, and the adsorbent column must be operated so that the test water passes through these pores. To accomplish this nonchannelized flow of water, the right combination of wetting, particle size, and column diameter and length is needed.

The methodology published by Junk et al. (*78*) and reviewed by others (*318*, *423*, *536*) included this wetting concept through an empirical approach that resulted in a recommendation of 10 to 1 for the length-to-diameter ratio of a column containing 5 g of the hydrophobic XAD-2 polymer with particle sizes of approximately 0.25 mm. The optimum column shape to ensure intimate water contact to all surfaces of a polymer having specified particle size and pore diameter could be calculated from hydrodynamics, but this calculation has not been done by any of the investigators. Even if these calculated results were available, they would not be rigidly applicable because the pore sizes of the available polymers cover a rather wide range, and the particles used to pack adsorbent columns are usually not very uniform in size.

In general, small particles consistent with adequate flow under gravity conditions should be used in columns that are long and slender as opposed to short squat columns. Lengths about 5–10 times greater than the diameter are needed to avoid channelization of the water flow.

Advantages. The advantages of the use of solid adsorbents are best discussed within the framework of the generalized methodology just presented. For this reason, the key word advantages are discussed here rather than at the end of this review. The advantages are considered relative to solvent extraction procedures for accumulating organic components from water. Others have discussed these same desirable features of solid adsorbents but with different emphasis (*78, 318, 423, 536*).

SIMPLICITY. No complicated equipment is needed for the implementation of solid adsorbents to accumulate organic compounds from water. The procedure is identical to gravity flow adsorption chromatography used for decades by many chemists to remove extraneous material from liquid samples in a process called sample cleanup. The simplicity of this procedure with the reduced sample manipulations minimizes solute losses and sample contamination.

By analogy to solvent extraction, the column containing the solid adsorbent corresponds to the separatory funnel containing the immiscible organic solvent. The transfer of the solute to the solid adsorbent occurs in an unattended operation requiring no manual effort or additional equipment such as the shakers used in solvent extraction or the distillation apparatus used in some of the automatic extraction devices. This simplicity allows for facile automation either off-line or on-line with the separation and detection procedure (*495, 512, 536*).

SPEED. The adsorption columns can accommodate high flow rates of up to 20 bed volumes per minute (*52, 78, 122, 129, 156, 167, 171, 175, 181, 200, 203, 211, 225, 231, 244, 258, 286, 319*). Multiple columns can be used to process many water samples simultaneously in an unattended operation. This feature also makes the methodology well suited to automation (*469, 491, 495, 500, 512, 529, 536*).

FIELD SAMPLING. The small, simple, adsorption columns are very convenient for field sampling to avoid the expensive and troublesome transport of water samples back to the laboratory for processing. Only the small columns, with the organic solutes preserved on the polymer surfaces at ambient temperatures, need to be carried to and from the laboratory. The cost differential for transporting small columns in comparison to transporting the containers and the water becomes greatest when large volumes of water are to be processed for analytical and bioassay purposes. The cost differential is appreciable even for small volumes of water when one considers water preservation by refrigeration and container costs in addition to transportation costs.

INEXPENSIVE. Sampling and organic accumulation costs are a significant fraction of the total expense of determining organic components present in water. Aside from the savings accrued by field sampling with polymers as described earlier, other savings are seen when polymer adsorption is compared to solvent extraction. For example, (1) greatly reduced sample manipulations decrease labor costs, (2) expensive solvent use is decreased, (3) the considerable manual or mechanical energy of a shaking process is not necessary, and (4) the adsorbent columns can be reused (*78, 423, 512*). Even if the columns are discarded after a single use, the cost of accumulation, using solid adsorbents, is estimated to be 5–10 times less than accumulation by solvent extraction on the basis of unpublished results from my laboratory.

SAFETY. Only a small amount of solvent is needed. The adsorbed solutes are eluted under gentle column flow conditions. In solvent extraction, larger volumes of solvent are needed and violent agitation is usually necessary. Thus, the fire and toxic hazards are much less for polymer accumulation. This safety feature compliments the decreased analytical costs and the lower solvent blanks associated with the use of smaller amounts of solvent.

INTIMATE CONTACT. In a properly designed and operated column, contact of the dissolved solutes with the solid adsorbent surfaces is assured.

In solvent extraction, this same contact is achieved either slowly over extended time periods of mild shaking or more rapidly through generation of a forced emulsion by violent agitation that consumes large amounts of energy.

SOLVENT CHOICE. Solvent extraction is limited to water immiscible solvents. Solid adsorbents do not have this limitation, so miscible solvents, desirable for subsequent analytical or bioassay purposes, can be used. For example, DMSO is preferred for mutagenicity screening and has been used to elute the adsorbed organic material (*211–213, 216, 235, 328*). For analytical purposes, acid, base, and neutral eluents can be employed for on-column fractionation of the adsorbed organic solutes (*78, 80, 196*).

CAPACITY. The high capacity of neutral macroporous polymers for hydrophobic organic solutes is essential for the convenient accumulation of large amounts of solutes from hundreds of liters of water. For molecular weights near 100, more than 1 g of solute can be adsorbed on 1 g of polymer without exceeding monolayer coverage. This capacity is rarely achieved in any kind of solvent extraction. Amounts sufficient for extensive analytical work and bioassay feeding experiments can be accumulated in an unattended fashion from large volumes of water having low concentrations or from smaller volumes having higher concentrations. Although breakthrough of hydrophilic components occurs frequently, the capacity for hydrophobic solutes is usually sufficient to accumulate the large amounts desired for difficult analytical characterizations and bioassay work.

EMULSION FREE. The solvent extraction of many water solutions, particularly waste waters and biofluids, is a prolonged and often frustrating procedure because of the formation of emulsions. This problem is obviated when solid adsorbents are used because phase separation is inherent to the procedure.

Disadvantages. The discussion of the advantages of polymer adsorption was relative to solvent extraction. This same perspective will be used in the discussion of the disadvantages. Most analytical chemists are familiar with solvent extraction procedures using solvents that are readily available in the desired purity. When the solvents are not pure enough for direct use in trace analytical work, the purification procedures are usually ones that are familiar to most bench chemists and technicians. These advantages of solvent extraction are the disadvantages of polymer adsorption.

FAMILIARITY. Many analytical chemists are not familiar with the methodology for polymer adsorption and are unwilling to abandon the familiar and traditional procedures of solvent extraction that are satisfactory for the trace analyses of many organic compounds.

AVAILABILITY. Many polymers have not been readily available, in small quantities, from the manufacturers; those that are easily obtained are generally not of sufficient purity for direct use in analytical work.

PURIFICATION. The impure polymers, received from the manufacturers, must be purified by using procedures that are slow and complicated. This situation is in contrast to the familiar procedures used to purify solvents.

SUMMATION. Those chemists who have embraced polymer adsorption successfully have either accepted or circumvented the three major disadvantages; some have pioneered new and novel means of purification; a few have prepared their own pure polymers; others have waited for the availability of commercial polymers pure enough to suit their analytical or bioassay needs.

The availability disadvantage is expected to persist for some time because of industrial and user considerations. The polymer industry caters to high-volume sales of low-purity polymers because the users are unwilling to pay the price associated with low-volume high-purity products. Consequently, purification is necessary before the polymers can be used for trace analytical work. It would be very much more desirable to have the polymers produced from high-purity starting materials under rigid production conditions than to purify the technical-grade products produced under less stringent conditions from impure starting materials.

Another factor that has delayed the production of polymers pure enough for trace analytical work is the uncertainty of the market. The past controversies over the best solvent for solvent extractions are minor compared to the predicted future controversies over the most useful polymers. The permutations of functionality, mixed functionality, and physical form generate such a large number of polymers that objective determination of the best will be much more difficult than past determinations of the best extraction solvent. In this uncertain atmosphere, the commitment to the production and marketing of relatively small amounts of ultrapure specialty polymers is a bold venture.

Adsorbent Characteristics

The characteristics of solid adsorbents that are pertinent to accumulation are (1) functionality, (2) surface area, (3) pore size, (4) particle size and shape, (5) chemical stability, and (6) solubility. The first three characteristics are the most important, and these are compared for several solid adsorbents as part of Table I. Certain features of this table will be explained before discussing the six pertinent characteristics of solid adsorbents.

The list of solid adsorbents is extensive, although not exhaustive. Some polymers such as PRP-1 (*281, 420–422*) and Hitachi Gel (*126, 418, 419*) are not listed because their use has not been documented either extensively in one laboratory or less extensively in many laboratories. Other polymers such as PTFE, polyethylene, and polypropylene have received little attention but are listed because the polymer functionality may well be of interest. Coated supports and one ion-exchange polymer are included for comparison purposes and are given only minor discussion in this review. Bonded phases are not polymeric, but they are included because they are porous, heavily used in clinical laboratories, and have great potential for future analytical use.

The polymers are loosely grouped according to chemical functionality with the styrene–divinylbenzenes (some XADs, Chromosorbs, and Porapaks) together with the ethylvinylbenzene–divinylbenzenes (Porapak Q and Synachrom).

Many different kinds of polyurethanes are available (*391*), and the exact form used in the published investigations is not specified. The physical form as foam or open pore is specified, and these are listed separately.

Table I. Characteristics of Polymeric Adsorbents Used for Water Analyses

Common Name of Adsorbent	*Type*	*Area (m^2/g)*	*Pore Size (Å)*	*Method Refs.*	*Supplier*
Chromosorb 102	styrene–DVB	350	90	357	Johns-Manville
Chromosorb 105	polyaromatic	650	500	68, 353	Johns-Manville
Chromosorb 106	polystyrene	750	—	68, 353	Johns-Manville
Ostion SP-1	styrene–DVB	350	85	109, 124	Laboratory Instr.
Synachrom	ethylvinylbenzene–DVB	570	45	425	Laboratory Instr.
XAD-1	styrene–DVB	100	200	76, 78	Rohm and Haas
XAD-2	styrene–DVB	350	90	78	Rohm and Haas
XAD-4	styrene–DVB	780	50	112, 386	Rohm and Haas
Porapak Q[a]	ethylvinylbenzene–DVB	735	70	68, 337	Waters Associates
XAD-7	methyl methacrylate	450	80	114, 191	Rohm and Haas
XAD-8	methyl methacrylate	140	250	329	Rohm and Haas
Spheron MD	methacrylate–DVB	320	—	427	Laboratory Instr.
Spheron SE	methacrylate–styrene	70	—	417	Laboratory Instr.
Tenax	diphenylphenylene oxide	20	720	345, 346	Applied Science
Polyurethane[b]	ester–open pore	0.6	—	87	varied
Polyurethane[b]	amide ester–foam	0.02	—	7, 83	varied
Polypropylene	propylene	—	—	415	varied
Polyethylene	ethylene	—	—	412, 414	varied
A-24[c]	anion exchange	28	—	449	Rohm and Haas
Teflon	tetrafluoroethylene	—	—	375, 377	varied
Chromosorb T	tetrafluoroethylene	5	—	378	Johns-Manville
Fluoropak 80	tetrafluoroethylene	3	—	378	Fluorocarbon Company
Unspecified[d]	coated supports	low	—	394	varied
Unspecified[e]	bonded phases	—[f]	—	68, 480	varied

NOTE: DVB denotes divinylbenzene.
[a] Other Porapaks (P to T) have surface areas from 300 to 700 m^2/g.
[b] Generally unspecified; many different types are available.
[c] Single example of many different kinds.
[d] Many possibilities are available.
[e] Most are porous silica with C_4 to C_{18} alkyl bonded groups.
[f] A wide range is available from about 100 to 600 m^2/g.

The area values are averages to the nearest 10 m^2/g of the values reported in technical bulletins and review articles. The pore sizes are also averages to the nearest 5 Å.

The suppliers and the methodology references are included for the benefit of those who wish to locate a specific polymer and the details of the methodology specific to each polymer. The suppliers and the listed references are not complete, and they do not necessarily reflect the preferences of this reviewer.

Some discussion of the six characteristics outlined at the beginning of this section is now in order.

Functionality. Surface affinity to different solutes is the most important characteristic of solid adsorbents. These affinities depend on the chemical functionality and the surface orientation of the polymer. Affinity can be estimated in the static mode by measuring adsorption isotherms for different organic solutes.

The results of these static measurements can then be used to rate the probable usefulness of different adsorbents. However, the isotherm results from static water solutions do not apply to dynamic column situations in which equilibrium conditions may not occur. A better approach is to generate frontal breakthrough curves that can then be used to estimate the use of different polymers for different solutes dissolved in water. Theoretical and experimental reports (*97, 143, 181, 286, 319–321, 537*) discuss details about affinity measurements. These details are not included in this review because affinity is discussed only qualitatively in the sections on Theoretical Considerations and Generalized Methodology. These qualitative discussions suggest that neutral polymers such as the styrene–divinylbenzenes are efficient for adsorbing neutral hydrophobic solutes from water solutions but have little affinity for polar and ionic solutes. If the polarity of the polymer is increased to that of the acrylates, the affinity for neutral hydrophobic components will suffer but the more polar solutes will be better adsorbed. In the absence of actual test results under dynamic column flow conditions, the simple "likes adsorb likes" concept is invoked.

Surface Area. Given a particular affinity of the adsorbent surface for the dissolved solute, the effectiveness of the solid adsorbent for accumulation is directly proportional to the surface area. This theoretical conclusion is supported by experimental results from many laboratories, only a few of which are cited (*143, 181, 207, 423*). The value of surface area was realized first in the early investigations of activated carbons (*1–4*) and later in the initial studies of synthetic polymers (*76, 78, 116*). The first polymer investigations relied on the work of Kun and Kunin (*539*), who introduced the macroporous polymerization process. Solid adsorbents that do not have this macroporosity are less effective accumulators. Intuitively, there is a temptation to achieve high surface areas for nonporous adsorbents by comminution. If one assumes cubical particles and a density of 1 g/cm^3 for a hypothetical nonporous solid, the surface area equation is

$$A = 6 \times 10^{-4}/D \tag{1}$$

where A is area in square meters per gram and D is the diameter (edge length) in centimeters. To achieve a moderate surface area of 6 m^2/g, the nonporous solid would have to be ground into cubes 1×10^{-4} cm or 1 μm in diameter. One-micrometer particles are already too small to handle conveniently in an adsorption column, yet 6 m^2/g is low compared to the surface areas of commercial macroporous polymers (*see* Table I). To more fully appreciate this macroporous property, consider the data calculated from eq 1 and listed in Table II for a range of surface areas that extends to 600 m^2/g, which is often used in adsorption chromatography.

Clearly the macroporous polymers are the only solids that have the physical characteristics useful for gravity flow columns. This conclusion is frequently not considered or discussed. With respect to the polymers listed in Table I, the most useful are those with the highest surface areas; however, these are subject to the caveat about pore size discussed next.

Pore Size. All other factors being equal, the pore size is inversely proportional to the surface area. For high-surface-area polymers, problems begin when the molecular diameters of the solutes and the pore diameters become comparable so that passage of the solutes through the pores is severely restricted. These comparable diameters

Table II. Particle Size Versus Area for Comminution of 1 g of a Solid of Density = 1.0 g/cm^3

		Area	
Number of Particles	*Particle Size*[a] *(cm)*	*Per Particle (cm^2)*	*Total (m^2/g)*
1	1	6	6×10^{-4}
10^3	10^{-1}	6×10^{-2}	6×10^{-3}
10^6	10^{-2}	6×10^{-4}	6×10^{-2}
10^9	10^{-3}	6×10^{-6}	6×10^{-1}
10^{12}	10^{-4}	6×10^{-8}	6
10^{15}	10^{-5}	6×10^{-10}	60
10^{18}	10^{-6}	6×10^{-12}	600

[a] A cube is assumed.

for the dissolved solute and the polymer pores can easily occur for many natural humic substances passing through high-surface-area polymers. When this situation happens, many of the large molecules are adsorbed only on the limited external surface, and poor efficiency is achieved. An additional complicating factor is the partial or complete blocking of the pores that prevents the permeation of small molecules. This blockage will happen long before the pores are blocked to the flow of water whose molecules are much smaller than most solutes. The resultant reduction in diameter of the openings to the pores further restricts the possible contact of other solutes to the internal surface and causes a concomitant loss in capacity.

Large pores are desirable but not always possible when high surface area is also demanded. Therefore, a compromise is necessary. This compromise is evident in the choice of XAD-8, with its lower surface area, over XAD-7 when humic acids are accumulated from environmental waters (*143, 167, 319-321*). The same, but more subtle, considerations apply to XAD-2 and XAD-4. The 780 m^2/g for XAD-4 will be available for the adsorption of intermediate weight organic solutes only if the water is relatively free of humic material that can block many of the pores. Therefore, a situation can exist for environmental waters for which the lower surface area XAD-2 is more effective than the XAD-4 polymer.

Some of the discrepancies in the reported recoveries of different solutes from various waters by different investigators who have used the same functional polymers from different manufacturers can be rationalized by considering the discussions of pore size and surface area. Even when the pore sizes and surface areas are specified, awareness of the uncertainty in their determination is needed. Two polymers having the same listed pore size and surface area can behave quite differently as accumulators of organic solutes; surface area does not specify surface orientation for adsorption; pore size is not uniform, so the quoted value is an average and experimentally uncertain number.

Particle Size and Shape. The polymerization process for producing macroporous synthetic polymers (*539*) leads to the formation of spherical particles whose size can be controlled within certain limits. The popular XAD polymers are usually sold with approximately 90% of the total weight encompassing smooth beads with 20-50-mesh sizes. Most users incorporate a suspension step to remove the fines in their purification of the polymer, but they do not remove the small number of particles larger than 20 mesh. The particle size and distribution vary with different polymer batches, and it is advisable to mechanically sieve polymer beads and choose only those within the 20-50-mesh size for preparation of the adsorption columns.

Some investigators grind the polymer beads prior to cleanup and packing into adsorption columns. Apparent reasons for this practice are to increase the surface area and to reduce polymer costs; both reasons are unsound. For possible increases in surface area, consider the data in Table II discussed earlier. Comminution even to very small particles provides a negligible increase in the surface area of macroreticular polymers. The value of grinding would be achieved only in those instances of a column operated in a theoretically unsound fashion. For example, grinding is advantageous in those cases in which very large molecules cannot penetrate the pores leading to adsorption only on the relatively small amount of external surface.

Grinding to reduce polymer costs is also questionable because of the time necessary to purify the polymers. As received, the smooth polymer beads are structurally stable. When these beads are ground, the structural stability is lost because the jagged edges of the ground particles are easily broken even under the mildest handling. The breaking of the edges exposes pockets of occluded impurities that otherwise would be inaccessible to the eluting solvent. Thus, the structural distortion of a polymer that has already been purified to a low blank level will result in a dramatic increase in the blank and thereby necessitate repurification. Time-consuming Soxhlet extractions will usually have to be repeated to return to the original low blank level. Several authors (*28, 33, 78, 170, 177, 223*) have alluded to this blank problem caused by mechanical ruptures and possibly even more subtle cracking from changes in eluting and storage solvents (*93*). Reasonable care should be exercised in the handling of the polymer beads and the packed adsorption columns; even greater care should be taken to preserve the physical integrity of the ground beads.

Chemical Stability. Chemical stability is just as important as the physical stability just discussed. In general, chemical deterioration of the polymers is no problem, and they can be stored at room temperature for years. However, the polymeric surfaces are subjected to an extreme variety of chemicals during the accumulation process. Some of these may react with the polymer. For example, reactions of styrene–divinylbenzene polymers and Tenax with the components of air and stack gases have been documented (*336, 344, 540*). The uptake of residual chlorine from water solutions has also been observed in my laboratory and elsewhere (*110, 271, 287*). Although the homogeneous nature of synthetic polymers should tend to reduce the number of these reactions relative to those that occur on heterogeneous surfaces of activated carbons, the chemical reaction possibility is real. In the development of methods for specific chemicals, the polymer stability should always be checked. On occasion, these checks may lead to favorable observations such as the stabilization of labile organophosphorus pesticides adsorbed onto XAD polymers (*103*, *121*, *122*).

Another stability that should be considered in the choice of solid adsorbents is their ability to tolerate a wide pH range without loss of accumulation efficiency. All of the neutral synthetic polymers are unaffected by extremes in pH; such is not the case with the bonded phases whose adsorption properties are stable only with pH values near 7.

A final stability feature is the thermal limit for the polymers. For gas chromatography, these limits range from 150 °C for the acrylates to more than 400 °C for Tenax. These limits are unrealistic for thermal desorption. For example, XAD-2 has an intolerably high blank at the temperature limit of 250 °C quoted for gas chromatography. High blanks at the temperature limits are generally true for all the other polymers. Tenax is the only one having extensive documentation of usefulness in thermal desorption (*21*, *26*, *53*, *82*, *162*, *178*, *240*, *317*, *338*, *343–345*, *348*, *350–353*, *356–364*, *366*, *369*, *370*, *372*, *373*).

Solubility. The solubility of the polymers has not been emphasized sufficiently in reviews and the original literature reports. The reason for this omission may be the extremely low solubility of all of the polymers, except Tenax, in the solvents normally used for elution and purification. The adverse effects of polymer solubility are (1) even very small amounts of dissolved polymer can interfere with the subsequent chemical analyses and bioassays, and (2) dissolution of some of the polymer can change the surface characteristics both externally and internally such that reuse involves a polymer of much different adsorption efficiency. An example of the first adverse effect would be deterioration of the analytical capillary column when deposition followed by decomposition of the polymer occurs on the front end of the capillary. An extreme example of the second adverse effect is the dissolution of large amounts of Tenax when benzene is used in Soxhlet desorption.

Polymer Purification

The polymers, as received from the suppliers, have impurities that must be removed before being used for trace organic determinations. The purification procedures used by different investigators are (1) solvent extraction in a Soxhlet, (2) solvent extraction in a shaker, (3) washing a column of the polymer with different solvents, (4) sonication, (5) vacuum degassing, and (6) purging with hot inert gases. Of these, Soxhlet extraction has been the most widely used and is the most effective. Different solvents have been employed by different investigators, and most have chosen to use a series of two or three 4–12-h extractions each with a solvent of different polarity. The usual sequence is to go from the more polar to less polar solvent. Popular solvents are water, acetonitrile, acetone, methanol, ethyl ether, and hexane. Sequences of acetonitrile–methanol–ether and methanol–acetone have been used most frequently for the XAD series of polymers after water washings to remove the fines and residual $NaCl-Na_2CO_3$ used in the production process. Wetting of the polymer, which is so essential for effective adsorption and elution, is also important in the solvent purification of the polymers. Although cycling of the solvent in Soxhlet extractions tends to ensure wetting, it is best to choose the solvents and the extraction thimble so that the solvent must pass through the pores of the polymer.

Purifications of macroporous polymers by thermal means, either under vacuum or in an inert gas stream, and sonication of a solvent–polymer mixture are not recommended because they cause physical distortions that expose previously inaccessible pockets of contamination.

The potential impurities for the synthetic polymers are enormous. The amounts can vary from substantial for those impurities associated with the polymerization process to negligible for some components adsorbed from the atmosphere.

Although atmospheric contamination of the polymer is generally minimal, it can be extensive because the dry polymer has a high affinity and capacity for almost all organic components. This situation is one of the reasons why the cleaned polymers are stored under an organic solvent such as methanol (78). With proper storage, the atmospheric contamination can be maintained well below interference levels. This situation is not the case for impurities from the polymerization process.

Many polymerization impurities originate from the use of impure technical-grade starting materials. For example, benzoic acid, a whole series of alkylbenzenes, styrene, naphthalene, and biphenyl have all been observed as contaminants in the XAD-2 and XAD-4 polymers. These impurities and others are believed to be trapped during the polymerization in very small nonreticular pockets. These pockets of impurity cause no problems as long as the molecular walls holding them in place are not ruptured. In many polymer uses, no ruptures occur and low blanks are achieved and maintained. If the molecular walls are ruptured by grinding or harsh handling of the polymer, some of the occluded impurities will be released. Apparently, contamination can be released even in the absence of harsh treatment; an interesting example is the alkane impurity caused by minor swelling and shrinking of XAD-2 during solvent changeover as reported by James et al. (93). Known impurities of XAD-2 and XAD-4 polymers are given in the box on page 221 along with the corresponding references.

The variable success for purifications in which the same solvents

Reported Impurities in XAD-2 and XAD-4 and Corresponding References

Divinylbenzenes: 28, 33.
Alkylbenzenes: 28, 33, 78, 170, 177, 223.
2-Pentenylbenzene: 28, 33.
Styrene: 28, 33, 170, 223.
Alkylstyrenes: 28, 33, 170.
Indenes: 28, 33, 170.
Alkylindans: 28, 33.
Naphthalene: 28, 33, 78, 170, 177.
Alkylnaphthalenes: 28, 33, 78, 170, 177.
Biphenyl: 170.
Ethyl acetate: 141.
Benzoic acid: 78.
Methyl benzoate: 28, 33.
Methyl *m*-ethylbenzoate: 28, 33.
Alkanes: 93, 223.
3-Phenylpentane: 28, 33.
Tetralin: 28, 33, 177.
Acetophenone: 28, 33.
p-Ethylacetophenone: 28, 33.
o-Ethylbenzaldehyde: 28, 33.

are used in different laboratories and the variation in the ultimate blank level for the same synthetic polymer used by different investigators may well be due to subtle differences in the procedures employed. Likewise, subtle differences in the methodology can lead to variable results for tests of the accumulation efficiency of solid adsorbents, as discussed in the following section.

Accumulation Tests

Although some investigators have tested adsorption and desorption separately, most have reported the efficiency of the total process. Separate efficiencies are useful for methods development and are of academic interest, but the total efficiency is primary for analytical and bioassay purposes.

Model solutions of a wide variety of organic solutes in different kinds of water have been prepared and tested at different concentration levels, flow rates, and pH by using different column types. The results of these tests have been well reviewed by Dressler (*423*) and Frei (*536*). Those reports present a discussion of these data for different solid adsorbents.

The extensiveness of the reported recovery tests varies from studies of a large number of components and conditions (*7, 17, 18, 25, 36, 60,*

68, 76, 78, 83, 87, 101, 104, 288, 289, 372, 386, 480) to very limited studies of one or just a few components. Including all reports, 133 references cite recoveries of different compounds from aqueous media. Some of these citations grouped by major polymer types are listed in the box below along with the corresponding references.

The recovery of most of the studied compounds was greater than 50%, and many compounds were recovered with greater than 90% efficiency. For the frequent occasions on which different investigators tested the same compound or compound types, the agreements in the recovery results are remarkable considering the newness of the methodology.

For those cases in which the recovery results do not agree, the discrepancies can be explained by considering the factors that influence the accumulation efficiency. These factors are (1) the spiking

Solid Adsorbents Used to Recover Organic Compounds from Aqueous Media and Corresponding References

XAD-1, XAD-2, and XAD-4: 6, <u>7</u>, <u>17</u>, <u>18</u>, <u>25</u>, 28, 33, 39, 52, 56, <u>60</u>, <u>68</u>, <u>76-78</u>, 81, <u>83</u>, 84, 85, <u>87</u>, 89, 90, 92, 95, 97, <u>101</u>, 102, 103, <u>104</u>, 107, 113, 114, 119, 120, 122, 123, 127, 129, 137, 140, 144, 149, 154-156, 158, 161-163, 165, 166, 171, 175, 179, 181, 183, 186, 193, 194, 200-202, 208, 211, 213, 214, 216, 218, 222, 224, 227, 237, 243, 246, 248, 250, 256, 257, <u>258</u>, 276, 285, 286, <u>288</u>, <u>289</u>, 290, <u>301</u>, 309, 311, 322, 341, 345, 347, 348, 350, 354, 357, 365, 366, <u>372</u>, 378, 380, 381, 385, <u>386</u>, 388, 394, 398, 406, 415, 416, 418, 434, 435, 449, 480, 482, 485, 487, 508, 517, 525, 527, 532, 535.

XAD-7, XAD-8: 6, <u>25</u>, 56, <u>68</u>, 89, <u>104</u>, 114, 165, 181, 211, 213, 218, 222, 224, 227, 286, 322.

Bonded phases: <u>68</u>, 161, 256, 341, 398, 418, <u>480</u>, 482, 485, 508, 517, 525, 532, 535.

Tenax: <u>36</u>, <u>68</u>, <u>101</u>, 162, <u>289</u>, 345, 347, 348, 350, 354, 357, 365, 366, <u>372</u>.

Polyurethanes: <u>7</u>, <u>83</u>, <u>87</u>, 104, 380, 381, 385, <u>386</u>, 388, 394, 398, 406.

Porapaks: <u>68</u>, 119, 341.

Chromosorbs: <u>68</u>, 357.

PTFE: 107, 378.

Polypropylenes: 415.

NOTE: The references to the more extensive recovery tests are underlined in each category. Usually these tests involve more than 10 compounds; at least three chemical classes; and at least two conditions such as flow rate, pH, water type, and concentration.

procedure, which can cause either high or low values; (2) the volume of water tested—lower values are obtained for the larger volumes; (3) the concentration level for the tests—recovery is directly proportional to concentration; (4) the eluting solvent and volume—low recoveries are obtained for poor solvents and volumes inconsistent with the column size; (5) wetting, which causes low recovery if not complete in both the adsorption and the elution step; (6) pH, which can lead to low recovery for ionic or ionizable compounds; (7) complexing of the model test compounds with constituents present in the test water, which can give either high or low values; and (8) operational conditions, such as flow rate, column dimensions, particle size, and particle size distribution, which influence recovery on the basis of theoretical considerations and experimental observations.

Spiking Procedure

The proper procedures for adding model test compounds to water, frequently called spiking, have been reviewed (*423*). Attention to details is necessary to avoid large errors in the results. In general, spiking of the model compounds must be done from a concentrated test solution whose solvent is miscible with water; the water must be free of suspended particles that can irreversibly adsorb the solutes; all glassware must be clean; and the spike level should not exceed the aqueous solubility.

Some investigators (*78, 346*) have tested recovery without spiking the water. Instead, a small amount of the concentrated test solution is applied (spiked) directly to the top of the polymer column, and then a large volume of water is passed through the column. If all of the applied model compound can be recovered by using the proposed eluting solvent, 100% efficiency is assumed. This procedure is fundamentally sound provided care is taken in applying the concentrated solution to the top of the column. This application must result in the uniform deposition of the model compound on the surface of the polymer to give coverage less than a monolayer.

Applications

The use of synthetic polymers and bonded phases to accumulate organic compounds from water for subsequent analytical and/or bioassay applications is summarized in this section.

Polymer Use. The use of different synthetic polymers for accumulating organic components from water is represented by the distribution chart shown in Figure 2. The popularity of the XAD resins is evident. The distribution was calculated from the review of more than

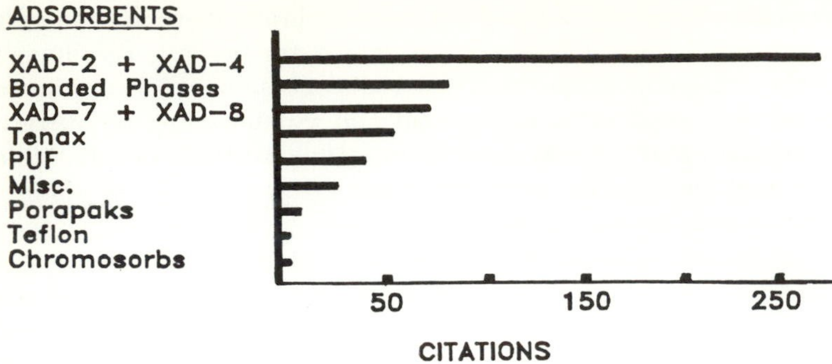

Figure 2. Comparative use of different synthetic polymers on the basis of number of literature citations.

350 references that described the application of synthetic polymers to different process and environmental waters. When the number of references to a specific polymer was too small to show separately on the chart, it was combined with the other low-use polymers and included under miscellaneous. This situation was the case for polyethylene, polypropylene, Ostion, Synachrom, Spheron, Hitachi Gel 3011, PRP-1, etc. The applications of carbon and the ion-exchange resins were not included.

The limited applications of Tenax and polyurethanes can be explained from theoretical considerations discussed earlier and because of practical problems that are more severe than those encountered in the use of the styrene–divinylbenzene and acrylate polymers.

The bonded phases compose an appreciable share of the total applications primarily because of an intense activity by clinical chemists in the determination of drugs and metabolites in biofluids. This activity is somewhat unique and deserves the separate discussion given in the following section.

Sample Cleanup. Many aqueous solutions in the biological sciences contain a complex mixture of solutes that interfere with the determination of target analytes that are present in appreciable amounts. In these cases, concentration is secondary to removing the interferences while maintaining the integrity of the analytes of interest. This situation arises frequently for the determination of drugs and metabolites in urine, blood, plasma, and tissues. Usually the sample is homogenized with aqueous acid, base, or salt, although urine is often processed directly. The procedure involves passing the urine or homogenized sample through a small column of solid adsorbent, washing the column with pure water, eluting the interference material with an appropriate solvent, and eluting the target analytes. The procedure can be used to achieve a small concentration factor ranging from 2 to 10, but this concentration is neither the primary intent nor does it compare with the 10^4 often obtained for other accumulation applications. The procedure is designed primarily to remove the interference compounds; XAD polymers are used heavily in clinical analyses for drugs and even for the isolation of enzymes from plant extracts (*138, 157, 292–294*). The extensive early use of XAD-2 is rapidly being replaced by bonded phases in those cases in which pH is not a problem. Representative recent references are given here (*469–476, 481, 483, 494*). Additional uses of bonded phases for environmental waters and biological and clinical samples can be found in the applications guide available from J. T. Baker. Bonded phase columns suitable for sample cleanup and in some cases general accumulation of organic compounds from nonbiological solutions include the following: SPE (J. T. Baker), Bond Elut (Analytichem Int.), SEP-PAK (Waters Associates), Cartridge (Alltech Associates), Cartridge (Chrompack), and SPICE (Rainin Instrument Company).

Dual Analytical–Bioassay. Many of the reports of polymer applications include dual analytical and bioassay features because these integrated studies help to focus attention on those compound classes and individual compounds that have the highest biological activity. These dual-purpose reports were counted in both categories to arrive at an application distribution of 85% analytical and 15% bioassay. The results given in these two report groups are discussed separately in the following sections.

ANALYTICAL. The data in the reports of the analytical applications cover a wide range of organic compounds accumulated from environmental, drinking, waste, and process waters; many drugs, metabolites, and other compounds have also been accumulated from aqueous biofluids. No satisfactory classification of compounds is available because of multiple functionalities, so arbitrary categories, which include nondescriptive designations such as plasticizers, pesticides, surfactants, and metabolites, were used to survey the literature for applications. References to representative applications within these categories are given on pages 226–227.

There are also isolated reports of the use of synthetic polymers for the determination of purine (*102, 245, 246*), warfarin (*523*), chlorophylls (*248, 514*), tetrachlorodibenzo-*p*-dioxin (*19, 79, 192*), and aldehydes and ketones (*45, 280, 340*).

BIOASSAY. The Rohm and Haas polymers—XAD-2, XAD-4, XAD-7, and XAD-8—have been used almost exclusively for the bioassay

Compounds Accumulated from Aqueous Media and Corresponding References

Pesticides: 3, 7a, 9, 29, 32, 34, 36, 40, 56, 79, 81, 83-85, 95, 100, 103, 111, 112, 120-122, 146, 172, 184, 188, 201, 203, 206, 221, 222, 224, 228, 236, 244, 249, 251, 261, 309-311, 317, 346, 354, 369, 372, 385, 386, 392, 394, 395, 398, 407, 412, 414, 416, 417, 452-454, 460, 480, 488, 501, 503, 505, 524, 529, 535.

Drugs (pharmaceuticals): 127, 129, 135, 137, 142, 147, 152, 156, 157, 166, 175, 186, 189, 193, 194, 200, 205, 226, 231, 239, 242, 243, 253, 258, 274, 300, 314, 322, 419, 420, 469-476, 483, 486, 487, 494, 500, 517, 522, 525, 530, 536.

Polycyclic aromatic hydrocarbons: 5, 11, 31, 45, 48, 79, 82, 84, 86-88, 96, 100, 107, 138a, 140, 161, 163, 172, 179-181, 288, 290, 302, 317, 345, 346, 349, 350, 362, 365, 369, 377, 380-382, 384, 415, 477, 482, 484, 492, 496, 497, 508, 535.

Phenolics: 1, 49, 90, 92, 114, 126, 158, 159, 164, 168, 173, 204, 207, 209, 225, 252, 268, 276, 279, 280, 285, 286, 293, 308, 404a, 404c, 417, 420, 422, 425, 427, 429, 430, 442, 449, 502, 511, 513.

Chlorinated hydrocarbons: 15, 16, 18, 27, 40, 63, 69-71, 94, 99, 101, 117-119, 264, 271, 276, 279, 280, 284, 302, 304, 307, 350, 360, 362, 372, 392, 407, 420, 422, 438, 459, 478, 479, 499, 508.

Polychlorinated biphenyls: 19, 34, 46, 83, 94, 101, 112, 116, 120, 146, 184, 202, 227, 251, 317, 346, 354, 379, 383, 386, 388, 389, 394, 396, 491.

Aliphatic acids: 114, 158, 190, 204, 214, 234, 238, 250, 252, 260, 288, 319, 323, 337, 375, 435, 439, 442, 444, 451, 455, 461, 463, 465-467, 502, 518.

Humic acids: 24, 65, 70, 71, 98, 115, 123, 143, 174, 320, 324, 326, 327, 329, 365, 434, 438, 443, 445, 456, 457.

Metabolites: 127, 128, 156, 244, 261, 298-300, 463, 470, 474, 475, 500, 501, 517, 518, 520, 523, 527, 532.

Miscellaneous ONPS[a] compounds: 67, 68, 75, 76, 78, 80, 125, 140, 190, 196, 207, 208, 252, 306, 311, 325, 339, 404b, 429, 436, 440, 451, 454, 458, 462, 519.

Hydrocarbons: 43, 64, 93, 136, 138a, 140, 148, 171, 288, 345, 347, 352, 375, 415, 427.

Amines, peptides, proteins: 165, 218, 252, 281, 283, 325, 376, 377, 432, 444, 463, 464, 495, 500, 521, 528, 534.

Other acids: 81, 181, 183, 185, 199, 254, 283, 302, 420, 468, 519, 526, 531.

Total organic chlorine: 59, 65, 119, 176, 272, 276, 279, 307, 370.

Alkaloids, steroids: 77, 254, 257, 263, 298, 521, 532, 533.

[a] ONPS refers to oxygen, nitrogen, phosphorus, and sulfur.

Compounds Accumulated from Aqueous Media—Continued

Aromatic acids: 102, 126, 207, 210, 238, 318, 337.

Nucleosides, nucleotides, conjugates: 77, 102, 156, 245, 246, 254.

Enzymes: 138, 151, 292–294, 515.

Plasticizers: 64, 146, 198, 393, 418, 484.

Surfactants: 7, 22, 109, 301, 333, 497.

Esters: 43, 64, 181, 239, 280.

Chelates: 35, 254, 400, 411, 489, 490.

applications. Of these four polymers, XAD-2 has the highest use. The most frequently employed bioassay procedure has been mutagenicity screening; many drinking waters and raw water sources for drinking waters have been actively investigated (*30, 54, 62, 74, 100, 105, 110, 150, 211–213, 216, 219, 230, 232, 233, 235, 266, 267, 269, 270, 277, 278, 288, 291, 296, 297, 312, 313, 316, 405*).

High-surface-area polymers can be used to accumulate large amounts of relatively insoluble organic compounds from very large volumes of water. In the adsorption step of the accumulation, the more soluble components are not recovered efficiently so that the accumulated solutes do not accurately reflect the proportions of different compounds present at trace levels in the water. Nevertheless, the very simple polymer approach can be used for many studies because the mutagenicity appears to reside in the hydrophobic fraction that the polymers accumulate efficiently. This conclusion is based on favorable bioassay comparison of the extracts accumulated by the XAD-2 method and the extracts collected by the more complicated and expensive freeze concentration method (*216, 233*).

For more complete accumulation of the organic components present in water, the combination use of reverse osmosis, solvent extraction, and synthetic polymers has been extensively studied by the Environmental Protection Agency (*177, 295*). The results of in-depth studies of styrene–divinylbenzene and acrylate polymers for accumulating the biologically active components from drinking, well, surface, and waste waters have been reported by Baird et al. (*296, 318a*), Loper et al. (*266*), Nestmann and co-workers (*100, 270, 273, 316*), and Kool and co-workers (*211–213, 216*). These studies have included comparisons to other accumulation techniques, mutagenic blank levels, and different eluting solvents, including DMSO (*211–213, 216, 235, 328*). Kopfler (*541*) and Loper (*291*) have recently reviewed the different procedures for accumulating organic compounds for bioassays.

Conclusions and Future Directions

The foremost conclusion is that synthetic polymers are applicable to a wide range of organic compounds present in many different kinds of aqueous samples, and the polymers are being used extensively. The extensive current and expected future use of polymers will force the availability of pure polymers from the commercial suppliers. The users will have to pay the price for this convenience in the same way they are now willing to pay a premium for the convenience of high-purity solvents.

A second important conclusion is the popularity of the styrene–divinylbenzene-based polymers. This popularity is partly because of precedent but mostly because of certain inherent advantages.

Third, the methodology is ideally suited to automation and on-line operations, so much more will occur in the future; clinical chemists responsible for drug analyses are already committed to automation.

A fourth conclusion, based on the advantages of the use of solid adsorbents, is the gradual replacement of solvent extractions with solid phase extractions. The movement toward this replacement is already evidenced by the commercial availability of several different cartridges of bonded phases and high-surface-area synthetic polymers.

A fifth conclusion is the tremendous advantages of field sampling with solid adsorbents, which will result in more extensive use of this approach for both analytical and bioassay purposes.

Finally, no single magic solid adsorbent that accumulates all organic components from water is available; no universal adsorbent is expected to emerge in the future. Instead, several different adsorbents and combinations of these adsorbents will be tested continuously for applicability to specific analytical and bioassay problems.

Acknowledgments

The author is indebted to long-time colleagues J. Richard, H. Svec, J. Fritz, M. Avery, C. Chriswell, R. Vick, and D. Witiak, who have worked at the Ames Laboratory, Iowa State University, in the development of the use of synthetic polymers. Appreciation is also expressed to V. Fassel and R. Fisher who provided administrative assistance and contributed to the professional atmosphere needed for completion of this review. Thanks go to J. Lemish and D. DoBell for extensive clerical help. This work was performed at the laboratories of the U.S. Department of Energy and supported under Contract No. W-7405-Eng-82, Division of the Office of Health and Environmental Research, Office of Energy Research and the Assistant Secretary for Fossil Energy, Division of Coal Utilization, through the Morgantown Energy Technology Center.

Literature Cited

1. Middleton, F. M.; Braus, H.; Ruchhoft, C. C. *J. Am. Water Works Assoc.* **1952,** *44,* 538.
2. Middleton, F. M.; Rosen, A. A. *Public Health Rep.* **1956,** *17(11),* 1125.
3. Rosen, A. A.; Middleton, F. M. *Anal. Chem.* **1959,** *31,* 1729.
4. Rosen, A. A.; Middleton, F. M. *Anal. Chem.* **1955,** *27,* 790.
5. Lagana, A.; Petronio, B. M.; Rotatori, M. *J. Chromatogr.* **1980,** *198,* 143.
6. Van Rossum, P.; Webb, R. G. *J. Chromatogr.* **1978,** *150,* 381.
7. Jones, P.; Nickless, G. *J. Chromatogr.* **1978,** *156,* 87.

7a. Paschal, D. C.; Bicknell, R.; Dresbach, D. *Anal. Chem.* **1977,** *49,* 1551.

8. Eichelberger, J. W.; Lichtenberger, J. J. *J. Am. Water Works Assoc.* **1971,** *63,* 25.
9. Kennedy, D. C. *Environ. Sci. Technol.* **1973,** 7, 138.
10. Sproul, O. J.; Ryckman, D. W. *J. Water Pollut. Control Fed.* **1961,** *23,* 1188.
11. Griest, W. H.; Caton, J. E. In *Handbook of Polycyclic Aromatic Hydrocarbons;* Bjorseth, A., Ed.; Marcel Dekker: New York, 1983; Chapter 3, p 95.
12. Tateda, A.; Fritz, J. S. *J. Chromatogr.* **1978,** *152,* 329.
13. Grob, K.; Grob, K., Jr.; Grob, G. *J. Chromatogr.* **1975,** *106,* 299.
14. Braus, H.; Middleton, F. M.; Graham, W. *Anal. Chem.* **1951,** *23,* 1160.
15. Feige, W. A.; Ruggiero, D. U.S. Environmental Protection Agency Report 600/52-82-027; U.S. Government Printing Office: Washington, DC, 1982.
16. Wood, P. R.; DeMarco, J. In *Activated Carbon Adsorption of Organics From the Aqueous Phase;* McGuire, M. J.; Suffett, I. H., Eds.; Ann Arbor Science: Ann Arbor, MI, 1980; Vol. 2.
17. Ewing, B. B.; Chian, E. S. K.; Cook, J. C.; Dewalle, F. B.; Evans, C. A.; Hopke, P. K.; Kim, J. H.; Means, J. C.; Milberg, R.; Perkins, E. G.; Sherwood, J. D.; Wadlin, W. H. U.S. Environmental Protection Agency Report 560/6-77-015; U.S. Government Printing Office: Washington, DC, 1977.
18. Junk, G. A.; Chriswell, C. D.; Chang, R. C.; Kissinger, L. D.; Richard, J. J.; Fritz, J. S.; Svec, H. J. *Z. Anal. Chem.* **1976,** *282,* 331.
19. Huckins, J. N.; Stalling, D. L.; Smith, A. *J. Assoc. Off. Anal. Chem.* **1978,** *61,* 32.
20. Buelow, R. W.; Carswell, J. K.; Symons, J. M. *J. Am. Water Works Assoc.* **1973,** *65,* 57.
21. Pellizzari, E. D.; Carpenter, B. H.; Bunch, J. E.; Sawicki, E. *Environ. Sci. Technol.* **1975,** *9,* 556.
22. Hinrichs, R. L.; Snoeyink, V. L. *Water Res.* **1976,** *10,* 79.
23. Suffet, I. H.; Brenner, L.; Coyle, J. T.; Cairo, P. R. *Environ. Sci. Technol.* **1978,** *12,* 1315.
24. Boening, P. H.; Beckmann, D. D.; Snoeyink, V. L. *J. Am. Water Works Assoc.* **1980,** *72(1),* 54.
25. Naipawer, R. E.; Potter, R.; Vallon, P.; Erickson, R. E. *Flavour Ind.* **1971,** *2(8),* 465.
26. Bertsch, W.; Chang, R. C.; Zlatkis, A. *J. Chromatogr. Sci.* **1974,** *12,* 175.
27. Symons, J. M.; Carswell, J. K.; DeMarco, J.; Love, O. T., Jr. "May Progress Report"; Drinking Water Research Division Municipal Environmental Research Laboratory: Cincinnati, OH, 1979.
28. Melton, R. G.; Coleman, W. E.; Slater, R. W.; Kopfler, F. C.; Allen, W. K.; Aurand, T. A.; Mitchell, D. E.; Voto, S. J.; Lucas, S. V.; Watson,

S. C. In *Advances in the Identification and Analysis of Organic Pollutants in Water;* Keith, L. H., Ed.; Ann Arbor Science: Ann Arbor, MI, 1981; Vol. 2, p 597.

29. Ermolaeva, L. P.; Ogoblina, I. P.; Il'icheva, I. A. *J. Anal. Chem. USSR* **1977,** *32(12),* 2429.
30. Glatz, B. A.; Chriswell, C. D.; Junk, G. A. In *Monitoring Toxic Substances;* Schuetzle, D., Ed.; ACS Symposium Series 94; American Chemical Society: Washington, DC, 1979; p 91.
31. Harrison, R. M.; Perry, R.; Wellings, R. A. *Water Res.* **1975,** *9,* 331.
32. Smith, G. E.; Isom, B. G. *Pestic. Monit. J.* **1967,** *1,* 16.
33. Melton, R. G.; Coleman, W. E.; Slater, R. W.; Kopfler, F. C.; Allen, W. K.; Aurand, T. A.; Mitchell, D. E.; Voto, S. J. U.S. Environmental Protection Agency Report 600/D-81-060; U.S. Government Printing Office: Washington, DC, 1981.
34. Petty, R. *Anal. Chem.* **1981,** *53,* 1548.
35. Vanderborght, B. M.; Van Grieken, R. E. *Anal. Chem.* **1977,** *49,* 311.
36. Bacaloni, A.; Goretti, G.; Lagana, A.; Petronio, B. M.; Rotatori, M. *Anal. Chem.* **1980,** *52,* 2033.
37. Chau, A. J. Y.; Babjack, L. J. *J. Assoc. Off. Anal. Chem.* **1979,** *62,* 107.
38. Babjack, L. J.; Chau, A. S. Y. *J. Assoc. Off. Anal. Chem.* **1979,** *62,* 1174.
39. Keith, L. H.; Garrison, A. W.; Allen, F. R.; Carter, M. H.; Floyd, T. L.; Pope, J. D.; Thurston, A. D., Jr. In *Identification and Analysis of Organic Pollutants in Water;* Keith, L. H., Ed.; Ann Arbor Science: Ann Arbor, MI, 1976; Chapter 22, p 329.
40. Kleopfer, R. D. In *Identification and Analysis of Organic Pollutants in Water;* Keith, L. H., Ed.; Ann Arbor Science: Ann Arbor, MI, 1976; Chapter 24, p 399.
41. Dunlap, W. J.; Shew, D. C.; Scalf, M. R.; Cosby, R. L.; Robertson, J. M. In *Identification and Analysis of Organic Pollutants in Water;* Keith, L. H., Ed.; Ann Arbor Science: Ann Arbor, MI, 1976; Chapter 27, p 453.
42. Garrison, A. W.; Pope, J. D.; Allen, F. R. In *Identification and Analysis of Organic Pollutants in Water;* Keith, L. H., Ed.; Ann Arbor Science: Ann Arbor, MI, 1976; Chapter 30, p 517.
43. Bruner, F.; Crescentini, G.; Mangani, F.; Petty, R. *Anal. Chem.* **1983,** *55,* 793.
44. Kleopfer, R. D.; Fairless, B. J. *Environ. Sci. Technol.* **1972,** *6,* 1036.
45. Reichert, J.; Kunte, H.; Engelhart, K.; Borneff, J. *Zentralbl. Bakteriol. Parasitenkd. Infektionskr. Hyg. Abt. 1: Orig. Reihe B* **1971,** *155(1),* 18.
46. Lawrence, J.; Tosine, H. M. *Environ. Sci. Technol.* **1976,** *10,* 381.
47. Buelow, R. W.; Carswell, J. K.; Symons, J. M. *J. Am. Water Works Assoc.* **1973,** *65,* 195.
48. Futoma, D. J.; Smith, S. R.; Smith, T. E.; Tanaka, J. In *Polycyclic Aromatic Hydrocarbons in Water Systems;* CRC: Boca Raton, FL, 1981.
49. Hoak, R. D. *Int. J. Air Water Pollut.* **1962,** *6,* 521.
50. Janardan, K. G.; Schaeffer, D. J.; Somani, S. M. *Bull. Environ. Contam. Toxicol.* **1980,** *24,* 145.
51. Hueper, W. C.; Payne, W. W. *Am. J. Clin. Pathol.* **1963,** *39,* 475.
52. Chriswell, C. D.; Ericson, R. L.; Junk, G. A.; Lee, K. W.; Fritz, J. S; Svec, H. J. *J. Am. Water Works Assoc.* **1977,** *69(12),* 669.
53. Krost, K. J.; Pellizzari, E. D.; Walburn, S. G.; Hubbard, S. A. *Anal. Chem.* **1982,** *54,* 810.
54. Glatz, B. A.; Chriswell, C. D.; Arguello, M. D.; Svec, H. J.; Fritz, J. S.; Grimm, S. M.; Thomson, M. A. *J. Am. Water Works Assoc.* **1978,** *70,* 465.

55. Marton, G.; Havas-Dencs, Mr. J.; Szokonya, L.; Illes, Z. *Hung. J. Ind. Chem.* **1981,** *9(3),* 251.
56. Koshima, H.; Onishi, H. *Bunseki Kagaku* **1983,** *32(8),* E251.
57. Weber, W. J., Jr.; van Vliet, B. M. *J. Am. Water Works Assoc.* **1981,** *73,* 420.
58. Marton, G.; Szokonya, L.; Havas-Dencs, Mr. J.; Illes, Z. *Hung. J. Ind. Chem.* **1981,** *9(3),* 263.
59. Kuhn, W.; Sontheimer, H. *Vom Wasser* **1973,** *15,* 65.
60. Grob, K. *J. Chromatogr.* **1973,** *84,* 255.
61. Janardan, K. G.; Schaeffer, D. J. *Anal. Chem.* **1979,** *51,* 1024.
62. Fiessinger, F.; Cabridenc, R.; Chouroulinkov, I.; Festy, B.; Lazar, P.; Richard, Y. In *Water Chlorination: Environmental Impact and Health Effects;* Jolley, R. L.; Brungs, W. A.; Cotruvo, J. A.; Cumming, R. B.; Mattice, J. S.; Jacobs, V. A., Eds.; Ann Arbor Science: Ann Arbor, MI, 1983; Vol. 4, Chapter 33, p 467.
63. Coyle, G. T.; Maloney, S. W.; Gibs, J.; Suffet, I. H. In *Water Chlorination: Environmental Impact and Health Effects;* Jolley, R. L.; Brungs, W. A.; Cotruvo, J. A.; Cumming, R. B.; Mattice, J. S.; Jacobs, V. A., Eds.; Ann Arbor Science: Ann Arbor, MI, 1983; Vol. 4, Chapter 30, p 791.
64. Lee, N. E.; Haag, W. R.; Jolley, R. L. In *Water Chlorination: Environmental Impact and Health Effects;* Jolley, R. L.; Brungs, W. A.; Cotruvo, J. A.; Cumming, R. B.; Mattice, J. S.; Jacobs, V. A., Eds.; Ann Arbor Science: Ann Arbor, MI, 1983; Vol. 4, Chapter 59, p 851.
65. Glaze, W. H.; Saleh, F. Y.; Kinstley, W. In *Water Chlorination: Environmental Impact and Health Effects;* Jolly, R. L.; Brungs, W. A.; Cumming, R. B., Eds.; Ann Arbor Science: Ann Arbor, MI, 1980; Vol. 3, Chapter 9, p 99.
66. Suffet, I. H.; Wicklund, A.; Cairo, P. R. In *Water Chlorination: Environmental Impact and Health Effects;* Jolley, R. L.; Brungs, W. A.; Cumming, R. B., Eds.; Ann Arbor Science: Ann Arbor, MI, 1980; Vol. 3, Chapter 66, p 757.
67. Golkiewicz, W.; Werkhoven-Goewie, C. E.; Brinkman, U. A. Th.; Frei, R. W.; Colin, H.; Guiochon, G. *J. Chromatogr. Sci.* **1982,** *21,* 27.
68. Anspach, G. L.; Jones, W. E., III; Kitchens, J. F. NTIS Report AD-A117-541; NTIS: Springfield, VA, 1982.
69. Robeck, G. G.; Love, O. T., Jr. NTIS Report PB83-168617; NTIS: Springfield, VA, 1983.
70. Snoeyink, V. L.; Chudyk, W. A.; Beckmann, D. D.; Boening, P. H.; Temperly, T. J. NTIS Report PB81-196776; NTIS: Springfield, VA, 1981.
71. Reed, G. D.; Zey, A. F. *J. Environ. Eng. Div. Am. Soc. Civ. Eng.* **1981,** *107,* 1095.
72. Coleman, W. E.; Melton, R. G.; Slater, R. W.; Kopfler, F. C.; Voto, S. J.; Allen, W. K.; Aurand, T. A. *J. Am. Water Works Assoc.* **1981,** *73(2),* 119.
73. Tabor, M. W. *Environ. Sci. Technol.* **1983,** *17,* 324.
74. Loper, J. C.; Tabor, M. W. In *Short-Term Bioassays in the Analysis of Complex Environmental Mixtures III;* Waters, M. D.; Sandhu, S. S.; Lewtax, J.; Claxton, L.; Chernoff, N.; Nesnow, S., Eds.; Plenum: New York, 1983, p 165.
75. Szachta, J. M. NTIS Report AD-A053-863; NTIS: Springfield, VA, 1978.
76. Riley, J. P.; Taylor, D. *Anal. Chim. Acta* **1969,** *46,* 307.
77. Bradlow, H. L. *Steroids* **1968,** *11,* 265.
78. Junk, G. A.; Richard, J. J.; Greiser, M. D.; Witiak, D.; Witiak, I. L.; Arguello, M. D.; Vick, R.; Svec, H. J.; Fritz, J. S.; Calder, G. V. *J. Chromatogr.* **1974,** *99,* 745.

79. Junk, G. A.; Richard, J. J.; Fritz, J. S.; Svec, H. J. In *Identification and Analysis of Organic Pollutants in Water;* Keith, L. H., Ed.; Ann Arbor Science: Ann Arbor, MI, 1976; Chapter 9, p 135.
80. Junk, G. A.; Richard, J. J. In *Identification and Analysis of Organic Pollutants in Water;* Keith, L. H., Ed.; Ann Arbor Science: Ann Arbor, MI, 1981; Chapter 19, p 295.
81. Mierzwa, S.; Witek, S. *J. Chromatogr.* **1977,** *136,* 105.
82. May, W. E.; Chesler, S. N.; Cram, S. P.; Gump, B. H.; Hertz, H. S.; Enagonio, D. P.; Dyszel, S. M. *J. Chromatogr. Sci.* **1975,** *13,* 535.
83. Musty, P. R.; Nickless, G. *J. Chromatogr.* **1976,** *120,* 369.
84. Osterroht, C. *J. Chromatogr.* **1974,** *101,* 289.
85. McNeil, E. E.; Otson, R.; Miles, W. F.; Rajabalee, J. M. *J. Chromatogr.* **1977,** *132,* 277.
86. Alben, K. *Environ. Sci. Technol.* **1980,** *14,* 468.
87. Navratil, J. D.; Sievers, R. E.; Walton, H. *Anal. Chem.* **1977,** *49,* 2260.
88. Olufsen, B. *Anal. Chim. Acta* **1980,** *113,* 393.
89. Burnham, A. K.; Calder, G. V.; Fritz, J. S.; Junk, G. A.; Svec, H. J.; Willis, R. *Anal. Chem.* **1972,** *44,* 139.
90. Hunt, G. T.; Clement, W. H.; Fause, S. D. In *Environmental Analysis;* Galen, E.; Ewing, W., Eds.; Academic: New York, 1977; p 57.
91. Tang, J. I. S.; Kawahara, F. K.; Yen, T. F. Presented at the American Chemical Society Symposium on Engine Combustion Deposits, Atlanta, GA, March 29, 1981; paper 2.
92. Vinson, J. A.; Burke, G. A.; Hager, B. L.; Casper, D. R.; Nylander, W. A.; Middlemiss, R. J. *Environ. Lett.* **1973,** *5(3),* 199.
93. James, H. A.; Steel, C. P.; Wilson, I. *J. Chromatogr.* **1981,** *208,* 89.
94. Harvey, G. R. Woods Hole Oceanographic Institution Report WHOI-72-86; Woods Hole Oceanographic Institution: Woods Hole, MA, 1972.
95. Sundaram, K. M. S.; Szeto, S. Y.; Hindle, R. *J. Chromatogr.* **1979,** *177,* 29.
96. Kveseth, K.; Sortland, B. *Chemosphere* **1982,** *11(7),* 623.
97. Harris, J. C.; Cohen, M. J.; Grozzer, Z. A.; Hayes, M. J. U.S. Environmental Protection Agency Report 600/52-80-193; U.S. Government Printing Office: Washington, DC, 1981.
98. Christman, R. F.; Johnson, J. D.; Norwood, D. L.; Lido, W. T.; Haas, J. R.; Pfaender, F. K.; Webb, M. R.; Bobenrieth, M. J. U.S. Environmental Protection Agency Report 600/52-81-016; U.S. Government Printing Office: Washington, DC, 1981.
99. Wallin, B. K.; Condren, A. J.; Walden, R. L. U.S. Environmental Protection Agency Report 600/52-81-043; U.S. Government Printing Office: Washington, DC, 1981.
100. Williams, D. T.; Nestmann, E. R.; LeBel, G. L.; Benoit, F. M.; Otson, R. *Chemosphere* **1982,** *11(3),* 263.
101. Picer, N.; Picer, M. *J. Chromatogr.* **1980,** *193,* 357.
102. Uematsu, T.; Kurita, T.; Hamado, A. *J. Chromatogr.* **1979,** *172,* 327.
103. Mallet, V. N.; Francoeur, J. M.; Volpe, G. *J. Chromatogr.* **1979,** *172,* 388.
104. Webb, R. G. U.S. Environmental Protection Agency Report 600/4-75/003; U.S. Government Printing Office: Washington, DC, 1975.
105. Kool, H. Quarterly Report No. 34; National Institute for Water Supply: Leidschendam, The Netherlands, 1983.
106. Smith, C. C. In *Application of Short-Term Bioassays in the Fractionation and Analysis of Complex Environmental Mixtures;* Waters, M.; Nesnow, S.; Huisingh, J.; Sandher, S.; Claxton, L., Eds.; Plenum: New York, 1978; p 227.

107. Griest, W. H.; Maskarinec, M. P.; Herbes, S. E.; Southworth, G. In *Analysis of Waters Associated With Alternative Fuel Production:* American Society for Testing and Materials: Philadelphia, 1980; p 167.
108. Rappaport, S. M.; Richard, M. G.; Hollstein, M. C.; Talcott, R. E. *Environ. Sci. Technol.* **1979,** *13,* 957.
109. Seidl, J. *Chem. Prum.* **1976,** *26,* 94.
110. Cheh, A. M.; Skochdopole, J.; Koski, P.; Cole, L. *Science* **1980,** *207,* 90.
111. Niederschulte, V.; Ballschmitter, K. Z. *Anal. Chem.* **1974,** *269,* 360.
112. Musty, P. R.; Nickless, G. *J. Chromatogr.* **1974,** *89,* 185.
113. Wun, C. K.; Walker, R. W.; Litsky, W. *Water Res.* **1976,** *10,* 955.
114. Stepan, S. F.; Smith, J. F. *Water Res.* **1977,** *11,* 339.
115. Mantoura, R. F. C.; Riley, J. P. *Anal. Chim. Acta* **1975,** *76,* 97.
116. Harvey, G. R.; Steinhauer, W. G.; Teal, J. M. *Science* **1973,** *180,* 643.
117. Glaze, W. H.; Henderson, J. E.; Bell, J. E.; Wheeler, V. A. *J. Chromatogr. Sci.* **1973,** *11,* 580.
118. Kissinger, L. D.; Fritz, J. S. *J. Am. Water Works Assoc.* **1976,** *68,* 435.
119. Glaze, W. H.; Peyton, G. R.; Rawley, R. *Environ. Sci. Technol.* **1977,** *11,* 685.
120. Coburn, J. A.; Valdmanis, I. A.; Chau, A. S. Y. *J. Assoc. Off. Anal. Chem.* **1977,** *60,* 224.
121. Berkane, K., Caissie, G. E.; Mallet, V. N. *J. Chromatogr.* **1977,** *139,* 386.
122. Mallet, V. N.; Brun, G. L.; MacDonald, R. N.; Berkane, K. *J. Chromatogr.* **1978,** *160,* 81.
123. Stuermer, D. M.; Harvey, G. R. *Nature* **1974,** *250,* 480.
124. Seidl, J. *Chem. Prum.* **1975,** *25,* 416.
125. Seidl, J. *Chem. Prum.* **1976,** *26,* 30.
126. Jahangir, L. M.; Samuelson, O. *J. Chromatogr.* **1980,** *193,* 197.
127. Fujimoto, J. M.; Haarstad, V. B. *J. Pharmacol. Exp. Ther.* **1969,** *165,* 45.
128. Stambaugh, J. E.; Feo, L. G.; Manthe, R. W. *Life Sci.* **1967,** *6,* 1811.
129. Mule, S. J.; Bastos, M. L.; Jukofsky, D.; Saffer, E. *J. Chromatogr.* **1971,** *63,* 289.
130. Holloway, P. W.; Popjak, G. *Biochem. J.* **1967,** *104,* 57.
131. Zaika, L. L.; Wasserman, A. E.; Monk, C. A.; Salay, J. J. *J. Food Sci.* **1968,** *33,* 53.
132. Zaika, L. L. *J. Agric. Food Chem.* **1969,** *17,* 893.
133. Technical Report 1738; Rohm and Haas Company: Philadelphia, PA.
134. Makino, I.; Sjovall, J. *Anal. Lett.* **1972,** *5,* 341.
135. Machata, G.; Vycudilik, W. *Arch. Toxicol.* **1975,** *33,* 115.
136. Zepp, R. G.; Wolfe, N. L.; Baughman, G. L.; Hollis, R. C. *Nature* **1977,** *267,* 421.
137. Fujimoto, J. M.; Wang, R. I. H. *Toxicol. Appl. Pharmacol.* **1970,** *16,* 186.
138. Croteau, R.; Burbott, A. V.; Loomis, W. D. *Biochem. Biophys. Res. Commun.* **1973,** *50,* 1006.
138a. Overton, E. B.; Mascarella, S. W.; McFall, J. A.; Laseter, J. L. *Chemosphere* **1980,** *9,* 629.
139. Wilks, A. D.; Pietryzk, D. J. *Anal. Chem.* **1972,** *44,* 676.
140. Shinohara, R.; Koga, M.; Shinohara, J.; Hori, T. *Bunseki Kagaku* **1977,** *26(12),* 856.
141. Suffet, I. H.; Brenner, L.; Radzuil, J. V. In *Identification and Analysis of Organic Pollutants in Water;* Keith, L. H., Ed.; Ann Arbor Science: Ann Arbor, MI, 1976; Chapter 23, p 375.
142. Stolman, A.; Pranitis, P. A. F. *Clin. Toxicol.* **1977,** *10,* 49.
143. Aiken, G. R.; Thurman, E. M.; Malcolm, R. L.; Walton, H. F. *Anal. Chem.* **1979,** *51,* 1799.

144. Pranitis, P. A. F.; Stolman, A. *J. Forensic Sci.* **1975,** *20,* 726.
145. Baum, R. G.; Saetre, R.; Cantwell, F. F. *Anal. Chem.* **1980,** *52,* 15.
146. Suffet, I. H.; Friant, S.; Marcinkiewicz, C.; McGuire, M. J.; Wong, D. T. L. *J. Water Pollut. Control Fed.* **1975,** *47,* 1169.
147. Missen, A. W.; Lewin, J. F. *Clin. Chim. Acta* **1974,** *53,* 389.
148. Burns, K. A.; Smith, J. L. *Aust. J. Mar. Freshwater Res.* **1980,** *31,* 251.
149. Missen, A. W. *Clin. Chem.* **1976,** *22,* 927.
150. Cheh, A. M.; Carlson, R. E. In *Advances in the Identification and Analysis of Organic Pollutants in Water;* Keith, L. H., Ed.; Ann Arbor Science: Ann Arbor, MI, 1981; Vol. 1, Chapter 29, p 457.
151. Shechter, I.; Bloch, K. *J. Biol. Chem.* **1971,** *246,* 7690.
152. Finkle, B. S. *Anal. Chem.* **1972,** *44(9),* 18A.
153. Rogers, I. H. *Pulp Pap. Mag. Can.* **1973,** *74(9),* 111.
154. Kullberg, M. P.; Gorodetzky, C. W. *Clin. Chem.* **1974,** *20,* 177.
155. Junk, G. A.; Svec, H. J. In *21st Annual Conference on Mass Spectrometry and Allied Topics;* San Francisco, 1973; p 308.
156. Sawada, H.; Hara, A.; Asano, S.; Matsumoto, Y. *Clin. Chem.* **1976,** *22,* 1596.
157. Wang, R. I. H.; Mueller, M. A. J. *Pharm. Sci.* **1973,** *62,* 2047.
158. Fox, M. E. In *Identification and Analysis of Organic Pollutants in Water;* Keith, L. H., Ed.; Ann Arbor Science: Ann Arbor, MI, 1976; Chapter 34, p 641.
159. Voss, R. H.; Wearing, J. T.; Wong, A. In *Advances in the Identification and Analysis of Organic Pollutants in Water;* Keith, L. H., Ed.; Ann Arbor Science: Ann Arbor, MI, 1981; Vol. 2, Chapter 53, p 1059.
160. Garrison, A. W.; Alford, A. L.; Craig, J. S.; Ellington, J. J.; Haeberer, A. F.; McGuire, J. M.; Pope, J. D.; Shackelford, W. H.; Pellizzari, E. D.; Gebhart, J. E. In *Advances in the Identification and Analysis of Organic Pollutants in Water;* Keith, L. H., Ed.; Ann Arbor Science: Ann Arbor, MI, 1981; Chapter 2, p 17.
161. Caton, J. E.; Barnes, Z. K.; Kubota, H.; Griest, W. H.; Maskarinec, M. P. In *Advances in the Identification and Analysis of Organic Pollutants in Water;* Keith, L. H., Ed.; Ann Arbor Science: Ann Arbor, MI, 1981; Vol. 2, Chapter 21, p 329.
162. Ryan, J. P.; Fritz, J. S. In *Advances in the Identification and Analysis of Organic Pollutants in Water;* Keith, L. H., Ed.; Ann Arbor Science: Ann Arbor, MI, 1981; Chapter 20, p 317.
163. Landrum, P. F.; Geisy, J. P. In *Advances in the Identification and Analysis of Organic Pollutants in Water;* Keith, L. H., Ed.; Ann Arbor Science: Ann Arbor, MI, 1981; Vol. 1, Chapter 22, p 345.
164. Haeberer, A. F.; Scott, T. A. In *Advances in the Identification and Analysis of Organic Pollutants in Water;* Keith, L. H., Ed.; Ann Arbor Science: Ann Arbor, MI, 1981; Vol. 1, Chapter 23, p 359.
165. Haeberer, A. F.; Scott, T. A. In *Advances in the Identification and Analysis of Organic Pollutants in Water;* Keith, L. H., Ed.; Ann Arbor Science: Ann Arbor, MI, 1981; Vol. 1, Chapter 27, p 433.
166. Vycudilik, W. *J. Chromatogr.* **1975,** *111,* 439.
167. Leenheer, J. A.; Huffman, E. W. D. *J. Res. U.S. Geol. Surv.* **1976,** *4(6),* 737.
168. Pietrzyk, D. J.; Chu, C. *Anal. Chem.* **1977,** *49,* 757.
169. Dolara, P.; Ricci, V.; Burrini, D.; Griffini, O. *Bull. Environ. Contam. Toxicol.* **1981,** *27,* 1.
170. Hunt, G. T.; Pangaro, N.; Zelenski, S. G. *Anal. Lett.* **1980,** *13,* 521.

171. Schnare, D. W. *J. Water Pollut. Control Fed.* **1979,** *51,* 2467.
172. Junk, G. A.; Richard, J. J.; Svec, H. J.; Fritz, J. S. *J. Am. Water Works Assoc.* **1976,** *68,* 218.
173. Glaze, W. H.; Henderson, J. E. IV *J. Water Pollut. Control Fed.* **1975,** *47,* 2511.
174. Christman, R. F.; Liad, W. T.; Millington, D. S.; Johnson, J. D. In *Advances in the Identification and Analysis of Organic Pollutants in Water;* Keith, L. H., Ed.; Ann Arbor Science: Ann Arbor, MI, 1981; Vol. 2, Chapter 49, p 979.
175. Kullberg, M. P.; Miller, W. L.; McGowan, F. J.; Doctor, B. P. *Biochem. Med.* **1973,** *7,* 323.
176. Glaze, W. H.; Henderson, J. E. IV; Smith, G. In *Identification and Analysis of Organic Pollutants in Water;* Keith, L. H., Ed.; Ann Arbor Science: Ann Arbor, MI, 1976; Chapter 6, p 247.
177. Lin, D. C. K.; Melton, R. G.; Kopfler, F. C.; Lucas, S. V. In *Advances in the Identification and Analysis of Organic Pollutants in Water;* Keith, L. H., Ed.; Ann Arbor Science: Ann Arbor, MI, 1981; Vol. 2, Chapter 46, p 861.
178. Gebhart, J. E.; Ryan, J. F.; Cox, R. D.; Pellizzari, E. D.; Michael, L. C.; Sheldon, L. S. In *Advances in the Identification and Analysis of Organic Pollutants in Water;* Keith, L. H., Ed.; Ann Arbor Science: Ann Arbor, MI, 1981; Vol. 1, Chapter 3, p 31.
179. Benoit, F. M.; Lebel, G. L.; Williams, D. T. *Int. J. Environ. Anal. Chem.* **1979,** *6,* 277.
180. Benoit, F. M.; Lebel, G. L.; Williams, D. T. *Bull. Environ. Contam. Toxicol.* **1979,** *23,* 774.
181. Grosser, Z. A.; Harris, J. C.; Levins, P. L. In *Polynuclear Aromatic Hydrocarbons;* Jones, P. W.; Leber, P., Eds.; Ann Arbor Science: Ann Arbor, MI, 1979; p 67.
182. Olsson, L.; Samuelson, O. *Chromatographia* **1977,** *10,* 135.
183. Pietrzyk, D. J.; Kroeff, E. P., Rotsch, T. D. *Anal. Chem.* **1978,** *50,* 497.
184. Rees, G. A. V.; Au, L. *Bull. Environ. Contam. Toxicol.* **1979,** *22,* 561.
185. Rosenfeld, J. M.; Muriek-Russell, M.; Phatak, A. *J. Chromatogr.* **1984,** *283,* 127.
186. Roerig, D. L.; Lewland, D.; Mueller, M.; Wang, J. I. H. *J. Chromatogr.* **1975,** *110,* 349.
187. Fritz, J. S. *Acc. Chem. Res.* **1977,** *10,* 67.
188. Richard, J. J.; Junk, G. A; Avery, M. J.; Nehring, N. L.; Fritz, J. S.; Svec, H. J. *Pestic. Monit. J.* **1975,** *9(3),* 117.
189. Bastos, M. L.; Jukofsky, D.; Saffer, E.; Chedekel, M.; Mule, S. J. *J. Chromatogr.* **1972,** *71,* 549.
190. Gustafson, R. L.; Albright, R. L.; Heisler, J.; Lirio, J. A.; Reid, O. T. *Ind. Eng. Chem. Prod. Res. Dev.* **1968,** *7,* 107.
191. Stepan, S. F.; Smith, J. F.; Flego, U.; Ruekers, J. *J. Water Res.* **1978,** *12,* 447.
192. Hopkins, S. J. *Manuf. Chem.* **1967,** *38(2),* 51.
193. Weissman, N.; Lowe, M. L.; Beattie, J. M.; Demetriou, J. A. *J. A. Clin. Chem.* **1971,** *17(9),* 875.
194. Pranitis, P. A. F.; Milzoff, J. R.; Stolman, A. *J. Forensic Sci.* **1974,** *19,* 917.
195. Robinson, J. L.; Marshall, M. A.; Draganjac, M. E.; Noggle, L. C. *Anal. Chim. Acta* **1980,** *115,* 229.
196. Burnham, A. K.; Calder, G. V.; Fritz, J. S.; Junk, G. A.; Svec, H. J.; Vick, R. *J. Am. Water Works Assoc.* **1973,** *65,* 722.

197. Chriswell, C. D.; Kissinger, L. D.; Fritz, J. S. *Anal. Chem.* **1976,** *48,* 1123.
198. Junk, G. A.; Svec, H. J.; Vick, R. D.; Avery, M. J. *J. Environ. Sci. Technol.* **1974,** *8,* 1100.
199. Leach, J. M.; Thakore, A. N. *J. Fish. Res. Board Can.* **1975,** *32,* 1249.
200. Gelbke, H. P.; Grell, T. H.; Schmidt, G. *Arch. Toxicol.* **1978,** *39,* 211.
201. Lebel, G. L.; Williams, D. T.; Griffith, G.; Benoit, F. M. *J. Assoc. Off. Anal. Chem.* **1979,** *62,* 241.
202. Lebel, G. L.; Williams, D. T. *Bull. Environ. Contam. Toxicol.* **1980,** *24,* 397.
203. Harvey, G. R.; Steinhauer, W. G. *J. Mar. Res.* **1976,** *34,* 561.
204. Paleos, J. *J. Colloid Interface Sci.* **1969,** *31,* 7.
205. Ibrahim, G.; Andryouskas, S.; Bastos, M. L. *J. Chromatogr.* **1975,** *108,* 107.
206. Daughton, G. G.; Crosby, D. G.; Garnos, R. L.; Hseih, D. D. *J. Agric. Food Chem.* **1976,** *24,* 236.
207. Pietrzyk, D. J.; Chu, C. H. *Anal. Chem.* **1977,** *49,* 860.
208. Kaczvinsky, J.; Saitoh, K.; Fritz, J. *Anal. Chem.* **1983,** *55,* 1210.
209. Puon, S.; Cantwell, F. F. *Anal. Chem.* **1977,** *49,* 1256.
210. Scoggins, M. W. *Anal. Chem.* **1972,** *44,* 1285.
211. Kool, H. J.; van Kreijl, C. F.; van Kranen, H. J.; de Greef, E. *Chemosphere* **1981,** *10,* 85.
212. Kool, H. J.; van Kreijl, C. F.; van Kranen, H. J. *Chemosphere* **1981,** *10,* 99.
213. Kool, H. J.; van Kreijl, C. F.; van Kranen, H. J.; de Greef, E. *Sci. Total Environ.* **1981,** *18,* 135.
214. Thompson, J. A.; Markey, S. P. *Anal. Chem.* **1975,** *47,* 1313.
215. Ames, B. N.; McCann, J.; Yamasaki, E. *Mutat. Res.* **1975,** *31,* 347.
216. van Kreijl, C. F.; Kool, H. J.; de Vries, M.; van Kranen, H. J.; de Greef, E. *Sci. Total Environ.* **1980,** *15,* 137.
217. Neeman, I.; Kroll, R.; Mahler, A.; Rubin, R. J. *Bull. Environ. Contam. Toxicol.* **1980,** *24,* 168.
218. Stuber, H. A.; Leenheer, J. A. *Anal. Chem.* **1982,** *55,* 111.
219. Maruoka, S.; Yamanaka, S. *Environ. Sci. Technol.* **1983,** *17(3),* 177.
220. Ehrhardt, M. *Deep Sea Res.* **1978,** **25,** 119.
221. Garrido-Segovia, J. J.; Monteoliva-Hernandez, M. *An. Edafol. Agrobiol.* **1981,** *40(9–10),* 1781.
222. Narang, A. S.; Eadon, G. *Int. J. Environ. Anal. Chem.* **1982,** *11(3–4),* 167.
223. Care, R.; Morrison, J. D.; Smith, J. F. *Water Res.* **1982,** *16(5),* 663.
224. Volpe, G.; Mallet, V. N. *Chromatographia* **1981,** *14(6),* 333.
225. Grieser, M. D.; Pietrzyk, D. J. *Anal. Chem.* **1973,** *45,* 1348.
226. Delbeke, F. T.; Debackere, M. *J. Chromatogr.* **1977,** *133,* 214.
227. Mukai, H.; Ozakio, K.; Kawada, K. *Nigata Ken Kogai Kenkyusho Kenkyu Hokoku* **1981,** *5,* 61.
228. Volpe, G. G.; Mallet, V. N. *Int. J. Environ. Anal. Chem.* **1980,** *8(4),* 291.
229. Shinohara, R.; Kido, A.; Eto, S.; Hori, T.; Koga, M.; Akiyama, T. *Esei Kagaku* **1980,** *26(2),* 84.
230. Maruoka, S.; Yamanaka, S. *Sci. Total Environ.* **1983,** *29(1–2),* 143.
231. Hux, R. A.; Mohammed, H. Y.; Cantwell, F. F. *Anal. Chem.* **1982,** *54,* 113.
232. Grim-Kibalo, S. M.; Glatz, B. A.; Fritz, J. S. *Bull. Environ. Contam. Toxicol.* **1981,** *26,* 188.
233. Maruoka, S.; Yamanaka, S. *Mutat. Res.* **1982,** *102,* 13.
234. Leach, J. M.; Thakore, A. N. *J. Fish. Res. Board Can.* **1973,** *30,* 479.
235. Sloof, W.; van Kreijl, C. F. *Aquatic Toxicol.* **1982,** *2,* 89.
236. Williams, D. T.; Benoit, F. M.; McNeil, E. E.; Otson, R. *Pestic. Monit. J.* **1978,** *12,* 163.

237. Cox, P. J.; Levin, L. *Biochem. Pharmacol.* **1975,** *24,* 1233.
238. Scoggins, M. W.; Miller, J. W. *Anal. Chem.* **1968,** *40,* 1155.
239. Cantwell, F. F. *Anal. Chem.* **1976,** *48,* 1854.
240. Sydor, R.; Pietrzyk, D. J. *Anal. Chem.* **1978,** *50(13),* 1842.
241. Tabor, M. W.; Loper, J. C. *Int. J Environ. Anal. Chem.* **1980,** *8,* 197.
242. Mohammed, H. Y.; Cantwell, F. F. *Anal. Chem.* **1978,** *50,* 491.
243. Bastos, M. L.; Jukofsky, D.; Mule, S. J. *J. Chromatogr.* **1973,** *81,* 93.
244. Volpe, G.; Mallet, V. N. *J. Chromatogr.* **1979,** *177,* 385.
245. Uematsu, T.; Sukadolnik, R. J. *J. Chromatogr.* **1976,** *123,* 347.
246. Zaika, L. L. *J. Chromatogr.* **1970,** *49,* 222.
247. Cantwell, F. F.; Puon, S. *Anal. Chem.* **1979,** *51,* 623.
248. Wun, C. K.; Rho, J.; Walker, R. W.; Litsky, W. *Water Res.* **1979,** *13,* 645.
249. Richard, J. J.; Fritz, J. S. *Talanta* **1974,** *21,* 91.
250. Richard, J. J.; Chriswell, C. D.; Fritz, J. S. *J. Chromatogr.* **1980,** *199,* 143.
251. Dawson, R.; Riley, J. P.; Tennant, R. H. *Mar. Chem.* **1976,** *4,* 83.
252. Leenheer, J. A.; Noyes, T. I.; Stuber, H. A. *Environ. Sci. Technol.* **1982,** *16,* 714.
253. Segar, G. A. *Effluent Water Treat. J.* **1969,** *9,* 433.
254. Al-Biaty, I. A.; Fritz, J. *Anal. Chim. Acta* **1983,** *146,* 191.
255. Hori, M. *Bull. Chem. Soc. Jpn.* **1969,** *42,* 2333.
256. Shackleton, C. H. L.; Whitney, J. O. *Clin. Chim. Acta* **1980,** *107,* 231.
257. Shackleton, C. H. L.; Sjovall, J.; Wisen, O. *Clin. Chim. Acta* **1970,** *27,* 354.
258. Miller, W. L.; Kullberg, M. P.; Banning, M. F.; Brown, L. D.; Doctor, B. P. *Biochem. Med.* **1973,** *7,* 145.
259. Bian, Y. *Huanjing Baohu* **1982,** *11,* 20.
260. Scheider, W.; Fuller, J. K. *Biochem. Biophys. Acta* **1970,** *221,* 376.
261. Still, G. G.; Mansager, E. R. *Pestic. Biochem. Physiol.* **1975,** *5,* 515.
262. Popjak, G.; Edmond, J.; Clifford, K.; Williams, V. *J. Biol. Chem.* **1969,** *244,* 1897.
263. Mattox, V. R.; Vrieze, W. *Fed. Am. Soc. Exp. Biol. Fed. Proc.* **1967,** Abstract 945.
264. Kringstad, K. P.; Ljungquist, P. O.; de Sousa, F.; Stromberg, L. M. In *Water Chlorination: Environmental Impact and Health Effects;* Jolley, R. L.; Brungs, W. A.; Cotruvo, J. A.; Cumming, R. B.; Mattice, J. S.; Jacobs, V. A., Eds.; Ann Arbor Science: Ann Arbor, MI, 1983; Vol. 4, Chapter 93, p 1311.
265. Spotte, S.; Adams, G. *Aquaculture* **1982,** *29(1-2),* 159.
266. Loper, J. C.; Tabor, M. W.; Miles, S. K. In *Water Chlorination: Environmental Impact and Health Effects;* Jolley, R. L.; Brungs, W. A.; Cotruvo, J. A.; Cumming, R. B.; Mattice, J. S.; Jacobs, V. A., Eds.; Ann Arbor Science: Ann Arbor, MI, 1983; Vol. 4, Chapter 86, p 1199.
267. Van Hoof, F. In *Water Chlorination: Environmental Impact and Health Effects;* Jolley, R. L.; Brungs, W. A.; Cotruvo, J. A.; Cumming, R. B.; Mattice, J. S.; Jacobs, V. A., Eds.; Ann Arbor Science: Ann Arbor, MI, 1983; Vol. 4, Chapter 87, p 1211.
268. Davis, W. P.; James, A. F. In *Water Chlorination: Environmental Impact and Health Effects;* Jolley, R. L.; Brungs, W. A.; Cotruvo, J. A.; Cumming, R. B.; Mattice, J. S.; Jacobs, V. A., Eds.; Ann Arbor Science: Ann Arbor, MI, 1983; Vol. 4, Chapter 54, p 791.
269. Cheh, A. M.; Carlson, R. E.; Hildedrandt, J. R.; Woodward, C.; Pereira, M. A. In *Water Chlorination: Environmental Impact and Health Effects;* Jolley, R. L.; Brungs, W. A.; Cotruvo, J. A.; Cumming, R. B.; Mattice, J. S.; Jacobs, V. A., Eds.; Ann Arbor Science: Ann Arbor, MI, 1983; Vol. 4, Chapter 88, p 1221.

270. Nestmann, E. R.; Otson, R.; Lebel, G. L.; Williams, D. T.; Lee, E. G. H.; Biggs, D. C. In *Water Chlorination: Environmental Impact and Health Effects;* Jolley, R. L.; Brungs, W. A.; Cotruvo, J. A.; Cumming, R. B.; Mattice, J. S.; Jacobs, V. A., Eds.; Ann Arbor Science: Ann Arbor, MI, 1983; Vol. 4, Chapter 82, p 1151.
271. Bean, R. M.; Riley, R. G.; Ryan, P. W. In *Water Chlorination: Environmental Impact and Health Effects;* Jolley, R. L.; Gorchev, H.; Hamilton, D. H., Eds.; Ann Arbor Science: Ann Arbor, MI, 1978; Vol. 2, Chapter 17, p 223.
272. Glaze, W. H.; Peyton, G. R. In *Water Chlorination: Environmental Impact and Health Effects;* Jolley, R. L.; Gorchev, H.; Hamilton, D. H., Eds.; Ann Arbor Science: Ann Arbor, MI, 1978; Vol. 2, Chapter 1, p 3.
273. Douglas, G. R.; Nestmann, E. R.; Betts, J. L.; Mueller, J. C.; Lee, E. G. H.; Stich, H. F.; San, R. H. C.; Brouzes, R. J. P.; Chimelanskas, A. J.; Paavila, H. D.; Walden, C. C. In *Water Chlorination: Environmental Impact and Health Effects;* Jolley, R. L.; Brungs, W. A.; Cumming, R. B.; Jacobs, V. A., Eds.; Ann Arbor Science: Ann Arbor, MI, 1980; Vol. 2, Chapter 76, p 865.
274. Osborne, D. N.; Gore, B. H. *J. Chromatogr.* **1973,** *77,* 233.
275. Neal, R. A. In *Water Chlorination: Environmental Impact and Health Effects;* Jolley, R. L.; Brungs, W. A.; Cumming, R. B.; Jacobs, V. A., Eds.; Ann Arbor Science: Ann Arbor, MI, 1980; Vol. 3, Chapter 87, p 1007.
276. Bean, R. M.; Mann, D. C.; Wilson, B. W.; Riley, R. G.; Lusty, E. W.; Thatcher, T. O. In *Water Chlorination: Environmental Impact and Health Effects;* Jolley, R. L.; Brungs, W. A.; Cumming, R. B.; Jacobs, V. A., Eds.; Ann Arbor Science: Ann Arbor, MI, 1980; Vol. 3, Chapter 33, p 369.
277. Tabor, M. W.; Loper, J. C.; Barone, K. In *Water Chlorination: Environmental Impact and Health Effects;* Jolley, J. L.; Brungs, W. A.; Cumming, R. B.; Jacobs, V. A., Eds.; Ann Arbor Science: Ann Arbor, MI, 1980; Vol. 3, Chapter 78, p 899.
278. Cheh, A. M.; Skochdopole, J.; Heilig, C.; Koski, M.; Cole, L. In *Water Chlorination: Environmental Impact and Health Effects;* Jolley, R. L.; Brungs, W. A.; Cumming, R. B.; Jacobs, V. A., Eds.; Ann Arbor Science: Ann Arbor, MI, 1980; Vol. 3, Chapter 70, p 803.
279. Bean, R. M.; Mann, D. C.; Neitzel, D. A. In *Water Chlorination: Environmental Impact and Health Effects;* Jolley, R. L.; Brungs, W. A.; Cotruvo, J. A.; Cummings, R. B.; Mattice, J. S.; Jacobs, V. A., Eds.; Ann Arbor Science: Ann Arbor, MI, 1983; Vol. 4, Chapter 27, p 383.
280. Kringstad, K. P.; Ljungquist, P. O.; deSousa, F.; Stromberg, L. M. *Environ. Sci. Technol.* **1981,** *15,* 562.
281. Lee, D. P.; Kindsvater, J. H. *Anal. Chem.* **1980,** *52,* 2425.
282. Pietrzyk, D. J.; Kroeff, E. P.; Rotsch, T. D. *Anal. Chem.* **1978,** *50,* 497.
283. Kroeff, E. P.; Pietrzyk, D. J. *Anal. Chem.* **1978,** *50,* 502.
284. Eriksson, K. E.; Kolar, M. C.; Kringstad, K. *Sven. Papperstidn.* **1979,** *82,* 95.
285. Edgerton, T. R.; Moseman, R. F.; Lores, E. M.; Wright, L. H. *Anal. Chem.* **1980,** *52,* 1774.
286. Harris, J. C.; Cohen, M. J.; Hayes, M. J. In *Proceedings of 2nd Symposium on Process Measurements for Environmental Assessment; 1981,* NTIS Report PB82-211574, p 41.
287. Cheh, A. M. NTIS Report PB83-208843; NTIS: Springfield, VA, 1983.
288. Epler, J. L.; Larimer, F. W.; Rao, T. K.; Burnett, E. M.; Griest, W. H.; Guerin, M. R.; Maskarinec, M. P.; Brown, D. A.; Edwards, N. T.; Gehrs,

C. W.; Milleman, R. E.; Parkhurst, B. R.; Ross-Todd, B. M.; Shriner, D. S.; Wilson, H. W., Jr. NTIS Report PB80-179328; NTIS: Springfield, VA, 1983.
289. Yamasaki, E.; Ames, B. N. *Proc. Natl. Acad. Sci.* **1977,** *74,* 3555.
290. Strup, P. E.; Wilkinson, J. E.; Jones, P. W. In *Carcinogenesis: Polynuclear Aromatic Hydrocarbons;* Jones, P. W.; Freudenthal, R. I., Eds.; Raven: New York, 1978; Vol. 3, p 131.
291. Loper, J. C. *Mutat. Res.* **1980,** *76,* 241.
292. Loomis, W. D. In *Methods in Enzymology;* Fleischer, S.; Packer, L., Eds.; Academic: New York, 1974; Vol. 31, p 528.
293. Loomis, W. D.; Lile, J. D.; Sandstrom, R.P.; Burbott, A. J. *Phytochemistry* **1979,** *18,* 1049.
294. Niemann, R. H.; Pap, D. L.; Clark, R. A. *J. Chromatogr.* **1978,** *161,* 137.
295. Kopfler, F. C.; Coleman, W. E.; Melton, R. G.; Tardiff, R. G.; Lynch, S. C.; Smith, J. K.; *Ann. N.Y. Acad. Sci.* **1977,** *298,* 20.
296. Baird, R.; Gute, J.; Jacks, C.; Jenkins, R.; Neisess, L.; Scheybeler, B.; Van Sluis, R.; Yanko, W. In *Water Chlorination: Environmental Impact and Health Effects;* Jolley, R. L.; Brungs, W. A.; Cumming, R. B.; Jacobs, V. A., Eds.; Ann Arbor Science: Ann Arbor, MI, 1980; Vol. 3, Chapter 80, p 925.
297. Loper, J. C.; Lang, D. R.; Schoeny, R. S.; Richmond, B. B.; Gallagher, P. M.; Smith, C. C. *J. Toxicol. Environ. Health* **1978,** *4,* 919.
298. Bradlow, H. L. *Steroids* **1977,** *30,* 581.
299. Derks, H. J. G. M.; Drayer, N. M. *Clin. Chem.* **1978,** *24,* 1158.
300. Thompson, J. A.; Holtzman, J. L. *J. Pharmacol. Exp. Ther.* **1973,** *186,* 640.
301. Kanabata, N.; Morigaki, T. *Environ. Sci. Technol.* **1980,** *14,* 1089.
302. Gardiner, J. Technical Report; Water Research Center: Medmenham, England, 1978; No. TR-89.
303. Thruston, A. D., Jr. *J. Chromatogr. Sci.* **1978,** *16(6),* 254.
304. Renberg, L. *Anal. Chem.* **1978,** *50(13), 1836.*
305. Kim, B. R.; Snoeyink, V. L.; Saunders, F. M. *J. Water Pollut. Control Fed.* **1976,** *48,* 120.
306. Walser, T. F. *Proc. S. Water Res. Pollut. Control Conf.* **1967,** *16,* 81.
307. Glaze, W. H.; Burleson, J. L.; Henderson, J. E. IV; Jones, P. C.; Kinstley, W. NTIS Report PB83-144444; NTIS: Springfield, VA, 1982.
308. Kawabata, N.; Ohira, K. *Environ. Sci. Technol.* **1979,** *13,* 1396.
309. Karoly, G.; Sebestyen, L.; Gyorfi, L. *Magy. Kem. Lapja* **1980,** *35(7),* 338.
310. Harris, R. L.; Huggett, R. J.; Slone, H. D. *Anal. Chem.* **1980,** *52,* 779.
311. Lebel, G. L.; Williams, D. T.; Benoit, F. M. *J. Assoc. Off. Anal. Chem.* **1981,** *64,* 991.
312. Cheh, A. M.; Carlson, R. E. *Anal. Chem.* **1981,** *53,* 1001.
313. Tabor, M. W.; Loper, J. C.; Barone, K. In *Water Chlorination: Environmental Impact and Health Effects;* Jolley, R. L.; Brungs, W. A.; Cumming, R. B.; Jacobs, V. A., Eds.; Ann Arbor Science: Ann Arbor, MI, 1980; Vol. 3, Chapter 78, p 899.
314. Davidow, B.; Quame, B.; Abell, L.; Lim, B. *Health Sci. Lab.* **1973,** *10(4),* 329.
315. Gustafson, R. L.; Paleos, J. In *Organic Compounds in Aquatic Environments;* Faust, S. J.; Hunter, J. V., Eds.; Marcel Dekker: New York, 1971; Chapter 10, p 213.
316. Nestmann, E. R.; Lebel, G. L.; Williams, D. T.; Kowbel, D. J. *Environ. Mutagen.* **1979,** *1,* 337.
317. Adams, J.; Menzies, K.; Levins, P. NTIS Report PB-268559: NTIS: Springfield, VA, 1977.

318. Walton, H. F. In *Chromatographic Analysis of the Environment;* Grob, R. L., Ed.; Marcel Dekker: New York, 1983; 2nd ed., Chapter 7, p 263.
318a. Baird, R. B.; Jacks, C. A.; Jenkins, R. L.; Gute, J. P.; Neisess, L.; Scheybeler, B. In *Chemical Water Reuse;* Cooper, W. J., Ed.; Ann Arbor Science: Ann Arbor, MI, 1981; Vol. 2, p 149.
319. Thurman, E. M.; Malcolm, R. L.; Aiken, G. R. *Anal. Chem.* **1978,** *50,* 775.
320. Thurman, E. M.; Malcolm, R. L. *Environ. Sci. Technol.* **1981,** *15,* 463.
321. Leenheer, J. A. *Environ. Sci. Technol.* **1981,** *15,* 578.
322. Hetland, L. B.; Knowlton, D. A.; Couri, D. *Clin. Chim. Acta* **1972,** *36,* 473.
323. Rogers, I. H.; Keith, L. H. In *Identification and Analysis of Organic Pollutants in Water;* Keith, L. H., Ed.; Ann Arbor Science: Ann Arbor, MI, 1976; Chapter 33, p 625.
324. Leenheer, J. A. *Acta Amazonica* **1980,** *10,* 513.
325. Leenheer, J. A.; Stuber, H. A. *Environ. Sci. Technol.* **1981,** *15,* 1467.
326. Oliver, B. G.; Thurman, E. M. In *Water Chlorination: Environmental Impact and Health Effects;* Jolley, R. L.; Brungs, W. A.; Cotruvo, J. A.; Cumming, R. B.; Mattrice, J. S.; Jacobs, V. A., Eds.; Ann Arbor Science: Ann Arbor, MI, 1983; Vol. 4, Chapter 16, p 231.
327. Colclough, C. A.; Johnson, J. D.; Christman, R. F.; Millington, D. S. In *Water Chlorination: Environmental Impact and Health Effects;* Jolley, R. L.; Brungs, W. A.; Cotruvo, J. A.; Cumming, R. B.; Mattice, J. S.; Jacobs, V. A., Eds.; Ann Arbor Science: Ann Arbor, MI, 1983; Vol. 4, Chapter 15, p 219.
328. de Greef, E.; Morris, J. C.; van Kreijl, C. F.; Morra, C. F. H. In *Water Chlorination: Environmental Impact and Health Effects;* Jolley, R. L.; Brungs, W. A.; Cummings, R. B.; Jacobs, V. A., Eds.; Ann Arbor Science: Ann Arbor, MI, 1980; Vol. 3, Chapter 79, p 913.
329. Malcolm, R. L.; Thurman, E. M.; Aiken, G. R. *Trace Subst. Environ. Health* **1977,** *11,* 307.
330. Hollis, O. L.; Hayes, W. V. *J. Gas Chromatogr.* **1966,** *4,* 235.
331. Sakodynskii, K.; Panina, L.; Klinskaya, N. *Chromatographia* **1974,** *7,* 339.
332. Cassidy, R. M.; Burteau, M. T.; Mislan, J. P.; Ashley, R. W. *J. Chromatogr. Sci.* **1976,** *14,* 444.
333. Krejci, M.; Roudna, M.; Vavrouch, Z. *J. Chromatogr.* **1974,** *91,* 549.
334. Creed, C. G. *Res. Dev.* **1976,** *27(9),* 40.
335. d'Aubigne, J. M.; Guiochon, G. *Chromatographia* **1970,** *3,* 153.
336. Trowell, J. M. *J. Chromatogr. Sci.* **1971,** *9,* 253.
337. Niederwieser, A.; Giliberti, P. *J. Chromatogr.* **1971,** *61,* 95.
338. Butler, L. D.; Burke, M. F. *J. Chromatogr. Sci.* **1976,** *14,* 117.
339. Vitzhum, O. G.; Werkoff, P. *J. Food Sci.* **1974,** *39,* 1210.
340. Neumann, M. G.; Morales, S. *J. Chromatogr.* **1972,** *74,* 332.
341. Graham, J. A.; Garrison, A. W. In *Advances in the Identification and Analysis of Organic Pollutants in Water;* Keith, L. H., Ed.; Ann Arbor Science: Ann Arbor, MI, 1981; Vol. 1, Chapter 26, p 399.
342. Johnson, J. F.; Barrall, E. M., II *J. Chromatogr.* **1967,** *31,* 547.
343. Pellizzari, E. D.; Bunch, J. E.; Berkley, R. E.; McRae, J. *Anal. Lett.* **1976,** *9,* 45.
344. Neher, M. B.; Jones, P. W. *Anal. Chem.* **1977,** *49,* 512.
345. Versino, B.; Knoppel, H.; De Groot, M.; Peil, A.; Poelman, J.; Schauenburg, H. *J. Chromatogr.* **1976,** *122,* 373.
346. Leoni, V.; Puccetti, G.; Grella, A. *J. Chromatogr.* **1975,** *106,* 119.
347. Katou, T.; Akiyama, K.; Suzuki, S. *Yokohama Kokuritsu Daigaku Kankyo Kagaku Kenkyu Senta Kiyo* **1981,** *7(1),* 11.

348. Pankow, J. F.; Isabelle, L. M.; Kristensen, T. J. *J. Chromatogr.* **1982,** *245,* 31.
349. Kadar, R.; Nagy, K.; Fremstad, D. *Talanta* **1980,** *27,* 227.
350. Pankow, J. F.; Isabelle, L. M.; Kristensen, T. J. *Anal. Chem.* **1982,** *54,* 1815.
351. Dowty, B. J.; Carlisle, D. R.; Laseter, J. L. *Environ. Sci. Technol.* **1975,** *9(8),* 762.
352. Marcelin, G. *J. Chromatogr.* **1979,** *174,* 208.
353. Murray, K. E. *J. Chromatogr.* **1977,** *135,* 49.
354. Leoni, V.; Puccetti, G.; Colombo, R. J.; Dovidio, A. M. *J. Chromatogr.* **1976,** *125,* 399.
355. Van Wijk, R. *J. Chromatogr. Sci.* **1970,** *8,* 418.
356. Zlatkis, A.; Lichlenstein, H. A.; Tishbee, A.; Bertsch, W.; Shumbo, F.; Liebich, H. M. *J. Chromatogr. Sci.* **1973,** *11,* 299.
357. Mieure, J. P.; Dietrich, M. W. *J. Chromatogr. Sci.* **1973,** *11,* 559.
358. Bertsch, W.; Anderson, E.; Holzer, G. *J. Chromatogr.* **1975,** *112,* 701.
359. Novotny, M.; Lee, M. L.; Bartle, K. D. *Chromatographia* **1974,** *7(7),* 333.
360. Pankow, J. F.; Isabelle, L. M.; Hewetson, J. P.; Cherry, J. A. *Ground Water* **1984,** *22,* 330.
361. Pankow, J. F. *J. High Resolut. Chromatogr. Chromatogr. Commun.* **1983,** *6,* 292.
362. Pankow, J. F.; Isabelle, L. M.; Asher, W. E. *Environ. Sci. Technol.* **1984,** *18,* 310.
363. Kopfler, F. C.; Melton, R. G.; Lingg, R. D.; Coleman, W. E. In *Identification and Analysis of Organic Pollutants in Water;* Keith, L. H., Ed.; Ann Arbor Science: Ann Arbor, MI, 1976; Chapter 6, p 87.
364. Bunn, W. W.; Deane, E. R.; Klein, D. W.; Kleopfer, R. D. *Water Air Soil Pollut.* **1975,** *4,* 367.
365. Shiaris, M. P.; Sherill, T. W.; Sayler, G. S. *Appl. Environ. Microbiol.* **1980,** *39,* 165.
366. Pankow, J. F.; Isabelle, L. M. *J. Chromatogr.* **1982,** *237,* 25.
367. Bartle, K. D.; Elstub, J.; Novotny, M.; Robinson, R. J. *J. Chromatogr.* **1977,** *135,* 351.
368. Daemen, J. M. H.; Dankelman, W.; Hendriks, M. E. *J. Chromatogr. Sci.* **1975,** *13,* 79.
369. Pankow, J. F.; Kristensen, T. J.; Isabelle, L. M. *Anal. Chem.* **1983,** *55(13),* 2187.
370. Sekerko, I.; Lechner, J. F. *Int. J. Environ. Anal. Chem.* **1982,** *11(1),* 43.
371. Smith, S. R.; Tanaka, J.; Collins, R.; Johnson, J. NTIS Report PB83-235259; NTIS: Springfield, VA, 1982.
372. Agostiano, A.; Caselli, M.; Provenzano, M. R. *Water Air Soil Pollut.* **1983,** *19(4),* 309.
373. Priestley, L. J., Jr.; Wilkes, B. E. *Microchem. J.* **1979,** *24,* 88.
374. Kirkland, J. J. *Anal. Chem.* **1963,** *35,* 2003.
375. Miget, R.; Kator, H.; Oppenheimer, C.; Laseter, J. L.; Ledet, E. J. *Anal. Chem.* **1974,** *46,* 1154.
376. Hjerten, S. *J. Chromatogr.* **1978,** *159,* 47.
377. Hjerten, S.; Hellman, U. *J. Chromatogr.* **1980,** *202,* 391.
378. Josefson, C. M.; Johnston, J. B.; Trubey, R. *Anal. Chem.* **1984,** *56,* 764.
379. Ross, W. D.; Jefferson, R. T. *J. Chromatogr. Sci.* **1970,** *8,* 386.
380. Basu, D. K.; Saxena, J. *Environ. Sci. Technol.* **1978,** *12,* 791.
381. Saxena, J.; Kozuchowski, J.; Basu, D. K. *Environ. Sci. Technol.* **1977,** *11,* 682.

382. Basu, D. K.; Saxena, J. *Environ. Sci. Technol.* **1978,** *12(7),* 795.
383. Gesser, H. D.; Chow, A.; Davis, F. C.; Uthe, J. F.; Reinke, J. *Anal. Lett.* **1971,** *4(12),* 883.
384. Saxena, J.; Basu, J. D.; Schwartz, D. J. In *Analytical Techniques in Environmental Chemistry;* Albaiges, J., Ed.; Pergamon: New York, 1980; p 119.
385. Uthe, J. F.; Reinke, J.; Gesser, H. *Environ. Lett.* **1972,** *3(2),* 117.
386. Musty, P. R.; Nickless, G. *J. Chromatogr.* **1974,** *100,* 83.
387. Lynn, T. R.; Rushneck, D. R.; Cooper, A. R. *J. Chromatogr. Sci.* **1974,** *12,* 76.
388. Bedford, J. W. *Bull. Environ. Contam. Toxicol.* **1974,** *12,* 622.
389. Bidleman, T. F.; Olney, C. E. *Bull. Environ. Contam. Toxicol.* **1974,** *11,* 442.
390. Braun, T.; Farag, A. B. *Talanta* **1975,** *22,* 699.
391. Moody, G. J.; Thomas, J. D. R. In *Chromatographic Separation and Extraction with Foamed Plastics and Rubbers;* Marcel Dekker: New York, 1982; Vol. 21.
392. Gesser, H. D.; Sparling, A. B.; Chow, A.; Turner, C. W. *J. Am. Water Works Assoc.* **1973,** *65,* 220.
393. Gough, K. M.; Gesser, H. D. *J. Chromatogr.* **1975,** *115,* 383.
394. Uthe, J. F.; Reinke, J.; O'Brodovich, H. *Environ. Lett.* **1974,** *6,* 103.
395. Bidleman, T. F.; Olney, C. E. *Science* **1974,** *183,* 516.
396. Murphy, T. J.; Rzeszutko, C. P. *J. Great Lakes Res.* **1977,** *3,* 305.
397. Tanaka, T.; Hiiro, K.; Kawahara, A. *Bunseki Kagaku* **1973,** *22,* 523.
398. Braun, T.; Farag, A. B. *Anal. Chim. Acta* **1978,** *99,* 1.
399. Bowen, H. J. M. *J. Chem. Soc.* **1970,** *A,* 1082.
400. Al-Bazi, S. J.; Chow, A. *Talanta* **1984,** *31(3),* 189.
401. Ahmed, S. M.; Beasley, M. D.; Efromson, A. C.; Hites, R. A. *Anal. Chem.* **1974,** *46,* 1858.
402. Moody, G. J.; Thomas, J. D. R. *Analyst* **1979,** *104,* 1.
403. Turner, B. C.; Glotfelty, D. E. *Anal. Chem.* **1977,** *49,* 7.
404. Mazurski, M. A. J.; Chow, A.; Gesser, H. D. *Anal. Chim. Acta* **1973,** *65,* 99.
404a. Kawabata, N.; Ohira, K. *Environ. Sci. Technol.* **1979,** *13(11),* 1396.
404b. Mourey, T.; Carpenter, A. P.; Siggia, S.; Lane, A. *Anal. Chem.* **1976,** *48(11),* 1592.
404c. Wojaczynska, M.; Kolarz, B. N. *J. Chromatogr.* **1980,** *196,* 75.
405. Schwartz, D. J.; Saxena, J.; Kopfler, F. C. *Environ. Sci. Technol.* **1979,** *13,* 1138.
406. Sukiman, S. *Radiochem. Radioanal. Lett.* **1974,** *18,* 129.
407. Wells, D. E.; Johnstone, S. J. *Water Air Soil Pollut.* **1978,** *9,* 271.
408. Hileman, F. D.; Sievers, R. E.; Hess, G. G.; Ross, W. D. *Anal. Chem.* **1973,** *45,* 1126.
409. Hansen, L. C.; Sievers, R. E. *J. Chromatogr.* **1974,** *99,* 123.
410. Yu, Y.; Tang, J.; Ye, M. *J. Radioanal. Chem.* **1983,** *76(2),* 275.
411. Lypka, G. N.; Gesser, H. D.; Chow, A. *Anal. Chim. Acta* **1975,** *78,* 367.
412. Weil, L.; Quentin, K. E.; Gitzowa, S. *GWF Wasser/Abwasser* **1972,** *113,* 64.
413. Winsten, W. A. *Anal. Chem.* **1962,** *34,* 1334.
414. Beyermann, K.; Eckrich, W. Z. *Anal. Chem.* **1973,** *265,* 1.
415. Rice, M. R.; Gold, H. S. *Anal. Chem.* **1984,** *56,* 1436.
416. Ahling, B.; Jensen, S. *Anal. Chem.* **1970,** *42,* 1483.
417. Brizova, E.; Popl, M.; Coupek, J. *J. Chromatogr.* **1977,** *139,* 15.

418. Ishii, D.; Hibi, K.; Asai, K.; Nagaya, M. *J. Chromatogr.* **1978,** *152,* 341.
419. Ishii, D.; Hibi, K.; Asai, K.; Nagaya, M.; Mochizuki, K. Mochida, Y. *J. Chromatogr.* **1978,** *156,* 173.
420. Lee, D. P. *J. Chromatogr. Sci.* **1982,** *20,* 203.
421. Iskandarani, Z; Pietrzyk, D. J. *Anal. Chem.* **1981,** *53,* 489.
422. Werkhoven-Goewie, C. E.; Boon, W. N.; Praat, A. J. J.; Frei, R. W.; Brinkman, U. A. Th.; Little, C. J. *Chromatographia* **1982,** *16,* 53.
423. Dressler, M. *J. Chromatogr.* **1979,** *165,* 167.
424. Kubelka, V.; Mitera, J.; Novak, J.; Mostecky, J. *Chem. Prum.* **1975,** *25,* 593.
425. Viden, I.; Kubelka, V.; Mostecky, J. Z. *Anal. Chem.* **1976,** *280,* 369.
426. Kubelka, V.; Mitera, J.; Novak, J.; Mostecky, J. *Chem. Listy* **1977,** *71,* 241.
427. Brizova, E.; Popl, M.; Coupek, J. *Chem. Prum.* **1977,** *27,* 352.
428. Loconto, P. R. *Abstracts of Papers,* The Pittsburgh Conference and Exposition on Analytical Chemistry and Applied Spectroscopy, Atlantic City, NJ; March 1984; Abstract 440.
429. Kunin, R. *Pure Appl. Chem.* **1976,** *46,* 205.
430. Yan, C.; Lu, M.; Gao, L.; Zhang, Y. *Huanjing Huaxue* **1983,** *2(2),* 39.
431. Yan, Y.; Lu, M.; Gao, L. *Haunjing Huaxue* **1982,** *1,* 5.
432. Sasagawa, T.; Ericsson, L. H.; Teller, D. C.; Titani, K.; Walsh, K. A. *J. Chromatogr.* **1984,** *307,* 29.
433. Junk, G. A.; Richard, J. J. *Water Qual. Bull.* **1981,** *6,* 40.
434. Miles, C. J.; Tuschall, J. R., Jr.; Brezonik, P. L. *Anal. Chem.* **1983,** *55,* 410.
435. Busch, H.; Hurlbert, R. B.; Potter, V. R. *J. Biol. Chem.* **1952,** *196,* 717.
436. Apffel, J. A.; Alfredson, T. V.; Majors, R. E. *J. Chromatogr.* **1981,** *206,* 43.
437. Dufka, O.; Malinsky, J.; Vladyka, J. *Chem. Prum.* **1971,** *21,* 459.
438. Rook, J. J.; Evans, S. *J. Am. Water Works Assoc.* **1979,** *71(9),* 520.
439. Baker, D. W. *J. Assoc. Off. Anal. Chem.* **1973,** *56,* 1257.
440. Leenheer, J. A. In *Contaminants and Sediments;* Baker, R. A., Ed.; Ann Arbor Science: Ann Arbor, MI, 1980; Vol. 2, Chapter 14, p 267.
441. Kunin, R.; Meitzner, E. F.; Bortnick, N. *J. Am. Chem. Soc.* **1962,** *84,* 305.
442. Richard, J. J.; Fritz, J. S. *J. Chromatogr. Sci.* **1980,** *18,* 35.
443. Christman, R. F.; Chassemi, M. *J. Am. Water Works Assoc.* **1966,** *58,* 723.
444. Koch, D. D.; Kissinger, P. T. *Life Sci.* **1980,** *26,* 1099.
445. Evans, S.; Maalman, T. F. *Environ. Sci. Technol.* **1979,** *13,* 741.
446. Painter, H. A. *Chem. Ind.* **1973,** *17,* 818.
447. Pitt, W. W., Jr.; Lolley, R. L.; Katz, S. In *Identification and Analysis of Organic Pollutants in Water;* Keith, L. H., Ed.; Ann Arbor Science: Ann Arbor, MI, 1976; Chapter 14, p 215.
448. Jolley, R. L.; Jones, G., Jr.; Pitt, W. W., Jr.; Thompson, J. E. In *Identification and Analysis of Organic Pollutants in Water;* Keith, L. H., Ed.; Ann Arbor Science: Ann Arbor, MI, 1976; Chapter 15, p 233.
449. Chriswell, C. D.; Chang, R. C.; Fritz, J. S. *Anal. Chem.* **1975,** *47,* 1325.
450. Anderson, C. T.; Maier, W. J. *J. Am. Water Works Assoc.* **1979,** *71,* 278.
451. Sirotkina, I. S.; Varshall, G. M.; Lure, Y. Y.; Stepanova, N. P. *J. Anal. Chem. USSR* **1974,** *29,* 1403.
452. Lores, E. M.; Bradway, D. E. *J. Agric. Food Chem.* **1977,** *25,* 75.
453. Verweij, A.; Degenhardt, C. E. A. M.; Boter, H. L. *Chemosphere* **1979,** *8(3),* 115.
454. Verweij, A.; Boter, H. L.; Degenhardt, C. E. A. M. *Science* **1979,** *204,* 616.
455. Aue, W. A.; Hastings, C. R.; Gerhardt, K. O.; Pierce, P. O.; Hill, H. H.; Mosemann, R. F. *J. Chromatogr.* **1972,** *72,* 259.
456. Slavinskaya, G. V.; Kuznetsova, N. S.; Zeleneva, L. A. *Teor. Prakt. Sorbtsionnykh Protsessov* **1981,** *14,* 93.

457. Vakulenko, V. A.; Kuznetsova, E. P. *Khim. Prom-St., Ser.: Prorvod. Pererab. Plastmass Sint. Smol.* **1981,** *7,* 27.
458. Heller, C. A.; Greni, S. R.; Erickson, E. D. *Anal. Chem.* **1982,** *54(2),* 286.
459. Semmens, M. J.; Staples, A.; Norgaard, G.; Hohenstein, G. *Environ. Technol. Lett.* **1983,** *4(8–9),* 343.
460. Plapp, F. W.; Casida, J. E. *Anal. Chem.* **1958,** *30,* 1622.
461. Gee, M. *Anal. Chem.* **1965,** *37,* 926.
462. Peppard, T. L.; Halsey, S. A.; *J. Chromatogr.* **1980,** *202,* 271.
463. Koch, D. D.; Kissinger, P. T. *Anal. Chem.* **1980,** *52,* 27.
464. Watkins, S. R.; Watson, H. F. *Anal. Chim. Acta* **1961,** *24,* 334.
465. Kuksis, A.; Prioreschi, P. *Anal. Biochem.* **1967,** *19,* 468.
466. Canvin, D. T. *Can. J. Biochem.* **1965,** *43,* 1281.
467. Reinbothe, H. *Pharmazie* **1957,** *12,* 732.
468. Matejka, Z.; Eliasek, J. *Water Res.* **1980,** *14,* 467.
469. Erni, F.; Frei, R. W. *J. Chromatogr.* **1978,** *149,* 561.
470. Voelter, W.; Kronbach, T.; Zech, K.; Huber, R. *J. Chromatogr.* **1982,** *237,* 475.
471. Lankelma, J.; Poppe, H. *J. Chromatogr.* **1978,** *149,* 587.
472. Roth, W.; Beschke, K.; Jauch, R.; Zimmer, A.; Koss, F. W. *J. Chromatogr.* **1981,** *222,* 13.
473. Yamatodani, A.; Wada, H. *Clin. Chem.* **1981,** *27,* 1983.
474. Niederwieser, A.; Staudenmann, W.; Wetzel, E. *J. Chromatogr.* **1984,** *290,* 237.
475. Santoni, Y.; Rolland, P. H.; Cano, J. P. *J. Chromatogr.* **1984,** *306,* 165.
476. Masoud, A. N.; Bueding, E. *J. Chromatogr. Biomed. Appl.* **1983,** *276,* 111.
477. Ogan, K.; Katz, E.; Slavin, W. *J. Chromatogr. Sci.* **1978,** *16,* 517.
478. Aue, W. A.; Kapila, S.; Hastings, C. R. *J. Chromatogr.* **1972,** *73,* 99.
479. Oyler, A. R.; Bodenner, D. L.; Welch, K. J.; Llukkonen, R. J.; Carlson, R. M.; Kopperman, H. L.; Caple, R. *Anal. Chem.* **1978,** *50,* 837.
480. Andrews, J. S.; Good, T. J. *Am. Lab.* **1982,** *14,* 70.
481. Lecaillon, J. B.; Rouan, M. C.; Souppart, C.; Febvre, N.; Juge, F. *J. Chromatogr.* **1982,** *228,* 257.
482. Eisenbeiss, F.; Hein, H.; Joester, R.; Naundorf, G. *Chromatogr. Newsl.* **1978,** *6(1),* 8.
483. de Jong, G. J. *J. Chromatogr.* **1980,** *183,* 203.
484. Sander, L. C.; Wise, S. A. *Anal. Chem.* **1984,** *56,* 504.
485. Saner, W. A.; Jadamec, J. R.; Sager, R. W. *Anal. Chem.* **1979,** *51,* 2180.
486. Erni, F.; Keller, H. P.; Morin, C.; Schmitt, M. *J. Chromatogr.* **1981,** *204,* 65.
487. Hageman, R. J.; Greving, J. E.; Jonkman, J. H.; De Zeeuw, R. A. *J. Chromatogr.* **1983,** *274,* 239.
488. Wachob, G. D. *Abstracts of Papers,* The Pittsburgh Conference and Exposition on Analytical Chemistry and Applied Spectroscopy, Atlantic City, NJ; March 1984; Abstract 992.
489. Watanabe, H.; Taguchi, S.; McLaren, J. W.; Berman, S. S.; Russell, D. S. *Anal. Chem.* **1981,** *53,* 738.
490. Sturgeon, R.; Berman, S.; Willie, S. *Talanta* **1982,** *29,* 167.
491. van Vliet, H. P. M.; Bootsman, T. C.; Frei, R. W.; Brinkman, U. A. Th. *J. Chromatogr.* **1979,** *185,* 483.
492. May, W. E.; Wasik, S. P.; Freeman, D. H. *Anal. Chem.* **1978,** *50,* 997.
493. Booth, R. L.; English, J. N.; McDermott, G. N. *J. Am. Water Works Assoc.* **1965,** *57,* 215.
494. Frei, R. W.; Lawrence, J. F.; Brinkman, U. A. Th.; Honigberg, I. *J. High Resolut. Chromatogr. Chromatogr. Commun.* **1979,** *2,* 11.

495. Frei, R. W. *Int. J. Environ. Anal. Chem.* **1978,** *5,* 143.
496. Euston, C. B.; Baker, D. R. *Am. Lab.* **1979,** *11(3),* 91.
497. Huber, J. F. K.; Becker, R. R. *J. Chromatogr.* **1977,** *142,* 765.
498. Kirkland, J. *Analyst* **1974,** *99,* 859.
499. Kummert, R.; Molnar-Kubica, E.; Giger, W. *Anal. Chem.* **1978,** *50,* 1637.
500. Krummen, K.; Frei, R. W. *J. Chromatogr.* **1977,** *132,* 429.
501. van Buuren, C.; Lawrence, J. F.; Brinkman, U. A. Th.; Honigberg, I. L.; Frei, R. W. *Anal. Chem.* **1980,** *52,* 700.
502. Horvath, C.; Melander, W.; Molnar, I. *Anal. Chem.* **1977,** *49,* 142.
503. Saner, W. A.; Gilbert, J. A. *J. Liq. Chromatogr.* **1980,** *3(11),* 1753.
504. Locke, D. C. *J. Chromatogr. Sci.* **1974,** *12,* 433.
505. de Jong, G.; Zeeman, J. *Chromatographia* **1982,** *15,* 453.
506. Aue, W. A.; Hastings, C. R. *J. Chromatogr.* **1969,** *42,* 319.
507. Aue, W. A.; Hastings, C. R.; Augl, J. M.; Norr, M. K.; Larsen, J. V. *J. Chromatogr.* **1971,** *56,* 295.
508. Symons, R. K.; Crick, I. *Anal. Chim. Acta* **1983,** *151(1),* 237.
509. Hastings, C. R.; Aue, W. A.; Augl, J. M. *J. Chromatogr.* **1970,** *53,* 487.
510. Kirkland, J. J. *Chromatographia* **1975,** *8,* 661.
511. Roessner, B.; Schwedt, G. Z. *Anal. Chem.* **1983,** *315(7),* 610.
512. Frei, R. W. *Anal. Proc.* **1980,** *17,* 519.
513. Werkhoven-Goewie, C. E.; Brinkman, U. A. Th., Frei, R. W. *Anal. Chem.* **1981,** *53,* 2072.
514. Eskins, K.; Dutton, H. J. *Anal. Chem.* **1979,** *51,* 1885.
515. Davis, G.; Kissinger, P. T. *Anal. Chem.* **1979,** *51,* 1960.
516. Little, J. N.; Fallick, G. J. *J. Chromatogr.* **1975,** *112,* 389.
517. Robert, J. *J. Liq. Chromatogr.* **1980,** *3,* 1561.
518. Ghebregzabher, M.; Rufini, S. T.; Castellucci, M. G.; Lato, M. *J. Chromatogr.* **1981,** *222,* 191.
519. De Vries, J. X.; Gunthert, W.; Ding, R. *J. Chromatogr.* **1980,** *221,* 161.
520. Allan, R. J.; Goodman, H. T.; Watson, T. R. *J. Chromatogr.* **1980,** *183,* 311.
521. Schauwecker, P.; Frei, R. W.; Erni, F. *J. Chromatogr.* **1977,** *136,* 63.
522. Dupont, D. G.; de Jager, R. L. *J. Liq. Chromatogr.* **1981,** *4,* 123.
523. Fasco, M. J.; Cashin, M. J.; Kaminsky, L. S. *J. Liq. Chromatogr.* **1979,** *2,* 565.
524. Dolphin, R. J.; Willmott, F. W.; Mills, A. D.; Hoogeveen, L. P. J. *J. Chromatogr.* **1976,** *122,* 259.
525. Gfeller, J. C.; Stockmeyer, M. *J. Chromatogr.* **1980,** *198,* 162.
526. Whitney, J. O.; Thaler, M. M. *J. Liq. Chromatogr.* **1980,** *3,* 545.
527. Winterlin, W.; Hall, G.; Hsieh, D. P. H. *Anal. Chem.* **1979,** *51,* 1873.
528. Falkowski, A. J.; Wei, R. *Anal. Biochem.* **1981,** *115,* 311.
529. Goewie, C. E.; Kwakman, P.; Frei, R. W.; Brinkman, U. A. Th.; Maasfeld, W.; Seshadri, T.; Kettrup, A. *J. Chromatogr.* **1984,** *284,* 73.
530. Lee, T. L.; Brooks, M. A. *J. Chromatogr.* **1984,** *306,* 429.
531. Powell, W. S. *Prostaglandins* **1980,** *20,* 947.
532. Luderer, J. R.; Riley, D. L.; Demers, L. M. *J. Chromatogr. Biomed. Appl.* **1983,** *273,* 402.
533. Muller, H.; Mrongovius, R.; Seyberth, H. W. *J. Chromatogr.* **1981,** *226,* 450.
534. Riggin, R. M.; Howard, C. C. *Anal. Chem.* **1979,** *51,* 210.
535. Wolkoff, A. W.; Creed, C. *J. Liq. Chromatogr.* **1981,** *4(8),* 1459.
536. Frei, R. W.; Brinkman, U. A. Th. *Trends Anal. Chem.* **1981,** *1,* 45.
537. Snyder, L. R. *Principles of Adsorption Chromatography;* Marcel Dekker: New York, 1968.

538. Frank, H. S.; Evans, M. W. *J. Chem. Phys.* **1945,** *13,* 507.
538a. Shinoda, K. *J. Phys. Chem.* **1977,** *81,* 1300.
538b. Tanford, C. In *The Hydrophobic Effect: Formation of Micelles and Biological Membranes;* John Wiley: New York, 1973; p 1.
539. Kun, K. A.; Kunin, R. *J. Polym. Sci.* **1968,** *6,* 2689.
540. Vick, R. D.; Richard, J. J.; Svec, H. J.; Junk, G. A. *Chemosphere* **1977,** *6,* 303.
541. Kopfler, F. C. In *Short-Term Bioassays in the Analysis of Complex Environmental Mixtures II;* Waters, M. D.; Sandhu, S. S.; Huisingh, J. L.; Claxton, L.; Nesnow, S., Eds.; Plenum: New York, 1981; p 141.

RECEIVED for review August 14, 1985. ACCEPTED December 16, 1985.

11

Potential Organic Contamination Associated with Commercially Available Polymeric Sorbents

Contaminant Sources, Types, and Amounts

Gary Hunt

Environmental Research and Technology, Inc., Concord, MA 01742

Polymeric sorbents are used extensively in the isolation and preconcentration of semivolatile trace organics from aquatic matrices. These synthetic materials contain measurable or significant quantities of one or more of the following types of chemical contamination: (1) residual monomers, (2) artifacts of the polymer synthetic pathway, and (3) chemical preservatives used to inhibit chemical or biological degradation. This chapter provides some insight into the chemistry of a number of commonly used polymeric sorbents. Particular focus is placed on the chemical identification of solvent-extractable semivolatile organic contaminants typically associated with each of the following types of polymeric sorbents as received from the manufacturer: Amberlite XAD resins, Ambersorb XE resins, and polyurethane foam.

POLYMERIC SORBENTS are frequently used in environmental analytical schemes for the isolation and/or preconcentration of trace organic contaminants from air and water matrices. Commercially manufactured polymeric sorbents such as Amberlite XAD resins, Ambersorb XE resins, Tenax (diphenyl-*p*-phenylene oxide), and polyurethane foam (PUF) have been used extensively for the collection of trace organic contaminants from ambient air, process streams (i.e., flue gas), and a variety of aquatic matrices including industrial effluents, ground water, surface water, and potable water supplies. Currently, these materials

0065–2393/87/0214/0247$06.00/0

are used extensively in the preconcentration of trace organics from waters for eventual use in biological testing. The use of solid sorbents in this manner permits concentration levels more compatible with current state-of-the-art biological testing procedures. Prior to the use of polymeric sorbents in actual isolation schemes, however, users should focus on a number of quality assurance issues including the impact of sorbent contaminants on subsequent biological testing. For instance, the presence of suspected carcinogens in sample extracts arising from the sorbent itself may bias the results of an eventual biological assay procedure.

The majority of these synthetic materials as received from the manufacturer can be expected to contain measurable or perhaps significant quantities of one or more of the following types of chemical contamination: (1) residual monomers, (2) artifacts of the polymer synthetic process (e.g., starting materials, catalysts, and byproducts), and (3) chemical preservatives to inhibit chemical or biological degradation. For these reasons, all sorbent materials should undergo rigorous cleanup procedures prior to use. Numerous sorbent preparation procedures that are capable of reducing contamination to levels compatible with the eventual end use of the sorbent are available in the literature. Users are cautioned, however, that because of the significant contamination often encountered, any cleanup procedure can at best only minimize these contaminants and oftentimes leave measurable quantities of organics characteristic of the respective polymer. In addition, although numerous cleanup procedures are available for establishing resin quality, the subsequent storage and handling of these materials can promote the reappearance of contamination. Users are, therefore, urged to familiarize themselves with the chemical nature of polymeric sorbent contaminants as well as proper cleanup, storage, and handling procedures.

This chapter provides some insight into the chemistry of a number of commonly used polymeric sorbents. Particular focus is placed on the chemical identification of contaminants typically associated with each of the following types of polymeric sorbents: Amberlite XAD resins, Ambersorb XE resins, and PUF. Emphasis is placed on the chemical speciation of solvent-extractable organic contaminants present in a number of these sorbents as received from the manufacturer. Both qualitative and quantitative data on a micrograms-per-gram (parts-per-million) basis are provided as determined by combined gas chromatography–mass spectrometry (GC–MS).

Experimental

Reagents. All reagents used in experimental procedures were of the highest grade commercially available. Methylene chloride, hexane, and ethyl ether used in the sorbent preparation procedure were distilled in glass quality manufactured by Burdick and Jackson.

d_{10}-Anthracene, which served as an internal standard in all GC–MS analyses, was obtained from KOR Isotopes.

Amberlite XAD-2 and XAD-4 and Ambersorb XE-340 and XE-348 resins were obtained through the courtesy of Rohm & Haas. Representative lots of each of the four resins were provided.

The PUF was purchased from Flexible Foam Products. The foam (type 1636) was purchased in sheets 4 ft × 4 ft and 3 in. in depth.

Sample Preparation. AMBERLITE XAD RESINS. Two lots of Amberlite XAD-2 resin and one lot of Amberlite XAD-4 resin were prepared for analysis. Aliquots of each resin sample (5–10 g) were extracted for approximately 24 h in a continuous Soxhlet apparatus containing methylene chloride[1]. Extracts were reduced to 2.0 mL by employing a rotary evaporator operating under reduced pressure. The use of a rotary evaporative concentration technique under reduced pressure will significantly reduce concentrations of volatile organics and selected lower boiling semivolatile species contained in the solvent extract. Hence, results provided for both the Amberlite and Ambersorb sorbent series more accurately represent the semivolatile contaminant chemistry characteristic of these sorbents. The volatile organic data reported here thus represents minimum values. Further details on the sample preparation procedures were reported previously (*1*).

AMBERSORB XE RESINS. Two lots of Ambersorb XE-340 and one lot of Ambersorb XE-348 were prepared for analysis. The sample size and preparation procedures were identical to those employed for the Amberlite XAD series. Further details pertinent to these analyses are contained in a previous publication (*2*).

PUF. Sorbent plugs were removed from the 4- × 4-ft sheets by employing a 4-in. template. Two such plugs were randomly removed, each measuring 4 in. in diameter and 3 in. in depth. Plug weights were taken to the nearest 0.1 g. One plug was extracted overnight (ca. 8 h) in a Soxhlet extractor containing an ethyl ether/hexane (5/95) solvent system. The second plug was extracted for the same period with methylene chloride. A third Soxhlet extractor designated as the procedural blank contained only the ethyl ether/hexane (5/95) solvent system. Upon completion of the extraction cycle, all extracts were reduced in volume to 2.0 mL in a Kuderna–Danish evaporative concentrator.

Gas Chromatography–Flame Ionization Detection and Gas Chromatography–Mass Spectrometry. AMBERLITE XAD RESINS. Extracts were analyzed for total chromatographable organics (TCO) with a gas chromatograph fitted with a flame ionization detector (GC–FID). The TCO data provided a quantitative distribution of organics boiling between 100 and 300 °C and divided them into discrete temperature intervals. Each of five boiling point ranges was established on the basis of the analysis of a C_7–C_{17} *n*-alkane mixture representing a temperature distribution of 100–300 °C. Quantitative values in units of micro-

[1] Methylene chloride was selected primarily on the basis of the following criteria: (1) It is commonly referred to as the universal solvent or the one used most frequently in the extraction of semivolatile organics sorbed on polymeric sorbent media. Hence, the contaminant chemistry associated with this solvent system would be of the most use to resin users. (2) The physical and chemical properties of methylene chloride make it ideally suited for the extraction of semivolatile organics sorbed on polymeric sorbent media.

grams per gram (parts per million) were established by comparison with an *n*-decane standard. All GC–FID chromatographic analyses were performed with a Tracor 560 gas chromatograph fitted with a dual FID unit. A summary of pertinent instrumental operating conditions is contained in a previous publication (*2*).

All GC–MS analyses were performed on a Hewlett-Packard 5985 quadrupole mass spectrometer. Pertinent chromatographic and mass spectral operating parameters are described elsewhere (*2*). Spectra were collected and recorded in the total ion mode. Individual component spectra were manually compared with U.S. Environmental Protection Agency–National Institutes of Health (USEPA–NIH) library spectra to provide component identifications. All quantitative data were provided as referenced to the d_{10}-anthracene internal standard.

AMBERSORB XE RESINS. All instrumental analyses were identical to those employed in the analysis of the Amberlite XAD resins as described previously (*2*).

PUF. Each of the three 2.0-mL PUF extracts was subjected to GC–MS analyses. All analyses were conducted with a Finnigan OWA 1020 GC–MS system fitted with an SE-54 fused-silica capillary column. All spectra were collected in the total ion mode. Component spectra, both raw and background subtracted, were manually compared against USEPA–NIH library spectra to permit component identifications. Quantitative data were again provided by using d_{10}-anthracene as the internal standard.

Results and Discussion

Amberlite XAD Resins. Amberlite XAD resins are synthetic adsorbents structurally composed of a styrene–divinylbenzene copolymer. Because of the polymeric production process, users are cautioned that these resins do contain significant quantities of preservatives and monomers as received from the manufacturer, Rohm and Haas (*3, 4*).

Results of the gas chromatographic analyses for each of the Amberlite XAD–polymeric sorbents are shown in Figure 1. Extractable organic concentrations in micrograms per gram (parts per million) are depicted for each of five boiling point intervals spanning from 100 to 300 °C. As shown, results for each of the two Amberlite sorbents are significantly higher than the corresponding values for each of the Ambersorb resins.

Further analyses of representative extracts of each of the Amberlite resins employing GC–MS indicated the presence of significant concentrations of a variety of aromatic hydrocarbons, including alkylated derivatives of benzene, styrene, naphthalene, and biphenyl. A more comprehensive listing of these contaminants, including their approximate concentrations in the sorbent matrix, is provided in Table I.

Although results for the two Ambersorb resins are in good qualitative agreement, the contaminant concentrations are consistently much higher for the Amberlite XAD–4 resin. This trend is directly attributable to the higher surface area of 725 m^2/g associated with the XAD–4

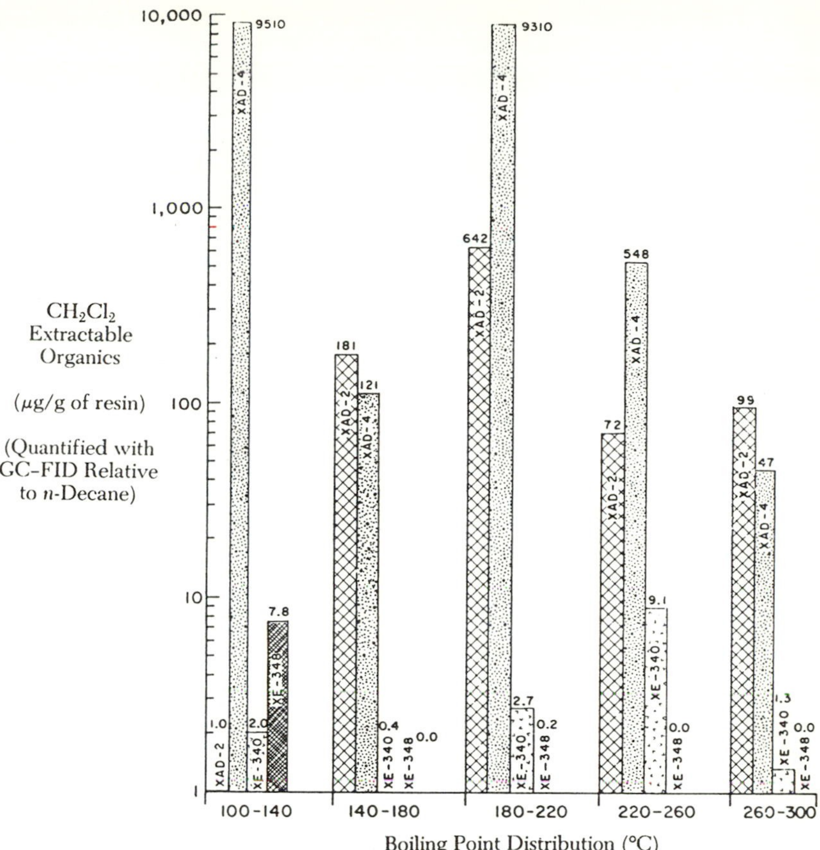

Figure 1. Comparison of extractable organics in four commercially available synthetic adsorbents.

sorbent beads as opposed to the 300-m^2/g area associated with the XAD-2 polymer (*3, 4*). At times, these properties have adversely affected the use of the Amberlite XAD-4 resin in environmental sampling schemes (*5, 6*). In both instances, the contaminants listed in Table I show a marked structural similarity to the characteristic parent structure of the Amberlite polymer repeating unit, as shown in Chart I. The predominance of benzene, styrene, naphthalene, and biphenyl derivatives suggests that the extractable contaminants are either residuals from the resin manufacturing process (e.g., starting materials or secondary byproducts) or artifacts from the degradation of the polymer itself during storage and handling, subsequent to the manufacturing process. In either case, prospective resin users are cautioned that XAD resins must undergo rigorous cleanup prior to use in actual environmental sampling regimes. The most widely accepted cleanup procedures are

Table I. Chemical Characterization and Quantitation of Organic Contaminants Extracted from Amberlite XAD Resins

	Concentration ($\mu g/g$)	
Compound Name	*XAD-2*	*XAD-4*
Methylbenzene (toluene)	nd	10000
Dimethylbenzene isomer	nd	53
Diethylbenzene isomer	102	nd
4-Ethyl-1,2-dimethylbenzene	d	d
1-Ethyl-2,3-dimethylbenzene	d	490
Triethylbenzene isomer	d	nd
Ethenylbenzene (styrene)	19	d
1-Ethenyl-4-ethylbenzene	320	5700
1-Ethenyl-3-methylbenzene	15	1800
1-Ethenyl-3,5-dimethylbenzene	d	nd
2-Ethenyl-1,3-dimethylbenzene	d	nd
1,4- or 1,3-Diethenylbenzene	45	5960
1- or 3-Methylindene	67	3470
Naphthalene	470	6870
1- or 2-Ethylnaphthalene	d	710
1- or 2-Methylnaphthalene	19	1020
1,1′-Biphenyl	69	1300
2- or 3-Methyl-1,1′-biphenyl	d	90
1,1′-Biphenyl, dimethyl isomer	d	nd
1,1′-Methylenebis[benzene]	d	nd
1,1′-Ethylidenebis[benzene]	130	55
1,1′-(1,2-Ethenediyl)bis[benzene]	d	67
1,1-Bis(*p*-ethylphenyl)ethane	26	nd

NOTE: d denotes component was detected but not quantitated; nd denotes component was not identified in the lots examined ($< 5\ \mu g/g$).
SOURCE: Adapted from reference 2.

those employing sequential solvent extraction in a Soxhlet apparatus (7, 8). As shown in Figure 2, a continuous extraction scheme employing a sequence of water, methanol, and methylene chloride can virtually eliminate chromatographable organic extractables associated with the sorbent matrix. Amberlite resin contamination is qualitatively consistent from lot to lot as received from the manufacturer. This finding is perhaps attributable to the patented synthetic process employed by Rohm and Haas in the manufacture of the Amberlite resin product line.

Ambersorb Resins. Chromatographable residues attributable to each of the Ambersorb resins, XE-340 and XE-348, as shown in Figure 1, are significantly lower than the values reported for each of the Amberlite resins. As shown in Figure 3, the majority of the components isolated from the Ambersorb XE-340 resin are readily classified as polycyclic aromatic hydrocarbons. (These are believed to be associated

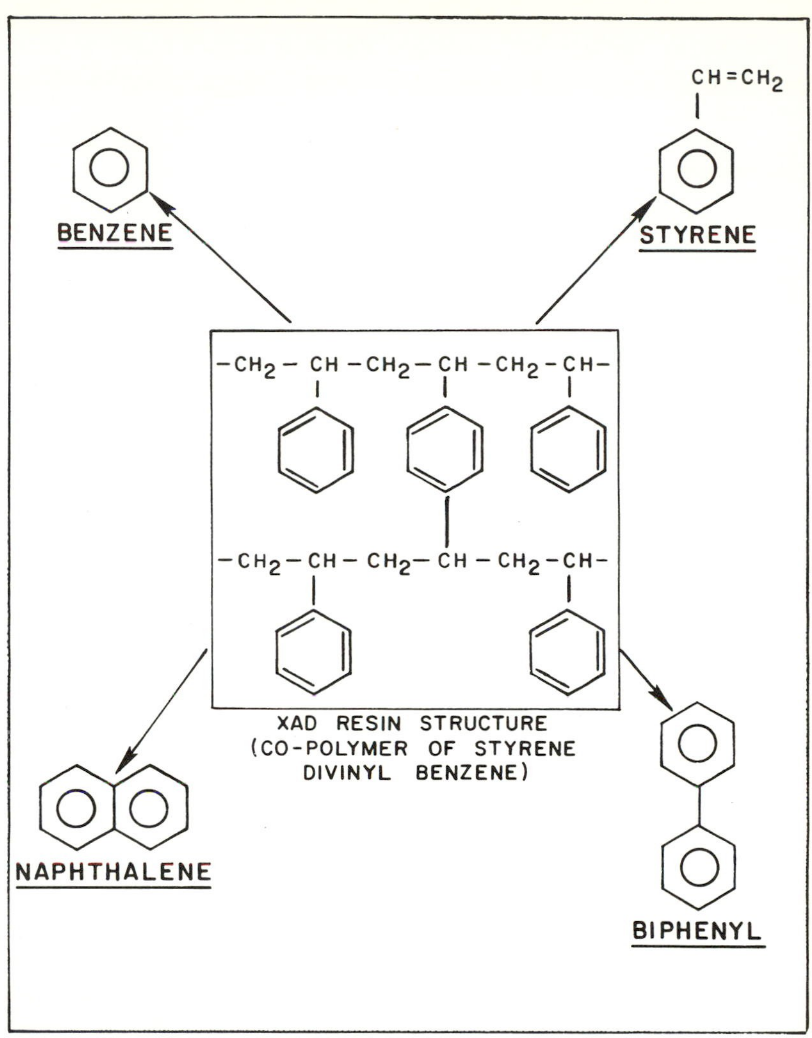

Chart I. Comparison of XAD-2 polymer repeating unit with parent structures of typical contaminant species. (Reproduced with permission from reference 1. Copyright 1980 Marcel Dekker.)

with the sorbent manufacturing process.) No discernible constituents were noted in the Ambersorb XE-348 extract (not shown) which closely approximated the accompanying laboratory method blank.

Both Ambersorb XE-340 and XE-348 are members of a carbonaceous polymer product line currently manufactured exclusively by Rohm and Haas. The chemical composition of these sorbents is generally regarded to be intermediate between that ascribed to either activated carbon or a purely polymeric sorbent (*9, 10*).

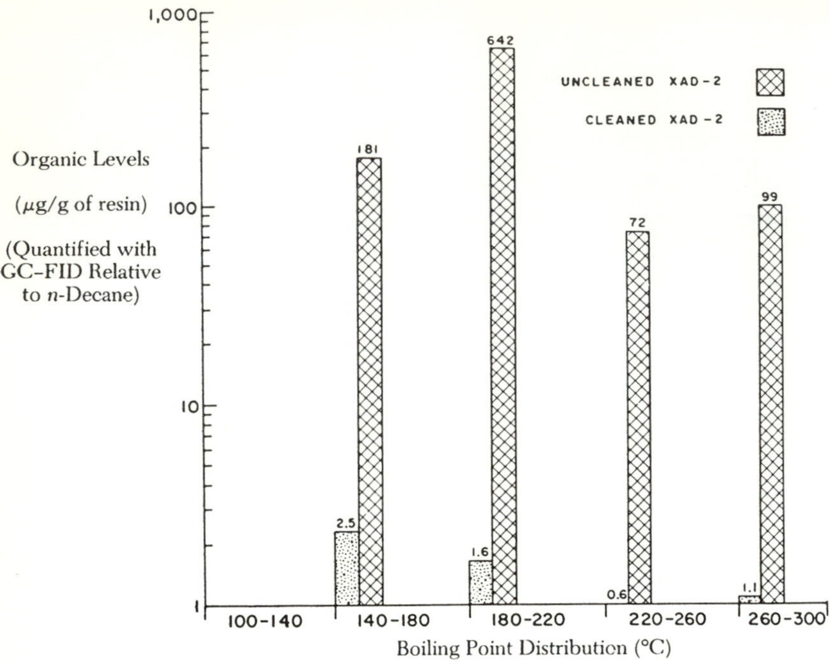

Figure 2. Comparison of extractable organic (CH_2Cl_2) levels in uncleaned and cleaned XAD-2 resin.

The patented Ambersorb manufacturing process relies on the carbonization of a macroreticular styrene-divinylbenzene-based copolymer (*11*). At the outset, the copolymer starting materials are sulfonated to render them infusible. The actual production process proceeds via a partial pyrolysis of the starting polymer at a prescribed temperature. When pyrolysis proceeds at a relatively low temperature (300–400 °C), the polymeric structure is retained and subsequently superimposed on the carbonaceous starting material. This process is consistent with that used in the actual manufacture of Ambersorb XE-340 resin. The fused-ring aromatic hydrocarbons noted during GC-MS analysis of the Ambersorb XE-340 extractable fraction may have been formed during this carbonization process. Perhaps they were formed via a free radical condensation involving the monomeric precursors (e.g., benzene and toluene) known to be associated with styrene-divinylbenzene copolymers. In fact, experimental data collected during this carbonization process have indicated that toluene and styrene account for the majority of the volatile hydrocarbon fraction released during the initial stages of the carbonization process (*11*). The actual Ambersorb synthetic process is assumed to proceed from an original styrene-divinylbenzene polymer backbone via a series of free radical combinations and electrolytic reactions. The postulated chemical structures for

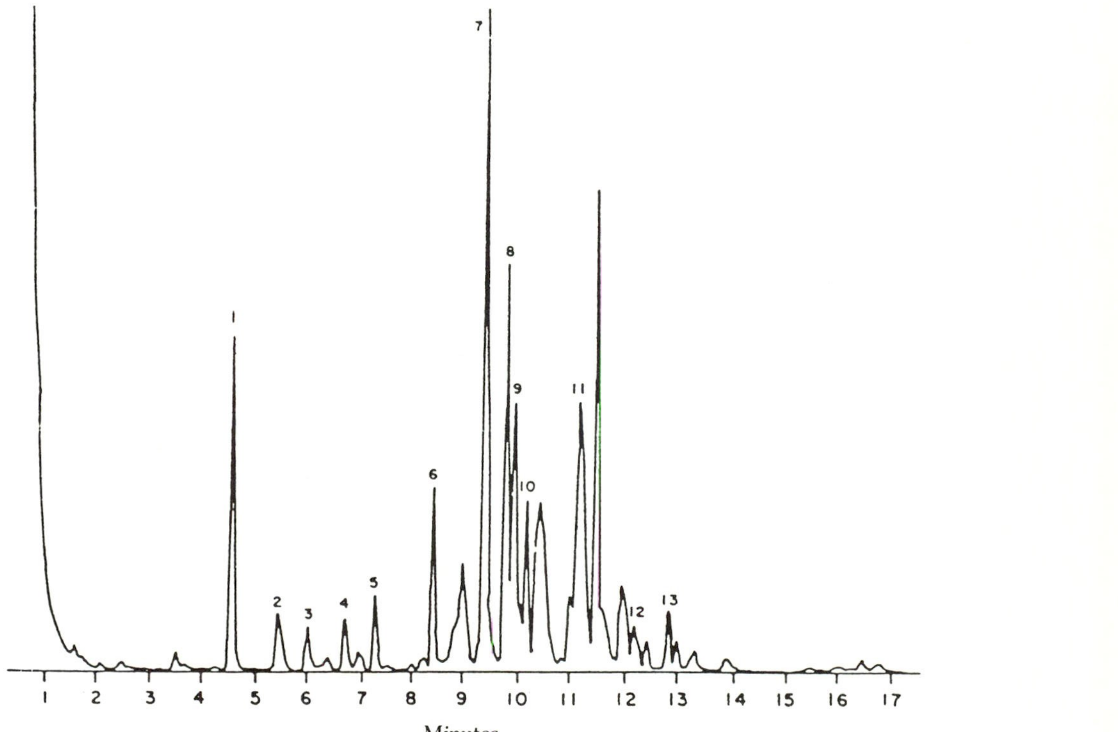

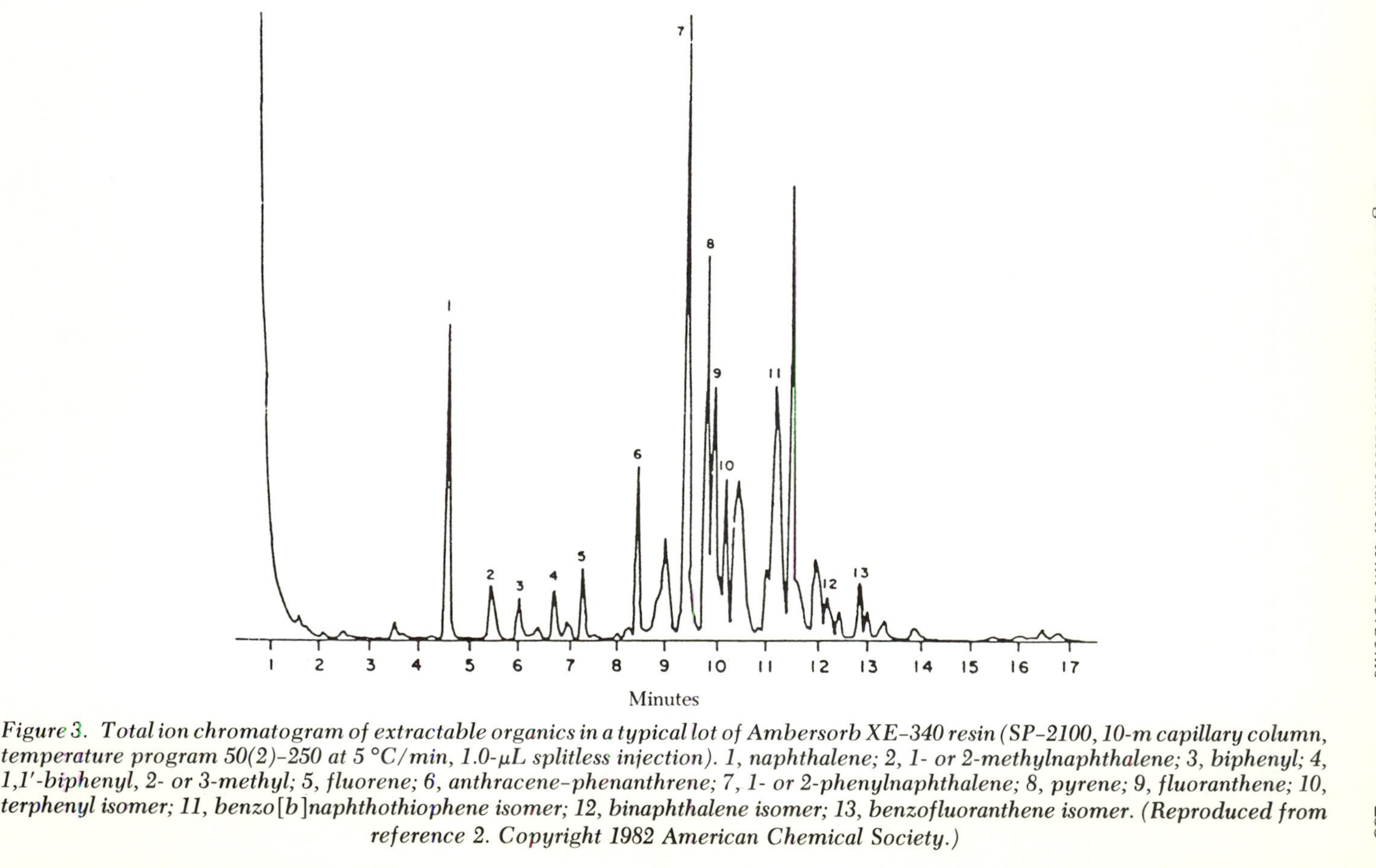

Figure 3. Total ion chromatogram of extractable organics in a typical lot of Ambersorb XE–340 resin (SP–2100, 10-m capillary column, temperature program 50(2)–250 at 5 °C/min, 1.0-μL splitless injection). 1, naphthalene; 2, 1- or 2-methylnaphthalene; 3, biphenyl; 4, 1,1′-biphenyl, 2- or 3-methyl; 5, fluorene; 6, anthracene–phenanthrene; 7, 1- or 2-phenylnaphthalene; 8, pyrene; 9, fluoranthene; 10, terphenyl isomer; 11, benzo[b]naphthothiophene isomer; 12, binaphthalene isomer; 13, benzofluoranthene isomer. (Reproduced from reference 2. Copyright 1982 American Chemical Society.)

some of these end products resulting from treatment of the starting copolymer at a variety of elevated temperatures under pyrolysis conditions (N_2) are shown in Scheme I. [Note the increased condensation and aromaticity of the end product in proceeding from Structure **I** to Structure **III.** Structure **I** is believed to represent the actual chemical structure of an Ambersorb XE-340 resin (*11*).] Similar reaction trends could account for the fused-ring aromatics isolated from the Ambersorb XE-340 product. Conversely, Ambersorb XE-348, which is produced via pyrolysis at a much higher temperature of 700 °C, shows no evidence of the presence of these same lower molecular weight fused-ring aromatics. At these higher temperatures, the polymeric properties of the sorbent are diminished, and the end product more closely approximates the properties of a carbonaceous sorbent.

PUF. Flexible PUF constitutes a generic class of polymeric sorbents that have been used extensively in recent years in a variety of environmental sampling applications. PUF, unlike other polymeric sorbents, which are generally manufactured by a single patented process, can be synthesized via any one of a number of patented processes. The most common synthetic pathways employ organic isocyanates (aliphatic or aromatic) and a polyol as starting materials, as shown in Scheme II. In addition, a number of chemical additives (*see* box) are generally introduced during the synthetic process to impart particular chemical or physical properties to the final foam product, as shown in Scheme III (*12*, *13*). High molecular weight halogenated organics, for example,

Ingredients Commonly Used in the PUF Manufacturing Industry

Diisocyanates

2,4-Toluene diisocyanate
2,6-Toluene diisocyanate
4,4′-Diphenylmethane diisocyanate
Hexamethylene diisocyanate

Blowing Agents (halogenated alkanes–water)

Trichlorofluoromethane
Methylene chloride

Fire Retardants

Tris(2-chloroethyl) phosphate
Tris(2,3-dibromopropyl) phosphate
Diammonium phosphate
Antimony oxides

Foam Stabilizers–Surface Active Agents

Silicone oils
Silicone–glycol copolymers

Cross-Linking Agents (triols–polyethers–alkanolamines)

Glycerol
Triethanolamine
Pentaerythritol

Catalyst(s) (tertiary amines)

Triethylenediamine
Triethylamine
Dimethylpiperazine
N,N-Dimethylcyclohexylamine

SOURCE: Adapted from references 12 and 13.

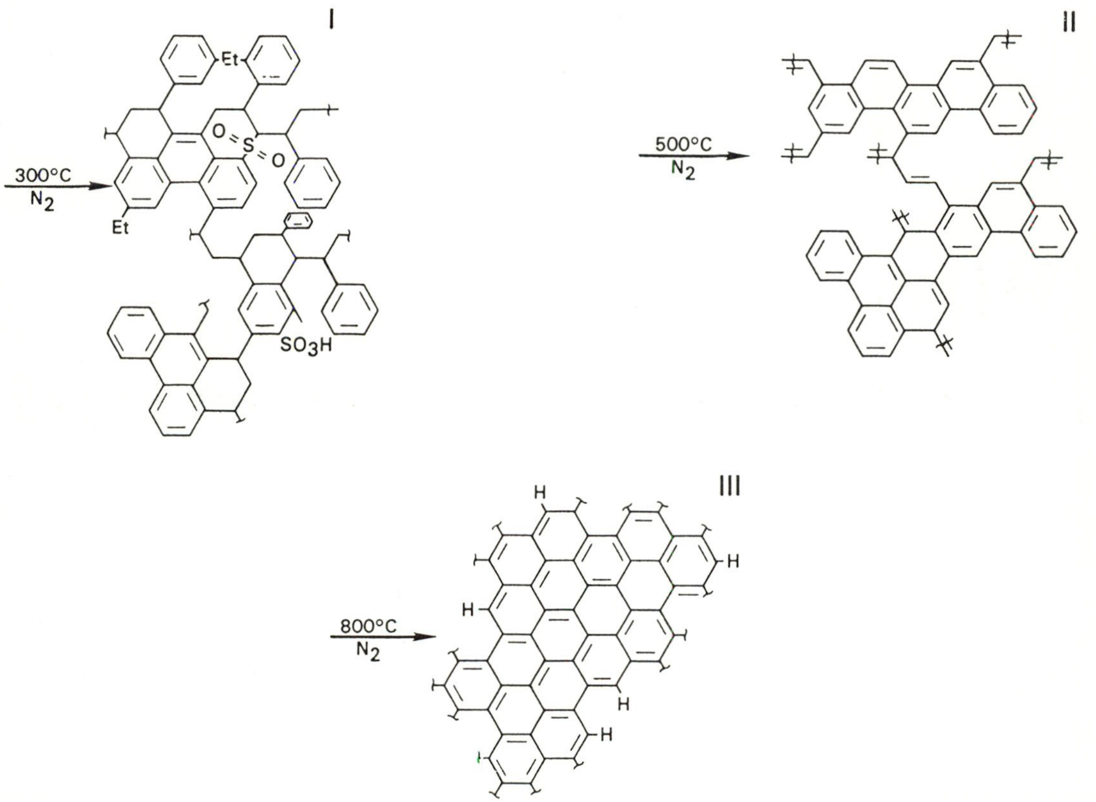

Scheme I. Chemical structures postulated for Ambersorb-type carbonaceous adsorbents (reaction products resulting from the thermal treatment of a styrene–divinylbenzene copolymer).

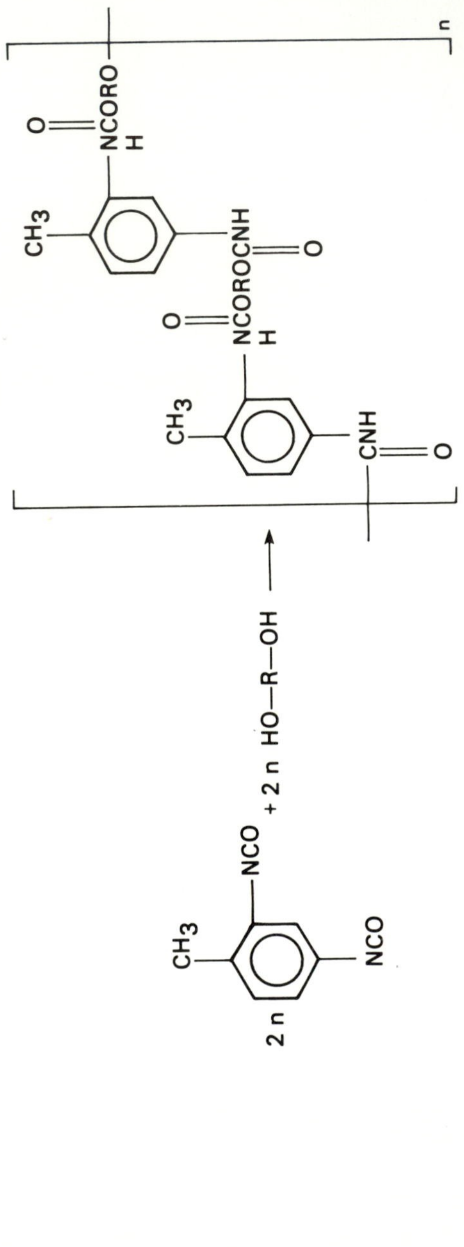

Scheme II. The synthesis of PUF: typical reaction sequence involving an aromatic isocyanate (TDI) and a triol.

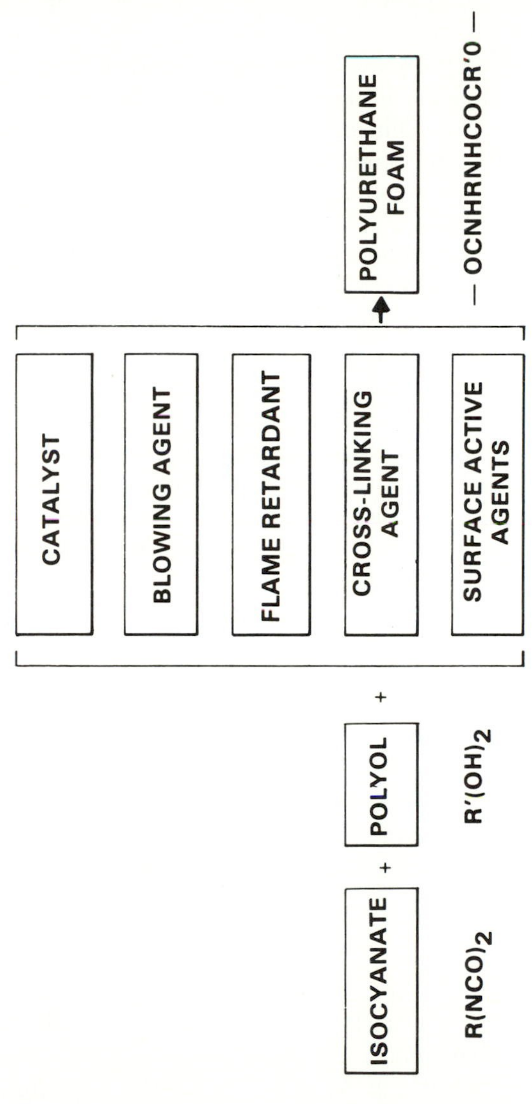

Scheme III. PUF manufacturing process: general synthetic pathway.

are typically added to impart fire-retardant properties to the foam (*13*). Some of the more commonly used chemical ingredients are listed in the box. Thus, PUF products can be expected to contain a number of these chemical additives as well as other synthetic artifacts as received from the supplier. Furthermore, the quality of flexible foam products will vary markedly from supplier to supplier as well as from manufacturer to manufacturer.

PUF, unlike other polymeric sorbents, tends to be a nonhomogeneous product containing a number of additives and artifacts in variable quantities from lot to lot. Our experience, however, indicates that these contaminants can be sufficiently reduced to permit trace organic analysis by employing a sequential solvent extraction procedure in conjunction with stringent quality control criteria prior to actual use. This observation is consistent with the experience of other investigators who have used flexible foams extensively in analytical environments.

A reconstructed ion chromatogram (GC–MS) containing extractable contaminants isolated from a typical lot of foam is shown in Figure 4. The qualitative composition of the extractable contaminants was provided by GC–MS. Contaminant profiles were identical for each of the two solvent systems employed, methylene chloride (100%) and ethyl ether/hexane (5/95). The contaminant chemistry shown here and again in Figure 5 in several instances is consistent with the manufacturing process data shown in the box, most notably the presence of residual toluene diisocyanate (starting materials, *see* Scheme II) and an aliphatic amine (possible reaction catalyst).

Because of the wide diversity in PUF manufacturing processes and likely contaminant chemistry, users are cautioned that sorbent quality control is more critical than with other synthetic polymers such as the Amberlite XAD series. Every effort should be made to procure PUF products consistently from the same manufacturer, preferably in each instance from the same production lot. Moreover, because of inconsistencies in manufacturing practices, first-time foam users should solicit the advice of other satisfied and experienced users in the selection of a sorbent supplier.

Conclusions

Chemical analyses of a variety of commercially available synthetic adsorbents indicate that significant quantities of extractable organics are present in these products as received from the manufacturer. The majority of these contaminants can be classified as either residual monomers, synthetic artifacts, or preservatives peculiar to the manufacturing process, packaging, storage, and handling of the sorbent itself. The manufacturing process is alleged to be the most significant contributor to the levels of extractable organics present in these materials.

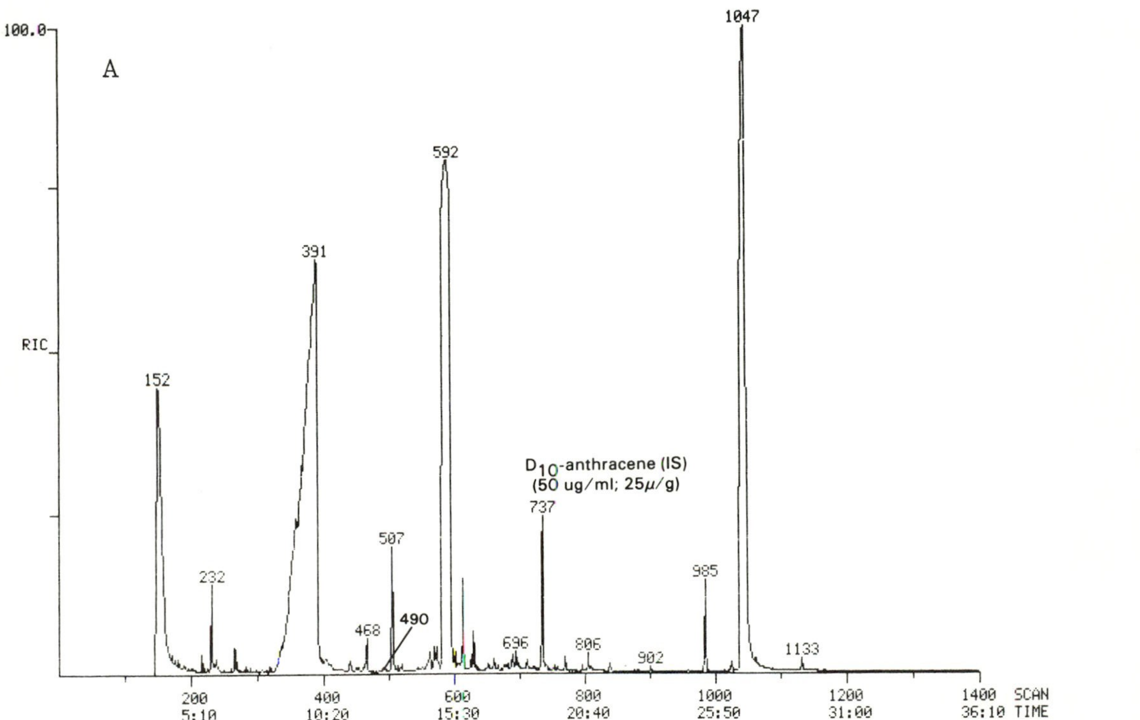

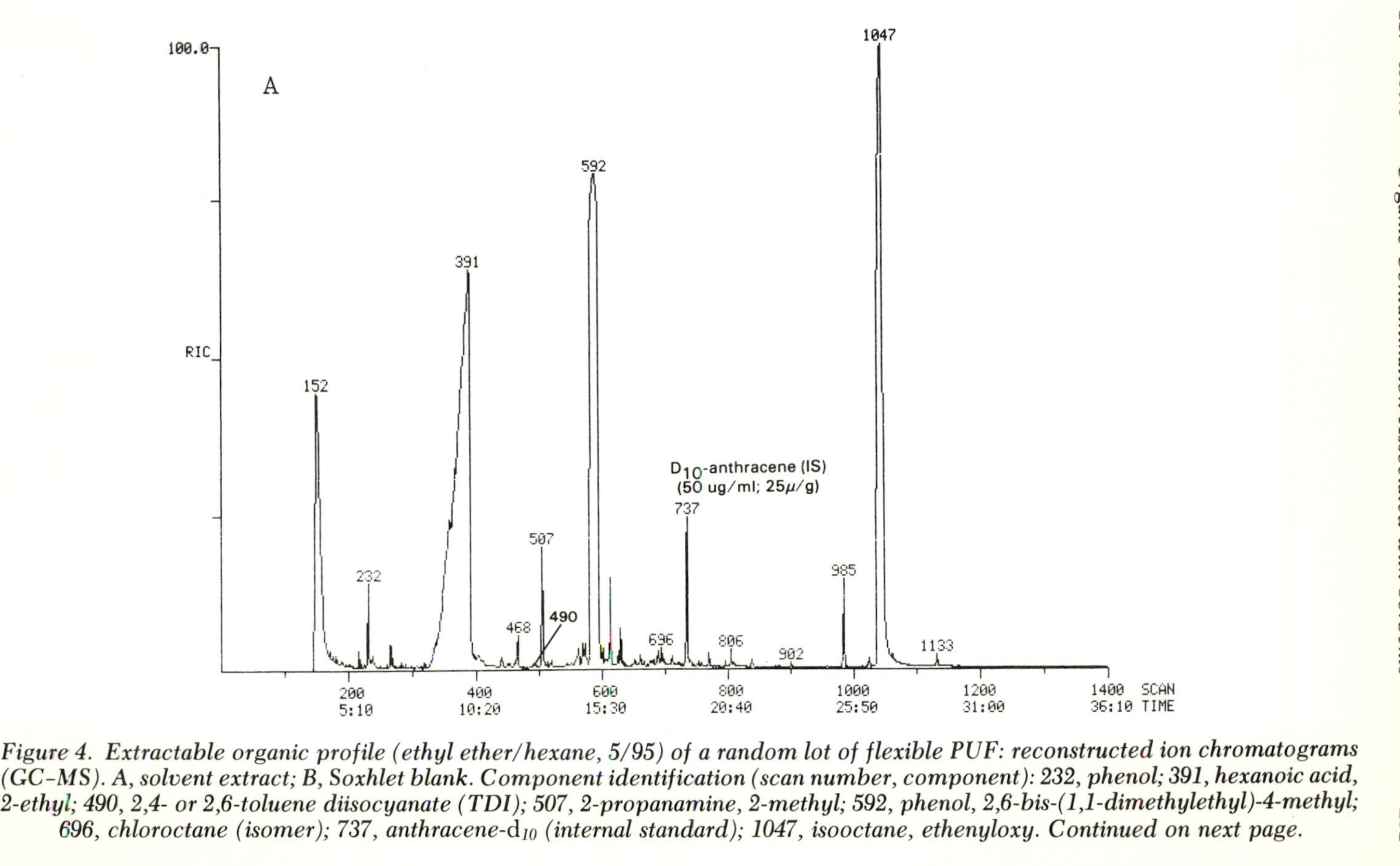

Figure 4. Extractable organic profile (ethyl ether/hexane, 5/95) of a random lot of flexible PUF: reconstructed ion chromatograms (GC–MS). A, solvent extract; B, Soxhlet blank. Component identification (scan number, component): 232, phenol; 391, hexanoic acid, 2-ethyl; 490, 2,4- or 2,6-toluene diisocyanate (TDI); 507, 2-propanamine, 2-methyl; 592, phenol, 2,6-bis-(1,1-dimethylethyl)-4-methyl; 696, chloroctane (isomer); 737, anthracene-d_{10} *(internal standard); 1047, isooctane, ethenyloxy.* *Continued on next page.*

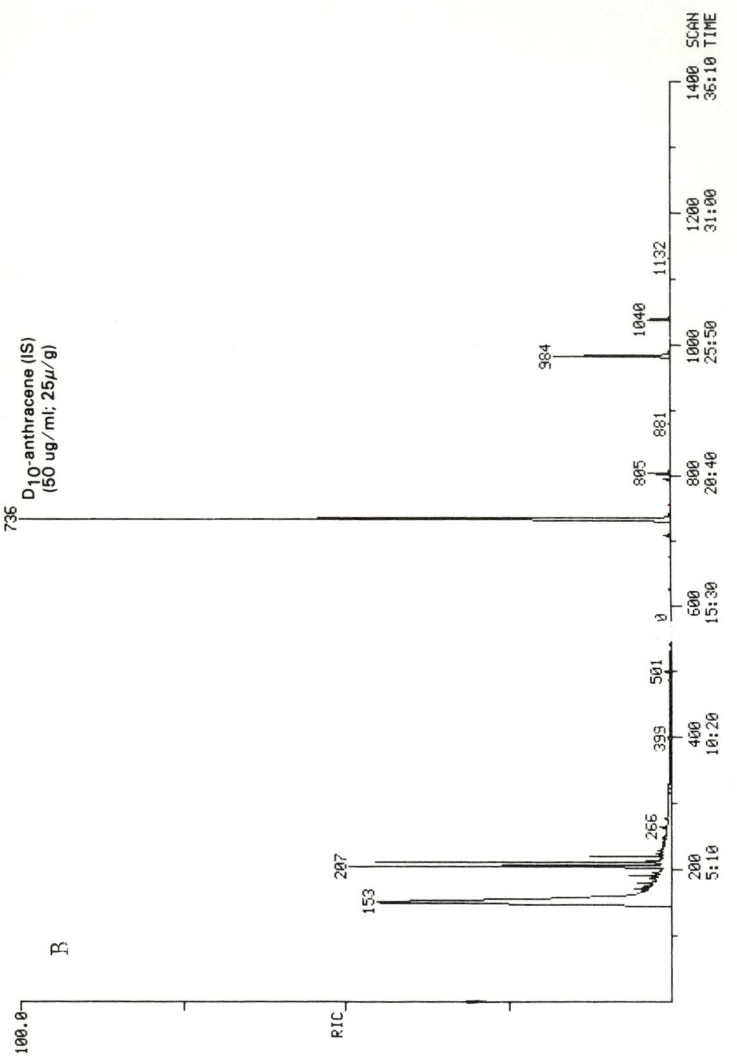
B
100.0
RIC
153
207
266
399
501
736
805
881
984
1040
1132
D10-anthracene (IS)
(50 ug/ml; 25μ/g)
200 5:10
400 10:20
600 15:30
800 20:40
1000 25:50
1200 31:00
1400 36:10
SCAN
TIME

Figure 4B.

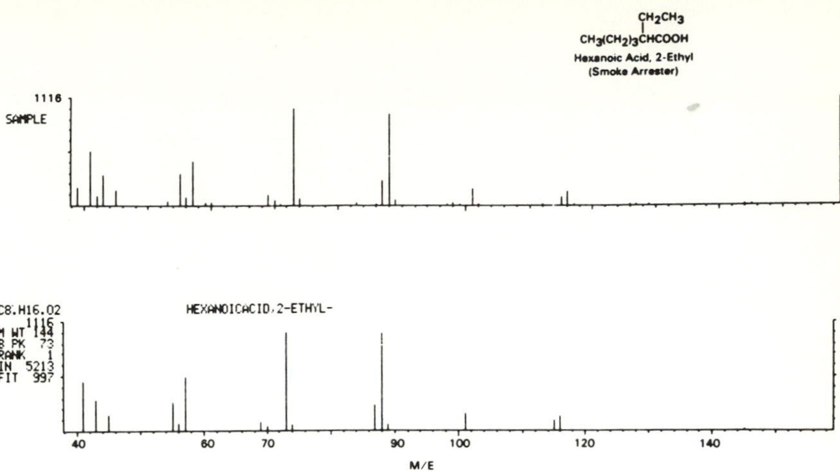

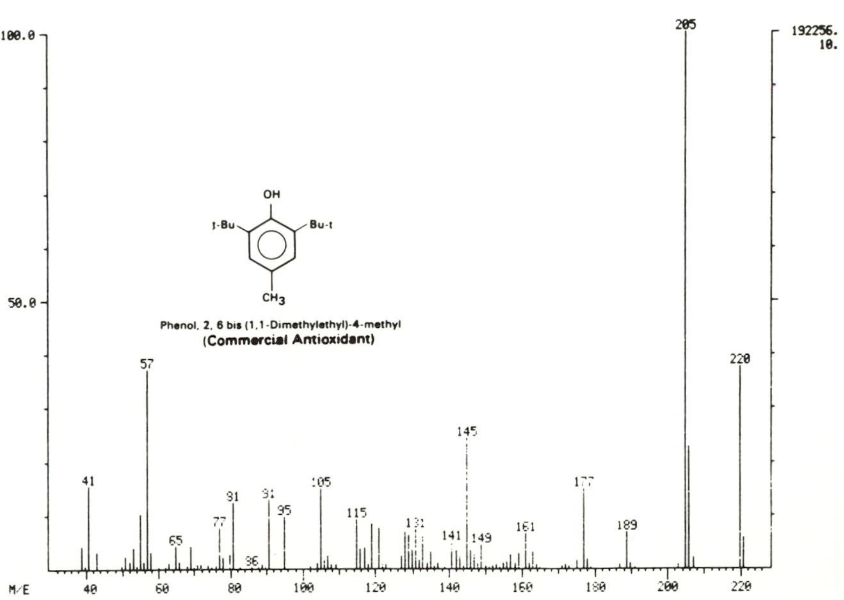

Figure 5. Mass spectral data: selected organic extractables (methylene chloride) isolated from a random lot of PUF.

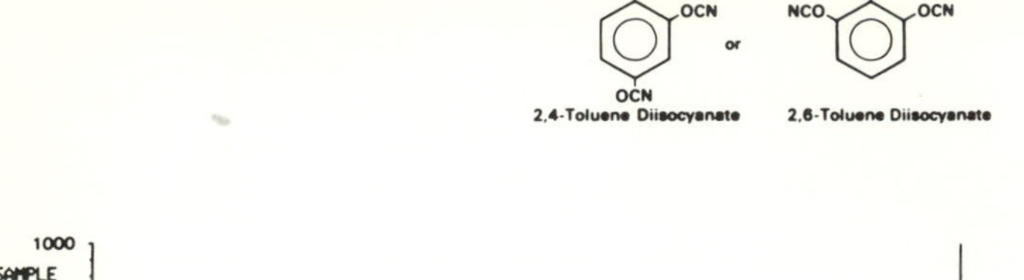

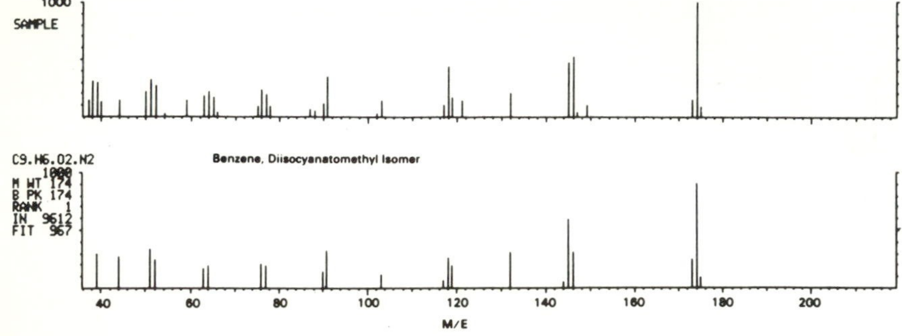

Figure 5. Continued.

Amberlite XAD-2 and XAD-4 resins, for example, contain significant quantities of alkyl derivatives of benzene, styrene, naphthalene, and biphenyl as received from the supplier. PUF products, on the other hand, generally contain numerous contaminants peculiar to one of the several patented commercial manufacturing processes. These include, but are not limited to, the following classes of chemical contaminants: isocyanate derivatives (e.g., toluene diisocyanates), alkyl amines, aliphatic acids, and brominated aromatics (e.g., fire retardants).

The native sorbent contamination should be reduced significantly prior to use of the collection media in actual aquatic sampling schemes. In practice, these materials cannot be removed entirely but can only be lowered sufficiently so as to permit subsequent analysis and to achieve the requisite detection limits. In all instances, users of polymer-type sorbents are urged to familiarize themselves with sorbent contaminant chemistry in order to recognize spurious data points arising from the use of improperly cleaned or mishandled resins. Furthermore, rigorous quality control measures, including field blanks, and sorbent contamination tolerance criteria should be instituted to both curtail and permit recognition of contamination arising from sorbent use.

We have found that rigorous sorbent pretreatment procedures (e.g., Soxhlet extraction and thermal desorption) in concert with a well-established quality control program will successfully control potential contamination effects arising from the sample collection media. Furthermore, a well-executed quality control program will permit identification of spurious data points attributable to media contamination when and if they do occur.

Literature Cited

1. Hunt, G. T.; Pangaro, N.; Zelenski, S. G. *Anal. Lett.* **1980,** *13(A7),* 521.
2. Hunt, G. T.; Pangaro, N. *Anal Chem.* **1982,** 54, 369.
3. *Amberlite XAD-2 Technical Bulletin;* Rohm and Haas: Philadelphia, 1978.
4. *Amberlite XAD-4 Technical Bulletin;* Rohm and Haas: Philadelphia, 1978.
5. Adam, J.; Menzies, K.; Levin, P. EPA Report No. 600/7-77-044, Arthur D. Little, Inc., 1977.
6. Thurman, E. M.; Aiken, G. R.; Malcolm, R. L. *Proceedings of the Fourth Conference on Sensing of Environmental Pollutants;* New Orleans, LA, 1978.
7. Lentzen, D. E. et al. *IERL/RTP Procedures Manual: Level 1 Environmental Assessment;* Environmental Protection Agency. U.S. Government Printing Office: Washington, DC, 1978; EPA-600/7-78-201.
8. Junk, G. A. et al. *J. Chromatogr.* **1974,** 99, 745.
9. *Ambersorb Carbonaceous Adsorbents;* Rohm and Haas: Philadelphia.
10. Neely, J. W. *Proceedings Division of Environmental Chemistry,* American Chemical Society National Meeting, Miami, FL; American Chemical Society: Washington, DC, 1978; paper 72, pp 204-205.
11. *Carbonaceous Adsorbents for the Treatment of Ground and Surface Waters;* Neely, J. F.; Isacoff, E. G., Eds.; Marcel Dekker: New York, 1983; pp 41-78.
12. Edwards, G. D.; Rice, D. M.; Soulen, R. L. U.S. Patent 3 297 597, 1967.
13. *Kirk-Othmer Encyclopedia of Chemical Technology;* John Wiley: New York, 1981; 3rd ed., Vol. 23.

RECEIVED for review August 14, 1985. ACCEPTED January 28, 1986.

12

An Evaluation of the Preparation of Resin Samplers for Broad Spectrum Analysis of Large-Volume Samples

J. Gibs[1], L. Brenner, L. Cognet[2], and I. H. Suffet

Environmental Studies Institute, Drexel University, Philadelphia, PA 19104

XAD resin cleaning by exhaustive extraction with a series of solvents was studied to observe if the resin cleaned in this manner was acceptable for use in broad spectrum capillary gas chromatographic (GC) analysis. Resin artifacts were always found in distilled water blanks after cleaning and after recleaning and reuse of the resin. The artifacts were identified by GC–mass spectrometry and compared to artifacts previously reported. Variations in artifacts were found between production lots. Several hypotheses are proposed for the existence of the resin artifacts after cleaning. Recommendations are given for storage, cleaning, blanking, and reusing resins for broad spectrum analysis of environmental samples.

RESIN SAMPLERS are used to isolate nanogram to microgram amounts of organic compounds from samples of natural and treated waters. Sampling of waters by this method can be categorized as either low-volume sampling (<10 L) or large-volume, broad spectrum sampling (>10 L). Table I indicates the sampling methods used during past environmental studies (*1–25*).

Low-volume sampling is typically used for quantitative analysis for specific compounds, such as pesticides, polychlorobiphenyls (PCBs), and organic acids. Large-volume sampling is used to simultaneously

[1] Current address: U.S. Geological Survey, Water Resources Division, Trenton, NJ 08628

[2] Permanent address: Lyonnaise des Eaux, Central Laboratory, 38 rue de Président Wilson, 78239 Le Pecq, France

0065–2393/87/0214/0267$07.75/0

Table I. Resin Cleanup and Elution Procedures for Resin Sampling

Sample Size and Compounds Studied	*Cleanup*	*Elution*	*Ref*
Model compounds and drinking water (<10 L and >10 L)	—	(1) 20 mL of 0.05 M $NaHCO_3$ (2) 20 mL of 0.05 M NaOH (3) 100 mL of methanol alternate method: (1) 100 mL of 1 M HCl (2) 100 mL of 0.05 M HCl (3) 100 mL of 0.05 M NaOH (4) 15 mL of ethyl ether	1
Broad spectrum analysis of drinking water (>10 L)	(A) 6–8× wash with ethyl ether (B) 6–8× wash with methanol	15–100 mL of ethyl ether (one elution)	2
Raw and treated drinking water (<10 L)	(A) 25 mL of methanol (B) 3 × 15 mL of water	20 mL of ethyl ether	3
Chlorinated pesticides in seawater (>10 L)	(A) wash with distilled water (B) 2 × 18-h Soxhlet extraction with acetonitrile (C) 10 bed volumes of distilled water	(1) 1 × 4 bed volumes of acetonitrile or ethyl ether (2) 1 × 500 mL of water (3) combine 1 and 2 (4) extract 3 with 2 × 50 mL of *n*-hexane	4
110 Model compounds (<10 L)	8-h sequential Soxhlet extraction with methanol, acetonitrile, and ethyl ether	(1) 2 × 10 mL of ethyl ether (10 min) (2) 5 mL of ethyl ether (10 min) (3) 30 mL of methanol (4) store in methanol	5
Model compounds in seawater (>10 L)	—	(1) wash with 2 L of distilled water (2) 8-h Soxhlet extraction with mixture of 150 mL of methanol and 100 mL of water	6

		(3) extract methanol–water mixture with 3 × 25 mL of *n*-hexane	
Model pesticides and well water (<10 L and >10 L)	*see* ref 5	3 × 15 mL of ethyl ether, 10 min between elutions, or 1 × 25 mL of ethyl ether, 10 min	7
EPA cleanup procedure for resin	(A) wash with 1 L of water (B) 8-h sequential Soxhlet extraction with water, methanol, and 2× methylene chloride alternate method: (A) rinse with 20 L of water (B) continuous extractor (20–40 mL/min) with methanol (10–20 h), then methylene chloride (10–20 h)	—	8
Organic acids (<10 L)	(A) wash 5× with 0.1 M NaOH (B) 24-h sequential Soxhlet extraction with methanol, acetonitrile, and ethyl ether	0.1 N NaOH	9
13 Model compounds (<10 L and >10 L)	wash successively (A) 1.5 L of acetone (B) 1.5 L of methanol (C) 1.5 L of methylene chloride or chloroform	(1) 10 mL of acetone (2) 40 mL of chloroform (3) combine 1 and 2 alternate method: (1) 15 mL of acetone (2) 40 mL of chloroform (3) combine 1 and 2	10
Model compounds in distilled water (<10 L)	(A) wash with tap water (B) Soxhlet extraction: water (8 h), methylene chloride (24 h) (C) store in methanol	continuous extractor: methylene chloride alternate method: Soxhlet extraction using methylene chloride or methanol (24 h)	11

Continued on next page.

Table I. Continued

Sample Size and Compounds Studied	*Cleanup*	*Elution*	*Ref*
Mutagens in natural samples (<10 L)	Soxhlet extraction	XAD-2 with benzene XAD-7 with petroleum ether	12
Mutagens in natural waters (>10 L)	(A) washed and Soxhlet extracted with acetone (16 h) (B) stored in acetone	(1) washed with 3 bed volumes of distilled water (2) residual water blown out with dry nitrogen (3) elution with 4 bed volumes acetone (140 mL)	13
Model compounds (<10 L and >10 L)	—	30 mL of ethyl ether	14
Model pesticides and river water (<10 L and >10 L)	—	ethyl ether or 10% acetone/90% ethyl ether	15
Broad spectrum analysis of chlorinated river and sea-water (>10 L)	*see* ref 5	(1) 125 mL of ethyl ether (2) concentrate and dissolve in benzene (3) fractionate by liquid chromatography	16
Model compounds and drinking water (>10 L)	sequential Soxhlet extraction with methanol, pyridine, and ethyl ether, 8 h each	*see* ref 5 and 7	17
Chlorinated pesticides and PCB model compounds and natural water samples	*see* ref 5	—	18
Broad spectrum analysis of drinking water (<10 L)	sequential Soxhlet extraction with methanol, acetonitrile, and ethyl ether	1 × 25 to 30 mL of ethyl ether	19

Chlorinated pesticides and phthalates in drinking water (>10 L)	(A) 250 mL of water (B) 250 mL of *n*-hexane (C) 250 mL of acetone (D) 250 mL of *n*-hexane (E) analyze by GC (F) if not clean, repeat D and E until clean	1 × 250 mL of *n*-hexane	20
9 Model compounds (<10 L)	(A) sequential Soxhlet extraction with methanol and ethyl ether (6 h) (B) 20 mL of ethyl ether (C) analyze B by GC (D) repeat B and C until clean (E) store in methanol	(1) 20 mL of ethyl ether (2) ethyl ether (10 min) (3) 2 × 20 mL of methanol	21
Nonpolar hydrocarbons in distilled water (>10 L)	sequential Soxhlet (4 h each) acetone, ethyl ether, and methanol	ethyl ether at 0.8 bed volumes/min for 2 bed volumes	22
Anthracene and *N,N*-dimethylthiourea (>10 L)	Soxhlet extraction with methanol and acetone	—	23
Dissolved organic carbon in natural seawater (>10 L and <10 L)	(A) rinse XAD–2 with water until free of chloride (B) sequential Soxhlet extraction with acetone, methanol, and water (24 h each) or (A) XAD–8 soaked in 0.1 M NaOH solution for 5 days, then same procedure as for XAD–2	(1) ammonium hydroxide solution (2) methanol, then ammonium hydroxide in methanol	24
Organic bases (<10 L)	Soxhlet extract: methanol, tetrahydrofuran, acetonitrile, acetone, ethyl ether, (10 h each), then dried at 50 °C overnight	15% methanol and 85% ethyl ether	25

isolate from a large sample as many trace organic compounds as possible. The extract from the large-volume sample can then be (1) chemically analyzed for as many compounds as possible (broad spectrum analysis) or (2) used as a medium for bioassay studies such as the Ames test.

Artifacts from the resin can interfere with the chromatographic analysis of the XAD resin extract. For example, the artifact may be a pollutant being studied, or coelution of the resin artifacts and compounds of interest may occur during capillary gas chromatographic (GC) analysis. Artifacts can also take part in competitive adsorption during sampling. This situation can cause sample breakthrough because certain compounds are preferentially collected.

Furthermore, many bioassay procedures can be affected synergistically or antagonistically by artifacts. In current practice, chemists use various solvents to clean the resin as completely as possible, run a resin blank, and chromatographically analyze the blank sample. Rarely are the artifacts identified. This procedure does not help the biologist who requires a blank that does not show a positive response in the bioassay test of the blank. Biologists usually clean the resin as the chemists do and complete a blank for the bioassay. Artifacts should be identified to assure that the bioassay results are in response to the compounds under study and not the artifacts in combination with the sample.

For these reasons, artifacts must be limited as part of a quality assurance program. As defined by the American Chemical Society Committee on Environmental Improvement (*26*), "the objective of a quality assurance program is to reduce measurement errors to agreed upon limits and to assure that the results have a high probability of being acceptable quality."

Cleaning and blanking procedures, therefore, must be carefully evaluated to assure reliable results from broad spectrum analyses of large volumes of water. Many questions arise about the reliability of present methods because researchers have based large-volume sampling procedures on results from low-volume procedures.

Early studies using resins for isolation and analysis of trace organics, such as pesticides, PCBs, and organic acids, from small volumes of water showed excellent recovery and the potential of easy application to environmental samples. Isotherm studies in distilled water were used to define the sampling parameters for quantitative analysis of these compounds. Later, studies using resin samplers for large-volume environmental samples were extrapolated from the early low-volume resin work of Junk et al. (*5, 14*) and Thurman et al. (*27*) (*see* Table I).

The cleaning and blanking of early large-volume samplers of Bean et al. (*16*), Harvey (*4*), McNeil et al. (*20*), and Osterroht (*6*) did not undergo quality assurance; observations and results from low-volume samplers were considered sufficient. In some cases, such as Van Rossum and Webb (*10*), a low-volume sampler was used to design a high-volume sampler.

The objective of this chapter is to evaluate and recommend procedures for cleaning, blanking, and reusing XAD resin samplers for broad spectrum analysis. Experimental studies were done to develop a methodology to minimize artifacts occurring after each procedure. The artifacts identified that reoccur during sampling processes are described. This work will help develop quality assurance procedures for large-volume XAD resin sampling programs.

Chemical and Physical Properties of XAD Resins

XAD-2 resin is a styrene-divinylbenzene copolymer that has a highly aromatic structure. It has a surface area of about 300 m^2/g, an average pore diameter of 90 Å, and a maximum pore diameter of 290 Å. XAD-8 resin is a methyl methacrylate copolymer and has a slightly more hydrophilic surface than XAD-2. It has a surface area of about 140 m^2/g, an average pore diameter of 250 Å, and a maximum pore diameter of 375–840 Å. Each exhibits different adsorption characteristics because of structural and surface chemistry differences. Both XAD resins have a significant number of micropores $<$25 Å in diameter. The adsorption area is distributed biomodally in XAD-8 and has maximums at 22.5 Å or less and at 375 Å. XAD-2 has a maximum at 22.5 Å or less and a slight increase at 240 Å (*11*). This result indicates that XAD-2 adsorption is initially more diffusion-limited than XAD-8.

The advantage of using these resins according to the manufacturer (Rohm and Haas) (*28*, *29*) is that adsorption is due to van der Waals forces. Therefore, simple elution or washing of the resin bed with an appropriate solvent and/or water of different pH will recover the adsorbed solute. This feature makes the use of resin appear attractive for sampling large volumes of water containing trace concentrations of organic compounds.

Experiments and Results

Resin Cleaning Procedure. Prior to use in a sampler, the resins must be cleaned to remove preservatives and the monomeric chemical residues that are byproducts of their production. The cleaning procedure recommended by Rohm and Haas is the following (*29*):

> The resin is backwashed with water to remove fines as well as the sodium chloride and sodium carbonate used to inhibit growth of bacteria and mold during storage. After backwashing, the resin remains covered with water and is then drained while methanol or acetone is

added, making sure that the resin is never dry. Four bed volumes of methanol (or acetone) is added at a rate of 4 bed volumes per hour. After the methanol rinse, 4 bed volumes of water is added at the same flow rate as the methanol to remove the methanol.

Junk et al. (*5*) found that the Rohm and Haas procedure (*29*) did not clean the resin sufficiently for use as an adsorbent for recovering nanogram-per-liter amounts of organic materials. Junk's resin preparation and cleaning procedure started with backwashing the resin with water to remove fines, sodium chloride, and sodium carbonate.

Table I shows that most investigators used exhaustive extraction with a series of solvents of decreasing polarity to remove resin contaminants. The procedure of Junk et al. (*5*) seems to have gained wide acceptance for preparing resins for sampling water for small amounts of organics.

This study used the Junk et al. (*5*) procedure for resin cleaning beginning with a water backwash. The remaining steps consisted of sequential Soxhlet extraction of 24-h duration for each of the following solvents: methanol, acetonitrile, and ethyl ether. This process was followed by a final rinse of 2 bed volumes of 1 N sodium hydroxide. Ethyl ether was replaced by whatever sample eluting solvent was to be used. Approximately 700 mL of wet resin was Soxhlet extracted with 900 mL of solvent. The solvent distillation rate was approximately 30 mL/min. After extraction, each of the unconcentrated solvents was analyzed by capillary GC with a flame ionization detector (FID) and GC–mass spectrometry (GC–MS).

Figure 1 and Table II show the results of GC analysis and GC–MS identification of XAD-2 and XAD-8 resin blanks, respectively. The compounds identified by GC–MS analysis (Table II) are suspected of being resin contaminants because they are monomers and polymer fractions of the resins. These compounds are compared in Table III to those previously identified in XAD-2 blanks by Hunt et al. (*30*), Melton et al. (*31*), Lochmueller and Jensen (*32*), Care et al. (*22*), and Fritz (*19*). Table III shows that each of the compounds identified in the XAD-2 resin blanks has been confirmed to be present by other workers. The compounds identified in the XAD-8 resin blanks were found by other workers also to be present in XAD-2 blanks.

Figure 1 shows that the final XAD-2 and XAD-8 resin Soxhlet extraction by ether contains 46 and 17 GC peaks, respectively. Of these GC peaks, 33 (XAD-2) and 12 (XAD-8) were observed within the first third of the chromatograms. The number of GC peaks in the final cleanup step was felt to be too many for broad spectrum sampling and analysis because the concentrates of the sample extracts are at least 200-fold more concentrated.

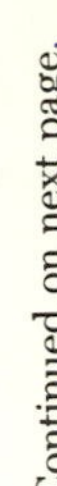

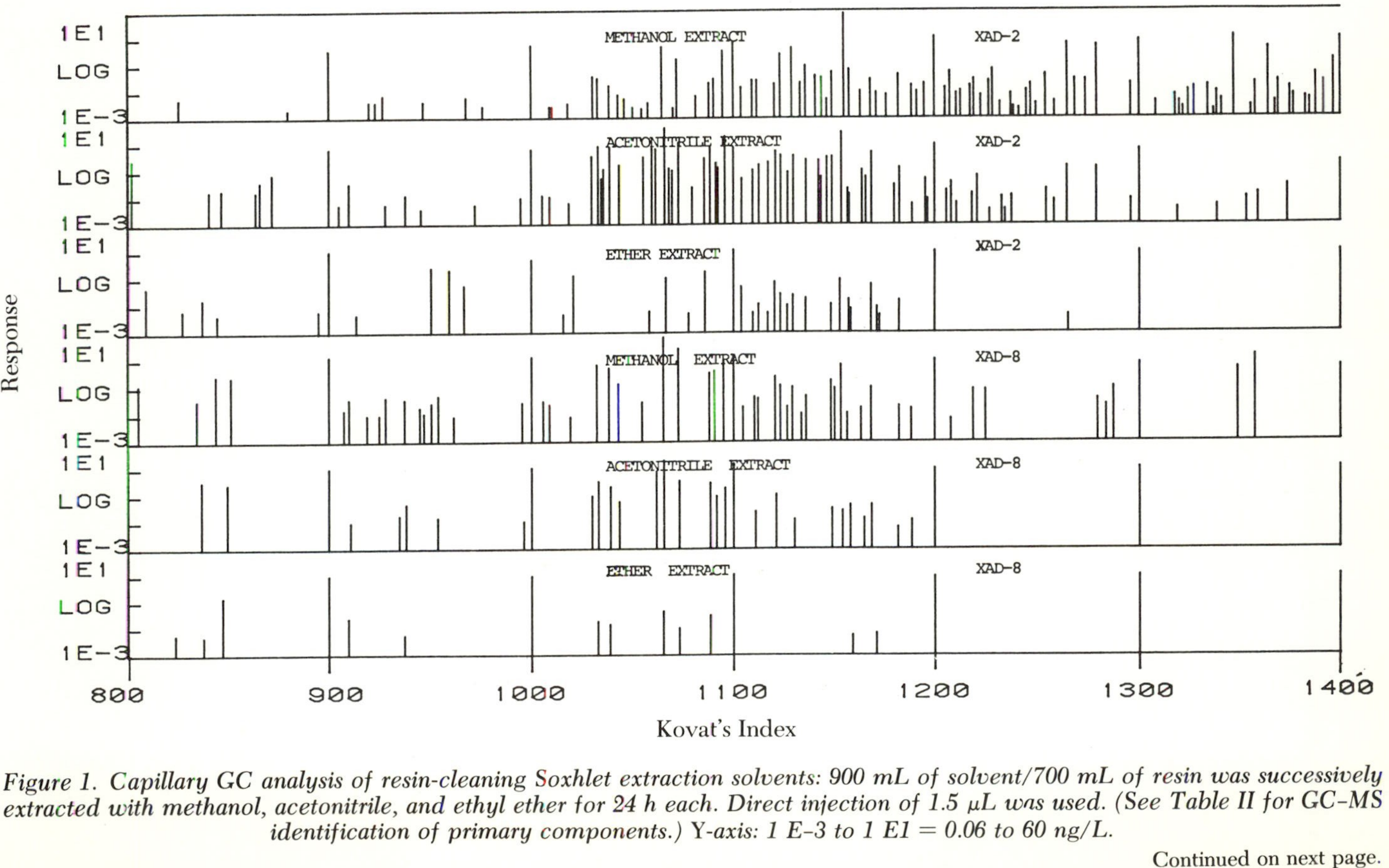

Figure 1. Capillary GC analysis of resin-cleaning Soxhlet extraction solvents: 900 mL of solvent/700 mL of resin was successively extracted with methanol, acetonitrile, and ethyl ether for 24 h each. Direct injection of 1.5 μL was used. (See Table II for GC-MS identification of primary components.) Y-axis: 1 E–3 to 1 E1 = 0.06 to 60 ng/L.

Continued on next page.

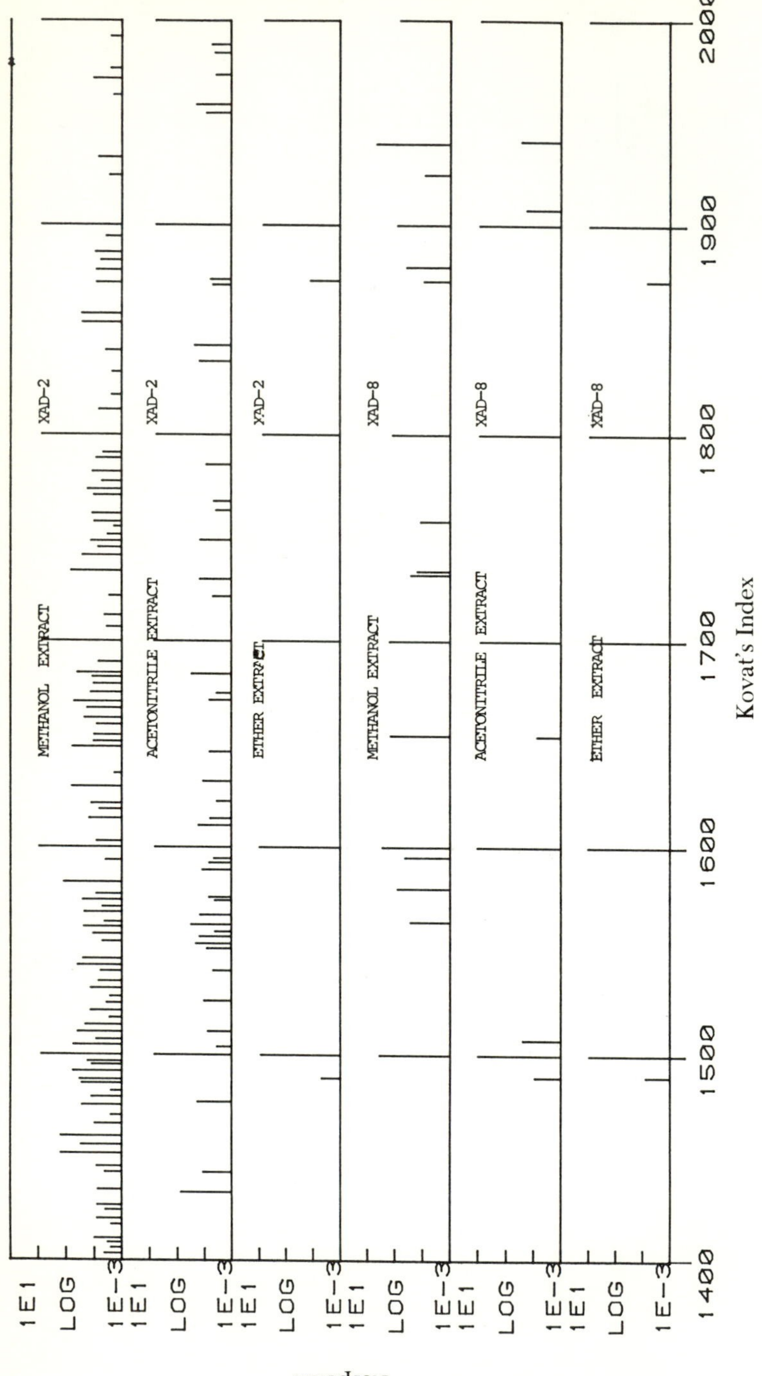

Figure 1. Continued.

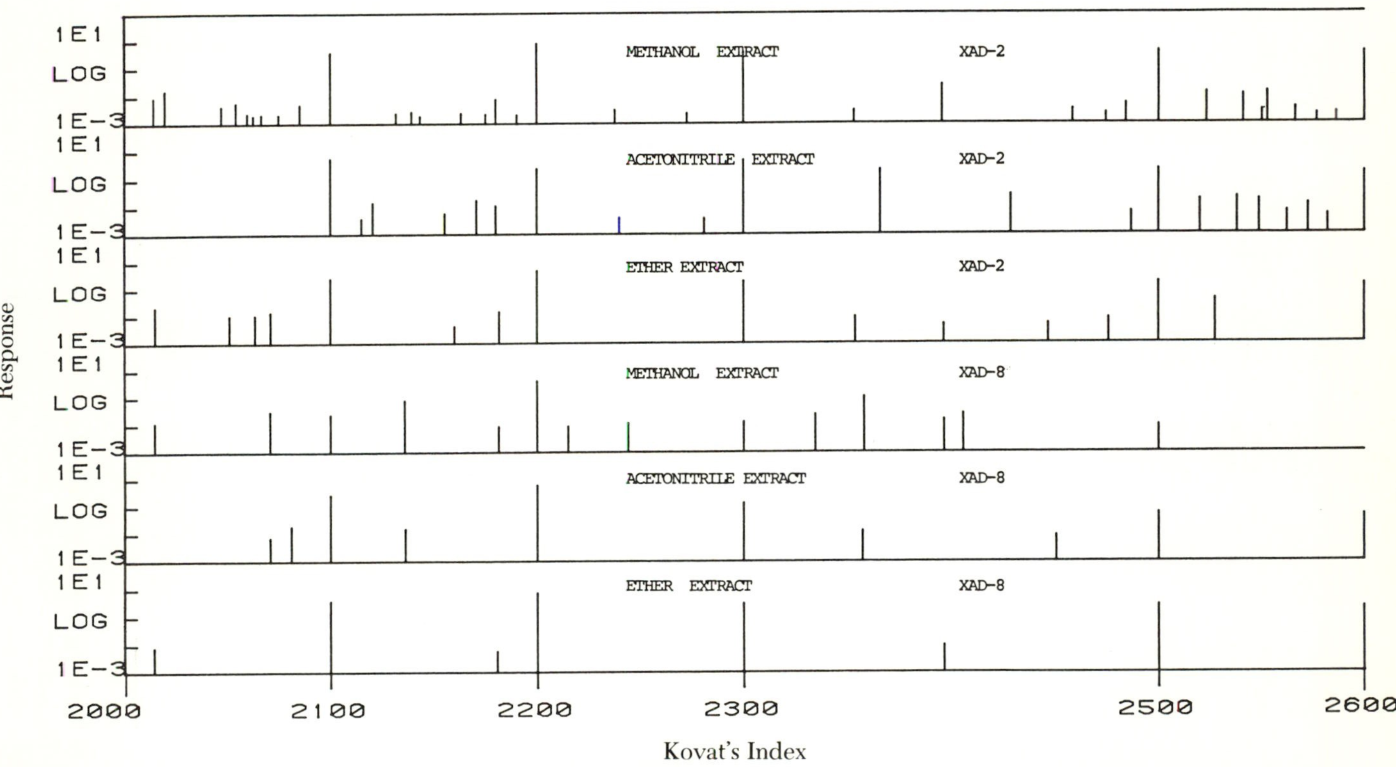

Figure 1. Continued.

Table II. XAD Resin Cleanup by Soxhlet Extractions: GC–MS Tentative Identifications by Computer Search

Compounds	XAD-8		XAD-2	
	CH_3OH	$CH_3C{\equiv}N$	CH_3OH	$CH_3C{\equiv}N$
Toluene	X[a]	X		
C6-Alcohol (e.g., 4-methyl-2-pentanol)	2[b]			
C2-Benzene	3	3		
Benzaldehyde	X			
C4-Benzene	2	2		
C2-Styrene	4	2		
Divinylbenzene	X			
Naphthalene	X		X	X
1-Undecanol	X			
Diethyl phthalate	X			
C17 HC branched	X			
Nonamide	X			
Dichlorobenzophenone	X			
N-(*p*-chlorophenyl)-4-chlorobenzamide	X			
Methyl benzoate[c]			X	
C2-Benzaldehyde[c]			X	
C2-Acetophenone[c]			X	
Methyl naphthalene			2	
Biphenyl			X	
C2-Naphthalene			X	
Methylstyrene				3
Methacrylic acid	X			
Unknown[c,d]	2			

NOTE: Ethyl acetate is a solvent impurity of ether. 900 mL of solvent/700 mL of resin was successively Soxhlet extracted for 24 h with CH_3OH, $CH_3C{\equiv}N$, and ether. No compounds were identified in the ether fractions. A sample was injected directly into the GC–MS. Identifications are tentative; no retention time data have yet been correlated. Identifications were completed by MS only.

[a] X indicates the presence of the corresponding compound.

[b] A number indicates the number of isomers present.

[c] Primary resin artifacts are always observed in resin samples. C2-methyl benzoate is occasionally found as an artifact in XAD-8 samples, although it was not found in these extracts.

[d] Unknown was identified by computer search as nitrooctanone. Chemical ionization MS does not confirm this identification.

SOURCE: Reproduced with permission from reference 39.

Therefore, an additional cleaning and quality assurance step was used. Columns of 100 mL of resin were prepared and eluted with 200 mL of ether. The ether was then concentrated to 1 mL by Kuderna–Danish evaporation and analyzed by a capillary GC with an FID. This analysis showed that many peaks were not removed after 200-fold concentration, and additional resin cleaning steps were needed. Alternate elution of the resin with successive single bed volumes of ether, methanol, and ether, including a 10-min residence time for each elution after draining the previous solvent, was added to the procedure. The last ether elution was concentrated to 1 mL for GC analysis. This process was repeated until the final ethyl ether elution had a reproducible minimum number and level of contaminant peaks on an FID chromatogram. Figure 2 shows the chromatogram of the final cleanup blank. The total number of GC peaks above 0.04 ppb on the XAD-2 and XAD-8 resins were 18 and 13, respectively. The total process of cleaning and quality assurance required 2 weeks per resin batch and large quantities of high-purity solvents.

Ultrasonic Cleaning of Resin. The 2 weeks required to clean a resin batch as just described is excessive. Low diffusion velocities of the resin contaminants moving from the resin pores to the bulk solvent may be causing the long time needed for Soxhlet extraction and column elution. An ultrasonic bath was tried to increase the diffusion velocities and thereby shorten the cleaning time. The manufacturer of the resin revealed that ion-exchange resins are commercially cleaned by using low-power ultrasonic generators and that an ultrasonic bath (BRANSONIC Model B-220) of 100 W should not cause the resin beads to fracture (33).

A comparative study was conducted by cleaning XAD-2 resin by ultrasonic and Soxhlet extraction procedures. Two batches of XAD-2 resins from the same lot of uncleaned resin were cleaned by backwashing with water and then were subsequently extracted with methanol, acetonitrile, and methylene chloride. The solvent-to-resin volume ratio for both methods was 5:1. The volumes of wet resin for the Soxhlet extraction and the ultrasonic bath were 100 mL and 50 mL, respectively. The ultrasonic cleaning procedure was done in a Teflon-lined screw-cap amber jar (250 mL) for 1 h in each solvent. The Soxhlet extraction sequence was 24 h in methanol, 24 h in acetonitrile, and 48 h in methylene chloride.

After the three solvent extractions, 50 mL of wet resin from each method was placed in separate columns and eluted three times with 1 bed volume of methylene chloride; a 20-min contact time between elutions was used. The third elution was collected and concentrated to 1 mL. A homologous series of normal hydrocarbons from C7 to C26 were added as internal standards. Two and one-half microliters of each concentrate was injected onto a 60-m × 0.32-mm i.d., DB-1 fused-silica GC capillary column (J & W Scientific).

Figure 3 shows a comparison between the two methods. The ultrasonically cleaned resin has 10 more contaminant peaks than the Soxhlet-extracted resin. However, the ultrasonic procedure achieves almost the same results as Soxhlet extraction in one-fourth the elapsed time.

Table III. XAD Resin Contaminants

Compounds	Resin Cleanup XAD-2 Blank (Drexel)[a,f]	XAD-2 Solvent Elutions: Ether[b] (μg/5.5 g of resin)	XAD-2 Solvent Elutions: Ether[c] Relative Amt. (range)	XAD-2 Solvent Elutions: CH_2Cl_2[d] (μg/g of resin)	XAD-2 Thermal Elution[e]	Resin Cleanup XAD-8 Blank (Drexel)[a,f]
Undecane		23–39				X
Toluene		0–9				
C2-Benzene		0.9–3.2	890–5600[g]			
C3-Benzene		0.4–1.3	150–420			
C4-Benzene			720–1820	102		X
C5-Benzene			70–1000			
C6-Benzene			X	X		
Styrene		1.3–5.9	690	19		
Methylindan			180–3040			
Methylstyrene	X		X			
C2-Styrene			400–5160	15–320	X	X
C3-Styrene					X	
Dimethylindan			450			
Tetralin			360			
Divinylbenzene			630–790	45		X
2-Pentenylbenzene			240			
1,1a,6,6a-Tetrahydrocycloprop-[a]indene			170			

Methyl benzoate	X	1720		X	
Acetophenone		90			
C2-Benzaldehyde	X	120			
Naphthalene	X	1390	470	X	X
C2-Methyl benzoate		50			
C2-Acetophenone	X	50			
Methylnaphthalene isomer	X	40–70	X		
Methylindene			67		
C2-Naphthalene	X		X		
Biphenyl	X		69		
Methylbiphenyl			X		
C2-Biphenyl			X		
1,1′-Methylenebis[benzene]			X		
1,1′-Ethylidenebis[benzene]			130		
1,1′-(1,2-Ethenediyl)bis[benzene]			X		
1,1-Bis(*p*-ethylphenyl)ethane			26		
Benzoic acid		—[g]		X	

NOTE: X indicates the presence of the corresponding compound.

[a] 900 mL of solvent/700 mL of resin was successively extracted with methanol, acetonitrile, and ethyl ether, 24-h Soxhlet each. All identifications are tentative. This information is from Table I.

[b] Cleaned resin (15 mL of methanol slurry) (5.5 g of dry resin eluted with 2 bed volumes of ethyl ether) (*22*).

[c] Cleaned resin eluted with ethyl ether. Amounts are relative to hexaethylbenzene standard at 0.2 ppb (*31*).

[d] Cleanup resin eluted with CH_2Cl_2 (*30*).

[e] Data are from reference 32.

[f] Additional compounds were found only in XAD-8 resin cleanup: C6 alcohol, benzaldehyde, undecanol, diethyl phthalate, C17 HC branched, nonamide, dichlorobenzophenone, *N*-(*p*-chlorophenyl)-4-chlorobenzamide, methacrylic acid, and an unknown compound identified by computer search as nitrooctanone (the unknown was not confirmed by chemical ionization MS). All identifications are tentative.

[g] Data are from reference 19.

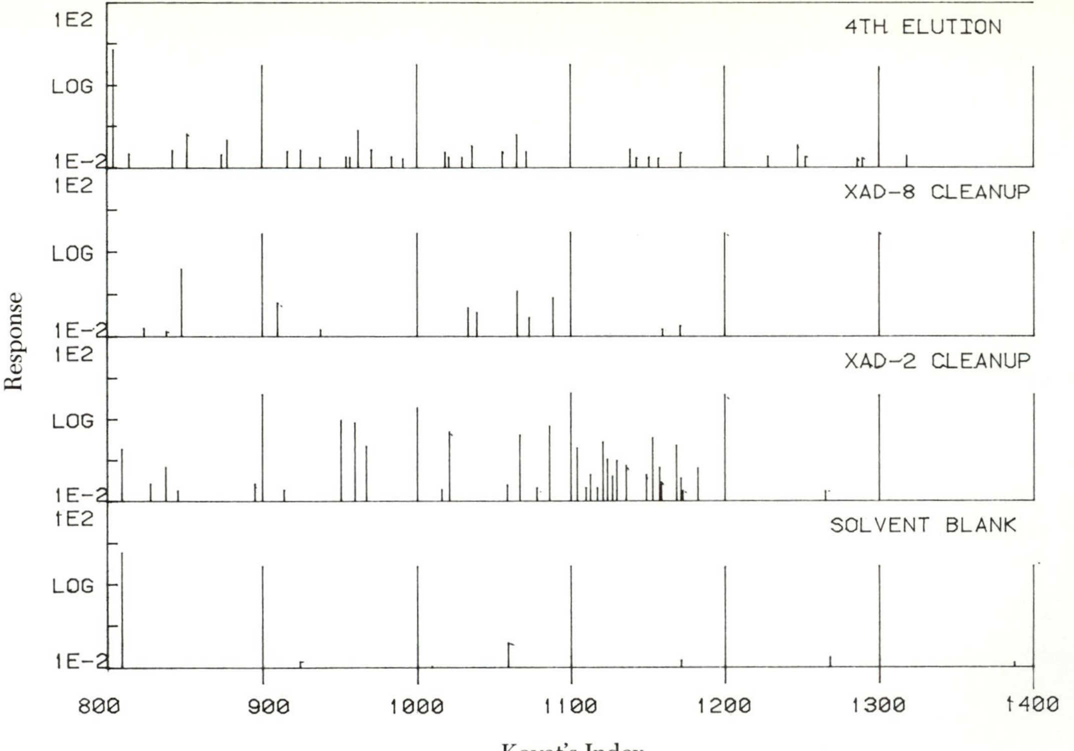

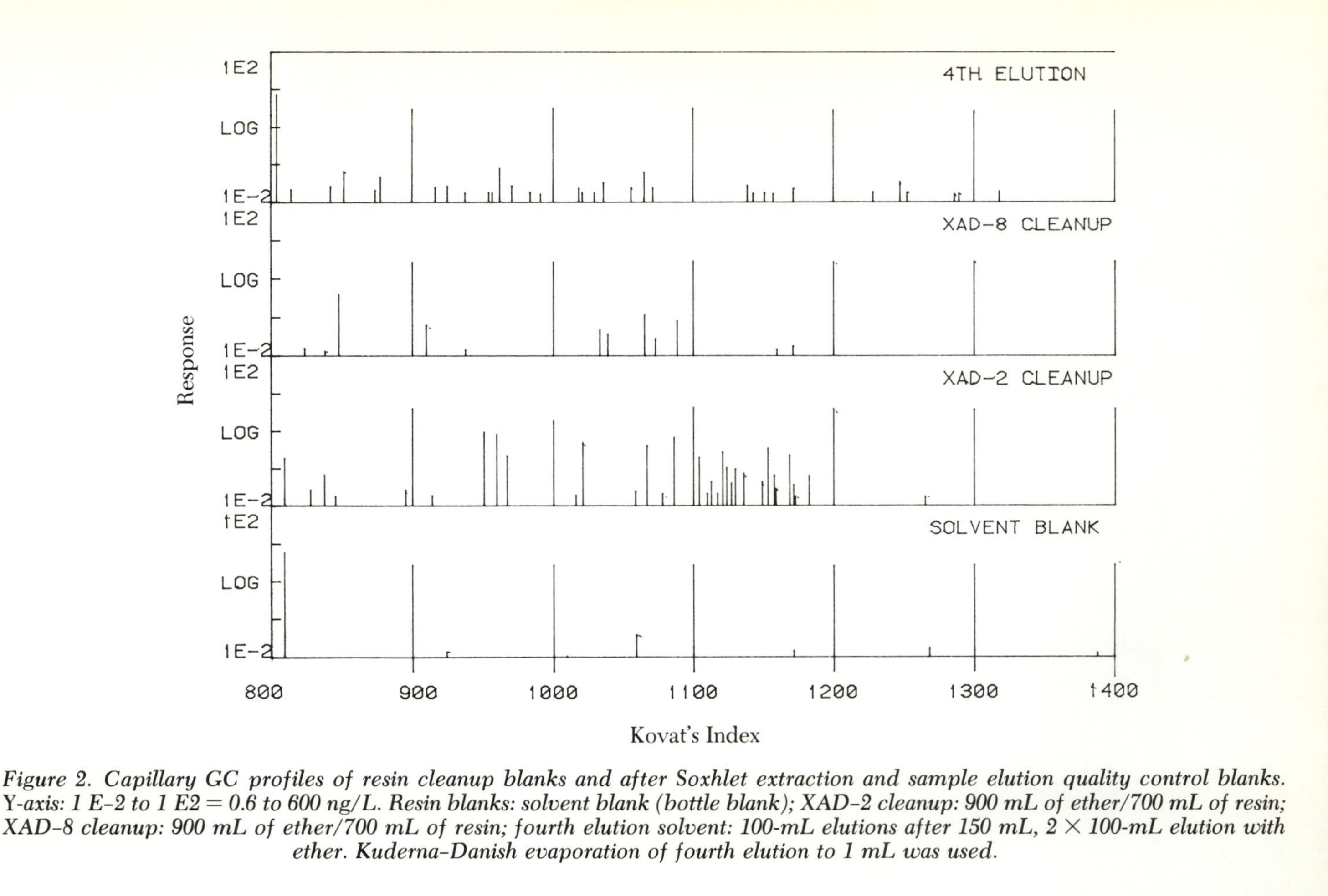

Figure 2. Capillary GC profiles of resin cleanup blanks and after Soxhlet extraction and sample elution quality control blanks. Y-axis: 1 E–2 to 1 E2 = 0.6 to 600 ng/L. Resin blanks: solvent blank (bottle blank); XAD–2 cleanup: 900 mL of ether/700 mL of resin; XAD–8 cleanup: 900 mL of ether/700 mL of resin; fourth elution solvent: 100-mL elutions after 150 mL, 2 × 100-mL elution with ether. Kuderna–Danish evaporation of fourth elution to 1 mL was used.

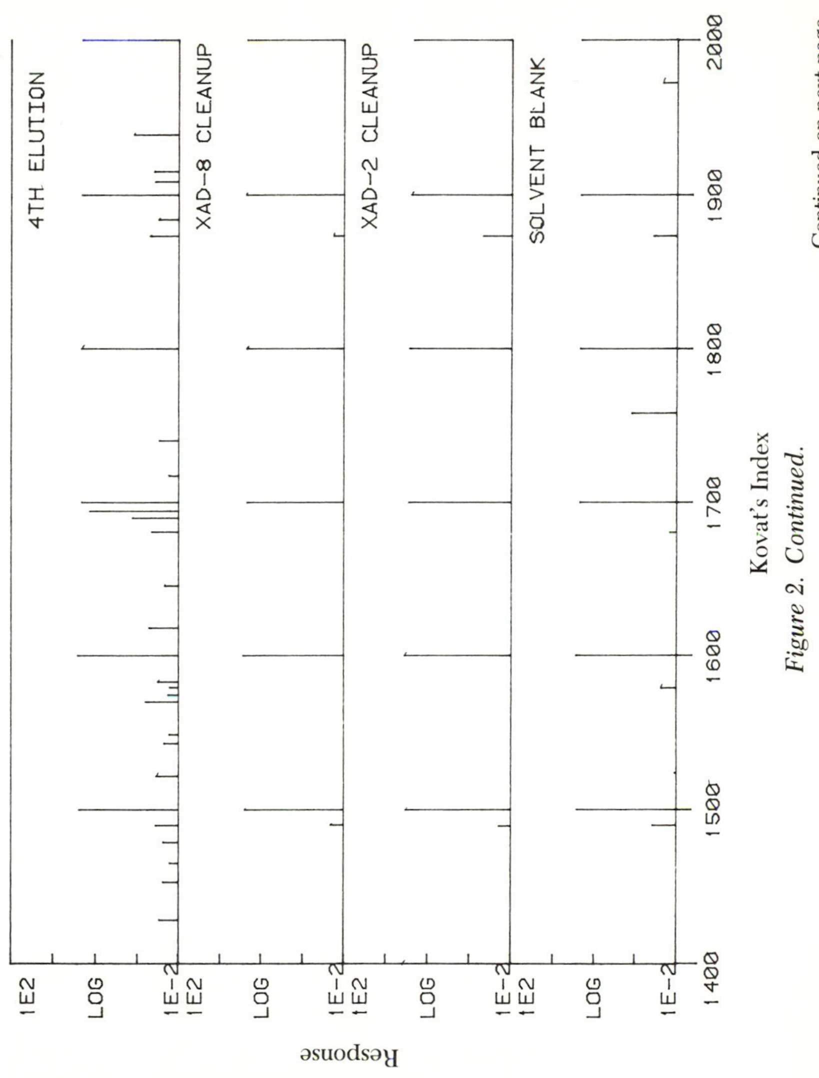

Figure 2. Continued.

Continued on next page.

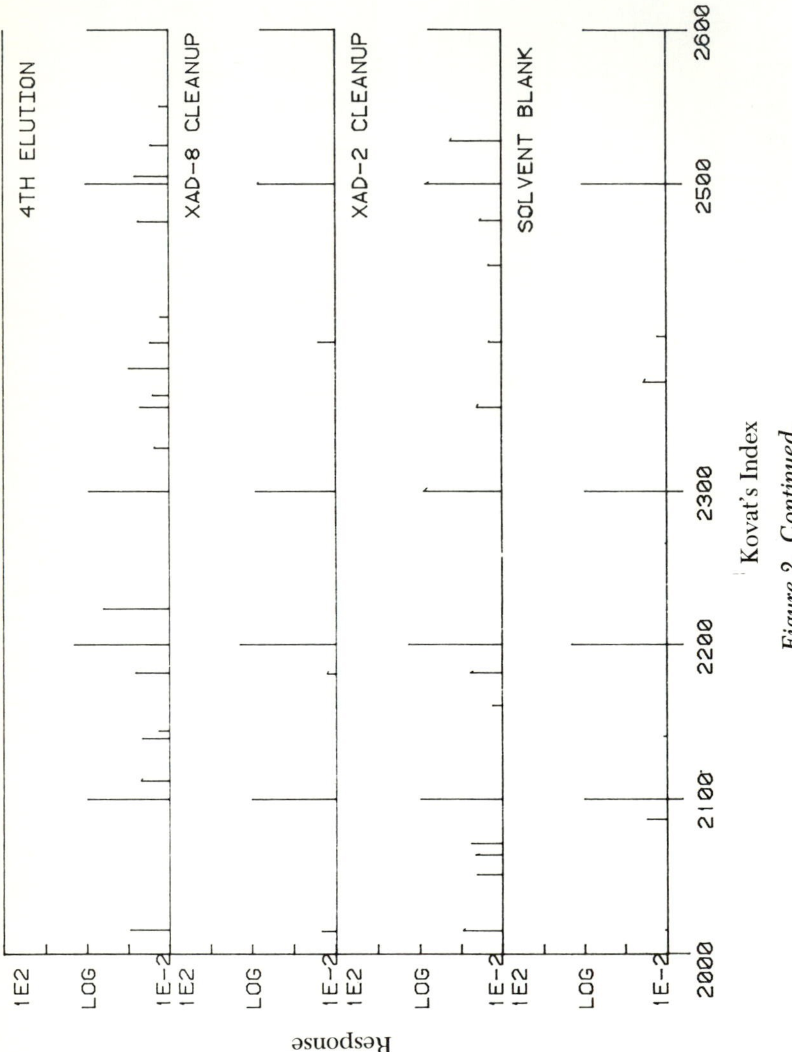

Figure 2. Continued.

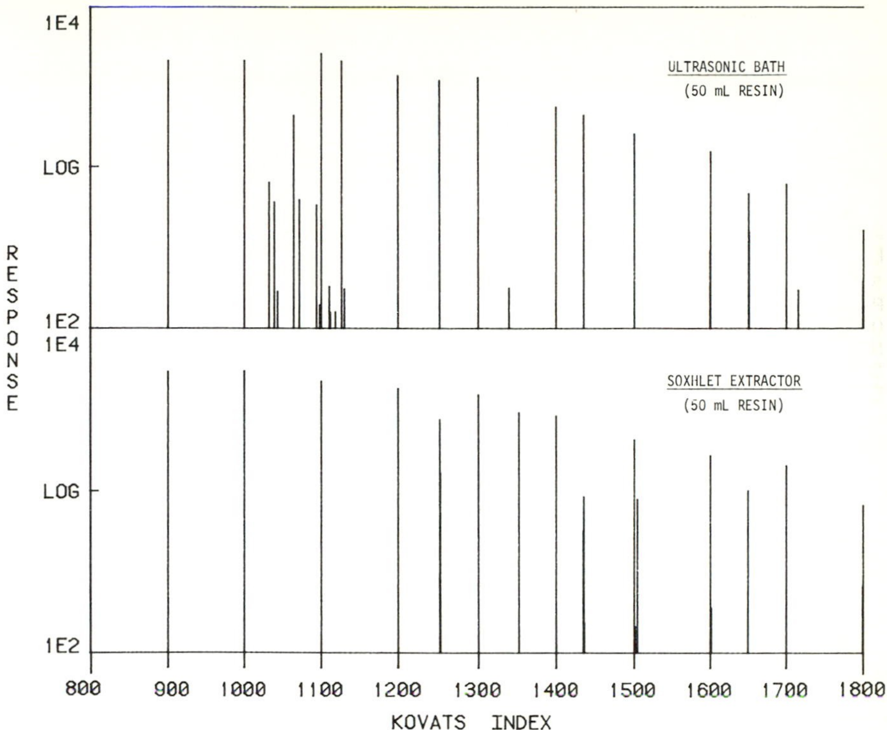

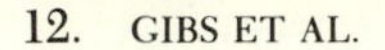

Figure 3. Capillary GC profile comparison of resin cleanup procedures by ultrasonic bath and Soxhlet extraction. Y-axis: 1 E2 = 0.005 ng/L to 1 E4 = 0.5 ng/L. The resin was extracted with methanol, acetonitrile, and methylene chloride. Fifty milliliters of resin was placed in a column and eluted with 2 bed volumes (50 mL) of methylene chloride. The third bed volume was concentrated to 1 mL, and 2.5 μL was injected into the GC. This sample represents a sample elution quality assurance control blank.

Blanking Procedures. A true blank would be pure water passed through cleaned resin by using the field procedure. An experiment was performed to individually blank the use of XAD-8 and XAD-2 resins. Resin columns of 100 mL (5 cm diameter) were prepared by the Soxhlet procedure, and CH_2Cl_2 was used as the elution solvent. Figures 4 and 5 show a series of three chromatograms for XAD-2 and XAD-8 resins, respectively. An explanation of the chromatograms follows:

1. The first chromatogram in each figure is the cleaned resin bed eluted with 100 mL of the CH_2Cl_2 concentrated to 1 mL. The CH_2Cl_2 elution of the column was not prewashed before water sampling began. This procedure was to be used in the field.

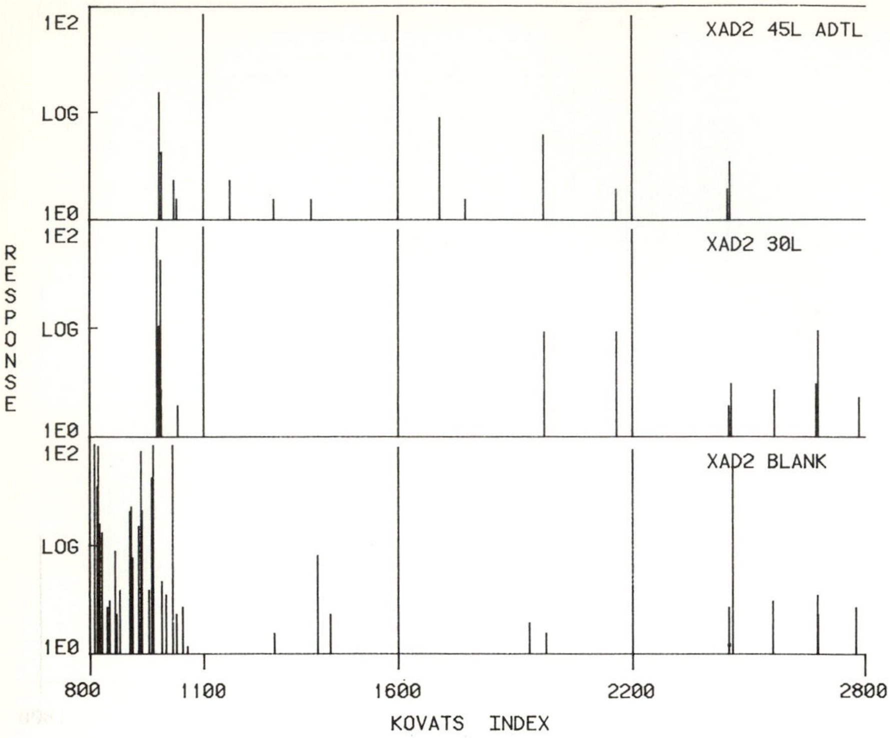

Figure 4. Capillary GC profile of 200-mL resin sampling blanks of XAD–2 column with Milli-Q water (pH 2.3). An initial methylene chloride elution of the clean resin (XAD–2 blank) was followed by a 30-L run of distilled water (XAD–2, 30 L) and then an additional 45 L of distilled water (XAD–2, 45 L additional). (See text for sampling detail.) The third bed volume was concentrated to 1 mL, and 2.5 μL was injected into the GC. Y-axis: 1 E0 = 0.067 ng/L to 1 E2 = 6.7 ng/L.

2. The second chromatogram in each figure shows the change in the blank after passing 30 L of Milli-Q (organic free) water (pH 2.3) through the resin bed at a flow rate of 1.25 L/h for 24 h. The resin bed was eluted with three 100-mL volumes of CH_2Cl_2, which were combined and concentrated to 1 mL for GC–FID analysis.
3. The third chromatogram in each figure represents another 45-L sample of Milli-Q water (pH 2.3) taken over a subsequent 4-day period at a flow rate of approximately 0.5 L/h. The resin bed was again eluted with three 100-mL volumes of CH_2Cl_2, which were combined and concentrated to 1 mL for GC–FID analysis.

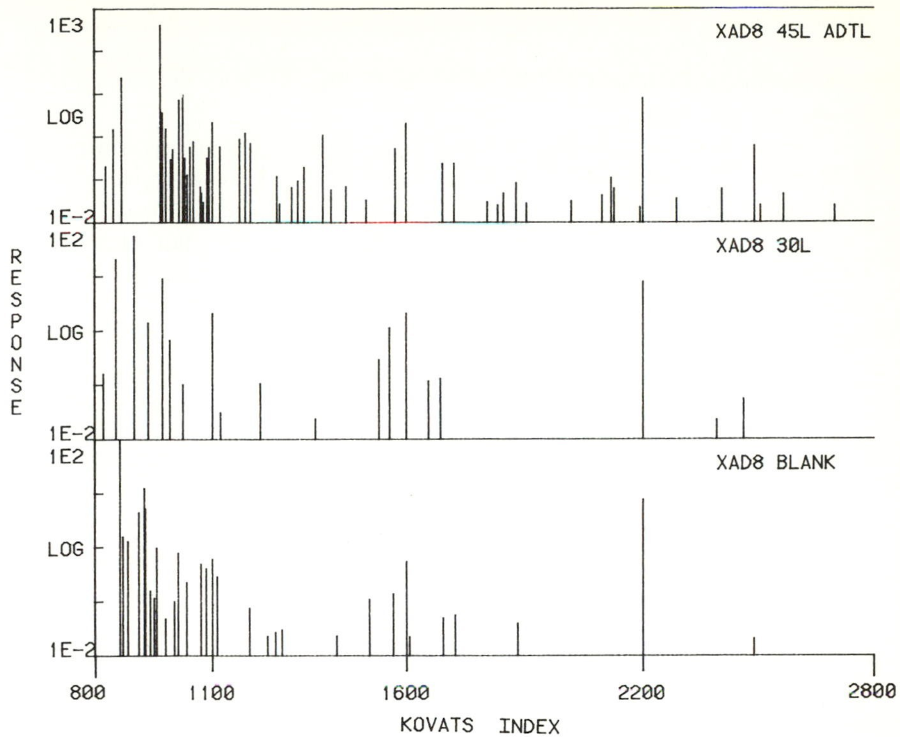

Figure 5. Capillary GC profile of a 200-mL resin sampling blank of an XAD-8 column with Milli-Q water (pH 2.3). An initial methylene chloride elution of the clean resin (XAD-8 blank) was followed by a 30-L run of distilled water (XAD-8, 30 L) and then an additional 45 L of distilled water (XAD-8, 45 L ADT'L). (See text for sampling detail.) The third bed volume was concentrated to 1 mL, and 2.5 μL was injected into the GC. Y-axis: 1 E-2 = 0.00067 ng/L to 1 E3 = 67 ng/L.

Figure 4 shows a capillary GC profile of an extract of an XAD-2 resin column ready for large-volume water sampling of trace organic chemicals (labeled as XAD-2 Blank). Few contaminants occur in the GC-FID profile of the XAD-2 Blank. These contaminants are probably the same contaminants that recur in all XAD-2 samples, namely methyl benzoate, C2-acetophenone, and C2-benzaldehyde (Table II). Other researchers also have shown the recurrence of naphthalene, phenanthrene, and pyrene (*16*), as well as toluene, styrene, C2-benzenes, undecane, and polysubstituted benzenes (22). Although these compounds recur in the blank, Figure 4 shows that XAD-2 extract maintains a low background of artifacts. Figure 4 (middle chromatogram) shows a decrease in peaks after 30 L of Milli-Q water. This decrease may be caused by washing the resin with water (*see* Kovat's index 800–1100), whereas the increase in peaks may be due to the impurities in Milli-Q water or the leaching of the resin.

Figure 5 shows that the XAD-8 resin extract has many more contaminants appearing after 30 L of sampling and even more after a subsequent 45 L. Thus, this batch of XAD-8 resin has a much higher background for sampling of trace organic compounds than the XAD-2 batch studied. In our experience, only the following compounds recur while sampling with XAD-8: two unknowns (*see* Table II, Unknown) and C2-methyl benzoate.

Resampling with the Same Resin. Resin cleaning was performed to produce the minimum number of GC-FID responses from the cleaning solvent. One cycle of exposure to a sample of water followed by elution of the columns caused the reappearance of the characteristic contaminants, which have been tentatively identified by GC-MS as methyl benzoate, C2-benzaldehyde, C2-acetophenone, and the unknown listed in Table II and shown in Figures 4 and 5. A similar phenomenon also has been reported by Bean et al. (*16*) and Care et al. (*22*).

An added elution step was used in this experiment to check the completeness of elution and the cleanliness of the resin. After sampling and elution of the resin (200 mL, 2 × 150 mL) with ethyl ether, the resin column was eluted with an additional 200 mL of ether, which was then concentrated to 1 mL for broad spectrum GC analysis. This fourth elution was a quality control blank for the subsequent reuse of the resin (Figure 2, top chromatogram).

Results from the fourth elution show that there are artifacts from the resins after elution with 2.5 bed volumes of solvent. These artifacts are found at very low levels—less than 0.3 parts per trillion in 150 L of water (*see* Figure 2).

Discussion

Artifacts. The following hypotheses have been presented to explain the presence of the artifacts occurring during sampling and elution of the resin after it is initially cleaned:

1. During sampling with water, the XAD resin beads fracture under the back pressure produced by pumping, and organic compounds are released (*16*).
2. The cleaning procedure does not remove the materials from the micropores in the resin, and they are slowly removed with each nonpolar elution (Rohm and Haas) (*34*).
3. It is hypothesized from this work that alternating expansion and contraction of the resin during aqueous sampling and

nonpolar solvent elution fractures the polymeric resin beads. Ethyl ether and CH_2Cl_2 are nonpolar solvents that, when added to the resin bed previously in water, cause a 10–15% change in the wet resin volume.

4. The contaminants are dissolved in the resin polymer and diffuse slowly out of the resin–solution interface (*35*).

The presence of the organic compounds originating from the resin (Tables II and III) and the GC peaks observed in Figures 4 and 5 in a sample collected on resin cause a problem in environmental chemical analysis. For quantitative analyses that use resins to isolate a specific group of chemicals, the usual procedure has been to ignore the presence of these compounds unless they interfere with the method. The observed phenomenon is a much more serious problem for broad spectrum analysis. The resin artifacts also become a problem when the sample is used for biological testing at the nanogram- to microgram-per-liter concentration range.

Recommended Solvent Selection Procedure. Resin contaminants have been identified only recently. Until 1977, only the following resin contaminants were identified: naphthalene, ethylbenzene, and benzoic acid (*36*). The resin cleaning procedure of Junk et al. (*5*), which uses a series of three solvents of decreasing polarity, removes the widest possible variety of organic contaminants from the resins. This method is necessary when many compounds must be removed.

Currently, much more is known about resin contaminants. The major contaminants of XAD-2 and XAD-8 resins have been identified by GC–MS (Tables II and III). It may now be possible to optimize the cleaning procedure by selecting solvents that effectively remove the known contaminants. Time and cost needed to prepare resins for sampler use could thus be reduced.

One approach that was considered is based upon the solubility parameter theory of Hildebrand and Scott (*36*, *37*). This approach attempts to determine the best solvent for cleaning resins as a function of the known resin contaminants. Hildebrand and Scott (*36*, *37*) developed the solubility parameter (δ) to describe the property of solvents:

$$\delta_x = (\Delta H_{\mathrm{vap}}/V_x)^{1/2} \tag{1}$$

where ΔH_{vap} is the heat of vaporization and V_x is the molar volume of compound x.

This scale ranges from 7.3 for *n*-hexane to 23.4 for water. Compounds with higher solubility parameters are generally more polar or hydrophilic than those with lower solubility parameters. The solubility parameter of Hildebrand and Scott (*36*, *37*) has been subdivided into

Table IV. Solubility Parameters for Selected XAD Resin Artifacts

	Solubility Parameters				*Cleanup Artifacts*[b]	
Compounds[a]	*Nonpolar*	*Polar*	*Hydrogen Bonding*	*Total*	*XAD-2*	*XAD-8*
Toluene	8.0	3.9	0.8	8.93		X
C6-Alcohol (4-methyl-1-pentanol)	5.57	4.24	4.95	8.57		X
C2-Benzene	8.06	3.63	0.0	8.84		X
	8.08	3.50	1.17	8.88		X
	8.29	3.65	0.0	9.06		X
	8.07	3.44	0.97	8.83		X
C4-Benzene	8.06	3.09	0.0	8.63		X
	8.14	3.18	0.0	8.74		X
	8.05	3.04	0.0	8.60		X
C2-Styrene (methyl)	8.21	3.78	0.0	9.04		X
n-Undecyl alcohol	7.51	3.29	5.45	9.85		X
Diethyl phthalate	7.32	5.34	4.17	9.97		X
Methyl benzoate	7.76	5.60	3.51	10.91	X	
Methylnaphthalene	8.73	4.60	0.0	9.9	X	
Methylstyrene	8.09	4.15	0.0	9.1	X	
Methacrylic acid	8.3	6.34	7.92	13.11		X
Undecane	7.80	0.0	0.0	7.80	X	
C3-Benzene						
Pseudocumene	8.17	3.46	0.0	8.87	X	
Isopropyl	7.92	3.36	0.0	8.60		
n-Propyl	7.94	3.36	0.0	8.62		
Styrene	8.22	4.46	0.0	9.35	X	
Tetralin (1,2,3,4-Tetrahydro-naphthalene)	8.69	3.80	0.0	9.48	X	
Acetophenone	7.86	5.83	4.03	10.58	X	
Average all XAD-8	7.82	3.89	1.92	8.94		
Average all XAD-2	8.11	3.86	0.76	9.01		

[a] Compounds are from Table II.
[b] X indicates the presence of the corresponding compound.
SOURCE: Adapted from reference 38.

polar, nonpolar, and hydrogen bonding components by Union Carbide (*38*).

Table IV shows the calculated solubility parameter (*38*) of some of the resin contaminants from Table III. The average total solubility parameters for the XAD–2 and XAD–8 organic contaminants (Table IV) are approximately the same. However, there are differences between XAD–2 and XAD–8 for the average hydrogen bonding and average nonpolar components of the solubility parameter.

Table V lists the common solvents used for cleaning XAD resins and their respective solubility parameters. Methanol and acetonitrile [which are used in the cleanup procedure of Junk et. al. (*5*)] have very different total solubility parameters than the compounds listed in Table IV. Thus, they should be less efficient for eluting the resin contaminants from the resin polymers. This situation explains the GC profile results, which show large numbers and high concentrations of contaminants after the successive 24-h Soxhlet extractions using methanol, acetonitrile, and ethyl ether (Figure 2).

An additional consideration in the cleaning of XAD resins is their flexible structure. The resin beads swell on contact with hydrophilic or polar solvents such as methanol (*29*). The nonpolar surface of the resin is repelled by polar solvents and attracted by nonpolar solvents. This effect causes the internal pore diameters to increase or decrease, respectively. The cleaning solvent or mixture of solvents must be polar to keep the internal pores open so that the contaminants will diffuse faster from the interior of the beads to the bulk solvent. However, the resin contaminants themselves are nonpolar, as shown in Table IV, and are not very soluble in polar solvents. The choice of an optimum resin-cleaning solvent should therefore be a compromise between diffusion and solubility. In addition, the solvent used after the water backwash must be miscible with water to remove the water from the resin pores.

Table V. Solubility Parameters of Resin Cleanup Solvents

Solvents	*Nonpolar*	*Polar*	*Hydrogen Bonding*	*Total*
Methanol	5.66	6.36	11.73	14.50
Acetonitrile	5.04	5.45	9.56	12.11
Ethyl ether	6.53	2.57	2.73	7.53
Acetone	6.37	4.79	5.39	9.62
Methylene chloride	6.55	5.71	4.70	9.88
n-Hexane	7.27	0.0	0.0	7.27
Cyclohexane	8.05	1.50	0.0	8.19
Isopropyl alcohol	6.86	4.79	7.81	11.45

SOURCE: Adapted from reference 38.

A single solvent probably will not have all the characteristics necessary for optimum resin cleaning. If a mixture of solvents is chosen for use in either the traditional Soxhlet extractor or the Environmental Protection Agency continuous extractor (*8*), the mixture must also form an azeotrope because both devices recycle their solvents by using distillation.

Recommended Blanking Procedure. The number of recurring resin impurities in resin elutions appears to be a variable related to a particular batch of resin, the cleaning procedure, the sequence of sampling and elution, and the reuse of the resin. A reproducible blank may never be possible between different production lots of resin for quantitative broad spectrum analysis of nanogram- to microgram-per-liter quantities of trace organic chemicals in large volumes of water. However, within a lot and with a blank using Milli-Q water to identify lot impurities, a reproducible set of recurring impurities may be obtained, and semiquantitative (order of magnitude) analysis appears viable (*39*). The concentrations of the resin impurities within a production lot may vary between large-volume samples and are a function of the exposure of the resin to the sample, resin elution, or reuse of the resin. Only compounds that have never been observed as resin impurities can be studied quantitatively by resin isolation methods.

Recommended Resin Reuse Procedure. Storing used resin in methanol between sampling runs is recommended (*5*). For good quality assurance, a study of storage in methanol after sampling and elution should be undertaken to test if adsorbed compounds that were not eluted are leaching from the used resin stored in methanol.

Conclusions

The conclusion of this study is that semiquantitative (order of magnitude) analysis is possible for broad spectrum analysis of resin lots cleaned by a consistent method and blanked with Milli-Q water. The concentrations of the resin impurities will vary as a function of the following:

1. The resin's exposure to a large-volume environmental sample with varying concentrations of chemicals.
2. The resin's elution procedure.
3. The resin's storage before use in methanol or sampling directly after cleaning with the eluting solvent.
4. The resin cleaning and reuse procedure.

This quality assessment study indicates that to use resins for broad spectrum analysis, the resin elution, storage, cleaning, and reuse procedures must be rigorously defined so that the artifacts produced do not interfere. Once the procedures have been defined, they must be strictly followed.

Literature Cited

1. Burnham, A. K.; Calder, G. V.; Fritz, J. S.; Junk, G. A.; Svec, H. J.; Willis, R. *Anal. Chem.* **1972,** *44,* 139–142.
2. Burnham, A. K.; Calder, G. V.; Fritz, J. S.; Junk, G. A.; Svec, H. J.; Vick, R. *J. Am. Water Works Assoc.* **1973,** *65,* 722–725.
3. Svec, H. J.; Fritz, J. S.; Calder, G. V. "Trace Soluble Organic Compounds in Potable Water"; Completion Report, Project A-064-1A; Department of Chemistry, Iowa State University, December 1973.
4. Harvey, G. R. *Adsorption of Chlorinated Hydrocarbons from Seawater by a Crosslinked Polymer;* U.S. Environmental Protection Agency. U.S. Government Printing Office: Washington, DC, 1973; EPA Report No. R2-73-177.
5. Junk, G. A.; Richard, J. J.; Grieser, M. D.; Witiak, D.; Witiak, D. L.; Arguelo, M. D.; Vick, R.; Svec, H. J.; Fritz, J. S.; Calder, G. V. *J. Chromatogr.* **1974,** *99,* 745-762.
6. Osterroht, C. *J. Chromatogr.* **1974,** *101,* 289–298.
7. Junk, G. A.; Richard, J. J.; Svec, H. J.; Fritz, J. S. *J. Am. Water Works Assoc.* **1976,** *68,* 218–222.
8. Lentzen, D. E.; Wagoner, D. E.; Estes, E. D.; Gulknecht, W. F. *Procedures for Preparation of XAD Resins Free of Extractable Organics;* U.S. Environmental Protection Agency. U.S. Government Printing Office: Washington, DC, 1978; EPA Report No. 600/7-78-201, Appendix B.
9. Thurman, E. M.; Aiken, G. R.; Malcolm, R. L. *Proceedings of the 4th Joint Conference on Sensing of Environmental Pollutants,* 1978.
10. Van Rossum, P.; Webb, R. G. *J. Chromatogr.* **1978,** *150,* 381–392.
11. Harris, J. C.; Cohen, M. J.; Grosser, Z. A.; Hayes, M. J. *Evaluation of Solid Sorbent for Water Sampling;* Arthur D. Little, Inc. U.S. Environmental Protection Agency. U.S. Government Printing Office: Washington, DC, 1980; *EPA Report* No. 600/2-80-193.
12. Dutka, B. J.; Jova, A.; Brechin, J. *Bull. Environ. Contam. Toxicol.* **1981,** *27,* 758–764.
13. Grabow, W. O. K.; Burger, J. S.; Hilner, C. A. *Bull. Environ. Contam. Toxicol.* **1981,** *27,* 424–429.
14. Junk, G. A.; Chriswell, C. D.; Chang, R. C.; Kissinger, C. D.; Richard, J. J.; Fritz, J. S.; Svec, H. J. Z. *Anal. Chem.* **1976,** *282,* 331–337.
15. Pfaender, F. K. *ESE Notes* **1976,** *12,* 4; Department of Environmental Science and Engineering, University of North Carolina, Chapel Hill.
16. Bean, R. M; Ryan, P. W.; Riley, R. G. *Proceedings of the Symposium on High Resolution Gas Chromatography of the American Chemical Society;* Academic: Chicago, 1978.
17. Chriswell, C. D.; Ericson, R. L.; Junk, G. A.; Lee, K. W.; Fritz, J. S.; Svec, H. J. *J. Am. Water Works Assoc.* **1977,** *69,* 669.
18. Coburn, J. A.; Valmandis, I. A.; Chau, A. S. Y. *J. Assoc. Off. Anal. Chem.* **1977,** *60,* 225–228.

19. Fritz, J. A. *Acc. Chem. Res.* **1977,** *10,* 67-72.
20. McNeil, E. E.; Otson, R.; Miles, W. F.; Rajabalee, F. J. M. *J. Chromatogr.* **1977,** *132,* 277-286.
21. Stephan, S. F.; Smith, J. F. *Water Res.* **1977,** *11,* 339-342.
22. Care, R.; Morrison, J. D.; Smith, J. F. *Water Res.* **1982,** *16,* 663-665.
23. McRae, T. G.; Gregson, R. P.; Quinn, R. J. *J. Chromatogr. Sci.* **1982,** *20,* 475-478.
24. Fu, T.; Pocklington, R. *Matrice Chem.* **1983,** *13,* 225-264.
25. Kaczvinsky, J. R., Jr.; Saitoh, K.; Fritz, J. S. *Anal. Chem.* **1983,** *55,* 1210-1215.
26. ACS Committee on Environmental Improvement *Anal. Chem.* **1980,** *52,* 2242-2249.
27. Thurman, E. M.; Malcolm, R. L. *Environ. Sci. Technol.* **1981,** *15,* 463-466.
28. Rohm and Haas Co. *Decolorization of Kraft Pulp Bleaching Effluents Using Amberlite XAD-8 Polymeric Adsorbent, IE-75;* Philadelphia, 1972.
29. Rohm and Haas Co. *Amberlite XAD-2, IE-89-65/78;* Philadelphia, 1982.
30. Hunt, G.; Pagano, N. *Anal. Chem.* **1982,** *54,* 369-379.
31. Melton, R.; Coleman, W. E; Slater, R. W.; Kopfler, F. C.; Allen, W. K.; Aurand, T. A.; Mitchell, D. E.; Voto, S. J.; Lucas, S. V.; Watson, S. C. In *Advances in the Identification and Analysis of Organic Pollutants in Water;* Keith, L., Ed.; Ann Arbor Science: Ann Arbor, MI, 1981; Vol. 2, Chapter 36.
32. Lochmueller, C. H.; Jensen, E. C. *XAD-2 Thermal Blank Determination;* U.S. Environmental Protection Agency. U.S. Government Printing Office: Washington, DC, 1981; EPA Report No. 600/82-8-047.
33. Rohm and Haas Co. Personal communication; Philadelphia, 1983.
34. Rohm and Haas Co. Personal communication; Philadelphia, 1982.
35. Gesser, Department of Chemistry, University of Manitoba, Winnipeg, Manitoba, Canada, Personal communication; 1985.
36. Hildebrand, J. H.; Scott, R. L. *The Solubility of Nonelectrolytes;* Dover: New York, 1964; 3rd ed.
37. Hildebrand, J. H.; Scott, R. L. *Regular Solutions;* Prentice Hall: Englewood Cliffs, NJ, 1962.
38. *Tables for Solubility Parameters;* Union Carbide Corp.: Tarrytown, NY, May 16, 1975.
39. Gibs, J.; Najar, B.; Suffet, I. H. In *Water Chlorination;* Lewis Publishers: Chelsea, MI, 1985; Vol. 5, p 110.

RECEIVED for review October 14, 1985. ACCEPTED May 14, 1986.

13

Isolation of Organic Acids from Large Volumes of Water by Adsorption on Macroporous Resins

George R. Aiken

U.S. Geological Survey, Denver Federal Center, Denver, CO 80225

Adsorption on synthetic macroporous resins, such as the Amberlite XAD series and Duolite A-7, is routinely used to isolate and concentrate organic acids from large volumes of water. Samples as large as 24,500 L have been processed on site by using these resins. Two established extraction schemes using XAD-8 and Duolite A-7 resins are described. The choice of the appropriate resin and extraction scheme is dependent on the organic solutes of interest. The factors that affect resin performance, selectivity, and capacity for a particular solute are solution pH, resin surface area and pore size, and resin composition. The logistical problems of sample handling, filtration, and preservation are also discussed.

THE CONCENTRATION OF DISSOLVED ORGANIC CARBON from most natural waters ranges from 1 to 20 milligrams of carbon per liter (mg C/L), of which approximately 75% are organic acids. To effectively study any of these organic acids, they must be isolated from other organic and inorganic species and concentrated. Often large volumes of water must be processed to obtain enough material to carry out the desired measurements. In the case of aquatic humic substances, a complete chemical characterization, including elemental analysis and the determination of structure, functional group content, and molecular weight, requires 500 mg of material. Further studies require more material. Adsorption chromatography on synthetic, macroporous resins has proven to be an effective method for this purpose. The use of the nonionic Amberlite XAD-8 and XAD-4 resins and the anion-exchange resin Duolite A-7 for isolating and concentrating organic acids from large volumes of water will be presented in this chapter.

XAD resins gained popularity as adsorbents for organic solutes from water during the last decade (*1*, *2*), and sorption of organic solutes on this resin series has been extensively studied (*3–5*). XAD-8, an acrylic ester resin, and XAD-4, a styrene–divinylbenzene resin, adsorb organic acids when the acids are in the uncharged state; desorption is favored when the acids are in the ionic form. The driving force for sorption of nonpolar organic solutes on these resins is the hydrophobic effect. The hydrophobic effect refers to the unfavorable entropy of solution caused by the ordering of water molecules around the organic solute (*6*). Differences in resin pore size, surface area, and chemical composition result in different capacity factors for the same solute on each resin.

The use of the anion-exchange resin Duolite A-7 for concentrating organic acids was reported as early as 1965 by Abrams and Breslin (*7*) and more recently by Leenheer (*8*). A-7 is a high-surface-area, macroporous, phenol–formaldehyde, weak-base resin. This resin combines weak-base, secondary-amine functional groups with the relatively hydrophilic phenol–formaldehyde matrix to effectively sorb and elute organic acids.

Processing large volumes of water through an isolation procedure presents a number of logistical problems. The procedure is best done on site so that problems associated with transporting large volumes of water can be circumvented. Every aspect of the job is magnified because of the large sample size. Many glass bottles or stainless steel drums are required to contain the sample, more filtration equipment is needed to provide sufficient quantities of filtered sample, the columns and quantities of resin used must be large enough to efficiently process the sample, and large quantities of reagents are required. Organizing and successfully completing the field collection of organic compounds from water is often a challenge.

The purpose of this chapter is to discuss the use of Amberlite XAD-8 and Duolite A-7 in the field to process large volumes of water for the extraction of organic acids. Schemes involving filtration, concentration, and isolation steps for extracting organic acids from water by using XAD-8 and Duolite A-7 have been devised (*9*, *10*). Some of the logistical problems of filtration and sample preservation are also discussed.

Experimental

Resin. Amberlite XAD-8 and XAD-4 were obtained from Rohm and Haas. These resins were cleaned according to the method of Thurman and Malcolm (*9*). Resin was stored in CH_3OH. Before going to the field, resin was packed in columns, rinsed with distilled water to remove CH_3OH, and sequentially rinsed three times with 0.1 N NaOH and 0.1 N HCl. Resin was transported to the field in 0.1 N HCl in 4-L glass jugs. Columns were packed on site, and resin again was rinsed with 0.1 N NaOH and 0.1 N HCl.

Duolite A-7 was obtained from Diamond-Shamrock and cleaned according to the method of Leenheer (*8*). This resin was packed and stored in large glass columns 10 cm i.d. × 120 cm long. The resin was rinsed with 3 N NH_4OH, followed by a distilled water rinse, until the specific conductance of the effluent was less than 10 μmho.

Reagents. All reagents used were reagent grade.

Instrumentation. Glenco glass columns, 3500 series, with Teflon end fittings and Teflon tubing, were used in the field, along with Cole-Parmer peristaltic pumps and Flotec positive-displacement pumps. A Beckman 915 B carbon analyzer was used to quantify influents, effluents, and eluates.

Procedure. The procedure followed with XAD-8 was that of Thurman and Malcolm (*8*). The Suwannee River (GA) sample was filtered through 0.45-μm Ag filters in Millipore, stainless steel, 142-mm, plate filter holders. The Laramie-Foxhills ground-water sample did not require filtration. Samples were processed on site within 24 h after collection to prevent alteration. All column eluates were adjusted to pH 2 upon collection and immediately refrigerated.

The procedure followed with Duolite A-7 was that of Leenheer (*8*). The sample was pumped directly from the source through Balston stainless steel cartridge filter holders fitted with 0.3-μm Balston glass-fiber filter tubes.

Filter flow rate was compared between 42-mm diameter, 0.45-μm Selas Ag filters and 0.86-cm × 5.78-cm, 0.3-μm Balston glass-fiber filter tubes, grade AA. Samples were pumped at a constant pressure of 30 psi of N_2 gas through each filter; filtrate was collected as a function of time.

Results and Discussion

XAD-8 Method. The choice of an appropriate adsorbent for the concentration of organic solutes in water is dependent on the solutes of interest. Table I presents k' (capacity factor) data for various organic acids on XAD-8 and XAD-4. XAD-4 is a styrene-divinylbenzene resin with a smaller pore size and larger surface area than XAD-8. The larger surface area (750 m^2/g) of the XAD-4 makes this resin very efficient for smaller organic acids; however, it is not as efficient for larger solutes. XAD-8 has been shown to be well suited for concentrating high molecular weight organic acids, such as fulvic acid, from water (*5, 11*). XAD-8 has a greater pore diameter, and larger molecules can diffuse into the beads more easily than on the XAD-4 (*5*). XAD-8 also elutes more efficiently than XAD-4 when fulvic acid is the solute of interest (*5*).

By using XAD-8, hydrophobic organic acids in a natural water can be isolated from the remainder of the organic solutes present, including the hydrophilic organic acids, by properly selecting the influent pH and the volume of sample to be processed for a given column size. The capacity factor, k', for a solute at 50% breakthrough of the influent concentration (Figure 1), is related to the volume of water processed by the following expression:

$$V_E = V_O (1 + k') \tag{1}$$

where V_E is the breakthrough volume and V_O is the void volume of the column. The breakthrough volume, V_E, is the volume of sample that must pass through the resin bed so that the concentration of solute in the column effluent is 50% of its concentration in the influent. By using this equation, a k' value, referred to as the k'_{cutoff}, is chosen to ensure 100% retention of the solutes of interest. Solutes with $k' < k'_{cutoff}$ will not be completely retained by the resin, whereas solutes with $k' > k'_{cutoff}$ will be 100% retained. The amount of water processed, therefore, is a function of the solutes of interest. Care should be exercised not to overtreat the XAD-8 column with too much sample because such treatment will result in biasing the isolate composition toward the more hydrophobic constituents.

The separation of retained and nonretained solutes is not sharp, and a certain amount of solute with $k' < k'_{cutoff}$ will be concentrated along with the solute of interest. When twice the breakthrough volume for a particular solute, $2V_E$, has passed through the column (Figure 1), the concentration of solute in the effluent is equal to the concentration of solute in the influent, C_o, and the solute is completely broken through. At this point, one half of the total amount of solute applied to the resin has been retained (hatched area of Figure 1). This value is the maximum

Table I. Capacity Factors of Selected Organic Acids on XAD-8 and XAD-4

Solute	*XAD-8*[a]	*XAD-4*[b]
High Molecular Weight		
North Carolina fulvic acid	604	332
Polyacrylic acid MW 2000	945	735
Polyacrylic acid MW 5000	1500	175
Polyacrylic acid MW 90,000	350	0
Low Molecular Weight		
Valeric acid	159	907
Benzoic acid	601	2038
Phenol	325	864
Butanoic acid	39	196
Cyclohexylcarboxylic acid	390	—
Heptanoic acid	960	—
Hexanoic acid	377	—
Phthalic acid	144	—
p-Toluic acid	1037	—

[a] Pore size = 250 Å; surface area = 140 m^2/g.
[b] Pore size = 50 Å; surface area = 750 m^2/g.

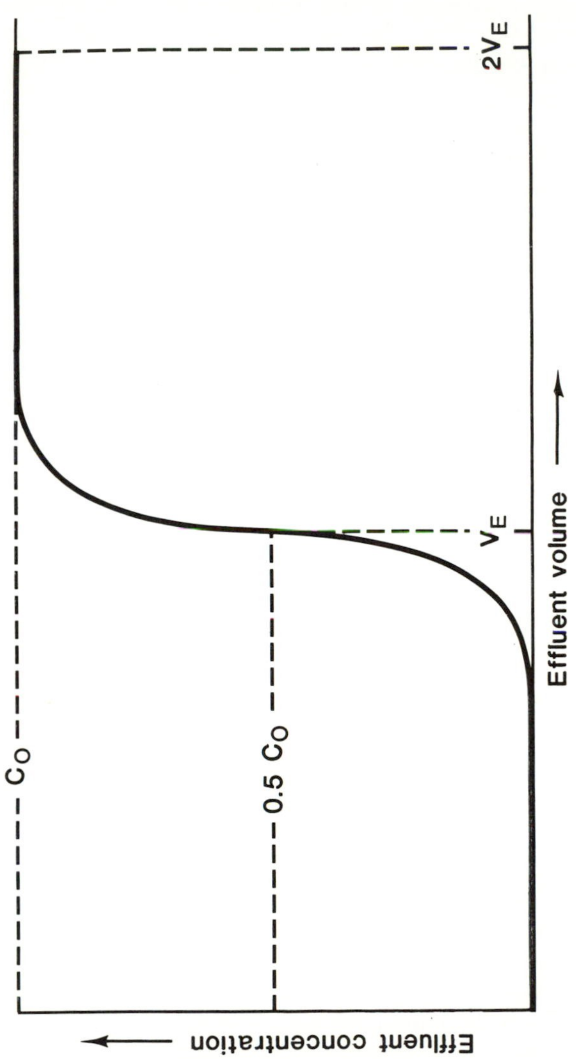

Figure 1. Frontal chromatography breakthrough curve.

amount of this solute that will be absorbed because the concentration of solute on the resin is now in equilibrium with the concentration of solute in the mobile phase.

An extraction method for isolating humic substances from water by using XAD-8 has been proposed by Thurman and Malcolm (9) (*see* box). Humic substances in natural waters represent almost the entire hydrophobic acid fraction. This method has been used to isolate 4.25 g of humic substances from 24,500 L of ground water from the Foxhills–Laramie aquifer and to obtain 500 g of humic material from 10,400 L of the Suwannee River (Table II). The sample from the Suwannee River was collected as a reference sample of aquatic humic substances by the International Humic Substances Society. In both of the examples cited, a k'_{cutoff} of 100 was used.

In processing volumes of water as large as in the two examples cited, large columns of resin are preferred. In both the Foxhills–Laramie and Suwannee River samples, columns with 8 L of resin and a void volume of 6 L were used. By using eq 1 with a k'_{cutoff} of 100, it was calculated that, for the Foxhills–Laramie sample, 460 L of water could

Extraction Process Using XAD–8 To Concentrate Aquatic Humic Substances

Step 1. Filter sample through 0.45-μm silver-membrane filter and lower pH to 2.0 with HCl.

Step 2. Pass acidified sample through column of XAD–8; aquatic humic substances adsorb to resin.

Step 3. Elute XAD–8 resin in reverse direction with 0.1 N NaOH; acidify immediately to avoid oxidation of humic material.

Step 4. Reconcentrate on smaller XAD–8 column until dissolved organic carbon (DOC) is greater than 500.

Step 5. Adjust pH to 1.0 with HCl to precipitate humic acid. Separate humic and fulvic acids by centrifugation. Rinse humic-acid fraction with distilled water until $AgNO_3$ test shows no Cl^-. Dissolve humic acid in 0.1 N NaOH and hydrogen saturate by passing solution through cation-exchange resin in H-form.

Step 6. Reapply fulvic acid fraction at pH 2 to XAD–8 column. Desalt fulvic acid by rinsing column with 1-void volume of distilled water to remove HCl and inorganic salts; elute fulvic acid in reverse direction with 0.1 N NaOH.

Step 7. H-saturate fulvic acid fraction by immediately passing 0.1 N NaOH eluate through cation-exchange resin in H-form. Continue cation-exchange process until final concentration of Na^+ is less than 0.1 ppm.

Step 8. Freeze-dry humic acid and fulvic acid fractions.

Table II. Amounts of Organic Acids Extracted from Various Natural Waters with XAD-8 by the Method of Thurman and Malcolm (9)

Sample	*DOC (mg C/L)*	*Volume of Water Processed (L)*	*Hydrophobic Acids Extracted (g)*
Suwannee River, Georgia	38[a]	10,400	500
Foxhills-Laramie ground water	0.7	24,500	4.25

[a] Average value from 12/1/82 to 2/15/83.

be passed through the column before eluting without saturating the column with fulvic acid. At a flow rate of 2 L/min, which is 15 times the bed volume of the resin, 460 L of sample was passed through the column in 4 h. Samples with low DOC values require multiple reconcentration steps on XAD-8 before the compounds of interest are sufficiently concentrated. In the case of the Foxhills-Laramie sample, the initial DOC of the water was 0.7 mg C/L. Of this initial DOC, 0.1 mg C/L was humic material. This sample required many reconcentration steps to concentrate the sample to a DOC of 500, as called for in the method. Samples having a higher concentration of the solute of interest, such as those from the Suwannee River, do not require further concentration steps before processing.

A practical consideration in using XAD-8 resin should be noted. Despite exhaustive cleaning of the resin before use, the resin bleeds on the order of 3 mg C/L in 0.1 N NaOH (*5*). A major component of the bleed is acrylic acid, which has a low capacity factor on XAD-8 at pH 2, and is separated to a large extent from the sample during successive reconcentration steps. However, a small amount of contamination from resin bleed may be present in the final product.

Duolite A-7 Method. The criteria necessary for an efficient anion sorbent of organic acids from natural waters are weak-base functional groups, macroporous structure, and a hydrophilic matrix that is negatively charged at pH 10 (*7*). Diamond-Shamrock's Duolite A-7, a phenol-formaldehyde weak-base resin, is an effective resin for isolating and concentrating organic acids from water (*8*). This resin has excellent elution characteristics; 100% recovery of colored organic solutes from water was reported when loading was limited to one-half to two-thirds of resin capacity (*7*). Decreased resin performance has been reported when the resin is loaded to capacity (*7*, *12*); however, Leenheer (*13*) has not found this situation to be the case in his studies.

An extraction method that concentrates and fractionates all the organic constituents in water has been outlined in detail by Leenheer and Noyes (*10*). As shown in Figure 2, a sample is pumped, without

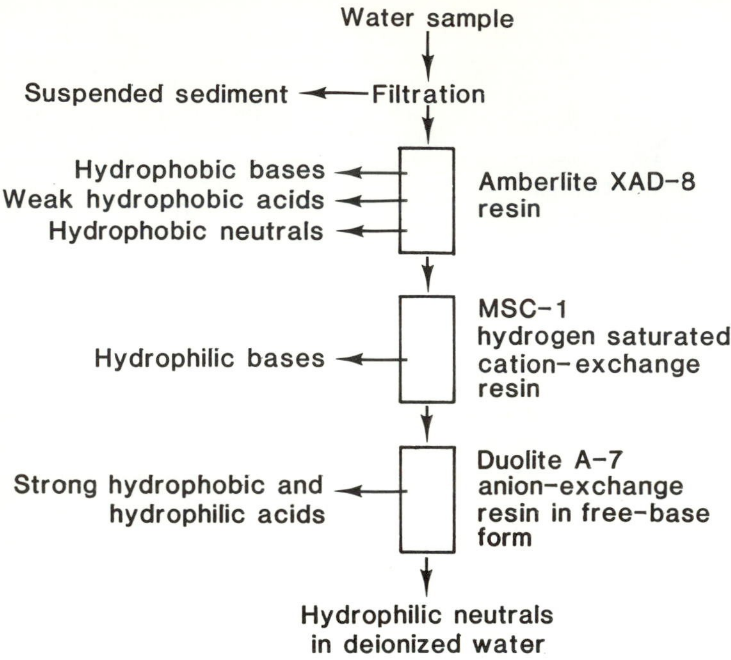

Figure 2. Fractionation of organic solutes in water by the method of Leenheer and Noyes (10).

prior pH adjustment, through a column array consisting of XAD-8, MSC-1 hydrogen-saturated cation-exchange resin, and A-7 resin in the free base form. Some very weak organic acids, such as phenols, sorb onto XAD-8 at pH 7. However, these constitute a small fraction of the acids present in a natural water sample. The majority of the organic acids in the sample, such as fulvic acid, sorb onto the A-7 resin. These acids are eluted from the column by adjusting the influent pH to 11.5.

In this method, the hydrogen-saturated cation-exchange resin, MSC-1, is essential for the collection of organic acids on the A-7 resin. Sorption of organic acids on the A-7 resin is by hydrogen bonding and ion exchange. Both of these mechanisms are pH dependent, and efficient sorption is favored at acidic pH values. The cation-exchange resin serves two functions: (1) All of the organic acids are in the hydrogen form after passing through the cation-exchange resin; and (2) the effluent from the cation-exchange resin is of low pH. XAD-8 is not a necessary part of the system, and the scheme can be simplified, if organic acids are the only solutes of interest, by omitting the XAD-8 resin.

All the organic acids present in a water sample, with the exception of the very weak organic acids, sorb onto the A-7 resin. In addition, all inorganic anions are concentrated on the A-7 resin. These inorganic anions are coeluted with the organic acids and are included in the column eluate. The organic acids can be further fractionated into hydrophobic acids and hydrophilic acids by adjusting the A-7 eluate to pH 2 and passing the solution through a column of XAD-8 and by using the k'_{cutoff} limits discussed earlier. At this stage, the hydrophobic acids can be desalted on XAD-8, hydrogen saturated, and freeze-dried in the manner described in steps 6, 7, and 8 of the box. Desalting the hydrophilic acids is a tougher problem to solve. Chromatographic and solvent-extraction methods are presently under study to improve the efficiency of this step.

The major disadvantage of the method involves the incorporation of resin bleed into the concentrated sample. Any organic acids that bleed from the acrylic ester XAD-8 resin as the sample passes through this resin will also be concentrated on A-7 with the sample. Generally, at pH values less than 10, this bleed will be low; however, it must still be considered as a contaminant and may be significant if the DOC of the original sample is low, as in a ground-water sample. Additional bleed contamination of the sample occurs in the elution of the phenol-formaldehyde A-7 resin. Bleed concentrations on the order of 30 mg C/L have been observed in the elution of A-7 at pH 11.5 (*10*). This bleed is predominantly aminophenols, formic acid, and formaldehyde. Formic acid and formaldehyde are volatile and are removable by evaporation, and the nonvolatile aminophenols can be removed by a hydrogen-saturated cation-exchange resin.

This system has been employed quite successfully to process water samples in excess of 1000 L (Table III). In each of the examples cited, a 10-L column of A-7 resin was used; with the exception of the Suwannee River sample, this quantity of resin was sufficient to process the entire sample without regenerating the resin. The method of Leenheer and Noyes (*10*) differs from the XAD-8 method in that the filtered sample does not require adjustment to pH 2 before concentration. This

Table III. Amounts of Organic Acids Extracted from Various Natural Waters with A-7 by the Method of Leenheer and Noyes (*10*)

Sample	*DOC (mg C/L)*	*Volume of Water Processed (L)*	*Hydrophobic and Hydrophilic Acids Extracted (g)*
White River, Colorado[a]	2.3	1091	1.498
Platte River, Colorado[a]	1.4	1136	1.085
Suwannee River, Georgia[b]	38.4	8102	490

[a] Data are taken from reference 10.
[b] Data are taken from reference 13.

feature allows for the incorporation of a series of in-line glass-fiber filtration tubes, which greatly speeds up the entire process. In addition, the large number of 12-gal glass jugs required to contain acidified filtrate in the XAD-8 method are no longer necessary. This method is more efficient than the XAD-8 method in both space and time. Suwannee River samples were processed by each method (Tables II and III). In 7 days, 8100 L of water was processed by the A-7 method, whereas in approximately 60 days, 10,400 L of water was processed by filtration through silver filters and the XAD-8 method. For a given sampling site, the effectiveness of the A-7 resin for concentrating organic solutes is dependent on the amounts and types of inorganic anions in the sample. Each of the examples cited above was a low-conductivity sample. For high-conductivity waters, the effectiveness of the A-7 resin decreases because of competition for adsorption sites.

Other Practical Considerations. FILTRATION. Filtration of the sample is important to separate the dissolved compounds of interest from suspended organic carbon; from biological organisms such as bacteria, algae, and microcrustaceans; and from clay minerals, all of which could interfere with subsequent analyses. Filtration through a 0.45-μm or smaller filter is the accepted practice, but this step can become prohibitive both in time and expense in a procedure of this magnitude.

Filters can be divided into two types: membrane (screen) filters and depth filters. Membrane filters, such as silver membrane filters, physically screen and retain particles on their surfaces. These filters have uniform pore sizes and are rated for absolute retention; all particles larger than the pore size are retained. Depth filters, such as glass-fiber filters, consist of a matrix of fibers that form a tortuous maze of flow channels. The particulate fraction becomes entrapped by this matrix. These filters do not have a uniform pore size, and it is not possible to rate them for absolute retention. They are rated according to nominal pore size, which is determined by the particle size that is retained by the filter to a predetermined percentage. This percentage is usually given as 98% retention; however, it can be as low as 90%.

Two types of filters are suitable for the study of organic acids in water: 0.45-μm Selas silver membrane filters and 0.3-μm Balston glass-fiber filter tubes (*11*). Silver membrane filters are advantageous in that they have uniform pore size and they do not alter the composition of dissolved organic carbon; however, they have slow flow characteristics and are expensive. On the other hand, glass-fiber filters have good flow characteristics and are economical; however, particles larger than the nominal pore size can pass the filter and slight sorption of certain organic compounds occurs. Both filters are inert with respect to organic acids in natural waters, with the exception of mercaptans on Ag filters (*11*). The glass-fiber filter tubes are supported by an inert screen, and water is pumped through the filter. Water from the Ogeechee River in Georgia, a river high in clays, was filtered through each filter type at 30 psi. The glass-fiber filters have much better flow characteristics than the silver filters (Figure 3). Even though the filters are of comparable size, the glass-fiber tubes have 2.4 times more surface area, on the basis of surface-area calculations from filter dimensions. In reality, the glass-fiber filters have an even greater surface area than the silver membrane filters because of the macroreticular nature of depth filters. Flow rate for the silver membrane filter dropped off rapidly, whereas flow rate on the glass-fiber filters dropped off gradually. The difference in flow rate resulted in 1500 mL of water being filtered by the glass-fiber filters compared to 200 mL of water being filtered by the silver membrane filters in 30 min of filtration time. The advantages of good flow characteristics are magnified many times in the number of filters used and the amount of time required for filtration.

SAMPLE PRESERVATION. Samples of naturally occurring organic acids are subject to both biological and chemical degradation. To avoid degradation of the sample, it is important to filter the sample immediately after collection. Filtration through a 0.45-μm or smaller filter effectively removes organisms as small as bacteria from the sample. After filtration, the sample should be processed as soon as possible.

Biological activity in the resin concentrates can be suppressed by the addition of a preservative. Silver has been found to be an effective bactericide in water at concentrations as low as 10 ppb (*14*) and can be added to the sample as $AgNO_3$. Adjustments of sample pH are another effective means to minimize biological degradation. *Pseudomonas* bacteria, responsible for the degradation of phenols, are destroyed at both high and low pH (*15*). High pH increases the oxidation rate of organic acids and should be avoided.

The organic compounds of interest are best protected from degradation by filtering, concentrating the sample on the resin, and eluting the sample immediately after collection. In each of the methods described in this chapter, the resin is eluted with a solution of high pH that can result in the oxidation of the compounds of interest. The eluate should be adjusted to pH 2 and refrigerated after it is collected. If delays in sample processing are anticipated, it is most important that the sample be filtered immediately. The sample should then be chilled if the delay will be more than a day before concentrating and eluting it from the resin. A sample that has been unfiltered and at room temperature for a few days is of little value and should be discarded.

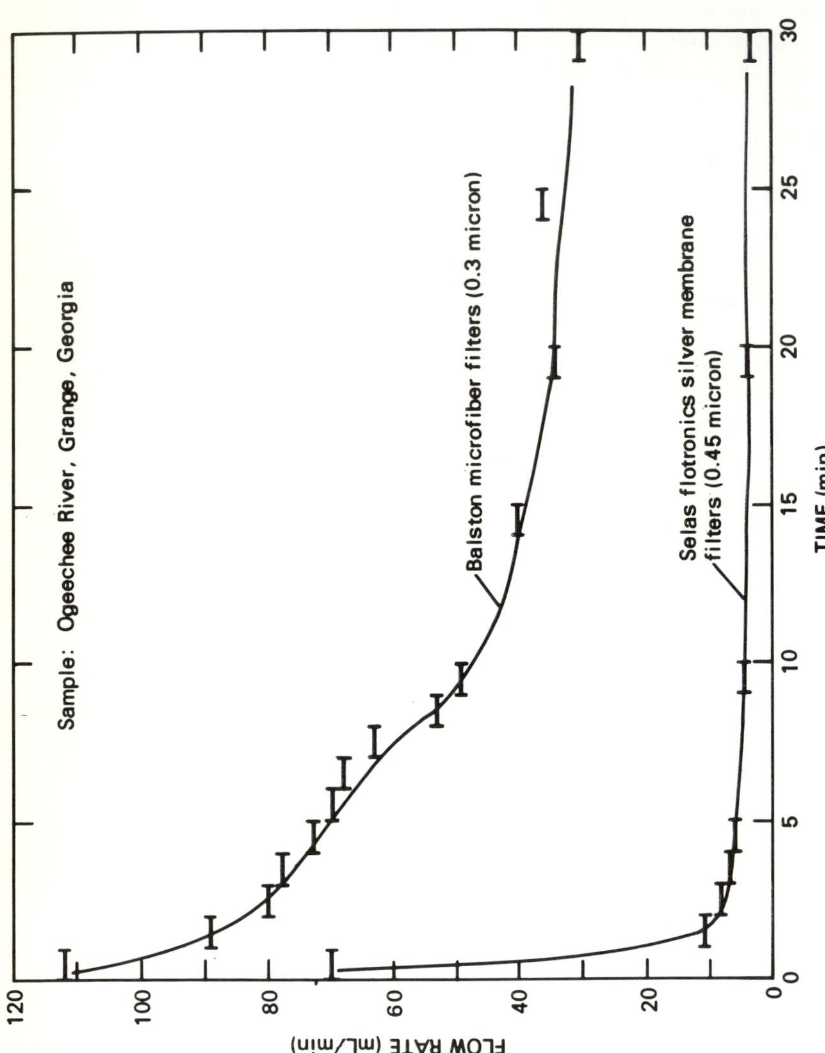

Figure 3. *Variation of flow rate with filtration time for silver membrane and glass-fiber filters at 30 psi for Ogeechee River water. (Reproduced with permission from reference 11. Copyright 1985 John Wiley.)*

Conclusions

Adsorption chromatography is an efficient way to isolate organic acids from large volumes of water. The nonionic, macroporous, Amberlite XAD-8 and the weak-base anion-exchange resin Duolite A-7 are two resins well suited for this purpose. These resins have been successfully used to extract organic acids from natural waters at sites where it was necessary to process thousands of gallons of sample.

Acknowledgments

The use of brand names in this report is for identification purposes only and does not imply endorsement by the U.S. Geological Survey.

Literature Cited

1. Grieser, M. D.; Pietrzyk, D. J. *Anal. Chem.* **1973,** *45,* 1348.
2. Junk, G. A.; Richard, J. J.; Grieser, M. D.; Witiak, D.; Witiak, J. L.; Arguello, M. C.; Vick, R.; Svec, H. J.; Fritz, J. S.; Calder, G. V. *J. Chromatogr.* **1974,** *99,* 745.
3. Pietrzyk, D. J.; Chu, C. H. *Anal. Chem.* **1978,** *50,* 497.
4. Thurman, E. M.; Malcolm, R. L.; Aiken, G. R. *Anal. Chem.* **1978,** *50,* 775.
5. Aiken, G. R.; Thurman, E. M.; Malcolm, R. L.; Walton, H. F. *Anal. Chem.* **1979,** *51,* 1799.
6. Tanford, C. *The Hydrophobic Effect: Formation of Micelles and Biological Membranes;* Wiley Interscience: New York, 1973.
7. Abrams, I. M.; Breslin, R. P. *Proc. Int. Water. Conf.* Pittsburgh, PA, 1965.
8. Leenheer, J. A. *Environ. Sci. Technol.* **1981,** *15,* 578.
9. Thurman, E. M.; Malcolm, R. L. *Environ. Sci. Technol.* **1981,** *15,* 463.
10. Leenheer, J. A.; Noyes, T. I. U.S. Geological Survey Water-Supply Paper No. 2230, 1984.
11. Aiken, G. R. In *Humic Substances in Soil, Sediment and Water: Geochemistry, Isolation and Characterization,* Aiken, G. R.; McKnight, D. M.; Wershaw, R. L.; MacCarthy, P., Eds.; John Wiley: New York, 1985; p 692.
12. Kunin, R.; Suffet, I. H. In *Activated Carbon Adsorption of Organics from the Aqueous Phase;* McGuire, M. J.; Suffet, I. H., Eds.; Ann Arbor Science: Ann Arbor, MI, 1980; Vol. 2, pp 425–442.
13. Leenheer, J. A. personal communication.
14. Woodward, R. L. *J. Am. Water Works Assoc.* **1963,** *55,* 881.
15. Afghan, B. K.; Belliveau, P. E.; LaRose, R. H.; Ryan, J. F. *Anal. Chim. Acta* **1974,** *71,* 355.

RECEIVED for review August 14, 1985. ACCEPTED December 13, 1985.

14

Use of Large-Volume Resin Cartridges for the Determination of Organic Contaminants in Drinking Water Derived from the Great Lakes

Guy L. LeBel, David T. Williams, and Frank M. Benoit

Monitoring and Criteria Division, Environmental Health Directorate, Health and Welfare Canada, Tunney's Pasture, Ottawa, Ontario, K1A OL2 Canada

A large-volume XAD-2 resin sampling cartridge was designed to sample approximately 1500 L of raw and treated drinking water to obtain sufficient organic water extracts for chemical analysis and biological testing. The concentration technique was evaluated for the extraction and subsequent chemical analysis of several target compounds in 1500 L of water samples fortified at 10 ng/L. Recoveries were greater than 70%; these recoveries were similar to earlier recoveries employing a smaller volume cartridge. The cartridges were used to obtain water extracts from six Great Lakes area drinking water supplies, and the extracts were analyzed for organophosphorus, organochlorine, polyaromatic hydrocarbon, and triaryl-alkyl phosphate compounds by gas chromatography-mass spectrometry (GC-MS). Some of the compounds were found at nanograms-per-liter concentrations in both raw and treated water samples. Analysis of fortified and field-water extracts by two GC-MS laboratories gave similar intra- and interlaboratory precision and accuracy data.

MACRORETICULAR RESINS, particularly the Amberlite XAD series, have been used extensively to isolate and concentrate trace organic compounds from drinking water (*1–8*). We have previously reported the use of an XAD cartridge for this purpose and have evaluated the system for the analysis of organophosphorus pesticides (OPs) (*4*), polynuclear aromatic hydrocarbons (PAHs) (*5*), phosphate triesters (TAAPs) (*6*), or-

0065-2393/87/0214/0309$06.00/0

ganochlorine compounds (OCs) (7), and polychlorinated biphenyls (PCBs) (8). The drinking water extracts were also tested for biological activity by using the *Salmonella*/mammalian-microsome assay (9, 10).

To provide sufficient material for chemical analysis and an extended battery of biological tests consisting of *Salmonella*/mammalian-microsome assay, sister-chromatid exchange (SCE), and micronucleus (MN) induction in Chinese hamster ovary (CHO) cells (*11*), it was necessary to modify and evaluate the XAD resin cartridge for sampling very large volumes (>1000 L) of raw and treated water.

This chapter presents the results of studies on cartridge design for evaluation using fortified water samples and the application of the modified cartridge to obtain residual organic compounds for chemical analysis of field samples of drinking water. The biological activity results have already been reported elsewhere (*11*).

Materials and Methods

XAD Resin Cartridge Design. The purification and sources of materials and sampling parameters have been described previously (*4*, *6*) for a small XAD-2 cartridge used for sampling up to 2000 L of water. A large-volume cartridge for sampling up to 2000 L of water was constructed from stainless steel tubing and fittings and had a bed volume of approximately 325 mL of XAD-2 resin. A cross section of the sampling cartridge is illustrated in Figure 1. The dimensions and sampling parameters of the two types of cartridges are shown in Table I. By using a relative flow rate similar to that of the small-volume cartridge, that is, 3 bed volume/min, the sampling rate for the large-volume cartridge was approximately 1 L/min; this rate allowed for the collection of a total sample volume of approximately 1500 L during a 24-h period.

Evaluation of Recovery of Organic Compounds with the LV-XAD-2 Cartridge. The recoveries of selected organic compounds

Table I. Parameters for XAD-2 Sampling Cartridges

Parameter	*Small-Volume Cartridge*	*Large-Volume Cartridge*
Length	40.5 cm	40.5 cm
Outer diameter	1.6 cm	3.8 cm
Inner diameter	1.4 cm	3.2 cm
Bed volume	45 mL	325 mL
Flow rate (3 bed vol/min)	140 mL/min	1000 mL/min
Total volume/24 h	200 L	1450 L
Eluent volume	300 mL	1000 mL

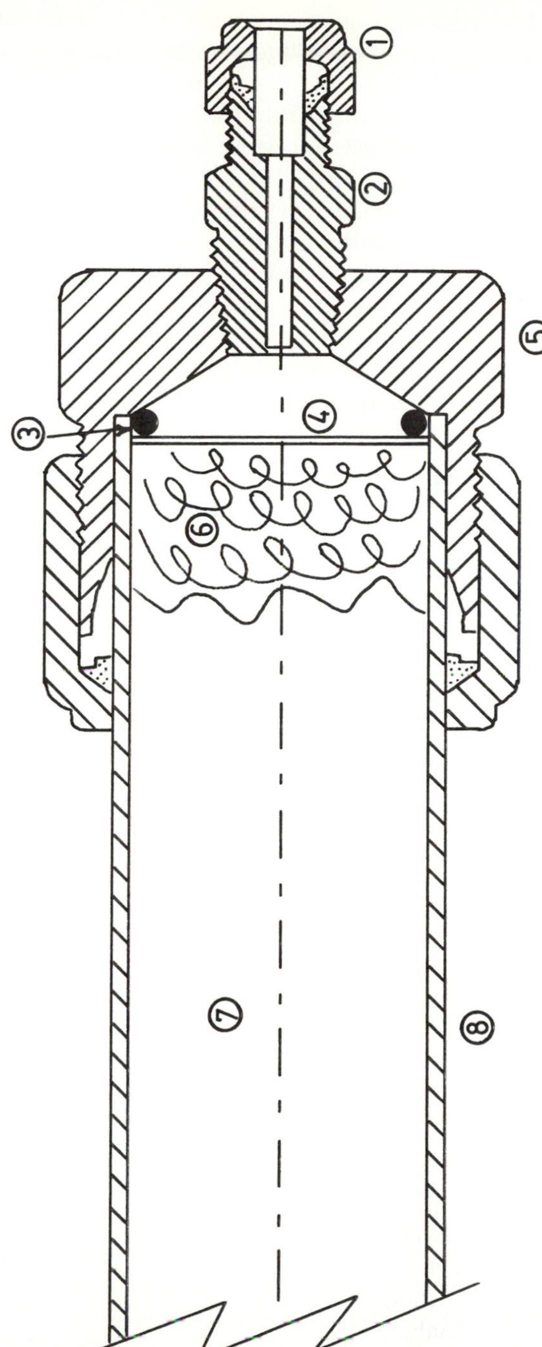

Figure 1. Cross section of sampling cartridge: 1, stainless steel swagelok nut, 0.64 cm ($\frac{1}{4}$ in.), no. SS-402-1; 2, stainless steel swagelok male connector, 0.95 cm ($\frac{3}{8}$ in.), NPT no. SS-400-1-6; 3, Teflon O-ring, 2.664 cm i.d., wall thickness 0.262 cm; 4, stainless steel screen, 40 mesh; 5, stainless steel swagelok cap (modified with female $\frac{3}{8}$ in. NPT through end), no. SS-2400-C; 6, Pyrex brand glass wool; 7, XAD-2 Amberlite resin; and 8, 304 stainless steel tube, 3.81 × 0.305 cm (1.5 × 0.120 in.), ASTM A269.

were determined by the direct fortification technique (*6*). For this technique, a known amount of organic compounds was added directly to the XAD resin column. A second cartridge was attached in series to the fortified cartridge to determine whether low recoveries of specific compounds were due to breakthrough. After 1500 L of tap water had passed through the cartridges, the residual water was drained from the cartridges, which were then eluted with 1 L of 15% acetone/hexane (v/v). The organic phase from the eluate was dried by filtration through purified anhydrous sodium sulfate and was then concentrated to a small volume. The aqueous phase of the eluate was extracted with methylene chloride to determine the loss of the more polar organic compounds. For organochlorine determinations, the extracts were further purified by Florisil (magnesium silicate) column chromatography (*12*) before analysis by capillary gas chromatography with an electron capture detector (GC–ECD).

Field Samples. Duplicate cartridges were connected in parallel to both raw and finished water outlets at each water processing plant after the taps were flushed for at least 15 min (*10*). The flow rate of water through each cartridge was set at 1 L/min, monitored, and, if necessary, adjusted at set intervals. After 24 h, the cartridges were disconnected, sealed with swagelok nuts, and brought back to the laboratory. The adsorbed organic compounds were eluted from the resin with 1 L of 15% acetone/hexane followed by acetone (350 mL). The eluates were each concentrated to 10 mL and stored at 4 °C. Blank extracts were obtained by elution of regenerated XAD-2 cartridges (*4*). Two aliquots (10% extract-aliquot) of the acetone/hexane eluates were used for gas chromatographic–mass spectrometric (GC–MS) and gas chromatographic (GC) analyses by two different laboratories. The remaining extracts were retained for biological testing (*11*).

Chemical Analysis of Extracts. The extracts were analyzed by capillary column GC–MS for OCs, TAAPs, and PAHs (*see* the list on page 313). The GC–MS parameters used at the two laboratories are shown in Table II. The identification and quantitation were all done by using automatic routines based on a mass spectra library created from authentic standards of the selected compounds. Compounds were located by searching the reconstructed ion chromatogram for each library entry within a narrow retention time window relative to the internal standard (anthracene-d_{10} or phenanthrene-d_{10}). Quantitation was achieved by comparison of characteristic ion areas in the field samples with ion areas of the internal standard. These ion areas were normalized by response factors established by comparison of ion ratios of a standard mixture of all 66 analytes at a concentration of 2.5 ng/μL.

The extracts were also analyzed for OPs and TAAPs by capillary GC equipped with a nitrogen–phosphorus selective detector (GC–NPD) (*6, 8*).

List of Organic Compounds Analyzed in XAD-2 Extracts

OCs	*PAHs*	*TAAPs*
1,4-Dichlorobenzene	Naphthalene	Triethyl phosphate
1,2-Dichlorobenzene	2-Methylnaphthalene	Tributyl phosphate
1,3,5-Trichlorobenzene	1-Methylnaphthalene	Tris(2-chloroethyl) phosphate
1,2,3-Trichlorobenzene	Biphenyl	Tris(1,3-dichloropropyl) phosphate
2,4,5-Trichlorotoluene	1-Ethylnaphthalene	Triphenyl phosphate
1,2,3,5-Tetrachlorobenzene	2-Ethylnaphthalene	Tributoxyethyl phosphate
1,2,3,4-Tetrachlorobenzene	2,6-Dimethylnaphthalene	2-Ethylhexyldiphenyl phosphate
Pentachlorobenzene	1,3-Dimethylnaphthalene	*o*-Isopropylphenyldiphenyl phosphate
Hexachlorobenzene	1,4-Dimethylnaphthalene	Tri-*o*-tolyl phosphate
γ-BHC	1,5-Dimethylnaphthalene	Tri-*m*-tolyl phosphate
α-BHC	2,3-Dimethylnaphthalene	*p*-*tert*-Butylphenyldiphenyl phosphate
β-BHC	1,2-Dimethylnaphthalene	Tri-2,4-xylyl phosphate
Aldrin	Bibenzyl	
Dieldrin	*cis*-Stilbene	*OPs*
Octachlorostyrene	*trans*-Stilbene	Dimethoate
p,p'-DDE	2,3,6-Trimethylnaphthalene	Diazinon
p,p'-DDT	2,3,5-Trimethylnaphthalene	Phosphamidon
Photomirex	Fluorene	Methylparathion
Mirex	3,3'-Dimethylbiphenyl	Ronnel
Hexabromobiphenyl	4,4'-Dimethylbiphenyl	Fenitrothion
Heptachlor epoxide	9-Fluorenone	Malathion
Oxychlordane	Phenanthrene	Dursban
γ-Chlordane	Anthracene	Parathion
α-Chlordane	2-Methylanthracene	Ruelene
α-Endosulfan	9-Methylanthracene	Supracide
β-Endosulfan	Anthraquinone	Ethion
	Fluoranthene	Trithion
	Pyrene	Imidan
		EPN
		Azinphos-methyl
		Phosalone

NOTE: OCs and PAHs were analyzed by GC-MS; TAAPs were analyzed by GC-MS and GC-NPD; and OPs were analyzed by GC-NPD.

Results and Discussion

Large quantities of organic extracts of Great Lakes raw and treated water were required for chemical analysis and for biological testing. To provide a sufficient amount of material for these tests, an XAD-2 cartridge was designed to allow the sampling of 1500 L of water, about 7 times the volume of our earlier studies (*10*). Water samples were collected during a 24-h period at a flow rate of 1 L/min, that is, 3 bed volume/min.

Table II. GC–MS Parameters for Analysis of LV–XAD Extracts

Parameter	*Lab A*	*Lab B*
System	Finnigan 4100–INCOS	Finnigan 4500–INCOS
Column	DB-5 15 m × 0.25 mm i.d.	DB-1 30 m × 0.25 mm i.d.
Sensitivity	10^{-7} A/V	10^{-8} A/V
Filament current	0.3 A	0.5 A
Electrometer voltage	1350 V	1350 –≥ 1450 V
Injection volume	3 μL splitless	2 μL splitless
Injection temperature	250 °C	250 °C
Oven temperature program		
Initial temperature	50 °C	50 °C
Hold time	0.1 min	0.1 min
Final temperature (level 1)	100 °C	180 °C
Rate (level 1)	15 °C/min	8 °C/min
Final temperature (level 2)	270 °C	285 °C
Rate (level 2)	5 °C/min	10 °C/min
Hold time	60 min total run	10 min
Internal standard	anthracene-d_{10}	phenanthrene-d_{10}
Quantitation (number of ions)	4 (average)	2 (average)

The recovery of selected OP and OC compounds was determined by fortifying the XAD-2 resin at a level equivalent to 10 ng/L, and 1500 L passed through the cartridge. The concentrated extracts from the fortified samples were analyzed by GC–NPD (Table III) and GC–ECD (Table IV). The results in Tables III and IV show similar recoveries for the small and the large cartridges. This finding indicates that proportional increases in the volumes of XAD-2 resin and water samples did not cause significant changes in recovery of the organic compounds. No OCs and only a small percentage of the more polar TAAPs (Table III) broke through the first column onto the second XAD-2 cartridge. Only tris(2-chloroethyl) phosphate (13.9%) was detected in the extract of the aqueous phase derived from solvent elution of the XAD-2 cartridge. Although the percentage recoveries were not determined for other organic compounds (i.e., PAHs, OC and OP pesticides, and PCBs), we believe from our previous studies (*4–8*) that the percentage recoveries would be similar for these other classes of compounds.

To field test the large-volume cartridges, the cartridges were used to accumulate the waterborne organic compounds in drinking water supplies from five selected Great Lakes areas: Sault Ste. Marie, Amherstburg, Fort Erie, Burlington, and Cornwall. Barrie, which is about 25 km from the Great Lakes and has ground water supply, was chosen as a control site. Duplicate organic concentrates from raw and treated water were obtained from each site once during the summer of 1982 and once during the winter of 1983. Aliquots of the extracts were

Table III. TAAP Percent Recoveries at 10-ng/L Fortification Level

TAAP	Large Volume (1500 L) 1st Cartridge	Large Volume (1500 L) 2nd Cartridge	Low Volume (200 L)[a]
Tributyl phosphate	112.9[b]	0.3	101.2
Tris(2-chloroethyl) phosphate	81.8 (13.9)[c]	5.5 (1.6)[c]	95.5
Tris(1,3-dichloropropyl) phosphate	96.8	—	nd
Triphenyl phosphate	97.8	—	107.4
Tributoxyethyl phosphate	101.6	1.8	141.0
2-Ethylhexyldiphenyl phosphate	95.5	—	nd
o-Isopropylphenyldiphenyl phosphate	97.0	—	nd
Tri-*o*-tolyl phosphate	94.6	—	97.0 (cresyl)
Tri-*m*-tolyl phosphate	92.2	—	nd
p-*tert*-Butylphenyldiphenyl phosphate	95.9	—	nd
Tri-2,4-xylyl phosphate	90.7	—	nd

NOTE: —means not found; nd means not determined.
[a] Data are taken from reference 6.
[b] Value is from a single determination.
[c] Value is for the aqueous phase (*see* text).

analyzed by two laboratories by GC–MS for 66 analytes (listed on page 313), which were selected on the basis of the results of earlier studies.

The results of analyses of the extracts by GC–MS are presented in Tables V–VII where the analytes detected are listed. The results presented in the tables are the average results of the duplicate samples and both laboratories when a compound was detected by both analysts.

Table IV. Percent Recoveries for OC Compounds at 10-ng/L Fortification Level

OC	Large Volume (1500 L)	Low Volume (200 L)
1,4-Dichlorobenzene	70.0	67.0
1,2-Dichlorobenzene	80.7	68.3
1,3,5-Trichlorobenzene	70.0	69.6
1,2,3-Trichlorobenzene	86.0	79.7
2,4,5-Trichlorotoluene	81.3	79.4
1,2,3,5-Tetrachlorobenzene	103.3	87.5
1,2,3,4-Tetrachlorobenzene	88.7	86.9
Pentachlorobenzene	76.7	90.8
Hexachlorobenzene	76.0	89.7
Octachlorostyrene	74.7	89.0
Photomirex	79.3	79.8
Mirex	72.0	68.4
Hexabromobiphenyl	74.0	61.9

NOTE: No OCs were found in the backup cartridge.

Table V. PAH Levels (ng/L) in Great Lakes Water as Determined by GC–MS

Compound	*Barrie*		*Cornwall*		*Burlington*		*Fort Erie*		*Amherstburg*		*Sault Ste. Marie*	
	S	W	S	W	S	W	S	W	S	W	S	W
Naphthalene	14.5	9.4	3.4	5.9	15.1	12.8	3.5	6.6	3.9	23.2	48.9	95.7
	16.8	8.4	40.5	4.2	9.8	4.1	5.2	9.9	5.2	20.0	101.0	23.7
Methylnaphthalene	—	—	0.9[b]	—	61.9	72.2	2.2	1.8	4.7	32.3	9.4	10.4
	—	—	1.5	1.3	5.3	2.9	3.1	4.4	1.1	17.5	14.7	2.6
Dimethylnaphthalene	—	—	—	—	63.1	232.4	—	—	1.8	19.5	—	—
	—	—	—	—	0.2	2.3	—	0.3[b]	—	3.9	5.0	—
Trimethylnaphthalene	—	—	—	—	23.2	99.8	—	—	—	2.7	—	—
	—	—	—	—	—	—	—	—	—	—	—	—
Biphenyl	—	—	—	—	1.6[b]	4.0	—	—	0.3[b]	1.8	0.8[b]	—
	—	—	—	—	—	—	—	—	0.3[b]	1.3	2.4	—
3,3′-Dimethylbiphenyl	—	—	—	—	—	5.7	—	—	—	—	—	—
	—	—	—	—	—	—	—	—	—	—	—	—
4,4′-Dimethylbiphenyl	—	—	—	—	—	5.7[b]	—	—	—	—	—	—
	—	—	—	—	—	—	—	—	—	—	—	—

Fluorene	0.6	0.4	—	—	5.4	18.9	—	—	0.4[b]	1.9	2.4	—
	0.1	0.4	—	—	0.8	—	0.9	—	0.7	1.6	5.6	0.9
9-Fluorenone	—	—	0.6[b]	—	—	—	—	0.7[b]	—	1.6	—	—
	—	—	1.1	0.4[b]	1.3	—	1.6	1.4	2.1	2.5	3.1	0.3[b]
Phenanthrene	1.0	—	0.5	—	13.1	33.7	1.3	—	0.7	2.4	6.5	—
	0.9	—	0.7	0.6	1.5	0.5	3.0	1.7	2.2	2.1	23.1	5.3
Anthracene	—	—	0.3[b]	—	—	1.2	—	—	—	—	—	—
	—	—	—	—	—	—	—	—	—	—	0.4	—
Methylanthracene	—	—	—	—	14.0	20.5	—	—	—	—	—	—
	—	—	—	—	—	—	—	—	—	—	3.0	—
Anthraquinone	—	—	—	—	—	—	—	—	—	1.0[b]	—	—
	—	—	—	—	0.3[b]	0.2[b]	—	0.3[b]	2.4	0.6	2.7	—
Fluoranthene	0.6[b]	0.3[b]	0.8[b]	1.0[b]	0.8[b]	2.6	0.5[b]	1.3[b]	—	4.2	4.9	1.2
	0.4[b]	0.2[b]	0.3	0.8[b]	0.4[b]	0.6	0.6	0.7	1.1[a]	0.7	9.8	2.2
Pyrene	—	—	0.3[b]	0.7[b]	0.7[b]	3.4	0.6[b]	0.8[b]	1.0	8.5	1.4	1.2[b]
	—	—	—	0.3[b]	—	0.5	0.3[b]	0.7	0.2[b]	0.8[b]	3.1	1.1[b]

NOTE: S indicates summer, and W indicates winter. For each compound, the first row gives levels for raw samples, and the second row gives levels for treated samples.
[a] Laboratory A.
[b] Laboratory B.

Table VI. TAAP Concentrations (ng/L) in Great Lakes Water As Determined by GC–MS

Compound	*Barrie*		*Cornwall*		*Burlington*		*Fort Erie*		*Amherstburg*	
	S	W	S	W	S	W	S	W	S	W
Triethyl phosphate	—	—	14.3	11.9	7.8	3.4	—	—	—	—
	—	—	7.0	5.3	11.6	8.2	—	—	—	—
Tributyl phosphate	—	—	8.3	4.5	2.3	2.8	1.4	0.9	1.5	—
	—	0.4	4.3	2.1	3.0	4.9	1.9	1.3	1.2	0.4
Tris(2-chloroethyl) phosphate	—	—	3.7	0.8	—	1.6[b]	1.3[b]	1.7[b]	—	—
	—	—	1.7	0.6	2.6	1.6[b]	1.3[b]	1.0	—	—
Tris(1,3-dichloropropyl) phosphate	—	—	—	—	—	—	—	—	—	—
	—	—	—	—	2.2[b]	—	—	—	—	—
Triphenyl phosphate	—	—	—	—	12.9	—	—	—	5.9	6.1
	—	—	—	—	—	—	—	—	6.0	6.5
Tributoxyethyl phosphate	—	—	9.8	—	54.4	16.6	—	—	—	10.7
	—	—	0.2	—	11.0	6.4	—	—	—	3.6

NOTE: S indicates summer, and W indicates winter. For each compound, the first row gives levels for raw samples, and the second row gives levels for treated samples. None of the TAAP compounds were detected in Sault Ste. Marie samples. *o*-Isopropylphenyldiphenyl phosphate was only detected by laboratory B in the treated, summer, Amherstburg sample at a concentration of 1.2 ng/L.
[a] Laboratory A (not shown).
[b] Laboratory B.

Table VII. OC Concentrations (ng/L) in Great Lakes Water as Determined by GC–MS

	Barrie		*Cornwall*		*Burlington*		*Fort Erie*		*Amherstburg*	
Compound	S	W	S	W	S	W	S	W	S	W
1,4-Dichlorobenzene	0.2	—	0.4	—	1.6	0.2	—	—	0.4	1.6
	0.4	—	0.2	—	1.9	1.9	—	0.4	0.6	1.0
1,2-Dichlorobenzene	—	—	0.6	—	—	—	—	—	0.6[b]	0.4[b]
	—	—	0.9	—	0.2	0.3	—	—	—	0.5[b]
α-BHC	—	—	2.2	0.8	—	—	0.8	1.4	1.8	1.4
	—	—	2.5	0.8	2.5	2.1	2.6	1.5	3.8	2.0

NOTE: S indicates summer, and W indicates winter. For each compound, the first row gives levels for raw samples, and the second row gives levels for treated samples. None of the OC compounds were detected in Sault Ste. Marie samples. Dieldrin was only detected by laboratory A in the raw, winter, Cornwall sample at a concentration of 13.8 ng/L.

[a] Laboratory A (not shown).

[b] Laboratory B.

However, when one laboratory reported detection of a compound in both replicates and that compound was not reported by the other laboratory, only the average for the one laboratory was reported. Single occurrences were not reported.

The field-sample extracts were also analyzed for TAAPs and OP pesticides by capillary GC–NPD; however, no OP pesticides were detected. The TAAP results are shown in Table VIII. The use of the NPD-specific detector permitted lower detection limits than GC–MS for the TAAP (8); however, all of the higher concentrations were in agreement with results obtained by GC–MS. Similar to our earlier surveys, triethyl phosphate levels were highest in the eastern section of the Great Lakes system.

To estimate the validity of the results obtained by GC–MS analysis of XAD extracts, eight fortified drinking water extracts were analyzed by two GC–MS laboratories. Several target-compound mixture solutions were prepared, and fortified extract was made by spiking various aliquots of the mixture solutions into analyte-free LV–XAD extracts. The extracts were fortified with at least two concentrations of target compounds (*see* the list below). By assuming a 150-L representative water extract, the analyte amounts would be equivalent to approximately 1–100 ng/L of water.

The results from the two laboratories (Tables IX and X) show an overall percentage deviation of 43.3% from the fortified values. No strong trends related to the parameters tested were evident, although laboratory A experienced poorer accuracy at the lower spiking levels.

Organics and Fortification Levels (ng/μL) Used in GC–MS Accuracy Study

OCs

1,2,3-Trichlorobenzene (6.0, 1.6)
2,4,5-Trichlorotoluene (6.0, 1.8)
1,2,3,4-Tetrachlorobenzene (3.0, 0.6)
Pentachlorobenzene (4.5, 1.2)
Hexachlorobenzene (1.5, 0.8)
γ-BHC (4.5, 0.6)
Octachlorostyrene (1.5, 0.4)
Heptachlor epoxide (12.0, 1.6)
γ-Chlordane (4.5, 0.9)
α-Chlordane (3.0, 0.8)
α-Endosulfan (6.0, 3.2)
p,p'-DDE (7.5, 1.5)
Dieldrin (1.5, 0.8)
Mirex (15.0, 2.0)

TAAPs

Triethyl phosphate (7.5, 4.0)
Tris(1,3-dichloropropyl) phosphate (30.0, 6.0)
Triphenyl phosphate (3.0 [4×], 0.8, 1.6, 0.4, 0.6)
o-Isopropylphenyldiphenyl phosphate (7.5, 1.0)
Tri-*o*-tolyl phosphate (4.5, 1.2)

PAHs

Naphthalene (1.5, 0.4)
2-Methylnaphthalene (4.5, 2.4)
Fluorene (9.0, 1.2)
9-Fluorenone (6.0, 1.2)

NOTE: Fortified extracts were 0.5 mL.

Table VIII. TAAP Concentrations (ng/L) in Great Lakes Water As Determined by GC–NPD

	Barrie		*Cornwall*		*Burlington*		*Fort Erie*		*Amherstburg*	
Compound	S	W	S	W	S	W	S	W	S	W
Triethyl phosphate	—	—	11.8	9.9	9.5	5.2	1.0	0.9	0.1	—
	—	—	11.2	10.1	11.2	7.0	3.3	0.6	1.1	0.3
Tributyl phosphate	—	—	9.5	4.1	5.3	3.5	1.6	0.3	1.6	—
	—	0.1	7.8	2.2	7.4	4.8	3.3	1.2	1.2	0.8
Tris(2-chloroethyl) phosphate	—	—	9.6	6.2	4.2	3.0	5.7	6.0	3.0	4.0
	—	—	6.3	3.2	6.2	4.4	9.1	4.6	6.5	4.9
Tris(1,3-dichloropropyl) phosphate	—	—	2.5	1.7	1.5	0.7	1.3	1.3	0.3	0.4
	—	—	1.2	0.9	2.1	1.5	2.2	1.2	0.5	0.4
Triphenyl phosphate	—	—	1.1	0.6	20.4	2.7	0.3	0.2	8.4	4.5
	—	—	0.9	1.1	1.2	0.9	1.3	0.3	8.6	8.0
Tributoxyethyl phosphate	0.4	1.8	16.4	4.4	73.8	19.8	3.5	2.8	5.8	8.1
	0.4	1.3	4.1	1.8	16.4	8.8	8.0	2.1	5.5	7.8
o-Isopropylphenyldiphenyl phosphate	—	—	—	—	—	—	—	—	1.8	0.7
	—	—	—	—	—	—	—	—	1.3	0.6

NOTE: S indicates summer, and W indicates winter. For each compound, the first row gives levels for raw samples, and the second row gives levels for treated samples. None of the TAAP compounds were detected in Sault Ste. Marie samples except for triphenyl phosphate, which was detected in the raw, summer sample at a concentration of 0.1 ng/L. All values are the mean of two replicate extracts.

Table IX. Accuracy of GC–MS Analysis of Fortified Extracts by Compound Groups

Compound	*No. of Determinations per Lab*	*Mean Percent Deviation from Fortified Concentration*		
		All	*Lab A*	*Lab B*
All	52	43.3	50.1	36.5
PAHs	8	25.0	17.1	32.9
TAAPs	16	45.8	49.4	42.1
OCs	28	47.1	59.8	34.3

Poorer accuracy is to be expected at the low levels because the uncertainty of the measurement rises as the minimum detection limit is approached. Laboratory B reported better overall accuracy than laboratory A, particularly for the OCs.

To estimate the precision of the analysis, five of the field-sample extracts were analyzed by GC–MS in duplicate at laboratory A. The results, gathered by compound classes, are shown in Table XI. The table also includes the results of a single analysis from laboratory B. The average percent deviation of the duplicate analyses from the mean value was 16%, whereas the average percent deviation between the results from laboratory B and the mean value from laboratory A was 22%. A comparison of the results from a single analysis of the treated water extracts (summer and winter) for class aggregate concentration in laboratories A and B (Table XII) showed an average percent deviation of 20%. Hence, the intra- and interlaboratory uncertainties in the results are comparable.

To examine the effect of seasons or water types on the concentrations of organic compounds, the aggregate sums for the summer and winter water extracts were tabulated (Tables XIII and XIV) for each water type. On the basis of the uncertainty data just presented, and with one exception (*see* Tables XIII and XIV, Burlington PAHs results), no apparent differences or trends in aggregate organic compound

Table X. Accuracy of GC–MS Analysis of Fortified Extracts by Level of Fortification

Level (ng/extract)	*No. of Determinations per Lab*	*Mean Percent Deviation from Fortified Concentration*		
		All	*Lab A*	*Lab B*
<500	12	47.8	65.3	30.3
500–5000	37	39.8	44.7	35.9
>5000	3	62.6	55.4	69.8

Table XI. Aggregate Concentrations (ng/L) in Great Lakes Water by GC–MS

Sample[a]	*Lab*	*Injection No.*	*PAHs*	*TAAPs*	*OCs*
Cornwall (R, S)	A	1	8.8	35.6	3.4
		2	3.6	35.3	3.5
	B	—	5.8	35.1	4.0
Cornwall (T, W)	A	1	12.9	10.2	2.0
		2	14.5	10.8	1.8
	B	—	5.3	6.1	.5
Fort Erie (T, W)	A	1	22.3	2.0	1.8
		2	32.4	.7	2.5
	B	—	22.6	2.1	2.8
Amherstburg (T, S)	A	1	15.0	10.7	4.2
		2	8.2	7.9	2.8
	B	—	13.1	9.4	3.9
Amherstburg (R, W)	A	1	125.1	22.3	5.4
		2	119.9	11.0	5.0
	B	—	106.1	15.4	3.5

[a] S indicates summer, and W indicates winter; R means raw sample, and T means treated sample.

concentrations could be found for the water types and seasonal variation. The aggregate organic compound concentrations for the treated water were also of the same magnitude as those found in our 1980 survey of Great Lakes water quality (*10*).

All the results have been corrected for blank contribution. The XAD-2 cartridge blanks for the evaluation stage and the field cartridge blanks were essentially free of all of the compounds analyzed. However, the field samples, as well as the field blanks, had significant amounts

Table XII. Comparison of Results (ng/L) from Two GC–MS Laboratories

		TAAPs		*OCs*		*PAHs*	
City	*Season*[a]	*Lab A*	*Lab B*	*Lab A*	*Lab B*	*Lab A*	*Lab B*
Barrie	S	—	—	0.5	0.3	20.0	16.2
	W	0.7	0.2	—	—	6.0	11.7
Cornwall	S	7.6	16.3	2.4	5.0	45.7	43.5
	W	7.8	8.2	0.4	1.2	5.0	4.3
Burlington	S	27.6	31.2	5.1	4.3	20.1	18.3
	W	19.4	21.5	5.5	3.1	10.0	12.9
Fort Erie	S	2.6	5.2	2.5	2.8	14.2	15.8
	W	2.7	2.1	1.4	2.4	16.2	22.7
Amherstburg	S	7.2	10.2	4.6	4.1	15.1	14.2
	W	9.0	12.0	2.0	4.6	40.5	60.6
Sault Ste. Marie	S	—	1.6	—	3.8	157.9	190.2
	W	—	0.4	1.7	2.1	26.7	44.9

[a] S indicates summer, and W indicates winter.

Table XIII. Aggregate Results (ng/L) for Treated Water (GC–MS)

Compond	*Barrie*		*Cornwall*		*Burlington*		*Fort Erie*		*Amherstburg*		*Sault Ste. Marie*	
	S	W	S	W	S	W	S	W	S	W	S	W
TAAPs	—	0.5	12.0	8.0	29.4	20.5	3.9	2.4	8.7	10.5	0.8	0.2
OCs	0.4	—	3.7	0.8	4.7	4.3	2.7	1.9	4.4	3.3	1.9	2.0
PAHs	18.1	8.9	44.6	4.6	19.2	11.5	15.0	19.5	14.7	50.5	174.1	35.8

NOTE: S indicates summer, and W indicates winter.

Table XIV. Aggregate Results (ng/L) for Raw Water (GC–MS)

Compond	*Barrie*		*Cornwall*		*Burlington*		*Fort Erie*		*Amherstburg*		*Sault Ste. Marie*	
	S	W	S	W	S	W	S	W	S	W	S	W
TAAPs	1.5	—	36.5	18.8	77.4	23.6	2.7	1.8	9.2	16.9	—	—
OCs	0.2	—	3.2	0.8	1.6	0.6	0.9	1.5	2.6	3.2	—	—
PAHs	16.5	10.2	5.4	7.1	202.4	518.4	8.0	10.1	13.1	99.7	74.4	110.1

NOTE: S indicates summer, and W indicates winter.

of diethylhexyl phthalate (DEHP) (0.5–20 μg/L of DEHP for an equivalent 150-L aliquot). This finding was traced to the inadvertent use of a short piece of Tygon tubing by the analyst during the extract preparation stage, although the protocol required the avoidance of all plastic/rubber components. This interference did not cause serious difficulties during the GC–MS analysis because the resolving power of the capillary column allowed the isolation of DEHP from the other analytes, and/or the data system could ignore the ions due to DEHP when generating reconstructed ion chromatograms.

Acknowledgments

The collection of samples and preparation of the XAD extracts were carried out under contract (H&W no. 885) by Concord Scientific Corporation, Downsview, Ontario. Part of the GC–MS analysis was performed under contract (H&W no. 1084) by Mann Testing Laboratories, Mississauga, Ontario.

Literature Cited

1. Junk, G. A.; Richard, J. J.; Greiser, M. D.; Witiak, D.; Witiak, J. L.; Arguello, M. D.; Vick, R.; Svec, H. J.; Fritz, J. S.; Calder, G. V. *J. Chromatogr.* **1974,** *99*, 745–762.
2. Burnham, A. K.; Calder, G. V.; Fritz, J. S.; Junk, G. A.; Svec, H. J.; Willis, R. *Anal. Chem.* **1972,** *44*, 139–142.
3. Van Rossum, P.; Webb, R. G. *J. Chromatgr.* **1978,** *150*, 381–392.
4. LeBel, G. L.; Williams, D. T.; Griffith, G.; Benoit, F. M. *J. Assoc. Off. Anal. Chem.* **1969,** *62*, 241–249.
5. Benoit, F. M.; LeBel, G. L.; Williams, D. T. *Int. J. Environ. Anal. Chem.* **1979,** *6*, 277–287.
6. LeBel, G. L.; Williams, D. T.; Benoit, F. M. *J. Assoc. Off. Anal. Chem.* **1981,** *64*, 991–998.
7. McNeil, E.; Otson, R.; Miles, W.; Rajabalee, F. J. M. *J. Chromatogr.* **1977,** *132*, 277–286.
8. LeBel, G. L.; Williams, D. T. *Bull. Environ. Contam. Toxicol.* **1980,** *24*, 397–403.
9. Nestmann, E. R.; LeBel, G. L.; Williams, D. T.; Kowbel, D. J. *Environ. Mutagenesis* **1979,** *1*, 337–345.
10. Williams, D. T.; Nestmann, E. R.; LeBel, G. L.; Benoit, F.M.; Otson, R. *Chemosphere* **1982,** *11*, 263–276.
11. Douglas, G. R.; Nestmann, E. R.; LeBel, G. L. *Environ. Health Perspect.* in press.
12. LeBel, G. L.; Williams, D. T. *J. Assoc. Off. Anal. Chem.* **1983,** *66*, 691–699.

RECEIVED for review August 14, 1985. ACCEPTED February 10, 1986.

15

Broad Spectrum Analysis of Resin Extracts: A Base Extraction Cleanup Procedure

J. Gibs[1] and I. H. Suffet

Environmental Studies Institute, Drexel University, Philadelphia, PA 19104

A base extraction procedure was developed to minimize the degradation of the performance of fused-silica capillary chromatographic columns used to analyze XAD resin extracts. The degradation of the capillary gas chromatographic column was apparently caused by acidic nonvolatiles called humic materials. The humic materials were absorbed on XAD resins and eluted by nonpolar solvents along with the nonpolar organic compounds of interest in the samples. The base extraction procedure removed approximately 84% of the humic materials present in the ether extract.

BROAD SPECTRUM GAS CHROMATOGRAPHIC CAPILLARY ANALYSIS of trace organic chemicals in water can be defined as a method that analyzes, at one time, the largest possible number of chemicals contained in a sample. Broad spectrum organic analyses are best applied to (1) samples needing minimum pretreatment and (2) samples containing a large molecular weight range of organic components.

The usefulness of broad spectrum analysis is based upon being able to observe the changes in water quality data represented by differences between chromatograms. Thus, the data analysis involves the interpretation of large quantities of chromatographic data at one time. Therefore, sample and instrument quality assurance becomes extremely critical for reliable comparison of interchromatographic data (*1*).

Broad spectrum chromatographic analysis (*2–8*) has been used to determine order of magnitude changes between samples by (1) evaluating peak-to-peak changes in gas chromatographic (GC) detector

[1] Current address: Water Resources Division, U.S. Geological Survey, Trenton, NJ 08628

0065–2393/87/0214/0327$06.00/0

response (*2, 4, 5, 8*), (2) noting visual impressions of the overall relative numbers of peaks and their areas (*6*), (3) noting changes in the response in parts of different chromatograms (*2–4, 6, 7*), and (4) counting peaks that have different concentration levels (*7*).

The change from packed to capillary GC columns made possible the separation of extremely complex mixtures of organic chemicals at nanograms-per-liter concentrations. Capillary GC has increased the number of peaks observed in a typical drinking water sample from the Delaware River at Philadelphia, PA, from 60 peaks with packed-column GC to more than 250 (*8*). This situation indicates the increased information now available with capillary GC analysis.

The increased ability of capillary GC to resolve organic species caused an unanticipated drawback for broad spectrum analysis. The amount of stationary phase on a capillary GC column is much less than on a packed column. This condition increases the likelihood of observing a decrease in chromatographic performance caused by the sample matrix. The altered stationary phase may cause reduced precision of retention times and peak areas. The changes in the chromatographic performance of the stationary phase are measured by the Grob general purpose test mix (*9*).

Part of the strategy of broad spectrum analysis is to minimize sample pretreatment (and potential artifact production). This strategy was satisfactory for packed-column GC analysis (*2*). However, minimal sample pretreatment has pitfalls. The problems involved in the broad spectrum capillary GC analysis of ethyl ether elutions of XAD macroreticular resins used for environmental sampling were described by Bean et al. (*10*). The five major chromatographic problems found were (1) unresolved GC peaks on a 30-m OV-101 capillary column with a flame-ionization detector (FID), (2) apparent decomposition of the stationary phase, (3) chromatographic column plugging at the injector end, (4) an injection port glass liner covered with black residue, and (5) an unfruitful GC–mass spectrometric (GC–MS) investigation because of a lack of clean spectra.

Bean et al. (*10*) assumed the cause to be "significant quantities of high molecular weight materials present in the sample". These workers used a two-step method to solve the problem: (1) gel permeation chromatography to remove materials having molecular weights (MW) greater than 800 (standardized versus polypropylene glycol, MW = 880) and (2) deactivated silica gel to fractionate compounds of MW <800 into a nonpolar fraction in *n*-hexane and a more polar fraction in 16% ethyl ether and 84% *n*-hexane. No data was published on the sample losses caused by this procedure.

Most environmental GC analyses involve a sample cleanup procedure. For example, Schomberg (*11*) recommends optimized sample cleanup by using liquid chromatography with size exclusion, ion exchange, etc. to enhance the performance of GC. The sample cleanup steps should remove the nonvolatile, temperature-unstable, and/or nonchromatographable organic species present in the sample matrix.

This chapter describes the simpler cleanup approach of base extraction for XAD resin extracts. Quality assurance procedures statistically define the benefits of this analytical approach for broad spectrum capillary GC. The extraction procedure was optimized by studying UV absorbance of the base extractant.

Experimental

Sampling. An XAD resin accumulator was used to collect samples at different locations in a water treatment plant. Previous studies used a 100-mL XAD-2 resin accumulator to isolate the trace organics from water (*2*). This arrangement was changed because of the findings of Aiken et al. (*12*) that the XAD-2 resin irreversibly adsorbed humic materials. Therefore, an intimately mixed 200-mL bed of 100 mL of XAD-2 and 100 mL of XAD-8 was used. The XAD-8 adsorbed the larger molecules that slowly desorbed from XAD-2. This design enabled the XAD-2 resin to maintain its adsorption capacity for nonpolar compounds throughout the sampling program.

The aqueous sample was transferred to the samplers by a 1/8-in. Teflon line from each of the sampling points. In this line was a stainless steel metering valve (Nupro Company) with an orifice of 0.055 in. The glass sampling column was 700 mm long × 50 mm i.d. with an overflow 200 mm from the top and a glass frit at the bottom. Below the glass frit was a stopcock shut-off valve. A peristaltic pump (Cole-Parmer Masterflex) was used to meter the flow of water (4.5 L/h) through the macroreticular resin bed. Total sample volume was 325 L over a 72-h period.

A mixture of 10% by weight sodium sulfite and phosphoric acid was added to the sampled water by a metering pump before it entered the resin bed. The solution was used to reduce chlorine in the water and lower the pH to less than 3.5. The lowered pH was to inhibit biological growth on the resin during water sampling and to facilitate the collection of phenols and organic acids. The addition of the reagent solution was accomplished by a pressurized tank from which the acid and reducing agent were transferred by a Teflon manifold with 316 stainless steel fittings to several individual samplers.

Resin Bed Elution. Once composite sampling was completed, the resin bed was kept wet. Redistilled ethyl ether was added three times at volumes of 150, 100, and 100 mL. Ethyl ether removes the compounds collected on the surfaces of the resin. Each elution was equilibrated for 10 min to permit the transfer of materials from the resin surfaces to the eluant. The first three ether elutions were collected together. The fourth elution of 100 mL was used as a quality control sample to see if the removal of adsorbed species was complete and to identify any impurities in the ether.

After the fourth elution was drained from the resin, 100 mL of methanol was added to remove the ethyl ether from the interior pores of the resin. The methanol also removed polar compounds, which include humic material, and allowed the resin to be easily wetted by water. The elutions following methanol were 200 mL of 0.1 N NaOH followed by two elutions of 200 mL of distilled water. The original elution procedure used was identical to that of Junk et al. (*13*) for equivalent bed volumes of solvent.

The 0.1 N sodium hydroxide should elute the polar materials that cannot be removed from the resin by a nonpolar solvent (ethyl ether) or a polar solvent (methanol). The materials removed by this elution should be primarily organic acids and non-ether-soluble materials. This elution is drained after 10 min, and the resin bed is subsequently rinsed with two washes of distilled water. The resin is stored in 200 mL of methanol until the next sampling period. The openings of sampler were covered with aluminum foil between sampling periods.

Resin Elution Concentration. The first three ethyl ether elutions were combined into a 500-mL glass bottle with a Teflon cap liner. The bottles and all the laboratory glassware used in the sample analysis were cleaned with 6.0 N HCl, acetone, and hexane and then baked at 200 °C to remove the solvent. The combined elutions were brought to the laboratory for evaporative concentration and analysis.

The concentration procedure was performed on the sample within 3 days of arrival at the laboratory. Upon arrival, all ether extract samples were stored at −10 °C in a freezer. The ether elution after the water was frozen out was further dried by passage through anhydrous sodium sulfate. The sample was then concentrated by a Kuderna–Danish evaporator to 2 mL. After concentration, 1.0 mL of the 2-mL concentrated sample was base extracted to remove the chemicals interfering with the capillary GC analysis. The base extraction procedure developed in this study is the following:

1. Make up a 0.2 N K_2CO_3 solution and adjust to pH 12 with 5 N KOH.
2. Extract the pH 12 solution with redistilled ethyl ether at a water-to-solvent volume ratio of 10:1 to remove trace impurities.
3. Cover the pH 12 solution with 10 mL of ethyl ether for storage.
4. Add 10 mL of the basic solution to 1 mL of the ether extract in a screw-cap test tube. Close the test tube with a Teflon–rubber septum cap.
5. Shake for 5 min.
6. Centrifuge at 500 g to separate the phases.
7. Freeze out the water phase.
8. Transfer the ether phase to a storage vial.
9. Add hexane with C–23 methyl ester internal standard to the storage vial up to 1.7 mL.
10. Check the pH of the base after extraction. (NOTE: If the pH is lower than 11.2, acidic compounds probably remain in the extract; therefore, either increase the water-to-solvent ratio, increase the pH of the base, or change the buffer for that sample and redo the procedure.)

Gas Chromatography. The gas chromatograph was a Carlo Erba 2150AC with an on-column injector similar to that of Grob (*14*) and a flame ionization detector. The operating conditions of the gas chromatograph detector were as follows: detector temperature, 275 °C; makeup gas for flame, prepurified H_2 (0.65 Kp/cm^2 or 9.6 psi) and zero-grade air (1.05 Kp/cm^2 or 15.4 psi). The oven temperature program was as follows: initial isothermal, 42 °C for 7 min; temperature program rate, 2.5 °C per min; and final isothermal, 295 °C for 15 min.

The hydrogen carrier gas velocity was 50 cm/s. The carrier gas velocity was measured by the methane peak retention time as recommended by Rijks (*15*).

The chromatographic column used was a wall-coated, open tubular column (WCOT) (J & W Scientific) with a DB-1 Durabond chemically bonded stationary phase that had a nominal film thickness of 0.25 μm. The column was 60 m long $\times$ 0.32 mm i.d. The DB-1 stationary phase has chromatographic properties similar to SE-30.

When the analysis began, the chemically bonded phase was not available. An SE-30 WCOT capillary column of the same physical dimensions as the column just described was used instead. The change was made because the chemically bonded phase was found to deteriorate more slowly than a conventional coated SE-30 column (*16*). This claim was substantiated during this study.

The signal from the electrometer was integrated by a Supergrater III (Columbia Scientific). The retention time and peak areas were subsequently converted into Kovat's indexes by the method of van Den Dool and Krantz (*17*) and relative detector responses by software on a Tektronix 4051 computer. The normalized chromatographic data was then plotted in spike profile format (Figure 1). The original computer plotting program used to produce reconstructed chromatograms similar to Figure 1 was written by Glaser et al. (*18*) in FORTRAN. This program (*18*) was modified and rewritten in BASIC so that chromatograms could be plotted in Kovat's indexes.

Results and Discussion

Chromatographic Problems from Interference. The overall reproducibilities of the GC retention times of the internal standards were poor for the ether elutions of the resin extracts (Figures 2 and 3). The retention time differences of the resin extracts were not uniformly distributed over the entire chromatogram. The first 38% (C-7 to C-12) and the last 18% (C-22 to C-26) showed standard deviations >1 min. The smallest standard deviation occurred at C-14, which was in the middle of the chromatogram. This pattern was not the same for the retention-time standard deviations of the internal standards alone in ether (*see* Figure 3). For the internal standards alone, the standard deviation of retention times was fairly uniform over the entire chromatogram. A significant bias was also seen in the mean retention times in the first 38% of the chromatograms (through the C-12 internal standard). Significant interferences appeared to exist in the GC analysis of concentrates from the ethyl ether elution from the macroreticular resin isolation method.

The interferences created other chromatographic problems besides poor retention-time precision. These problems included the presence of a large rising and falling base line (a hump between C-10 and C-17), an unexpected rise in the level of noise from the flame ionization detector, and contamination of the stationary phase of the chromatographic column. These problems were especially severe for extracts of 150-L samples of chlorinated drinking water. The contamination of the

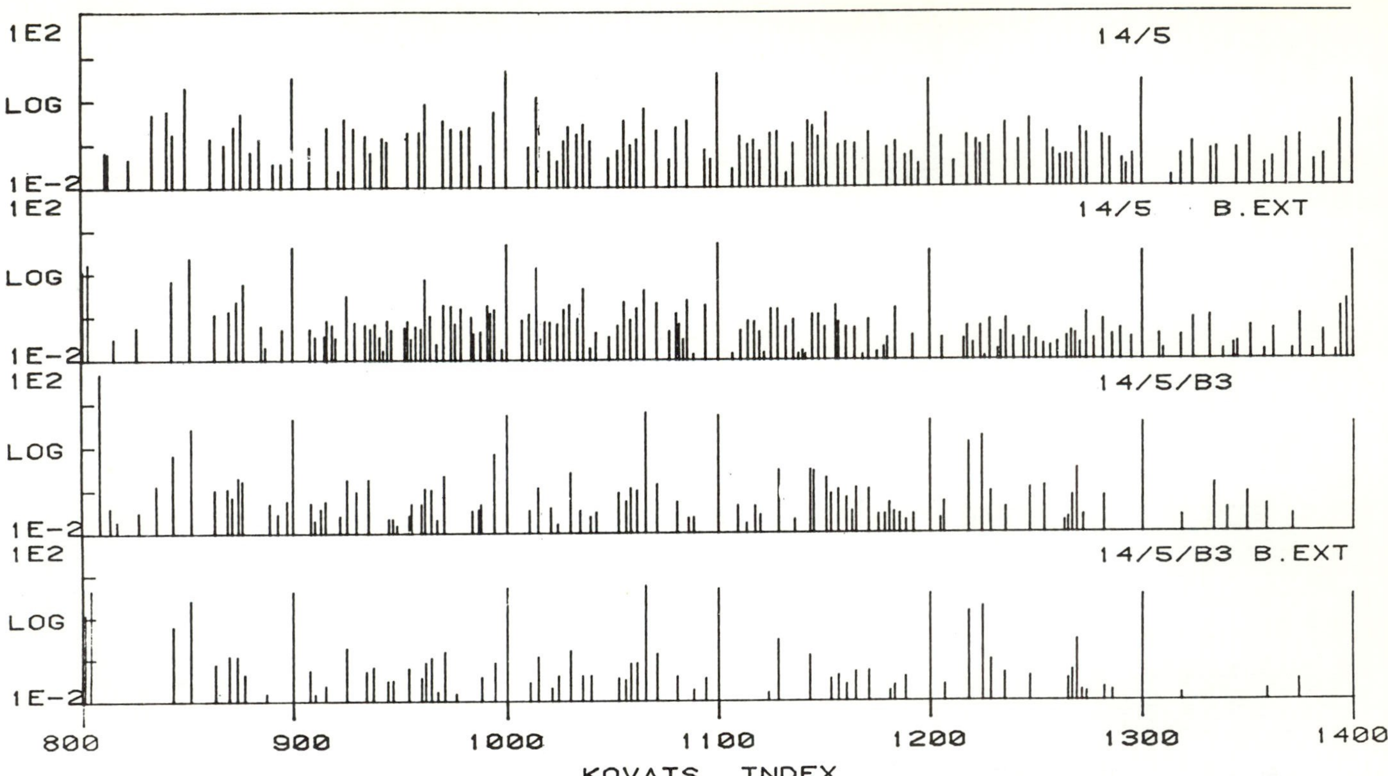

Figure 1A. Effect of base extraction on GC capillary chromatographic profiles.

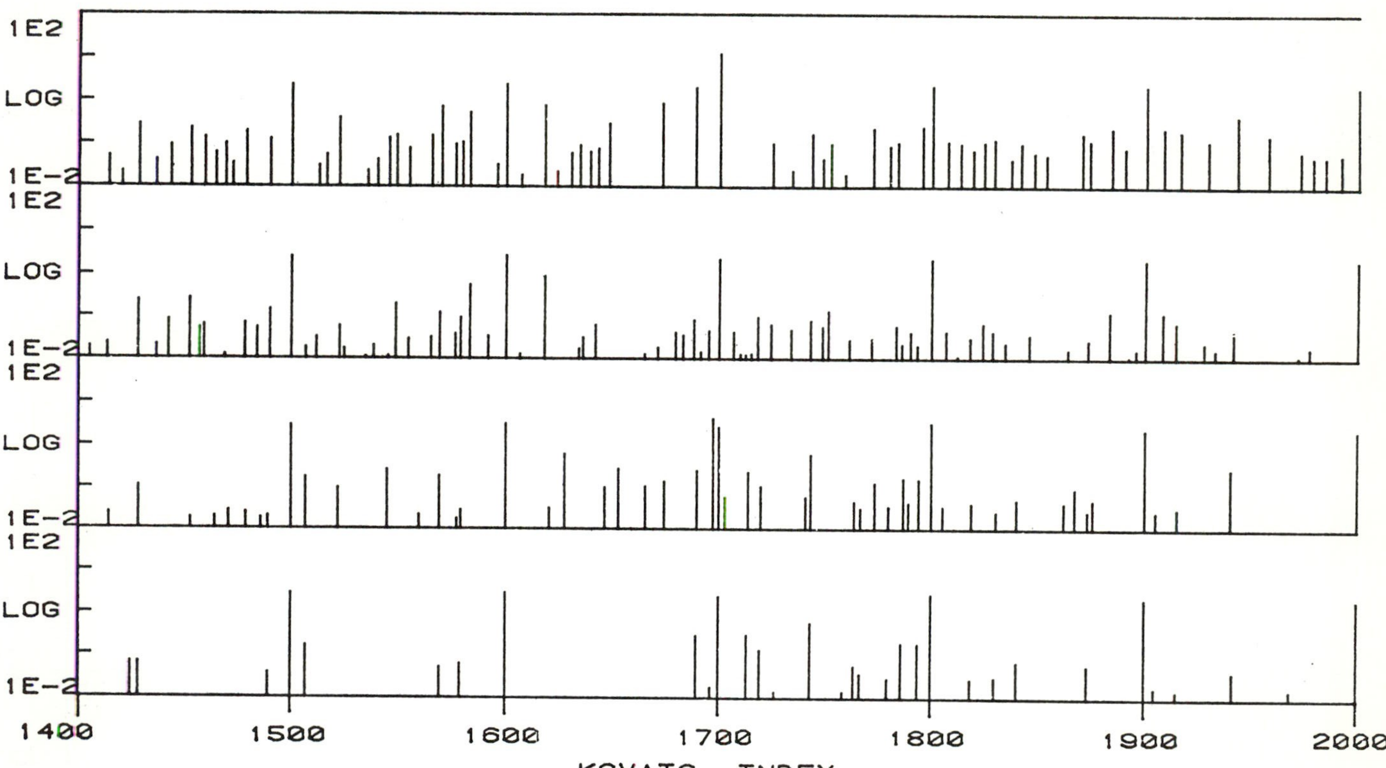

Figure 1B. Effect of base extraction on GC capillary chromatographic profiles.

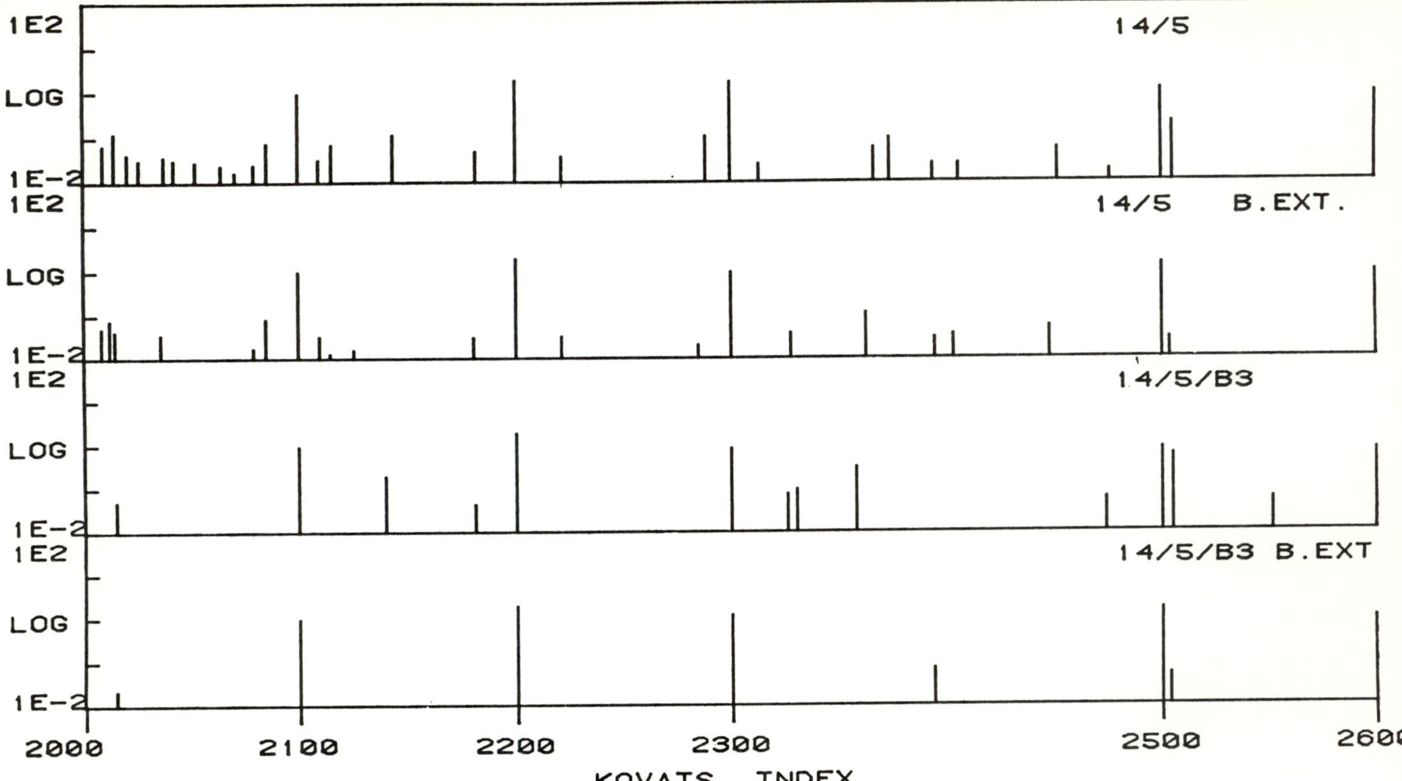

Figure 1C. Effect of base extraction on GC capillary chromatographic profiles.

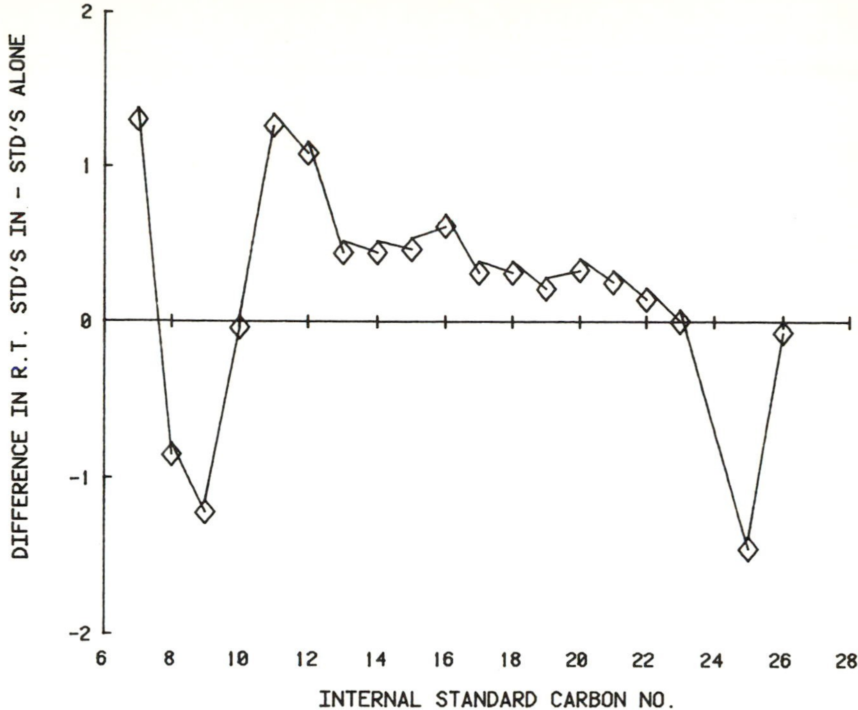

Figure 2. Changes in mean retention time due to sample matrix.

column stationary phase was determined by a Grob column-performance test mixture (*9*).

The interferences appeared to be similar to those of Bean et al. (*10*). This observation suggested that high molecular weight, non-chromatographable materials were present in the ether concentrate from the XAD resin accumulator. The dramatic effects of the interferences were not anticipated because previous XAD-2 resin accumulator analyses by packed-column GC reported by Suffet et al. (*2*) and Yohe et al (*5*) showed acceptable results.

Evaluation of GC Capillary Column. The Grob test was used to evaluate GC capillary column performance (*9*). Degradation of the GC columns occurred after one or two capillary GC analyses of the ether extracts. The methyl esters, used to measure the Trenzahl (TZ) number and effective stationary-phase film thickness, showed severe tailing and reduced areas. The tailing increased with increasing carbon number. This finding indicated that the stationary phase was being altered by the XAD resin extract. The area of the acid peak (2-ethylhexanoic acid) in the Grob test mix decreased markedly and almost disappeared. This

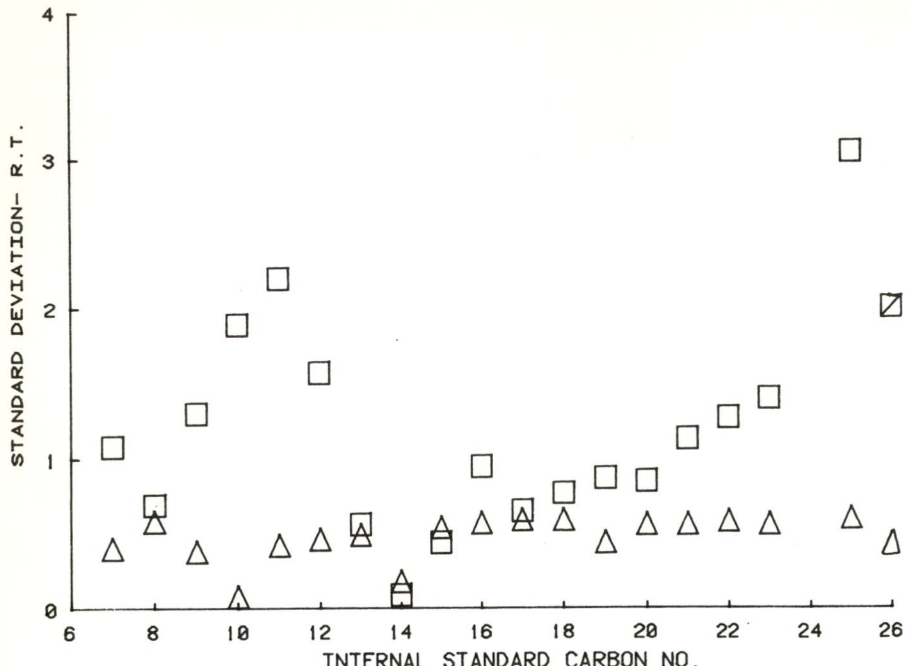

Figure 3. Sample matrix effect on reproducibility of internal standard retention times: □, standards in sample; Δ, standards alone. (Reproduced with permission from reference 26. Copyright 1985 Lewis Publishers.)

result indicated the presence of acidic adsorption sites. The other components of the Grob test mix did not change appreciably. The acidic adsorption sites indicated by the Grob test mix did not affect the peak shape and relative areas of the *n*-alkane homologous series used as internal standards on the SE-30 or DB-1 capillary columns. Therefore, the test mix was a much more sensitive measurement of the condition of the capillary column stationary phase.

GC Capillary Column Cleanup. Several GC capillary column cleanup procedures were tried with moderate success. These included breaking off the first meter of column and treating the column with various on-column silylation and derivatizing agents. The most successful of these agents were Methelute (*N,N,N*-trimethylbenzenaminium hydroxide) and trimethylsilyldiethylamine (Pierce Chemical Company). Fluorinated silylation agents degraded the stationary phase by causing active sites on the fused-silica column. The on-column chemical treatments took place at oven temperatures held at 300 °C. The reaction was completed after 4 h or when the base line returned to its previous preinjection level. Injections of on-column cleanup agents frequently had to be repeated to get acceptable capillary GC analysis as measured by the Grob test mix.

The capillary column cleanup procedures were time-consuming and seemed to increase the fragility of the fused-silica column. Thus, this approach is not the best when many samples need to be analyzed each week and each sample analysis takes approximately 2 h to complete.

Humic Substance Interference. The probable interferences within the sample matrix were thought to be high molecular weight, nonchromatographable materials called humic substances. This conclusion was reached on the basis of (1) the results of the Grob column-performance text mix, (2) the type of successful on-column chemical cleanup, and (3) a review of the literature regarding the nonchromatographable constituents of natural waters.

Bean et al. (*10*) used size-exclusion chromatography to remove materials of MW >800. This method of cleanup was rejected because size-exclusion chromatography might introduce new artifacts and cause additional delays in sample analysis time. Because the matrix problem appeared to have acidic components, it was decided that a base extraction procedure might remove these materials.

Optimization of Base Extraction. The base extraction, sample cleanup procedure was optimized by taking into account the following facts:

1. The material to be removed was probably humic substances, which may be considered diprotic acids having a $pK_1 = 4.8$ and a $pK_2 = 10.5 + 0.3$ (*19, 20*).
2. The reaction to remove the interferences from the ether sample was a simple acid–base reaction.
3. The amount of acidic materials in the sample varied considerably from sample to sample.
4. The amount of acidic materials in any ethyl ether extract was unknown.
5. Humic materials absorbed in the UV-vis spectral region because of an aromatic structure containing phenolic and carboxylic functional groups.

The following decisions were made to achieve consistent removal of interferences:

1. The base should be a strong buffer. A comparison of the titration of 0.2 N KOH and 0.2 N K_2CO_3 adjusted to pH 12 shows that it is easier to exceed the acid-neutralizing capacity of the KOH compared with the buffered solution of K_2CO_3 (*see* Figure 4).

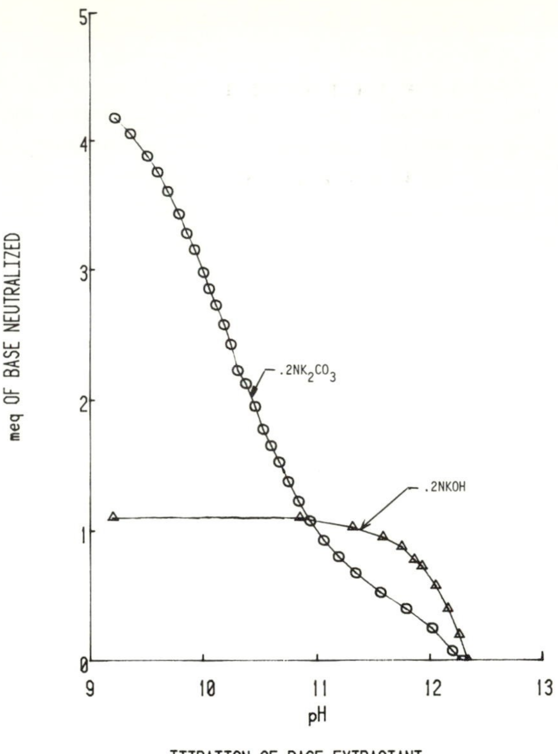

Figure 4. Buffer capacity of two bases.

2. The liquid–liquid extraction should have a constant volume ratio of base to ethyl ether. The volume ratio should be kept constant in order to keep the fraction of the sample solute extracted into the base a constant (*21*).

K_2CO_3 was chosen as the buffer for the base because it is stable and has no interferences in the UV-vis spectrum. The extraction procedure was optimized by diluting the aqueous extract to twice its original volume and scanning it in a 5-mm cell from a wavelength of 190 nm to 500 nm with a Perkin–Elmer 559 UV–vis spectrophotometer with background correction.

The UV spectrum showed that the basic solution maximum absorbance wavelength changed from 225 nm before extraction to 240 nm after extraction. In addition, the shape of the spectrum also changed. This result indicated that a variety of organic compounds with differing absorbance maxima were being removed from the ethyl ether sample. The total absorbance (A_T) of the base extractant at any wavelength may be represented as $A_T = \Sigma A_i$, where A_i is the absorbance of a constituent in the mixture.

Thus, the area under the spectrum would be a better measure of the overall efficiency of the extraction procedure than the absorbance at a single wavelength. The area was integrated by the trapezoidal area rule over the wavelength range 210–280 nm. This integration avoided the low transmittance at wavelengths below 200 nm.

The parameters to be optimized for the extraction of the concentrated ethyl ether elution are (1) base-to-solvent volume ratio, (2) pH, and (3) number of extraction steps.

Figure 5 shows a series of spectra for pH 11 base extracts of ethyl ether with different volume ratios. These curves are not corrected for dilution due to the change in volume of base. The increase in the ratio of the volume of base to the volume of ethyl ether decreases the spectral area; for example, increasing the volume ratio by threefold decreases the spectral area by 29%. If no additional acidic materials were extracted, the expected decrease in spectral area would be 67%. A further increase in the volume ratio to fivefold decreases the spectral area by 58% as opposed to the expected 80% by dilution. Thus, adding more volume of base removes more acidic materials from the ethyl ether resin extract.

The effects of the pH of the base (10.5, 11, 12) and two-step extractions were also investigated at a base-to-ether volume ratio of 9:1. As the number of milliequivalents of base increased, the absorbance also increased. Sixteen percent of the total extractable material from the two-step extraction (9:1) at pH 11 was left in the ether after the first extraction step.

A comparison of packed-column 10% SE-30 gas chromatograms before and after base extraction is shown in Figure 6 for pH 12. The rise and fall of the base line hump in the later portion of the chromatogram was dramatically reduced. This result was verified by observing the same phenomenon with chromatograms at pH 11 and 11.5.

The cleanup procedure used was discussed in the section entitled Resin Elution Concentration. The extraction procedure was not the optimum procedure that would result from the observations. The data indicated that a base extraction at pH >12 and a base-to-solvent ratio of 15:1 with multiple extraction steps were needed to maximize removal of the matrix interferences. However, the extraction efficiency must be balanced against minimal sample destruction and operational ease.

Therefore, pH 12 was chosen to minimize base hydrolysis of organic compounds. The 10:1 volume ratio was used to assure sufficient recovery of the ether phase from a larger volume ratio, and only one extraction step was used to minimize loss of volatiles during the cleanup method.

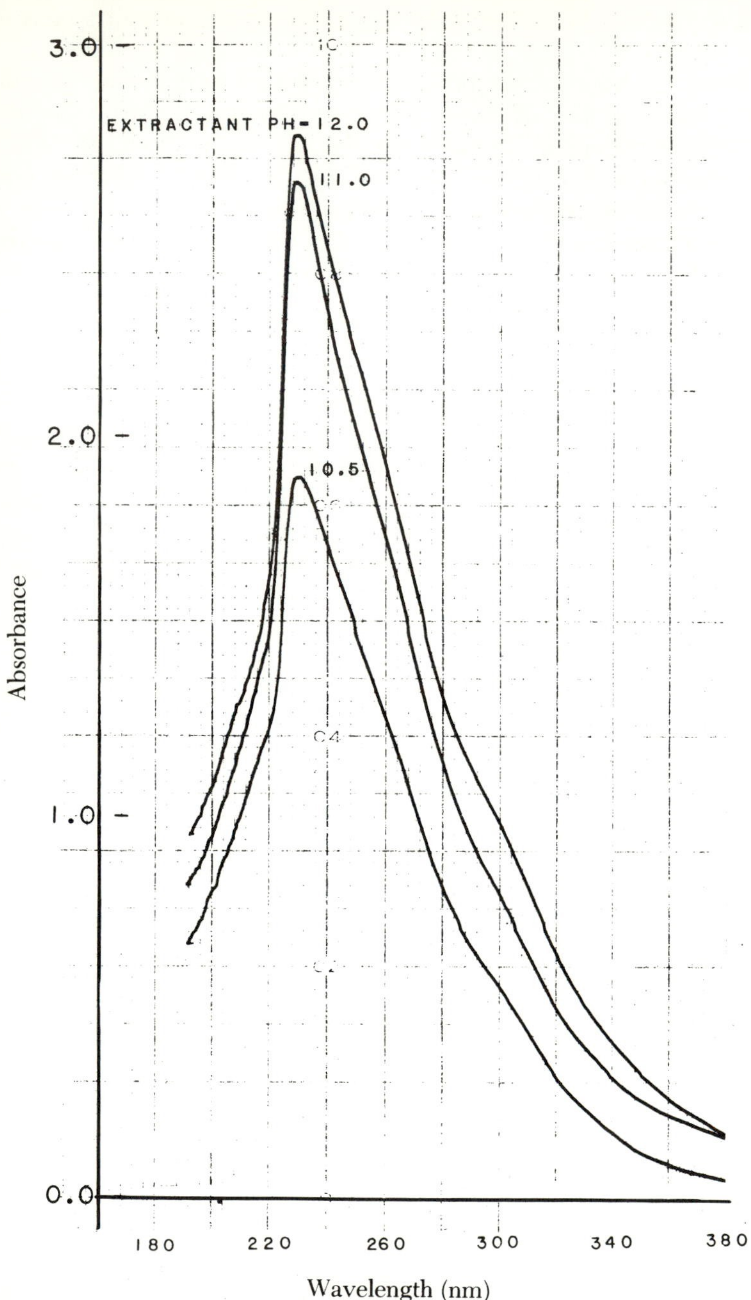

Figure 5. UV-vis absorbance spectra of aqueous extract of ether macroreticular resin elution.

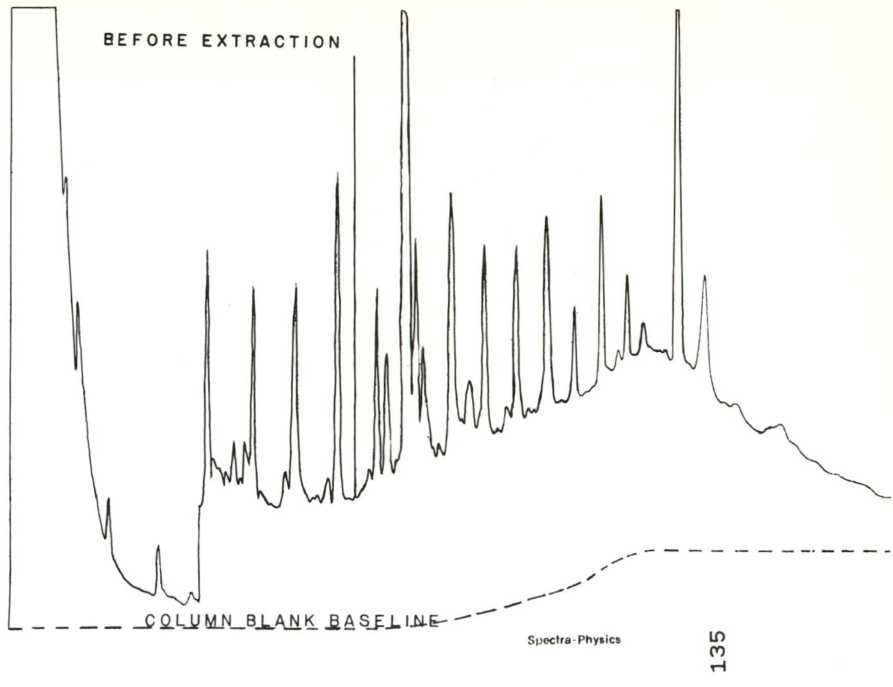

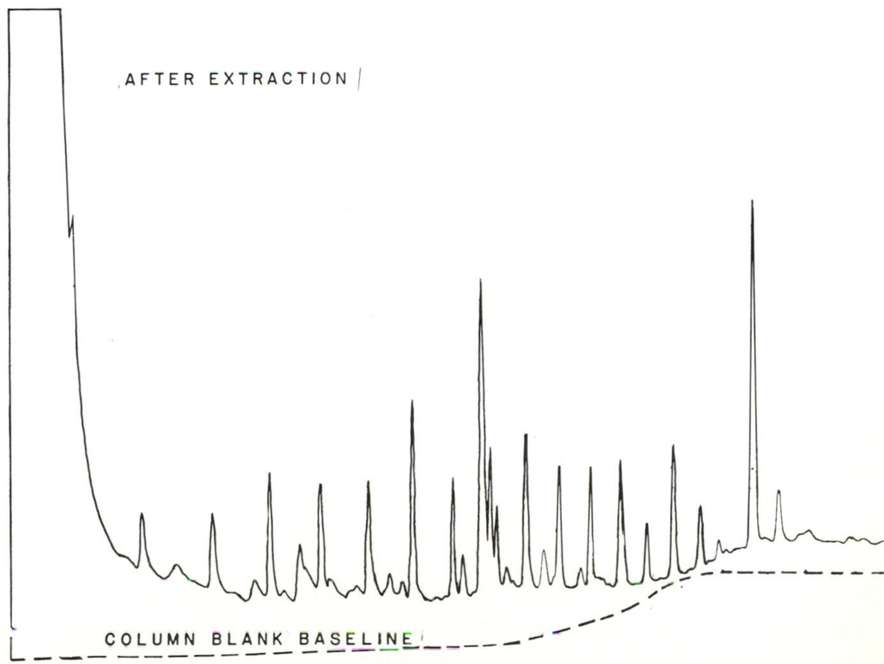

Figure 6. Effect of pH 11.5 extraction on chromatogram base line.

Results After Base Extraction. Base extraction was found to eliminate the observed chromatographic hump (as shown in Figure 6) and improve peak resolution and detection. On the negative side, the procedure could remove compounds of interest or add contaminants to sample. To study the effects in more detail, a sample was chromatographed on a capillary column before and after base extraction (Figure 1). Computer reconstructed plots of the sample capillary chromatograms before and after base extraction were used to observe significant changes in the number and response of peaks. The detector response in Figure 1 was plotted logarithmically, and the minimum response was 0.01 (0.6 ppt). The comparison showed fluctuations in the number and response of peaks after base extraction primarily for the small peaks below 0.6 ppt. The more concentrated GC peaks were not greatly affected by base extraction. The decreasing response of peaks does not appear to be due to the physical manipulation of the sample during the procedure. Chemical explanations are more likely, possibly including the conversion to new compounds or the loss due to solubility changes, that is, the hump materials dissolving into the aqueous base solution.

The increase or appearance of peaks has several possible explanations. Addition of contaminants by the base during extraction is minimal, as proven by GC analysis of blank ether extractions of the buffer used in the procedure. The base could convert compounds and thus account for some of the new peaks seen. However, the hump obscures the normal column base line, and the removal of the hump by base extraction may cause previously undetected or unintegrated compounds to be measured. The hump-associated retention times have more detector noise than the other retention times in the chromatograms. The integrator usually defines the detector noise as a peak unless the sensitivity of the integrator is decreased (22).

Although qualitative improvement in the packed-column GC was observed (Figure 6), the improvement in identifying the capillary column GC peaks by their Kovat's indexes was determined by using the Fisher *F* test. Table I shows the standard deviations of the retention times of the normal alkane internal standards from packed-column GC analysis before and after a base extraction. The Fisher *F* test ($P = 0.05$) showed that only the standard deviation of the retention time of the C-14 hydrocarbon of the base-extracted sample was not significantly smaller. The better precision of internal standard retention time improved the ability to compare GC peaks by their retention times and, in turn, the Kovat's indexes throughout the whole chromatogram. Such comparisons are essential to the interpretation of broad spectrum analysis.

The base extraction cleanup procedure changed the initial GC profile data contained in the original ethyl ether extract. Specifically,

Table I. Retention Time Standard Deviations of Internal Alkane Standards of Composite XAD Resin Water Samples With and Without Cleanup

	Standard Deviation (min)		
Compound	*With Cleanup* ($n = 7$)	*Without Cleanup* ($n = 4$)	F^a
C-8	0.143 (n = 6)	0.69	23.28
C-9	0.114	1.30	130.04
C-10	0.120	1.89	248.06
C-11	0.131	2.20	282.03
C-12	0.139	1.57	127.58
C-13	0.136	0.56	16.96
C-14	0.140	0.09	0.413
C-15	0.135	0.44	10.62
C-16	0.134	0.95	50.26
C-17	0.138	0.65	22.19
C-18	0.136	0.77	32.06
C-19	0.123	0.87	50.03
C-20	0.126	0.85	45.51
C-21	0.129	1.13	76.73
C-22	0.133	1.27	91.18
C-23	0.119	1.40	138.41
C-25	0.136	3.05	502.95
C-26	0.132	2.01	231.87

NOTE: All compounds, except C-8 and C-14, exceeded the critical value of F 0.05 (df = 3, 6) = 4.76. C-8 exceeded the critical value of F 0.05 (df = 3, 5) = 5.41, and C-14 did not exceed the critical value.

[a] F = (standard deviation without cleanup)^2/(standard deviation with cleanup)^2.

acidic sample constituents were removed in the cleanup process. A measure of the total amount of acidic components in a sample can be estimated by the amount of base neutralized during the extraction procedure. The following assumptions are necessary to make this estimate:

1. The compounds are soluble in water at a pH between 2.9 and 3.5, which was the pH of the water sample.
2. The compounds can be sorbed by the XAD-2 and XAD-8 resin at a pH between 2.9 and 3.5.
3. The compounds are soluble in ethyl ether, which was used to elute the resin column.

The amount of base neutralized was computed from the measurement of the pH of the buffer after the cleanup step. This pH value was entered on the acid titration curve of the buffer (*see* Figure 4), and the

Table II. Acid Neutralization of Base Extract

Sample Location	*Sampling Period*[a]		
	8/1/80	*9/9/80*	*9/30/80*
Chlorinated rapid sand filter effluent	1.0	0.45	0.70
Chlorinated ozonator effluent	0.84	0.65	0.31
Chlorinated GAC effluent	0.16	0.13	0.65
Chlorinated ozonated GAC effluent	0.20	0.295	0.04
Nonchlorinated rapid sand filter effluent	0.44	—	0.11
Nonchlorinated ozonator effluent	0.34	0.20	0.0
Nonchlorinated ozonated GAC effluent	0.21	0.285	0.0
Nonchlorinated GAC effluent	0.21	0.26	0.0
Lab still blank	0.18	0.0	—
Transportation blank	0.22	0.0	—

Note: GAC denotes granular activated carbon.
[a] Values are the milliequivalents of base neutralized during the sampling period ending on the given date.

corresponding number of milliequivalents of acid was determined. An example of the results of this procedure is shown in Table II.

Loss of Compounds During Base Extraction. There are no perfect cleanup methods, and certain constituents in the sample extracts may react or be lost during the base extraction step. Gas chromatographic–mass spectrometric (GC–MS) analysis of base extracts of resin eluates revealed that base hydrolysis was occurring for at least one class of compounds, namely, phthalates (*23*). The two isopropylidene sugars undergoing losses were identified by GC–MS as the diacetone sugars of L-sorbose and D-xylose. The phthalate derivatives undergoing base hydrolysis were dimethyl, diethyl, and dibutyl phthalates. Rhoades et al. (*24*) showed that dimethyl phthalate and benzyl phthalate undergo significant storage losses at pH 10. The isopropylidene sugars are moderately polar compounds that probably were back extracted into the base. The back extraction occurred because isopropylidene sugars are moderately soluble in water, that is, 14 g of 1,2,3,5-di-*O*-isopropylidene-D-glucofuranose dissolves in 100 g of boiling water (*25*).

The loss of compounds in any method limits the usefulness of the approach for broad spectrum screening. The loss of compounds during base extraction occurred either by dissolution back into the water or by reaction with base. These losses are deemed to be an acceptable trade-off for the following reasons:

1. Improved accuracy and reproducibility for the nonpolar synthetic organic chemicals that are being studied.
2. More productive use of analytical equipment because the capillary column is not degraded rapidly.

3. One source of organics in drinking water is studied.
4. Detection of compounds having a low concentration in a sample is enhanced by a significantly lower base line noise level.

Conclusions

The removal of chromatographic analysis interference increases the reliability of the broad spectrum approach to organic analysis. The improved retention-time precision and lower background noise level make it possible to use statistical significance-level testing of the broad spectrum data (*26*).

The sample cleanup procedure is efficient and easy to perform. The cleanup procedure removes approximately 84% of the non-chromatographable materials. The losses from the sample affect only a small number of the synthetic organic chemicals of interest in the field of drinking water treatment. The compounds undergoing the largest losses were phthalates and diacetone sugars.

The use of UV-absorbance measurements to optimize the base extraction was demonstrated. UV measurements over a wide wavelength range were demonstrated to be necessary because many humic materials being removed may not have the same maximum absorbance wavelength.

Acknowledgments

Research support was provided by the U.S. Environmental Protection Agency under Contract No. 806256-02, J. Keith Carswell, Project Officer. We wish to thank M. E. Post for her advice and help in developing the laboratory procedures and B. Najar for significantly contributing to the laboratory analyses.

Literature Cited

1. Gibs, J. Ph.D. Thesis, Drexel University, 1983.
2. Suffet, I. H.; Brenner, L.; Coyle, J. T.; Cairo, P. R. *Environ. Sci. Technol.* **1978**, *12*, 1315.
3. Suffet, I. H. In *Activated Carbon Adsorption of Organics from the Aqueous Phase, Vol. 2*; McGuire, M. J.; Suffet, I. H., Eds.; Ann Arbor Science: Ann Arbor, MI, 1980; Chapter 24, p 539.
4. Yohe, T. L.; Suffet, I. H.; Coyle, J. T. In *Activated Carbon Adsorption of Organics from the Aqueous Phase, Vol. 2*; McGuire, M. J.; Suffet, I. H., Eds; Ann Arbor Science: Ann Arbor, MI, 1980; Chapter 2, p 37.
5. Yohe, T. L.; Suffet, I. H.; Cairo, P. R. *J. Am. Water Works Assoc.* **1981**, *73*, 402.
6. Stevens, A. A.; Seeger, D. R.; Slocum, C. J.; Domino, M. M. *J. Am. Water Works Assoc.* **1981,** *73*, 548.

7. Van Rensberg, J. F. J.; Hassett, A.; Theron, S.; Wiecher, S. G. *Prog. Water Technol.* **1980**, *12*, 537.
8. Coyle, G. T.; Maloney, S. W.; Gibs, J.; Suffet, I. H. In *Water Chlorination: Environmental Impact and Health Effects*; Jolley, R. et al., Eds.; Ann Arbor Science: Ann Arbor, MI, 1983; Vol. 4, Chapter 30, p 42.
9. Grob, K., Jr.; Grob, G.; Grob, K. *J. Chromatogr.* **1978**, *156*, 1.
10. Bean, R. M.; Ryan, P. W.; Riley, R. G. *Proc. Symp. High Resolut. Gas Chromatogr. Am. Chem. Soc.* 1977, Aug. 28–Sept. 2, Chicago, IL; Academic: New York.
11. Schomberg, G. *Chromatogr. Electrophor.* **1980**, *10*, 327.
12. Aiken, G. R.; Thurman, E. M.; Malcolm, R. L.; Walton, H. F. *Anal. Chem.* *1979*, *51*, 1803.
13. Junk, G. A.; Richard, R. J.; Grieser, M. D.; Witiak, D.; Witiak, J. L.; Arguello, M. D.; Vick, R.; Sveck, H. J.; Fritz, J. S.; Calder, G. V. *J. Chromatogr.* **1974**, *99*, 745.
14. Grob, K. *J. High Resolut. Chromatogr. Chromatogr. Commun.* **1978**, *1*, 263.
15. Rijks, J. A. Ph.D. Thesis, Technical University of Enidhoven, The Netherlands, 1973.
16. *High Resolution Chromatography Products*; J & W Scientific: Rancho Cordova, CA, 1981.
17. Van Den Dool, H.; Krantz, P. D. *J. Chromatogr.* **1963**, *11*, 463.
18. Glaser, E. R.; Silver, B. L.; Suffet, I. H. *J. Chromatogr. Sci.* **1977**, *15*, 22.
19. MacCarthy, P.; Peterson, M. J.; Malcolm, R. L.; Thurman, E. A. *Anal. Chem.* **1979**, *51*, 2041.
20. Perdue, E. M. *Geochim. Cosmochim. Acta* **1978**, *42*, 1351.
21. Suffet, I. H. *J. Agric. Food Chem.* **1973**, *21*, 288.
22. Environmental Protection Agency. *Problems with Chromatographic Integrators: Process Measurements Review*; Industrial Environmental Research Laboratory: Research Triangle Park, NC, 1979; Vol. 2, No. 1, p 45.
23. Neukrug, H. M.; Smith, M. G.; Coyle, J. T.; Santo, J. P.; McElhaney, J.; Suffet, I. H.; Maloney, S. W.; Chostowski, P. C.; Pipes, W. O.; Gibs, J.; Bancroft, K. *Removing Organics from Drinking Water by Combined Ozonization and Adsorption;* U.S. Environmental Protection Agency. U.S. Government Printing Office: Washington, DC, 1981; EPA-600152-83-048.
24. Rhoades, J. W.; Thomas, R. E.; Johnson, D. E.; Tillery, J. B. *Determination of Phthalates in Industrial and Municipal Waste Waters*; U.S. Environmental Protection Agency. U.S. Government Printing Office: Washington, DC, 1981; EPA-600/S4:81-063.
25. *Dictionary of Organic Compounds*; Buckingham, J., Ed.; Chapman-Hall: New York; 1982.
26. Gibs, J.; Suffet, I. H.; Najar, B. In *Water Chlorination: Environmental Impact and Health Effects, Vol. 5*, Jolly, R.; Bull, R. J.; Davis, W. P.; Katz, S.; Roberts, M. H. Jr.; Jacobs, V. A., Eds.; Lewis: Ann Arbor, MI, 1985; Chapter 88, p 1099.

RECEIVED for review October 17, 1985. ACCEPTED May 14, 1986.

NOVEL METHODS TO ISOLATE AND FRACTIONATE ORGANIC CHEMICALS IN WATER SAMPLES

16

Solvent Extraction Using a Polymer as Solvent with an Amperometric Flow-Injection Detector

You-Wei Feng and Calvin O. Huber

Department of Chemistry and Center for Great Lakes Studies, University of Wisconsin—Milwaukee, Milwaukee, WI 53201

Solvent extraction offers unique advantages among separation techniques. A system based on extraction into a polymer [poly(vinyl chloride)] as solvent was examined here because of possible advantages in speed, simplicity, sample size, solvent handling, etc., especially when coupled with flow injection and an amperometric detector. Solutes examined included salicylic acid and 8-hydroxyquinoline. The apparatus typically consisted of 0.8-mm i.d. × 170-cm coiled tubing that could be connected directly to the injection loop of a flow-injection amperometric detector system containing a nickel oxide electrode.

LIQUID EXTRACTION FOR THE SEPARATION and enrichment of organic compounds in aqueous samples has been used successfully. Automated solvent extraction with flow-injection analysis has been reported (*1*).

The handling and disposal problems associated with the use of liquid solvent extractors have resulted in increased attention to the separation and preconcentration of organic compounds in water by collection in synthetic polymers followed by elution with an organic solvent (*2*). For example, selective collection and concentration of organic bases on methylacrylic ester resin from dilute water samples have been reported (*3*). Such collection techniques are especially well-suited to flow-injection measurement techniques. In this study, ionizable organic analytes such as salicylic acid and 8-hydroxyquinoline (oxine) were extracted into a polymer and then back extracted by an aqueous solution. Amperometric measurement using a flow-injection technique was employed to monitor the process.

0065-2393/87/0214/0349$06.00/0

The detection step involves electrochemical oxidation at a nickel electrode. This electrode has been applied to measurements of glucose (*4*), ethanol (*5*), amines, and amino acids (*6, 7*). The reaction mechanism involves a catalytic higher oxide of nickel. The electrolyte solution consists of 0.1 M sodium hydroxide containing 10^{-4} M nickel as suspended nickel hydroxide to ensure stability of the electrode process. The flow-injection technique offers the advantages of convenience and speed in solution handling and ready maintenance of the active electrode surface.

Experimental

The polymer extractor consisted of 0.8-mm i.d. poly(vinyl chloride) (PVC) tubing (Norton Company) typically 170 cm in length and coiled about a 1- or 2-cm diameter rod. The outflow of the extractor tube was transferred to the detector by direct connection or by collection and transfer (*see* Figure 1).

The continuous-stream flow-injection system (Figure 2) consisted of a gravity-feed electrolyte reservoir, a sample injection valve (Rheodyne, Model 50) fitted with a 30 μL-sample loop, and a flow-through electrochemical detector cell. The channel diameter of the Teflon tubing for the stream was 0.8 mm. The tubing length from injector to detector was 10 cm.

The detector cell was a three-electrode system consisting of a flow-through nickel working electrode, a saturated calomel reference electrode (SCE), and a stainless steel outlet tubing counter electrode. The tubular-type electrode cell housing was constructed of molded Teflon, which was machined to provide the channels and to accommodate the fittings. The working electrode area was

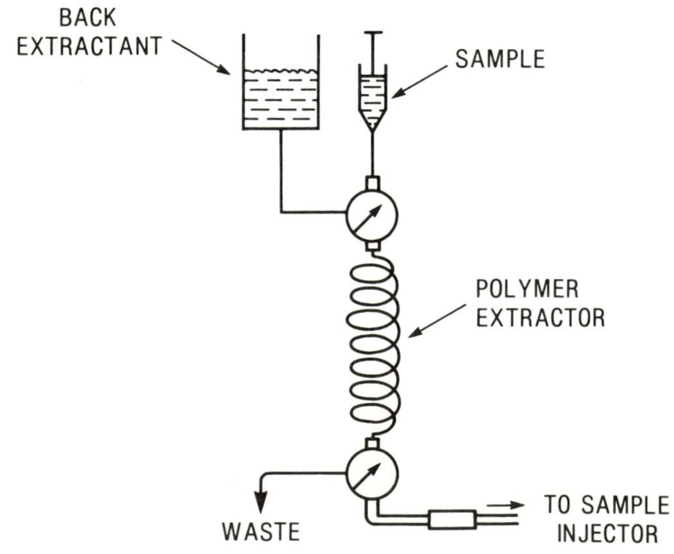

Figure 1. Extraction apparatus.

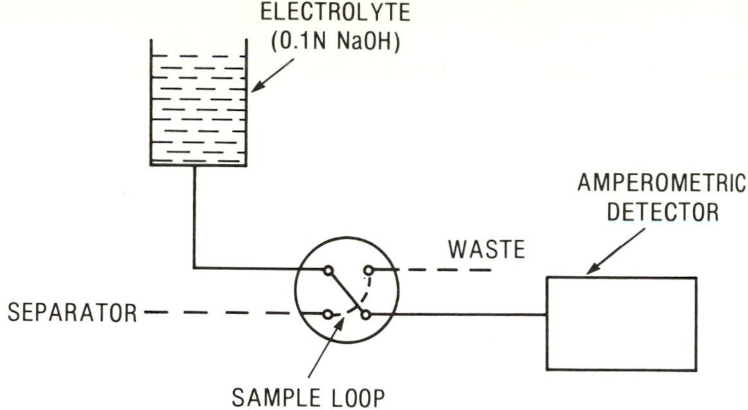

Figure 2. Flow-injection detector apparatus.

about 1 mm^2. The SCE salt bridge was located downstream from the working electrode. The working electrode applied potential was +0.56 V versus SCE. The circuitry consisted of a two op-amp based potentiostat and current-to-voltage converter with offset. It was powered by a ±15-V dc power supply. A potentiometric recorder was used for readout.

Results and Discussion

Detector Characteristics. The applied potential that produced the largest analytical signal was 0.56 V versus SCE. The decrease in analytical signal at potentials more positive than 0.56 V suggested a decrease in the active surface area of the electrode due to competitive solvent oxidation at the active sites.

The analytical signal increased with flow rate up to 1.4 mL/min, and no further increases at higher flow rates occurred.

A linear relationship between peak current and salicylate concentration in 0.1 M NaOH carrier electrolyte was found for the range 8×10^{-6} M to 1×10^{-2} M with a sensitivity of 2.5 μA/mM.

The linear range for salicylate in acetate buffer solution (0.025 M NaAc/HAc, pH 4.8) was found to be 2.0×10^{-5} to 5.0×10^{-3} M with a sensitivity of 2.2 μA/mM.

Polymer Selection. The polymer was selected on the basis of observations using salicylic acid–salicylate as analyte. The following organic polymers were examined: polystyrene, methyl methacrylate–ethyl acrylate, Teflon, silicone rubber, PVC, and polyester. Ten-millimolar salicylic acid in 0.01 M HCl was first extracted for 30 s and then back extracted with 0.1 M NaOH. Peak currents for back extractants (nA) were as follows: PVC, 1780; methyl methacrylate–ethyl acrylate, 380; Teflon, 90; polyester, 410; polystyrene, 290; and silicone rubber, 1420. PVC tubing afforded the most efficient extraction and was used for the remainder of the study. The selection of an optimum polymer depends upon analyte species.

Separation Dynamics. The mass transport dynamics of the system were examined with respect to extraction, back extraction, and time interval between extraction and back extraction. Concentration gradients limited by diffusion rate must be considered; that is, equilibrium with homogenous concentrations of analyte does not describe this system.

Extraction time effects are summarized in Figure 3. For extraction times greater than about 2 min, the extraction rate is apparently limited by concentration gradient establishment in one of the phases.

When the concentration decrease of the extraction sample rather than the back extracted amount was observed, the extent of extraction also leveled within 2 min, as shown by the following data: For 0.1 mM salicylic acid, the peak currents (μA) were 0.180, 0.116, 0.043, and 0.045 after extraction times of 0, 1.0, 2.0, and 4.0 min, respectively; for 1.0 mM salicylic acid, the peak currents (μA) were 1.79, 1.14, 0.90, and 0.91

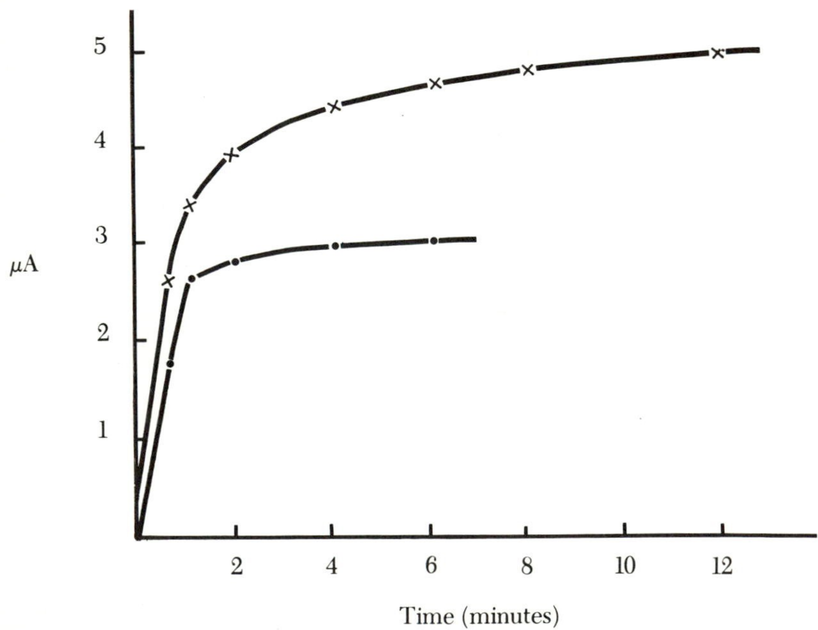

Figure 3. Extraction and back extraction time effects: (●), extaction times followed by 0.50-min back extractions; (×), back extraction times after 1.0-min extractions. Extraction was from 10 mM salicylic acid in 0.01 m HCl, and back extraction was into 0.1 M NaOH.

after extraction times of 0, 1.0, 2.0, and 4.0 min, respectively. Extraction leveling data were obtained over a salicylic concentration range of 0.1–5 mM with an extraction efficiency of 30–50%. These results indicate that within 2 min a steady-state concentration profile is reached. When the extraction step is done with continuous flow, the amount of sample extracted per minute increases as the flow rate increases but levels at higher flow rates. This result indicates eventual mass transfer control of extraction efficiency.

For back extraction rate studies, an extraction time of 1.0 min was used, and the back extraction time was varied up to 12 min, as shown in Figure 3. The time dependence in the back extraction process indicates that a diffusion rate to or from the interface is regulating the process. The gradient in the aqueous solution is very steep because of ionization of the analyte; thus, diffusion of analyte molecules in the polymer to the interface limits the back extraction rate. This rate limitation supports the view of a solvent extraction rather than a surface adsorption mechanism. The longer time dependence for back extraction (~12 min) versus that for extraction (~2 min) can be attributed to the collapse of the concentration gradient in the polymer on proceeding to back extraction. Back extraction for analytical applications was by continuous flow of 0.1 M NaOH with detector injection of the initial portion of the back extract. This concentration gradient broadens further during the period between extraction and back extraction while the extraction tubing contains air. Figure 4 shows results of experiments in which this time interval was varied. As predicted, the rate of back extraction is decreased for larger time intervals. The rapid drop during the first 2 or 3 min in Figure 4 suggests that diffusion coefficients in the polymer are comparable to those in liquids. The analytical procedure accordingly requires both short and constant intervals between extraction and back extraction.

Between samples, the residual solute in the polymer is removed by continued exposure to back extractant. This process requires about 2 min for 1 mM sample solutions. As expected from diffusion laws, this time period for residual removal varies with the square of the sample concentrations. Higher sampling rates require keeping analyte concentrations below about 1 mM.

Plasticizer. The polymer tubing extractor contains the plasticizer 1,2-bis(ethylhexyl) phthalate. Longer exposures to 0.1 M NaOH indicated that electroactive hydrolysis products of plasticizer, probably ethyl and hexyl alcohols, were produced, as suggested by the following peak currents given in nanoamperes (the times of the peak currents are given in parentheses): 30 (2 min), 7 (1 min), 5 (30 s), and 2 (15 s). The analytical procedure presented here uses back extraction times of less

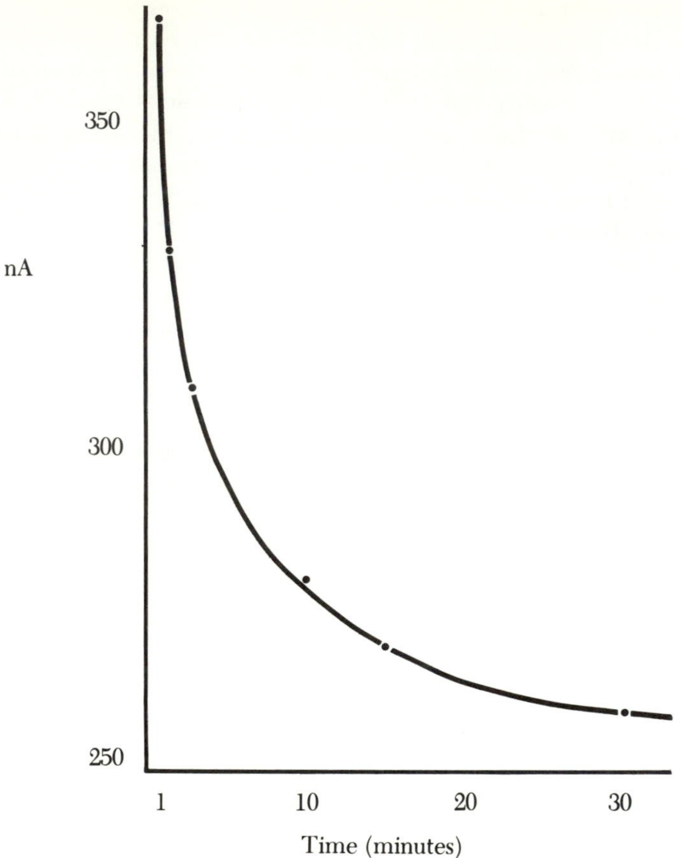

Figure 4. Effect of time interval between extraction and back extraction. Extraction was from 1 mM salicylic acid in 0.01 M HCl, and back extraction was into 0.01 M NaOH.

than 30 s so that the plasticizer contribution to analytical signals is negligible. Blank determinations further supported the conclusion that, for the prescribed procedure, plasticizer contribution is insignificant.

To determine whether the plasticizer was functioning as solvent, attempts were made to extract the analyte (salicylic acid) with liquid diethyl phthalate. The distribution ratio for this extraction was negligibly small. This finding indicated that the PVC polymer rather than the phthalate ester plasticizer served as the solvent.

Evaluation of Distribution Parameters. The fraction of salicylic acid extracted ($\%E$) was found to be proportional to the equilibrium fraction (α_o) of the un-ionized neutral form present in the pH range 1.65–3.01, as shown in Table I. To analytically describe the partition

Table I. Relation of the Fraction of Salicylic Acid Extracted (%E) to the Equilibrium Fraction (α_o) of the Un-ionized Neutral Form in the pH Range 1.65–3.01

pH	%E	α_o	E/α_o
1.65	44.8	0.95	0.47
1.88	38.9	0.92	0.42
2.23	34.8	0.85	0.41
2.49	33.1	0.75	0.44
2.93	28.8	0.53	0.55
3.01	26.3	0.48	0.54

process, it was assumed that mass transport into the polymer was limited by diffusion of solute molecules in the bulk aqueous solution toward the interface. The mass exchange near the interface reached steady state in about 2 min. Assuming that the effective volumes of the thin layers of water and polymer phases were equal, one can estimate the apparent distribution coefficient (K_D) in terms of the observed capacity ratio (k'):

$$K_D = (n_o/V_o)/(n_w/V_w) = n_o/n_w = k' \tag{1}$$

where n is the number of molecules of un-ionized salicylic acid; V is the volume; and subscripts o and w designate polymer and water phases, respectively. In the water phase, the fraction of un-ionized salicylic acid can be stated as

$$\alpha_o = n_w/n_s \tag{2}$$

where n_s is the sum of salicylic acid and salicylate species in the water phase. The total solute species (n_t) in both phases is

$$n_t = n_o + n_s \tag{3}$$

Eq 1–3 can be combined to yield a linear relationship in reciprocal parameters:

$$1/n_o = (1/n_t) + (1/n_t k' \alpha_o) \tag{4}$$

The peak current (i_p) of the analytical signal after back extraction is proportional to the amount extracted:

$$i_p = kn_o \tag{5}$$

Combining eq 4 and 5 gives

$$1/i_p = (1/kn_t) + (1/k'kn_t\alpha_o) \tag{6}$$

By using experimentally obtained data for 1 mM salicylic acid, a plot of reciprocal analytical signal versus reciprocal α_o yielded a linear relationship for the pH range 1.65–3.01. This result supported the solvent extraction model. The corresponding estimate of capacity ratio and distribution coefficient using this treatment was 8.5.

Selectivity. Several additional analytes were examined. Extraction efficiencies from 1 mM solutions for a 2-min extraction time were as follows: glucose, 2%; salicylic acid, 34%; oxine, 75%; phenol, 22%; and *p*-cresol, 79%. 8-Hydroxyquinoline and *p*-cresol are significantly more extractable than salicylic acid. 8-Hydroxyquinoline can be back extracted either by basic solution (pH >12) or acidic solution (pH <3) because of its amphiprotic behavior (pK_1 = 5.0; pK_2 = 9.90). This experiment was performed by using either 0.1 M NaOH or 0.01 M HCl with essentially the same results. The extraction step should be performed at pH 7–10.

Separation selectivity was demonstrated by extraction of salicylic acid from pH 2 solutions in the presence of a 60-fold excess of ethanol and sixfold excesses of barbituric acid and caffeine. No measurable interference was observed. Experiments showed that the principal selectivity is in the extraction rather than the back extraction step. This finding indicated that the polar and/or ionic nature of these interferences prevents retention by polymer. Preconcentration of analyte was examined by means of extraction from a flowing stream and back extraction into a minimum volume. The extractor tube length was 4.3 m, and the sample was 10 mL of 1 mM oxine with an extraction time of 2.0 min. The back extractant was 80 μL of 0.2 M NaOH. A sevenfold increase in concentration was observed. Enhanced preconcentration can be expected with smaller tubing diameter-to-length ratios, larger sample volumes, and repetitive use of back extractant.

Salicylic acid extractions from inorganic electrolyte solutions of 0.1 ionic strength yielded about 10% more extraction efficiency than those from water solutions, probably attributable to a salting-out effect.

The pK_a for salicylic acid is 3.0; thus, back extraction into a solution of even moderate pH is predicted. Experimentally, successful determinations of salicylic acid were performed by back extraction with 0.025 M acetate buffer (pH 4.7).

Precision and Linearity. Detector response for salicylate was linear for the range from 10^{-5} M to 10^{-2} M; the relative standard deviation of 3% was based on eight repetitive measurements of 0.5 mM solutions, and the sensitivity was 2.5 μA/mM. The corresponding coupled separation–detection system yielded a linear range from 2×10^{-5} M to 2×10^{-3} M, a relative standard deviation of 4%, and a sensitivity of 1.2 μA/mM. Such results suggest further applications of the technique.

Conclusions

Analytical separation of several organics from water by PVC polymer is feasible. A solvent extraction model describes the separation dynamics and pH dependence. Selectivity via pH control of the extraction step and preconcentration of analyte can be accomplished. These results suggest that other polymer solvent extraction schemes can be developed by using this approach. The flow-through amperometric technique provides a well-suited detector component for the technique.

Literature Cited

1. Shelly, D. C.; Rossl, T. M.; Warner, I. M. *Anal. Chem.* **1982,** *54,* 87–91.
2. Rice, M. R.; Gold, H. S. *Anal. Chem.* **1983,** *55,* 1436–1440.
3. Stuber, H. A.; Leenheer, J. A. *Anal. Chem.* **1983,** *55,* 111–115.
4. Schick, K. G.; Mageani, V. G.; Huber, C. O. *Clin. Chem.* **1978,** *24,* 448.
5. Morrison, T. N.; Schick, K. G.; Huber, C. O. *Anal. Chim. Acta* **1980,** *120,* 75.
6. Hui, B. S.; Huber, C. O. *Anal. Chim. Acta* **1982,** *134,* 211.
7. Kafil, B.; Huber, C. O. *Anal. Chim. Acta* **1982,** *139,* 347–352.

RECEIVED for review August 14, 1985. ACCEPTED April 22, 1986.

17

Evaluation of Bonded-Phase Extraction Techniques Using a Statistical Factorial Experimental Design

R. E. Hannah, V. L. Cunningham, and J. P. McGough

Smith Kline & French Laboratories, Philadelphia, PA 19101

Isolation techniques using bonded-phase silicas (bonded-phase extraction) were evaluated as an alternative to liquid–liquid extraction methods. The relative importance of four variables on extraction efficiencies for bonded-phase isolation techniques was evaluated by using a statistical 2^4 factorial experimental design (4 variables at 2 levels). Extraction efficiencies were based on percent recoveries for a 27-component synthetic test mixture containing a variety of organic compounds typical of those likely to be found in samples of interest. The experimental variables that were identified and included in the design were sample pH, nonpolar solid-phase extraction strength, polar-phase extraction strength, and conditioning solvent concentration. This 2^4 extraction factorial design allowed more information to be obtained with relatively few runs by varying several parameters at once. The application of statistical methods permitted the evaluation of data quality and also the determination of variable interactions.

ENVIRONMENTAL PROCESS ANALYSIS requires the characterization of chemical process and waste streams in order to evaluate their environmental abuse potential and treatability characteristics. An integral part of this analysis, as well as environmental fate determinations, is the isolation of organic compounds and metabolic products from very complex matrices such as waste water effluents, process streams, biological reactors, and fermentation broths. Generally, the organics involved are fairly polar, water-soluble compounds that must be ex-

0065–2393/87/0214/0359$06.00/0

tracted from aqueous solutions. Although liquid–liquid extraction methods usually work well for nonpolar organics, they are not always selective enough to separate the compounds of interest from all matrix interferences and frequently produce poor extraction efficiencies for more polar analytes.

Bonded silicas are now widely used in chromatographic separation techniques. Disposable extraction columns containing these materials are capable of extracting organic compounds from various sample matrices very efficiently. Extraction systems using bonded-phase silicas have been successfully applied to a wide range of sample preparation problems involving biological and biomedical studies (*1–7*); petrochemical analysis (*8*); food and cosmetics analysis (*9–12*); and environmental analysis involving raw water, drinking water, and waste water matrices (*13–16*). In most of these applications, the goal is to isolate a specific compound or class of compounds from the sample. More recently, especially in the environmental area, solid-phase extraction techniques are being applied to much broader compound groups such as priority pollutants (*15*) and hazardous organic compounds (*16*). With these broader applications in mind, an experimental design was set up to evaluate bonded-phase extraction techniques as general survey methods for organics in aqueous matrices.

Factorial Experimental Design

In a factorial experiment, a fixed number of levels are selected for each of a number of variables. For a full factorial, experiments that consist of all possible combinations that can be formed from the different factors and their levels are then performed. This approach allows the investigator to study several factors and examine their interactions simultaneously. The object is to obtain a broad picture of the effects of the selected experimental variables and detect major trends that can determine more promising directions for further experimentation. Advantages of a factorial design over single-factor experiments are (1) more than one factor can be varied at a time to allow the examination of interaction effects and (2) the use of all experimental runs in evaluating an effect increases the efficiency of the experiment and provides more complete information.

In this study, a factorial experiment was set up to evaluate the effects of four variables at two levels on extraction efficiencies by using bonded-phase isolation techniques and a 27-component synthetic test mixture. The compounds studied and the respective mass ions used for quantitation are given in the box. The compounds in the mix vary greatly in water solubility and volatility and, in general, represent a wide class of organic compounds typical of those present in environmental samples. To maximize solute recoveries, the procedure was

Test Mixture Components and Quantitation Mass Ions

Compound	*Mass Ion*
Isooctane	99
Acetone	43
Tetrahydrofuran	42
Ethyl acetate	61
Methyl ethyl ketone	72
Isopropyl alcohol	45
Ethanol	45
Methylene chloride	84
Methyl isobutyl ketone	100
Tetrachloroethylene	164
Toluene	65
Dodecane	85
Cyclohexanone	55
Dimethylformamide	73
Cyclohexanol	57
1,2-Dichlorobenzene	146
1-Octanol	84
Dimethyl sulfoxide	63
1,3,5-Trichlorobenzene	180
1,2,4-Trichlorobenzene	180
Nitrobenzene	123
Phenol	94
4-Methylphenol	108
1,2,3,4-Tetrahydroisoquinoline	132
2,4-Dichlorophenol	162
2,5-Dichlorophenol	162
4-Chlorophenol	128

designed to use two extraction column types of different polarities connected in series. The first column or primary column was a nonpolar phase, and the sorbent for the second column was more polar in nature. Also, in half the experimental runs, the samples were fortified with 500 ppm of methanol. The hydrophobic nonpolar phases required conditioning with methanol prior to sample loading to wet the phase surface. This facilitated contact between the aqueous sample matrix and the hydrophobic phase surface and maximized extraction efficiency. Methanol, acting as a bridge solvent, was added directly to the sample prior to extraction to determine what additional effect, if any, it had on extraction efficiency.

Four experimental variables were selected: sample pH, primary column type, secondary column type, and methanol concentration. By using each of the four variables at two levels, the complete arrangement of experimental runs became a $2 \times 2 \times 2 \times 2$ or 2^4 factorial design requiring 16 runs. Table I represents the design matrix; the high and

Table I. 2^4 Factorial Design Matrix

Test Run Number[a]	*Variable Levels*[b]			
	pH	*C1*	*C2*	*BS*
1	−	−	−	−
2	+	−	−	−
3	−	+	−	−
4	+	+	−	−
5	−	−	+	−
6	+	−	+	−
7	−	+	+	−
8	+	+	+	−
9	−	−	−	+
10	+	−	−	+
11	−	+	−	+
12	+	+	−	+
13	−	−	+	+
14	+	−	+	+
15	−	+	+	+
16	+	+	+	+

[a] Test runs were made in random order to eliminate possible bias.
[b] pH: low level (−) 2.0, high level (+) 8.0; C1 nonpolar phase extraction type: low level (−) C18, high level (+) C8; C2 polar phase extraction type: low level (−) cyano, high level (+) diol; BS bridge solvent concentration in sample: low level (−) 0 ppm, high level (+) 500 ppm.

low levels of each of the variables are coded as plus and minus signs. For the quantitative variables—pH and methanol concentration—a plus sign represents the high level. For the qualitative variables—primary column type and secondary column type—it does not matter which level is associated with the plus sign, as long as the designations are consistent. Table II represents the design matrix with the variable levels uncoded.

Experimental

Chemicals and Standard Solutions. Cyclohexanone, cyclohexanol, 1,3,5-trichlorobenzene, 1,2,4-trichlorobenzene, phenol, 4-methylphenol, 4-chlorophenol, 1,2,3,4-tetrahydroisoquinoline, 1-chlorohexane, 1-chlorododecane, and 1-chlorooctadecane were obtained from Aldrich. Acetone, tetrahydrofuran, ethyl acetate, toluene, dimethyl sulfoxide, and methanol were obtained from J. T. Baker. Distilled-in-glass isooctane, methylene chloride, ethyl ether, and pentane were obtained from Burdick and Jackson. Analytical standard kits from Analabs provided methyl ethyl ketone, isopropyl alcohol, ethanol, methyl isobutyl ketone, tetrachloroethylene, dodecane, dimethylformamide, 1,2-dichlorobenzene, 1-octanol, nitrobenzene, 2,4-dichlorophenol, and 2,5-dichlorophenol. All chemicals obtained from the vendors were of the highest purity available and were used without further purification. High-purity water

Table II. 2^4 Factorial Design Matrix

Test Run Number[a]	Variable Levels[b]			
	pH	*C1*	*C2*	*BS*
1	2	C18	cyano	0
2	8	C18	cyano	0
3	2	C8	cyano	0
4	8	C8	cyano	0
5	2	C18	diol	0
6	8	C18	diol	0
7	2	C8	diol	0
8	8	C8	diol	0
9	2	C18	cyano	500
10	8	C18	cyano	500
11	2	C8	cyano	500
12	8	C8	cyano	500
13	2	C18	diol	500
14	8	C18	diol	500
15	2	C8	diol	500
16	8	C8	diol	500

[a] Test runs were made in random order to eliminate possible bias.
[b] *See* Table I for description of variable levels.

was obtained by passing deionized carbon-filtered water supplied by the in-house system through a second system (Hydro Ultra Water Systems) containing mixed-bed, high-capacity, ion-exchange cartridges and an activated carbon cartridge. The water then passed through a final carbon polishing filter and was ultrafiltered with a 0.2-μm filter.

Individual stock solutions of the test compounds were prepared in methanol at a concentration of 50 mg/mL. A standard test mixture was prepared by adding 100-μL aliquots of each of the individual stock solutions to 500 mL of ultrapure water to give a concentration of 10 ppm per component. Samples requiring the addition of 500 ppm of methanol (conditioning solvent) were prepared by adding 6.3 μL of methanol to 10-mL aliquots of the aqueous standard mix just prior to extraction. The pH of the samples was adjusted with either 6 M HCl (for pH 2 samples) or 6 M NaOH (for pH 8 samples). To separate ionic strength effects from pH effects, the ionic strength of the samples was held constant.

Gas chromatographic–mass spectrometric (GC–MS) calibration standard mixes for quantitation were prepared in ethyl ether at concentrations of 20, 50, and 75 ppm. Internal standard spiking solution containing 1-chlorohexane, 1-chlorododecane, and 1-chlorooctadecane was prepared from individual stock solutions in methanol of each component. Two hundred microliters of each solution were added to ethyl ether and diluted to 2 mL. Forty microliters of this internal standard mix was added to the column extracts before diluting to 2 mL to yield a final concentration of 100 ppm per internal standard component.

Equipment and Instrumentation. Solid-phase extraction columns were obtained from J. T. Baker. Octadecyl (C18) and octyl (C8) 1-mL low-displacement columns were used as the primary extraction columns. Cyano and diol

3-mL columns were used as the secondary extraction columns. A Finnigan 1020 GC–MS was used to analyze the column extracts. The autoquantitation software available with this system was used to generate calibration curves and perform the necessary calculations.

Procedures. SORBENT CONDITIONING. To facilitate partitioning of organics in aqueous solution onto hydrophobic nonpolar sorbents, the sorbents must first be conditioned with methanol to increase their wettability. This solvation of the solid phase is necessary to provide efficient and reproducible extractions. The conditioning process was carried out according to the instructions accompanying the extraction columns.

SAMPLE LOADING. Prior to sample loading, the conditioned primary and secondary columns were connected in series by using column adapter connectors (J. T. Baker). The sorbents in each column were not allowed to dry out. A 10-mL aliquot of the 10-ppm aqueous standard mixture adjusted according to the treatment number of the design matrix (Table II) was loaded onto the column at a rate of 1.0–1.5 mL/min under an air pressure of 10 lb/in.^{2} applied at the top of the column. Air pressure was used instead of a vacuum to minimize losses of the more volatile components.

SORBENT ELUTION. The loaded extraction columns were disconnected and eluted with solvents individually. The elution procedure was as follows: First, 1 column bed volume of pesticide-grade pentane was added to the sorbent to displace any residual water. Then, 3 column bed volumes of ethyl ether were added. The eluting solvents were collected in a 2-mL volumetric flask. The eluates from the primary and secondary columns were combined in the volumetric flask. The internal standard solution was added, and the final extract volume was adjusted to 2 mL. Air pressure was again used to push the solvents through the sorbents to avoid volatility losses.

Analytical Methods. All of the sorbent extracts were analyzed by using GC–MS. The solutions were chromatographed on a 30-m $\times$ 0.32-mm i.d. Supelcowax 10 capillary column with a film thickness of 1 μm (Supelco). The Supelcowax 10 is a Carbowax PEG 20 M bonded-phase capillary column. The instrument conditions were as follows: injector, 250 °C; separator oven, 250 °C; column over initial, 40 °C programmed to 250 °C at 6 °C/min; linear velocity of helium carrier gas, 35 cm/s at 40 °C; column head pressure, 6 lb/in.^{2}; mass range, 33–333 amu scanned in 1-s intervals. Two-microliter aliquots were injected in the split mode at a 10-to-1 split ratio. A typical chromatogram appears in Figure 1.

Quantitative results were produced for each compound on the basis of internal standard method calculations. A three-point calibration curve was generated for each compound by using peak areas of a quantitation ion extracted from the mass spectrum of the compound. The ion was selected on the basis of it being a uniquely characteristic mass of the compound. The use of extracted ion quantitation produces more accurate results than total ion-current quantitation in cases in which two or more components are not completely resolved chromatographically. This situation is generally the case in complex mixture analysis. The quantitation ions selected for each of the compounds in the mix are listed in the box.

The Supelcowax 10 column has the ability to accept water injections. Direct aqueous injections of the 10-ppm standard water mix were performed to verify

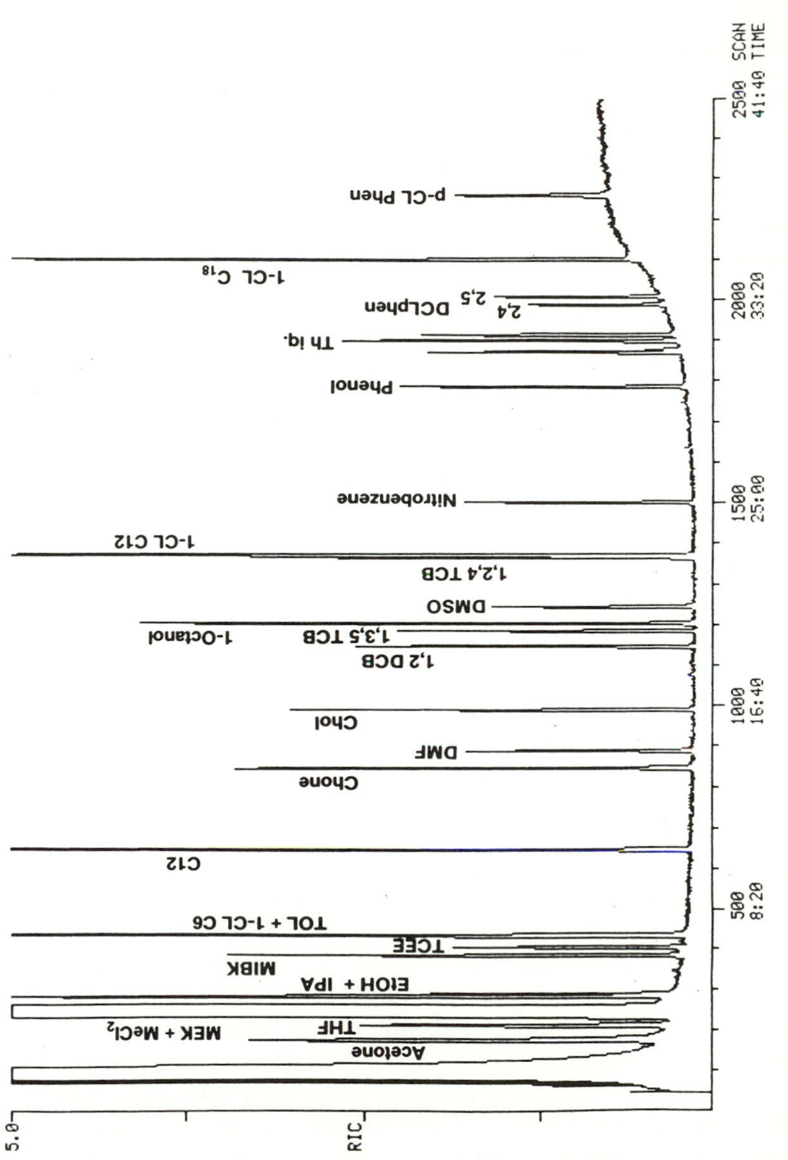

Figure 1. GC–MS chromatogram of the standard mixture.

the integrity of the sample solutions prior to extraction. A chromatogram of an aqueous injection of the mixture appears in Figure 2. Except for a base line upset at approximately 8 min, all features of the chromatogram appear very much like the ethyl ether standard mixture chromatogram in Figure 1.

Results and Discussion

The percent recoveries of 20 compounds from all 16 experimental runs are listed in Table III. The results for the seven remaining compounds in the mixture are not listed because either they were not recovered at all in any of the extractions (dimethylformamide and dimethyl sulfoxide), or because determinate system errors were discovered in the experimental protocol and rendered the data for those compounds unreliable.

Main effects were calculated to determine the influence of each variable on the extraction efficiency for each compound. An *effect* is the change in response for a variable as you move from the low to high level of that variable over all conditions of the other variables. The *main effect* of a variable is the difference between the average of the high-level responses and the average of the low-level responses. A cubic representation of responses for methyl isobutyl ketone is illustrated schematically in Figure 3. The values at each vertex of the cube represent the response (percent recovery) measured at that combination of variable levels. The values followed by a minus sign represent the response when variable B (primary column type) was low (C18). The values followed by a plus sign represent the response when variable B was high (C8). Schematically, the change in responses in going from the low to high level of variable D (methanol concentration) can be determined by comparing the response values on the bottom plane of the cube (low) to the top plane of the cube (high). Likewise, the effect of changing variable A (pH) is determined by comparing the left and right faces of the cube, and the effect of variable C (secondary column type) is determined by comparing the responses on the front and back faces of the cube.

To manually calculate the effects of a 2^4 factorial design for 20 different compounds would be quite time-consuming. Fortunately, more rapid methods can be used, one being Yate's algorithm (*17*). This algorithm is applied to the responses after they have been arranged in standard order. A factorial design is in standard order when, as in Table I, the first column consists of alternating minus and plus signs, the second column consists of successive pairs of minus and plus signs, the third column consists of alternating sets of four, etc. By using the Yate's algorithm contained in RS1 software (Bolt, Beranek, and Newman) mounted on a VAX system (Digital), analysis tables were generated for each compound. Table IV contains the results for *p*-cresol. Column 0

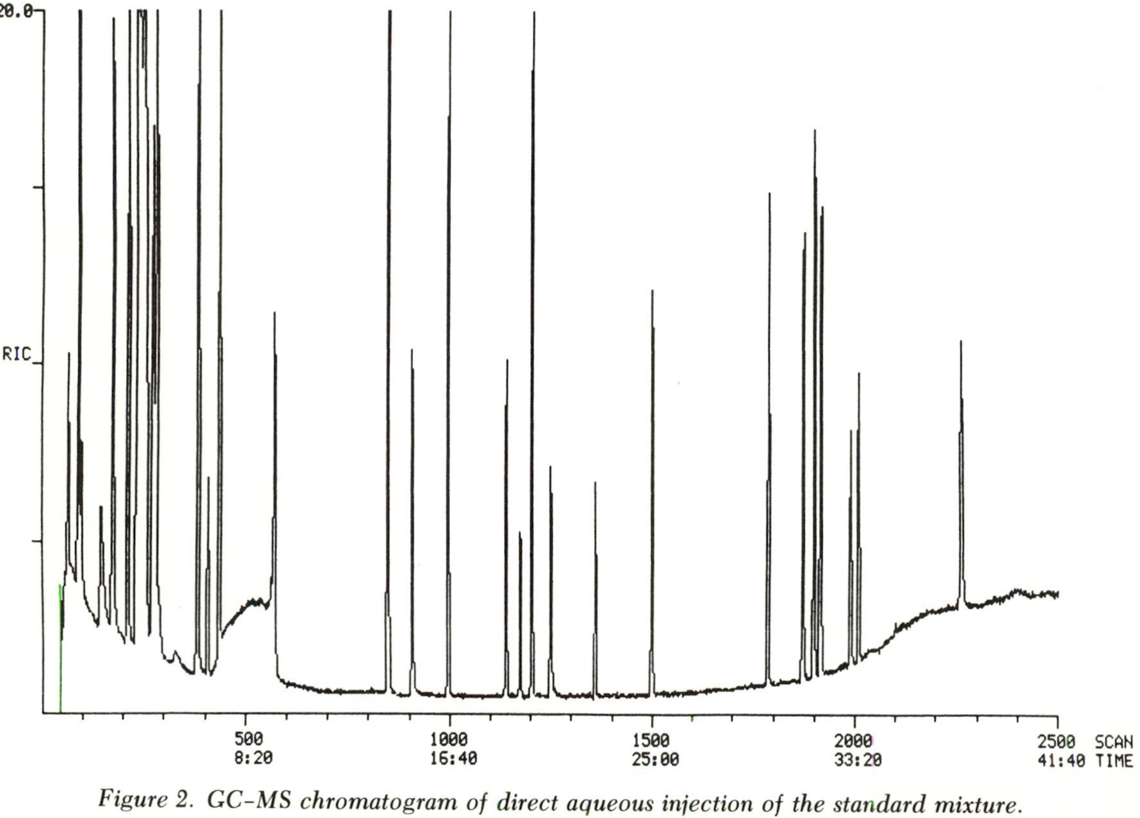

Figure 2. GC–MS chromatogram of direct aqueous injection of the standard mixture.

Table III. Percent Recoveries of Compounds from Experimental Runs

Test Run Number	*Toluene*	*Dodecane*	*135-TCB*	*124-TCB*	*TCEE*	*$MECL_2$*	*CHOL*	*ACET*	*1-Octanol*	*NO_2 Benzene*
1	20	14	21	24	12	5	56	18	58	60
2	0	11	17	18	5	13	38	10	46	42
3	4	10	12	14	6	20	38	10	40	46
4	8	13	14	15	4	10	58	22	54	60
5	30	22	26	30	22	50	54	10	68	70
6	0	8	8	7	1	8	42	13	48	48
7	7	12	16	18	6	10	52	16	54	54
8	3	14	17	19	6	13	72	15	50	64
9	30	17	25	29	17	40	58	16	64	66
10	11	12	17	20	7	19	53	16	62	60
11	24	19	30	34	19	45	64	18	66	72
12	1	12	19	22	5	13	50	14	48	48
13	0	8	11	11	3	10	34	9	42	42
14	0	8	8	8	1	10	50	15	56	54
15	36	21	24	28	17	28	60	16	66	66
16	0	12	9	8	2	9	52	15	54	56

Table III. Continued

Test Run Number	*MIBK*	p-*Cresol*	*CHONE*	*DCB*	*THIQ*	p-*CLPHEN*	*THF*	*Phenol*	*24DCLPHEN*	*25DCLPHEN*
1	54	56	52	3	8	62	24	26	72	64
2	28	11	36	2	42	12	11	3	32	28
3	28	42	34	2	8	43	8	33	50	46
4	44	18	56	2	54	22	22	6	42	40
5	70	52	52	4	8	60	26	24	74	68
6	38	8	38	1	42	13	12	2	30	26
7	50	52	52	2	11	54	18	38	62	58
8	60	14	66	2	52	21	26	4	40	37
9	60	56	54	4	8	63	24	28	76	70
10	58	12	48	2	48	19	20	6	54	50
11	60	60	58	4	8	68	30	50	78	72
12	30	17	44	2	44	20	14	6	38	32
13	25	40	30	1	6	44	5	19	52	46
14	44	10	44	1	48	16	14	3	36	30
15	60	56	60	4	12	60	30	42	72	64
16	46	12	48	1	50	15	18	2	24	21

NOTE: Column headings key: 135-TCB: 1,3,5-trichlorobenzene; 124-TCB: 1,2,4-trichlorobenzene; TCEE: tetrachloroethylene; $MECL_2$: methylene chloride; CHOL: cyclohexanol; ACET: acetone; NO_2 benzene: nitrobenzene; MIBK: methyl isobutyl ketone; CHONE: cyclohexanone; DCB: 1,2-dichlorobenzene; THIQ: 1,2,3,4-tetrahydroisoquinoline; *p*-CLPHEN: 4-chlorophenol; THF: tetrahydrofuran; 24DCLPHEN: 2,4-dichlorophenol; 25DCLPHEN: 2,5-dichlorophenol.

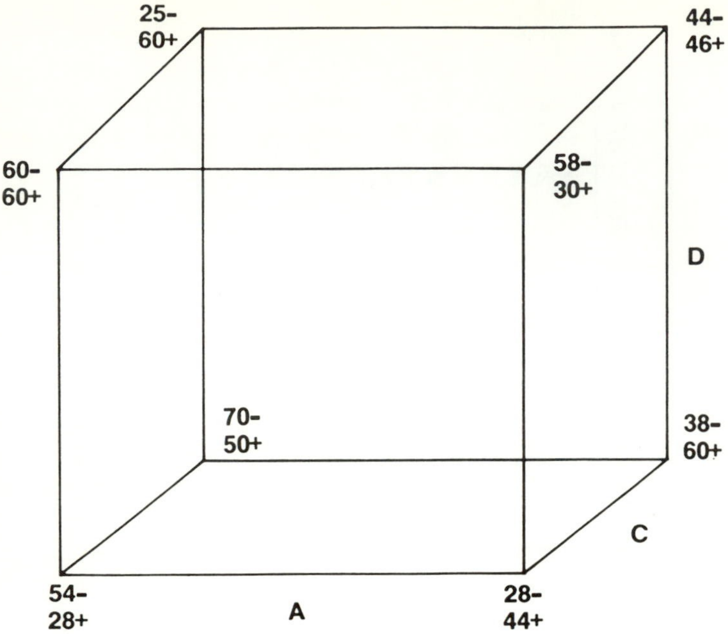

Figure 3. Cubic representation of responses for methyl isobutyl ketone. A = pH 2, 8; C = cyano, diol; and D = no solvent, solvent.

Table IV. Yates Analysis for p-Cresol

0		*1*	*2*	*3*	*4*
1	(1)	56	516	—	—
2	A	11	−312	−39.00	6084.00
3	B	42	26	3.25	42.25
4	AB	18	14	1.75	12.25
5	C	52	−28	−3.50	49.00
6	AC	8	0	0.00	0.00
7	BC	52	22	2.75	30.25
8	ABC	14	−30	−3.75	56.25
9	D	56	10	1.25	6.25
10	AD	12	−10	−1.25	6.25
11	BD	60	28	3.50	49.00
12	ABD	17	−40	−5.00	100.00
13	CD	40	−26	−3.25	42.25
14	ACD	10	26	3.25	42.25
15	BCD	56	−4	−0.50	1.00
16	ABCD	12	0	0.00	0.00

NOTE: Column designations are as follows: 0, treatment number and variables; 1, response mean (percent recovery); 2, sum of response; 3, effect calculation; and 4, mean square. Each treatment mean has one degree of freedom. (*See* text for more detail.)

lists the treatment number and variables that were at the plus or high level. Row 1 represents the response when all variable levels were low. When only one variable is listed in column 0, the effect in that row is due to that single variable. When more than one variable is listed, the effect in that row is due to the interactions between the variables. Column 1 lists the responses (percent recoveries) when the variable or variables in Column 0 are at the high level and all other variables are at the low level. Column 2 lists the sum of the recoveries for variables at the high level minus the sum of the recoveries for variables at the low level. Column 3 lists the effect calculation, which is the average of the responses when the variable level is high minus the average of the responses at low level. The greater the absolute value of the calculated effect for a given treatment, the greater the likelihood that that variable or combination of variables (interaction) has a statistically significant influence on the response. Each treatment mean has one degree of freedom, as shown in Column 4. Because the sum of the differences between each mean and the grand mean must be zero, the set of 16 experiments has a total of $n - 1$ or 15 degrees of freedom. Column 5 lists the mean square associated with each treatment, or the square root of the residual sum of squares (RSS). The ratio of RSS values follows an F distribution and is a measure of the statistical significance of the effect. For example, the ratio of the square of the largest mean square value divided by its respective degrees of freedom ($6084^2/1$) to the square of the next highest value divided by its degrees of freedom ($100^2/1$) gives an F value of 3701.5. Comparison of this value with those given in an F distribution table indicates that the effect is statistically significant at almost the 99% level. The actual F value for the 99% level at these degrees of freedom is 4052 (*18*).

Any experimental design that is intended to determine the effect of a parameter on a response must be able to differentiate a real effect from normal experimental error. One usual means of doing this determination is to run replicate experiments. The variations observed between the replicates can then be used to estimate the standard deviation of a single observation and hence the standard deviation of the effects. However, in the absence of replicates, other methods are available for ascertaining, at least in a qualitative way, whether an observed effect may be statistically significant. One very useful technique used with the data presented here involves the analysis of the factorial by using half-normal probability paper (*19*).

One of the fundamental premises of statistics is that, in the absence of real effects, the errors associated with a set of experimental observations are random variable with a normal probability distribution and have a mean (μ) equal to 0 and a variance equal to σ^2. A typical normal probability distribution is illustrated in Figure 4.

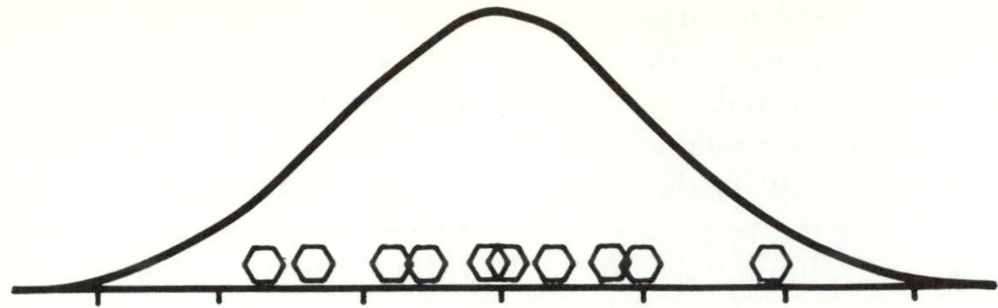

Figure 4. Normal probability distribution.

If the percentage probability of the occurrence of some value of *X* is plotted against *X*, a sigmoidal cumulative normal curve is obtained, as shown in Figure 5.

Normal probability paper is obtained by adjusting the vertical in such a way that the plot of *P* versus *X* is a straight line. Thus, data that follow a normal probability distribution will produce a straight line when plotted on normal probability paper, as shown in Figure 6.

Half-normal plots are a modification in which the absolute value of *X* is used rather than the actual value itself (*20*). This technique has the benefit of a plot whose *X* axis starts at 0.

Use of Half-Normal Plots with Factorial Data. The application of this method to the factorial data is straightforward. If, for any given compound, the data from the factorial experiment occurred simply as the result of random variation about a fixed mean, and the changes in the levels of the variables had no real effect at all on the percent recovery, then the 15 main effects and interactions, representing 15

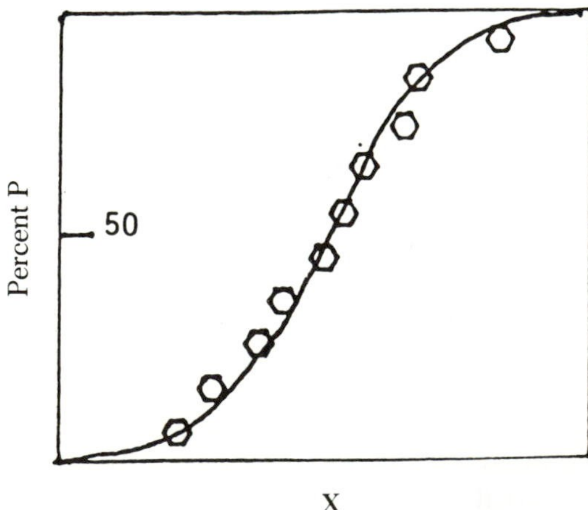

Figure 5. Plot of cumulative percent probability versus X.

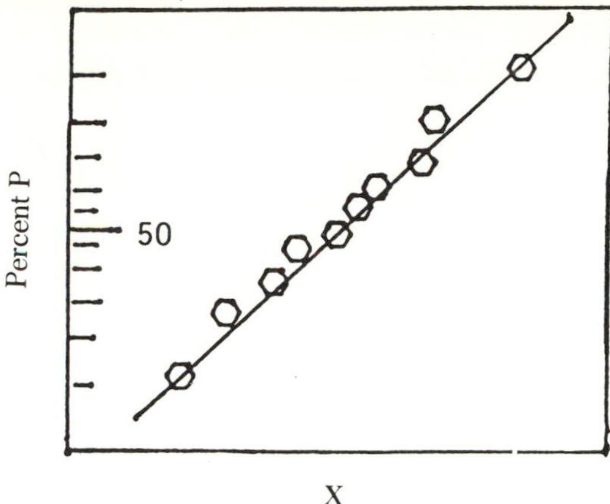

Figure 6. Plot of cumulative percent probability versus X plotted on normal probability paper.

contrasts between pairs of averages each containing eight observations, would be roughly normal and distributed about zero. They would therefore plot on half-normal probability paper as a straight line. The data for methylene chloride, shown in Figure 7, clearly represent this type of behavior. Therefore, none of the variations in the experimental design had a statistically significant effect on the recovery of methylene chloride. Ten compounds from the mix fell into this category on the basis of half-normal analysis. They include 1,2,4-trichlorobenzene, 1,3,5-trichlorobenzene, dodecane, cyclohexanone, cyclohexanol, 1-octanol, tetrahydrofuran, 1,2-dichlorobenzene, nitrobenzene, and methylene chloride.

Conversely, examination of the half-normal probability plot for *p*-cresol, Figure 8, shows that although 14 of the 15 data points fall approximately on a straight line through the origin, one point does not. This aberrant point relates to the main effect A, which is the pH level. Here, then, the effect of pH on the recovery of *p*-cresol is statistically significant. Indeed, pH was expected to have a significant effect on extraction efficiency in consideration of the pK_a of *p*-cresol (10.17). Five compounds had half-normal probability plots that possessed one significant outlier, which was the effect due to pH. The compounds were phenol, *p*-chlorophenol, *p*-cresol, 1,2,3,4-tetrahydroisoquinoline, and toluene.

The real power of the use of half-normal probability plots, however, comes with data that are likely to have embedded outliers. These data profoundly distort the half-normal plots, as illustrated with the data for methyl isobutyl ketone shown in Figure 9. The plot shows neither normal random error nor significant effects cleanly. Thus, this

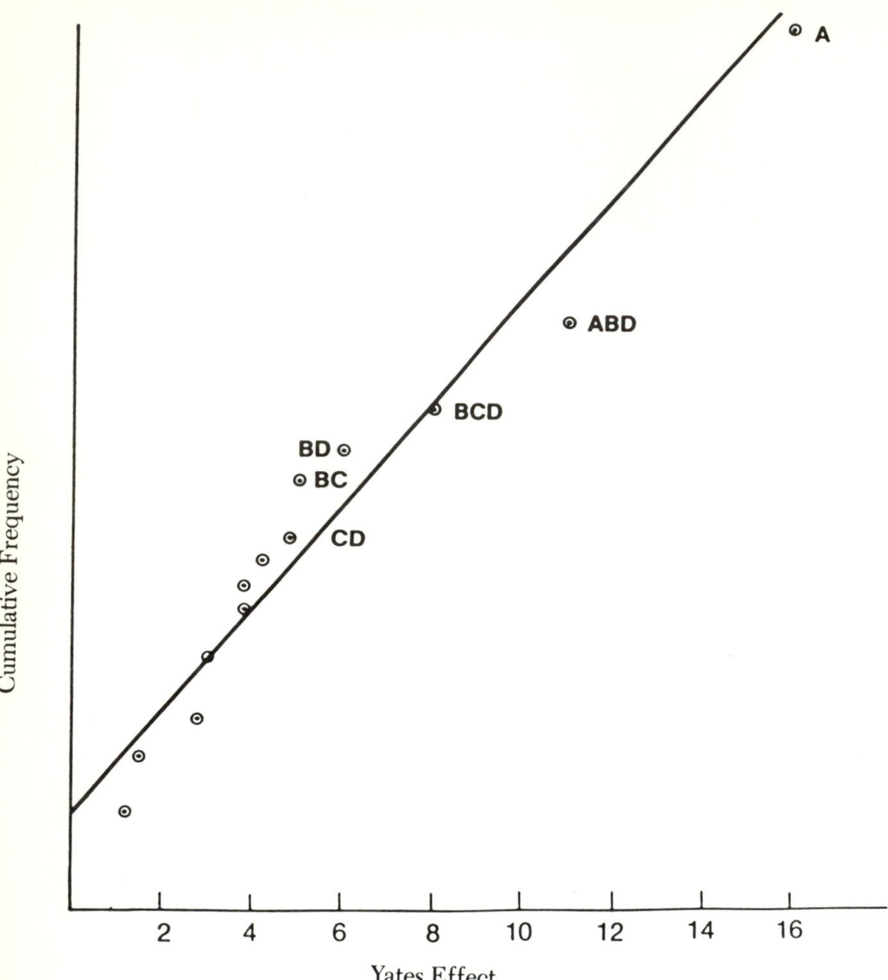

Figure 7. Half-normal probability plot for methylene chloride.

plot serves as a diagnostic tool and focuses attention on the need to repeat all or part of the factorial. Five compounds in all—methyl isobutyl ketone, 2,4-dichlorophenol, 2,5-dichlorophenol, acetone, and tetrachloroethylene—produced distorted half-normal plots.

As a result of this anomalous data, part of the main factorial experiments was repeated by using a 2^3 design that included the same parameters and conventions as the original factorial, except that one factor, the methanol bridge solvent concentration, was excluded from the design. This smaller factorial consisted of eight experiments that were run in duplicate. The design matrix is depicted in Table V.

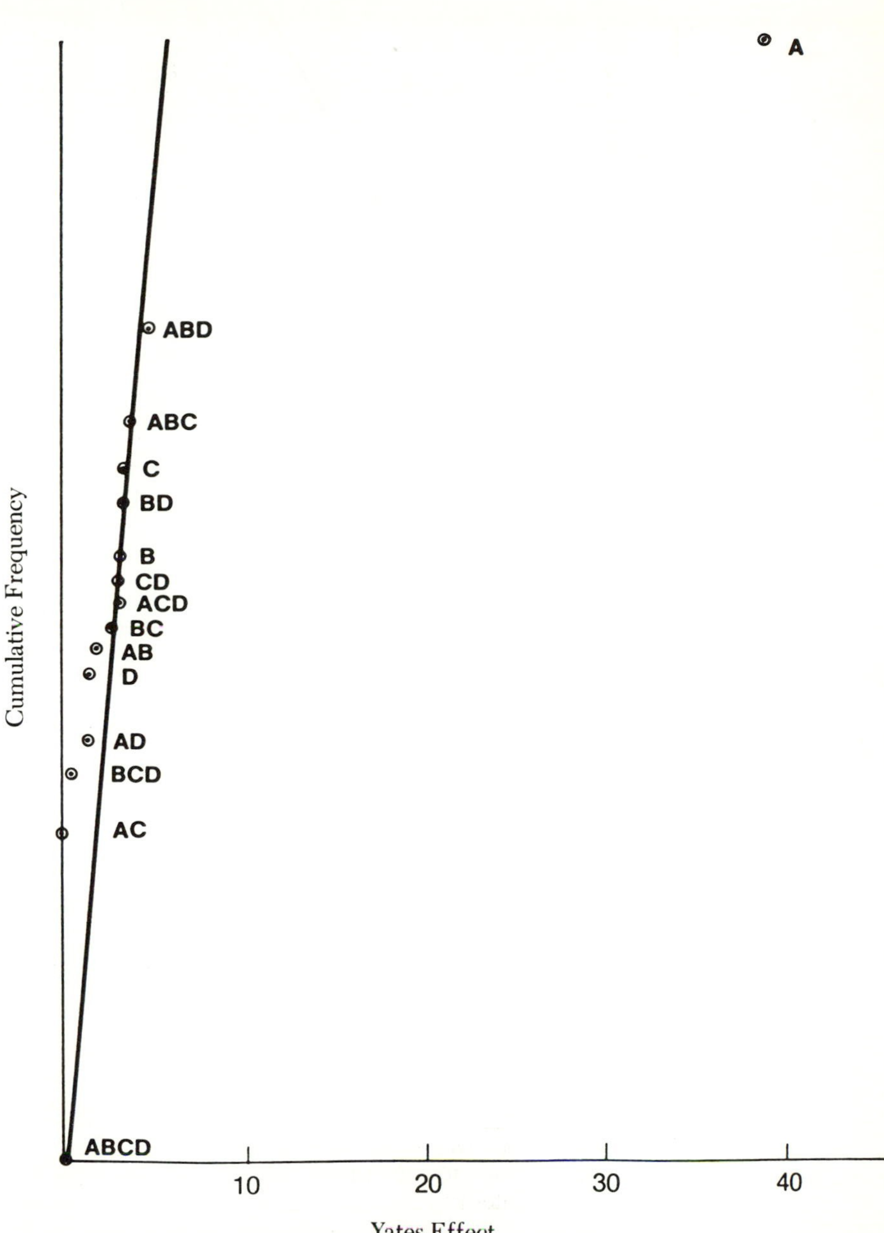

Figure 8. Half-normal probability plot for p-*cresol.*

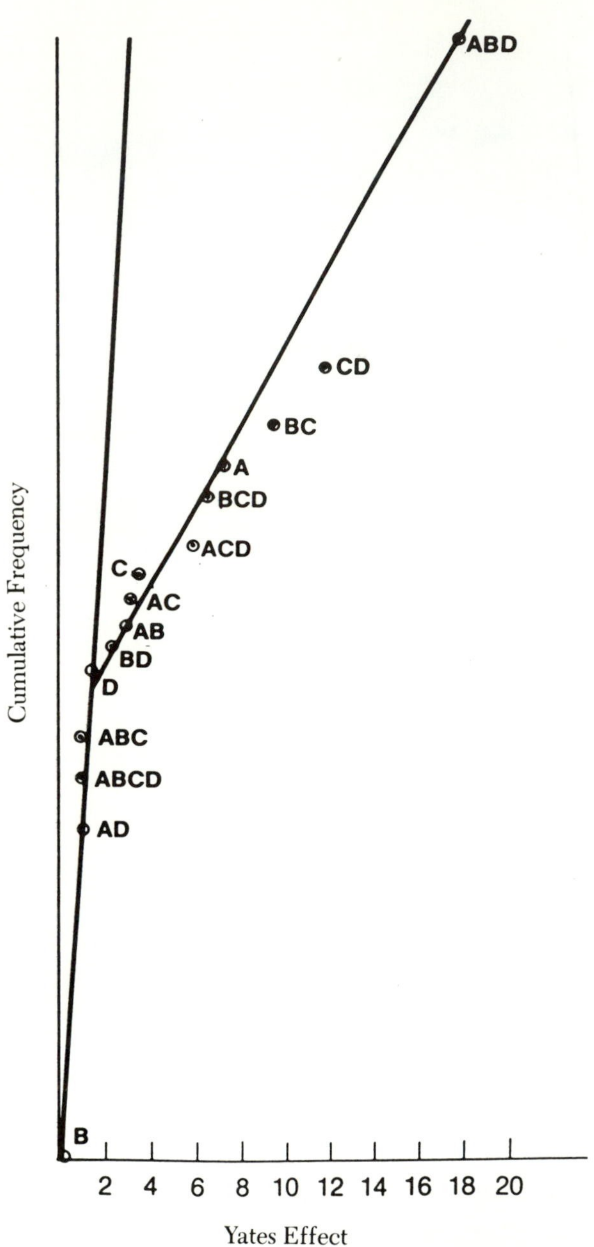

Figure 9. Half-normal probability plot for methyl isobutyl ketone.

Table V. 2^3 Factorial Design Matrix

Test Run Number[a]	*Variable Levels*[b]		
	pH	*C1*	*C2*
1	2	C18	cyano
2	8	C18	cyano
3	2	C8	cyano
4	8	C8	cyano
5	2	C18	diol
6	8	C18	diol
7	2	C8	diol
8	8	C8	diol

[a] Test runs were made in random order to eliminate possible bias.
[b] *See* Table I for description of variable levels.

Compound recovery data for duplicate runs differed by 2–15%, depending on the compound. Half-normal probability plot analysis of the new data for the anomalous compounds indicated none of the distortion encountered earlier. Results for acetone and tetrachloroethylene now indicated only random variation with no significant outliers. Results for 2,4-dichlorophenol and 2,5-dichlorophenol indicated a significant pH effect. A significant interaction effect (AB) was detected between variables pH and primary column type for the dichlorophenols and also for methyl isobutyl ketone. This interaction effect indicates that at approximately low pH (pH 2), compound recoveries for dichlorophenols will be greater when a C18 phase is used as the primary column. The half-normal plot for 2,5-dichlorophenol is shown in Figure 10. In examining data for all the compounds from the 2^3 replicate factorials, this interaction consistently appears for phenolic compounds.

Conclusions

Analysis of data from the factorials indicates that pH has a consistently significant effect on compound recoveries. A summary of the effect of pH level on compounds used in the study is given in Table VI. There is also an interaction between pH and primary column sorbent type for some compounds. This interaction suggests that at low sample pH, a C18 column will produce the best extraction efficiencies for phenolic compounds. The effect of adding methanol to the sample before extraction clearly produced odd results when the recovery data from the 2^4 factorial was analyzed by using half-normal plots. This effect will be studied in future work. Additionally, different elution solvents will be examined as well as new sorbent phases as they become available.

In general, many of the analytes examined in this study can be isolated by using alternate methods such as purge and trap, liquid–liquid

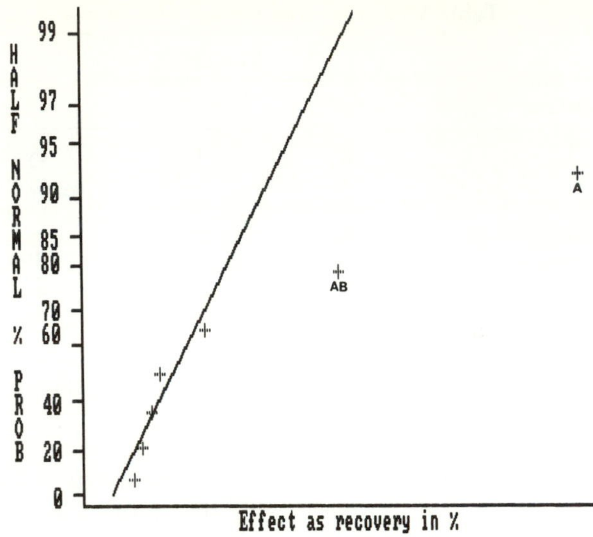

Figure 10. Half-normal probability plot for 2,5-dichlorophenol.

Table VI. Summary of pH Effect on Compound Recovery

Compound	*Mean*	*Low pH Mean*	*High pH Mean*
Toluene	11	19	3
Dodecane	13	15	11
1,3,5-Trichlorobenzene	17	21	14
1,2,4-Trichlorobenzene	20	24	15
Tetrachloroethylene	9	13	4
Methylene chloride	19	26	12
Cyclohexanol	52	52	52
Acetone	15	14	15
1-Octanol	55	57	52
Nitrobenzene	57	60	54
Methyl isobutyl ketone	47	51	43
p-Cresol	33	52	13
Cyclohexanone	49	49	48
1,2-Dichlorobenzene	3	3	2
1,2,3,4-Tetrahydroisoquinoline	29	9	48
4-Chlorophenol	37	57	17
Tetrahydrofuran	19	21	17
Phenol	22	33	11
2,4-Dichlorophenol	52	67	37
2,5-Dichlorophenol	47	61	33

NOTE: All values are percent recoveries.

extraction, and steam distillation. However, this study has shown that as a general survey method, solid-phase extraction techniques are effective in isolating and concentrating a variety of organic compounds from aqueous matrices at trace levels. These techniques generally require less sample manipulation and are much quicker and more economical than traditional extraction procedures. The use of two-level experimental factorial designs has been demonstrated to be extremely useful for measuring the effects of variables on a response. Factorial designs are economical, easy to use, and can provide a great deal of valuable information. Also there are powerful diagnostics for easily testing the quality of experimental data.

Literature Cited

1. Mills, J. *Biochem. Soc. Trans.* **1985,** *13(6),* 1073–1077.
2. Lato, M. *Lab (Milan)* **1984,** *11(5),* 413–416.
3. Karege, F. *J. Chromatogr.* **1984,** *311(2),* 361–368.
4. Canfell, C.; Binder, S.; Khayam-Bashi, H. *Clin. Chem.* **1982,** *28,* 25.
5. Ford, B.; Vine, J.; Watson, T. R. *Anal. Toxicol.* **1983,** *7,* 116–118.
6. Good, T. J. *J. Chromatogr. Sci.* **1981,** *19,* 562–566.
7. Frayn, K. N.; Maycock, P. F. *Clin. Chem.* **1983,** *29(7),* 1426–1428.
8. *Baker-10 SPE Applications Guide;* J. T. Baker Chemical: Phillipsburg, NJ, 1984; Vol. 1, p 28.
9. Reid, S. J.; Good, T. J. *J. Agric. Food Chem.* **1982,** *30(4),* 775–778.
10. Fox, A.; Morgan, S.; Hudson, J.; Zhu, Z. T.; Lau, P. *J. Chromatogr.* **1983,** *256,* 429–438.
11. Hilker, D. M.; Clifford, A. J. *J. Chromatogr.* **1982,** *231,* 433–438.
12. Baker-10 SPE Applications Guide; J. T. Baker Chemical: Phillipsburg, NJ, 1984; Vol. 1, p 56.
13. *Analytichem Applications Note E2;* Analytichem International: Harbor City, CA, 1984.
14. Dimson, P. *LC* **1983,** *1(4),* 236–237.
15. Chladek, E.; Marano, R. S. *J. Chromatogr. Sci.* **1984,** *22(8),* 313–320.
16. Rostad, C. E.; Pereira, W. E.; Ratcliff, S. M. *Anal. Chem.* **1984,** *56(14),* 2856–2860.
17. Yates, F. *Imp. Bur. Soil Sci. Tech. Commun.* **1937,** *35.*
18. Neter, J.; Wasserman, W. *Applied Linear Statistical Models: Regression, Analysis of Variance, and Experimental Design;* Richard D. Irwin: Homewood, IL, 1974.
19. Box, G. E.; Hunter, W. G.; Hunter, J. S. *Statistics for Experimenters: An Introduction to Design, Data Analysis, and Model Building;* John Wiley: New York, 1978.
20. Daniel, C. *Applications of Statistics to Industrial Experimentation;* John Wiley: New York, 1976.

RECEIVED for review August 14, 1985. ACCEPTED December 26, 1985.

18

Use of Gel Permeation Chromatography To Study Water Treatment Processes

A. Bruchet, Y. Tsutsumi, J. P. Duguet, and J. Mallevialle

Lyonnaise des Eaux Central Laboratory, 38 rue du President Wilson 78230 Le Pecq, France

The efficiency of a water treatment process is often evaluated by using nonspecific and specific parameters. Most of the time, specific determinations involve extraction techniques followed by gas chromatography (GC) or high-performance liquid chromatography. Consequently, determinations are limited to the study of volatile and semivolatile organics. This chapter presents a gel permeation technique used to study higher molecular weight or more polar compounds. The gel permeation chromatography and pyrolysis GC–mass spectrometry were used in a pilot study at Vigneux, south of Paris, to determine the efficiency of the combination of ozone and granular activated carbon unit processes.

EVALUATION OF A WATER TREATMENT PROCESS usually involves the measurement of nonspecific parameters (e.g., dissolved organic carbon [DOC], total organic halogens [TOX], UV absorbance, fluorescence). In addition, the development of chromatographic methods (gas and liquid phase) and extraction techniques (e.g., resin adsorption, continuous liquid–liquid extraction) has expanded analysis to many volatile and semivolatile nonpolar compounds. However, the majority of the organic carbon is generally found in the form of humic and fulvic acids (*1*) together with other biological macromolecules. These compounds are not well-characterized and are not amenable to analysis with current organic analytical techniques.

These humic and fulvic acids contribute substantially to the carbon balance in a unit process. They can also modify the way micropollutants

0065–2393/87/0214/0381$06.00/0

act in a process by association (*2*), and they may be precursors for volatile halogenated compounds (*3, 4*). Several attempts to characterize these dissolved organic compounds have resulted in general models for humic acid (*5, 6*).

Two techniques are generally used to separate organic compounds by molecular weight: ultrafiltration on membranes and gel permeation chromatography (GPC). Use of ultrafiltration to separate humic acids from natural surface water and to study the effects of ozonation on these humic acids has been previously reported (*7*). The method was shown to be quite useful but very time-consuming. Recently, GPC using Sephadex gels has been applied in the field of sewage treatment (*8*) and in the natural environment on lake or ground waters (*9, 10*). Also, new silica gels for high-pressure liquid chromatography (HPLC) have been developed, and their ability to separate natural organic compounds in water has been demonstrated (*11, 12*).

This study reports two gel permeation techniques used to evaluate water at various stages of (potable water) treatment. The water samples were taken from a surface-water pilot treatment plant that employed several levels of ozone treatment. Selected fractions of the permeate were subsequently analyzed by pyrolysis gas chromatography–mass spectrometry (GC–MS).

Materials and Methods

The source of water samples was a 4-m^3/h pilot plant on the Seine River located upstream from Paris, France. The background organic concentration ranged from 2 to 3 mg/L. The process, shown in Figure 1, included an upflow solids contact clarifier (Pulsator, Degremont, Rueil Malmaison, France) followed by rapid sand filtration (RSF). The effluent of the RSF was then split into four lines, which received various levels of ozonation followed by granular activated carbon (GAC) adsorption. Postchlorination (0.2 mg/L residual after 1 h) was used for bacterial control.

Gel Permeation Chromatography. The molecular weight distributions were determined by gel permeation chromatography (GPC) on Sephadex G25 (Sephadex Pharmacia, Uppsala, Sweden). Conditions were as follows: column size, 0.25 × 90 cm; eluant, water; flow rate, 100 mL/h. Water samples were first concentrated by rotary evaporation to obtain a total organic carbon (TOC) concentration between 100 and 200 mg/L. Ten milliliters of these concentrates was then injected into the Sephadex column. Fractions were collected following chromatographic separation for analysis. UV absorbance at various wavelengths, TOC (Dohrman DC80, Envirotech Corp.), and fluorescence (excitation wavelength = 320 nm, emission wavelength = 405 nm) were measured on the fractions collected.

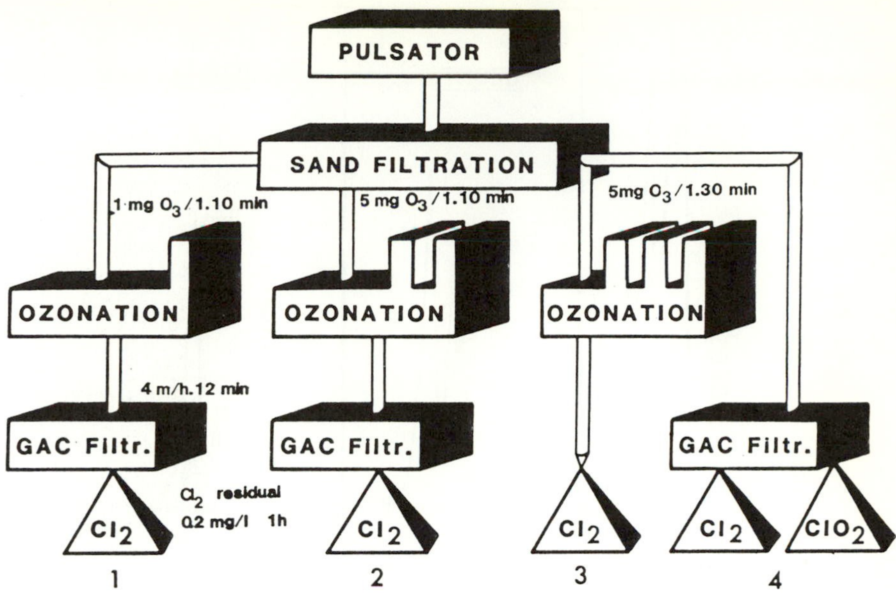

Figure 1. Diagram of Vigneux pilot plant.

High-pressure GPC was performed on a Du Pont 8800 chromatograph equipped with two TSK G 2000 SW columns (7.5 × 300 mm each column) in series. The chromatograph was operated at 1 mL/min, and 5 mM phosphate buffer was used as the eluent. Detection was by UV at 254 nm.

Pyrolysis GC–MS Analysis. Flash pyrolysis was performed by using a pyroprobe 100 (Chemical Data Systems) temperature-control system. Samples were pyrolyzed from 150 to 750 °C with a temperature program of 20 °C/ms and a final hold for 20 s. After pyrolysis, the fragments were separated on a 25-m CP WAX 57 fused silica capillary column (temperature program: 25–220 °C at 3 °C/min), followed by MS on a R 10–10 C (Ribermag, Rueil-Malmaison, France) operated at 70 eV and scanned from 20 to 400 *m/z*.

Results and Discussion

Size Exclusion Chromatography. Examples of HPLC chromatograms obtained for line I by using TSK G 2000 SW are shown in Figure 2. The chromatograms came from rapid sand-filtered water (RSW), ozonated water, and GAC-filtered water. In each chromatogram, three peaks were observed in less than 15 min. Because no exclusion peak occurred, apparently no UV-absorbing compounds having molecular weights (MW) >10,000 were present.

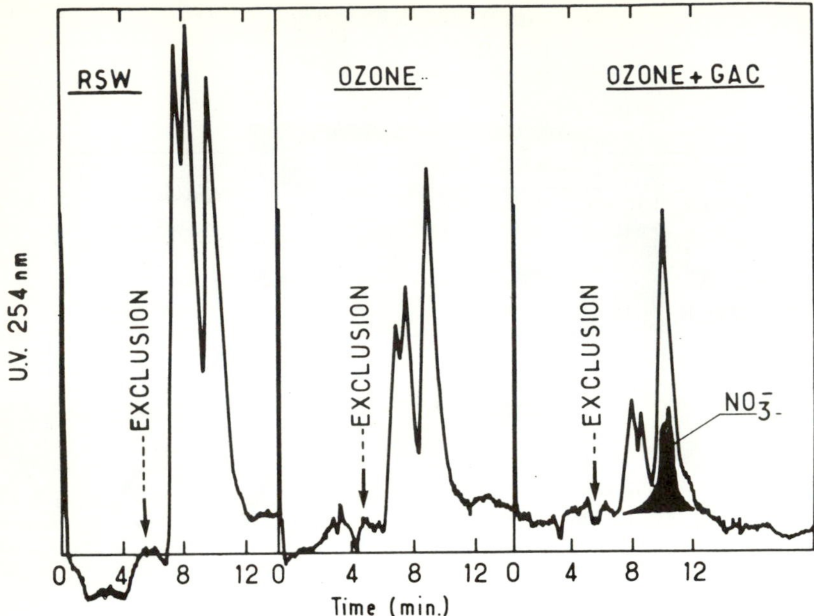

Figure 2. HPLC chromatograms obtained for line I by using TSK G 2000 SW.

The predominant effect of ozone and subsequent GAC contact is seen in the first two peaks. These peaks represent intermediate molecular weights and undergo a dramatic reduction in concentration. The third peak is only slightly diminished by the treatment process. The use of UV excitation at 254 nm (UV254) introduces potential interferences from inorganic materials. Several solutes (nitrate, chloride, phosphate, sulfate) were analyzed to determine their contribution to the UV254 response. Only nitrate was found in sufficient quantity to substantially contribute to the peaks observed. This situation is shown graphically in Figure 2 in the ozone + GAC chromatogram. Nitrate, at a concentration of 20 mg/L, accounted for 30% of the third peak.

This method proved to be both easy to operate and rapid. However, some drawbacks may limit its applicability. Silanol groups on the gel may adsorb materials irreversibly and can bleed organic carbon into the system. Because of the low volumes of sample injected, TOC balances using this type of technique are not possible.

Size-exclusion chromatography with Sephadex gel was used to study changes in the organic matrix as it passed through the treatment process at Vigneux. The results obtained are presented in Figures 3–7. Points where significant differences were observed are indicated by arrows.

RAPID SAND FILTRATION. A comparison between raw water and sand-filtered water is shown in Figure 3. Three main fractions are typically recovered with Sephadex G25 from low-TOC surface waters. These fractions will be referred to as G1, G3, and G5, which correspond to the fraction numbers by order of elution. The apparent MWs for these fractions are as follows: >5000 daltons for G1, 1000–5000 daltons for G3, and <1000 daltons for G5. During RSF, G5 exhibits a decrease in TOC, fluorescence, and UV260, whereas G1 shows a decrease only in TOC. The distribution by MW of organic carbon after sand filtration is as follows: >5000 daltons, 17% (G1); 1000–5000 daltons, 29% (G3); and <1000 daltons, 54% (G5).

Another important observation is the low magnitude of the ratios UV260/TOC and fluorescence/TOC for the largest size fraction. This finding is not consistent with the behavior of the polyhydroxyaromatic core of humic acids (*13*).

OZONATION. The effect of ozonation at various doses and contact times is shown in Figure 4. As usual, arrows are used on the figure to

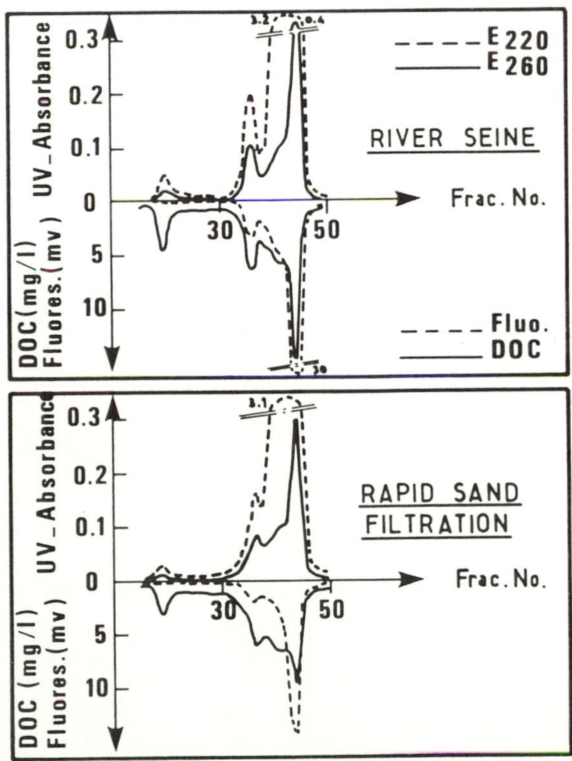

Figure 3. Comparison between raw water and sand-filtered water.

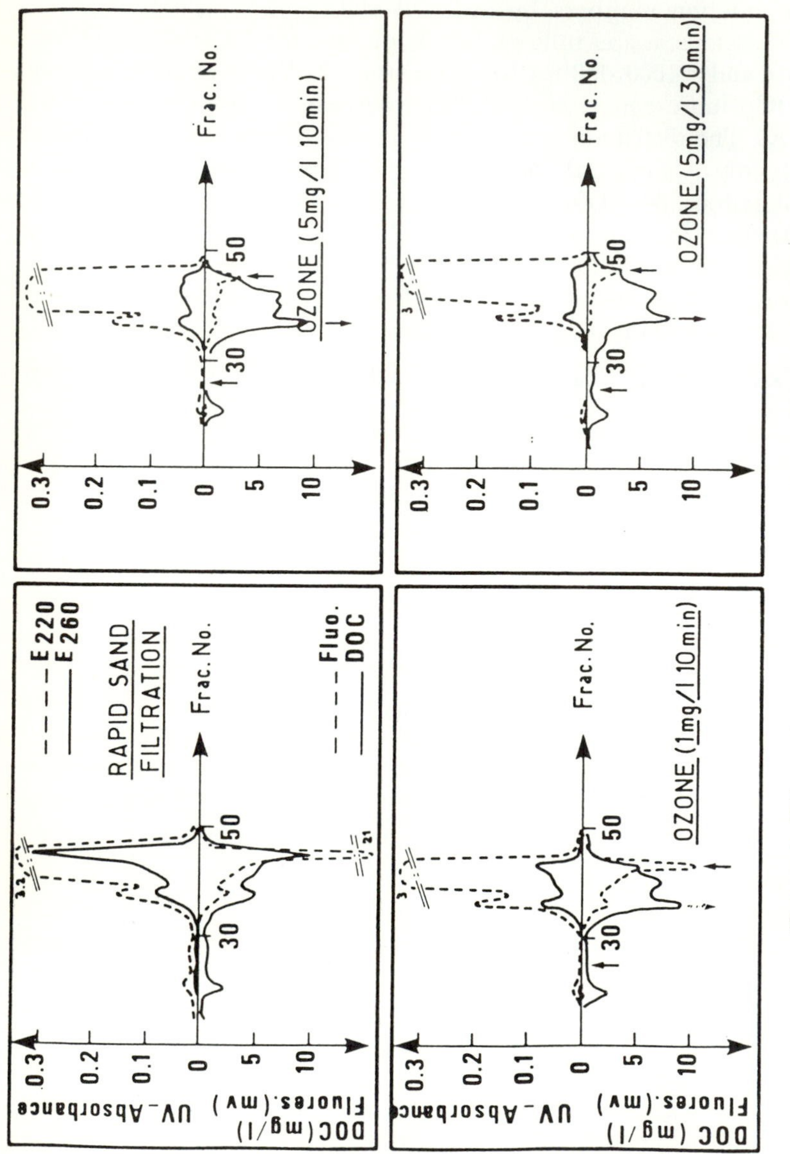

Figure 4. Effect of ozonation at various doses and contact times.

indicate where changes in any parameter occur across the unit operation of ozonation. UV260 and fluorescence are greatly reduced for fraction G5, and the reduction increases with ozone dose and contact time. This finding is interpreted as destruction of aromatic rings and unsaturated bonds.

A slight decrease in TOC is observed in the high molecular weight (HMW) fraction (G1), but increasing the ozone dose and contact time does not bring about complete destruction of this fraction. Thus, a portion of the HMW fraction is apparently composed of organics very resistant to ozonation, such as sugars and proteins (*14*). The most surprising result was that transformation of compounds does not increase the TOC of the lowest MW fraction (G5), which shows a decrease.

On the other hand, the intermediate fraction (G3) shows an increase in TOC. This shift could be partially due to transformation of the HMW to fraction G3. However, the increase in TOC for fraction G3 seems to be too great to be explained solely by migration from G1 to G3. Thus, some of the TOC may have shifted from G5 to G3, either because of the formation of polar oxidation byproducts with elution volumes matching those of G3 (under the experimental conditions used, certain carboxylic acids such as acetic or citric acid would be eluted in fraction G3) or possibly by the mechanism of oxidative polymerization (*15, 16*).

GAC ADSORPTION. Figure 5 shows the evolution of parameters for GAC filtration. In this case, significant reductions in the parameters had already occurred for GAC preceded by ozonation. GAC further reduced the TOC of the low (G5) and intermediate (G3) fractions, whereas the HMW fraction remained relatively unchanged. In the case of GAC filtration without preozonation, the UV260 was reduced compared with the RSW, but not to the same degree as was observed for ozonation at 5 mg/L.

DISINFECTION. Figure 6 shows the distribution of parameters after disinfection. The addition of chlorine or chlorine dioxide after GAC filtration (without preozonation) did little to affect the parameters reported here. A slight decrease was observed for fluorescence and UV260 for fractions G3 and G5, both for chlorine and chlorine dioxide.

The effect of the various treatments on the distribution of TOC within the three molecular weight fractions is summarized in Table I. Arrows are used to show whether the TOC increased, decreased, or remained unchanged. The only unexpected result was the increase in the intermediate MW TOC after ozonation.

Pyrolysis GC–MS. Pyrolysis GC–MS was used to obtain general information on the composition of each Sephadex fraction. This tech-

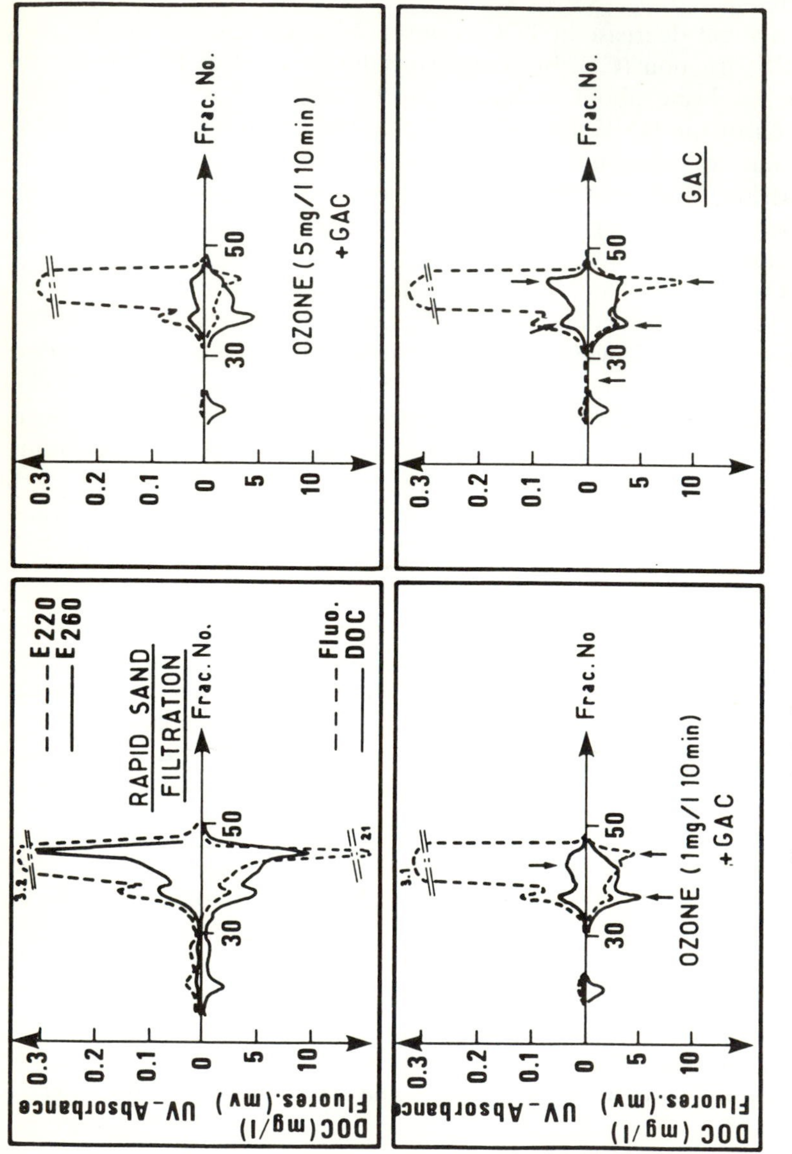

Figure 5. Evolution of parameters for GAC filtration.

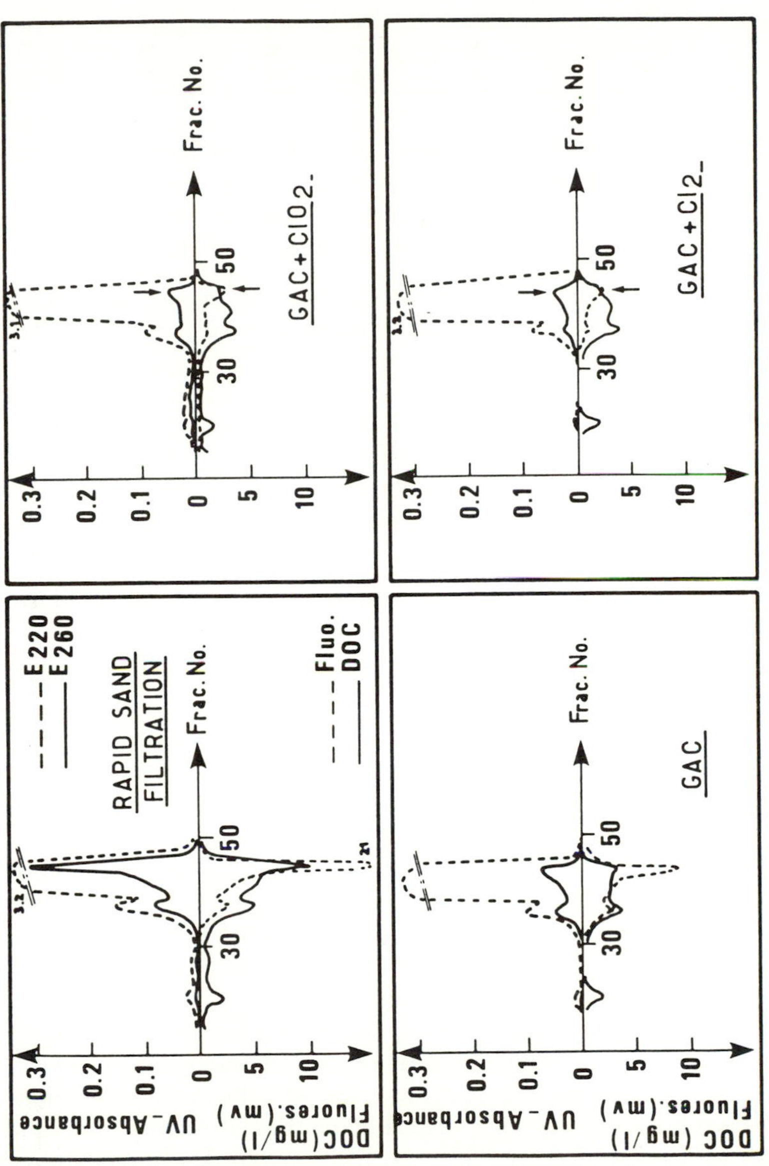

Figure 6. Distribution of parameters after disinfection.

Table I. Evolution of the Sephadex Fractions (TOC)

Fraction (MW)	*Sand Filtration*	O_3	*GAC*	Cl_2,ClO_2
>5000	↘	↘	↘	→
1000–5000	↘	↗	↘	→
<1000	↘	↘	↘	→

nique has proven useful because each biological macromolecule undergoes typical and specific fragmentation (*17–19*). For example, polyhydroxyaromatics give rise to phenol derivatives, proteins yield pyrrole and indole derivatives, and sugars result in furan derivatives.

Rapid Sand Filtration. Pyrochromatograms of fractions G1, G3, and G5 collected from sand-filtered water are presented in Figures 7–9.

G1 (MW >5000). Most of the pyrolysis fragments found in this fraction are derived from the three general classes just mentioned. Pyrrole and methylpyrrole originate from proteinaceous material such as polypeptides as well as from single amino acids such as proline and hydroxyproline. A quantitative relationship between amino acid hydrolyzable content and pyrrole abundance was established by Bracewell (*20*) for some Scottish brown forest soils, and such a correlation probably could be established for water.

The phenols and cresols observed on the pyrochromatogram could arise from tyrosine-containing proteins, which yield phenol and *p*-cresol (*17, 21*) during pyrolysis. However, because of (1) the similarity in concentration of *m*- and *p*-cresol and (2) the low *p*-cresol/phenol ratio found here (≃0.1 expressed as the ratio of the peak areas) compared with the higher values observed with natural proteins such as bovine serum albumin (BSA) (*21*) for which this ratio was close to 1, it is more likely that these phenolic peaks originated from polyhydroxyaromatic compounds.

The relatively low yield of phenol generated (compared with acetamide, for instance) indicates that humic substances are not major constituents of this higher MW fraction.

Furfural and methylfurfural are highly characteristic of sugars. 2-Cyclopenten-1-one has been described as a characteristic product of polycarboxylic acids (*22*). However, it has also been detected during furnace pyrolysis of D-glucose (*17*) and in this laboratory during the pyrolysis of Dextran, along with two other cyclic ketones tentatively identified as 2- and 3-methyl-2-cyclopenten-1-one. Thus, possible polysaccharide origin is suggested. Acetol (1-hydroxy-2-propanone) also arises from sugars and may become the main pyrolysis fragment if salts are present (*23*).

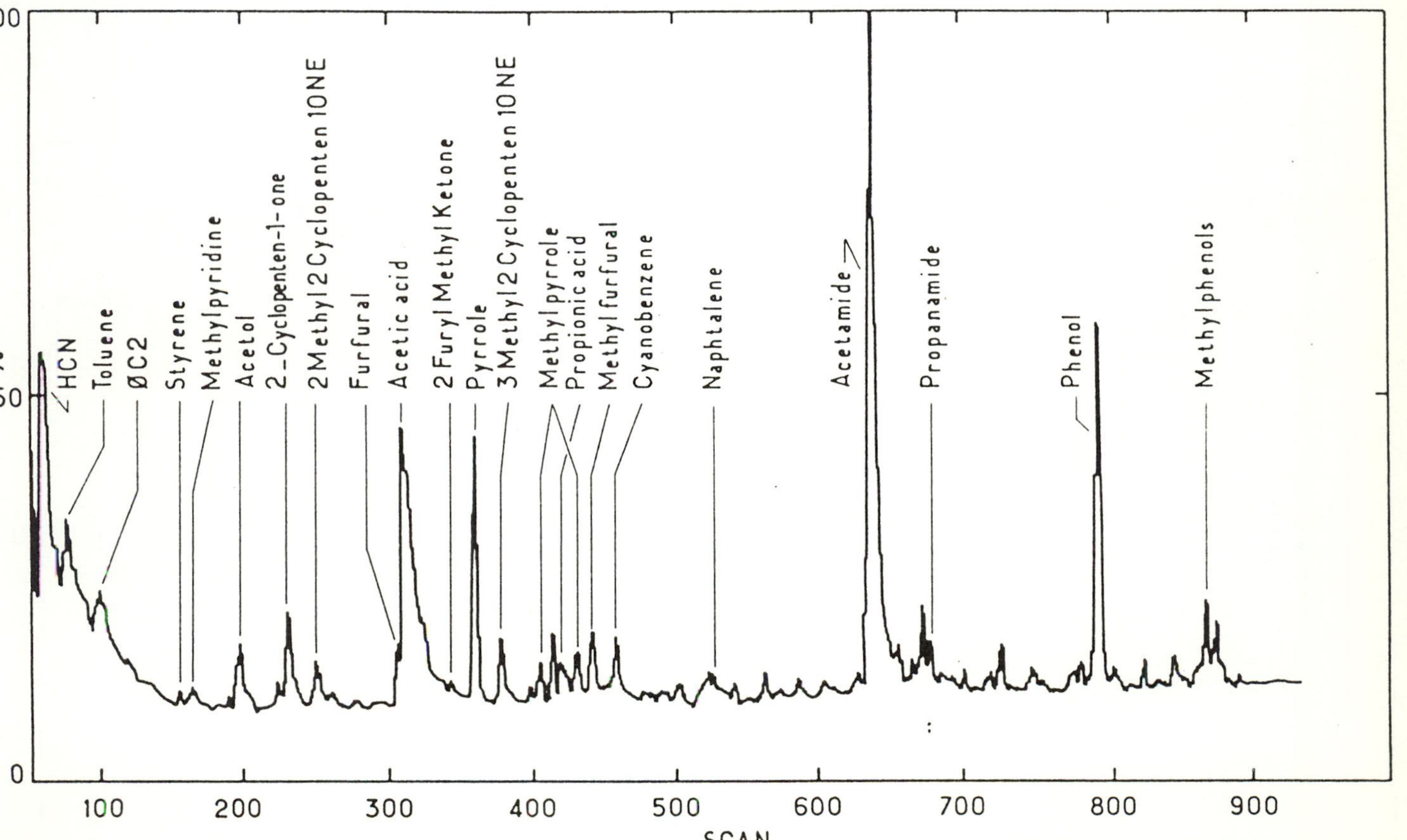

Figure 7. Pyrochromatogram of fraction G1 (MW >5000) collected from sand-filtered water.

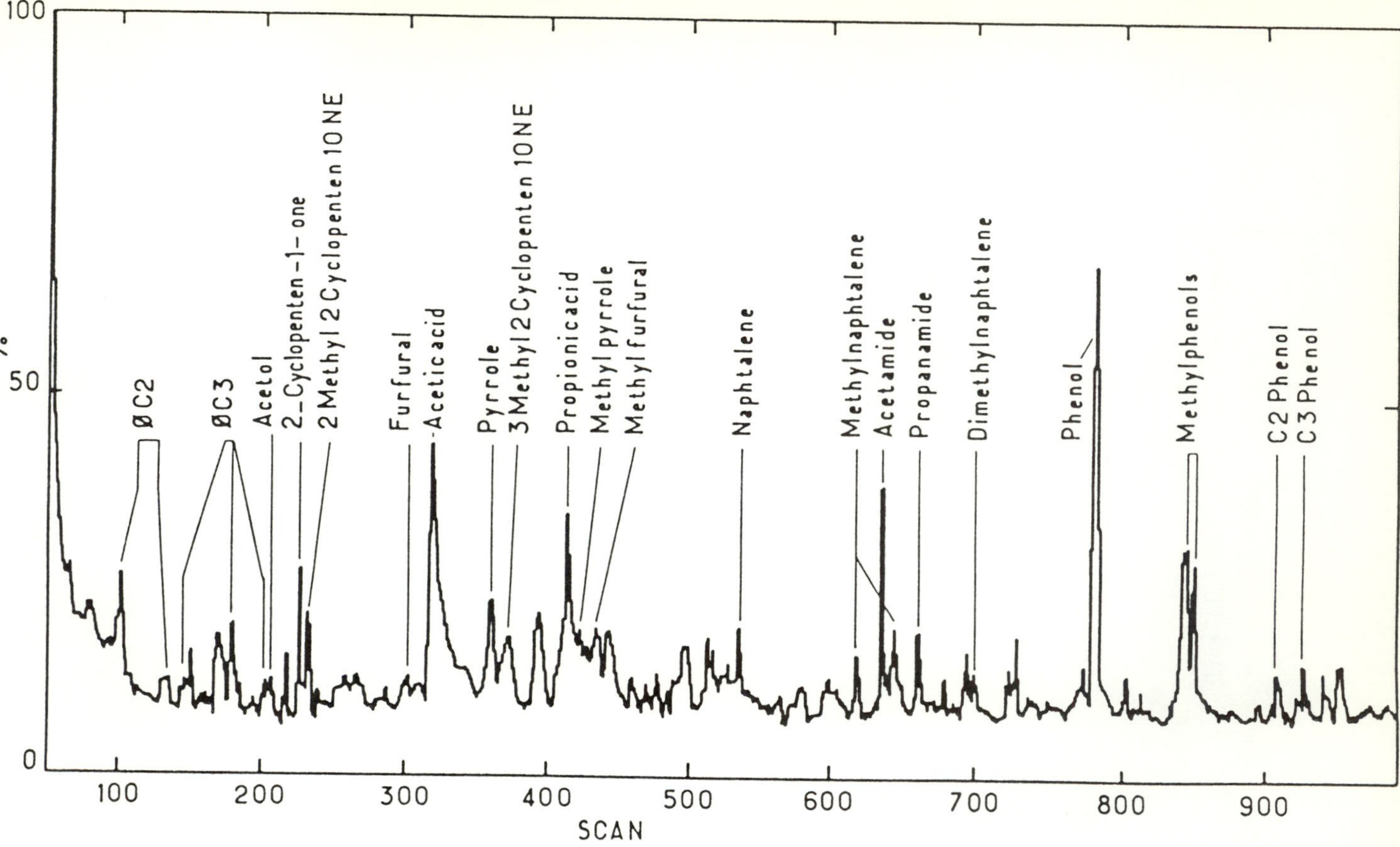

Figure 8. Pyrochromatogram of fraction G3 (MW 1000–5000) collected from sand-filtered water.

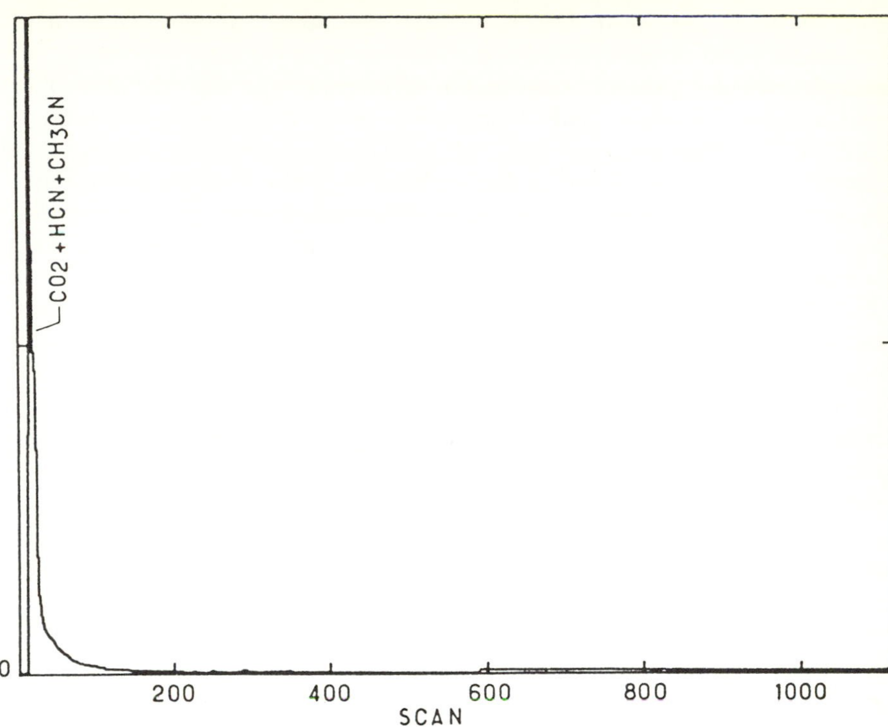

Figure 9. Pyrochromatogram of fraction G5 (MW <1000) collected from sand-filtered water.

The predominant acetamide peak is considered by many authors (*17, 24, 25*) to be characteristic of decomposition products of *N*-acetylamino sugars. These materials could enter the water sample from natural processes because they form a major constituent of microbial cell walls (e.g., chitin peptidoglycan).

Other compounds on the pyrochromatogram may result from a variety of sources. Aromatic compounds such as toluene, naphthalene, and benzonitrile are often found during pyrolysis of natural organic materials from water or soil (*18*). Their exact origin is uncertain because they may arise from the breakdown of humic aromatic structures or from the cyclic condensation of aliphatic chains containing electrophilic substituents, as in the case of poly(vinyl chloride) (*26*). Large yields of acetic acid may occur during the pyrolysis of amino sugars.

In summary, the principal components of the G1 fraction appear to belong to a class of nonpolyphenolic structures. *N*-Acetylamino sugars probably compose a major portion of this fraction.

G3 (MW 1000–5000). The compounds found in fraction G3 (shown in Figure 8) were very similar to those detected in fraction G1. Thus, the two fractions appear to largely contain the same compounds from a structural point of view, and the differences lie in the proportions of the macromolecules present. One major difference was that the phenolic peaks were dominant. This observation is consistent with the higher UV/TOC or fluorescence/TOC ratio observed in this fraction. For the same reasons discussed earlier, these phenolic peaks originate from polyhydroxyaromatics (and not tyrosine-containing proteins). This finding indicates that humic or fulvic acids are a major constituent of this fraction.

G5 (MW <1000). In Figure 9, the pyrochromatogram for the lowest molecular weight fraction does not contain any detectable organic compounds except the simple degradation byproducts carbon dioxide, cyanide, and acetonitrile. This result demonstrates the presence of carbonaceous and nitrogenous compounds. The absence of pyrolysis products is most likely because the lowest fraction is composed of easily pyrolyzed molecules. Furthermore, an absence of aromatic structures is indicated because aromatic rings are not destroyed under these conditions (*18*). The absence of nonvolatile aromatic structures (volatile compounds have been lost during sample preparation, which includes two steps of evaporation) is unusual because of the high fluorescence and UV levels observed for this fraction, which are normally associated with aromatics.

This apparent contradiction may be related to UV absorbance contributed by mineral forms. The gel permeation technique used in this laboratory has been observed to concentrate mineral components. For example, nitrate in this fraction has been found to exceed 1 g/L. Furthermore, some mineral forms such as ferrous iron have been observed to absorb at 260 nm. Chelates have been found to quench fluorescence. Thus, the concentrations of all minerals and their contributions to UV absorbance or fluorescence quenching should be carefully examined. High concentrations of metallic ions also may play a catalytic role during pyrolysis and further contribute to the absence of pyrolysis products (27).

OZONATED WATER. After ozonation (1 mg/L for 10 min), the pyrolysis of fraction G1 (Figure 10) showed a drastic reduction in the phenol peak and no significant change in the peaks of the other typical fragments. The pyrrole/acetamide and furfural/acetamide ratios were practically unmodified and thus indicated that little change occurred in the sugars, amino sugars, and peptides. Moreover, because all samples were prepared in roughly the same manner, the pyrochromatograms were directly comparable. Thus, ozone removes only the phenolic compounds and does not affect the other macromolecules, which are resistant to oxidation by ozonation. Thus, after ozonation, the G1 fraction appears to be almost entirely composed of sugars, amino sugars, and proteins.

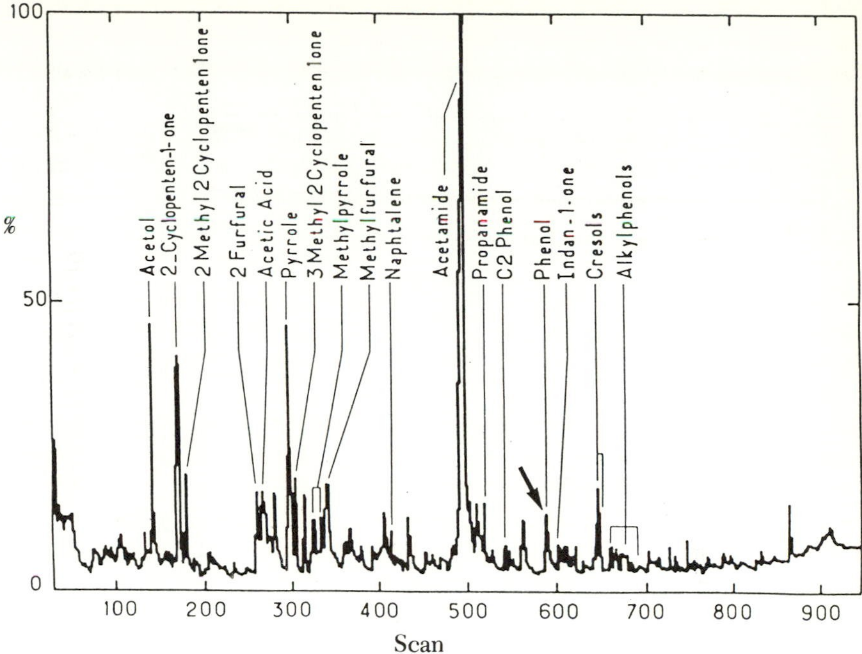

Figure 10. Pyrochromatogram of fraction G1 (MW >5000) collected from ozonated water (1 mg/L for 10 min).

Surprisingly, the G3 fraction (Figure 11) shows almost the complete opposite behavior. The phenol peak becomes predominant, and this change is expressed in Figure 12, where the relative surface areas (response) of three typical fragments (base peaks for pyrrole, phenol, and 2-cyclopenten-1-one) are shown for line I. Similar results were obtained for line II (ozonation: 5 mg/L for 10 min) where a high carboxylic content was also found (Figure 13).

This unexpected result may be related to the increase in TOC on fraction G3 and may be further evidence of the polymerization phenomenon discussed earlier. However, this hypothesis must be carefully considered because of our limited knowledge of pyrolysis mechanisms. The possibility of phenol formation during the thermal fragmentation process from elimination reactions followed by cyclization of polyconjugated chains has been suggested by Bracewell (22) and should be investigated.

Conclusions

The use of GPC in combination with pyrolysis GC–MS has proven to be a very powerful tool to evaluate the efficiency of water treatment processes. The measurement of nonspecific parameters after separation

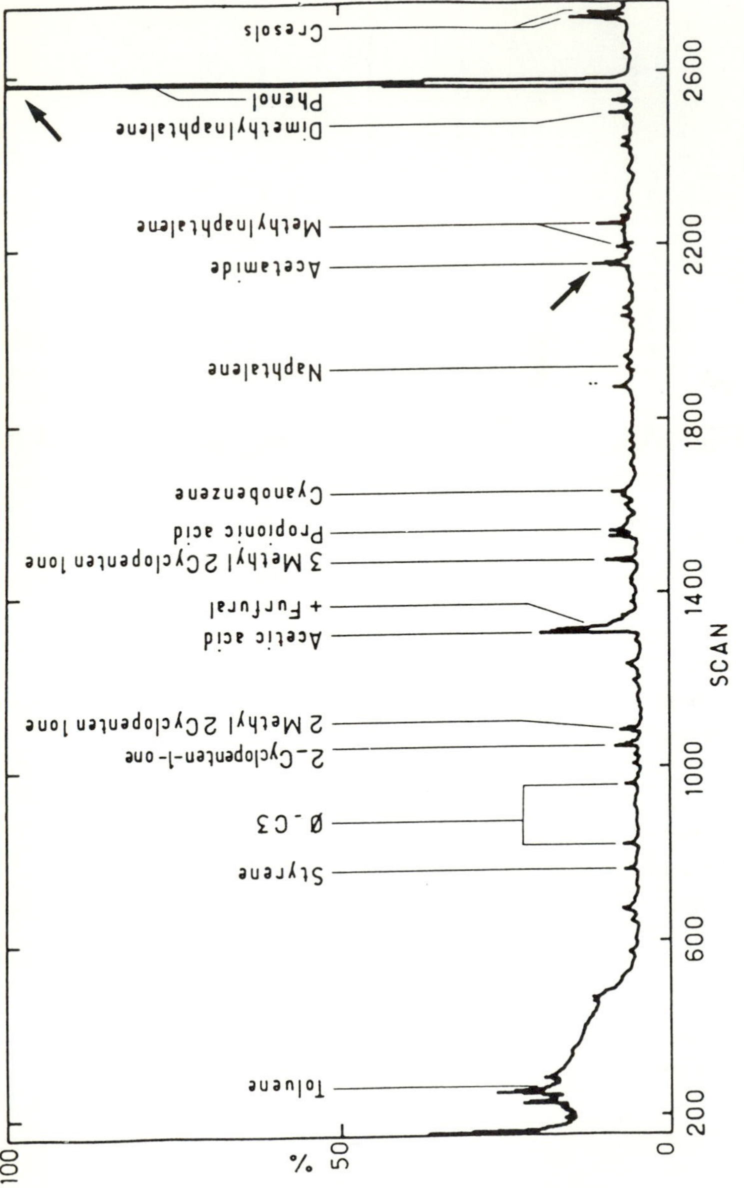

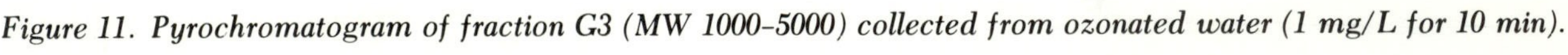

Figure 11. Pyrochromatogram of fraction G3 (MW 1000–5000) collected from ozonated water (1 mg/L for 10 min).

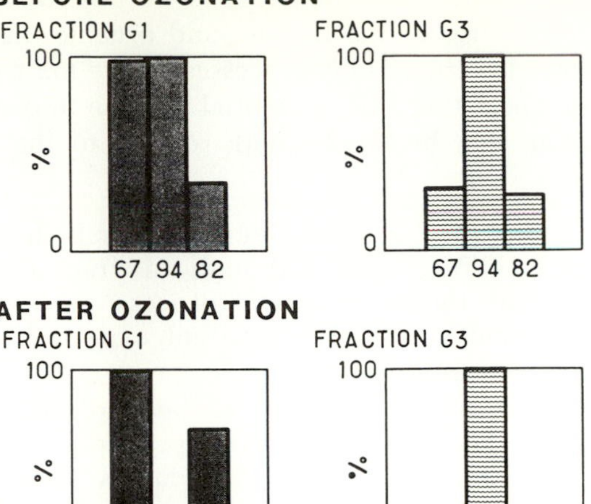

Figure 12. Relative evolution of typical pyrolysis fragments: 67 = pyrrole, 94 = phenol, and 82 = 2-cyclopenten-1-one.

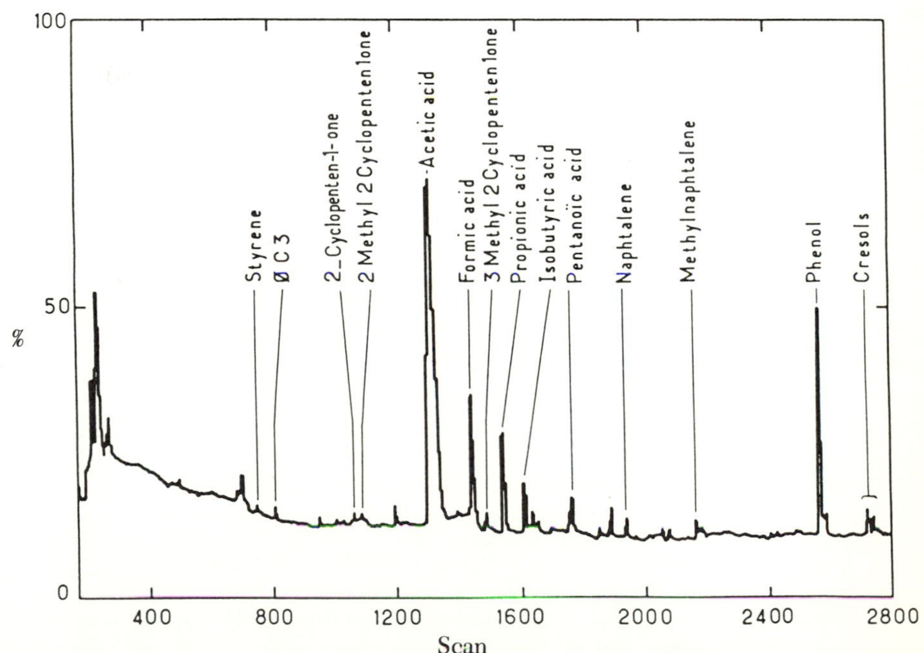

Figure 13. Pyrochromatogram of fraction G3 (MW 1000–5000) collected from ozonated water (5 mg/L for 10 min).

by Sephadex GPC allows mass balances to be determined and allows insight into the structures of the background organic matrix as well as the effects of water treatment unit processes on the matrix (*28*).

Beyond demonstrating the potential of the technique, several specific conclusions can be drawn with respect to the Vigneux pilot plant:

1. Only 15% of the TOC in sand-filtered water is due to compounds having MWs >5000. Although this fraction has been decreased during the pretreatment process, the very HMW compounds found in soils (*29*) are not expected in waters because they are nonsoluble.
2. Humic material does not appear to be a prominent component of this HMW fraction. The principal components appear to be proteins and sugars, particularly *N*-acetylamino sugars, which probably originate from the degradation of microbial cell walls.
3. These proteins and amino sugars are practically unaffected by the ozonation process. A supplementary example of this situation can be found elsewhere (*21*) and suggests that the proteins and amino sugars may represent a large portion of nonadsorbable organic carbon in water.
4. Humic substances dominate the intermediate MW (1000–5000) range (G3), which represents 29% of the total organic carbon and thus contributes substantially to the carbon balance.
5. Polymerization may occur during ozonation, as suggested by the increase in the TOC of the intermediate MW fraction (G3) and the increase in the phenolic compounds found in this fraction.

Acknowledgments

We thank the Ministere de l'Environnement and the Agence Financiere de Bassin Seine Normandie for their financial aid and contribution to this work.

Literature Cited

1. McCarty, P. L. *J. Environ. Eng. Div. Am. Soc. Civ. Eng.* **1980,** *106,* 1.
2. Carter, C. W.; Suffet, I. H. *Environ. Sci. Technol.* **1982,** *16(11),* 735.
3. Rook, J. J. *J. Water Treatment Exam.* **1974,** *23,* 254.
4. Christman, Russel F.; Norwood, Daniel L.; Millington, David S.; Johnson, J. Donald; Stevens, Alan A. *Environ. Sci. Technol.* **1983,** *17(10),* 625.
5. Gjessing, E. T. *Physical and Chemical Characteristics of Aquatic Humus;* Ann Arbor Science: Ann Arbor, MI, 1976.
6. Schnitzer, M.; Kahn, S. U. *Humic Substances in the Environment;* Marcel Dekker: New York, 1972.

7. Mallevialle, J. In *Procedes d'Oxydation Appliques au Traitement de l'Eau Potable;* Kuhn, H.; Sontheimer, H., Eds.; Engler-Bunte Institute: Karlsruhe, West Germany, 1979.
8. Tambo, N.; Kamei T. *Jpn. Water Works Assoc.* **1976,** *502,* 2–24.
9. Eisenreich, S. J.; Armstrong, D. E. *Environ. Sci. Technol.* **1977,** *11,* 497–501.
10. Gloor, G.; Hans, L. *Anal. Chem.* **1979,** *51(6),* 645.
11. Glaze, W. H.; Jones, P. C.; Saleh, F. W. In *Advances in the Identification and Analysis of Organic Pollutants in Water;* Keith, L. H., Ed.; Ann Arbor Science: Ann Arbor, MI, 1981; pp 371–382.
12. Gloor, R.; Leidner, H.; Wuhrmann, K.; Fleischmann, T. H. *Water Res.* **1981,** *15,* 457–462.
13. Mallevialle, J.; Rousseau, C.; Suchet, M. *Tech. Sci. Munic. Spec. Hydrol.* **1979,** *74,* 182.
14. Mallevialle, J.; Schmitt, E.; Bruchet, A. *Journees Information Eau de Poitiers* 30 Sept. ler Oct., 1982.
15. Chrostowski, P. C. Ph.D. Thesis, Drexel University, 1981.
16. Duguet, J. P.; Dussert, B.; Mallevialle, J.; Fiessinger, F. Presented at the 13th Biomedical International Conference IAWPRC, Rio de Janeiro, Aug. 17–22, 1986.
17. Irwin, W. J. *Analytical Pyrolysis: A Comprehensive Guide;* Chromatographic Science Series Vol. 22; Marcel Dekker: New York, 1982.
18. Bracewell, J. M.; Robertson, G. W. *Anal. Proc.* **1981,** *18,* 532.
19. Saiz-Jimenez, C. *Geoderma* **1979,** *22,* 25.
20. Bracewell, J. M.; Robertson, G. W. *J. Anal. Appl. Pyrolysis* **1984,** *6,* 19.
21. Mallevialle, J.; Bruchet, A.; Schmitt, E. Presented at the 1983 AWWA Water Quality Technology Conference, Norfolk, VA. Dec. 1983.
22. Bracewell, J. M.; Robertson, G. W.; Welch, D. I. *J. Anal. Appl. Pyrolysis* **1980,** *2,* 239.
23. Richards, G. N.; Shafizadeh, F.; Stevenson, T. T. *Carbohydr. Res.* **1983,** *117,* 322.
24. Windig, W.; Haverkamp, J.; Kistemaker, P. G., *Anal. Chem.* **1983,** *55,* 81.
25. Bracewell, J. M.; Robertson, G. W. *J. Soil Sci.* **1984,** *35.*
26. O'Mara, M. M. *J. Polym. Sci., Part A-1* **1970,** *8,* 1887.
27. Rolin, A.; Richard, C.; Masson, D.; Deglise, X. *J. Anal. Appl. Pyrolysis* **1983,** *5,* 151.
28. Bruchet, A.; Tsutsumi, Y.; Duguet, J. P.; Mallevialle, J. Presented at the 5th Conference on Water Chlorination Environmental Impact and Health Effects, Williamsburg, VA, June 3–8, 1984.
29. Butler, J. H. A.; Ladd, J. N. *Aust. J. Soil Res.* **1969,** *1,* 229–239.

RECEIVED for review August 14, 1984. ACCEPTED May 2, 1986.

Mutagen Isolation Methods

Fractionation of Residue Organic Compounds from Aqueous Environmental Samples

M. Wilson Tabor[1] and John C. Loper[1,2]

Department of Environmental Health[1] and Department of Microbiology and Molecular Genetics[2], University of Cincinnati Medical Center, Cincinnati, OH 45267

A general preparative procedure, based on high-performance liquid chromatography (HPLC) and the Salmonella *microsome mutagenicity assay, has been developed for the isolation of mutagenic components from samples of complex mixtures of residue organics obtained from environmental waters. This procedure features preliminary HPLC separation to characterize the sample; preparative-scale HPLC separation with mutagenic bioassay of the fractions; further HPLC separation of bioactive fractions, employing different elution techniques; and chemical-biological characterization of isolated mutagenic components. Results of the application of this approach to residue organics from drinking water, ground water, and waste water are presented along with chemical characterization, via high-resolution mass spectrometry, of the constituents of a mutagenic subfraction isolated from waste water residue organics.*

TRACE LEVELS OF ANTHROPOGENIC ORGANIC COMPOUNDS in environmental waters have heightened concern as to the possible human health impact of chronic exposure to these contaminants via routes such as drinking water. One response to this concern in the United States has been the enactment of a series of federal acts, including the Safe Drinking Water Act of 1974, PL 93-523, and the Clean Water Act of 1977, PL 95-217. Although the implementation of this federal legislation has yielded improvements in the nation's drinking water and national waterways regarding substances such as toxic metals (*1*), a recent review (2) concluded that "further studies of the identities, carcinogenicity,

0065-2393/87/0214/0401$06.00/0

mutagenicity, mode of formation, and practical methods of removal are needed for the organic contaminants". The goal of our studies has been to understand the biohazardous components of complex mixtures of organics from aqueous samples in terms of their prevalence, concentration, chemical structure, origin, and mechanism of biological effects. One specific focus of this research effort has been the isolation, identification, and chemical-biological characterization of mutagenic components from aqueous environmental samples.

The first step required for this process was the application of methods for the isolation of residue organics from aqueous samples, for example, drinking water, ground water, and waste water (*3, 4*), by using columns of XAD resins to concentrate the organics. Residue organics have been defined as those organics adsorbed by XAD-2/XAD-7 resins under the conditions described and recovered by the solvent elution method employed in the U.S. Environmental Protection Agency (USEPA) Interim Protocol for drinking water (*5*). In the past, these organics were referred to as nonvolatile residue organics (*6–10*) on the basis of previous descriptions of organics in drinking water (*11*). Following isolation, the residue organics can be tested for mutagenicity and other biological end points via a variety of short-term bioassay procedures (*12*) and/or fractionated for the isolation of mutagenic components for chemical-biological characterization.

The need for the fractionation of residue organics isolated from water samples to assess mutagenicity and to identify the biohazardous constituents has been discussed (*8, 13*). On the basis of this need, a biological approach to the chemical fractionation and separation of residue organics is the method of choice for the isolation of biohazardous compounds from complex mixtures of residue organics. We introduced a general method based upon high-performance liquid chromatography (HPLC) for the fractionation of residue organics from drinking water. This method assessed the polarity distribution of constituents by using an analytical HPLC separation followed by preparative-scale HPLC separation of the residue organics for mutagenicity testing and compound identification (*6, 7*). The strategy for this method is summarized in Figure 1. This approach led to the first identification of a previously unidentified mutagen from drinking water residue organics (*14*).

The use of HPLC for the separation of residue organics from aqueous samples for mutagenicity testing has been extended to studies of many other types of water samples from different parts of the world. Baird and co-workers (*15, 16*) have used such an approach in their studies of mutagenicity of residue organics from drinking water, river water, storm runoff, reclaimed waste waters, and other waste waters. Jolley and co-workers (*17, 18*) have applied HPLC to the separation

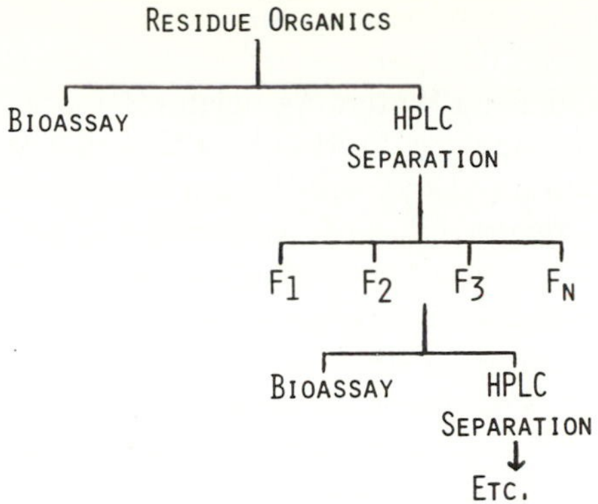

Figure 1. Schematic of the strategy for the coupled bioassay-analytical fractionation of residue organics isolated from aqueous environmental samples.

and waste water effluent residue organics to study the relationship of mutagenicity to disinfection methods and to fractionate residue organics from other polluted waters for bioassay of separated fractions. In a study of genotoxic components of oil shale retort process water, Strniste et al. (*19*) first separated these samples by a classical liquid-liquid extraction scheme followed by HPLC separation of the extracts for bioassays. One research group (*20*) reported the use of HPLC to fractionate drinking water residue organics for mutagenicity assessment. Our research group has made extensive use of HPLC for the fractionation of drinking water and waste water residue organics (*3, 4, 6–10, 21, 22*). Residue organics from samples of ground and river water and from samples of various stages in the processing of both drinking waters and of diverse waste waters have been fractionated in these studies. This approach has provided a better assessment of the mutagenic potential of such residue organics. The purpose of this chapter is to present the specifics of our HPLC approach as applied to residue organics from a variety of aqueous samples, particularly as the method is used in the isolation of mutagenic constituents for compound identification. Examples of the separation of residue organics from a variety of aqueous environmental samples will be used to illustrate the applicability of the approach.

Methods

Instrumentation. HPLC separations were performed on a Waters Associates system consisting of two Model 590 pumps, a Model U6K injector, a Model 680 automated gradient controller, and both Model 440 fixed wavelength (254 nm) and Model 480 variable wavelength absorbance detectors. Chromatograms were displayed on a Fisher Recordall Series 5000 two-pen recorder. For preparative-scale (milligram level) separations, a Waters Associates Z-module radial compression column (RCM) unit was used. The RCM unit was fitted with an 8-mm × 10-cm column packed with 10-μm silica particles bonded with octadecylsilane for reverse-phase separations; for normal-phase separations, similar RCM columns packed with 10-μm silica particles were used. For analytical-scale (microgram level) separations, prepacked 3.9-mm × 30-cm columns (Waters Associates), packed with 10-μm silica particles bonded with octadecylsilane for reverse-phase separations or with 10-μm silica particles for normal-phase separations, were used.

Mass spectral (MS) determinations were performed on a Kratos MS 80 high-resolution mass spectrometer as described previously (*3, 13*).

Sample Description. Finished drinking water I (DWI) was from a city that draws its raw water from a river polluted by chemicals from numerous industrial, municipal, and agricultural sources. To prepare finished drinking water, the raw water is treated by a series of settling, coagulation, and flocculation steps, and the final product is chlorinated to a residual level of 1–2 mg/L. Finished drinking water II (DWII) was from a city that draws its raw water from a network of streams and rivers that principally drain wilderness regions. The raw water is settled in a series of reservoirs and then chlorinated to a residual level of 2–4 mg/L for distribution. Finished drinking water III (DWIII) was from a city that draws raw ground water from a major U.S. aquifer system. The raw water is treated by pH adjustment to >pH 10 with lime to remove minerals, followed by a series of settling steps. The final product is chlorinated to a free chlorine residual of 1–2 mg/L, a process that lowers the pH to 9.0. Because this aquifer is recharged in part by stream bank infiltration from a river subject to multiple points of contamination from industrial, municipal, and agricultural sources, residue organics were isolated from both the river water (RW) and the raw ground water (GW) for mutagenesis–separation assessment. In addition to these drinking waters and raw waters, both influent waste water (IWW) and effluent waste water (EWW) from a municipal sewage treatment plant, heavily polluted (>80% by volume) by industrial discharges, were examined.

Reagents. Organic solvents for HPLC separations—methylene chloride, methanol, isopropyl alcohol, hexane, and acetonitrile—were obtained as HPLC grade from Fisher Scientific. Type I water for HPLC and for the preparation of other aqueous solutions was purified as described previously (*7*). All HPLC solvents were filtered through a 0.45-μm Millipore membrane filter (Millipore Corporation) and degassed by 15 min of sonication while under reduced pressure immediately prior to use. Sample-enrichment purification cartridges, SEP-PAKS (Waters Associates), packed with silica particles bonded with octadecylsilane, were activated and used according to methods described previously (*6, 7, 9*) for the processing of HPLC fractions for mutagenesis testing. All other chemicals were of reagent grade and were used without further purification.

Isolation of Residue Organics from Waters via XAD Chromatography. Residue organics were isolated from the water samples via XAD chromatographic procedures developed in our laboratory. Drinking water and ground water samples were processed via the XAD procedure described in publications (*3, 9, 10, 21, 22*) and detailed in the Interim Protocol developed for the USEPA (*5*). Waste water samples were processed via a modification of the XAD procedure (*4*).

Biological Analysis. Tester strains TA98 and TA100 for the *Salmonella* microsomal mutagenicity tests were provided by B. Ames. Characteristic properties of the bacterial tester strains were verified for each fresh stock, and their mutagenicity properties were verified again by using positive and negative controls as part of each experiment, as recommended (*23*). Mutagenesis tests requiring metabolic activation used polychlorinated biphenyl mixture Aroclor 1254, induced rat liver 9000 × *g* supernatant fraction, S9 (Litton Bionetics). Mutagenesis assays, without (−S9) and with (+S9) metabolic activation, were conducted as described previously (*3, 7, 10, 24*). The detection of mutagenic activity in experimental samples was based upon a dose-dependent response exceeding the zero-dose, spontaneous control value by at least twofold; that is, the ratio of total revertant colonies per plate to spontaneous colonies per plate was ≥2. In some situations involving HPLC subfractions in which the amount of sample was limiting, semiquantitative determinations of mutagenesis were made as described previously (*3, 10*). All recoveries of bioactivity from concentrated or fractionated residue organic samples were based upon an expression of mutagenesis per liter equivalent, representative of the original water sample. Typical mean revertant colony counts, ± standard error, obtained from spontaneous plates and positive control plates from our laboratory for the time period of the experiments described herein, were reported recently (*3, 21*).

HPLC Separations. SAMPLE PREPARATION FOR INJECTION. Isolated residue organics were dissolved in water/acetonitrile solvent mixtures for reverse-phase HPLC separations as follows: sample was dissolved in a minimum volume of acetonitrile and diluted with water until a volume equal to the acetonitrile was added or until the solution became cloudy; in the latter case, additional acetonitrile was added, about 10% of the total volume or until the solution became clear again. For normal-phase HPLC separations, residue organics were dissolved in methylene chloride/hexane solvent mixtures as follows: sample was dissolved in methylene chloride, minimum volume, then diluted with an equal volume of hexane by using the same procedure and criteria followed for the acetonitrile/water sample solutions.

Analytical-Scale Separations. A preliminary analytical reverse-phase HPLC separation of an aliquot of the residue organics was conducted to characterize the sample of residue organics according to the polarity of the components. The separation was accomplished by injecting 25 μL of a 1-μg/μL solution of the residue organics into the HPLC unit operating under the following conditions: flow rate, 2.0 mL/min; initial mobile phase, water. Following sample injection, the column was washed with water for 5 min or until no more 254-nm absorbing components were eluted; that is, there were no peaks >5% of full-scale absorbance setting, usually 0.5 or 1.0 absorbance units full scale. After the water wash, a linear mobile-phase gradient of 45-min duration was initiated from 100% water to 100% acetonitrile. On completion of the gradient, the column was washed with acetonitrile for 10 min or until no more UV-absorbing components were eluted. In general, results from this HPLC separation indicate the composition of the sample of residue organics. For example, if the majority, >75%, of the sample components eluted at a solvent composition of 80% water:20% acetonitrile or later, the sample would be ready for separation via preparative-scale reverse-phase HPLC to collect fractions for mutagenesis testing. However, if the majority, >75%, of the sample eluted before the 80% water:20% acetonitrile solvent composition, then the sample would be examined via an analytical-scale normal-phase HPLC separation prior to scale-up to the preparative level.

To run a preliminary analytical-scale normal-phase HPLC separation, 25 μL of a 1-μg/μL solution of residue organics was injected into the HPLC unit fitted with a normal-phase column. The column was washed with hexane flowing at 2.0 mL/min for 5 min or until no more UV-absorbing components were eluted. Following the hexane wash, a linear gradient of 30-min duration was initiated from 100% hexane to 100% methylene chloride. On completion of the gradient, the column was washed with methylene chloride until no more UV-absorbing components were eluted. Following this wash, a linear mobile-phase gradient of 30-min duration was initiated from 100% methylene chloride to 100% isopropyl alcohol. On completion of this second gradient, the column was washed with isopropyl alcohol for 10 min or until no more

UV-absorbing components were eluted. The results of this HPLC separation identified those samples most suited for preparative-scale normal-phase separation into subfractions for mutagenesis testing.

INITIAL PREPARATIVE-SCALE HPLC SEPARATIONS. On the basis of the results of the analytical scale HPLC separation, one of the following preparative-scale HPLC separations was used for the initial fractionation of the complex mixture of residue organics for mutagenesis testing. For these separations, mobile-phase flow rates of 2.0 mL/min were used, and sample sizes from 40 to 80 mg were loaded per fractionation run. Each preparative-scale run involved the collection of a series of fractions, each of which contained components eluted with a specific combination of isocratic and linear-gradient mobile-phase compositions. In each case, the respective fraction for each mobile-phase composition was collected until no more UV-absorbing components were eluted, then the mobile phase composition was changed along with the collection vessel for the collection of the next fraction.

The initial preparative-scale reverse-phase HPLC fractionation involved the collection of the following five fractions based on mobile-phase composition: Fraction A, 100% water; Fraction B, a 5-min linear gradient from 100% water to 75% water:25% acetonitrile, then isocratic until no more UV-absorbing components are eluted; Fraction C, a 5-min linear gradient to 50% water:50% acetonitrile, then isocratic until no more UV-absorbing components are eluted; Fraction D, a 5-min linear gradient to 25% water:75% acetonitrile, then isocratic as before; and Fraction E, a 5-min linear gradient to 100% acetonitrile, then isocratic until no more UV-absorbing components are eluted. The fractions were processed for mutagenesis testing and/or further separation according to the following procedures: The aqueous solution fraction, Fraction A, was concentrated via gentle evaporation under a stream of dry nitrogen according to procedures described previously (*4*). Fractions B–E were processed via our previously published procedure employing reverse-phase SEP-PAK cartridges (*6*, *7*, *9*).

In this procedure, HPLC subfractions B–E were diluted with two volumes of water, and this solution was passed through a C18 SEP-PAK, previously activated as follows: Slowly pass 10 mL of acetonitrile through the SEP-PAK according to the instructions from the manufacturer; the SEP-PAK then is washed with 20 mL of Type I water. Following the slow passage of the HPLC subfraction through the SEP-PAK, 5 mL of air is gently passed through the SEP-PAK to remove the residual solvent. The residue organics are eluted from the SEP-PAK by the slow passage of 5 mL of methylene chloride through the cartridge. The volumes of the methylene chloride concentrates are recorded, and these samples are stored in Teflon-capped amber vials at −20 °C until mutagenic testing. These solutions were then processed as described in the following paragraphs for the preparation of dimethyl sulfoxide (DMSO) solutions for bioassay.

For preparative-scale normal-phase HPLC separations, the following fractions were collected: Fraction A, isocratic hexane wash until no more UV-absorbing components eluted; Fraction B, a 5-min linear gradient from 100% hexane to 65% hexane:35% methylene chloride, then isocratic until no more UV-absorbing components are eluted; Fraction C, a 5-min linear gradient from 65% hexane:35% methylene chloride to 35% hexane:65% methylene chloride, then isocratic until no more UV-absorbing components are eluted; Fraction D, a 5-min linear gradient to 100% methylene chloride, then isocratic until no more UV-absorbing components are eluted; Fraction E, a 5-min linear gradient from 100% methylene chloride to 65% methylene chloride:35% isopropyl alcohol, then isocratic until no more UV-absorbing components are eluted; Fraction F, a 5-min linear gradient from the previous solvent to 35% methylene chloride:65% isopropyl alcohol, then isocratic until no more UV-absorbing components are eluted; and Fraction G, a 5-min linear gradient from the previous solvent to 100% isopropyl alcohol, then isocratic until no more UV-absorbing components are eluted. These fractions were processed for mutagenesis testing and/or further HPLC separations by gentle evaporation under a stream of nitrogen (*4, 10*), described as follows.

The volume of each HPLC subfraction was reduced by using a micro Snyder apparatus. Usually a measured aliquot of an extract is concentrated at one time, rather than concentrating the whole extract. The sample is gently heated by using an N-Evap bath (Organomation). As the solution was concentrated, some constituents came out of the solution. In those cases, a small volume of acetone was added to keep the components in solution until sufficient evaporation had occurred to azeotrope the remaining HPLC solvent from the subfraction. Usually three to four additions of acetone were required. The final volumes of the acetone concentrates were recorded, and these samples were stored in Teflon-capped amber vials at −20 °C until mutagenesis testing. At the time of bioassay, an aliquot of the residue solution was removed from the sample vial; typically, this aliquot was adjusted to the necessary bioassay volume with DMSO.

ADDITIONAL PREPARATIVE-SCALE HPLC SEPARATIONS. After mutagenesis assessment of the HPLC fractions from the initial preparative-scale separation just discussed, those fractions containing mutagenic constituents are further separated on HPLC by employing the following strategy: For example, if the mutagenic constituents were found to be in Fraction D from an initial reverse-phase HPLC preparative-scale separation, that is, a mobile-phase composition of 25% water:75% acetonitrile, a solution of the bioactive fraction would be injected into the HPLC unit. The separation would be accomplished on a reverse-phase column by using an initial mobile phase of 40% water:60% acetonitrile. Following a wash of 5 min, or until no more UV-absorbing constituents are eluted (collected as Fraction D-1), a linear gradient of 20-min duration to a mobile-phase composition of 20% water:80% acetonitrile would be run. During this gradient, fractions would be collected at 5-min intervals, that is, four fractions, Fractions D-2 through D-5. At the completion of the gradient, the collection vessel would be changed for the collection of Fraction D-6, and the column would be washed isocratically with the latter mobile phase until no more UV-absorbing constituents eluted or for a time period of 10 min. The six fractions, D-1 through D-6, would be processed for this hypothetical case via the SEP-PAK procedure (*6*, *7*, *9*) for mutagenesis testing as described earlier. The same general approach would apply for an additional HPLC separation of mutagenic fractions from either initial normal or reverse-phase HPLC preparative-scale separations.

If further fractionation of mutagenic subfractions is required to isolate constituents from fractions collected during the second separation, a similar HPLC strategy may be employed (*7*).

Results and Discussion

The purpose of this chapter is to present a general approach for the isolation of mutagenic constituents from residue organics of aqueous environmental samples. The isolated mutagens then can be characterized chemically as to their identity (*14*) or characterized biologically-biochemically as to their mechanism of mutagenesis (*25*, *26*) or other properties. The overall strategy for this approach to isolation is outlined in Figure 1. This strategy of a coupled bioassay-chemical fractionation procedure was proposed by Loper and Lang (*27*), based largely upon their mutagenicity and carcinogenicity studies of residue organics isolated by Kopfler et al. (*28*) as part of a USEPA five-city study (*24*) of drinking water organics. Originally, the procedure was developed by using reverse-phase HPLC as a method for the chemical fractionation of the complex mixture of residue organics (*7*, *9*). However, it was found subsequently (*10*) that residue organics from some drinking water samples would require a combination of reverse-phase and normal-phase HPLC separations to isolate mutagens for compound identification. Therefore, the more general approach described herein was developed.

Mutagenic Activity of Residue Organics from a Variety of Aqueous Samples. Samples of residue organics from differing aqueous environmental sources have been found to contain a wide variation not only in the total amount of mutagenic activity per liter equivalent of water but also in tester strain specificity and requirements for metabolic activation (*8, 13, 28*). Results of the mutagenic activities for residue organics from a variety of aqueous samples are summarized in Table I. In general, from these data and those reported by others, mutagenic activity of residue organics on a per liter equivalent basis was found to be the greatest for waste waters from industrially contaminated municipal sewage treatment plants (e.g., IWW and EWW, Table I) (*4, 29, 30*), whereas the lowest mutagenic activity appears to be associated with waters such as raw ground waters, drinking waters from ground sources (e.g., GW, Table I) (*3*), settled river water as a drinking water source sampled prior to disinfection (*21*), and domestically contaminated municipal treatment plant waste waters prior to disinfection [(*31*); Tabor and Loper, unpublished]. These data suggest a wide variation in the chemical constituents of the residue organics from different waters. Preliminary analytical-scale HPLC separations of the residue organics are required to assess the distribution of organic constituents and are useful in planning preparative-scale HPLC separations for mutagenic compound isolation.

Analytical-Scale HPLC Separations. Reverse-phase HPLC chromatography favors the distribution of the semi- and nonpolar constituents of a sample of residue organics, whereas normal-phase HPLC chromatography favors the distribution of semipolar constituents (*32*). This approach is illustrated in Figure 2 by the chromatograms of residue organics from a waste water sample separated by both reverse-

Table I. Mutagenic Activity for Residue Organics Isolated from a Variety of Aqueous Environmental Samples

	Net Revertants per Liter Equivalent[a]			
	TA98		*TA100*	
Sample	*−S9*	*+S9*	*−S9*	*+S9*
Drinking Water I (DWI)	93	36	328	149
Drinking Water II (DWII)	124	48	218	—
Drinking Water III (DWIII)	56	30	184	—
Ground Water (GW)	39	—	—	—
River Water (RW)	53	67	—	—
Influent Waste Water (IWW)	90	12000	—	—
Effluent Waste Water (EWW)	2800	3800	—	—

NOTE: — indicates the response was not equal to or greater than 2× the spontaneous rate.
[a] Net revertants = total revertants − spontaneous revertants, i.e., controls. Representative average spontaneous rates ± standard deviation ($n > 15$) over the time period for these experiments were TA98 − S9, 15 ± 3; TA98 + S9, 28 ± 6; TA100 − S9, 122 ± 10; TA100 + S9, 126 ± 14.

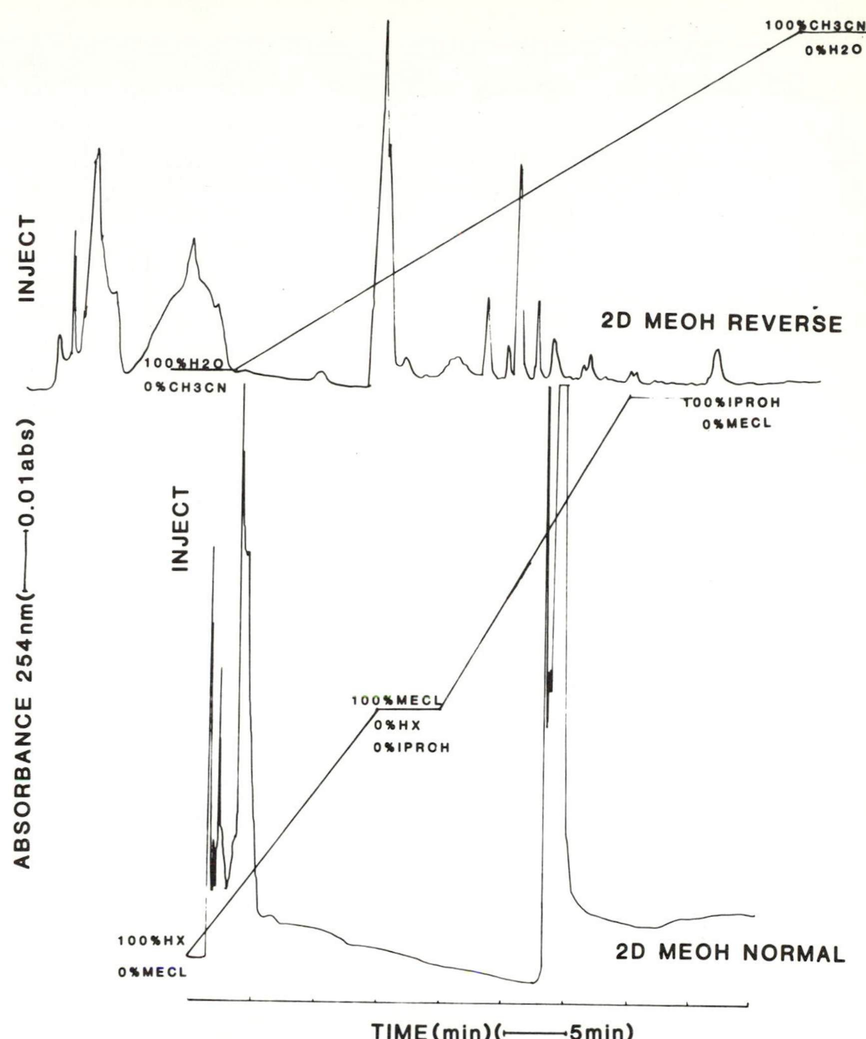

Figure 2. Absorbance (254 nm) profile of the analytical-scale HPLC separation of 25 µg of residue organics isolated from an industrially impacted influent waste water. The separations were accomplished via reverse-phase HPLC (top) and normal-phase HPLC (bottom). Mobile phases used in these separations were water (H_2O), acetonitrile (CH_3CN), methylene chloride (MECL), hexane (HX), and isopropyl alcohol (IPROH), as indicated.

phase and normal-phase analytical-scale HPLC. These chromatograms show that this sample of residue organics consists of two groups of constituents, that is, nonpolar and somewhat polar.

However, the typical HPLC distribution of constituents found in drinking water residue organics is as shown in the chromatogram of a

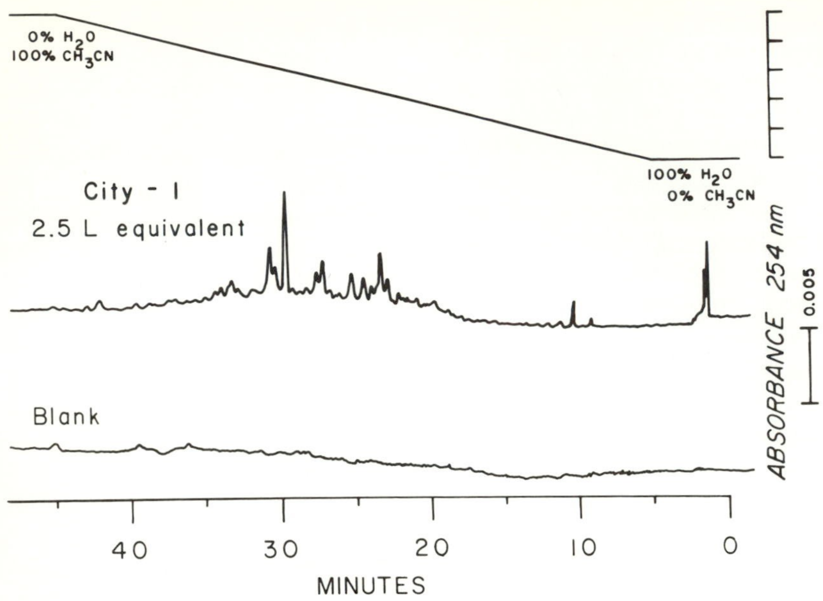

Figure 3. Absorbance (254 nm) profile of the analytical-scale reverse-phase HPLC separation of 2.5-L equivalents of residue organics isolated from finished drinking water I.

reverse-phase HPLC separation in Figure 3. Here, the majority of components are eluted in the semipolar to nonpolar range, that is, with a solvent composition less polar than 60% water:40% acetonitrile. Similar results have been observed for samples of mutagenic residue organics from a variety of drinking waters isolated via XAD chromatography, reverse osmosis, or similar methods (*3, 6, 7, 10, 15, 16, 20, 21*). These results are not unexpected in that Yamasaki and Ames (*33*) noted that organic compounds, expected to exert biological effects and capable of passing through biological membranes, are most probably lipophilic compounds of low polarity. The same argument was presented recently by Neal (*34*), wherein he added that the toxic components of residue organics from drinking water are likely to be <500 molecular weight. From these arguments, Neal proposed the first priority for compound isolation, in order to conduct a more extensive toxicological evaluation, should be lipid-soluble components in this molecular weight range (*35*).

Preparative-Scale HPLC Separations. Results of the analytical-scale HPLC separations are used to develop an approach in the scale-up of the HPLC separations for the preparation of subfractions of the residue organics in quantities suitable for mutagenesis testing and compound isolation. If the analytical-scale separation results indicate the constituents of the residues are principally semi- to nonpolar, then reverse-phase HPLC would be the chromatographic mode of choice to be used for a preparative-scale separation. However, if the results indicate the constituents of the residues are principally polar to semipolar, then normal-phase HPLC would be the chromatographic mode of choice to be used.

The strategy, summarized in Figure 1, employed in the first preparative-scale HPLC separation, Step I, is to separate the residue organics into discrete groupings of organics according to polarity by using a combination of linear gradient and isocratic solvent elutions. Fractions are collected for each distinct solvent elution employed in the separation. Aliquots of each fraction are bioassayed for mutagenicity.

An example of the preparative-scale HPLC separation of residue organics isolated by the XAD procedure from 120 L of drinking water, source DWI, is illustrated in Figure 4. Fractions were collected, as indicated, and bioassayed for direct-acting mutagenesis. The results of the mutagenesis assays for each fraction, Figure 5, indicate that the majority of the mutagenic components from this sample of residue organics eluted in a relatively nonpolar fraction, Fraction 3.

Also in this experiment, proportionate amounts, by volume, of each fraction were combined and then bioassayed. The mutagenic activity for this mixture, Mix 1–4 in Figure 5, when compared to the arithmetic sum of bioassay results for the individual fractions collected, Sum 1–4 in Figure 5, indicates the presence of components in the residue organics mixture that are antagonistic to the mutagenic components isolated in each fraction. The results of this experiment illustrate a further application of the preparative-scale separation of residue organics: In addition to its use for mutagen isolation, the procedure permits a better assessment of the mutagenicity of residue organics. Such assessments are valuable in cases in which mutagenesis testing results for the initial bioassay of the isolated residue organics show toxicity or another form of nonlinear dose responses rather than a linear dose response, whereas the mutagenesis testing results for the initial bioassay show no mutagenic activity. In these cases, the bioassay results could be due to the presence of antagonistic and/or toxic components present in the original mixture of isolated residue organics. The effects of such components have been observed by this HPLC technique for some residue organics isolated from drinking water, for example, Figure 5, but not all (*3, 10*). In addition to drinking water residue organics, Alfheim and co-workers (*36, 37*) observed toxicity for residue organics isolated from wood-combustion emissions but were able to accurately assess the mutagenicity of these samples following HPLC fractionation of the residue organics.

Isolation of Mutagens. Applying the general preparative-scale HPLC separation procedure outlined in the previous section (Step I),

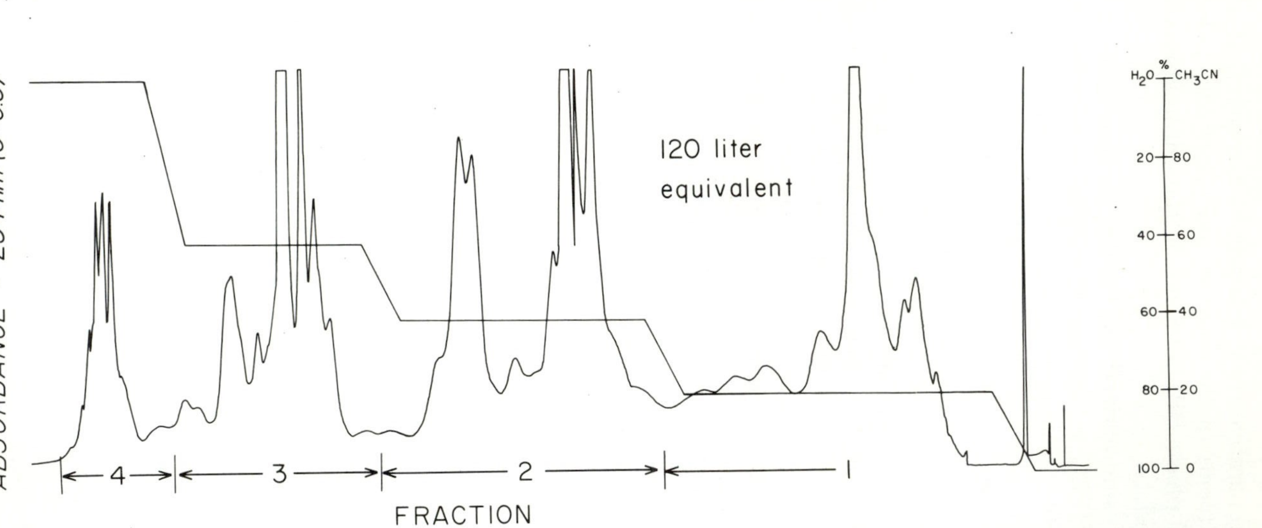

Figure 4. Absorbance (254 nm) profile of the preparative-scale reverse-phase HPLC separation of 120-L equivalents of residue organics isolated from finished drinking water I.

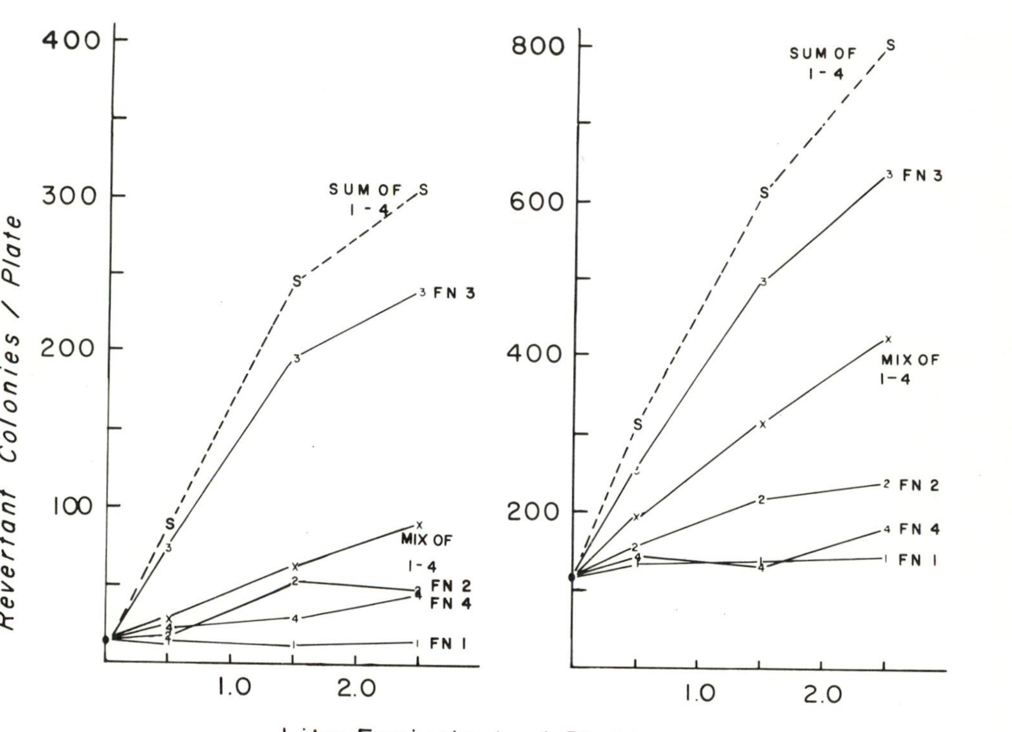

Figure 5. Direct-acting mutagenic activity of HPLC fractions (1–4, Figure 4) obtained by sequential step elution of residue organics isolated from finished drinking water I. "MIX" indicates the mutagenic activity measure for an aliquot of residue organics proportionately reconstituted from the separate fractions; "SUM" is the arithmetic sum of the net revertant colonies for each dose for the four fractions.

one generally finds the bulk of the mutagenic activity of a given mixture of residue organics to have eluted in one or possibly two of the broad fractional cuts. Therefore, the next and subsequent preparative-scale HPLC separations (Steps II, III, etc.) of the active fraction(s) from Step I will involve one of two approaches. The first approach uses the same chromatographic mode, that is, reverse phase or normal phase, used in Step I; the second approach involves the alternate chromatographic mode. In either case, the general strategy of HPLC fractionation followed by bioassay of an aliquot of the collected fractions is applicable (Figure 1).

The application of the same preparative chromatographic mode for Step II as employed in Step I was described by Tabor and Loper (*7*) for the isolation of a TA100 promutagen from drinking water residue organics. The promutagen subsequently was identified as 3-(2-chloroethoxy)-1,2-dichloropropene (*14*). In the first step, bioactivity was isolated in the 50:50 water:acetonitrile fraction. The Step II HPLC fractionation was accomplished by a shallow linear solvent gradient that started at a solvent composition 5% less in acetonitrile, that is, the strong solvent, and went to a composition 20% greater in strong solvent than the isocratic solvent composition used for elution of the mutagenic fraction in Step I. (NOTE: A shallow linear gradient is defined as a 1% solvent composition change per minute at a mobile-phase flow rate of approximately one column volume per minute.) This approach should be generally applicable to the separation of bioactive fractions from Step I that originally eluted at solvent compositions >20% strong solvent.

The application of the second approach, use of a chromatographic mode for Step II alternate to that used in Step I, is illustrated by the isolation of an S9-dependent mutagenic subfraction from residue organics isolated from an effluent waste water sample, EWW in Table I. In this isolation, Step I involved the separation of the residue organics via normal-phase preparative HPLC. The resulting chromatogram for this separation is shown in Figure 6. Bioassay of aliquots of the collected fractions showed that >80% of the S9-dependent TA98 mutagenesis had been eluted by 100% methylene chloride in Fraction 5, Figure 6, and no activity for TA98 was detected in the later eluting components. Because this mutagenic fraction eluted in a semipolar region, preparative-scale reverse-phase HPLC was applied for the Step II separation. The separation involved a series of isocratic and linear gradient elutions using combinations of water and acetonitrile, as shown in Figure 7. Bioassay of the collected fractions showed that >90% of the S9-dependent TA98 mutagenesis had been eluted via an isocratic wash of 50:50 water:acetonitrile into Fraction 4, Figure 7.

Identification of Mutagens in Waste Water—EWW. Attempts to further separate this fraction into discrete components via HPLC were

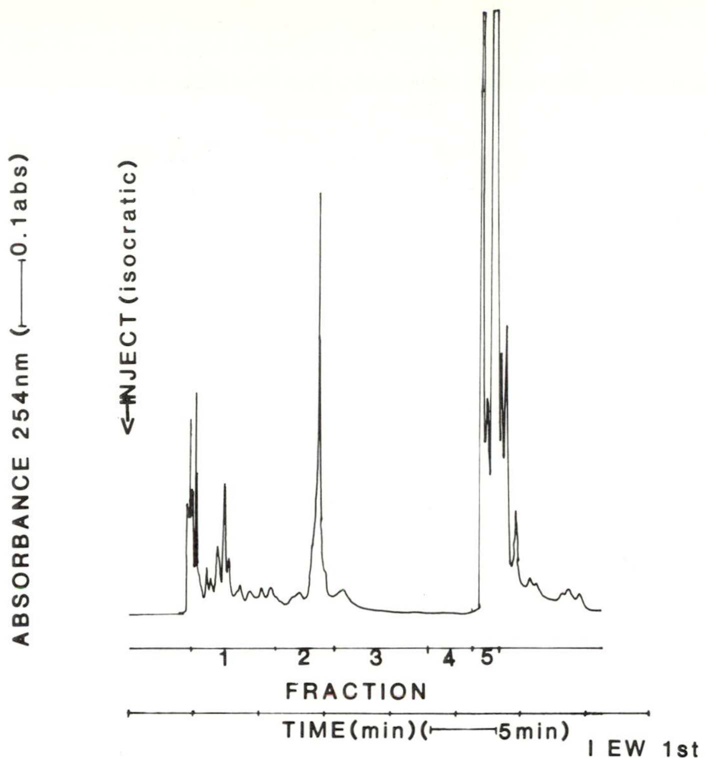

Figure 6. Absorbance (254 nm) profile of the preparative-scale normal-phase HPLC separation of 5-L equivalents of residue organics from effluent waste water I. The bulk of the S9-dependent TA98 mutagenesis was eluted in Fraction 5.

unsuccessful; however, the sample was found to be suitable for preliminary analysis via MS. This procedure showed the sample contained at least three components. The first was found to have a molecular ion at 135.0135 mass units and a characteristic ion at 108.0010 mass units, that is, loss of HCN. From these and other ions in the spectra, this compound was tentatively identified as benzothiazole. Comparison of this spectra with that obtained on an authentic sample of benzothiazole confirmed the identification. Although Kinae et al. (*38*) have reported benzothiazole to be an S9-dependent mutagen for the *Salmonella* tester strain, TA1537, our studies indicate this compound does not account for the majority of mutagenic activity in this sample.

A second component in this mutagenic subfraction was shown to have an empirical formula of $C_8H_{10}O_2S$. The compound was tentatively identified as the ethyl ester of benzenesulfonic acid. This structural assignment was made from exact mass analysis and interpretation of the following ions: a molecular ion at 170.0400 mass units; a characteristic

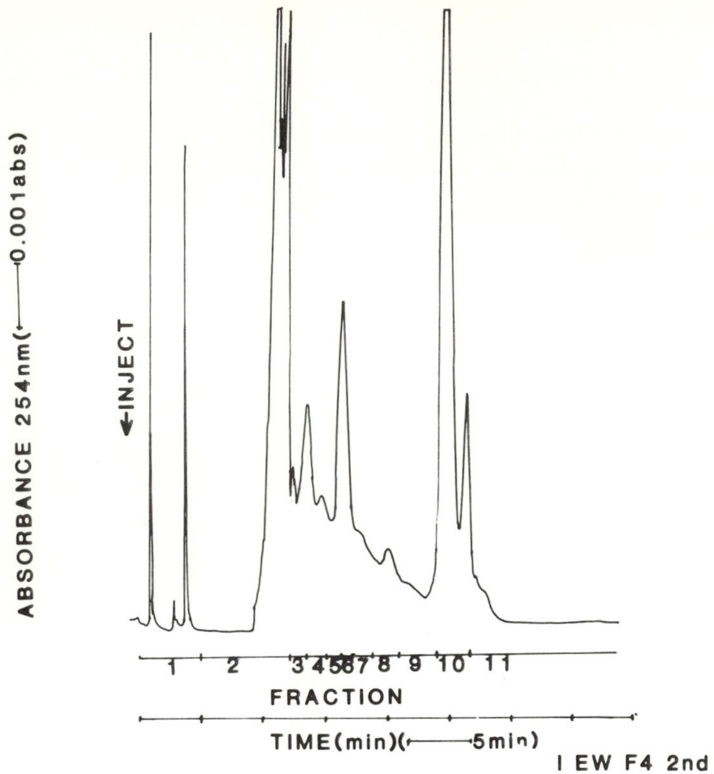

Figure 7. Absorbance (254 nm) profile of the preparative-scale reverse-phase HPLC separation of the mutagenic Fraction 5, Figure 6. The bulk of the S9-dependent TA98 mutagenesis was eluted in Fraction 4.

ion at 141.0029, loss of ethyl; a characteristic ion at 93.9042, loss of phenyl; and other characteristic ions in the spectrum. The mass spectrum of this compound was compared to the spectrum obtained for phenylethyl sulfone, which was run under identical conditions. Although the two spectra were similar, they did not match. Therefore, a tentative structural assignment is made at this time, and confirmation studies are in progress. The possibility of the benzenesulfonic acid ethyl ester being the predominant S9-dependent mutagen in this sample is under investigation.

The third major component detected in this mutagenic sample has not yet been characterized chemically. Thus, although characterization of this mutagenic subfraction has not been completed, two candidates have been identified as possibly contributing to the S9-dependent mutagenesis in this sample.

Conclusions

A basic assessment of the mutagenic activity of organic residues from water entails the application of the *Salmonella* test to both the parent residue mixture and to major subfractions obtained by reverse-phase and/or normal-phase HPLC. Such analyses overcome the presence in particular residue mixtures of components that are antagonistic or toxic for the bioassay. When applied to residues obtained by various procedures, the method reveals a wide range of mutagenic activity among environmental waters, depending upon the source of the water sample, the apparent types of natural and anthropogenic compounds the water contains, and the water-processing and disinfection procedures used for treatment. Biologically, this range of activity emcompasses mutagenic components that are differentiated on the basis of tester strain specificity and the effects of the absence or presence of microsomal activation. Chemically, the various mutagenic components are separable by HPLC across the entire range of polar, midpolar, relatively nonpolar, and nonpolar compounds. Through combinations of specific mutagenic bioassays plus sequential analytical and preparative HPLC fractionations, this method has been shown to be applicable for the isolation of mutagens from residues of a variety of environmental waters.

Acknowledgments

Technical assistance by S. MacDonald, L. Rosenblum, B. Myers, R. Hutchenson, and M. Niemi is gratefully acknowledged. Our appreciation is extended to K. Jayasimhulu for assistance in MS analysis, to R. Trosset for assistance in literature searching, and to R. Jones, R. Miller, and G. Hicks for assistance in sample acquisition. This research was supported by grants from the USEPA (CR808603 and CR810792), the Virginia Environmental Endowment (81-103), and the National Science Foundation (PCM-8219912). This chapter has not been subjected to USEPA review and therefore does not reflect the views of that agency, and no official endorsement should be inferred.

Literature Cited

1. *Environmental Quality, The 13th Annual Report of the Council on Environmental Quality;* Council on Environmental Quality. U.S. Government Printing Office: Washington, DC, 1982.
2. Crump, K. S.; Guess, H. A. *Drinking Water and Cancer: Review of Recent Findings and Assessment of Risks;* Prepared for Council on Environmental Quality, Springfield, VA. National Technical Information Service. U.S. Department of Commerce. Contract No. EQ10AC018, 1980.
3. Tabor, M. W.; Loper, J. C. *Int. J. Environ. Anal. Chem.* **1985,** *19*, 281–318.

4. Tabor, M. W.; Loper, J. C.; Myers, B. L.; Rosenblum, L.; Daniels, F. B. In *Short-Term Genetic Bioassays in the Evaluation of Complex Environmental Mixtures;* Waters, M. D.; Sandhu, S. S.; Lewtas, J.; Claxton, L.; Chernoff, N.; Newnow, S., Eds.; Plenum: New York, 1985; Vol. 4, pp 269–288.
5. Tabor, M. W. In *Development of Procedures for the Preparation of Environmental and Waste Samples for Mutagenicity Testing;* Pearson, G. J.; Williams, L. R., Eds.; Office of Research and Development. U.S. Environmental Protection Agency: Las Vegas, NV, 1985; Section 3, pp 67–108; USEPA-600/4-85/058.
6. Tabor, M. W.; Loper, J. C.; Barone, K. In *Water Chlorination: Environmental Impact and Health Effects;* Jolley, R. L.; Brungs, W. A.; Cummings, R. B.; Jacobs, V. A., Eds.; Ann Arbor Science: Ann Arbor, MI, 1980; Vol. 3, pp 899–912.
7. Tabor, M. W.; Loper, J. C. *Int. J. Environ. Anal. Chem.* **1980,** *8,* 197–215.
8. Loper, J. C. *Mutat. Res.* **1980,** *76,* 241–268.
9. Loper, J. C.; Tabor, M. W. In *Short-Term Bioassays in the Analysis of Complex Environmental Mixtures;* Waters, M. D.; Sandher, S. S.; Husingh, T. L.; Claxton, L.; Nesnow, S., Eds.; Plenum: New York, 1981; Vol. 2, pp 155–165.
10. Loper, J. C.; Tabor, M. W.; Miles, S. K. In *Water Chlorination: Environmental Impact and Health Effects;* Jolley, R. L.; Brungs, W. A.; Cotruvo, J. A.; Cummings, R. B.; Mattice, J. S.; Jacobs, V. A., Eds.; Ann Arbor Science: Ann Arbor, MI, 1983; Vol. 4, pp 1199–1210.
11. *Drinking Water and Health;* National Academy of Sciences: Washington, DC, 1977; Vol. 1, p 492.
12. Hoffman, G. R. *Environ. Sci. Technol.* **1982,** *16,* 560A–572A.
13. Loper, J. C. In *Water Chlorination: Environmental Impact and Health Effects;* Jolley, R. L.; Brungs, W. A.; Cummings, R. B.; Jacobs, V. A., Eds.; Ann Arbor Science: Ann Arbor, MI, 1980; Vol. 3, pp 937–945.
14. Tabor, M. W. *Environ. Sci. Technol.* **1983,** *17,* 324–328.
15. Baird, R.; Gute, J.; Jacks, C.; Jenkins, R.; Neisess, L.; Scheybeler, B.; Van Sluis, R.; Yanko, W. In *Water Chlorination: Environmental Impact and Health Effects;* Jolley, R. L.; Brungs, W. A.; Cumming, R. B.; Jacobs, V. A., Eds.; Ann Arbor Science: Ann Arbor, MI, 1980; pp 925–935.
16. Nellor, M. H.; Baird, R. B.; Smyth, J. R. *Health Effects Study, Final Report;* County Sanitation Districts of Los Angeles County: Whittier, CA, 1984.
17. Jolley, R. L.; Cumming, R. B. *Ozone Sci. Eng.* **1979,** *1,* 31–37.
18. Cumming, R. B.; Lee, N. E.; Lewis, L. R.; Thompson, J. E.; Jolley, R. L. In *Water Chlorination: Environmental Impact and Health Effects;* Jolley, R. L.; Brungs, W. A.; Cumming, R. B.; Jacobs, V. A., Eds.; Ann Arbor Science: Ann Arbor, MI, 1980; Vol. 3, pp 881–898.
19. Strniste, G. F.; Bingham, J. M.; Spall, W. D.; Nickols, J. W.; Okinaka, R. T.; Chen, D. J. C. In *Short-Term Bioassays in the Analysis of Complex Environmental Mixtures;* Waters, M. D.; Sandhu, S. S.; Lewtas, J.; Claxton, L.; Chernoff, N.; Nesnow, S., Eds.; Plenum: New York, 1983; Vol. 3, pp 139–151.
20. Kool, H. J.; van Kreijl, C. F.; Verlaan de Vries, M. In *Organic Pollutants from Water: Sampling and Analysis;* Suffet, I. H.; Malaiyandi, M., Eds.; Advances in Chemistry 214; American Chemical Society: Washington, DC, 1986; Chapter 29.
21. Loper, J. C.; Tabor, M. W.; Rosenblum, L.; DeMarco, J. *Environ. Sci. Technol.* **1985,** *19,* 333–339.
22. Loper, J. C.; Tabor, M. W.; Rosenblum, L. In *Water Chlorination: Environ-*

mental Impact and Health Effects; Jolley, R. L.; Bull, R. J.; Davis, W. P.; Katz, S.; Roberts, M. H.; Jacobs, V. A., Eds.; Lewis: Chelsea, M. I., 1985; Vol. 5, pp 1329–1339.

23. Ames, B. N.; McCann, J.; Yamasaki, E. *Mutat. Res.* **1975,** *31,* 347–364.
24. Loper, J. C.; Lang, D. R.; Schoeny, R. S.; Richard, B. B.; Gallagher, P. M.; Smith, C. C. *J. Toxicol. Environ. Health* **1978,** *4,* 919–938.
25. Distlerath, L. M.; Loper, J. C.; Tabor, M. W. *Biochem. Pharmacol.* **1983,** *32,* 3739–3748.
26. Distlerath, L. M.; Loper, J. C.; Tabor, M. W. *Environ. Mutagen.* **1985,** *7,* 303–312.
27. Loper, J. C.; Lang, D. R. In *Application of Short-Term Bioassays in the Fractionation and Analysis of Complex Environmental Mixtures;* Waters, M. D.; Nesnow, S.; Huisingh, J.; Sandhu, S. S.; Claxton, L., Eds.; Plenum: New York, 1978; pp 513–528.
28. Kopfler, F. C.; Coleman, W. E.; Melton, R. G.; Tardiff, R. G.; Lynch, S. C.; Smith, J. K. *Ann. N.Y. Acad. Sci.* **1977,** *298,* 20–30.
29. Hopke, P. K.; Plewa, M. J.; Johnson, J. B.; Weaver, W.; Wood, S. G.; Larson, R. A.; Hinesly, T. *Environ. Sci. Technol* **1982,** *16,* 140–147.
30. Hopke, P. K.; Plewa, M. J.; Stapleton, P. L.; Weaver, D. L. *Environ. Sci. Technol.* **1984,** *18,* 909–916.
31. Loper, J. C. In *Organic Pollutants from Water: Sampling and Analysis;* Suffet, I. H.; Malaiyandi M., Eds.; Advances in Chemistry 214; American Chemical Society: Washington, DC, 1986; Chapter 28.
32. Tabor, M. W. In *Chromatography: Theory and Practice;* Kaplan, L. A.; Pesce, A. J., Eds.; C. V. Mosby: St. Louis, MO, 1984; Chapter 4, pp 74–99.
33. Yamaski, E.; Ames, B. N. *Proc. Natl. Acad. Sci. U.S.A.* **1977,** *74,* 3555-3559.
34. Neal, R. A. *Environ. Sci. Technol.* **1983,** *17,* 113A.
35. Neal, R. A., unpublished data.
36. Alfheim, I.; Becher, G.; Hongslo, J. K.; Ramadhl, T. *Environ. Mutagen.* **1984,** *6,* 91–102.
37. Alfheim, I.; Ramdahl, T. *Environ. Mutagen.,* **1984,** *6,* 121–130.
38. Kinae, N.; Kawashima, H.; Kawane, R.; Saitou, M.; Saitou, S.; Tomita, I. *J. Pharmacobio. Dyn.* **1981,** *4,* 55–63.

RECEIVED for review August 14, 1985. ACCEPTED December 18, 1985.

COMPARISON OF ISOLATION METHODS TO COLLECT LARGE AMOUNTS OF ORGANIC CHEMICALS FROM WATER

20

A Comparison of Seven Methods for Concentrating Organic Chemicals from Environmental Water Samples

F. C. Kopfler, H. P. Ringhand, and R. G. Miller

Chemical and Statistical Support Branch, Toxicology and Microbiology Division, Health Effects Research Laboratory, U.S. Environmental Protection Agency, Cincinnati, OH 45268

Because there are no quantitative analytical techniques for the complex organic matter that occurs in chlorinated water, direct determination of the efficiency of techniques for isolating this matter is not possible. Seven methods capable of isolating gram quantities of organic matter from water samples were evaluated by determining the ability of each to recover a set of model compounds possessing a wide variation in polarity, functional groups, water solubility, and molecular weight. No single method appeared to be superior overall, on the basis of the recovery of the model solutes, but some methods could be eliminated from field application for the present time because the adsorbents required were not commercially available. Field application of two methods was undertaken, and the samples collected were tested in several bioassays.

ESTIMATING THE HEALTH RISK associated with organic matter in potable water is a major objective of the U.S. Environmental Protection Agency (USEPA). Such estimates are made by using various biological tests in which animals or lower organisms are exposed to the organic contaminants in drinking water at sufficiently high concentrations to ensure that the lack of a positive response provides a desired margin of safety. Estimates can be made by assessing each chemical contaminant separately or by testing directly a concentrate of the aqueous sample of interest. A direct evaluation of the water sample is generally not possible because the concentration of organic matter in most drinking water sources is less than 10 mg/L. Because the organic matter in surface waters is a complex mixture of natural and anthropogenic substances that defies complete analytical characterization, the direct toxicological evaluation of organic concentrates offers a practical alternative.

Two general classes of methods can be functionally defined for preparing concentrates of organic substances. Concentration methods involve the removal of water (e.g., lyophilization, freeze concentration, vacuum distillation, reverse osmosis [RO], and ultrafiltration) and result in a more highly concentrated aqueous solution of organic contaminants. Isolation methods are those methods in which the organic substances are physically removed from the aqueous solution, for example, adsorption onto a solid substrate followed by desorption (*1*).

Evaluation Conditions

The approach taken by the Health Effects Research Laboratory of the USEPA to determine the most acceptable method for producing sample concentrates for biological testing was to solicit proposals for practical methods that could be used in the field to either isolate the organic matter from water or concentrate it 50-fold or more. Because potable waters are ever-changing sources of organic matter, these sources were considered unsuitable for use in any methods evaluation. Therefore, a decision was made to evaluate various approaches under standardized conditions. Consequently, the approach taken was to have each research group (Table I) determine the efficiency of its method in recovering a set of model solutes from organic-free water. Each group was responsible for obtaining its own organic-free water and model compounds. The humic acid was supplied by the USEPA. The model solutes shown in Table II reflect a wide range of physical and chemical properties and were selected for the most part from the list of consensus voluntary

Table I. Research Groups Involved in Methods Evaluation

Institute	*Principal Investigator*	*Method Evaluated*
Drexel University	I. H. Suffet	RO–CLLE[a]
Arthur D. Little	D. J. Ehntholt	LLE–SCF CO_2[b]
Gulf South Research Institute	J. K. Smith	RO–Donnan dialysis
Georgia Institute of Technology	E. S. K. Chian	Solid adsorbents (XAD-8, AG MP-50, GCB)
University of Illinois	J. B. Johnston	Solid adsorbents (PTFE, MSC-1, A-162)
Envirodyne Engineers, Inc.	D. C. Kennedy	Solid adsorbent (quaternary XAD-4)
Los Angeles County Sanitation District	R. B. Baird	Solid adsorbents (MD-1, MP-50, XAD-2, XAD-7)

[a] Continuous liquid–liquid extraction.
[b] Liquid–liquid extraction and supercritical fluid CO_2.

Table II. Percent Recovery of Model Compounds by Different Procedures

Model Compound	*Concentration in Test Solution* ($\mu g/L$)	*XAD-8, AG MP-50, GCB* n = 5 *100 L*	*Quaternary XAD-4* n = *1* *500 L*	*RO-Donnan Dialysis* n = *1* *500 L*	*SCF* CO_2 n = 3 *10 L*	*PFTE, MSC-1, A-162* n = 3 *8 L*	*MP-1, MP-50, XAD-2, XAD-7* n = 5 *500 L*	*RO-CLLE* n = 2 *12.5 L*
Stearic acid	50	7.8 ± 9.1	9.6	3.2	nd	98.4 ± 6.6[a]	36.3 ± 14.0	na
Trimesic acid	50	47.6 ± 19.8	22.3	nd	nd	22.9 ± 13.9	nd	na
2,4-Dichlorophenol	50	11.3 ± 11.3	48.4	52.0	26.3 ± 8.5	51.5 ± 25.6	76.2 ± 10.8	74 ± 8[b]
Quinaldic acid	50	nd	96.4	nd	nd	na	31.8 ± 9.0	na
Isophorone	50	44.4 ± 3.3	55.2	ns	28.0 ± 5.6	25.3 ± 9.0	77.0 ± 15.3	88 ± 12
Biphenyl	50	57.0 ± 14.5	52.4	8.0	16.3 ± 11.9	89.2 ± 8.1[a]	81.6 ± 16.4	50 ± 17
1-Chlorododecane	5	95.2 ± 3.0	42.0	49.0	25.0 ± 11.5	70.4 ± 8.0[a]	76.5 ± 33.7	19 ± 3
2,6-Di-*tert*-butyl-4-methylphenol	50	6.8 ± 11.6	45.2	2.0	30.7 ± 13.6	91.6 ± 10.3[a]	37.8 ± 23.0	na
2,4′-Dichlorobiphenyl	50	70.1 ± 10.3	46.9	0.7	45.0 ± 4.6	79.1 ± 3.5[a]	72.4 ± 9.2	na
2,2′,5,5′-Tetrachlorobiphenyl	5	65.0 ± 14.8	28.7	16.0	30.3 ± 7.2	84.7 ± 4.6[a]	75.1 ± 12.6	na
Anthraquinone	50	59.9 ± 13.8	45.3	5.0	31.6 ± 29.3	109 ± 9.0[a]	74.6 ± 22.9	na
Phenanthrene	1	38.7 ± 6.9	44.6	nt	14.3 ± 3.2	na	na	na
Bis(2-ethylhexyl) phthalate	50	65.3 ± 21.2	15.6	nd	29.7 ± 4.7	94.3 ± 13.5[a]	65.7 ± 34.9	na
Glucose	50	na	na	16	nd	na	nd	na
Furfural	50	nd	0.6	0	3.0 ± 2.6	35.6 ± 12.9	6.6 ± 1.8	na
Quinoline	50	61.1 ± 25.7	21.7	12.0	4.3 ± 7.5	67.4 ± 35.8	79.9 ± 26.5	25 ± 5
5-Chlorouracil	50	27.8 ± 10.7	41.7	45.0	9.3 ± 4.0	na	7.5 ± 32.3	na
Caffeine	50	39.3 ± 32.3	0.6	53.0	6.0 ± 5.6	8.4 ± 1.2	41.8 ± 4.7	na
Glycine	50	4.8 ± 1.9	nd	nd	nd	na	3.3 ± 3.1	na
Chloroform	50	nd	43.5	21	nd	na	67.9 ± 22.8	na
Methyl isobutyl ketone	50	3.3 ± 2.7	25.6	41	4.7 ± 0.6	na	26.8 ± 15.7	na
Humic acid	2000	34.0 ± 4.6	40.7	54	1.0 ± 0	na	na	na

NOTE: nd denotes not detected; na denotes not analyzed.
[a] Recovered only from PFTE.
[b] $n = 4$, without humic acid present.

reference compounds recommended by the Council on Environmental Pollutants (2). The solute concentrations listed in Table II reflect the predominance of humic substances among the organic contaminants of drinking water.

After developing acceptable analytical techniques and conducting laboratory-scale recovery studies, each investigator was to use his method to isolate or concentrate at least 50-fold the complete mixture of model solutes from 500 L of organic-free water (i.e., distilled water or equivalent containing 70 ppm of $NaHCO_3$, 120 ppm of $CaSO_4$, and 47 ppm of $CaCl_2 \cdot 2H_20$). The inorganic matrix was used in an attempt to simulate Cincinnati potable water (3) and to minimize the effect that differences in water sources may have on the efficiency of a particular method.

Methods Evaluated

Concentration Method. The concentration procedure that was developed and evaluated was a RO–Donnan dialysis system (*4*). The initial objective during method development was to conduct membrane-screening tests to evaluate the suitability of various RO and ion-exchange membranes. The four membranes considered for final evaluation on the basis of solute rejection, chlorine stability, and artifact production were the cellulose acetate and FT–30 (Film Tec) RO membranes, the Nafion cation-exchange membrane, and the IONAC MA 3475 anion-exchange membrane.

The method basically involved repetitive batch concentration whereby 167-L portions of water containing the model solutes were forced by applied pressure through a semipermeable membrane against the osmotic pressure gradient. The RO system was operated in a recirculating mode so that the portion of the water not forced through the RO membrane was recycled back to the batch concentration tank along with the inorganic salts and model solutes rejected by the membrane. Because the inorganic salts are concentrated along with the model solutes, a 4-h Donnan softening cycle was used to reduce the concentration of calcium ions by exchanging them with sodium ions and thus preventing inorganic salt precipitation.

Isolation Methods. One of the isolation methods evaluated was a liquid–liquid extraction procedure using supercritical fluid (SCF) CO_2 as the extraction solvent (*5*). The effectiveness of SCF CO_2 as an extraction solvent compared to gaseous and liquid CO_2 is associated with the marked increase in the density of CO_2 at its critical temperature and pressure, resulting in increased solvating power. The extraction unit that was evaluated was an open system consisting of a SCF CO_2

source, a stainless steel extraction vessel, a pressure-reduction valve for CO_2 removal, and a U-tube trap system for solute collection.

The other isolation methods evaluated employed solid adsorbents to isolate the model solutes (*6–8*). The first of these used XAD-4, a macroreticular, polystyrene-divinylbenzene resin (Rohm and Haas), into which trimethylamine groups had been introduced (*9*). The purpose of the resulting quaternary ammonium functional groups was to allow more efficient adsorption of acidic compounds without an appreciable loss of capacity for hydrophobic compounds. This feature is important because the vast majority of the organic matter in potable water is neutral or acidic in nature (*10*).

Desorption of the model compounds isolated on the quaternary XAD-4 (QXAD-4) resin column was accomplished by sequential elution with ethyl ether, methanol, ethyl ether, 0.1 N HCl/ether, 0.1 N HCl/methanol, and saturated HCl/methanol. The acidified organic solvents were used for acidic solute removal.

Evaluations of an integrated adsorption system were also conducted. In this system, by varying the pH conditions, the dissolved organics (model compounds) are separated into fractions by isolation onto Amberlite XAD-8, AG MP-50 cation-exchange resin, and graphitized carbon black. The procedure is based on the separation of organic solutes into hydrophobic and hydrophilic neutral, acidic, and basic fractions.

Another adsorption system evaluated was a high-volume, high-pressure, macroporous-resin-based concentrator system designed to provide a 10,000-fold concentrate. This system used four stainless steel columns in series. Columns one through four were filled with AG MP-1 (Bio-Rad), AG MP-50 (Bio-Rad), XAD-2 (Rohm and Haas), and XAD-7 (Rohm and Haas), respectively. Unlike the other adsorption processes, a pump system was employed for both the adsorption and desorption phases of the evaluation. The use of acetonitrile as the elution solvent permitted UV monitoring of the eluant.

Combination Methods. Combination methods refer to those methods that employ both concentration and isolation methodologies. The modified parfait-distillation method that was evaluated used a series of adsorbents coupled with vacuum distillation to recover unadsorbed solutes (*11*). Porous polytetrafluoroethylene (PTFE) was used to adsorb the hydrophobic neutrals, and in field sampling it serves as a filter for particulate removal. Cation- and anion-exchange resins were then used to remove the ionized organic substances and to deionize the sample prior to the concentration of the nonadsorbed hydrophilic neutrals by vacuum distillation. Organic constituents were selectively desorbed from the ion-exchange resins by using organic solvents containing either HCl or NH_3.

Direct solvent extraction of large-volume water samples with immiscible organic solvents is generally not employed in the preparation of organic concentrates for biological testing. However, volume limitations can be overcome by the use of a continuous liquid–liquid extraction (CLLE) apparatus in combination with a separate preconcentration technique such as RO (*12*). Although a RO preconcentration step was not actually performed, all test evaluations of the CLLE procedure were performed by assuming a prior 15-fold concentration by RO. Consequently, evaluations were performed by using higher concentrations of model solutes and inorganic salts. Use was made of a 1:10 solvent-to-water ratio and a solvent recycle feature to maximize solvent efficiency and to minimize solvent artifacts.

Results and Discussion

The general goal of this overall effort, supported in part or full by the Health Effects Research Laboratory, was to develop an efficient method for preparing a 50-fold or greater concentrate that was representative of the organic constituents present in potable water. Ideally, the procedure should avoid chemical transformations, be artifact free, have high capacity, and result in minimal losses. In addition, the concentrates should be in a solvent or easily exchanged to a solvent system compatible with in vivo and in vitro biological test systems. Efforts to determine the most acceptable method for preparing representative concentrates on the basis of recovery data and mass balance determinations were hampered by the fact that the different investigators varied the starting volumes, the final concentration factor, and the total number of model solutes quantified because of their differing capabilities and problems unique to the individual method.

Model Compound Studies. Two major problems that each of the investigating teams faced were how to effectively spike a large volume of water with a variety of model solutes of differing solubility and how to analyze the model solutes for mass balance determinations. Because aqueous solutions are not amenable to direct gas chromatographic (GC) or GC–mass spectrometric (GC–MS) analyses, direct analytical testing of the aqueous influent or effluent in adsorption techniques and of the aqueous phase in either RO or liquid–liquid extractions (LLE) was not possible. Therefore, LLE became an integral part of all the mass balance determinations of the influent or feed stock. Additional analytical problems were associated with glucose, glycine, trimesic acid, quinaldic acid, humic acid, and 5-chlorouracil, none of which could be extracted by conventional LLE; thus, mass balance determinations were extremely difficult, if not impossible.

However, recovery data were more easily obtained because of the increased concentration of solutes and in some instances the incorporation of an organic solvent in the desorption process that was compatible with GC analyses. The results of the different recovery studies are listed in Table II. The original reports should be read for details and complete recovery data (*4–8, 11, 12*). Even though the recovery data indicate the shortcomings of preparing a representative concentrate of the organic contaminants in potable water, they do demonstrate certain trends concerning similar solutes and permit a limited comparison of methods.

In general, methods using solid adsorbents, compared with methods using RO and LLE, yielded higher recoveries for the majority of solutes. This result was expected because a variety of ionic and nonionic adsorbents are now available for the specific recovery of acidic, basic, or neutral solutes. Like all concentration–isolation procedures, the use of solid adsorbents has certain limitations. Major problems generally associated with adsorbents are the potential introduction of artifacts, the separation of solutes into multiple fractions with different solvent properties, and the incompatibility of the various solvent systems with toxicological test systems. Possibly because of the extensive cleanup procedures used by each research group, artifact production was shown not to be a major factor. Even though fractionation is generally desirable for analytical purposes, the resultant increased costs of biological testing associated with multiple samples are often prohibitive. Consequently, one is faced with recombining the fractions into a common solvent or solvent mixture suitable for subsequent health-effects testing. None of the solvent systems used in the adsorption procedures under evaluation were considered compatible for direct toxicological testing.

Model solutes recovered at levels in excess of 50% by two or more methods were 2,4-dichlorophenol, isophorone, biphenyl, 1-chlorododecane, 2,4′-dichlorobiphenyl, 2,2′,5,5′-tetrachlorobiphenyl, anthraquinone, bis(2-ethylhexyl) phthalate, and quinoline. Except for anthraquinone and quinoline, these solutes are hydrophobic. Porous PTFE effectively recovered stearic acid in addition to each of the hydrophobic solutes mentioned earlier. The higher recoveries by porous PTFE are not attributed to a greater adsorptive capacity but rather to the ease with which the hydrophobic solutes can be desorbed by conventional elution processes.

Each of the methods had trouble recovering the more highly water-soluble or volatile model compounds such as trimesic acid, furfural, glucose, glycine, caffeine, and methyl isobutyl ketone. Quinaldic acid, an amphoteric substance of moderate water solubility, was also poorly recovered by each method, except for the QXAD-4 procedure.

Field Application. Toward the end of the method evaluation period, the requirement arose to produce samples of organic materials for toxicological testing from a pilot drinking water treatment plant. The toxicologists wanted two types of samples: (1) a concentrated aqueous solution of the organic material that could be used as drinking water for the experimental animals and (2) a highly concentrated sample in an organic solvent for in vivo and in vitro testing. There were five streams at the plant: one stream that was not disinfected and four streams that received either chlorine, chloramine, ozone, or chlorine dioxide. There was also a carbon treatment step followed by redisinfection on the chlorinated stream, so there were seven sampling points. Two thousand gallons of water from each were required to provide the amount of residue needed for the desired sensitivity of the toxicological assays. Four sets of samples were to be taken over a 1-year period, so the total volume of water to be processed was 56×10^3 gal (21.2×10^4 L). Because of the magnitude of this project, the practicality of the methods had to be considered as well as the ability to recover certain classes of chemicals. None of the methods were superior in recovering all of the model compounds. Some methods required special adsorbents that were not commercially available (QXAD-4) or were not available in sufficient quantity (powdered Teflon or graphitized carbon black) for this project. As already discussed, some of the adsorption methods produced samples in several incompatible solvent systems or in solvents not suitable for use in biological tests.

The method chosen to provide an aqueous concentrate was RO. The RO procedure provided 50-fold aqueous concentrates containing almost all of the organic carbon (*4*). However, the removal of salt that was required to achieve a 400-fold concentrate without precipitation and without forming a hypertonic solution resulted in substantial losses of organic carbon (*13*).

The method chosen to provide a sample concentrate in an organic solvent was an adsorption method using XAD-8 and XAD-2 resins with a subsequent acetone elution. The system used for the adsorption method is illustrated in Figure 1. The adsorption columns contained 5 L of XAD resin that was cleaned by exhaustive extractions with methylene chloride, acetone, and methanol prior to use. The XAD-8 resin was placed before the XAD-2 resin to remove humic material, which is reported to bind irreversibly to XAD-2 (*14*). Two thousand gallons of water was passed through each set of columns at 9–11 bed volumes/h. Immediately prior to the water entering the columns, HCl was added to bring the water to pH 2. Samples were taken for total organic carbon (TOC) analysis before and after each resin column to monitor the efficiency of removal. Columns were eluted separately with 3 bed volumes of acetone followed by 3 bed volumes of acetone containing HCl.

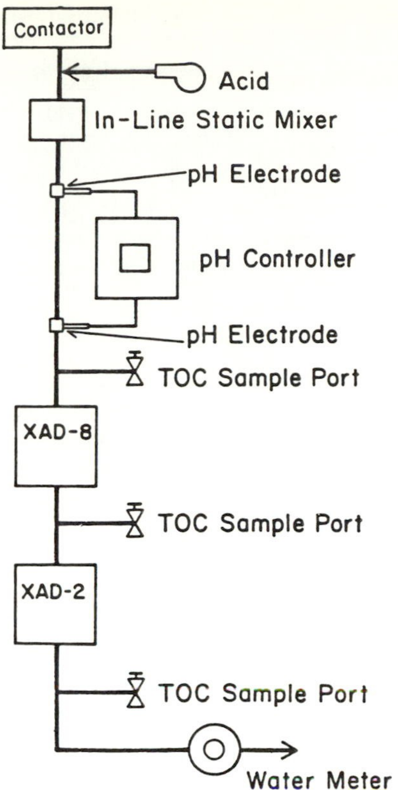

Figure 1. Schematic of adsorption process used for sample collection.

Table III contains the results of TOC analysis of one of the samples taken during the collection of organics from the chlorine-treated streams. The TOC content of the sample at zero time is elevated because the methanol used to clean and store the resin was not completely removed by rinsing the columns with 50 gal of the sample stream prior to beginning the sample collection. The results obtained thereafter indicate that within the first 500 gal of sample, either equilibrium was reached or only a select fraction of the organic chemicals was removed. Although the actual amount of organic carbon removed was different for each stream, the results for each followed this pattern.

Similar results have been observed for granular activated carbon columns used in water treatment and have been attributed to either biological or chemical oxidation of TOC within the column (*15*). Because only 67 h was required to collect the samples, it is unlikely that biological oxidation was responsible for the significant losses in TOC; decreased TOC levels most probably resulted from the majority of the

Table III. Typical TOC Data by Resin Process

TOC Sampling Point	*Gallons*				
	0	*500*	*1000*	*1500*	*2000*
Influent	3.6	3.5	3.4	3.4	3.4
After XAD-8	11.0[a]	2.2	2.7	2.7	2.5
After XAD-2	30.0[a]	1.6	2.0	2.1	1.9

NOTE: Values are TOC levels in milligrams per liter.
[a] Resin was contaminated with MeOH even after prerinse with 50 gal of water.

organic material in the water being highly polar and only the nonpolar fraction being adsorbed. This view is strengthened by the fact that the least efficient recovery was observed in the ozonated water. Ozonation produces more polar organic byproducts than other disinfectants, and this sample contained no residual ozone to cause oxidative degradation within the column.

If constant loading is assumed during the collection period, the percent of organic carbon adsorbed can be calculated (Table IV). Results varied from 30% for the ozonated water to 50% for the chloraminated sample. The residues obtained after evaporating the acetone eluates were analyzed for carbon content. The amount of organic carbon contained in the residues of the acetone eluates of each column was calculated and compared to the organic carbon content of the volume of the water sample passed through the columns. The actual fraction of organic carbon recovered varied from 11% to 39%. The eluates obtained with acidified acetone could not be used because of condensation products of acetone, which formed during storage.

Table IV. Organic Carbon Recovery with XAD Columns from Drinking Water Pilot Plant Samples

TOC Sampling Point	*Not Disinfected*	*Ozone*	*Chlorine Dioxide*	*Chloramine*	*Chlorine*
Influent[a]	3.4	3.3	3.3	3.4	3.4
After XAD-8[a]	2.5	2.8	2.2	2.3	2.5
After XAD-2[a]	1.8	2.3	1.8	1.7	1.9
Apparent percent adsorbed	47	30	45	50	44
Percent recovered with acetone	30	11	24	39	39

[a] Values are the average of the organic carbon recoveries in milligrams per liter obtained at 500, 1000, 1500, and 2000 gal during a typical run.

To estimate the size column required to isolate more of this polar material, we used information published by Thurman et al. (*16*), who described an empirical relationship between aqueous solubility of organic compounds and capacity factors on XAD-8 resin. They defined *capacity factor* as the mass of solute sorbed on the resin divided by the mass of solute present in the void volume of the column at the 50% breakthrough point. On the basis of their data, a substance having a capacity factor of 1000 will have a solubility of about 1×10^{-3} mol/L. Examples of compounds having solubility in this range are methyl benzoate, 2,4,6-trichlorophenol, and chlorobenzene.

By using Thurman's empirical formula, it was calculated that a column of 45 L of XAD-8 would be required to completely retain these compounds. This size column would require 135 L of solvent for elution. Collecting five samples at once would require a pilot plant for distillation and extraction comparable in magnitude with the water treatment pilot plant being sampled. Therefore, a decision was made to use smaller columns at the risk of losing more soluble organic compounds.

Conclusion

An ideal method for recovering sufficient representative samples of organic matter from water samples for toxicological testing has not yet been developed. However, of the seven methods evaluated, the use of solid adsorbents was the most efficient and showed the greatest potential in concentrating organics from potable water for biological testing.

Literature Cited

1. Kopfler, F. C. In *Short-Term Bioassays in the Analysis of Complex Environmental Mixtures II;* Waters, M. D.; Sandhu, S. S.; Huisingh, J. C.; Claxton, L.; Nesnow, S., Eds.; Plenum: New York, 1981; pp 141–153.
2. Keith, L. H. *Environ. Sci. Technol.* **1979,** *13,* 1469–1471.
3. Durfor, C. N.; Becker, E. *Public Water Supplies of the 100 Largest Cities in the United States;* U.S. Geological Survey: 1962; Water Supply Paper No. 1812.
4. Lynch, S. C.; Smith, J. K. In *Organic Pollutants in Water: Sampling and Analysis;* Suffet, I. H.; Malaiyandi, M., Eds.; Advances in Chemistry 214; American Chemical Society: Washington, DC, 1986; Chapter 21.
5. Ehntholt, D. J.; Eppig, C.; Thrun, K. E. In *Organic Pollutants in Water: Sampling and Analysis;* Suffet, I. H.; Malaiyandi, M., Eds.; Advances in Chemistry 214; American Chemical Society: Washington, DC, 1986; Chapter 23.
6. Ben-Poorat, S.; Kennedy, D. C.; Byington, C. H. In *Organic Pollutants in Water: Sampling and Analysis;* Suffet, I. H.; Malaiyandi, M., Eds.; Advances in Chemistry 214; American Chemical Society: Washington, DC, 1986; Chapter 25.

7. Giabbai, M.; Chian, E. S. K.; Reuter, J. H.; Ringhand, H. P.; Kopfler, F. C. In *Organic Pollutants in Water: Sampling and Analysis;* Suffet, I. H.; Malaiyandi, M., Eds.; Advances in Chemistry 214; American Chemical Society: Washington, DC, 1986; Chapter 22.
8. Baird, R. B.; Jacks, C. A.; Neisess, L. B. In *Organic Pollutants in Water: Sampling and Analysis;* Suffet, I. H.; Malaiyandi, M., Eds.; Advances in Chemistry 214; American Chemical Society: Washington, DC, 1986; Chapter 26.
9. Richard, J. J.; Fritz, J. S. *J. Chromatogr. Sci.* **1980,** *18,* 35–38.
10. Leenheer, J. A. *Environ. Sci. Technol.* **1981,** *15,* 578–587.
11. Johnston, J. B.; Josefson, C.; Trubey, R. In *Organic Pollutants in Water: Sampling and Analysis;* Suffet, I. H.; Malaiyandi, M., Eds.; Advances in Chemistry 214; American Chemical Society: Washington, DC, 1986; Chapter 24.
12. Baker, R. J.; Suffet, I. H. In *Organic Pollutants in Water: Sampling and Analysis;* Suffet, I. H.; Malaiyandi, M., Eds.; Advances in Chemistry 214; American Chemical Society: Washington, DC, 1986; Chapter 27.
13. Smith, J. K.; Lynch, S. C.; Kopfler, F. C. In *Chemistry in Water Resuse;* Cooper, W. J., Ed.; Ann Arbor Science: Ann Arbor, MI, 1981; pp 191–203.
14. Thurman, E. M.; Akin, G. R.; Malcolm, R. L. *Proc. 4th Joint Conference on Sensing of Environmental Pollutants;* American Chemical Society: Washington, DC, 1978; pp 630–634.
15. Yohe, T. L.; Suffet, I. H.; Cagle, J. T. In *Activated Carbon Adsorption of Organics from the Aqueous Phase;* McGuire, M. J.; Suffet, I. H., Eds.; Ann Arbor Science: Ann Arbor, MI, 1980; Vol. 2, pp 27–69.
16. Thurman, E. M.; Malcolm, R. L.; Aiken, G. R. *Anal. Chem.* **1978,** *50,* 775–779.

RECEIVED for review October 31, 1985. ACCEPTED February 21, 1986.

21

Evaluation of Reverse Osmosis To Concentrate Organic Contaminants from Water

Stephen C. Lynch and James K. Smith

Gulf South Research Institute, New Orleans, LA 70186

Reverse osmosis for concentrating trace organic contaminants in aqueous systems by using cellulose acetate and Film Tec FT-30 commercial membrane systems was evaluated for the recovery of 19 trace organics representing 10 chemical classes. Mass balance analysis required determination of solute rejection, adsorption within the system, and leachates. The rejections with the cellulose acetate membrane ranged from a negative value to 97%, whereas the FT-30 membrane exhibited 46-99% rejection. Adsorption was a major problem; some model solutes showed up to 70% losses. These losses can be minimized by the mode of operation in the field. Leachables were not a major problem. Actual recoveries are reported for a field trial in which 9500 L was concentrated to 190 L.

PUBLIC CONCERN over trace chemical contamination of drinking water continues to increase each year despite extensive efforts to define and rectify the problem. The major focus of water contamination has shifted from the microbiological problems of several decades ago to the trace organic and inorganic chemical problems of today. Solutions can be developed only as potential classes of problem compounds, such as trihalomethanes, are identified. For example, significant advances have been made toward correcting and reducing levels of trihalomethanes because the problem was studied and defined and solutions were brought forth. Many types of chemical contamination have not yet evolved to this point.

In vivo and in vitro toxicity testing methods are used to assess potential adverse health effects of chemical contaminants. These methods have been used to confirm many suspected substances as toxic and carcinogenic. To date, only a small fraction of the organic makeup of most drinking waters has been elucidated and tested. Broad spectrum

0065-2393/87/0214/0437$08.00/0

concentration techniques are now employed to help identify organic components and to prepare samples for toxicity testing. These techniques are necessary because of the very low levels (parts per billion) of the organics and the low sensitivity of the toxicity assays to these compounds. Reverse-osmosis (RO) concentration is one method used to prepare drinking water concentrates for toxicity testing. Other methods that have been evaluated for concentrating organics in drinking water include adsorbent resins (e.g., XADs), vacuum distillation, supercritical fluid (CO_2) extraction, and parfait-distillation. The value of the RO concentration method lies primarily in the ability to concentrate a broad range of different chemical compounds, originally present at trace levels in natural waters, to levels that are of value in toxicological evaluations. The process has the advantage of low temperature, no extracting solvent, and no phase change. This method has not been developed as a low-detection-level analytical procedure.

Reverse Osmosis Concentration

This concentration method uses a polymeric semipermeable membrane and principles of RO to effect separation of water from the organics in drinking water samples. In this process, a water sample is recirculated past the semipermeable membrane while hydraulic pressures exceeding the osmotic pressure are maintained. Water is transported through the membrane under these conditions (permeation). The concentration of solutes continues to build as water is removed from the system.

Maximized recovery of as many different organic compounds as possible is a primary goal of the RO concentration process. Potential losses of compounds can occur through (1) the permeate water stream, (2) volatile headspaces, (3) adsorption and absorption onto system components, and (4) binding or coprecipitation to or with other compounds such as humic acids or insoluble inorganic salts. Control of most if not all of these factors can be attained through the manipulation of process variables.

Maximum recoveries can be significantly affected by permeation losses. Control of permeation losses must be achieved through the proper choice of membrane type. A membrane with high rejection of most organics must be selected. The membrane rejection (R) for a solute is defined in terms of the permeate solute concentration (C^P) and feed solute concentration (C^F):

$$R = [1 - (C^P/C^F)] \qquad (1)$$

Recovery of organic solutes can be predicted by the following equation for RO concentrations operated in batch modes (*1*):

$$M_f/M_i = (C_f^F V_f^F)/(C_i^F V_i^F) = (V_i^F/V_f^F)^{R-1} \qquad (2)$$

where M_f and M_i are the final and initial masses, respectively; C_f^F and C_i^F are the final and initial concentrations, respectively; V_f^F and V_i^F are the final and initial volumes, respectively; and R is the membrane rejection.

The exponential nature of the recovery response is represented graphically in Figure 1. The necessity for selecting membranes with high organic rejections is quite apparent. For example, in a 50-fold concentration, the recovery of a compound will not exceed 70% unless membrane rejection of that solute exceeds 0.9 (90% rejection).

Asymmetric cellulose acetate membranes were developed in the early 1960s by Loeb and Sourirajan (2). For more than a decade, cellulose acetate and its blends were the only commercially available RO membranes. Improved membranes (with respect to operating pH, biodegradation, compaction, and organic compound rejection) were developed in the early 1970s (3). These membranes used aromatic

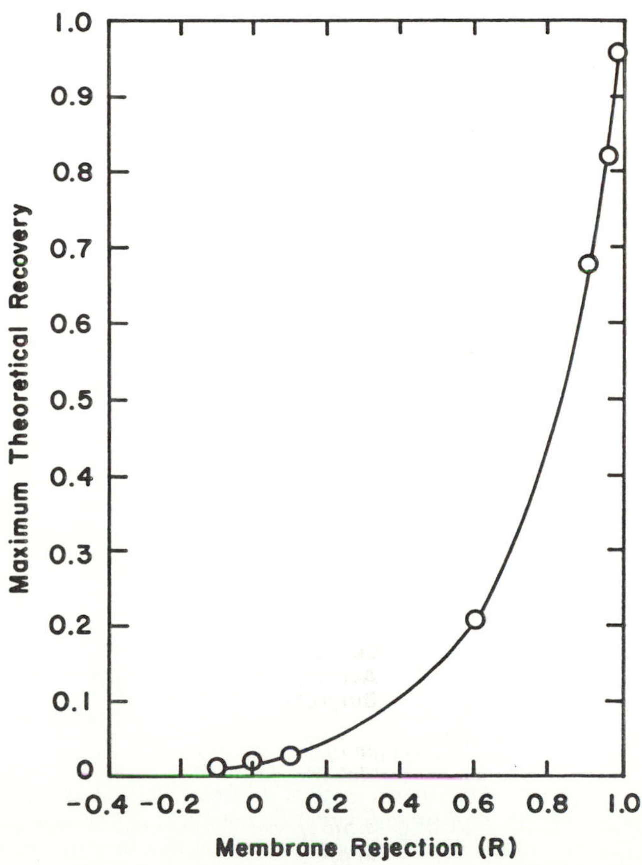

Figure 1. Theoretical recovery versus rejection for 50× concentration.

polyamide polymer. Both cellulose acetate and the improved polyamide were used in early RO concentrations (*4*). In 1977, the first thin-film composite membranes started to appear on the market (*5*). The composite membranes offered potential for enhanced RO recovery of organics.

Improved organic rejections result from the chemical nature of the newer polymeric materials in the nylon and thin-film composite membranes. Data by Chian and Fang (*6*) represented in Figure 2 illustrate different membrane rejections for a variety of organic com-

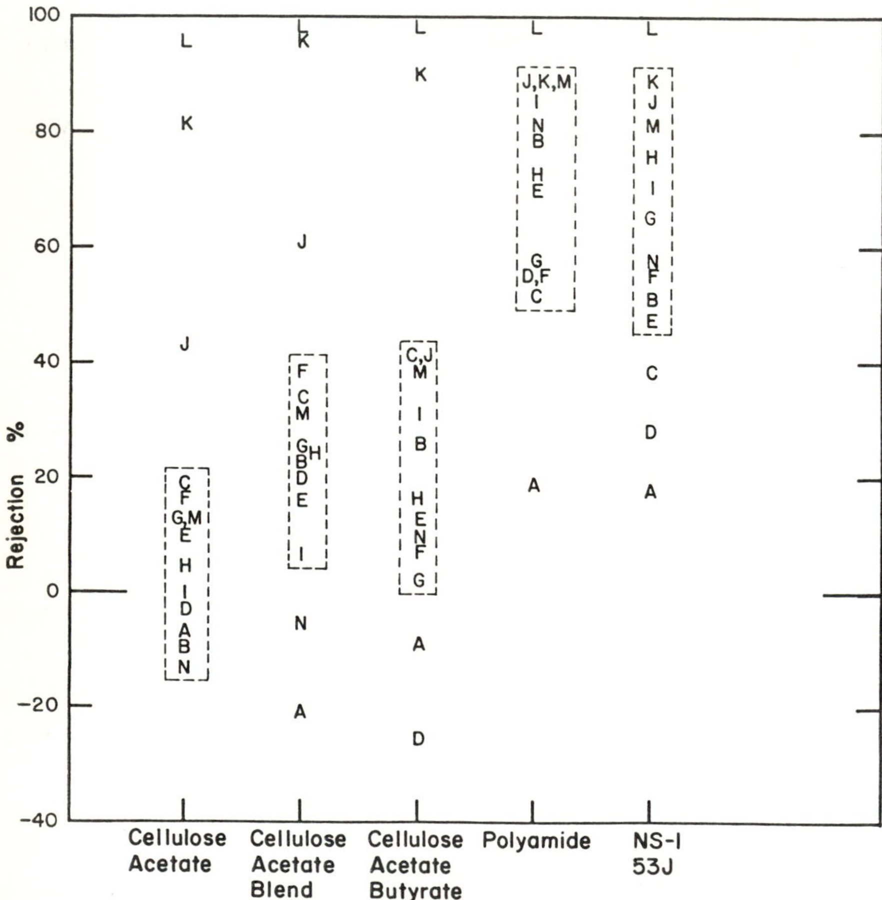

Figure 2. Rejection performance of five different RO membranes (6). A, methanol; B, aniline; C, formaldehyde; D, methyl acetate; E, acetic acid; F, urea; G, ethanol; H, acetone; I, hydroquinone; J, isopropyl alcohol; K, glycerol; L, sodium chloride; M, ethyl ether; N, phenol. Conditions: pressure, 40.8 atm; temperature, 24 °C; feed, 0.30 gal/min.

pounds and membrane types. Solute concentrations were maintained at the parts-per-million level for the majority of the compounds studied. Several points are of interest in Figure 2. First, sodium chloride salt rejection cannot be used as a guide for potential organic rejection. However, membranes with low salt rejection are unlikely to exhibit acceptable rejection of organics. The three cellulose acetate membranes demonstrated markedly lower overall rejection of organics than the newer nylon and composite membranes. Several compounds were even preferentially transported through the cellulose acetate membrane, as indicated by negative rejection values. Newer membranes not based on cellulose acetate offer potential for improved recovery through their enhanced rejection of organic compounds.

Proper control and design of process variables can improve recoveries significantly. Volatile chemicals can be better retained with suitable equipment closures and regulation of operating temperatures. Materials of construction and conditioning of the process equipment may well compensate for compounds that can be lost because of adsorption–absorption. Proper operation of process variables such as pH and appropriate use of water softening (by a membrane process such as Donnan softening) can prevent inorganic chemical precipitation and consequent losses of organic compounds due to entrapment, coprecipitation, and degraded system performance (*7*).

Factors as subtle as mode of operation can significantly affect the ultimate organic recovery (*1*). Figure 3 compares predicted recoveries for concentration processes operated in a batch mode (where the total volume to be reduced is processed in discrete sequential batches) and a less efficient continuous mode (where the total volume to be reduced is processed continuously by maintaining a working volume with makeup rate equal to permeate rate). A 90% membrane rejection is assumed. Batch processing is the preferred procedure for large-volume samples and higher concentration levels at which the mode becomes critical.

In RO concentration, as with all separation methods, complete recovery of all compounds is not possible. RO offers several distinct advantages for the concentration of organics in water for the analysis and assessment of health effects. These advantages are the following:

1. A sample can be concentrated without removing the solutes from the aqueous matrix.
2. No phase changes are required.
3. It is a low-temperature process.
4. Large volumes can be processed.
5. High concentration factors can be achieved.
6. The system has been demonstrated in the field.

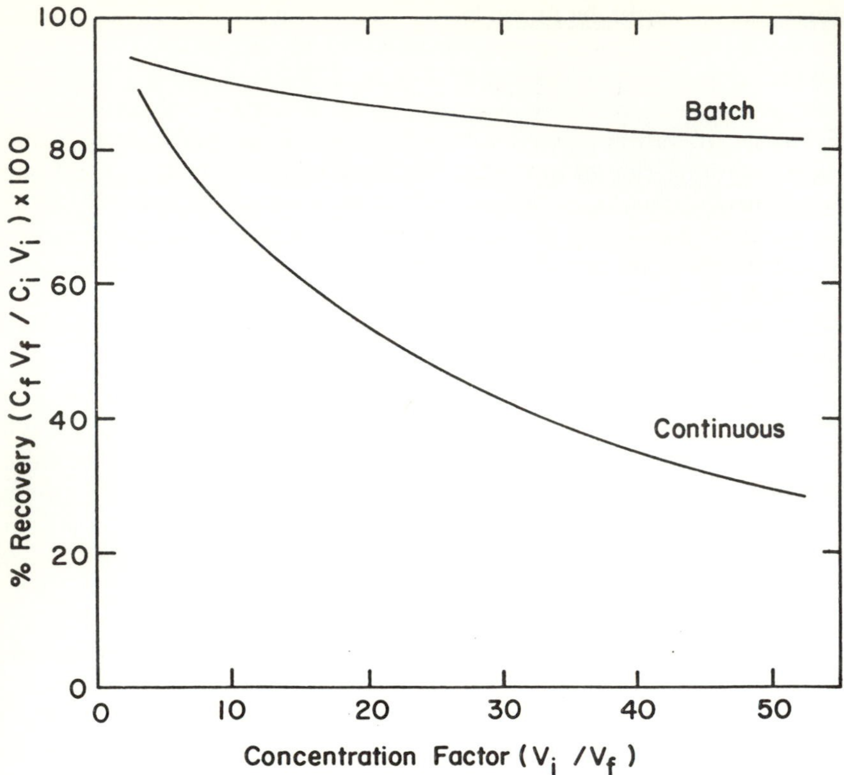

Figure 3. Theoretical recoveries for batch and continuous processing modes (membrane rejection = 90%).

7. A broad spectrum of organics is recovered.
8. Inorganic ions can be removed concurrently with organic concentration.

Factors Affecting Reverse-Osmosis Concentration

To provide a more comprehensive understanding of the use of RO as a method for concentrating organics from water supplies, several factors must be addressed:

1. What recoveries of specific organic compounds can be expected in a RO concentration process? What recoveries are actually observed?
2. What are the membranes' inherent rejections of different classes of organic compounds that may be found in water at trace levels? Can measurable improvements in the rejection of organics be expected from newer membrane materials?

3. What are the magnitude and probable cause of loss of organic compounds during the concentration process?
4. What materials are added to concentrates through leaching of system components, and does flushing remove these compounds from the system?
5. What effect do disinfectants, such as chlorine, have on leachable organic compounds?

This chapter provides some insight into factors affecting RO concentration of organics present at trace levels in water. The information was compiled primarily from data generated in a study comparing alternate methodologies for preparing organic drinking water concentrates (*8*), which was sponsored by the U.S. Environmental Protection Agency (USEPA) Health Effects Research Laboratory.

Experimental

Model Compounds. A spectrum of organic compounds incorporating varied chemical functionalities (acidic, amine, aldehyde, hydroxyl, aromatic, etc.) and aqueous solubilities was selected to evaluate the RO process. Membrane-screening and concentration test series using two starting levels of model solutes were performed in this study. Table I lists specific model compounds, target concentrations, and methodologies developed for the analysis of the compounds. Higher levels (2500 and 250 ppb) were used in the early screening test, whereas the low levels (50 and 5 ppb) were starting solute concentrations in actual concentration experiments.

All evaluation studies were performed by using trace quantities of the organic compounds spiked into a synthetic tap water matrix. The inorganic makeups of the synthetic tap waters used for the screening test and concentration test series are shown in Table II. Levels of inorganic salts used in the screening test, in which concentrations were not allowed to build up, were near the saturation point for the least soluble salt, calcium sulfate (i.e., 5 times the base levels). This condition simulates the worst case with respect to any potential organic salting-out effects that might exist.

Analytical methodologies listed in Table I were optimized for the conditions present in samples generated during this study (i.e., low concentrations of model solutes, high inorganic salt level, high humic acid backgrounds, and presence of acetone carrier solvent). Calibration curves were constructed to establish response linearity and detection limits. Interferences were investigated and sensitivities were maximized to provide reliable analysis of the model compounds. Despite these efforts, limits of detection and low concentrations in the water matrix and permeated water samples resulted in inadequate analytical definition for some organic solutes.

Membrane-Screening Test. Membrane-screening tests were performed to evaluate (1) the adequacy of flushing procedures, (2) the rejection of model compounds by the RO membrane, and (3) losses of model compounds to system components via adsorption, volatilization, solubility artifacts, or other phenomena.

The apparatus shown in Figure 4 was used to complete the membrane-

Table I. Model Compounds and Conditions Used in Evaluation of the RO Concentration Method

Chemical Class	*Model Compound*	*Concentration (μg/L)*		*Method of Detection*[b]
		Screening Test[a]	*Concentration Test*[a]	
Acids	trimesic acid	2500	50	HPLC
	stearic acid	2500	50	GC, GC–MS
	humic acid	2000	2000	HPLC
	glycine	2500	50	fluorescence
Aldehydes	furfural	2500	50	HPLC
Amines	quinoline	2500	50	GC, GC–MS
	caffeine	2500	50	GC, GC–MS
	5-chlorouracil	2500	50	HPLC
Carbohydrates	glucose	NT[c]	50	liquid scintillation
Chlorobiphenyls	2,4′-dichlorobiphenyl	250	5	GC, GC–MS
	2,2′,5,5′-tetrachlorobiphenyl	2500	50	GC, GC–MS
Esters	bis(2-ethylhexyl) phthalate	2500	50	GC, GC–MS
Hydrocarbons	1-chlorododecane	250	5	GC, GC–MS
	biphenyl	2500	50	GC, GC–MS
Ketones	isophorone	2500	50	GC, GC–MS
	anthraquinone	2500	50	GC, GC–MS
	methyl isobutyl ketone	2500	50	GC, GC–MS
Phenols	2,4-dichlorophenol	2500	50	GC, GC–MS
	2,6-di-*tert*-butyl-4-methylphenol	2500	50	GC, GC–MS
Trihalomethanes	chloroform	2500	50	GC

[a] See text for description of these test series.
[b] HPLC denotes high-performance liquid chromatography.
[c] NT means not tested.

Table II. Inorganic Makeup of Synthetic Tap Water Used in Laboratory Test

	Concentration (mg/L)	
Ion	*Screening Test*	*Concentration Test*[a]
Na^+	96	19.2
Ca^{2+}	241	48.1
HCO_3^-	254	50.8
SO_4^{2-}	424	84.7
Cl^-	114	22.7
Total	1130	226

[a] Salt levels at beginning of concentration test are given.

screening studies. Of particular interest is the equipment shown in the right-hand portion of the illustration, the RO circulation loop. The system (55-gal stainless steel reservoir, high-pressure pumps, RO membrane, and associated control instrumentation) was operated on total recycle for the screening test. Permeate and reject flows from the RO membrane were returned to the reservoir during tests. The permeate would normally be discarded during an actual RO concentration. No solute buildup (either inorganic or organic) was expected when operating in this mode. Operation of the system on total recycle provided a convenient evaluation of compound losses (adsorptive or volatile), membrane rejections, and leachable substances.

A routine screening test first involved a series of flush–discard cycles followed by a period (3–4 h) of total recycle with nonspiked synthetic tap water. Analysis of water at the end of the recirculation period provided an assessment of flush efficiencies and an estimate of leachable organic substances.

Membrane rejection of model compounds and losses of organics were determined after spiking and mixing the model compounds into the synthetic tap water. The spiked water was sampled at the beginning and end of an extended period (4 h) of total recirculation to measure loss of solutes. Analysis of this data yielded estimates of combined losses to adsorption, volatilization, precipitation, or other unknown mechanisms. Analysis of a permeate sample collected at the end of the recirculation period provided an assessment of membrane rejection of solutes. The 4-h equilibration period was chosen because actual field use of the method incorporates 50-gal batch cycles requiring approximately 4-h process time. There are indications that low molecular weight unsaturated compounds may require longer equilibration periods before a steady state condition is established (*9*). Longer equilibrations were not examined in this study.

Two different RO membrane types were evaluated in this study. The first was a standard cellulose acetate based asymmetric membrane. The second type, a proprietary cross-linked polyamine thin-film composite membrane supported on polysulfone backing, was selected to represent potentially improved (especially for organic rejection) membranes. Manufacturer specifications for these membranes are provided in Table III. Important considerations in the selection of both membranes were commercial availability, high rejection (sodium chloride), and purported tolerance for levels of chlorine typically found in drinking water supplies. Other membrane types having excellent potential for organic recovery were not evaluated either because they were not commercially

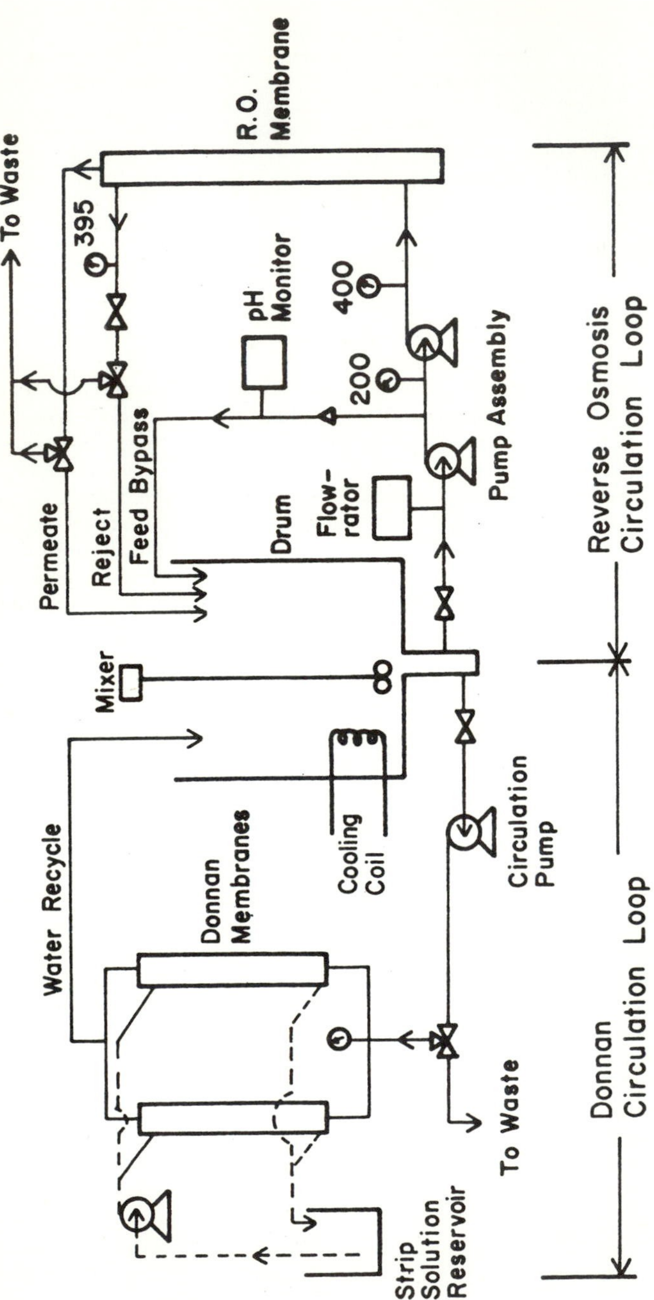

Figure 4. Major components of an RO concentration–Donnan dialysis organic recovery system.

Table III. Manufacturers, Suppliers, and Performance Specifications for Membranes Evaluated in Laboratory Tests

Membrane	*Data*	*Specifications*	*Observed Performance*
Cellulose acetate	manufacturer	Osmonics, Inc.	—
	model	OSMO-112-97-CA	—
	salt rejection	94-97%	97%
	feed concentration	1000 ppm of NaCl	500 ppm (mixed salts)
	pressure	400 lb/in.^{2}	400 lb/in.^{2}
	module flux	0.38 L/min	0.38 L/min
	temperature	25 °C	26 °C
	percent conversion	10%	4%
FT-30 polysulfone thin-film composite	manufacturer	Film Tec Corporation	—
	model	BW30-20-26	—
	salt rejection	98%	99.2%
	feed concentration	10,000 ppm of NaCl	500 ppm (mixed salts)
	pressure	350 lb/in.^{2}	400 lb/in.^{2}
	module flux	0.38 L/min	0.94 L/min
	temperature	25 °C	26 °C
	percent conversion	5%	4%

available or because they did not possess the required chemical resistance to chlorine.

Membrane Concentration Test. Process potential was demonstrated by concentrating 500 L of synthetic tap water spiked with trace levels of the model compounds. A 50× volumetric concentration was achieved by reducing the sample volume from 500 to 10 L. The recovery of model compounds and membrane rejection of compounds were evaluated, and the location of system losses was approximated.

The experimental design is presented schematically in Figure 5. Concentration experiments were performed by using the system shown in Figure 4. It was necessary to maintain the concentration of inorganics below precipitation levels. This condition was achieved by using a secondary membrane process, Donnan softening. The original 500-L sample was concentrated in three 167-L batches to an approximate concentration factor of 12.6×; then the sample was composited and further processed. During this step, 460 L of permeate water was discarded. At this point, the Donnan softening was performed. After a fixed Donnan softening period adequate to prevent precipitation, the intermediate concentrate was further processed to 10 L and evaluated for model compound recoveries. RO feed and permeate samples were collected at various volumetric concentration factor stages during the process. The RO feed samples were representative of the concentrate developing in the process reservoir.

Results and Discussion

Model compounds, selected by the Health Effects Research Laboratory at the USEPA, represent a broad spectrum of chemical classes and include several compounds of current environmental interest. Spike

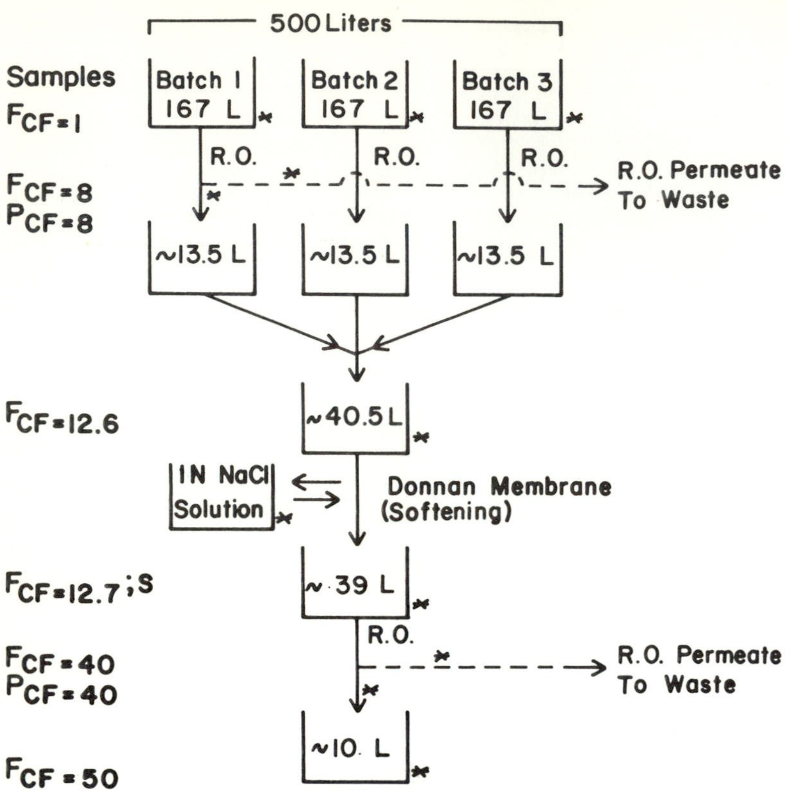

Figure 5. General experimental design and sampling regime (*) for the RO concentration runs.

levels are representative of those that might be found in more severely contaminated drinking waters. Many model compounds selected have very limited water solubility and require the use of cosolvents to facilitate the preparation of test waters. Experimental objectives were established to assist in comparisons of various concentration methods.

The importance of proper RO membrane selection has already been discussed. A review of commercially available RO membranes revealed five different basic membranes that could provide organic recovery. Cellulose acetate and cellulose acetate blends, aromatic polyamide, polyamide thin-film composite, cross-linked polyimine thin-film composite (FT-30), and polybenzimidazole were available when this work was performed. Only the first four types were commercially available. All membranes were available with excellent salt rejection (>97% sodium chloride). Two types of membranes, cellulose acetate and FT-30, have shown short-term (<2-months intermittent use) resistance to levels of chlorine typically found in tap water. This factor is important in concentrating drinking water samples because of chlorine's frequent use and presence as a disinfectant. The FT–30 membrane should not be used on chlorinated waters because the ultimate lifetime of the membrane may be affected. Both cellulose acetate and FT–30 composite membranes were studied in the RO concentration experiments. No deterioration in membrane performance was observed during the reported study period.

Cellulose acetate membrane was studied because of its past use in concentrate preparation and the need to better define its performance for specific organic recovery. Cellulose acetate continues to be widely used for a variety of industrial and commercial water purification applications. Cellulose acetate was not expected to perform at the level of the more highly cross-linked and inert thin-film composite membrane.

Membrane Rejection. Both cellulose acetate and FT–30 composite membranes were evaluated for rejection of solutes. Sodium chloride rejections were confirmed and listed in Table III. Typical organic rejections of model compounds are listed in Tables IV and V for cellulose acetate and FT–30 composite membranes, respectively. Rejections were measured during screening and concentration tests; solute levels were in the parts-per-billion range. Measurement of feed and permeate stream solute concentrations provided the necessary information to calculate solute rejection. Eq 1 was used to calculate rejection values.

Permeate solute levels were often below analytical detection limits. When this situation occurred, the rejection values shown in Tables IV and V were calculated by using detection limits, and the values were stated as minimums (i.e., rejection >90%). At very low concentrations, interferences and matrix effects became critical. Proper attention to analytical detail should be a primary concern of studies at the low parts-per-billion levels.

Several points should be made in regard to the data presented in Tables IV and V.

1. Rejections of all solutes studied were substantially higher for the FT–30 composite membrane (Table V) than for the cellulose acetate membrane (Table IV). This trend is predictable on the basis of the cross-linked nature and chemical inertness of the barrier layers.
2. Negative rejections were consistently measured for several compounds by using the cellulose acetate membrane system. Compounds of this nature must possess a strong affinity for the membrane material and have relatively high transport rates through the membrane.

Table IV. Model Compound Rejection for the Cellulose Acetate Membrane from Screening and Concentration Test

		Typical Rejections (Percent)[a]				
Chemical Class	*Model Compound*	*(589–85) No Humics*	*(618–04) No Humics*	*(618–30) With Humics*	*(618–41) No Humics*	*Average Rejection*
Acids	trimesic acid	—	95	>90	—/—	95
	stearic acid	33	>22	—	—/—	44
	humic acid	—	—	—	—/—	—
	glycine	—	97	—	—/—	97
Aldehydes	furfural	—	−90	−520	—/—	neg
Amines	quinoline	38	9	0	<0/>17	21
	caffeine	9	78	73	67/67	71
	5-chlorouracil	—	40	46	32/44	41
Carbohydrates	glucose	—	—	—	96/96	96
Chlorobiphenyls	2,2′,5,5′-tetrachlorobiphenyl	—	—	<80	—/—	—
Esters	bis(2-ethylhexyl) phthalate	64	—	—	—/—	64
Hydrocarbons	1-chlorododecane	—	55	60	—/78	64
	biphenyl	−28	−8	−52	−31/−64	neg
Ketones	isophorone	68	73	97	74/61	69
	anthraquinone	12	25	15	−11/−378	—
	methyl isobutyl ketone	—	—	—	17/−20	—
Phenols	2,4-dichlorophenol	−158	−5	−20	−20/−57	neg
	2,6-di-*tert*-butyl-4-methylphenol	38	94	72	>38/>62	61
Trihalomethanes	chloroform	−123	−80	−210	−29/<−20	neg

NOTE: Some original target compounds (*see* Table I) are not included because of inconsistent or incomplete data.
[a] All rejections indicated by greater than (>) were obtained by using the analytical limit of rejection in eq 1. Actual rejection may approach 100%. — means cannot be determined from available data; neg means negative value.

Table V. Model Compound Rejection for the FT-30 Composite Membrane from Screening and Concentration Test

Chemical Class	*Model Compound*	*Typical Rejections (Percent)*[a]					*Average Rejection*
		(589-81) No Humics	*(618-09) No Humics*	*(618-34) With Humics*	*(618-51) No Humics*	*(618-57) With Humics*	
Acids	trimesic acid	—	93	>97	—/—	—/—	96
	stearic acid	65	>43	—	—/—	—/>50	71
	humic acid	—	—	—	—/—	>92/>99	98
	glycine	—	>97	—	—/>70	>42/>71	78
Aldehydes	furfural	—	−26	—	61/36	>21/—	53
Amines	quinoline	88	96	97	99/98	99/>99	97
	caffeine	>99	>99	>98	45/>99	98/99	99
	5-chlorouracil	—	97	71	>86/>97	>48/>95	88
Carbohydrates	glucose	—	—	—	99/99	99/99	99
Chlorobiphenyls	2,2′,5,5′-tetrachlorobiphenyl	—	—	—	>38/23	—/>83	46
Esters	bis(2-ethylhexyl) phthalate	54	—	—	>81/>88	>92/>88	94
Hydrocarbons	1-chlorododecane	>57	>40	—	83/>96	>90/>96	87
	biphenyl	88	>40	93	>95/>97	>96/—	91
Ketones	isophorone	>99	>90	>88	>91/99	—/>64	96
	anthraquinone	89	>80	>89	>98/87	>96/>90	93
	methyl isobutyl ketone	—	—	—	95/99	>99/99	98
Phenols	2,4-dichlorophenol	79	96	>98	91/96	98/94	93
	2,6-di-*tert*-butyl-4-methylphenol	>92	91	>95	>83/>95	>90/>17	96
Trihalomethanes	chloroform	22	84	88	>88/91	>88/90	90

NOTE: Some original target compounds (*see* Table I) are not included because of inconsistent or incomplete data.

[a] All rejections indicated by greater than (>) were obtained by using the analytical limit of rejection in eq 1. Actual rejection may approach 100%. — means cannot be determined from available data.

3. No consistent adverse or beneficial effect was observed for tests in the presence or absence of humic acid.

The overwhelming conclusion supported by data is the superiority of the FT-30 composite membrane for the majority of organic compounds tested. From arguments presented earlier, improved recovery of organic compounds on the basis of these higher rejection properties would be expected. Data from selected literature sources (*6, 10–20*) on membrane rejections of organics in water at parts-per-million levels were reviewed. Results are presented by chemical class in Table VI. Data are compiled for cellulose acetate and a cross-linked NS-1-type composite membrane. Differences in the rejection of various compound classes by the two membrane types determined at higher solute levels are similar to those observed and reported here at parts-per-billion levels.

Leaching Studies. A determination of the adequacy of RO system flushing and a definition of substances that could leach from the system are important to the use of the sampling method. Both of these questions were addressed in these studies.

The RO system was flushed immediately prior to each screening test. The flush sequence consisted of (1) two 3-h, 95-L cycles with organic-free water; (2) four 30-min, 34-L cycles with unspiked synthetic tap water; and (3) one additional 34-L rinse with synthetic tap water. Adequacy of the flush regime was assessed by chemical analysis of unspiked synthetic tap water that was recirculated through the system

Table VI. Organic Solute Rejections for Cellulose Acetate and Cross-linked NS-1 Composite Membranes

	Percent Rejection	
Solute	*Cellulose Acetate*	*NS-1*
Acids	10–99	48–89
Aldehydes	2–78	37–95
Amines	−9–88	31–99
Carbohydrates	37–99	nd
Chlorobiphenyls	nd	nd
Esters	9–50	23–97
Hydrocarbons	42–99	nd
Ketones	5–67	76–96
Phenols	−37–81	57–83
Trihalomethanes	nd	nd

NOTE: nd means not determined.
SOURCE: Adapted from references 6 and 10–20.

(on total recycle) following the flush procedure. All screening tests incorporated this flush evaluation.

Examination of the system blank samples revealed sporadic occurrences of the model compounds used for method evaluation. No pattern was found to these occurrences; this result indicated that after flushing, contamination from the manufacture or repeated reuse of the membrane was not significant and the flushing regimes used in the study were adequate. A detailed examination of the system blanks using gas chromatographic–mass spectrometric (GC–MS) methodology revealed no major sources of contamination from any of the membranes in the absence of chlorine. In the presence of chlorine, a small increase in the number of organic compounds was observed for both membrane types, but this effect was not considered major.

Concentration and Recovery of Solutes. The RO method was evaluated by using small-scale concentrations and selected model organic solutes. Similar concentrations were performed by other researchers by using alternate sampling methods as part of a comparison study. The concentration provided a 50-fold volume reduction (500 L down to 10 L). Field applications of the RO method usually involve sample volumes of 2000–8000 L. No steps were taken to condition membranes and equipment prior to the laboratory tests. This laboratory performance evaluation was conducted, in many respects, as a worst case exercise.

The experimental plan for the concentration was described earlier in Figure 5. Information regarding membrane solute rejections, solute recoveries, and magnitudes of losses was obtained from data generated during the concentrations and in earlier screening experiments. Sample volumes, solute levels, and data from earlier experiments were used to estimate solute masses in various streams associated with the concentration process. Figure 6 shows a block diagram with four sections assigned to major areas where organic mass could be located at the end of the concentration. The object of the concentration is to maximize organics in the final concentrate (M_f^F). Estimates of solute in various locations were made, and mass balance calculations were completed.

The following equations were used to calculate the various mass fractions indicated in Figure 6. Solute masses in the initial sample (M_i^F), final concentrate (M_f^F), and salt-exchange bath (M_f^S) are straightforward products of measured solute concentrations and solution volumes:

$$M_i^F = (C_i^F)(V_i^F) \quad (3)$$

$$M_f^F = (C_f^F)(V_f^F) \quad (4)$$

$$M_f^S = (C_f^S)(V_f^S) \quad (5)$$

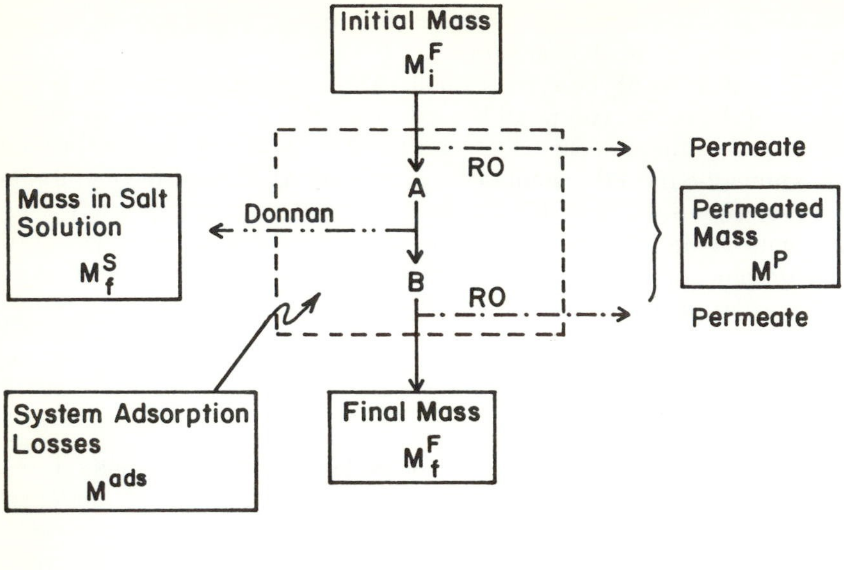

$$M_i^F = M_f^F + M_f^S + M^P + M^{ads}$$

Figure 6. Component streams used in estimating system mass balances.

The mass of solute permeated, M^P, was estimated by using eq 6, first by assuming a maximum concentration (i.e., $R = 1.0$), and then by assuming an expected concentration (i.e., R = experimentally determined value; *see* Tables IV and V).

$$C_f = C_i \ (V_i/V_f)^R \tag{6}$$

The differences in the masses predicted by using these two concentrations provided an estimate of the permeation losses.

The last category created to complete mass balance calculations was designated adsorption losses, M^{ads}, because the type of compounds most affected in this group suggested an adsorption mechanism rather than other potential loss mechanisms (e.g., volatilization or precipitation). Early experiments on temperature effects also suggested adsorption–absorption was involved rather than volatile losses. Adsorption data generated during screening tests were used to estimate material loss via this mechanism. Table VII shows these losses as experienced during screening tests of both membrane types. The lack of specific losses, which could be attributed to the two different types of membranes, suggests that other system components (e.g., gaskets, seals, backings,

Table VII. Percent Solute Adsorption Losses (Gains) Experienced During Screening Test

Chemical Class	*Model Compound*	*Typical Losses (Percent)*[a] CA (589–85) No Humics	CA (618–04) No Humics	CA (618–30) With Humics	FT-30 (618–09) No Humics	FT-30 (589–81) No Humics	FT-30 (618–34) With Humics	*Average Loss*
Acids	trimesic acid	—	+9.5	−4.5	0	—	−11	−1.5
	stearic acid	−36	−85	>−84	−74	−70	−70	−70 to −73
	glycine	—	+19	—	+5.9	—	—	+12
Aldehydes	furfural	—	+203	−98	+213	—	—	[b]
Amines	quinoline	−27	−22	−4.8	−39	−24	−6.3	−21
	caffeine	—	0	−4.5	0	+4.7	−6.3	−1.2
	5-chlorouracil	—	−3.8	−3.7	0	—	−4.8	−3.1
Chlorobiphenyls	2,2′,5,5′-tetrachlorobiphenyl	—	−50	>−50	>−63	—	−29	−48 to −70
Esters	bis(2-ethylhexyl) phthalate	−83	−93	>−85	>−61	−84	−75	−80 to −90
Hydrocarbons	1-chlorododecane	>−75	−31	−64	−29	−84	>−17	−50 to −68
	biphenyl	−73	−65	−74	−96	−83	−71	−77
Ketones	isophorone	−8.7	+29	+5.9	−35	0	−8.9	−2.9
	anthraquinone	−47	−68	−82	−55	—	−73	−65
	methyl isobutyl ketone	—	—	—	—	—	—	—
Phenols	2,4-dichlorophenol	−33	−27	−23	+29	−19	−14	−14.5
	2,6-di-*tert*-butyl-4-methylphenol	−46	—	−83	−73	−75	−71	−70
Trihalomethanes	chloroform	−24	−17	−68	0	−38	−35	−30

NOTE: Some original target compounds (*see* Table I) are not included because of inconsistent or incomplete data.
[a] Positive values suggest gains in solute levels during recirculation period. CA denotes cellulose acetate; — means insufficient data to calculate values.
[b] Data were too scattered to assess an average percent loss.

support, and other wetted surfaces) may play a primary role in these losses. All values are expressed as percentages of the initial solute mass.

The adsorption losses (%) shown in Table VII were used to calculate the amount of solute taken up by a freshly flushed system. Field application of the RO concentration method incorporated conditioning periods in which membranes and other system components were exposed to the sample (and its concentrates) to satisfy and minimize adsorptive solute loss.

Tables VIII, IX, and X present mass balance data for three concentration tests. Two tests (one with each membrane type) were completed without humic acid, and the third was a repeat of the FT-30 composite membrane test including humic acid. Referral to Figure 6 will help clarify the first six columns in Tables VIII–X. Column 1 (M_i^F) and the sum of the components (column 6) were used to calculate mass accountability values (column 7). Although scattered, accountability of mass was good, considering the complexity of the system and the trace levels at which this work was performed.

Final sample mass (M_f^F) and initial concentrate mass, adjusted for expected adsorption losses (M_i^F), were used to calculate total mass recovery (*see* last column in Tables VIII–X). These values reflect efforts to correct recoveries for adsorption losses.

Ultimate recovery of specific organic compounds was calculated from the initial sample mass (M_i^F) and final concentrate masses (M_f^F) listed in columns 1 and 2 (Tables VIII–X). Calculated recovery values are listed in Table XI. Maximum potential recoveries are projected by using eq 2 (and experimentally determined solute rejections).

The sum of all organic compounds added to this system was <0.5 g. To gain some perspective, mass of the RO element alone is estimated to be greater than 700 times (~350 g) the organic level. Measures should always be taken to compensate for the demand for organics that must be expected from system components.

In summarizing Tables VIII–XI, several points should be highlighted: (1) Recoveries were higher, as expected, for the FT-30 composite membrane than for the cellulose acetate membrane. (2) Compounds exhibiting negative rejections with the cellulose acetate membrane were not recovered. (3) Many compounds were not recovered at their predicted levels in tests with either type of membrane.

Rejections of solute compounds measured during the concentrations generally corresponded well with those measured in the screening test. Rejection values determined in the concentration test are presented in Tables IV and V along with values measured in the screening test. Rejection measurements made at 8- and 40-fold volume reductions are reported with the value determined at the lower concentration factor listed first (e.g., 87/83, *see* Tables IV and V).

Table VIII. Mass Balance Data and Estimates of Solute Recovery (Test Run 618–41): Cellulose Acetate Membrane Concentration without Humics

Chemical Class	*Model Compound*	M_i^F *(mg)*	M_f^F *(mg)*	M^{ads} *(mg)*	M^P *(mg)*	M_f^S *(mg)*	*Sum (mg)*[a]	*Mass Accountable (%)*	$\hat{M}_i^F$ *(mg)*	*Total Recovery (%)*
Acids	humic acid				. . . compound not present . . .					
Amines	quinoline	25[b]	0.05	5.3	24.1	0.05	29.5	118	19.7	0.3
	caffeine	27.5	6.1	0.3	22.2	0.3	28.9	105	27.2	22
	5-chlorouracil	28[b]	1.6	0.8	25.2	<0.5	27.6–28.1	99–100	27.2	5.9
Carbohydrates	glucose	20	4.8	1.0[c]	3.0	0.1	8.9	45	19	25
Chlorobiphenyls	2,4′-dichlorobiphenyl				. . . compound not present . . .					
	2,2′,5,5′-tetrachlorobiphenyl	2.5	nd	1.2–1.8[d]	2.5	nd	3.7–4.3	148–172	0.7–1.3	0
Hydrocarbons	1-chlorododecane	5	0.3	2.5–3.4[d]	3.8	<0.05	6.6–7.5	132–150	1.6–2.5	12–19
	biphenyl	21.5	0.09	16.6	20.5	<0.05	37.2	173	4.9	1.8
Ketones	isophorone	26	2.2	0.8	18.3	0.08	21.3	82	25.2	8.7
	anthraquinone	28	0.2	18.2	27.2	<0.05	45.6	163	0.8	25
	methyl isobutyl ketone	22	0.5	4.4[c]	21.6	0.1	26.6	121	17.6	2.8
Phenols	2,4 dichlorophenol	15	0.2	2.3	15	0.09	17.6	117	12.7	1.6
	2,6-di-*tert*-butyl-4-methylphenol	24[b]	0.12	16.8	18.8	<0.05	35.7	149	7.2	1.7
Trihalomethanes	chloroform	24	0.1	7.2	24	<0.1	31.4	131	16.8	0.6

NOTE: Some original target compounds (*see* Table I) are not included because of inconsistent or incomplete data. Corrections for losses to adsorption have been factored into these calculations of overall solute recovery.

[a] Sum = $M_f^F + M^{ads} + M^P + M_f^S$.

[b] These initial masses are estimated from the amount of solute added to the system and are not calculated from the analyzed solute concentration and initial volume.

[c] Because of limited data to calculate adsorptive loss percentages, values have been assumed primarily on the basis of solute water solubility and secondly on the basis of solute volatility (e.g., glucose = −5% and methyl isobutyl ketone = −20%).

[d] Ranges of data. These result from percentage adsorption loss ranges, such as the range −50% to −68% for 1-chlorododecane, as seen in Table VII.

Table IX. Mass Balance Data and Estimates of Solute Recovery (Test Run 618–51): FT-30 Membrane Concentration without Humics

Chemical Class	*Model Compound*	M_i^F *(mg)*	M_f^F *(mg)*	M^{ads} *(mg)*	M^P *(mg)*	M_f^S *(mg)*	*Sum (mg)*[a]	*Mass Accountable (%)*	$\hat{M}_i^F$ *(mg)*	*Total Recovery (%)*
Acids	humic acid				. . . compound not present . . .					
Amines	quinoline	25	4.8	5.3	1.0	1.3	12.4	50	19.7	24
	caffeine	25	13.9	0.3	1.0	1.1	16.3	65	24.7	56
	5-chlorouracil	34[b]	22.5	1.0	13	1.5	38	112	33	68
Carbohydrates	glucose	18	4.4	0.9[c]	1.0	0.2	6.4	36	17.1	26
Hydrocarbons	1-chlorododecane	4.5	1.6	2.3–3.1[d]	0.7	<0.05	4.6–5.4	102–120	1.4–2.2	73–114
	biphenyl	25	1.6	19.3	4.0	0.2	25.1	100	5.7	28
Ketones	isophorone	25	18.6	0.8	6.0	1.3	26.7	107	24.2	77
	anthraquinone	25	2.1	16.3	8	0.2	26.6	111	8.7	24
	methyl isobutyl ketone	20	12.1	4.0[c]	2.9	1.3	20.3	102	16	76
Phenols	2,4-dichlorophenol	20	10.7	2.9	3.0	0.5	17.1	86	17.1	63
	2,6-di-*tert*-butyl-4-methylphenol	24[b]	1.1	16.8	8.4	0.32	26.6	111	7.2	15
Trihalomethanes	chloroform	22.5	4.6	6.8	8.5	0.7	20.6	92	15.7	29

NOTE: Some original target compounds (*see* Table I) are not included because of erratic or incomplete data. Corrections for losses to adsorption have been factored into these calculations of overall solute recovery.

[a] Sum = $M_f^F + M^{ads} + M^P + M_f^S$.

[b] Initial masses are estimated from the amount of solute added to the system and are not calculated from the analyzed solute concentration and initial volume.

[c] Because of limited data to calculate adsorptive loss percentages, values have been assumed primarily on the basis of solute water solubility and secondly on the basis of solute volatility (e.g., glucose = −5% and methyl isobutyl ketone = −20%).

[d] Ranges of data. These result from percentage adsorption loss ranges, such as the range −50% to −68% for 1-chlorododecane, as seen in Table VII.

Table X. Mass Balance Data and Estimates of Solute Recovery (Test Run 618–57): FT–30 Membrane Concentration with Humics

Chemical Class	Model Compound	M_i^F *(mg)*	M_f^F *(mg)*	M^{ads} *(mg)*	M^P *(mg)*	M_f^S *(mg)*	*Sum (mg)*[a]	*Mass Accountable (%)*	$\hat{M}_i^F$ *(mg)*	*Total Recovery (%)*
Acids	stearic acid	28	0.9	19.6	22.6	<0.3	43	155	8.4	11
	humic acid	900	486	45	72	28	631	70	855	57
Amines	quinoline	30	3.6	6.3	3	0.1	13	43	23.7	15
	caffeine	30	16	0.4	2.0	1.1	19.5	65	29.6	54
	5-chlorouracil	27	12.1	0.8	10	0.8	23.7	88	26.2	46
Carbohydrates	glucose	26	4.2	1.3[b]	2.0	0.2	7.7	30	24.7	17
Chlorobiphenyls	2,2′,4,4′-tetrachlorobiphenyl	1.5[c]	0.24	0.7–1.1[d]	1.2	<0.05	2.1–2.5	140–167	0.4–0.8	30–60
Hydrocarbons	1-chlorododecane	3.5	1.7	1.8–2.4[d]	0.8	<0.05	4.3–4.9	123–140	1.1–1.7	100–154
	biphenyl	15	1.2	11.6	2.7	0.1	15.6	104	3.4	35
Ketones	isophorone			. . . compound not added — lab accident . . .						
	anthraquinone	30	1.5	19.5	7.0	0.2	28.2	94	10.5	14
	methyl isobutyl ketone	21	8.6	4.2[b]	2.0	1.3	16.1	77	16.8	51
Phenols	2,4-dichlorophenol	20	10.3	3	3	0.4	16.7	84	17	61
	2,6-di-*tert*-butyl-4-methylphenol	5	0.1	3.5	1.6	<0.05	5.2	104	1.5	6.7
Trihalomethanes	chloroform	22	4.6	6.6	8	0.7	19.9	90	15.4	30

NOTE: Some original target compounds (*see* Table I) are not included because of inconsistent or incomplete data. Corrections for losses to adsorption have been factored into these calculations of overall solute recovery.

[a] Sum = $M_f^F + M^{ads} + M^P + M_f^S$.

[b] Because of limited data to calculate adsorptive loss percentages, values have been assumed primarily on the basis of solute water solubility and secondly on the basis of solute volatility (e.g., glucose = −5% and methyl isobutyl ketone = −20%).

[c] Initial masses are estimated from the amount of solute added to the system and are not calculated from the analyzed solute concentration and initial volume.

[d] Ranges of data. These result from percentage adsorption loss ranges, such as the range −50% to −68% for 1-chlorododecane, as seen in Table VII.

Table XI. Model Solute Mass Recoveries Experienced in a 50× Concentration

Chemical Class	Model Compound	Cellulose Acetate		FT-30		
		Actual	Maximum[a]	Actual w/o Humic	Actual with Humic	Maximum[a]
Acids	stearic acid	[b]	>6	[b]	3.2	86
	humic acid	[c]	[c]	[c]	54	92
Amines	quinoline	0.2	3.7	19	12	89
	caffeine	22	32	56	53	96
	5-chlorouracil	5.7	10	66	45	63
Carbohydrates	glucose	24	86	24	16	96
Esters	bis(2-ethylhexyl) phthalate	[b]	[b]	[b]	[b]	79
Hydrocarbons	1-chlorododecane	6.0	24	36	49	60
	biphenyl	0.4	~0	6.4	8.0	70
Ketones	isophorone	8.5	30	74	LA[d]	86
	anthraquinone	0.7	3	8.4	5.0	76
	methyl isobutyl ketone	2.3	~0	61	41	92
Phenols	2,4-dichlorophenol	1.3	~0	54	52	76
	2,6-di-*tert*-butyl-4-methylphenol	0.5	22	4.6	2.0	86
Trihalomethanes	chloroform	0.4	~0	20	21	67

NOTE: Some original target compounds (*see* Table I) are not included because of inconsistent or incomplete data. All values are percentages.

[a] Theoretical maximum recovery based on calculations using eq 2.

[b] Insufficient data to calculate a value.

[c] Not present in test matrix.

[d] LA indicates lab accident.

As expected, compounds demonstrating consistent negative rejections with cellulose acetate membranes (dichlorophenol, biphenyl, furfural, chloroform) were not recovered to any extent. Compounds with the best rejections (>90%) were the better recovered substances. The FT-30 composite membrane clearly demonstrated superior performance to the cellulose acetate membrane for organic rejection, concentration, and recovery. Sodium chloride rejection was no indicator of potential organic rejection.

The use of this process for the preparation of samples for analysis and bioassay is related to the predictability of the concentration and recovery of specific organics and the representative recovery of the majority of organics in the stream being evaluated. A summary of the degree of concentration and recovery with the FT-30 membrane system and selected low molecular weight organics examined is shown in Table XII. The concentration of all organics was achieved. The recovery of these solutes appeared to be compromised by adsorption losses, low solubility, complex matrices, methods of spiking, and analytical detection limits. Recoveries of individual solutes were reproducible on the test matrices studied.

The results of field trials, based on total organic carbon (TOC) levels, show excellent recovery of the majority of organics present (*21*). Recovery data from a typical field RO concentration are shown in Table XIII. These results reflect high membrane rejections and recoveries found with higher molecular weight organics. Reductions in the total amount of adsorbed material in relation to the total sample

Table XII. Comparison of Volumetric Concentration, Organic Compound Concentration, and Trace Organic Mass Recovery Obtained in a Laboratory RO Concentration with the FT-30 Composite Membrane

Chemical Class	*Model Compound*	*Volumetric Concentration*	*Organic Compound Concentration*	*Organic Mass Recovery (%)*
Acids	stearic acid	50×	1.6×	3.2
	humic acid	50×	27×	54
Amines	quinoline	50×	6×	12
	caffeine	50×	27×	53
	5-chlorouracil	50×	22×	45
Carbohydrates	glucose	50×	8.1×	16
Hydrocarbons	1-chlorododecane	50×	24×	49
	biphenyl	50×	4×	8
Ketones	anthraquinone	50×	2.5×	5
	methyl isobutyl ketone	50×	21×	41
Phenols	2,4-dichlorophenol	50×	26×	52
Trihalomethanes	chloroform	50×	11×	21

NOTE: Some original target compounds (*see* Table I) are not included because of inconsistent or incomplete data.

Table XIII. RO Field Concentration Experience in Jefferson Parish, Louisiana, in October 1983 (9500 L Concentrated to 190 L)

Pilot Stream	*Initial TOC (mg/L)*	*Final TOC (mg/L)*	*Concentration Factor*[a]	*Percent Recovery*
1	3.32	152	48	96
2	3.29	198	59	102
3	3.17	161	53	96
4	3.32	149	46	97
5	2.55	126	50	98

[a] Concentration factor = starting volume/final concentrate volume.

were achieved by better total equilibration and conditioning under field conditions.

Conclusions

The primary use of the RO method lies in preparing concentrates from large volumes of sample in which the compounds of interest are too dilute to be of value. In vivo and in vitro toxicity testing methods are one such application. In many of these studies, the quantity and concentration requirements dictate a very large initial sample volume. Membrane processes such as ultrafiltration and RO also show great potential for the enrichment of valuable biotechnology products; the concentrates might be used as prepared or as pretreated products for further processing. The RO concentration method has successfully been used for the preparation of samples for health effects evaluations. The concentration method advantages of rapid large-volume processing, low-temperature conditions, and lack of phase change cannot be matched by any other process.

Results of this study confirm the expected improved recoveries of trace organics with membranes more selective and more highly crosslinked than the classical cellulose acetate membrane. Improved recoveries were predicted from literature data reported for similar membrane types. In light of these results, cellulose acetate should no longer be considered for applications such as these. Further improvements in recovery can be expected as developmental membranes with more highly selective barriers are brought into commercial use. Each new membrane type considered for use on disinfected waters should be evaluated for sensitivity to common disinfectants (oxidants). Both decreased selectivity and potentially troublesome chemical breakdown products should be considerations under these conditions. Although the cellulose acetate and FT–30 composite membranes did not prove to be particularly sensitive to chlorine, many commercially available membranes are known to be affected. This situation is especially true of the polyaramide and polyamide composite structures.

Recovery levels for individual organic compounds were, as expected, less than the values predicted by using the membrane solute rejections. Differences in the actual recovery and the theoretical recovery were partially due to adsorption losses. Mass balance analysis still indicated a deficiency in some cases. Rectification of the inconsistencies in these data was complicated by the limited water solubility of compounds chosen for study, the necessity of using a cosolvent in spiking, and, in particular, the limitations of the analytical procedures at these extremely low concentrations.

Excellent recoveries, based on TOC analysis, have been achieved in field concentrations with large-volume samples. This result is not totally unexpected because the majority of the organics comprising TOC measurements are higher molecular weight materials, and equilibration of the system with organics in the sample is practiced. Laboratory and field studies show that the majority of all organics in aqueous solutions can be concentrated by using highly cross-linked RO membrane systems. The amount of concentration for classes of low molecular weight organics is less than the volumetric concentration and varies between classes of materials. Correlation of the organic levels in the final concentrate with the concentration in the organic sample appears feasible for a fixed matrix and single solutes. Additional studies are needed to better define the relationship of classes of compounds and their concentration and recovery. RO has been demonstrated as an excellent process for the concentration and recovery of the bulk of organic material present in a water stream.

Symbols

F	feed stream
P	permeate stream
S	salt stream
R	membrane rejection
C^P	permeate stream concentration
C^F	feed stream concentration
C_i^F	initial concentration in feed stream
C_f^F	final concentration in feed stream
C_f^S	final concentration in salt stream
V_i^F	initial feed stream volume
V_f^F	final feed stream volume
V_f^S	final salt bath volume
CF	concentration factor (V_i^F/V_f^F)

M_i^F	initial mass in feed stream
M_f^F	final mass in feed stream
M_f^S	final mass in salt bath
M^P	mass in permeate stream
M^{ads}	mass adsorbed

Acknowledgments

We would like to express our appreciation for research funding by the Health Effects Research Laboratory of the USEPA (Contract No. 68-03-2999 and Contract No. 68-03-3164). We would like to thank Paul Ringhand, Frederick Kopfler, and Robert Miller for their valuable technical assistance and project guidance. The assistance of the following Gulf South Research Institute laboratory scientists is sincerely appreciated: Laurence Rando, William Yauger, Dennis Catalano, Steve Ruiz, and Lynn Kupfer. The significant typing and editing contributions of Brenda Zimny, Shirley Roper, and Kay Magyari are recognized.

Although the research described in this chapter has been funded wholly or in part by the Health Effects Research Laboratory of the USEPA, it has not been subjected to USEPA review and therefore does not necessarily reflect the views of the USEPA, and no official endorsement should be inferred. Mention of trade names or commercial products does not constitute endorsement or recommendation for use.

Literature Cited

1. Smith, J. K.; Cabasso, I.; Klein, E.; Kopfler, F. C. Presented at the 2nd National Conference on Complete Water Reuse, AIChE, Chicago, IL, May 1965.
2. Loeb, S.; Sourirajan, S. In *Sea Water Demineralization by Means of an Osmotic Membrane;* Advances in Chemistry 38; American Chemical Society: Washington, DC, 1963; p 17.
3. Richter, J. W.; Hoehn, H. H. U.S. Patent 3 567 632, 1971.
4. Smith, J. K.; Lynch, S. C. EPA Reports and Samples for EPA Contract Nos. 68-02-2194, 68-03-2367, and 68-03-2464; submitted to Health Effects Research Laboratory; NTIS, 1973–1978.
5. Wrasidlo, W. J. U.S. Patent 4 005 012, 1977.
6. Chian, E. S. K.; Fang, H. H. P. In *Evaluation of New Reverse Osmosis Membranes for the Separation of Toxic Organic Compounds from Wastewater;* Annual Report to the U.S. Army Medical Research and Development Command, Department of the Army: Washington, DC, 1973; Contract No. DADA17-73-C-3025.
7. Smith, J. K.; Lynch, S. C.; Kopfler, F. C. In *Chemistry in Water Reuse;* Cooper, W. J., Ed.; Ann Arbor Science: Ann Arbor, MI, 1981; Vol. 2, Chapter 10, pp 191–203.
8. Lynch, S. C.; Smith, J. K.; Rando, L. C.; Yauger, W. L. *Isolation or Concentration of Organic Substances From Water—An Evaluation of Reverse*

Osmosis Concentration; Report to EPA Health Effects Research Laboratory; NTIS, September 1983; Contract 68-03-2999.

9. Sorg, T. J.; Love, O. T., Jr. *Reverse Osmosis Treatment to Control Inorganic and Volatile Organic Contamination;* Report to EPA DWRD, MERL, OEET, ORD; NTIS, June 25, 1984.
10. Matsuura, T.; Sourirajan, S. *J. Appl. Polym. Sci.* **1973,** *17,* 1043–1071.
11. Matsuura, T.; Sourirajan, S. *J. Appl. Polym. Sci.* **1973,** *17,* 3683–3708.
12. Matsuura, T.; Sourirajan, S. *J. Appl. Polym. Sci.* **1971,** *15,* 2905–2927.
13. Lonsdale, H. K.; Cross, B. P.; Graber, F. M.; Milstead, C. E. *J. Macromol. Sci. Phys.* **1971,** *B5(1),* 167–188.
14. Anderson, J. E.; Hoffman, S. J.; Peters, C. R. *J. Phys. Chem.* **1972,** 76, 4006–4011.
15. Matsuura, T.; Sourirajan, S. *J. Appl. Polym. Sci.* **1972,** *16,* 1663–1688.
16. Wendt, R. P.; Toups, R. J.; Smith, J. K.; Leger, N.; Klein, E., *I&EC Fundam.* **1971,** *10,* 406–411.
17. Sourirajan, S. In *Reverse Osmosis;* Academic: New York, 1970; p 28.
18. Deinzer, M.; Melton, R.; Mitchell, D. *Water Research* **1975,** *9,* 799–805.
19. Kamizawa, C.; Masuda, H.; Ishizaka, S. *Bull. Chem. Soc. Jpn.* **1972,** *45, 2967–2969.*
20. Cadotte, J. E.; Kopp, C. V.; Cobain, K. E.; Rozelle, L. T. *Progress Report on In Situ-Formed Condensation Polymers for Reverse Osmosis Membranes;* North Star Research Institute, Minneapolis, MI. Office of Saline Water Report. U.S. Department of the Interior; June 1974, p 92.
21. Lynch, S. C.; Smith, J. K.; McTopy, J. *Collect and Concentrate Drinking and/or Waste Waters by Reverse Osmosis—Jefferson Parish Disinfection Pilot Plant: October 1983;* Report to EPA Health Effects Research Laboratory; January 1984; Contract No. 68-03-3164.

RECEIVED for review August 14, 1985. ACCEPTED March 26, 1986.

Evaluation of an Integrated Adsorption Method for the Isolation and Concentration of Trace Organic Substances from Water

M. F. Giabbai[1,2], E. S. K. Chian[1], J. H. Reuter[3], H. P. Ringhand[4], and F. C. Kopfler[4]

[1]School of Civil Engineering and [3]School of Geophysical Sciences, Georgia Institute of Technology, Atlanta, GA 30332
[4]Health Effects Research Laboratory, U.S. Environmental Protection Agency, Cincinnati, OH 45268

A scheme for the isolation and concentration of dissolved trace organic substances from water for toxicological and chemical characterization was evaluated. The principle behind this scheme consists of the separation of organic solutes into fractions by adsorption onto different adsorbents (i.e., XAD-8 resin, AG MP-50 cation-exchange resin, and Carbopack B graphitized carbon black) under varying pH conditions. Test solutions containing 22 model organic substances along with inorganic salts were used to monitor process performance. High-resolution gas chromatography and high-performance liquid chromatography were employed for the quantitation of each model compound. The isolation-fractionation scheme proved to be effective for 16 out of 22 model compounds; average recoveries varied between 30% and 90%.

THE HIGH COMPLEXITY AND DILUTED FORM in which organic compounds occur in natural and drinking waters require that isolation, concentration, and fractionation procedures be employed to achieve a suitable sample for chemical and toxicological characterization. The use of these methods in analytical schemes has thus far allowed the identification of several hundred trace organic substances in drinking water

[2]Current address: EnvironScience Laboratories, Inc., Atlanta International Industrial Park, Atlanta, GA 30316

0065-2393/87/0214/0467$06.00/0

(*1*). Compounds thus identified may then be subjected to various in vivo and in vitro test systems to assess biological activity. Because of the large number of compounds and the possibility of additive or synergistic effects, it is impossible to test all combinations. Alternatively, a primary biological screen may be used to identify those natural and drinking water concentrates or fractions that contain biologically active components. Chemical identification of the bioactive substances is then attempted (*2*). Currently, the alternative approach is preferred for the risk assessment of the nonvolatile fraction of organic substances present in water because of analytical limitations.

In either approach, the selection of isolation (e.g., solvent extraction, adsorption on carbon and synthetic resins) and concentration (e.g., lyophilization, vacuum distillation, reverse osmosis, ultrafiltration) methods is of paramount importance in properly assessing the potential toxicity of waterborne organics. A comprehensive literature review on the development and application of these and other methods to biological testing has recently been published by Jolley (*3*).

Several attempts to improve the percent recovery of organic constituents from water have been pursued by sequentially arranging different methods in a multistep scheme. Baird et al. (*4*) experienced 80–90% removal of organic carbon by using a series of ion-exchange and macroreticular resins to prepare organic residues from large volumes of water for chemical and toxicologic testing. Amberlite XAD–4/8 columns in series with activated carbon columns were used by Van Rossum and Webb (*5*) to process 1000 L of tap water. Organic pollutants were subsequently identified by gas chromatographic–mass spectrometric (GC–MS) analysis of the solvent-eluted fractions. Recently, Leenheer (*6*) proposed a comprehensive analytical scheme whereby the dissolved organic carbon (DOC) in natural waters may be separated into operationally defined fractions on the basis of their adsorption onto different substrates (macroreticular adsorbents and ion-exchange resins) under varying pH conditions. Recovery of input and size of the individual fractions has been evaluated in terms of total organic carbon (TOC) analysis.

If specific classes of compounds or specific components are to be biologically tested, suitable concentration methods can be designed. On the other hand, to estimate the overall hazard associated with the organic constituents of a water sample, a viable alternative consists in the use of different methods in a sequential scheme. In both cases, several critical areas of concern must be considered before applying such methods to real water samples: (1) The aqueous organic concentrate prepared by the selected concentration scheme has to be representative of the original water sample with regard to the relative abundance of the individual components. (2) The transformation of organic constituents between preparation of concentrates and biological testing and/or chemical analysis must be avoided. (3) The effect of humic material, which constitutes the bulk of the organic fraction of natural and drinking waters, on the recovery of trace solutes has to be taken into account. (4) The introduction of artifacts and constituents' alteration by the concentration methods must be kept to a minimum. (5) The co-recovery of toxic inorganic constituents must be evaluated. (6) The potential effect of chlorine residual on the material used in the concentration scheme (e.g., resin, membranes) must be assessed. Moreover, the development and application of several different concentration schemes require that a strict comparison be based on the recovery of selected model organic substances representative of a wide range of chemical classes, functional group contents, and molecular weights.

These considerations, as well as the necessity for a comprehensive approach toward the isolation, concentration, and fractionation of trace components of DOC in water, have led to the evaluation of a fractionation scheme wherein selected organic compounds having different functionalities and sorption parameters were separated and concentrated. Test solutions containing 22 model compounds at parts-per-billion (micrograms-per-liter) concentration levels were chosen as the basis for process evaluation. The criteria used in selecting the model compounds were to provide (1) a variety of functional groups; (2) a range of physical properties such as volatility, solubility, polarity, and molecular weight; (3) known water pollutants; and (4) halogenated derivatives. The majority of the compounds were taken from a list of consensus voluntary reference compounds (*7*). The inclusion of humic acid and inorganic salts in the test solutions was an attempt to simulate drinking water. The proposed fractionation scheme was initially evaluated on a laboratory scale and was subsequently adapted for processing several hundred liters of aqueous test solutions.

Experimental

Resin and Carbon Adsorbents. Amberlite XAD-8 was obtained from Rohm and Haas as an industrial-grade preparation in 20–50-mesh size beads. The cation-exchange AG MP-50, 20–50 mesh size, was supplied by BioRad Laboratories. Leenheer's (*6*) procedures for cleanup, preparation, and storage of the resins were followed. In addition, XAD-8 was Soxhlet extracted with methylene chloride immediately after the acetone and hexane extractions. Glass columns (200 × 13 mm i.d.) with Teflon stopcocks were packed with ~15-mL bed volumes of resin. The graphitized carbon black (GCB) Carbopack B, 100–120 mesh particle size, was purchased from Supelco. Acetone, methylene chloride, and organic-free water (OFW) were used to wash the carbon prior to column packing. Because of the fragile nature of this material, care was taken to avoid excessive mechanical stress during its handling. Two hundred milligrams of GCB was packed in a glass column (200 × 5 mm i.d.), as recommended by Bacaloni et al. (*8*).

Reagents. The organic model compounds were purchased from Aldrich Chemical Company, Alfa Products, Fluka Chemical Company, and Analabs;

purities varied from 96% to 99%, as specified in the manufacturers' literature. Inorganic salts, hydrogen peroxide (50% solution), mineral acids, and bases were obtained from Fisher Scientific Company. Organic solvents for purifying and extracting and for analytical operations were distilled-in-glass grade (Burdick–Jackson). Heptafluorobutyric anhydride and trifluoroacetic anhydride were obtained from PCR Research Chemicals; acetyl chloride was supplied by Mallinckrodt; Diazald was supplied by Aldrich; and *N*,*O*-bis(trimethylsilyl)trifluoroacetamide was supplied by Pierce Chemical Company. The humic acid used in these experiments was provided by the Health Effects Research Laboratory of the U.S. Environmental Protection Agency, and it had been prepared from a commercial-grade humic acid (Fluka). OFW was prepared by passing tap water through a series of treatments in the following sequence: a Millipore no. 360 activated carbon cartridge (Continental Water Systems Company), a Millipore no. 300 deionizer cartridge, and a glass column (60 cm $\times$ 2.5 cm i.d.) packed with 50 g of 16–30-mesh size filtrasorb F-400 virgin activated carbon (Calgon Company). Hydrogen peroxide was added to the stream, which then passed through UV light sources (Modified Model H-50, Ultraviolet Technology). This process resulted in water having an average TOC concentration of 27 $\pm$ 15 ppb and a hydrogen peroxide residue of $<$100 ppb, produced at a rate of 72 L/day.

Instrumentation. TOC was analyzed by a Dohrmann DC-54 ultra-low-level carbon analyzer (Envirotech Company). A Hewlett Packard 5830-A gas chromatograph (GC) equipped with a capillary injection system and a flame ionization detector (FID) was employed for the quantitative analysis of each model compound. Separation of the organics was accomplished on a glass capillary column (0.3 mm i.d. $\times$ 30 m) deactivated according to Grob (*9*) and coated by the static method (*10*) with SE-54 silicone gum phase, 0.2 μm film thickness. 5-Chlorouracil was analyzed on a Perkin–Elmer series 3 high-pressure liquid chromatograph (HPLC) equipped with a Rheodyne injection system, a LC 65-T variable-wavelength UV-detector, and a Lichrosorb C_{18} reverse-phase column. The operating conditions of HPLC were as follows: H_2O:CH_3OH (90:10) solvent system under isocratic conditions, a flow rate of 4 mL/min, and photometrical monitoring at 254 nm. Analytical confirmation of the recovery of the model compounds and tentative identification of organic impurities introduced during the handling of the test solution were accomplished by means of a Finnigan 4023 mass spectrometer (MS) interfaced with a Hewlett Packard 5830-A GC. The MS conditions were as follows: electron impact ionization mode; electron multiplier, 1500 V; electron energy, 70 eV; emission current, 0.5 mA; mass range, 45–450 amu; and scan rate, 0.95 s/decade. The chromatographic conditions were identical to those employed in the GC–FID analysis. Humic acids were quantified at 450 nm by means of a Beckman spectrophotometer. Details of the analytical procedures for the determination of the model compounds are published elsewhere (*11*).

Preparation of Model Compound Test Solutions. Individual stock solutions containing 500 mg/L were prepared by dissolving quinaldic acid, glycine, and glucose in OFW; 5-chlorouracil in 2 N NH_4OH; phenanthrene, 1-chlorododecane, 2,4′-dichlorobiphenyl, and 2,2′,5,5′-tetrachlorobiphenyl in hexane; and the remaining compounds in methanol. Humic acid stock solution was prepared in 0.02 N NaOH. The composition of the test solutions is reported in Table I. Test solutions (500 mL) were prepared by adding salts and the required volumes of stock solutions in OFW. Phenanthrene, 1-chlorododecane,

Table I. Model Compounds, Test Solution Composition, and Water Solubility

Compound	*Concentration*[a]	*Water Solubility*[b]
Trimesic acid	50	—
Stearic acid	50	i
Quinaldic acid	50	—
Humic acid	2000	—
Glycine	50	s
Furfural	50	s
Quinoline	50	i
Caffeine	50	δ
5-Chlorouracil	50	—
Glucose	50	s
2,4′-Dichlorobiphenyl	50	i
2,2′,5,5′-Tetrachlorobiphenyl	5	i
Bis(2-ethylhexyl) phthalate	50	—
1-Chlorododecane	5	i
Biphenyl	50	i
Phenanthrene	1	i
Isophorone	50	δ
Anthraquinone	50	i
Methyl isobutyl ketone	50	—
2,4-Dichlorophenol	50	δ
2,6-Di-*tert*-butyl-4-methylphenol	50	—
Chloroform	50	i
$NaHCO_3$	70	—
$CaSO_4$	210	—
$CaCl_2{\cdot}2H_2O$	47	—

[a] All concentrations are given in micrograms per liter except those for $NaHCO_3$, $CaSO_4$, and $CaCl_2{\cdot}2H_2O$, which are given in milligrams per liter.
[b] Data are taken from references 24 and 25; i = insoluble, s = soluble, δ = slightly soluble, and — = no data.

2,4′-dichlorobiphenyl, and 2,2′,5,5′-tetrachlorobiphenyl were spiked by sequentially exposing these compounds to solvents of increasing polarity (hexane × acetone × OFW). Blowing dry with a gentle N_2 stream and sonication were used to remove the solvent and to aid solubilization, respectively. The humic acid stock solution was added as the last component.

Isolation–Fractionation Scheme. Figure 1 illustrates the isolation–fractionation scheme devised and evaluated in this study. Step 1: The test solution was first acidified to pH 2 and passed through the XAD-8 column by gravity flow at a rate of 15 bed volumes/h. The last portion of the test solution remaining in the column was displaced from the resin by 1 bed volume of 0.01 N HCl rinse, which was combined with the original test solution. Step 2: The hydrophobic acid fraction was desorbed with 0.25 bed volumes of 0.1 N NaOH followed by 1.5 bed volumes of OFW. Step 3: The test solution effluent from the XAD-8 (pH 2) was adjusted to pH 10 with 1 N NaOH and recycled through the XAD-8 column at a flow rate of 15 bed volumes/h. Following the sample, 2.5 bed volumes of OFW were used to rinse the XAD-8 column. The rinse was com-

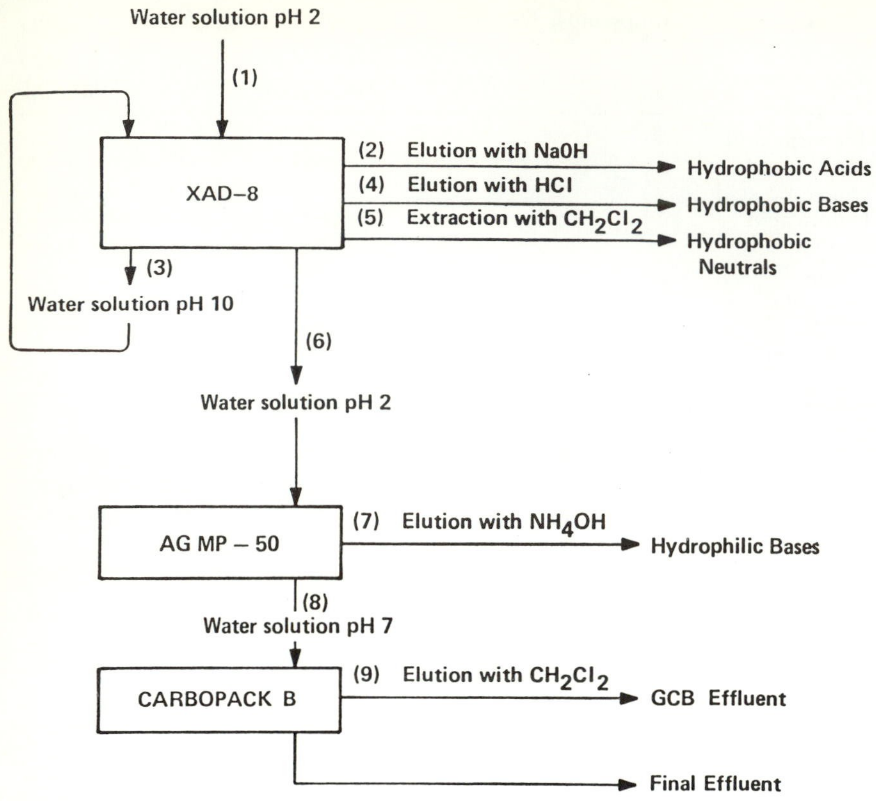

Figure 1. Flow scheme of isolation–fractionation. (Reproduced with permission from reference 11. Copyright 1983 Elsevier.)

bined with the test solution effluent. Step 4: The hydrophobic base fraction was eluted with 0.25 bed volumes of 0.01 N HCl followed by 1.5 bed volumes of 0.1 N HCl. Step 5: Finally, the XAD–8 resin was transferred from the column to a separatory funnel and extracted with three 50-mL aliquots of methylene chloride to desorb the hydrophobic neutral fraction. Step 6: The column effluent (pH 10), with its remaining dissolved hydrophilic substances, was readjusted to pH 2 with concentrated HCl and then passed through the AG MP–50 cation-exchange column at a flow rate of ~15 bed volumes/h. Step 7: The hydrophilic base fraction was desorbed by elution with approximately 0.8 bed volumes of 1 N NH_4OH. Step 8: Finally, the test solution effluent was adjusted to pH 7 and processed through the Carbopack B column at a flow rate that allowed a contact time of approximately 0.5 min. Step 9: The GCB was extracted with methylene chloride.

Analytical Procedures. HYDROPHOBIC NEUTRAL FRACTION. The hydrophobic neutral fraction, which was desorbed in methylene chloride, was concentrated to an appropriate volume (1 mL) in a Kuderna–Danish apparatus. Then, under a stream of N_2 and after addition of the internal standard (i.e., hexamethylbenzene), this fraction was analyzed by GC–FID and GC–MS.

Hydrophobic Acid Fraction. Known amounts of surrogate compounds (i.e., undecanoic acid and 3-quinolinecarboxylic acid) were added to 1 or 2 mL of the hydrophobic acid fraction, and the water was removed under a stream of N_2 at room temperature. The residue was acidified with approximately 0.3 mL of 6 N HCl and brought to dryness under a stream of pure N_2. Finally, it was redissolved in approximately 1 mL of ethyl ether by carefully stirring with a glass rod to help dissolve any acids. The solution was subsequently methylated with gaseous diazomethane (*12*). Diazomethane was generated by adding 15 drops of aqueous NaOH solution (35%) to a solution of Diazald in methanol (~1 mg in 10 mL). Diazomethane gas was bubbled under N_2 pressure (flow rate of ~40–60 mL/min) into the ethereal solution containing the acids for approximately 10–20 s. After addition of hexamethylbenzene, the solution was analyzed by GC. For every batch of hydrophobic acid samples, a standard solution consisting of trimesic, stearic, quinaldic, and surrogate acids was prepared in OFW (concentration of 50 μg/mL), dried, and methylated according to the method just described. This standard solution served as the basis for the quantitative evaluation of the samples. Humic acid was measured in this fraction by spectrophotometry at 430 nm. Standards (10–40 mg/L) and samples were analyzed at identical pH values.

Hydrophobic Base Fraction. The hydrophobic base fraction was adjusted to pH 10. A 50-μL aliquot of the aqueous solution was subjected to HPLC analysis to test for the presence of 5-chlorouracil. The remaining aqueous solution was solvent extracted with methylene chloride. The extract was first concentrated in a Kuderna–Danish apparatus, then under a stream of N_2, and analyzed by GC–FID and GC–MS.

Hydrophilic Base Fraction. An aliquot (1–2 mL) of the hydrophilic base fraction was dried under a stream of N_2, acidified with HCl, and analyzed for glycine after derivatization with isoamyl alcohol, acetyl chloride, and heptafluorobutyric anhydride according to the procedure described by Burleson et al. (*13*). An aliquot (1–2 mL) of the same hydrophilic base fraction was analyzed for quinaldic acid following the procedure mentioned for the hydrophobic acid fraction. The remaining portion of the hydrophilic base fraction was extracted at pH 10 with 50 mL of methylene chloride. The extract was concentrated to 1 mL and analyzed by GC–FID and GC–MS.

Carbopack B Fraction. The Carbopack B column was eluted with 100 mL of methylene chloride. The effluent was concentrated to 1 mL and directly analyzed by GC–FID and GC–MS.

Final Effluent. The final effluent (*see* Figure 1) was solvent extracted with methylene chloride and analyzed by GC.

Details of the analytical procedures for the determination of the model organic compounds are published elsewhere (*11*).

Results and Discussion

Preliminary experiments resulted in the formation of a precipitate due to the presence of inorganic salts when the pH of the test solution was raised to 10 for the first passage through the XAD-8 column to isolate the hydrophobic base fraction. Initially, desalting the solution with a cation-exchange resin (i.e., AG-50-X8, Na^+ form) was tried before processing through the fractionation scheme. However, several model compounds were lost presumably by adsorption on the nonionic lattice of the resin. Precipitate formation was avoided when the sequence of adsorption onto the XAD-8 column was reversed: the test solution was first adjusted to pH 2 for the adsorption of the hydrophobic acids and neutrals and then adjusted to pH 10 for the isolation of the hydrophobic bases. Therefore, this approach was ultimately adopted in this study. The final effluent of the test solution was readjusted to pH 2 before passing through the cation-exchange resin Ag MP-50 to isolate the hydrophilic bases. Six experiments were conducted under these conditions. The results are expressed as average percent recoveries in Table II.

Malcolm et al. (*14*) and Thurman et al. (*15*) noticed that the adsorption of solutes onto XAD-8 macroreticular resin could be predicted by means of a linear correlation between the logarithm of the capacity factor and the inverse of the logarithm of the water solubility of each compound. Their investigation, however, was limited to approximately 20 selected organic compounds in individual aqueous solutions. By comparing the results shown in Table II and the water solubility properties of each model compound used in this study (*see* Table I), it appears that the predictive model could serve for a first estimate of the recovery of multisolute solutions at trace levels. However, low recoveries and the erratic behavior of several compounds included in this study suggest that additional factors need to be considered.

It appears that 2,4-dichlorophenol, which was expected primarily in the hydrophobic acid fraction, does not follow the predictive model (*see* Table II). Solute-solute interactions may be responsible for this unexpected behavior, and the fact that 2,4-dichlorophenol was partially recovered in the hydrophilic base fraction suggests an adsorptive affinity to the styrene-divinylbenzene lattice of the cation-exchange resin. The relatively poor recovery of 1-chlorododecane and 2,2′,5,5′-tetrachlorobiphenyl in the hydrophobic neutral fraction (*see* Table II) may be attributed to difficulties encountered in solubilizing them in water and to subsequent losses by adsorption onto glass walls and Teflon tubing, although precautions against solubilization problems had been taken during the preparation of the test solution (*see* Experimental section). Because of the low concentrations, no attempts were made to verify adsorption losses. Bis(2-ethylhexyl) phthalate was found in several fractions (*see* Table III), a fact that may indicate nonspecific adsorption onto both macroreticular and ion-exchange resins. The small concentrations of methyl isobutyl ketone (MIBK), 5-chlorouracil, and quinaldic acid recovered in their respective fractions made the quantitative analysis of these compounds unreliable. MIBK was detected

Table II. Average Percent Recovery of Model Compounds from Resin Scheme

Compound	*Mean Recovery ± Standard Deviation*				
	OA	*OB*	*ON*	*IB*	*EF*
Stearic acid	32.4[a]				
Trimesic acid	41.8[b]				
2,4-Dichlorophenol				13.8 ± 11.1	23.6 ± 8.9
Quinaldic acid				NQ	
Isophorone			80.8 ± 18.5		
Biphenyl			82.7 ± 5.8		
1-Chlorododecane			33.8 ± 6.8		
2,6-Di-*tert*-butyl-4-methylphenol			50.2 ± 8.6		
2,4′-Dichlorobiphenyl			74.2 ± 5.3		
2,2′,5,5-Tetrachlorobiphenyl			44.4 ± 22.1		
Anthraquinone			58.0 ± 13.3		
Phenanthrene			77.8 ± 13.3		
Bis(2-ethylhexyl) phthalate	1.8 ± 1.5[c]		37.6 ± 7.9	2.3[b]	9.2 ± 4.2
Furfural					38.3
Quinoline		22.1 ± 10.6			
5-Chlorouracil				NQ	
Caffeine			16.4 ± 5.4	3.7[b]	25.2 ± 6.2
Glycine				55.5 ± 19.6[c]	
Humic acids	88.1 ± 6.5				
Chloroform					
Methyl isobutyl ketone		NQ			

NOTE: OA = hydrophobic acid (XAD-8); OB = hydrophobic base (XAD-8); ON = hydrophobic neutral (XAD-8); IB = hydrophilic base (AG MP-50); EF = final effluent (solvent extraction); NQ = found but not quantitated.
[a] Three values.
[b] Two values.
[c] Four values.
SOURCE: Reproduced with permission from reference 11. Copyright 1983 Elsevier.

Table III. Percent Recovery of Model Compounds on Carbopack B without Inorganic Salts and at pH 7

Compound	*Percent Recovery*	
	Desorbed from GCB	*Extracted from Water after GCB*
2,4-Dichlorophenol	115.2	NF
Quinoline	97.5	NF
Isophorone	16.3	92.4
1-Chlorododecane	51.2	NF
2,4′-Dichlorobiphenyl	48.6	0.9
2,2′,5,5′-Tetrachlorobiphenyl	54.1	3.7
Anthraquinone	92.1	NF
Bis(2-ethylhexyl) phthalate	51.1	64.3
Phenanthrene	114.0	NF
Caffeine	92.1	NF
Furfural	NF	26.0
Methyl isobutyl ketone	6.7	65.5

NOTE: NF indicates not found.
SOURCE: Reproduced with permission from reference 11. Copyright 1983 Elsevier.

primarily in the hydrophobic neutral fraction, whereas 5-chlorouracil and quinaldic acid were found at very low concentrations in the hydrophilic base fraction (*see* Table II). The behavior of quinaldic acid confirms the findings of Leenheer and Huffman (*16*), who used test solutions spiked at the milligrams-per-liter level through a similar fractionation scheme. Quinaldic acid was recovered in the hydrophilic base fraction after the test solution had gone through the XAD-8 column under acidic and alkaline conditions. This result supported the suggestion of the amphoteric behavior of this compound. Chloroform could not be detected in the hydrophobic neutral fraction probably because it was lost by volatilization.

That several model organic compounds were only partially or incompletely retained by the resins prompted us to investigate the use of Carbopack B as an alternative or complementary adsorbent. Test solutions without humic acids were used to verify the sorptive–desorptive behavior of several model compounds under the experimental conditions proposed by Bacaloni et al. (*8*), except that the compounds were desorbed with methylene chloride. The results of duplicate experiments are given in Table III. Isophorone and MIBK were not effectively retained by Carbopack B, whereas bis(2-ethylhexyl) phthalate was almost equally distributed between the aqueous phase and the carbon. The relatively poor recovery of 1-chlorododecane, 2,4′-dichlorobiphenyl, and 2,2′,5,5′-tetrachlorobiphenyl may be ascribed to sorptive losses onto reservoir glass wall, whereas furfural may be inefficiently desorbed. Phenanthrene, quinoline, anthraquinone, and, in particular, caffeine and 2,4-dichlorophenol were quantitatively recovered.

In consideration of the results presented in Tables II and III, it was decided to integrate the resins and carbon columns in a single scheme as shown in Figure 1. The test solution was adjusted to pH 7 immediately following the AG MP-50 column and processed through the Carbopack B column. The results from one experiment conducted under these conditions practically confirmed the overall process performance anticipated from the individual experiments (*see* Table IV). Trimesic acid, stearic acid, and humic acids were found in the hydrophobic acid fraction; quinoline was primarily quantitated in the hydrophobic base fraction, whereas a small amount of it was also detected in the hydrophobic neutral fraction. Isophorone, biphenyl, 1-chlorododecane, 2,6-di-*tert*-butyl-4-methylphenol, 2,4′-dichlorobiphenyl, 2,2′5,5′-tetrachlorobiphenyl, anthraquinone, and phenanthrene were re-

Table IV. Average Percent Recovery of Model Compounds from Integrated Adsorption Scheme

	Percent Recovery				
Compound	*OA*	*OB*	*ON*	*IB*	*GCB*
Stearic acid	31.3				
Trimesic acid	39.7				
2,4-Dichlorophenol				11.6	23.1
Quinaldic acid				NQ	
Isophorone			75.6		
Biphenyl			66.8		
1-Chlorododecane			40.1		
2,6-Di-*tert*-butyl-4-methylphenol			49.3		
2,4′-Dichlorobiphenyl			70.1		
2,2′,5,5′-Tetrachlorobiphenyl			55.6		
Anthraquinone			62.3		
Phenanthrene			60.1		
Bis(2-ethylhexyl) phthalate			39.1		12.3
Glucose (NA)					
Furfural			NF		NF
Quinoline		31.4	3.5		
5-Chlorouracil				NQ	
Caffeine			11.3	2.3	31.3
Glycine				44.5	
Humic acids	83.1				
Chloroform			NF		
Methyl isobutyl ketone			NF		

NOTE: OA = hydrophobic acid (XAD-8); OB = hydrophobic base (XAD-8); ON = hydrophobic neutral (XAD-8); IB = hydrophilic base; GCB = Carbopack B; NQ = found but not quantitated; NA = not analyzed; NF = not found.

covered exclusively in the hydrophobic neutral fraction. Bis-(2-ethylhexyl) phthalate was primarily quantitated in the hydrophobic neutral and Carbopack B fractions and partially found in the final effluent. As expected, caffeine was primarily recovered in the Carbopack B fraction, although a small amount was also detected in the hydrophobic neutral and hydrophilic base fractions. Glycine, quinaldic acid, and 5-chlorouracil were detected in the hydrophilic base fraction. Furfural was quantitated only in the final effluent fraction. This result confirmed that neither the resins (i.e., XAD-8, AG MP-50) nor the carbon (i.e., Carbopack B) was able to effectively isolate it from the water solutions. The results of integrated experiments and those obtained from the separate resin and carbon experiments (*see* Tables II and III) allow us to conclude that the proposed isolation-fractionation scheme (*see* Figure 1) is effective for recovering 16 out of 22 model compounds under this study at recoveries ranging from 30% to 90%.

One of the major concerns over the use of synthetic resins for the isolation of trace organic compounds is the potential contamination of the isolated samples, which is a major limitation particularly when attempting to collect organic concentrates for biological testing. During these experiments, the GC-FID trace of the fractions generated from the resin scheme revealed the presence of organics other than those of the selected model compounds. The hydrophobic neutral fraction was the relatively more contaminated. The bulk of the impurities appeared to be in small quantities, except for two or three major ones, whose amount was comparable to that of the recovered model compounds. Attempts to confirm the origin of the contaminants were pursued by GC-MS analysis of each isolated fraction and of the methylene chloride extract of a similar aliquot of the test solution. A list of the tentatively identified contaminants is shown in the box. The detection of several compounds of high volatility in both resin fraction and solvent extract can be ascribed to contributions from the lab environment (e.g., refrigerator for stock solution storage) and/or to the humic acid solids used to prepare the test solution. Chlorocyclohexene is an impurity commonly found in the best grade methylene chloride commercially available, whereas phthalates are widespread contaminants because of their large use as plasticizers.

Conclusions

An isolation-fractionation scheme for the separation of trace organic solutes from natural and drinking waters has been developed. This process involves the separation of a number of organic solutes into several fractions on the basis of their sorptive characteristics onto different adsorbents under varying pH conditions. The specific adsor-

Tentatively Identified Artifact Contaminants from Resin Fractionation Scheme

ON

Chlorocyclohexene
Phenol
5-Amino-2,4-(1*H*,3*H*)-pyrimidinedione
N,4-Dimethylbenzenesulfonamide
Phthalate
Phthalate
Bromoform
Xylene
Ethylbenzene
Chlorobenzene
4-Methyl-3-penten-2-one
Dibromochloromethane

EF

1-(4-Hydroxyphenyl)ethanone
Dichlorocyclohexane
Chlorocyclohexanol
Phenol
Chlorocyclohexene
Tetrachloroethane
Bromoform
Ethylbenzene
Chlorobenzene
3-Methylenepentanone
Dibromochloromethane

Solvent-Extracted Test Solutions, pH 2 and 10

Chlorobenzene
1-Cyclohexene
2-Cyclohexen-1-one
Trichloropropene
Chlorocyclohexanol
Phthalate
Phenol
Toluene
Trichloroethane
Ethylbenzene
Xylene
Bromoform

NOTE: ON = hydrophobic neutral and EF = final effluent.

bents evaluated include Amberlite XAD-8 and AG MP-50 resins and Carbopack B GCB. Of the 22 model compounds used in this study, 16 have been separated at recoveries varying from 30% to 90%. The recovery of the model compounds on XAD-8 resin appears to be controlled by their water solubility properties, except for the high-volatile compounds (i.e., chloroform), which may be lost by volatilization during sample handling. Highly polar solutes, present as cations in acidic water solutions, can be effectively recovered on AG MP-50. Meanwhile, nonionic solutes, which have water solubilities not suitable for adsorption on XAD-8 and a strong affinity for the graphite structure of Carbopack B, are effectively recovered on Carbopack B. Still, the poor recovery of several compounds cannot be fully explained, and further experiments are required to elucidate the fate of these compounds on the scheme.

When the samples and resins were handled properly, contaminants introduced throughout the isolation–fractionation scheme were found to be minimal. Therefore, it is felt that the proposed process can be properly scaled up to handle large quantities of water for the preparation of concentrates for biological and chemical characterization.

However, because several classes of organic compounds cannot be recovered effectively, the investigation of other supplemental isolation–concentration methods is warranted. For example, the highly volatile purgeable organic compounds (i.e., chloroform, MIBK) may first be analytically identified and quantitated and then spiked at a level that would be expected in the concentrate for the toxicologic study. Other methods, such as reverse osmosis or freeze-drying processes, can be used as an integral part of the proposed isolation–fractionation scheme to concentrate the highly polar, water-soluble compounds (e.g., glucose, furfural).

Acknowledgments

This research was supported by the Health Effects Research Laboratory of the U.S. Environmental Protection Agency under Contract No. 68-03-3000. The excellent technical assistance of Z. Geskin, P. May, B. Ghosh, and J. S. Kim is gratefully appreciated.

Literature Cited

1. Lin, D. C. K.; Melton, R. G.; Kopfler, F. C.; Lucas, S. V. In *Advances in the Identification and Analysis of Organic Pollutants in Water;* Keith, L. H., Ed.; Ann Arbor Science: Ann Arbor, MI, 1981; Vol. 2, pp 861-906.
2. Tabor, M. W.; Loper, J. C.; Barone, K. In *Water Chlorination: Environmental Impact and Health Effects;* Jolley, R. L.; Brungs, W. A.; Cumming,

R. B.; Jacobs, V. A., Eds.; Ann Arbor Science: Ann Arbor, MI, 1980; Vol. 3, pp 899–913.

3. Jolley, R. L. *Environ. Sci. Technol.* **1981,** *15(8),* 874–880.
4. Baird, R.; Cute, J.; Jacks, C.; Jenkins, R.; Neisess, L.; Scheybeler, B.; Van Sluis, R.; Yanko, W. In *Water Chlorination: Environmental Impact and Health Effects;* Jolley, R. L.; Brungs, W. A.; Cumming, R. B.; Jacobs, V. A., Eds.; Ann Arbor Science: Ann Arbor, MI, 1980; Vol. 3, pp 925–935.
5. Van Rossum, P.; Webb, R. G. *J. Chromatogr.* **1978,** *150,* 381–392.
6. Leenheer, J. A. *Environ. Sci. Technol.* **1981,** *15(5),* 1578–1587.
7. Keith, L. H. In *Advances in the Identification and Analysis of Organic Pollutants in Water;* Keith, L. H., Ed.; Ann Arbor Science: Ann Arbor, MI, 1981; Vol. 2, pp 1165–1170.
8. Bacaloni, A.; Goretti, G.; Lagana, A.; Petronio, B. M.; Rotatori, M. *Anal. Chem.* **1980,** *52,* 2033–2036.
9. Grob, K. *J. High Resolut. Chromatogr. Chromatogr. Commun.* **1980,** *3,* 493.
10. Giabbai, M.; Shoults, M.; Bertsch, W. *J. High Resolut. Chromatogr. Chromatogr. Commun.* **1978,** *1,* 277.
11. Giabbai, M.; Roland, L.; Ghosal, M.; Reuter, J. H.; Chian, E. S. K. *J. Chromatogr.* **1983,** *279,* 373–382.
12. Schlenk, H.; Gilerman, J. L. *Anal. Chem.* **1960,** *32,* 1412.
13. Burleson, J. L.; Peyton, G. R.; Glaze, W. H. *Environ. Sci. Technol.* **1980,** *14(11),* 1354–1359.
14. Malcolm, R. L.; Thurman, E. M.; Aiken, G. R. In *Trace Substances in Environmental Health;* Hemphill, D. D., Ed.; University of Missouri: Columbia, MO, 1977; Vol. 11, pp 307–314.
15. Thurman, E. M.; Malcolm, R. L.; Aiken, G. R. *Anal. Chem.* **1978,** *50(6),* 775–797.
16. Leenheer, J. A.; Huffman, E. W. D. *J. Res. U.S. Geol. Surv.* **1976,** *6,* 737–751.

RECEIVED for review August 14, 1985. ACCEPTED April 7, 1986.

23

Isolation of Organic Compounds Present in Water at Low Concentrations Using Supercritical Fluid Carbon Dioxide

Daniel J. Ehntholt, Christopher Eppig, and Kathleen E. Thrun
Arthur D. Little, Inc., Cambridge, MA 02140

The use of supercritical fluid carbon dioxide to extract low levels of organic substances from water was investigated for 23 different compounds. In general, compounds that were volatile and/or not highly soluble in water were readily extracted under the conditions used. Compounds of higher water solubility did not show evidence of extraction. In addition, those materials that tended to precipitate or form more soluble species under acidic conditions were not extracted.

THE USE OF BIOLOGICAL TESTS is one approach to understanding and evaluating the possible toxicological effects of the consumption of organic substances found in drinking waters. Many of these tests using experimental animals or organisms require concentration levels of the organic compounds that are significantly higher than those normally found in drinking waters. Although hundreds of organic compounds have been identified and quantified in samples of natural waters, much of the organic matter present cannot readily be characterized by using currently available analytical protocols. Without such prior qualitative and quantitative identification of the substances, they cannot be purchased or synthesized for use in the preparation of the concentrated solutions required for health-effects testing.

Direct concentration of the organic materials from aqueous samples offers an attractive alternative that circumvents the analytical problems associated with the identification and quantification of wide varieties of species present at trace levels. A number of techniques have been studied for their use in effecting such concentrations. These have

0065–2393/87/0214/0483$06.00/0

included the use of reverse osmosis, solid sorbents, and liquid–liquid extractions (*1, 2*). Serious problems may, however, be encountered in the use of concentrated solutions prepared by these methods because of inadvertent contamination of the sample. For example, membrane techniques may introduce impurities from the membrane and may not selectively isolate organic substances from inorganic species. Collection of organics on sorbents followed by recovery with organic solvents also poses a number of problems. Concerns have been expressed over the large blank contributions of resins, the possible interactions of the organic substances concentrated with the solvents (or impurities present in them) used for desorption, and the presence of traces of solvent in the prepared sample. The liquid–liquid extraction techniques using organic solvents yield concentrations of organic substances in media that may be undesirable for animal feeding studies. For example, immediate concern can be expressed about the use of benzene or the halogenated one- and two-carbon compounds, which are known or suspect carcinogens, if long-term biological tests are to be performed.

This study has examined the possible use of supercritical fluid carbon dioxide for the concentration and/or isolation of specified organic compounds present in waters at trace levels. This type of direct extraction using a nontoxic, nonhazardous solvent such as carbon dioxide represents a new concept for extracting trace levels of organic compounds from water. Solubility phenomena in supercritical fluids were reported by Hanney and Hogarth (*3*) as early as 1879. They found that inorganic salts such as cobalt chloride and potassium iodide could be dissolved in supercritical ethanol and ether. Furthermore, they found that the solubility level increased as the pressure increased. In the early 1900s, Bucher (*4*) studied the solubilities of a number of organic materials in supercritical carbon dioxide. His results showed that the concentration of organic species such as naphthalene, phenanthrene, phenols, and other aromatics dissolved in supercritical carbon dioxide was many times that which would be expected from the normal increase in vapor pressure due to external pressure. Other supercritical fluid solubility studies of the early 1900s were directed to similar considerations of solution thermodynamics, multiphase equilibria, etc. Booth and Bidwell (*5*) presented an excellent review of the developments during this period.

During the 1940s, a large amount of solubility data was obtained by Francis (*6, 7*), who carried out measurements on hundreds of binary and ternary systems with liquid carbon dioxide just below its critical point. Francis (*6, 7*) found that liquid carbon dioxide is also an excellent solvent for organic materials and that many of the compounds studied were completely miscible. In 1955, Todd and Elgin (*8*) reported on phase equilibrium studies with supercritical ethylene and a number of low-vapor-pressure organic materials such as fatty acids and high molecular weight alcohols. They found, as did previous investigators, that the solubility levels of the organic species were orders of magnitude higher than those predicted by vapor pressure considerations. Their findings led them to write "The magnitude . . . of solubility . . . is sufficient to consider the gas as an extracting medium, that is, fluid liquid or fluid solid extraction, analogous to liquid–liquid extraction and leaching . . .thus, compression of a gas over mixture of compounds could selectively dissolve one compound, permitting it to be removed from the mixture." This study was the first published reference to potential extraction process applications of solubility in supercritical fluids.

A few years later, Elgin and Weinstock (*9*) reported that a number of organic–water mixtures could be separated into organic-rich and water-rich phases by using supercritical ethylene, and they presented process concepts for separating such mixtures. Since Todd and Elgin's paper (*8*) in 1955, descriptions of a number of process applications of supercritical fluid solubility have appeared in the literature. Much of the effort reported has been directed to the extraction of edible (*10*) and essential (*11*) oils and other food and beverage products such as spices (*12*), coffee (*13*), and hops (*14*) using either supercritical or near-critical liquid carbon dioxide. The attributes of carbon dioxide, such as its low cost and absence of safety hazards and toxicity problems, were ideally suited for food applications. In the mid- and late-1960s, some developmental activity was directed toward supercritical fluid chromatography (*15*) and to extractions of fuels, namely, supercritical fluid extraction of coal (*16*), petroleum (*17*), and lignite (*18*).

Starting in about 1975, developmental efforts at a number of industrial and academic laboratories increased markedly in both the United States and Europe. The resurgent research activity was motivated by a number of factors:

1. Increased scrutiny of certain industrial solvents because of associated health and safety problems.
2. Increasing costs of traditional but energy-intensive separation processes such as distillation and evaporation.
3. Increasingly stringent pollution control legislation that increased costs of traditional extraction processes.
4. Identification of certain key areas in which supercritical fluid processing could be technically, as well as economically, superior to traditional separation processes.

Supercritical fluid separation processes operate at pressures ranging from 1000 to 4000 lb/in.^{2}, pressures that might be considered high, especially in the foods and essential oils industries. However, because of the factors just listed, supercritical fluid extraction has become economically attractive irrespective of the pressure requirements. Efforts to date have resulted in the development of several supercritical fluid processes in use throughout the world. Several large pilot plants for coal (*19*), coal ash (*20*), and asphalt (*21*) separation are in operation in the United States, United Kingdom, and Russia. Two commercial plants came on stream in 1979 for the extraction of beverage products: one in Germany for coffee decaffeination (*22*), and the other in Australia for hops extraction (*23*). Theoretical and practical efforts leading to several of these developments were summarized at a symposium devoted to "Extraction with Supercritical Gases" (*22*).

These efforts and others suggested there might be process advantages for using supercritical fluid carbon dioxide to extract low concentrations of organics from water. The U.S. Environmental Protection Agency (USEPA) therefore sponsored the work described herein, which was designed to evaluate such a process; the work was initiated in late 1980 and completed in 1983. Twenty-three organic substances were selected as representative of classes of compounds usually encountered in drinking waters. In addition, calcium and sodium salts were added to the spiked aqueous samples (as well as method blanks) to simulate inorganic salt concentrations found in Cincinnati drinking water. Lead nitrate was also studied to determine whether supercritical fluid carbon dioxide might extract and/or concentrate metal salts.

The ultimate goal of this program was the study of extraction feasibility for large volumes (~500 L) of drinking waters. However, to facilitate sample preparation steps and subsequent analyses, initial supercritical fluid carbon dioxide extraction studies were conducted by using small groups of similar compounds. These studies used 400-mL sample volumes. Subsequent 10-L extractions were run with aqueous samples containing all 23 organic substances as well as calcium and sodium salts present at the levels of interest in this effort.

Experimental

Most of the extraction studies carried out under this contract were conducted on 400-mL aqueous samples in a stainless steel extractor (extractor volume was approximately 600 mL) operated at about 2500 lb/in.^{2} (i.e., 173 bar) and 45 °C. Supercritical conditions are achieved for carbon dioxide at pressures >1070 lb/in.^{2} (i.e., 73.8 bar) and temperatures >31.1 °C. In our tests, approximately 300 standard liters of CO_2 was typically passed through the aqueous solutions into the traps. The trapping system usually consisted of a set of three sequential glass U-tubes maintained at −76 °C by a dry ice–acetone bath. Operation at this temperature precludes clogging by solid CO_2 but may be responsible for the loss of some extracted organic materials, as noted later. A scaled-up series of 10-L extractions were also carried out during this program. The apparatus used was similar to that used in the small-scale work but had an internal volume of 15 L. The traps were stainless steel impingers having a volume capacity of approximately 1 L.

The experimental apparatus used to perform the supercritical fluid carbon dioxide extractions is shown diagrammatically in Figure 1. Carbon dioxide provided from supply cylinder 1 is compressed by diaphragm compressor 2 and heated to the desired extraction temperature in heat exchanger 3. The pressurized, temperature-adjusted, carbon dioxide feed flows through the high-pressure fluid inlet line 4 to vessel 6, which contains the aqueous solution to be extracted. The extraction vessel is wrapped with electrical heating tape to regulate the extraction temperature, which is measured with thermocouple 7.

The supercritical carbon dioxide extract stream is passed from the extractor vessel outlet through pressure reduction valve 8, where the pressure is reduced to atmospheric pressure and the extracted organic substance is precipitated in collection device 9. The atmospheric pressure carbon dioxide then flows from the collection device through a rotameter 10 and dry gas meter 11, which measure CO_2 flow rate and total volume, respectively, to the vent 12.

To enhance the CO_2–aqueous phase interfacial area and facilitate contact by dispersion of the CO_2 as fine bubbles, a plug of silanized glass wool was placed in the bottom of the extraction vessel. After charging the vessel with 400 mL of aqueous feedstock solution, the vessel was slowly pressurized to the extraction pressure and simultaneously heated to the desired temperature. Carbon dioxide was then passed through the aqueous phase at a velocity of slightly more than 10 cm/min (about 10 standard liters/min at 1 atm and 70 °F). After a predetermined amount of carbon dioxide (typically 300 standard liters) flowed through the sample, the system was depressurized and the extracted aqueous raffinate (stream) was drained through valve 5 into a collection vessel. The extractor and all lines were of stainless steel construction; traps for the small-scale extractions and containers for feedstock and raffinate were all glass.

Results and Discussion

Analytical Methods. Twenty-three organic substances were selected for study during this effort. These compounds and the concentration levels at which they were investigated in this program are shown in Table I. All of the aqueous solutions for evaluation by using supercritical fluid CO_2 were prepared by spiking a small aliquot of the organic compounds dissolved in acetone into a distilled, deionized, water sample containing 70 ppm of $NaHCO_3$, 120 ppm of $CaSO_4$, and 47 ppm of $CaCl_2 \cdot H_2O$.

Special care must always be exercised in the study of parts-per-billion concentrations of organics in water to ensure minimal losses due to sample degradation, adsorption or absorption to process materials, and other similar losses. These issues were addressed by dividing the measurement procedures into two parts: (1) sample preparation and (2) analytical method, or finish. Because well-defined sample preparation steps were not available from the literature for the quantitative determination of parts-per-billion concentration levels of most of the model organic compounds in water, a considerable amount of effort was placed on the development of appropriate procedures for such measurements. In particular, each method was developed with the intent to have a procedure that could verify the presence of appropriate concen-

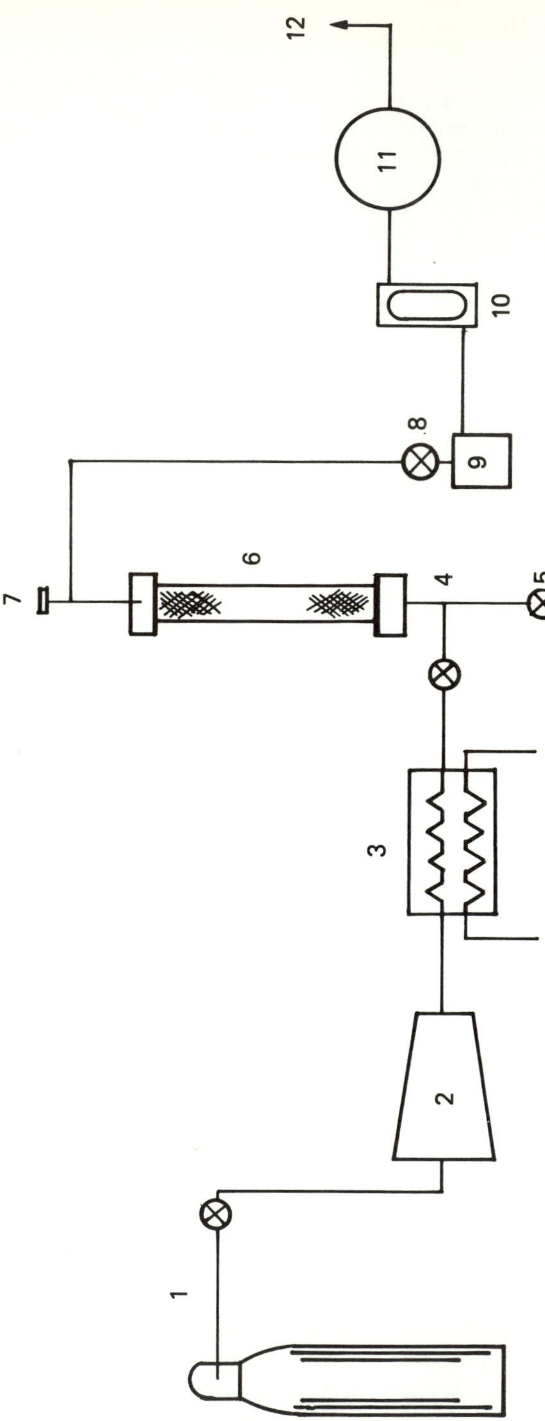

Figure 1. Schematic representation of supercritical fluid extraction apparatus.

Table I. Organic Substances Selected for Study

Group	*Compound*	*Concentration Level (μg/L)*	*Sample Preparation and Analysis*[a]
1	1-Chlorododecane	5	microextraction, GC–FID
	2,2′,5,5′-Tetrachlorobiphenyl	5	
	Biphenyl	50	
	Bis(2-ethylhexyl) phthalate	50	
	2,4′-Dichlorobiphenyl	50	
2	Crotonaldehyde	50	DNPH derivatization, HPLC–UV
	Furfural	50	
	Isophorone	50	
	Methyl isobutyl ketone	50	
3	Anthraquinone	50	microextraction, GC–FID
	Quinoline	50	
	Caffeine	50	
4	2,4-Dichlorophenol	50	microextraction, GC–FID
	2,6-Di-*tert*-butyl-4-methylphenol	50	
5	Quinaldic acid	50	sorbent extraction, CH_2N_2, GC–FID
	Trimesic acid	50	
	Stearic acid	50	
6	Glucose	50	evaporation, derivatization, GC–FID
	Glycine	50	
7	Chloroform	50	purge and trap, GC–ECD[a]
8	Phenanthrene	1	HPLC–UV–fluorescence
9	5-Chlorouracil	50	HPLC–UV
10	Humic acid	2000	HPLC–UV

[a] Abbreviations are defined in the text.

trations of the organics in the feedstock and also monitor their concentrations after extraction by carbon dioxide, that is, to quantify concentrations of organics in the raffinate (effluent aqueous stream), which might be as low as one-fiftieth of the starting concentration level. For some groups of compounds, this effort involved an extensive study of sample preparation steps. However, the goal of developing accurate and reproducible methods for studying the concentrations of the model compounds in water was met for the compounds in Table I. Table I identifies the sample preparation and analytical technique used to quantify each compound in this study.

To accurately determine the levels of compounds present in the aqueous solutions before and after CO_2 extraction, a sample prepartion step involving concentration of the compounds was necessary for most samples. In the case of Groups 1, 2, 3, 4, and 8, microextraction techniques with organic solvents were used (*24*). For Group 5 acids, resin concentration was used. Group 6 was concentrated by evaporation; Group 7 (chloroform), Group 9 (5-chlorouracil), and Group 10 (humic acids) samples were analyzed as received. Derivatization was used to enhance the detection limits for three of the groups. The formation of

2,4-dinitrophenylhydrazone (DNPH) derivatives of the Group 2 aldehydes and ketones permitted their determination by an HPLC method after microextraction. Methyl esters of the Group 5 acids were formed by using diazomethane and were subsequently detected by gas chromatography–flame ionization detection (GC–FID). Glucose and glycine (Group 6) were quantified after treatment with a hydroxylamine hydrochloride solution in pyridine, followed by *N*-trimethylsilylimidazole. The trimethylsilyl–glucose–oxime derivative and the glycine–trimethylsilyl derivative thus formed were analyzed by using GC–FID methods.

Analytical methods for monitoring the compounds were developed or modified to permit the quantification of all 23 compounds of interest. As noted earlier, the compounds were initially studied in small-scale extractions by groups. This approach assured minimal interferences in the analyses conducted during the initial supercritical fluid carbon dioxide extractions. Table II summarizes the data on the recovery of organics from aqueous samples containing the compounds of interest at concentration levels listed in Table I when the sample preparation techniques and analytical methods described were used. For each experimental run, blank and spiked aqueous samples were carried through the sample prepration and analytical finish steps to ensure accurate and reproducible results. Analyses of sodium, calcium, and lead content were also conducted on selected samples by using standard atomic ab-

Table II. Summary of Recovery Data (Microextraction–Derivatization Techniques)

Analyte	*No. of Experiments*	*Mean Recovery*	*Standard Deviation*	*Coefficient of Variation*
2,4′-Dichlorobiphenyl	6	96	18.1	18.8
2,2′,5,5′-Tetrachlorobiphenyl	6	87	17.9	20.6
Bis(2-ethylhexyl) phthalate	6	85	22.5	26.5
1-Chlorododecane	3	131	7.8	6
Biphenyl	6	98	11.8	12.0
Furfural	3	61.5	16.5	26.8
Crotonaldehyde	3	83.7	12.0	14.3
Isophorone	3	82.4	10.7	12.9
Methyl isobutyl ketone	3	63.1	28.1	44.4
Anthraquinone	3	66.7	19.3	66.7
Quinoline	4	70.8	19.9	28.1
Caffeine	4	86.1	10.6	12.3
2,4-Dichlorophenol	7	86	19.8	23.1
2,6-Di-*tert*-butyl-4-methylphenol	7	76	10.5	13.9
Quinaldic acid	3	9.1[a]	0.7	7.7
Trimesic acid	3	95.8[a]	3.3	3.4
Stearic acid	3	78.1[a]	8.8	11.2
Chloroform	12	80	—	—
Phenanthrene[b]	6	69.2	9.7	14.0

NOTE: All values are percentages.
[a] Recovery from XAD-7 resin.
[b] Data are taken from reference 26.

sorption spectroscopic methods for inorganic cations present in drinking waters (25).

All of the techniques developed were amenable to the analysis of the concentrated compounds in collection devices (traps). Because these compounds were expected to be present in neat form in the traps, it was not anticipated that the sample preparation aspect of the analysis would pose any difficulty (i.e., it was assumed that any adsorption to glass walls, etc., that might occur would affect only a proportionately very small amount of the collected sample). As discussed later, this situation did not prove to be the case for some of the compounds studied. In addition, the difficulties inherent in trapping trace quantities of organics in the effluent CO_2 stream were not obvious during the early stages of this program.

Small-Scale Extractions of Organic Compounds. Table III details the experimental results obtained for the small-scale supercritical fluid carbon dioxide extraction of the organic compounds. The compounds investigated, nominal spiking levels, and number of experiments performed are listed in the first three columns. The mean recoveries for each of the three U-tube traps connected in series are then presented along with the total recoveries obtained from all three traps. The quantity of compound recovered from the raffinate (effluent) solution after CO_2 extraction is contained in the last column. Although four of the five Group 1 compounds spiked into the aqueous samples could be recovered from the traps, only 20% to 31% of the total mass of each compound could be accounted for when the amounts in the traps and raffinate were summed. Losses may be due to incomplete trapping because of the low, but measurable, vapor pressures of these compounds and the large volume of CO_2 passed through the traps.

Results from the two experiments conducted on the aldehydes and ketones are listed in Table III as separate sets of data to illustrate the care that must be exercised in conducting and evaluating these runs. Both CO_2 extractions were performed under similar conditions. However, in the second run, the U-tube traps were contacted with the 2,4-dinitrophenylhydrazine derivatizing solution for longer periods of time. This modification in the analytical procedure permitted higher total mass accountabilities in the second experiment, ranging from 64.9% for isophorone to 28.7% for methyl isobutyl ketone. The recoveries from the raffinate for each of these compounds remained relatively constant. This result suggested that the trap recoveries in the first case were artificially low.

In the case of the Group 3 compounds anthraquinone, caffeine, and quinoline, it seems likely that the low trap recoveries and high residual concentrations of quinoline and caffeine in the raffinate were due to the

Table III. Small-Scale (400 mL of Aqueous Samples) Extractions

Compound	Concentration Level (μg/L)	No. of Experiments	Mean Trap Recoveries (%)				Mean Raffinate Recoveries (%)[a]
			Trap 1	Trap 2	Trap 3	Total[a]	
1-Chlorododecane	5	1	0	0	0	0	20.7
2,2′,5,5′-Tetrachlorobiphenyl	5	3	18.7	0	0	18.7 (18.0)	12.0 (20.8)
Biphenyl	50	1	9.1	11.0	3.3	23.4	3.8
Bis(2-ethylhexyl) phthalate	50	3	11.3	0	0	11.3 (3.6)	15.4 (17.5)
2,4′-Dichlorobiphenyl	50	3	15.7	4.6	0	20.3 (2.6)	8.5 (10.8)
Crotonaldehyde	50	1	0.8	0	0.2	1.0	25.1
		1	7.0	0.8	0	7.8	31.0
Furfural	50	1	3.7	0	0.1	3.8	43.4
		1	8.3	0.7	0	10.8	22.3
Isophorone	50	1	0	0	1.7	1.7	17.8
		1	39.2	0.7	0.5	40.4	24.5
Methyl isobutyl ketone	50	1	0	0	0	0	7.4
		1	15.6	1.5	0.2	17.3	11.4
Quinoline	50	2	1.7	1.7	0	3.4 (3.7)	46.1 (14.3)
Caffeine	50	2	0	0	0	0	81.4 (11.4)
Anthraquinone	50	2	56.0	14.3	14.3	84.6 (38.3)	21.4 (30.2)
2,4-Dichlorophenol	50	3(SCF CO_2)[b]	35.8	9.6	0	45.4 (15.0)	28.0 (14.0)
		1(liquid CO_2)	15.3	16.8	8.2	40.3	13.4
2,6-Di-*tert*-butyl-4-methylphenol	50	3(SCF CO_2)	26.6	6.0	0	32.7 (3.0)	0
		1(liquid CO_2)	8.7	10.7	6.6	26.0	6.8
o-Bromophenol	50	3(SCF CO_2)	24.6	17.0	0	41.7 (15.6)	31.6 (9.6)
		1(liquid CO_2)	10.8	12.3	10.8	33.9	19.5
Stearic acid	50	1	25	2.5	20	47.5	22
Quinaldic acid	50	1	0	0	0	0	85
Trimesic acid	50	1	0	0	0	0	0 (91)
Phenanthrene	1	1	58	36	3	97	0
Stearic acid	50	1	25	2.5	20	47.5	22
Quinaldic acid	50	1	0	0	0	0	85
Trimesic acid	50	1	0	0	0	0	91
5-Chlorouracil	50	1	0	0	0	0	96

[a] Standard deviations are given in parentheses.
[b] SCF denotes supercritical fluid.

low pH of the extracting media (pH 3) and the resulting poor solubility of these nitrogenous compounds in the CO_2 effluent stream. Anthraquinone was recovered in good yields from the same extractions.

Table III also presents our data for the extraction of Group 4 phenols from aqueous solutions. The *o*-bromophenol was added as an internal standard when some initial recovery problems were noted for the 2,6-di-*tert*-butyl-4-methylphenol; results for its extraction are also reported here. The three phenols show good recoveries in the traps and overall good mass recoveries. One experiment was conducted under liquid CO_2 extraction conditions (temperature = 30 °C and pressure = 1500 lb/in.^{2}) in an attempt to compare the relative efficiencies of the two states of CO_2 for phenol extraction. Unfortunately, the phenols showed evidence of substantial breakthrough from the trapping system. The experiment does, however, demonstrate that liquid CO_2 is also a good extractant for phenols present in water at parts-per-billion concentration levels.

Group 5 acids present at the 50-ppb level in aqueous solutions containing the specified inorganic salts were also extracted. The feedstock solution (pH 7.2) and the raffinate (pH 4.5) were analyzed by concentration on XAD-7 resin, elution by methanol, methylation by diazomethane, and, finally, measurement of peak areas by GC–FID. Comparison of the peak areas of the raffinate with the peak areas in the original solution indicated a 78% reduction for stearic acid, 15% reduction for quinaldic acid, and 9% reduction for trimesic acid. The three traps, including the glass wool plugs and connecting tubes, were also analyzed for acid content. Only stearic acid was detected.

The Group 6 compounds glucose and glycine were not tested at the small-scale level because previous work suggested that they would not be soluble in supercritical fluid CO_2. Additionally, chloroform was not tested at this level because it is undoubtedly extractable but would pose significant trapping problems because of its relatively high vapor pressure.

The Group 8 compound phenanthrene was tested in a 1-ppb (micrograms-per-liter) solution and showed virtually complete recovery, as detailed in Table III. The Group 9 compound 5-chlorouracil, on the other hand, did not exhibit any extraction. Recovery of the compound in the raffinate (at the 50-ppb level) was quantitative; this result indicated that little, if any, was extracted.

Several extractions were also conducted on 2.0-mg/L humic acid solutions. These solutions were prepared by dissolving a known quantity of humic acid (Fluka, further purified by USEPA pesonnel) in 0.20 M sodium hydroxide followed by dilution with water to a 0.02 M sodium hydroxide solution. Subsequent neutralization to pH 7.0 with 0.100 M HCl and dilution with water containing the salts noted earlier gave a

2.0-mg/L humic acid solution for extraction studies. Three supercritical fluid carbon dioxide extractions were carried out on this solution; no humic acid could be detected in the traps.

Table IV lists extraction conditions tried in these runs. Although the analyses of feedstock solutions showed the presence of 97.7% to 104.7% of the expected concentration levels of humic acids, analyses of raffinate solutions showed lower humic acid concentrations. The raffinate obtained after the CO_2 extraction indicated that the humic acids were present at 39.4% to 44.9% of the feedstock levels (as measured by high-performance liquid chromatography–UV spectroscopy (HPLC–UV). This result suggested that the acidic conditions present in the extractor caused some precipitation and resultant loss of material. Cleaning of the extractor after these experiments indicated that a dark organic material had precipitated on the walls of the extractor. In addition, an experiment was conducted on the neat humic material (in the absence of water) to verify that the organic substance was not dissolved or carried over into the trapping system. In that case, approximately 500 mg of humic acid material was treated sequentially with 100 standard liters of carbon dioxide (2500 lb/in.^{2} at 45–50 °C) and 50 standard liters of carbon dioxide (4000 lb/in.^{2} at 38–41 °C). There was no apparent collection of material in the traps (visually or gravimetrically), and there was only a very slight loss ($<5\%$ gravimetrically) of the starting mass. That loss may have been due to sample handling procedures. This experiment thus supported the observation of nonextraction of humic materials in the tests noted earlier.

Trapping Device Experiments. The system used to collect the organic compounds extracted from the aqueous stream was, in most cases, a series of glass U-tubes held at −76 °C. That temperature represented a practical lowest limit to prevent deposition of solid carbon dioxide. During the course of this program, it became evident that, for many compounds, complete mass balances were not being achieved. The trapping system appeared to be a likely source of such losses because many of the compounds studied had a finite vapor pressure at −76 °C. An effluent CO_2 stream saturated with these

Table IV. Extraction Conditions for 2.0-ppm Humic Acid Solutions

Experiment Number	*Temperature (°C)*	*Pressure (lb/in.^{2})*	*CO_2 Volume (standard liters)*
1	45	2400 ± 100	300
2	46 ± 3	2450 ± 50	340
3	46 ± 3	2200 ± 50	580

NOTE: Sample volume was 400 mL for all experiments.

compounds, even at −76 °C, might, therefore, be responsible for these losses if enough CO_2 was passed through the traps. Other possible sources of compound losses include entrainment in the CO_2 stream or adsorption to the trap walls, with inefficient rinsing of the trap before analysis.

By using data from the small-scale extractions of dichlorophenol as an example, the maximum theoretical amount that can be collected at −76 °C can be calculated to be 77%. Actual experimental values show recovery to be about 62% for the three small-scale supercritical fluid carbon dioxide extractions of dichlorophenol. These data support the suggestion that the vapor pressure of the compound being trapped is an extremely critical physical constant when large volumes of CO_2 relative to the aqueous sample volume are being used for the extraction process.

Several experiments were conducted during the program to better define the magnitude of such losses and to suggest a design for trapping systems that might mitigate vapor-phase loss of the extracted organics. Three alternative approaches for the collection of the organics were identified:

1. The use of a U-tube sequence operated at less than 1 atm and temperatures below −78 °C. The use of such a reduced-pressure trap would allow lower trap temperatures without the collection of solid CO_2 and would also increase the superficial linear velocity of the gas as it passes through the trap. This type of a system appears feasible but was not investigated during this program because other concepts seemed to offer comparable or better collection efficiencies.
2. The use of a collection device containing a solid sorbent such as granular activated charcoal (GAC), Tenax, or an XAD resin to concentrate the organics in the CO_2 effluent stream. Such a system would require subsequent desorption of the organics with the possibility of concomitant contamination and was, therefore, not a major focus of our study. A limited effort conducted to determine whether vapors of some extracted organics were being lost through the U-tube traps indicated that this situation was the case.
3. The collection of the total CO_2 effluent as a solid or liquid and subsequent isolation of the entrained organic compounds. Such a trapping system would minimize losses of organic substances due to the effects of poor thermal transfer possible in the other trap systems. In addition, if the effluent could be collected as liquid CO_2 (e.g., at 900 lb/in.^{2} and 20 °C), several additional benefits were deemed possible. For example, minimal loss of the organic compounds should occur because their concentration in the gas distilled off is

related to their partial pressure above the liquid CO_2 solution. The operation of a trap at 900 lb/in.^{2} would also facilitate the use of a recycling system for the CO_2. Several experiments designed to perform a preliminary evaluation of this type of trapping system were also promising.

During these experiments, we also investigated more efficient trap rinse procedures. For example, the three U-tube traps normally used in these experiments were rinsed sequentially with CH_2Cl_2, then CH_2Cl_2–base (3–5 drops of 6 N NH_4OH in 1 mL of CH_2Cl_2), instead of only a CH_2Cl_2 rinse (as in our small-scale experiments).

These studies demonstrated that sequential washing of the trap system with CH_2Cl_2 and basic CH_2Cl_2 resulted in higher mass recoveries of the compounds of interest. To maximize concentration levels in the rinse solvents for subsequent analysis, such sequential washings had not been used in the small-scale experiments conducted during this program. Many of the results for trap recoveries presented earlier may thus represent minimal values. This finding does not affect the raffinate values.

Small-Scale Extractions of Inorganic Species. Because the presence of high levels of inorganic salts interferes with animal feeding studies, it was also of interest to demonstrate whether or not salts could be carried over and concentrated by this extraction technique. We therefore examined the trapping system after several blank runs. In these cases, solutions containing the specified levels of salts (i.e., $NaHCO_3$, $CaSO_4$, and $CaCl_2$), as well as $PbNO_3$ at the 25-ppb level, were extracted by supercritical fluid CO_2. The results for sodium, calcium, and lead are shown in Table V and indicate that salts of these ions were not collected in the traps.

Chlorine Residual Extraction. A 400-mL solution of distilled water containing 70 ppm of sodium bicarbonate, 120 ppm of calcium sulfate, and 47 ppm of calcium chloride was extracted under typical operational conditions. Similarly, a solution containing all of these materials plus a 2-ppm chlorine residual (prepared with NaOCl) was extracted. Analysis of the U-tube traps, feedstock, and raffinate solutions in each case (blank and chlorine residual samples) showed that no new chlorinated compounds were formed by the presence of a chlorine residual.

Ten-Liter Extractions. After the conduct of the small-scale experiments just described, a series of four 10-L extractions were run as follows:

Run 1: 10 L of distilled, deionized water spiked with the 23 organic analytes and specified levels of inorganic salts.

Table V. Extraction of Aqueous Salt Solutions

Analyte	*Sodium Analysis*[a] *(mg)*	*Calcium Analysis*[b] *(mg)*	*Lead Analysis*[c] *(μg)*
Distilled Water			
Total traps	0.15	0.03	0.1
Raffinate	<0.02	0.03	<0.4
Solution 1			
Total traps	0.08	0.06	0.4
Raffinate	8.0	15.2	2.0
Solution 2			
Total traps	0.03	0.01	0.1
Raffinate	8.0	16.0	7.2
Solution 3			
Total traps	0.11	0.04	0.2
Raffinate	8.8	16.0	12.0

[a] Total sodium expected in solutions = 7.7 mg.
[b] Total calcium expected in solutions = 13.2 mg.
[c] Total lead expected in solutions = 15.6 μg.

Run 2: 10 L of distilled, deionized water spiked only with calcium and sodium inorganic salts.

Runs 3 and 4: replicates of Run 1.

On the basis of scale-up considerations from 0.4-L runs, each sample extraction was conducted at 1950 ± 50 lb/in.^{2} and 37–45 °C, lasted about 110 min, and involved passing approximately 11,200 standard liters of carbon dioxide through the aqueous solution. Because pressure–flow rate excursions might occur in the large-scale apparatus and lead to the rupture of glass traps, a series of three stainless steel impingers maintained at −76 °C were used to collect the organics present in the effluent carbon dioxide stream.

In addition, the earlier extraction experiments showed that quantitative removal of organics from the traps was a problem for some compounds, perhaps because of the small quantities of collected organics on the large trap surfaces. Therefore, a trap rinse sequence that maximized the dissolution of any of the types of organic compounds that could be recovered from the trap was used. The solvents used also needed to be compatible with any derivatization–sample preparation steps necessary before analysis. Thus, at the conclusion of each experiment, the three traps were rinsed sequentially with methylene chloride, methylene chloride–base (5 N NH_4OH), and Milli-Q water. The first methylene chloride trap rinse yielded some aqueous phase extract (approximately 20 mL), which was added to the Milli-Q rinse. Aliquots of the trap rinses, raffinate, and feedstock were analyzed according to methods previously developed for the groups of interest. Table VI summarizes the results obtained from these runs. In general, the types

Table VI. Summary of 10-L Extraction Results

Group	Compound	Nominal Spike (μg/L)	Percent Recovered from Traps			Percent Recovered from Raffinate		
			Run 1	*Run 3*	*Run 4*	*Run 1*	*Run 3*	*Run 4*
1	1-Chlorododecane	5	38	21	16	*	*	*
	2,2′,5,5′-Tetrachlorobiphenyl	5	34	35	22	*	*	*
	Biphenyl	50	30	11	8	*	*	*
	Bis(2-ethylhexyl) phthalate	50	28	35	26	*	*	*
	2,4′-Dichlorobiphenyl	50	49	46	40	*	*	*
2	Crotonaldehyde	50	5	3	1	2	2	3
	Furfural	50	5	4	*	*	*	*
	Isophorone	50	23	34	27	*	*	*
	Methyl isobutyl ketone	50	4	5	5	4	7	2
3	Anthraquinone	50	24	64	7	29	65	*
	Quinoline	50	13	*	*	58	35	*
	Caffeine	50	7	11	*	103	77	34
4	2,4-Dichlorophenol	50	23	36	20	*	*	*
	2,6-Di-*tert*-butyl-4-methylphenol	50	46	26	20	*	*	*
5	Quinaldic acid	50	*	*	*	83	83	100
	Trimesic acid	50	*	*	*	98	89	65
	Stearic acid	50	*	*	*	37	18	26
6	Glucose	50	*	*	*	NA	NA	NA
	Glycine	50	*	*	*	NA	NA	NA
7	Chloroform	50	NA	NA	NA	*	*	*
8	Phenanthrene	1	18	13	12	*	*	*
9	5-Chlorouracil	50	7	14	7	14	11	13
10	Humic acid	2000	1	1	1	x	x	x

NOTE: NA = not analyzed; * = none detected; x = none detected, but brown precipitate was recovered from the extractor upon cleaning.

of compounds that were extracted and trapped were the same as those found in the small-scale experiments. In particular, the hydrocarbons and phenols were collected in the traps, whereas more polar compounds such as the acids, glucose, and glycine were not detected in the trapping system. The mass balances for some types of materials (e.g., 5-chlorouracil and the humic acid) were poorer in the 10-L extraction; however, these runs were the first that contained all 23 compounds at the same time, and the extractions were also conducted for a longer period of time. It is possible that the interactions between compounds under the acidic extraction conditions account for the lower total recoveries in certain cases. For example, the absence of humic materials in the raffinate and the observation of a brown organic material (humics) upon cleaning the extractor indicate that this material was precipitated out of the solution.

The total organic carbon (TOC) contents (mg/L) of selected 10-L raffinate samples were as follows: Run 1, 22; Run 2, 16; Run 3, 18; and Run 4, 15. The TOC contents (mg/L) of selected 10-L feedstock samples were as follows: Sample 1, 160; and Sample 2, 155. The high feedstock values are due primarily to the amount of organic solvent (acetone) added as part of the spiking solutions for the 23 organics. These results indicate approximately 90% extraction efficiency of organic content from the aqueous feedstock and probably represent good indicators of the overall extraction efficiency of this process for volatile organic species such as acetone.

Conclusions

This study demonstrated the use of supercritical fluid carbon dioxide for the isolation and concentration of certain types of organic compounds present in water at low concentration levels. In general, compounds that were volatile and/or not highly soluble in water were readily extracted under the operating conditions used. However, the subsequent efficient trapping of these compounds was not a trivial problem. Those compounds that exhibited greater solubility in water (e.g., trimesic acid and 5-chlorouracil) did not show evidence of extraction; in addition, those materials that tended to precipitate (humic acid) or form more soluble species (caffeine) under acidic conditions were not extracted. Experiments were also conducted to determine whether or not inorganic salts such as sodium bicarbonate, calcium sulfate, calcium chloride, or lead nitrate (added to several solutions as surrogates for possible toxic metal salt contaminants) were extracted. Results indicated that the inorganics were not isolated or concentrated.

The extraction conditions used in the study were optimized on the basis of approximately 70% extraction of typical phenolic compounds.

Additional treatment with supercritical fluid carbon dioxide would likely increase the extraction efficiency of the process but might present additional trapping (recovery) problems.

On the basis of this work, it appears that the supercritical fluid carbon dioxide extraction of organic compounds present in water at low levels may be useful only in the case of volatile organic species. Even for these compounds, however, if efficient recovery of the compounds is desired and appropriate, certain limitations of existing trapping techniques remain to be overcome.

Acknowledgments

The suggestions, support, and encouragement of Paul Ringhand of the Health Effects Research Laboratory of the USEPA are gratefully acknowledged. In addition, the efforts of R. Bruni, R. Cruz-Alvarez, L. Fiebig, L. Guilmette, M. Randel, A. Tucci, D. Wilson, and C. Wong were essential to the completion of this work.

The research described in this chapter was funded by the USEPA through Contract No. 68–03–3001 to Arthur D. Little, Inc. (ADL Project No. 85474). Although these results have been reviewed by the Health Effects Research Laboratory, USEPA, and approved for publication, approval does not signify that the contents necessarily reflect the views and policies of the USEPA, nor does mention of trade names or commercial products constitute endorsement or recommendations for use.

Literature Cited

1. Karasek, F. W.; Clement, R. E.; Sweetman, J. A. *Anal. Chem.* **1981,** *53(9),* 1050A.
2. Jolley, R. L. *Environ. Sci. Technol.* **1980,** *15(8),* 874.
3. Hannay, J. B.; Hogarth, H. *Proc. R. Soc. London* **1879,** *29,* 324.
4. Buchner, E. H. Z. *Physik. Chem.* **1906,** *54,* 665.
5. Booth, H. S.; Bidwell, R. M. *Chem. Rev.* **1949,** *44,* 447.
6. Francis, A. W. *J. Phys. Chem.* **1954,** *58,* 1099.
7. Francis, A. W. *Liquid-Liquid Equilibriums;* Wiley-Interscience: New York, 1963.
8. Todd, D. B.; Elgin, J. C. *AIChE J.* **1955,** *1,* 20.
9. Elgin, J. C.; Weinstock, J. H. *J. Chem. Eng. Data* **1959,** *4(1),* 3.
10. Br. Patent 1 356 749.
11. Aleksandrov, L. G.; Popova, S. A.; Serdyuk, V. I. *Tr. Krasnodar. Nauchno Issled. Inst. Pishch. Promsti* **1973,** *6,* 146.
12. Br. Patent 1 336 511.
13. Br. Patent 1 346 134.
14. Br. Patent 1 388 581.
15. Giddings, J. C.; Myers, M. N.; King, J. W. *J. Chromatogr Sci.* **1969,** *7,* 276.
16. Bartle, K. O.; Martin, T. G.; Williams, D. F. *Fuel* **1975,** *54,* 226.
17. Irani, C. A.; Funk, E. W. In *Recent Developments in Separation Science;* Li., N., Ed.; CRC: Cleveland.

18. Tugrul, T.; Olcay, A. *Fuel* **1978,** *57*, 415.
19. Maddocks, R. R.; Givson, Jr. *Chem. Eng. Prog.* **1977,** *59*.
20. Adams, R. M.; Knebel, A. H.; Rhodes, D. E. *Chem. Eng. Prog.* **1979,** 44.
21. Paul, P. F. M.; Wise, W. S. *The Principles of Gas Extraction;* Mills and Boon: London, 1971.
22. *Extraction with Supercritical Gases;* Schnieder, G. M.; Stahl, E.; Wilke, G., Eds.; Verlag Chemie: Weinheim, West Germany, 1980.
23. Harold, F. V.; Clark, B. J. *Brew. Dig.* **1979,** 45.
24. Thrun, K. E.; Simmons, K. E.; Oberholtzer, J. E. *J. Environ. Sci. Health* **1980,** *A15(5)*, 485.
25. *Methods for Chemical Analysis of Water and Wastes;* U.S. Environmental Protection Agency. U.S. Government Printing Office: Washington, DC, 1979; No. 600/4-79-020.
26. Reunanen, M.; Droneld, R. *J. Chromatogr. Sci.* **1982,** *20*, 449–454.

RECEIVED for review August 14, 1985. ACCEPTED March 31, 1986.

24

Recovery of Trace Organic Compounds by the Parfait–Distillation Method

James B. Johnston[1], Clarence Josefson[2], and Richard Trubey[3]
Institute for Environmental Studies, University of Illinois, Urbana, IL 61801

The parfait-distillation method uses a sequential series of adsorbents to remove contaminants from water and vacuum distillation to recover unadsorbed materials. This method recovers a wide range of neutral, cationic, anionic, and hydrophobic contaminants. The first adsorbent, porous polytetrafluoroethylene (PTFE), removed humic acid and a broad range of hydrophobic compounds. PTFE was followed by Dowex MSC-1 and then Duolite A-162 ion-exchange resins. A synthetic hard water spiked parts-per-billion concentrations with 20 model compounds was used to evaluate the method. Poorly volatile, neutral, water-soluble species (glucose); cationic aromatics; and most hydrophobic compounds were recovered quantitatively. Model amphoterics were removed from the influent but were not recovered from the adsorption beds. The recovery of model acids and bases ranged from 22% to 70% of the amount applied.

THE NEED TO EVALUATE HEALTH RISKS associated with organic contaminants in surface and potable waters has prompted basic research into methods to recover and concentrate these substances. The most intensively studied contaminants are hydrophobic or volatile substances that are readily recovered by methods such as gas stripping and solvent extraction. However, these compounds account for only a small fraction of the organic material in a typical water (*1*, *2*). The majority of

[1] Current address: Smith Kline & French Laboratories, Chemical Engineering Department, Swedeland, PA 19479
[2] Current address: Millikan University, Department of Chemistry, Decatur, IL 62522
[3] Current address: E. I. du Pont de Nemours and Company, Agricultural Products Department, Wilmington, DE 19898

0065–2393/87/0214/0503$08.50/0

compounds in water are not well-known, both because methods to recover them have not been developed and because automated means of identification, such as gas chromatography–mass spectrometry (GC–MS), are not applicable.

A previous exploratory study attempted to recover the soluble, poorly volatile subclass of organic compounds in water (*3*). It used a set of sequential adsorbents. Silica gel, the first adsorbent, filtered out particulate matter and adsorbed some hydrophobic compounds. The next adsorbent was a cation-exchange bed that recovered cations and amphoteric substances, and the last adsorbent was an anion-exchange bed. The effluent from this series of adsorbents contained the neutral compounds. The eluates from each bed and the effluents were then concentrated under vacuum. This system, the parfait method, was demonstrated to recover parts-per-billion concentrations of several known mutagens in amounts sufficient to be detected by bioassay.

For the method to be most useful for the recovery of unknown toxicants in water, certain inherent difficulties of the original parfait method had to be overcome. Also, the method had to be evaluated by using a set of known compounds representative of major chemical groups that could contaminate water.

The original parfait method rested on the use of vacuum distillation–lyophilization to concentrate the poorly volatile species in water. It might be expected that the removal of water under vacuum should be simple and straightforward. Vacuum distillation and lyophilization do indeed recover the poorly volatile contaminants from unfractionated surface waters. However, the compounds are often obtained in an intractable, insoluble form. These intractable precipitates are believed to form when bicarbonate dissociates under vacuum to form metal carbonate precipitates that trap organic polymers and lipids (*4, 5*). The parfait method prevents the formation of these precipitates by removing metal ions on an acidic cation-exchange bed.

To protect the cation-exchange bed from particulate matter, the original parfait system had a bed of silica gel ahead of the cation-exchange bed. To neutralize the acid released from the cation bed, as well as to remove anions, a strongly basic anion-exchange bed was added after the cation exchanger.

In the original parfait system, the silica gel created problems by acting as a weak cation exchanger and by leaching silicic acid to the anion-exchange bed and into the final effluent. These silicate residues interfered with the recovery of compounds from these sources.

The modified parfait method developed here replaced silica gel with porous polytetrafluoroethylene (PTFE, Teflon). Properly wetted Chromosorb T, an aggregate of aqueous dispersion–polymerized PTFE, is an efficient adsorbent for many hydrophobic substances in water

(6). In addition, it is an exceptionally stable material that does not contaminate subsequent adsorption beds or eluates.

The remainder of the modified parfait column consisted of an MSC-1 cation-exchange and an A-162 anion-exchange bed. The elution conditions for these beds were modified to minimize contamination of eluates and to selectively desorb organic anions and cations. With the modified protocol, 20 model compounds (Table I), selected by the U.S. Environmental Protection Agency (USEPA) Health Effects Research Laboratory (HERL), were used to evaluate the recovery efficiency of the method. Recoveries were determined in the presence of 2 ppm of a humic acid supplied by HERL.

Experimental

Reagents. Sigma Chemical Company supplied quinaldic acid, trimesic acid, glycine, caffeine, 5-chlorouracil, ethidium bromide (EB), and fluorescamine. Analabs supplied stearic acid, 2,4′-dichlorobiphenyl, and 2,2′,5,5′-tetrachlorobiphenyl. Eastman supplied phenanthrene. Aldrich supplied pyrene,

Table I. Compounds Used To Assess the Performance of the Parfait Recovery Method

Compounds	*Concentration (μg/L)*	*Salient Character*[a]
Quinaldic acid	50	S, AM, PV
Trimesic acid	50	S, A, PV
Stearic acid	50	SS, A, PV
Glycine	50	S, AM, PV
Furfural	50	S, N, V
Quinoline	50	PS, N, V
Caffeine	50	SS, N, PV
5-Chlorouracil	50	SS, N, PV
Glucose	50	S, N, PV
2,4′-Dichlorobiphenyl	50	PS, N, PV
2,2′,5,5′-Tetrachlorobiphenyl	50	PS, N, PV
Bis(2-ethylhexyl) phthalate	50	PS, N, PV
1-Chlorododecane	50	PS, N, V
Biphenyl	50	PS, N, V
Isophorone	50	PS, N, V
Anthraquinone	50	PS, N, PV
Methyl isobutyl ketone	50	SS, N, V
2,4-Dichlorophenol	50	SS, N, V
2,6-Di-*tert*-butyl-4-methylphenol	50	S, N, PV
Phenanthrene	1	PS, N, PV
Humic acid	2000	—

[a] S = water soluble, SS = slightly water soluble, PS = poorly water soluble, C = cation, A = anion, AM = amphoteric, N = neutral, V = volatile, and PV = poorly volatile.

furfural, quinoline, glucose, bis(2-ethylhexyl) phthalate, biphenyl, isophorone, anthraquinone, methyl isobutyl ketone, 2,4-dichlorophenol, 2,6-di-*tert*-butyl-4-methylphenol (BHT), and acenaphthene. Fluka supplied 1-chlorododecane and produced the humic acid supplied by HERL. Inorganic reagents were supplied by Baker. All reagents were of the highest purity supplied by the manufacturers and were used without purification. Hexane, acetone, dichloromethane, and methanol were Burdick and Jackson distilled-in-glass.

Materials. Porous Teflon (Chromosorb T, 30–60 mesh, 7–8 m^2/g total surface area) was manufactured by Johns-Manville and supplied by Alltech Associates; Diamond Shamrock supplied Duolite A-162; Sigma supplied MSC-1.

Instrumentation. Vacuum distillation of parfait column effluents was performed on an FTS Systems model FD-20-84, high-capacity, corrosion-resistant, freeze-drying apparatus modified as described in the text.

Basic Procedures. The synthetic hard water used in all of the parfait column studies consisted of 0.070 g of $NaHCO_3$, 0.156 g of $CaSO_4 \cdot 2H_2O$, and 0.047 g of $CaCl_2$ per liter. This solution was prepared in 8-L batches just before use. The final pH was 7.2.

For parfait column studies, standard solutions (400 μg/mL) of each organic solute except phenanthrene were prepared in a suitable solvent (acetone, water, or methanol), and 1.00 mL of the standard was added to 8 L of the synthetic hard water with rapid stirring. This yielded a final organic solute concentration of 50 μg/L and a solvent concentration of less than 126 mg/L. In the runs simultaneously involving 16 compounds, four different mixtures of solutes were prepared so that the final solvent concentration was 500 mg/L. Phenanthrene was spiked to a final concentration of 1.0 μg/L, rather than 50 μg/L. Solutions were not degassed before application to a parfait column. No adjustments of pH were necessary after adding solutes to the synthetic hard water.

Stock solutions of 16.0 mg/mL of Fluka humic acid were prepared by adding 1 M NaOH dropwise to a water slurry of the solid in a volumetric flask until the acid was dissolved. This concentrate was diluted with the synthetic hard water to a final concentration of 2.0 mg/L.

The pH of the humic acid solutions behaved as expected for polymers containing weakly acidic groups. The pH of the 16-mg/mL stock, as detected by a combination glass-reference electrode, changed slowly and did not stabilize. A typical value for the pH after 30 min of stirring was 4.5. When the humic solution was supplemented with 0.5 M KCl to aid the equilibrium of protons with titratable groups, the pH came to a stable value after a few minutes. A pH of 3.9 was then recorded. If the humic acid solution was progressively diluted to 20 μg/mL in 0.5 M KCl, the equilibrium pH rose to 6.4. The pH of the humic acid diluted to 2.0 μg/mL in the synthetic hard water was stable and did not differ detectably from the pH of the water before addition of the humate.

Modified Parfait Protocol. Teflon adsorbents were cleaned by 24-h extraction with acetone in a Soxhlet apparatus and stored in acetone. Cleaned Chromosorb T was used to fill a 150-mL silanized addition funnel equipped with a 24–40 male joint on the bottom. A plug of silanized glass wool supported the bed just above the stopcock. The Teflon bed was compressed slightly with a large stirring rod to keep the top of the bed in place while the bed was washed with at least 6 bed volumes (900 mL) of high-purity water. The washed Teflon was kept covered with water and used immediately.

Large quantities of ion-exchange resin MSC-1, a strongly acidic macroporous cation-exchange resin, or Duolite A-162, a strongly basic macroporous anion-exchange resin, were cleaned at one time.

A resin to be cleaned was packed into a 3-L capacity addition funnel and washed first with two volumes of high-purity water. The MSC-1 resin was then washed sequentially with 2 bed volumes of 2.0 N HCl, 5 bed volumes of high-purity water, 2 bed volumes of 1.5 N NaOH, and 5 bed volumes of high-purity water; these HCl, water, NaOH, and water washes were then repeated. The same procedure was carried out with A-162, except that the order of acid and base was reversed. The resins were then extracted for at least 16 h in methanol in a Soxhlet apparatus. The final cleanup steps for MSC-1 consisted of a sequential elution with 2 bed volumes of methanol, 1 bed volume of 1.0 N NH_3 in 12% methanol/88% methylene chloride (made as described later), 2 bed volumes of 12% methanol/88% methylene chloride (solvent 1), and 3 bed volumes of methanol. The same steps were used with A-162 except that HCl replaced NH_3 in the second wash. The resins were considered to be clean at this point, and they were stored in methanol.

The NH_3 or HCl solutions in solvent 1 were prepared by passing the anhydrous gas through the methanol/methylene chloride and protecting the solution from water vapor by using anhydrous $CaSO_4$.

The ion-exchange resins were loaded in a methanol slurry into parfait column addition funnels over a small plug of silanized glass wool. Thirty milliliters of MSC-1 was loaded into a 150-mL funnel, and 65 mL of A-162 was loaded into a 60-mL funnel, filling it and preventing the resin from floating during later elution steps. The total exchange capacity of each bed was at least 50 meq. The resins were drained of methanol, 2 bed volumes of high-purity water was passed over the bed, and then 2 bed volumes of 2.0 N HCl (MSC-1) or 1.5 N NaOH (A-162) was passed through to charge the exchange sites. This process was followed by at least 5 bed volumes of high-purity water until the pH of the effluent (measured with pH paper) was near neutrality. From the initial water wash after the methanol was drained off, the beds were kept covered with solvent (i.e., they were not allowed to run dry). The addition funnels were then connected to each other in order: Teflon was above MSC-1, which was above A-162. A 10-L glass carboy containing the synthetic hard water and any test solute was put in place above the Teflon bed. A siphon flow through 4-mm-i.d. Teflon tubing was started by applying a gentle air pressure to the open hole of a two-holed stopper through which the tubing passed. The other end of the Teflon tubing passed through and beyond a one-holed rubber stopper fitted to the top of the porous Teflon bed of the parfait column. Thus, the water contacted only glass or Teflon surfaces before encountering the first parfait bed.

Flow through the column was initiated, and the column effluent was collected directly into a silanized 12-L round-bottom flask at a nominal flow rate of 15 mL/min, adjusted with the bottom stopcock. The entire parfait setup is illustrated in Figure 1. At the end of a run, 500 mL of high-purity water was passed through the train of adsorbents, which were allowed to drain completely by gravity.

The Teflon and ion-exchange beds were eluted with the eluents as follows: The addition funnels containing each bed were separated from the parfait column, and a separate, empty, addition funnel was loosely mounted on top to supply eluents. The bottom stopcock was closed, and the first eluent was slowly added until air escaped the bed and the bed was covered by 0.5 cm of liquid. The upper addition funnel was then firmly seated, and elution began by opening the bottom stopcock. Flow was adjusted to less than 3 mL/min with

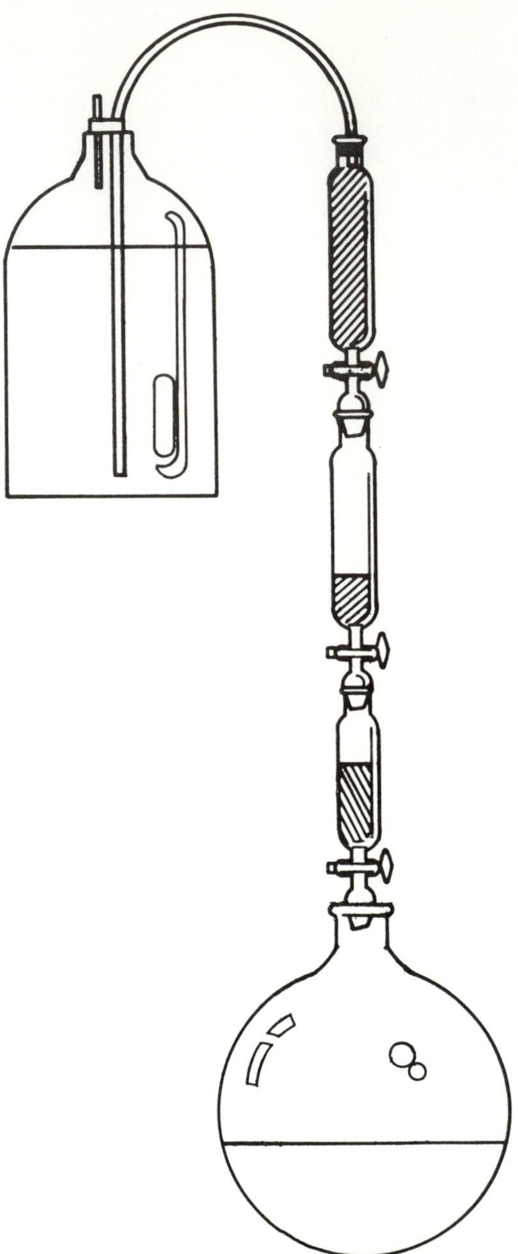

Figure 1. The modified parfait column.

the bottom stopcock. New eluents were added to the upper addition funnel as it ran dry. The adsorbent bed in the lower funnel was kept constantly covered with 1-5 mm of eluent.

The Teflon bed was eluted with 4 bed volumes of methylene chloride, which was collected directly into a separatory funnel. The water layer that formed in the eluate was separated and discarded. This methylene chloride became parfait fraction 1 (F1).

The addition funnels containing the ion-exchange beds were dried and eluted with a series of solvents. To dry each bed, 2 bed volumes of absolute methanol was passed through, followed by 2 bed volumes of absolute ethanol, followed by another 2 bed volumes of methanol. Each bed was then washed with 2 bed volumes of solvent 1. This washing diminished the shock to the resins of the next solvents. The ethanol, methanol, and solvent 1 eluates were discarded.

The MSC-1 bed was eluted with 150 mL of 0.33 N NH_3 in solvent 1, 4 bed volumes of 0.1 N NH_3 in solvent 1, and then 4 bed volumes of methanol. The combined eluates became parfait fraction 2 (F2).

The A-162 bed was eluted first with 150 mL of 0.33 N HCl in solvent 1 (F3) and then with 4 bed volumes of 0.10 N HCl in solvent 1 (F4), 4 bed volumes of methanol (F5), and 4 bed volumes of 0.10 N HCl in methanol (F6).

Each of these six organic fractions was reduced to a volume of about 5 mL in a Kuderna-Danish concentrator under a three-ball Snyder column and then to a volume of 1.0 mL under a gentle stream of dry nitrogen. These concentrates, or their methylated derivatives adjusted to the same concentration, were analyzed as described in the next section.

The aqueous effluent from the parfait column was exposed to vacuum in a silanized 12-L flask on a high-capacity freeze dryer. The effluent was stirred with a magnetically coupled stirring bar and heated by immersion of the lower half of the flask in a 30 °C constant-temperature water bath. The rate of distillation was controlled to about 8 L/24 h with a needle-valve air bleed. Distillation was continued to a volume of about 500 mL. Then the concentrate was transferred to a smaller flask, and the volume was further reduced to 10 mL on a conventional rotary evaporator.

Analytical Methods. Gas chromatography (GC) with a flame ionization detector was used to separate and quantify all solutes evaluated on parfait columns except glycine, glucose, phenanthrene, and 5-chlorouracil. Acidic compounds were methylated prior to chromatography with at least a 100-fold molar excess of diazomethane in ethyl ether, prepared from Diazald by using the Aldrich procedures. All concentrates were chromatographed on 3% Carbowax 20M on 60–80-mesh, acid-washed, dichlorodimethylsilane-treated Chromosorb W. The Carbowax packing is insensitive to water and thus permits direct injection of the vacuum distillate of the parfait column effluent to detect isophorone and caffeine in fraction 7. Generating conditions for chromatography on Carbowax were as follows: detector, 220 °C; injection port, 210 °C; and column, 100–200 °C at 8 °C/min. The principal difficulty with the Carbowax packing was that it could not be used at temperatures above 200 °C. All GC was initialized to a nitrogen carrier gas flow rate of 30 mL/min at 100 °C. Columns were frequently repacked, especially if there was noticeable deterioration in peak resolution.

In typical experiments, 4-μL aliquots of concentrates were injected. Recoveries were determined from the areas of peaks, and each determination was calibrated with standards prepared from the same stock solutions used to spike the column influent. If the identity of a peak determined by retention time alone was uncertain, that extract was spiked with a known amount of standard and rechromatographed to identify the compound and to enable an internal calibration of its amount. The standards for acidic solutes were always methylated at the same time as each experimental concentrate.

Reconstruction experiments showed that methylation efficiency was the same in the solvent used to prepare the standard and that obtained from experimental parfait column concentrates. Chromatography was usually completed within 36 h of final concentration or methylation of a sample. Samples were stored in Teflon-lined crimp-sealed vials at −20 °C.

Glycine was analyzed by the fluorescence of its fluorescamine derivative with excitation at 366 nm and emission at 480 nm (7). A standard working curve prepared simultaneously with the analyte permitted quantitation.

Glucose was analyzed by the glucose oxidase method (*8–10*) by using reagents obtained from Sigma. Unknowns were quantified by comparing absorbances at 500 nm with those of standards prepared at the same time.

Phenanthrene was analyzed directly by fluorescence with excitation at 365 nm and emission at 394 nm and quantified by comparison to a standard curve. 5-Chlorouracil was analyzed directly by absorbance at 270 nm. Attempts to determine 5-chlorouracil as the pyrolysis product of its tetramethylammonium salt (*11*) were unsuccessful. Hypochlorite was determined by the ferrous-*N*,*N*-diethyl-*p*-phenylenediamine titration method (*12*).

Results and Discussion

Experiments Determining Parfait Column Composition and Elution Protocol. MODEL COMPOUNDS. The compounds selected by HERL for evaluation of the parfait recovery method were intended to span a broad range of physical characteristics. Especially important were water solubility and volatility. Table I categorizes qualitatively the relative water solubility and volatilities of the model compounds and describes their ionic states in dilute aqueous solutions at neutral pH. Because water solubility and volatility can be observed for all compounds, that is, there are no truly insoluble or nonvolatile compounds, qualitative descriptions such as soluble, slightly soluble, poorly soluble, volatile, or poorly volatile are used. Water solubility behavior is usually more familiar to chemists than relative volatility, especially the volatility of solutes in highly dilute aqueous solutions. The dividing line for solutes characterized in this study as volatile or poorly volatile has been drawn arbitrarily between biphenyl (bp 256 °C) and 2,4′-dichlorobiphenyl (bp >300 °C). Boiling points were used to assign these classifications, although the related property, fugacity, is the actual property of interest. In terms of ionic character, the list is deficient by its lack of cationic model compounds.

LOSSES DURING VACUUM DISTILLATION AND LYOPHILIZATION. The original parfait method used vacuum distillation to concentrate aqueous eluates from each of the parfait beds and the aqueous effluent

Table II. Retention of Biphenyl, Phenanthrene, and Pyrene in Aqueous Solutions during Vacuum Distillation and Lyophilization

	Freeze-Drying		*Vacuum Distillation*	
Compound[a]	*Volume (mL)*	*Compound (%)*	*Volume (mL)*	*Compound (%)*
Biphenyl	4.2	87	5.0[b]	26
(4 μg/mL)	2.0	47	4.9	10
	0.5	6	2.0	6
	—	—	0.5	4
Phenanthrene	3.2	48	4.5	48
(4 μg/mL)	1.0	32	1.5	19
	0.5	13	0.5	13
Pyrene	5.0[b]	100	5.0[b]	100
(0.14 μg/mL)	0.5	62	0.5	19

NOTE: Values refer to the amount remaining from a 5-mL sample.
[a] Solutions were prepared on a generator column (*13*).
[b] Sample was exposed to vacuum for 60 s to degas.

from the parfait column itself. The modified parfait method used this method to concentrate only the aqueous column effluent.

To explore the potential for loss of moderately volatile compounds during vacuum distillation, solutions of biphenyl (4 μg/mL), phenanthrene (4 μg/mL), and pyrene (0.14 μg/mL) were freeze-dried or vacuum-distilled for varying times, and the amount of compound remaining in the flask was determined. The results (Table II) show that vacuum distillation resulted in greater losses than lyophilization and that the ease of loss varied considerably. Biphenyl was so readily lost that the mere degassing of its solution resulted in a loss of about two-thirds of the original compound (*see* Figure 2).

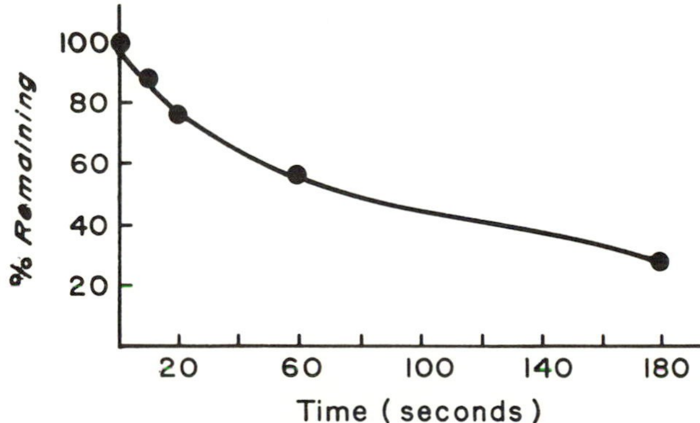

Figure 2. Loss of biphenyl during the degassing of an aqueous solution under vacuum.

To be certain that the compounds were actually escaping from the flask, rather than being adsorbed onto glass, another experiment was conducted with biphenyl (4 μg/mL), quinoline (15 μg/mL), and quinaldic acid (15 μg/mL). One hundred milliliters of each of these solutions was frozen in a 500-mL round-bottom flask and freeze-dried. A trap chilled with dry ice and propanol was placed in line between the flask and the mechanically refrigerated trap of the freeze dryer. Compounds in the flask and the dry ice trap were quantified after approximately half of the water had sublimed from the flask. The results (Table III) showed that the losses of quinoline were accounted for by the material in the trap and that initially the lost biphenyl was recovered from the trap. The biphenyl was so volatile that 10–40% of it apparently escaped from the trap during subsequent lyophilization of half of the original solution. The quinaldate was not lost from the flask. (It will be shown in later sections that D-glucose can be recovered quantitatively in the vacuum distillate of parfait column effluents.) Thus, the losses of aromatic hydrocarbons were due to volatilization under vacuum, especially during the degassing phase of vacuum distillation.

These few experiments suggested that an intuitive concept of relative volatility is inadequate to describe the behavior of dilute hydrophobic solutes under vacuum in aqueous solution. For example, the classification used in Table I, where compounds having boiling points over 300 °C at 1 atm are described as poorly volatile, inadequately predicts the losses of phenanthrene (bp 340 °C) in Table II. Also, relative volatility offers no prediction of the rapid loss of biphenyl during degassing (Tables II and III, Figure 2) and the relatively slow further loss of this compound during subsequent lyophilization (Tables II and III).

These experiments point out the need to avoid vacuum concentration during the recovery of hydrophobic compounds such as the

Table III. Loss of Biphenyl, Quinoline, and Quinaldic Acid from 100 mL of Aqueous Solution during Lyophilization

	Retained in Flask		
Compound	*Volume (mL)*	*Compound (%)*	*Recovered in Trap (%)*
Biphenyl[a]	55, 53	45, 31	44, 29
(4 μg/mL)	100[b]	45	54
Quinoline (15 μg/mL)	52	19	80
Quinaldic acid (15 μg/mL)	50, 52	100, 100	ND

NOTE: ND denotes not detected.
[a] Solution was prepared on a generator column (*13*).
[b] Sample was exposed to vacuum for 60 s.

polycyclic aromatic hydrocarbons. However, they suggest that this recovery method is suitable for compounds that interact strongly with water such as quinaldic acid and glucose. These experiments predict that any hydrophobic compounds reaching the parfait column effluent, fraction F7, will be lost on workup.

ADSORPTION ONTO SILANIZED GLASS. The original parfait column was composed entirely of glass, fiberglass, and Teflon. Preliminary studies showed that fiberglass prefilters, such as those originally used to support the parfait beds, can adsorb appreciable amounts of aromatic hydrocarbons. Silanized glass was tested to see if it had less affinity for such compounds. The breakthrough of caffeine (15 mg/L), pyrene (0.14 mg/L), and EB (10 mg/L) from 5-mL beds of silanized glass wool and silanized 60–80-mesh glass beads was determined (Figure 3). Only caffeine showed no appreciable adsorption. Both pyrene and EB were removed essentially quantitatively from the first two column volumes of solution, and pyrene in the effluent did not reach 50% of the influent concentration until 5 column volumes had been collected. Despite the adsorptivity of these media, the adsorbed compounds could be recovered in acetone washings. Thus, silanized glass and fiberglass were considered acceptable for use in the modified parfait method.

THE FIRST PARFAIT BED. The functions of the first parfait bed are to filter particulate matter from the sample and to adsorb hydrophobic compounds that would otherwise adhere to the polystyrene–divinylbenzene backbone of the ion-exchange resins. Chromosorb T, a porous PTFE originally designed as a stationary-phase support for GC, provides both of these functions. When properly wetted,

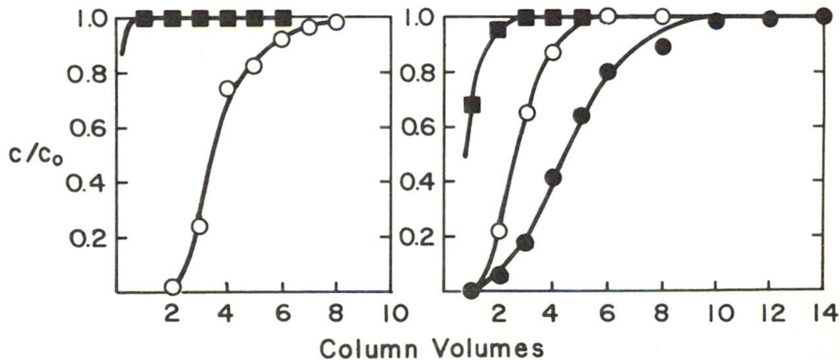

Figure 3. Adsorption of solutes to silanized glass wool and silanized glass beads. Aqueous solutions of caffeine (■), 15 mg/L; ethidium bromide (○), 10 mg/L; or pyrene (●), 0.14 mg/L were passed through 5-mL beds of silanized glass wool (left) or silanized glass beads (right).

Table IV. Recovery of Solutes from Chromosorb T in CH_2Cl_2

Solute	*Recovery*	*Mean ± S.D.*
2,4-Dichlorobiphenyl	83, 76, 89	83 ± 6
Anthraquinone	43, 53, 95	64 ± 27
Stearic acid	56, 53, 85	65 ± 18
Bis(2-ethylhexyl) phthalate	50, 62, 75	62 ± 13
2,6-Di-*tert*-butyl-4-methylphenol	50, 74, 77	67 ± 15

NOTE: All values are percentages. S.D. denotes standard deviation.

Chromosorb T shows the direct correlation of capacity with the octanol–water partition coefficient expected of hydrophobic adsorbents (*6*). Its overall affinity for hydrophobic water contaminants is similar to that of Amberlite XAD-8 (Rohm and Haas), but it has additional affinity for humic acid and for water-soluble, cationic, aromatic dyes. Chromosorb T is more easily cleaned than the XAD resins, and it is more inert, contributing essentially no contaminants to eluates. Table IV illustrates the adsorption and recovery of a series of hydrophobic test solutes at 50 ppb in 8 L of synthetic hard water on a 50-mL bed of Chromosorb T.

ION-EXCHANGE RESINS. A variety of ion-exchange resins (Table V) were considered for use in the parfait column. The resins eventually chosen were selected because they carry strongly acidic or basic exchange groups, they are macroporous, they have relatively higher exchange capacities than otherwise equivalent alternatives, and they are lower in cost. MSC-1 was finally chosen over AG MP-50 primarily on the basis of cost, and A-162 was preferred over other macroporous anion-exchange resins primarily because of its higher exchange capacity. A-162 gave the impression of cleaner eluates than AG MP-1, but this

Table V. Polystyrene Ion-Exchange Resins Considered for Use in the Parfait System

Resin	*Supplier*	*Type*[a]	*Manufacturer's Stated Capacity (meq/mL) Wet*
AG MP-1	BioRad	SB, MA	—
AG MP-50	BioRad	SA, MA	1.7
IR 120	Mallinckrodt	SA, MI	1.9
IR 252	Mallinckrodt	SA, MA	—
IRA 400	Mallinckrodt	SB, MI	1.4
IRA 900	Mallinckrodt	SB, MA	1.0
Duolite A-162	Diamond Shamrock	SB, MA	1.0
Dowex MSC-1	Sigma	SA, MA	2.0

[a] SA = strong acid, SB = strong base, MA = macroporous, and MI = microporous.

observation was not tested rigorously in side-by-side trials. Satisfactory parfait columns could probably be constructed with AG MP-50 in place of MSC-1 and with AG MP-1 in place of A-162.

ELUENTS FOR ION-EXCHANGE RESINS. Any formulation of the parfait method must solve the problem of recovering the organic solutes on the ion-exchange beds and separating them from the ions in the eluent. Because high-ionic-strength eluents are typically used to recover exchanged ions, the problem of separating traces of organic solutes from the eluting ions can be a difficult one. In the original parfait method, the separation was accomplished by using a volatile buffer, 2 M triethylammonium bicarbonate, pH 7.5, to elute the bed. When such an eluate was lyophilized, the eluting ions equilibrated with their neutral forms, which volatilized as triethylamine and CO_2 and left any poorly volatile solutes from the sample in the residue.

Volatile buffers were reconsidered for the modified method. Triethylamine was ruled out primarily because it could not be obtained in high purity and because the secondary and primary amines contaminating it could potentially react with solutes present in the water sample. Preliminary evidence of reaction between ethidium bromide and triethylammonium bicarbonate was obtained, but the reaction product was not characterized. The components of volatile buffers that appeared acceptable on the basis of chemical purity were ammonia, acetic acid, and formic acid. A few exploratory experiments were conducted involving the elution by ammonium formate and ammonium acetate of EB or quinaldic acid exchanged onto AG MP-50 or IRA 900. These experiments showed that 1 M ammonium formate in water was a very poor eluent, but that EB could be eluted from AG MP-50 with 1 M ammonium formate in methanol. Elution was essentially complete with 6 bed volumes of the methanolic eluent, whereas neither methanol alone nor aqueous 1 M ammonium formate was able to elute this solute. This situation pointed out the necessity for a counterion to displace exchanged solutes and, additionally, indicated that the displaced solute be highly soluble in the eluting solvent.

UV-absorbing material, considered to be quinaldic acid, was readily eluted by 1 M ammonium acetate in methanol. However, controls for this experiment showed that a considerable amount of UV-absorbing material could be eluted from columns that had not been loaded with quinaldate; therefore, firm conclusions about the suitability of ammonium acetate could not be made. Glacial acetic acid was also briefly considered in this series of experiments; results were similar to those obtained with ammonium acetate. Later in the study, after MSC-1 and A-162 were chosen as the ion-exchange resins, the choice of an eluting solvent was reinvestigated. By this time, the liberation of contaminating materials from the resin was recognized as the most serious question to be addressed.

The approach tested was to neutralize the H^+ or OH^- counterions on the bed, thereby converting weak organic acids or bases to their neutral form, and to elute with organic solvents in which neutral organic compounds should be soluble. This approach was expected to reduce the exposure of the resins to high-ionic-strength eluents that might liberate contaminants from the resin support matrix. The approach should recover organic solutes in a neutral, organic-soluble form separate from inorganic ions, and the eluted solutes should be easily concentrated in a Kuderna-Danish apparatus, in anticipation of analysis by GC. Full advantage depended upon the exclusion of water from the eluting solvent.

Experiments began by using 12% methanol in methylene chloride (solvent 1) containing anhydrous HCl or NH_3 gas. In general, a concentrated stock of HCl or NH_3 in solvent 1 was prepared, and its concentration was determined by titration. The stock was diluted with fresh solvent 1 to obtain the desired acid or ammonia concentration. All tubing used in this apparatus must be Teflon, or phthalate esters and other contaminants will occur in the eluates, and the solvents must be protected from moisture by admitting air only through drying tubes containing $CaSO_4$.

The protocol finally developed was performed as follows: The resins were gently dehydrated by washing with a series of alcohols. Exchanged solutes were retained during these washings because no mobile counterions were provided in the alcohol. The H^+ or OH^- counterions associated with the resin were neutralized by washing with 1 bed equivalent (50 meq) of HCl or NH_3 in solvent 1, and any neutralized organic acids or bases were eluted in solvent 1 and methanol eluents. Most of the contaminating material released by the ion-exchange resins was found to elute in the alcohol washings and from the A-162 bed in the 0.33 N HCl wash. The A-162 resin was more difficult to elute than the MSC-1 resin. The elution of A-162 was optimized by using 400 μg each of *o*-phthalic acid and trimesic acid in 8 L of synthetic hard water. Respective recoveries of 72% and 43% of these acids were possible with the final protocol.

INTERFERENCES. A number of interferences were identified during the development of the elution protocols. Material adsorbed to the injection port and the head of the chromatographic column packings was liberated and gave artifactual peaks following the injection of concentrated parfait column eluates. This result was particularly true of the acidic eluate of the A-162 column. Frequent cleaning of the injection port and repacking of the first few centimeters of the column mitigated this problem.

Analysis by GC–MS of the contaminating peaks in eluates from a blank control column revealed ethyl hexadecanoate, methylbutyl phthalate, stearic acid, methyl 4-methyldodecanoate, dioctyl phthalate (two peaks), and an alkyl benzene. An eighth peak corresponded to an isomer of geranyl citronellal [$(H_3C)_2C{=}CHCH_2CH_2C(CH_3){=}CHCH_2CH_2C(CH_3){=}CHCH_2CH_2C(CH_3)CH_2CHO$]. The A-162 eluate showed bis(2-ethylhexyl) phthalate.

These experiments pointed to plasticizers in the eluting solvents. Although plastic tubing had been avoided in all of the parfait column operations, a short section of Tygon tubing had been used to connect the anhydrous NH_3 and HCl bottles to the eluting solvent reservoirs. When this tubing was replaced with Teflon tubing, the profile from F2 improved, although it still contained some contaminants. Figures 4, 5, and 6 show the GC profiles of various eluates of blank parfait columns, solute standards, and solutes recovered from a spiked column. These figures show the results typical of the best chromatograms obtained during the study. Among them, the least acceptable profiles were obtained from F3, the concentrate of the most acidic methanol–methylene chloride solvent, the one that induces the greatest shock to the A-162 support matrix. The only precaution that was found to reduce contamination of the GC profile of this eluate was the use of a freshly packed Carbowax phase for chromatographic support.

Concentrates of parfait columns were routinely assayed within 48 h of preparation. Slow evaporation of F1 was observed in some experiments, even when Supelco Teflon-lined septum vials were used. Individual concentrates were rechromatographed at various times, and it appeared that profiles could not be reliably reproduced after several days, even when the concentrates were stored at −20 °C. Certain compounds, particularly BHT and furfural, were found to change their profiles in a matter of days, probably because of decomposition.

Because interfering peaks could not be completely removed from parfait eluates, blank (unspiked) columns were routinely run in parallel with each spiked sample, and only peaks that could be unequivocally ascribed to the spiked solute were quantified.

INFLUENCE OF HYPOCHLORITE ON PARFAIT COLUMNS. One potential use of the parfait method is the recovery of organic matter from drinking water. To test for the interaction of chlorine disinfectant with column components or eluents, the influence of 2 ppm of hypochlorite was assessed in an unspiked control column. Each eluate was assayed for hypochlorite by using the ferrous *N*,*N*-diethyl-*p*-phenylenediamine titrimetric method (*12*). No hypochlorite was detected. Each eluate was also analyzed by GC and found to be virtually identical to a blank column without hypochlorite run simultaneously.

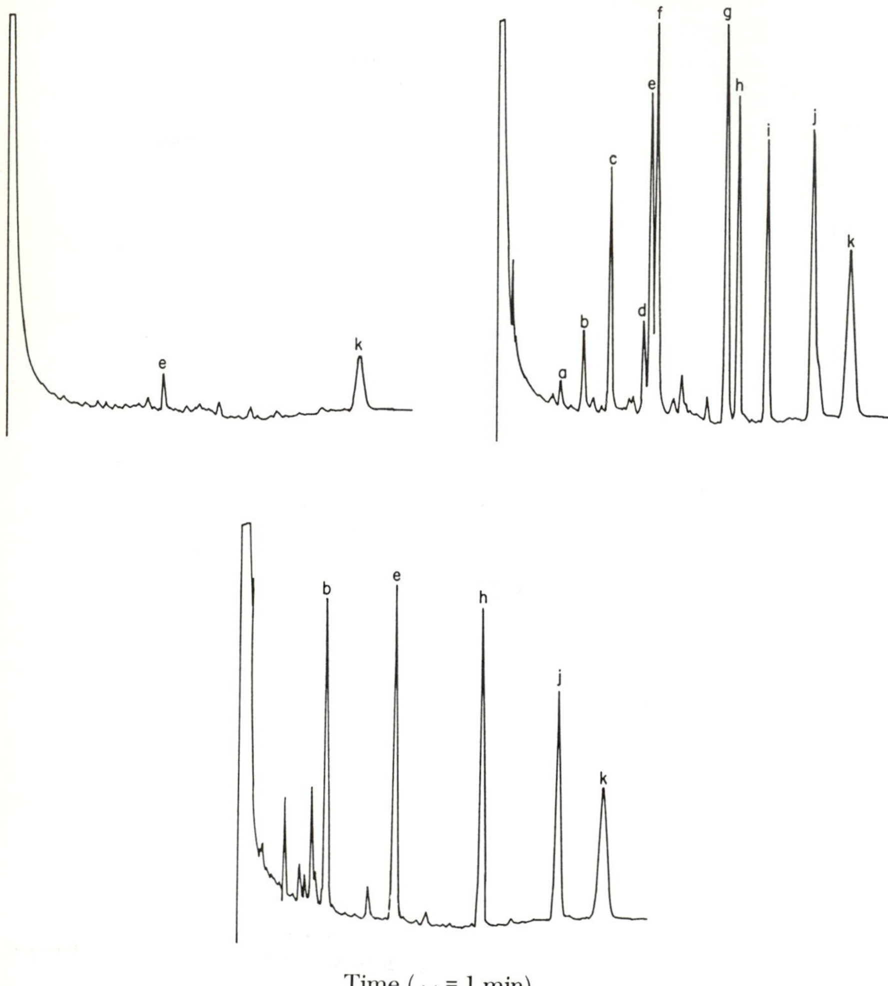

Figure 4. Gas chromatograms of parfait fraction F1 from an unspiked control column (top left), column 120 (top right), and one of the standards used to calibrate the chromatogram (bottom). Designated solutes are a, furfural; b, isophorone; c, 1-chlorododecane; d, quinoline; e, BHT; f, biphenyl; g, 2,4-dichlorobiphenyl; h, stearic acid; i, 2,2′,5,5′-tetrachlorobiphenyl; j, anthraquinone; and k, bis(2-ethylhexyl) phthalate.

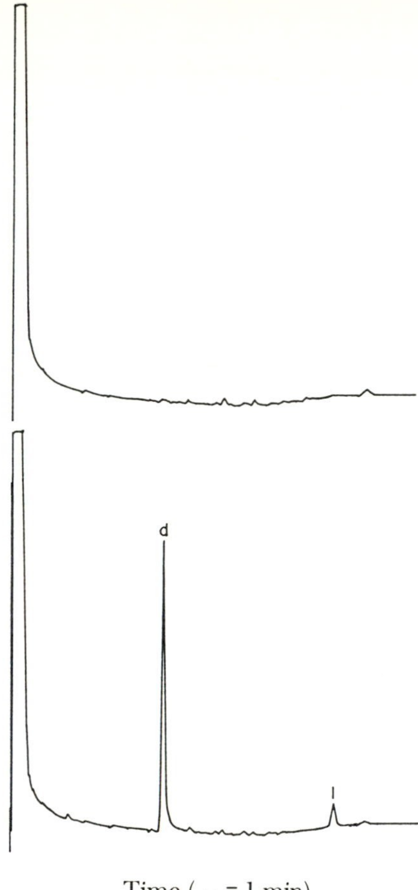

Figure 5. Gas chromatograms of parfait fraction F2 from an unspiked control column (top) and column 120 (bottom). Designated compounds are d, quinoline; and l, caffeine.

Recovery of Test Solutes. The recovery of test solutes is reported in the data from columns 102–123 (Tables VI–IX). Other than the suite of solutes tested on each column, the columns differ as follows: Columns 102–109 used a 50-mL Teflon bed, columns 110–120 used a 150-mL bed, columns 109–120 included 2.0 ppm of humic acid in the influent, and column 102 used 8 bed volumes of methanol for fraction 5, rather than the usual 4 bed volumes. Columns 117, 118, and 120 were replicate runs including 16 test solutes at once, excluding only those solutes that could not be analyzed reliably in the presence of the solutes included. Column 119 was an unspiked control column run with humic acid and used to identify contaminants in the eluates of columns 117,

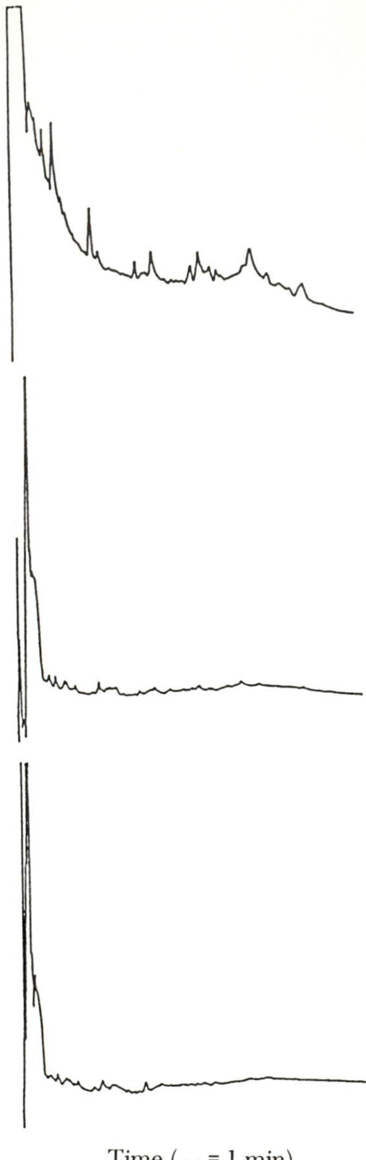

Time (⌴ = 1 min)

Figure 6. Gas chromatograms of parfait column fractions F3 (top), F4 (middle), and F5 (bottom) from an unspiked control column.

Table VI. Recovery of Compounds Found Exclusively on Teflon

Compound[a]	Percent Recovery on Column Number[b]						Mean ± S.D.[c]
	106	*108*	*109*	*117*	*118*	*120*	
Stearic acid	105.0	—	—	106.0	93.6	95.6	98.4 ± 6.6
2,4′-Dichlorobiphenyl	73.8	—	—	81.5	75.1	80.6	79.1 ± 3.5
2,2′,5,5′-Tetrachlorobiphenyl	—	69.0	77.0	85.1	80.0	89.1	84.7 ± 4.6
Bis(2-ethylhexyl) phthalate	61.6	—	—	104.0	100.0	78.8	94.3 ± 13.5
1-Chlorododecane	—	29.7	37.6	70.6	62.4	78.3	70.4 ± 8.0
Biphenyl	—	49.0	53.5	90.1	80.7	96.8	89.2 ± 8.1
Anthraquinone	65.8	—	—	107.0	102.0	119.0	109.0 ± 9.0
2,6-Di-*tert*-butyl-4-methylphenol	129.0	—	—	103.0	89.1	82.8	91.6 ± 10.3

NOTE: Column 112 had a 72.3% recovery of 2,4′-dichlorobiphenyl, and column 114 had a 47.8% recovery of methyl isobutyl ketone.
[a] All compounds were exposed to the parfait column at 50 ppb in synthetic hard water.
[b] Columns 109–120 were also spiked with 2.0 mg/L of humic acid. Columns 106–109 had a 50-mL bed of Chromosorb T; columns 112–120 had a 150-mL bed of Chromosorb T.
[c] Means only include data from columns 117, 118, and 120; S.D. denotes standard deviation.

118, and 120. All seven fractions from these columns were analyzed by GC after methylation. If a chromatographable compound was not detected in a fraction from these columns, it is not reported. Only fraction 7, the vacuum distillate of the column effluent, was analyzed by the glucose oxidase method.

HUMIC ACID. Humic acid did not contribute detectable impurities to the eluates of blank parfait columns. This result was apparently due to the insolubility of humate in the organic solvents used to elute the Teflon and ion-exchange beds and the inability of the humate to volatilize in the GC. Humic acid did, however, distribute itself throughout the parfait column, as indicated by the observation of color entering the column effluent, F7. When 16 mg of humate in 8 L of synthetic hard water was passed through a parfait column having the Teflon bed divided into three sequential 50-mL beds, 8.9%, 5.0%, and 2.9% of the total humate were found in the aqueous phases that separated upon elution of these beds, as indicated by absorbance at 200 nm. The column effluent from this experiment contained 5.1% of the humate applied. The majority of the humate applied was found as color adsorbed to PTFE, and it did not elute into methylene chloride. Conditions to elute it from PTFE were not explored.

The most remarkable finding regarding humic acid was its failure to adversely affect the recovery of the test solutes. The results in Tables VI and VII show no significant interference by humate in the recovery of the test solutes, except perhaps for caffeine and 2,4-dichlorophenol. With 2,2′,5,5′-tetrachlorobiphenyl, bis(2-ethylhexyl) phthalate, 1-chlorododecane, biphenyl, and anthraquinone, the columns containing

Table VII. Percent Recoveries of Test Solutes by Fraction

Trimesic Acid *Column No.*	*F3*	*F4*	*F5*	*F6*	*Total*
102	7.6	6.8	41.0	15.2	70.6
104	4.8	4.3	19.2	6.9	35.2
106	7.1	7.1	19.0	6.9	35.2
112	4.9	6.1	40.2	24.4	75.6
117	ND	1.2	14.9	9.3	25.4
118	3.1	2.5	20.5	9.3	35.4
120	TR	1.5	6.4	ND	7.9
Mean ± S.D.	1.0 ± 1.8	1.7 ± 0.7	13.9 ± 7.1	6.2 ± 5.4	22.9 ± 13.9

Isophorone *Column No.*	*F1*	*F2*	*F3*	*F7*	*Total*
114	35.8	1.5	0.6	5.0	42.9
117	20.3	TR	TR	TR	20.3
118	19.8	TR	TR	TR	19.8
120	35.7	ND	ND	ND	35.7
Mean ± S.D.	25.3 ± 9.0	—	—	—	25.3 ± 9.0

Caffeine *Column No.*	*F1*	*F2*	*F7*	*Total*
108	15.9	33.4	51.1	100.4
109	15.3	32.5	35.0	82.8
117	ND	7.4	26.4	33.8
118	ND	9.8	33.0	42.8
120	ND	8.1	ND	8.1
Mean ± S.D.	ND	8.4 ± 1.2	19.8 ± 17.5	28.2 ± 18.0

2,4-Dichlorophenol Column No.	*F1*	*F2*	*F3*	*F4*	*F5*	*F6*	*Total*
108	2.3	12.9	31.6	34.5	4.6	TR	85.9
109	1.7	8.9	7.0	5.8	5.8	4.7	33.2
117	3.9	ND	28.1	18.8	10.9	14.1	75.8
118	5.5	ND	10.9	18.0	7.8	11.7	53.9
120	ND	ND	19.8	4.0	1.0	ND	24.8
Mean ± S.D.	3.1 ± 2.8	ND	19.6 ± 8.6	13.6 ± 8.3	6.6 ± 5.1	8.6 ± 7.5	51.5 ± 25.6

Furfural Column No.	*F1*	*F3*	*F4*	*F5*	*F6*	*Total*
114	ND	11.7	26.3	ND	ND	38.0
117	29.5	1.0	ND	ND	ND	30.5
118	6.1	16.3	ND	ND	3.7	26.1
120	22.8	10.6	9.9	7.0	TR	50.3
Mean ± S.D.	19.5 ± 12.0	9.3 ± 7.7	3.3 ± 5.7	2.3 ± 4.0	1.2 ± 2.1	35.6 ± 12.9

Quinoline Column No.	*F2*	*Glucose Column No.*	*F7*
MSC-1 #2	57.0	106	56.4
104	80.8	107	119.0
112	46.6	110	97.5
117	46.3	117	89.4
118	51.3	118	67.8
120	111.0	120	105.0
Mean ± S.D.	69.4 ± 35.8	Mean ± S.D.	87.4 ± 18.7

NOTE: TR = trace; ND = not detected. All means include only columns 117, 118, and 120.

humic acid averaged a higher recovery than those lacking it. However, this result was probably due to the use of a 150-mL bed of Teflon in the later columns rather than to the presence of the humic acid. The tetrachlorobiphenyl, chlorododecane, and biphenyl were also tested on a pair of columns using a single 50-mL bed of Teflon and lacking (column 108) or containing (column 109) humic acid. A high recovery was seen in this pair of columns, but it was not as high as that seen in the columns using the 150-mL Teflon bed and containing humate (columns 112–120).

A similar comparison can be made with caffeine on columns 108 and 109, where recovery decreased about 20% in the presence of the humate. Caffeine recovery can also be compared on column 109, which contained humate and had a 50-mL Teflon bed, and columns 117, 118, and 120, which contained humate and had a 150-mL Teflon bed. Caffeine was recovered from column 109 fraction F-1 but not from columns 117, 118, and 120. Moreover, the total recovery of caffeine from columns 117, 118, and 120 was less than half that from column 109.

These results could be explained if the humic acid retained on the Teflon bed reduced the elution efficiency of caffeine into the methylene chloride eluent. This situation might occur either by an interaction of humate and caffeine adsorbed onto Teflon or during extraction of the residual aqueous phase recovered from methylene chloride elution of the Teflon. Substituted xanthines such as caffeine show an anomalously high adsorption to Teflon, relative to their water solubility, possibly because their water solubility is enhanced through self-association in aqueous solutions (*6, 14, 15*). Humic acids might stabilize or absorb such complexes, reducing partition of caffeine from water to methylene chloride. Aromatic compounds including phthalate esters and DDT have been found to associate with dissolved humic acids (*16, 17*), although it is not known if caffeine behaves in this way. Simple comparisons of the partition of caffeine between methylene chloride and water or methylene chloride and humate solutions would test this possibility.

The only other effect possibly attributed solely to humic acid involved 2,4-dichlorophenol, the most broadly distributed compound tested in this study. In columns 108 and 109, the humic acid apparently decreased the ease of elution of the chlorophenols because lower overall recoveries were obtained from column 109, which included the humate, than from column 108, which was humate-free. Also, recovery was detected in F6 of column 109 but not in F6 of column 108. This result suggests that humate enhanced binding of the phenol to the column. The reproducibility of 2,4-dichlorophenol recovery among the various parfait fractions was poor, as illustrated by the results from replicate columns 117, 118, and 120. Because of the variability, the differences in

2,4-dichlorophenol recovery in columns 108 and 109 might reflect the reproducibility of the method rather than the influence of the humate.

GLYCINE AND QUINALDIC ACID. Two solutes were not accounted for satisfactorily in this study: quinaldic acid and glycine. Most of the data concerning these two compounds cannot be interpreted unequivocally. Analytical problems, impurities eluting from the ion-exchange resins, and sources of loss unrecognized at the time of analysis account for this situation.

Quinaldic acid quantified by GC showed quite low apparent recoveries. A low yield of chromatographable product from methylation and interference with the chromatography by water probably accounted for this result. For example, when a standard solution containing equal weights of benzoic, trimesic, phthalic, and quinaldic acids was methylated and chromatographed, the peak areas obtained for quinaldic acid were much smaller than those obtained for the other acids. In one experiment, the methylated product from an original 400 ng of benzoic acid gave a peak area of 1.04×10^6 units, 400 ng of trimesic acid gave 1.29×10^6 units, but 400 ng of quinaldic acid in the same solution gave only 1.01×10^5 units.

Although these peak areas were sufficient to quantify quinaldic acid on Carbowax columns, residual water present in the extracts of the ion-exchange resins interfered in previous experiments using water-sensitive polydimethylsiloxane stationary phases. The change to a polyethylene glycol stationary phase eliminated the water interferences, and good linear calibration curves for methyl quinaldate were then obtained.

It was a surprise, therefore, when the recovery studies detected only traces of quinaldic acid (columns 112 and 122). The traces occurred in fractions F5 and F6, the methanolic eluates of the anion-exchange resin. More than 95% of the quinaldic acid was not found.

Biodegradation was rejected as an explanation for the loss of the quinaldic acid. First, it was recalled that eluates of parfait beds consisted of anhydrous neutral acidic or basic organic solvents, all inimical to microbial life. Second, it was noted that no other solutes, except glycine, were lost from the parfait system in substantial amounts, including readily metabolized solutes such as stearic acid, caffeine, and glucose. In columns 112 and 122, substantial recoveries of the solutes accompanying quinaldic acid were obtained. In preliminary experiments, benzoic and phthalic acids accompanying quinaldate were recovered in substantial amounts, but the quinaldate was not. Biodegradation was most likely in F7 because this fraction did not contact bacteriostatic agents, unlike the fractions obtained from the other parfait column beds. However, glucose, the most readily biodegraded solute tested in these experiments, was recovered quantitatively from fraction F7. To explain the loss of quinaldate through biodegradation, it would be necessary to postulate the selective biodegradation of this compound in the presence of equally available and susceptible materials, often under physical conditions inhibitory to microbial activity. This possibility seemed unlikely.

It is possible that amphoterics could be lost from the parfait method if they were soluble in the alcohols used to dehydrate the ion-exchange beds. Usually compounds can be eluted from an ion-exchange bed only if they are provided with a mobile counterion or if they can be converted to a neutral species by titration with acid or base. Amphoterics, however, can be associated with an ion-exchange bed without actually exchanging for the bed's original mobile counterion. For example, if quinaldic acid enters a cation-exchange bed in the H^+ form, the low pH prevalent there may both neutralize its carboxylic acid function and protonate its aromatic nitrogen. These occurrences would create a cation associated with the stationary anion of the ion-exchange bed but not actually exchanged for the original proton counterion. When the bed is neutralized by ammonia, the protonated quinaldinium ion will be converted to anionic quinaldate, and it will elute if it is soluble in the mobile phase. But if the bed is first washed with an ion-free solvent such as ethyl alcohol, the quinaldinium can, in principle, dissociate into neutral quinaldic acid and a proton, and the neutral acid may be eluted in the alcohol. Quinaldic acid is soluble in alcohol. Thus, it is at least possible that solubility in the dehydrating alcohols accounts for the virtually quantitative disappearance of quinaldic acid from the parfait system. If this hypothesis is correct, collection, concentration, and analysis of the dehydrating and conditioning solvents used in the standard protocol will recover the quinaldic acid. However, this prediction has not been tested experimentally.

The behavior of glycine in the parfait method also poses some questions. Table VIII shows the recovery of glycine from both nonstandard and standard parfait columns in which it was tested. In all but the last column described, the glycine was quantified as its fluorescamine derivative. In columns prior to number 104, no blank control columns were prepared, and all fluorescamine-active material is reported as glycine. From columns 104 onward, control columns were processed in parallel by using unspiked synthetic hard water, and the values reported as glycine represent the differences in fluorescamine-active material in corresponding fractions from spiked and unspiked columns. Control columns run with columns 104–107 showed no appreciable fluorescamine-active material in the column effluent, but color produced by the blank column's MSC-1 eluates corresponded to the recovery of about 50 μg (12%) of the glycine applied. In column 110, humic acid was also present, and the apparent recovery of glycine in the column

Table VIII. Recovery of Glycine on Parfait Column

Column No.	*Adsorbent: Eluant*[a]	*Percent Recovered*[b]
28, 29, 31[c]	Teflon: ethyl ether	TR, TR, TR
	AG MP-50: 95% methanol;	TR, TR, TR
	1 M ammonia in 95% methanol;	11, 10, 14
	1% ammonium acetate in 95% methanol	3, 1, 3
	AG MP-1: 95% methanol	TR, TR, TR
	1 M acetic acid in 95% methanol;	TR, 2, 2
	1% ammonium acetate in 95% methanol	2, 2, 3
	effluent	14, 23, TR
	total	30, 38, 22 (30 ± 8)
67, 68, 70[c]	Teflon: methylene chloride	TR, TR, NA
	MSC-1: 1 M ammonium hydroxide;	43, 19, 21
	acetone, then methylene chloride	TR, 2, NA
	IRA 400: 3 M ammonium hydrate;	TR, ND, NA
	acetone, then methylene chloride	TR, TR, NA
	effluent	4, ND, 1
	total	47, 21, 22 (30 ± 15)
104, 107	F2	5.9, 9.0
	F7	ND, ND
	total	5.9, 9.0
106	F2;	13.1
	4 bed volumes of water[d]	2.7
	F7	ND
	total	15.8
110	F2;	17.0
	5 bed volumes of 1 M ammonium acetate[d]	10.0
	F7	6.5
	total	33.5
123[e]	F1	0.052
	F2	0.289
	F3-6	0.835
	F7	0.130
	Teflon	1.51
	MSC-1	9.21
	A-162	0.001
	total	12.027

[a] F1-F7 are fractions from the standard parfait column.
[b] TR = trace; NA = not analyzed; ND = not detected. Entries correspond to each respective column number.
[c] Nonstandard parfait columns.
[d] Further elution of MSC-1 after standard elution yielded fraction F2.
[e] One-tenth scale standard parfait column using ^{14}C-glycine. Eluates and drying-conditioning washes from individual ion-exchange resins were pooled; Teflon, MSC-1, and A-162 denote radioactivity on these beds after elution (*see* text).

effluent may actually represent differences in color produced by humic material reacting with fluorescamine.

The most revealing results were for columns 106 and 110, in which water or ammonium acetate was used to elute the MSC-1 bed after the standard parfait elution had been performed. In column 110, an additional 10% of the applied glycine was recovered. This result suggested that a substantial amount of the glycine was still present unrecovered on this bed.

In column 110, it is also theoretically possible that glycine complexed with the added humic acid and that it was sequestered in the aqueous phase of the Teflon eluate and bound to the Teflon bed. To test this explanation, a 1/10-scale parfait column was constructed and 4 μCi of ^{14}C-glycine, 40 μg total, was applied in 800 mL of synthetic hard water (column 123). In this experiment, the alcohol and solvent 1 conditioning washes were combined with the standard eluates of each bed before counting. These solutions were not concentrated before counting. Quench correction was by the channels ratio method.

The data showed almost no recovery of glycine in the eluates. Furthermore, only 10.7% of the glycine could be accounted for on the Teflon and MSC-1 beds after elution. However, the data for the MSC-1 and Teflon beds are only minimum values because the correction for quench by the channels ratio method is not reliable in heterogeneous systems. To estimate the actual amount of glycine on the MSC-1 resin, decreasing weights of the resin were counted, the apparent efficiency from the channels ratio was used to quantify the isotope detected, and the percentage of the applied isotope was calculated. As the weight of resin counted was decreased from 786 to 13 mg, the observed counts per weight of resin rose sharply, and the calculated amount of glycine detected rose from 1.3% to 9.2% of the isotope applied. This increase in counts per milligram of resin was not a simple function and thus complicated attempts to estimate the limiting counts per unit weight. Therefore, the reported yield of 9.21% of the applied isotope, which is the value calculated from the counting of 13 mg of MSC-1, should be considered a minimum value.

Although a mass balance for glycine was not achieved, several conclusions can be drawn from these experiments. First, the parfait column removed glycine essentially quantitatively from the water (only 0.13% of the isotope was recovered in the column effluent). Second, the glycine was bound strongly to the cation-exchange bed and was eluted poorly by the standard protocol, as evidenced by the results from columns 106, 110, and 123. The results from columns 28–31 and 67–70 are difficult to interpret because of the possibility that material eluting from the column beds reacted with fluorescamine. Also, these experiments shed little light on the possibility that glycine reacts with humic acids, which are known to bind glycine (*18*). The only column run with humic acid, column 110, showed the familiar pattern of partial recovery of glycine from the cation-exchange bed and a small additional recovery in the column effluent. If the effluent value was genuine glycine, then the humic acid may have been responsible for mobilizing the glycine to the effluent. In the control column for this experiment, however, the background color was half that in the spiked column. The difference in these values was the basis for the glycine reported in the table. The validity of the results rests upon the reproducibility of the humic acid level in the effluents in these columns. However, because only a small percentage of the humate applied appeared in the effluent, high reproducibility of humate levels seems unlikely.

Recovery of Other Solutes on Standard Parfait Columns. The porous Teflon bed gave the most reproducible recoveries of test solutes. Table VI identifies the solutes found exclusively on Teflon and shows their recoveries. The influence of increasing the Teflon bed volume from 50 to 150 mL was mentioned earlier. From column 106, it is evident that BHT and stearic acid were the two most strongly adsorbed solutes. The least well-recovered solutes in this group were 1-chlorododecane and methyl isobutyl ketone; however, both of these solutes are so readily volatilized that losses during concentration of eluates must be considered a likely source of their low recovery.

Glucose was recovered quantitatively in the column effluent, whether humic acid was present or absent. Presumably, all other poorly volatile, neutral, hydrophilic solutes would also be recovered in this fraction.

Among compounds that adsorbed to Teflon and to at least one other bed, isophorone, caffeine, furfural, phenanthrene, and 2,4-dichlorophenol must be mentioned. Isophorone and phenanthrene were recovered essentially only from Teflon, minor amounts being found in other fractions. The finding of phenanthrene in F7 must be considered provisional because its assay was by fluorescence and the value reported was based on a very small signal difference between the F7 of the control and experimental columns. Caffeine was adsorbed to Teflon and the cation-exchange bed and was found in the column effluent. As mentioned earlier, caffeine's recovery was decreased slightly when humic acid was included (column 109 compared with 108) and decreased strikingly when the larger Teflon bed and 15 other solutes were included (columns 117, 118, and 120). In columns 117, 118, and 120, caffeine was no longer recovered from Teflon; its absence from F7 of column 120 is unexplained and is inconsistent with the previous four trials reported in the table.

Three compounds recovered from parfait columns were also previously tested for breakthrough from 5-mL Teflon beds (*6*). The capacity factors for these compounds and their recoveries from the Teflon bed of a parfait column showed a rough correlation. Phenanthrene, which was tested in the parfait column only in the presence of humate, was recovered essentially quantitatively from the 5-mL Teflon column and had a capacity factor of 368. About 15% of the caffeine applied to a parfait column in the absence of humate could be recovered from Teflon, and caffeine showed a capacity factor of 22. Only about 2% of the 2,4-dichlorophenol applied to parfait columns could be recovered on Teflon; its capacity factor was 5.6. It may therefore be anticipated that compounds following the inverse correlation of solubility with capacity factor and having a capacity factor greater than about 20 should be detectably absorbed to the Teflon bed of a parfait column. Simply increasing the volume of the Teflon bed may also increase the absolute recovery of adsorbable solutes that have modest values of k'. For this reason, a 150-mL bed of Teflon per 8 L of water may not be the ideal bed size; a larger bed may be better.

Unrecovered Solutes. Although not all of the applied solutes were recovered in parfait column eluates, reasonable suggestions can be made about the locations of the missing compounds. For example, the unrecovered methyl isobutyl ketone, 1-chlorododecane, and chlorobiphenyls were surely lost from the F1 eluate by vaporization during concentration.

Trimesic acid, on the other hand, appeared to be so strongly adsorbed to the anion-exchange resin that it was incompletely recovered in the eluate. As discussed earlier, glycine also appeared to bind tightly and to elute incompletely by the standard protocol. Both of these compounds probably could be eluted with aqueous eluents of high ionic strength, but this process would create the problem of recovering the solute from the eluate. Volatile buffers may offer a solution to this problem.

Isophorone, 5-chlorouracil, and quinoline were probably lost by vaporization from the column effluent, F7. Each of these solutes is sufficiently water-soluble to not adsorb to Teflon, but each is volatile enough to be lost during the vacuum concentration of F7. Recall that only 20% of 15 mg of quinoline was recovered during lyophilization of a solution to half of its original volume (Table III). Isophorone and 5-chlorouracil were detected in F7, and at least part of the unrecovered fraction of each of these compounds must have been lost during the vacuum concentration of this effluent. (Eluates of the anion-exchange bed were not analyzed for 5-chlorouracil, which also could have been present on this bed; *see* Table IX.) Quinoline was not detected in F7,

Table IX. Percent Recovery of Phenanthrene and 5-Chlorouracil

Fraction	*Phenanthrene*	*5-Chlorouracil*
F1	98.8	3.0
F2	ND	15.8
F7	1.2	10.0
Total	100.0	28.8

NOTE: ND denotes not detected.

but its water solubility and weak basicity suggest that any of it that escaped the cation bed would have gone to the effluent where it would have vaporized.

Furfural is such a reactive, volatile, and water-soluble compound that its loss could be due to several causes. Oxidation and volatilization from eluates, especially from the column effluent, are the most likely sources of loss. Humic acid seemed to affect the recovery of only two solutes: caffeine and 2,4-dichlorophenol.

Future Use of the Parfait Method. The object of this study was to understand the behavior of a broad range of chemical types in this recovery method. On the basis of this evaluation and similar work with other recovery methods, a synthesis of methods is to be proposed that would provide a comprehensive recovery of chemical contaminants in water.

The study of the parfait method reported here shows that it does not recover all classes of trace contaminants in water with equal efficiency. Volatile compounds are readily lost in this method. Contamination of eluates during elution of the ion-exchange beds is also a major problem. Even if this contamination were acceptable, the elution of these beds is not complete, as illustrated by the behavior of trimesic acid and glycine. Porous Teflon, on the other hand, offers a means to quantitatively and cleanly recover a set of water contaminants, albeit a set that was not the primary objective of the original parfait method.

Strengths of the method include its ability to quantitatively recover neutral, nonvolatile solutes such as glucose and its exceptional recovery of aromatic cations and humic acids on Chromosorb T. The exceptional affinity of aromatic cations for porous PTFE suggests its use to recover pollutants such as paraquat and diquat. The affinity of humate for porous PTFE should also be studied more intensively to learn whether other humic and fulvic acids, or fractions thereof, are adsorbed.

The original concept of the parfait method was to use vacuum distillation to concentrate the nonvolatile contaminants in water. The complicating factors in the current modified parfait method come mainly from the ion-exchange beds, which are necessary to remove the metal cations from water so that intractable precipitates will not form during vacuum concentration. The intractable nature of such precipitates is believed to come from the interaction of metal carbonates with lipids and humic substances during concentration. Presumably, the intractability of these precipitates would be reduced or abolished by removing the lipids and humics before concentration, rather than by removing the metals by ion exchange. This study showed that porous Teflon will remove these materials, a result suggesting that vacuum concentration might still be used as a comprehensive method to concentrate water-soluble, nonvolatile compounds if a water sample is first treated on a large bed of porous Teflon. Such an approach would yield three fractions: one containing water-soluble components including ionic and neutral compounds, one containing humic acids and any adsorbed materials, and one containing organic-soluble materials eluted from the PTFE–humic fraction.

Acknowledgments

Although the study described in this article was funded wholly by the USEPA under cooperative agreement 807126 with HERL, it has not been subjected to the Agency's required peer and administrative review and therefore does not necessarily reflect the views of the Agency and no official endorsement should be inferred. Mention of trade names and commercial products does not constitute endorsement or recommendation for use.

Literature Cited

1. Shackelford, W. M.; Keith, L. H. U.S. Environmental Protection Agency Report 600/4-76-062; U.S. Government Printing Office: Washington, DC, 1976.
2. Rosen, A. A. In *Identification and Analysis of Organic Pollutants in Water;* Keith, L. H., Ed.; Ann Arbor Science: Ann Arbor, MI, 1977; pp 3–14.
3. Johnston, J. B.; Verdeyen, M. K. In *Chemistry in Water Reuse;* Cooper, W. J., Ed.; Ann Arbor Science: Ann Arbor, MI, 1981; Vol. 2, pp 171–189.
4. Otsuki, A.; Wetzel, R. G. *Limnol. Oceanogr.* **1973,** *18*, 490–493.
5. Suess, E. *Geochim. Cosmochim. Acta* **1970,** *34*, 157–168.
6. Josefson, C. M.; Johnston, J. B.; Trubey, R. *Anal. Chem.* **1984,** *56*, 764–768.
7. Coppola, E. D.; Hanna, J. G. *J. Chem. Educ.* **1976,** *53*, 322–323.
8. *The Enzymatic Colorimetric Determination of Glucose;* Sigma Chemical Company: St. Louis, 1979; Technical Bulletin No. 510.
9. Fales, F. W. In *Standard Methods of Clinical Chemistry;* Seligson, D., Ed.; Academic: New York, 1963; Vol 4, pp 101–111.
10. Washko, M. E.; Rice, E. W. *Clin. Chem.* **1959,** *7*, 542–545.

11. Robb, E. W.; Westbrook, J. J., III *Anal. Chem.* **1963,** *35,* 1644-1647.
12. *Standard Methods for the Examination of Water and Wastewater;* American Public Health Association: Washington, DC, 1976; 14th ed., pp 329-332.
13. May, W. E.; Waslk, S. P.; Freeman, D. H. *Anal. Chem.* **1978,** *50,* 997-1000.
14. Guttman, D.; Higuchi, T. *J. Am. Pharm. Assoc.* **1957,** *46,* 4-10.
15. Cesaro, A.; Russo, E.; Tessarotto, D. *J. Solution Chem.* **1980,** *9,* 221-235.
16. Carter, C. W.; Suffet, I. H. *Environ. Sci. Technol.* **1982,** *16,* 735-740.
17. Matsuda, K.; Schnitzer, M. *Bull. Environ. Contam. Toxicol.* **1971,** *6,* 200-204.
18. Lytle, C. R.; Perdue, E. M. *Environ. Sci. Technol.* **1981,** *15,* 224-228.

RECEIVED for review August 14, 1985. ACCEPTED January 9, 1986.

25

Evaluation of a Quaternary Resin for the Isolation or Concentration of Organic Substances from Water

Shaaban Ben-Poorat, David C. Kennedy, and Carol H. Byington

Envirodyne Engineers, Inc., St. Louis, MO 63146

A synthetic resin (Amberlite XAD-4 quaternary) was evaluated as an adsorbent for the concentration-isolation of 22 specific organic solutes at micrograms-per-liter levels. Adsorption and desorption processes were first developed and tested on a laboratory scale and then adapted for a pilot-scale model. Studies determining the effect of humic substances and inorganic salts on the adsorption-desorption of model compounds were also performed. The effect of 2 ppm of chlorine residual on the generation of chlorinated organic compounds was also studied. XAD-4 quaternary resin in hydroxide form was efficient in recovering the majority of model compounds. Mass balances indicated accountability was generally higher in bench-scale experiments. Statistical evaluation of pilot-scale studies suggested that the presence of humic substances affected the concentration of model compounds.

THE FIELD OF SEPARATION SCIENCE has made great strides in recent years in developing techniques for isolating, separating, and concentrating organic species. One impetus for these advances has been the search for sensitive and accurate analytical methods for trace organics. A second impetus has been the need for effective concentration and isolation techniques for preparing biologically active substances for biomedical investigations; the objective of this project was directed toward this application.

Within the realm of analytical separation systems, by far the most fruitful approach has been the use of solid sorbent techniques. Although other approaches have been studied (reverse osmosis, solvent extraction,

0065-2393/87/0214/0535$06.25/0

foam separation, etc.), none are so versatile and offer so much potential for selectivity, concentration, and field use as adsorption techniques. This project dealt only with the investigation of adsorbents for the isolation of organic substances for toxicological testing. Furthermore, it was limited to the investigation of newly developed synthetic sorbents such as the polymeric XAD-4 quaternary anion-exchange resin adsorbent rather than traditional activated carbons.

Our goal for this project was to develop a system for sampling 500 L (or more) of drinking water that might contain 1–50 ppb of organic compounds. Mass balances for each compound were determined to reveal the unrecovered amount of each compound. The mass balance determinations were required to determine whether recovery losses were the result of volatilization, adsorption, and/or chemical transformation.

Synthetic sorbents are known to contain artifacts in the resin that could be eluted during desorption of the organic compounds concentrated on the resin. Therefore, separate experiments using XAD-4 quaternary resin (OH^- form) were also performed to evaluate the presence of artifacts, either those arising from the interaction of chlorine with the resin or those from the resin itself.

Experimental

Preparation of Model Compound Test Solutions. Test solutions of the model compounds selected by the U.S. Environmental Protection Agency (USEPA) are shown in the box on page 537. These compounds, which were used in bench-scale and pilot-scale studies, were prepared by diluting the required volume(s) of stock solution with organic-free water containing an inorganic salt matrix. The salt matrix consisted of 77 ppm of $NaHCO_3$, 120 ppm of Ca_2SO_4, and 47 ppm of $CaCl_2 \cdot 2H_2O$. During the experiments, some precipitation of salt occurred in the reservoir prior to passing the water through the column. The pH was, therefore, adjusted from approximately 8.5 to 7.0 with 1 N HCl to correct the problem.

Bench-Scale Column and Resin Preparation. The apparatus used during bench-scale studies was modeled after work done by Junk et al. (*1*). The adsorption–desorption column was 37 cm long × 1 cm i.d. In most bench-scale studies, this column was filled with 13 cm of XAD-4 quaternary resin (OH^- form) of 40–80 mesh. This amount was equivalent to approximately 10 cm^3 of wet resin. A diagram of the apparatus used for the isolation–concentration of organics from water is shown in Figure 1. Figure 2 shows the eluant concentration apparatus designed by Junk and notes the changes. Glassware shown in Figures 1 and 2 was manufactured by Southwestern Glass.

The resin cleanup procedure employed during this project used Soxhlet extraction with the same solvents used for elution. This method assured that any impurities were eluted prior to the actual adsorbent studies. Combining the best recommendations from previous studies, the following purification procedure was used in our laboratory:

Adsorption System Model Organic Compounds

Acids	*Amount (μg/L)*	*Amines*	*Amount (μg/L)*
Quinaldic acid	50	Quinoline	50
Trimesic acid	50	Caffeine	50
Stearic acid	50	5-Chlorouracil	50
Humic acid	2000		
Glycine	50	*Esters*	
Carbohydrates		Bis(2-ethylhexyl) phthalate	50
Glucose	50	*Chlorobiphenyls*	
Aldehydes		2,4′-Dichlorobiphenyl	50
		2,2′,5,5′-Tetrachlorobiphenyl	5
Furfural	50		
		Ketones	
Hydrocarbons			
		Isophorone	50
1-Chlorododecane	5	Anthraquinone	50
Biphenyl	50	Methyl isobutyl ketone	50
Polyaromatic Hydrocarbons		*Phenols*	
		2,4-Dichlorophenol	50
Phenanthrene	1	2,6-Di-*tert*-butyl-4-methylphenol	50
Trihalomethanes			
Chloroform	50		

1. A slurry of the resin was made with distilled water.
2. The resin was stirred gently.
3. The resin fines were removed by decantation.
4. Steps 1–3 were repeated three times.
5. The resin was rinsed with methanol three times.
6. The resin was then purified by sequential solvent extractions with methanol, acetonitrile, and ethyl ether in a Soxhlet extractor for 8 h per solvent.
7. The purified resin was stored in glass-stoppered bottles under water to maintain its purity.
8. The resin was not allowed to dry because cracking results in the release of impurities from the interior of the resin.

One other solvent, methylene chloride, was also used for resin cleanup but only during the resin blank study. The XAD-4 quaternary resin was stored in the chloride form under water rather than under methanol because our consul-

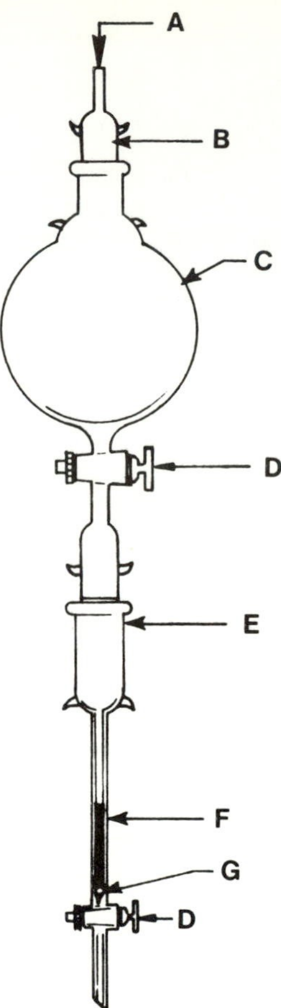

Figure 1. Apparatus for extracting organic solutes from water. A, pure inert gas pressure source; B, cap; C, 2-L reservoir; D, polytetrafluoroethylene stopcock; E, 24/40; F, 1.0-cm i.d. × 37-cm long glass tube packed with 13 cm of resin; G, silanized glass wool plug.

tants (G. A. Junk and J. S. Fritz) noted that XAD-4 quaternary resin is not stable when stored under methanol.

The XAD-4 quaternary resin used in these studies was prepared by the Ames Laboratory in Ames, Iowa. This resin had been used in studies by the Ames group for the adsorption and selective separation of acidic material in waste waters. For this study, the resin was chosen for its effectiveness in concentrating anionic material from solution. At the same time, it was thought that sufficient sites would be available to effectively adsorb neutral organic compounds from water. The resin was basically an XAD-4 macroreticular cross-linked polystyrene into which a trimethylamine group was introduced. The resin was stored in the chloride form but was converted to the hydroxide form before use in the resin sorption experiments.

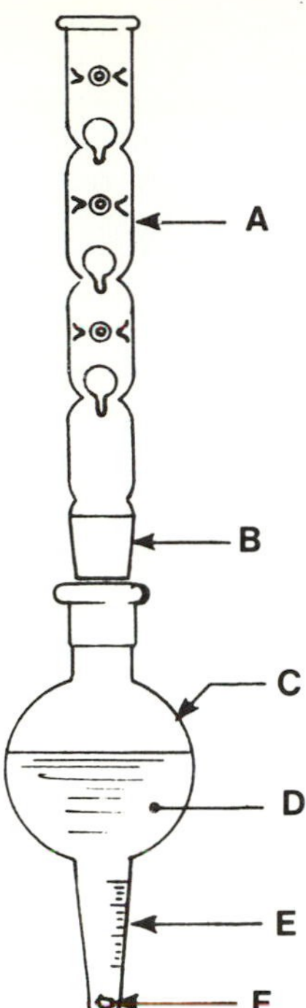

Figure 2. Concentration apparatus. A, Snyder distillation column; B, 24/40; C, 100-mL vessel; D, eluant solvent; E, graduated and calibrated taper; F, small boiling chip.

The bisulfite (HSO_3) form of the XAD-4 quaternary resin was used for the evaluation of several problem compounds and was cleaned in the same manner as the OH^- form.

Carbonaceous resins (i.e., XE-340) were cleaned by a somewhat different procedure, as follows:

1. The resin was extracted with methylene chloride for 24 h in a Soxhlet extractor.
2. The resin was dried in an oven at 104 °C overnight.
3. The clean resin was stored in a clean glass-stoppered bottle.
4. Before application, the resins were hydrated by soaking in methanol for 1 h.

Column Adsorption–Desorption Scheme. The XAD-4 quaternary resin was changed from the chloride to the hydroxide form by passing 75 mL of 0.1 N NaOH through the column, followed by organic-free Millipore water, until the pH of the effluent water was neutral (approximately 100 mL).

In a typical bench-scale study, five columns were set up under the strict quality control–quality assurance (QA-QC) requirements of this project: one column for the control of reagents and glassware, one column for the control of the resin blank, and triplicate columns for model compound studies.

When the reservoir (with a one-hole connector) was filled with water, salts were added and the compounds were spiked. The pH was adjusted with 1 N HCl to 7.0. The reservoir was capped and swirled for a few seconds. The upper and lower stopcocks were opened and pressure was applied by using high-pressure, ultrapure, nitrogen gas. The flow rate was controlled by the nitrogen cylinder pressure and by using the fine adjustment on the resin column stopcock. Two flow rates were studied and are described during the presentation of results. The initial flow rate used was 120–150 bed volumes/h (20–25 mL/min). A lower flow rate of approximately 48 bed volumes/h was studied after breakthrough of quinaldic acid was observed.

When most of the sample had passed through the column and the liquid level was at the top of the resin, the reservoir was washed with 25 mL of water, which was drained through the column. The effluent water was preserved and stored in a cooler for future extraction.

In general, desorption was accomplished in the same column with a series of solvents for the elution of neutral, basic, and acidic compounds. The following steps (or combinations of these steps) were used to desorb different classes of model compounds:

1. Six 25-mL portions of distilled ethyl ether. The first 25-mL portion of distilled ether was added to the resin column, and the column was agitated to free the resin. The column was then allowed to stand for 5–10 min before draining. The second through sixth 25-mL portions of ether eluants were used for desorption by gravity flow without agitation and were combined with the first aliquot. In some experiments, additional 25-mL aliquots of ether were used to elute the column.
2. Two 25-mL portions of methanol. Used for better desorption of caffeine and the removal of any remaining water from Step 1.
3. Two 10-mL portions of ether. Used for the removal of residual methanol from Step 2, which could interfere in the methylation of acidic compounds.
4. Two 25-mL portions of 0.1 N HCl/ether. Used for the removal of acidic compounds.
5. Two 25-mL portions of 0.1 N HCl/methanol. Used for the removal of acidic compounds.
6. Two 25-mL portions of saturated HCl/methanol. Used for the removal of humic substances.

Saturated HCl in methanol and ether was prepared by bubbling anhydrous, electronic grade, HCl gas through three containers of each solvent in series. A caustic trap was added after the third container to neutralize the excess HCl gas. The normality of each batch was determined by titration against a known sodium hydroxide 10 N solution. The 0.1 N HCl/ether was prepared by dilution in ether, and the normality was checked by standardization against a known 0.1 N sodium hydroxide solution. Fresh 0.1 N HCl/ether was prepared in most cases to ensure the correct normality.

One additional step (added late in the desorption scheme) consisted of rinsing the column with 0.1 N HCl in water to remove lead compounds. This step was necessary because lead nitrate at 25 ppb would precipitate, and lead was removed by the resin acting as a filter. This step may not be necessary in real field concentration–isolation of trace organic compounds.

Analytical Methods Development

Certain criteria and assumptions were used in the development of analytical methods. It was assumed that 1 L of water spiked with model compounds would be used in the laboratory-scale experiments. It was also assumed that the extracts taken for analysis could be reduced to 1 mL prior to analysis. The objective of the analyses would be to permit measurement at all values in excess of 1-2% of the concentration levels specified (i.e., for those compounds specified at 50 μg/L, one should be able to detect and measure concentrations in the range 1-50 μg/L).

The list on page 542 summarizes the analytical methods used for the 22 model compounds. Several problem compounds (including acids) were analyzed by a variety of methods that are described later. Linearity curves were prepared for each compound and for every type of detector or analytical method used throughout the project.

The analytical methods shown on page 542 were continuously modified because several resins showed different characteristics during the concentration of the model compounds. In addition, the extraction of several model compounds from the effluent water (or the analysis of the reservoir rinse) necessitated changes in the analytical methods.

In the following section, the analytical methods used for the evaluation of XAD-4 quaternary resin are described. For analytical purposes, the compounds were grouped as shown in the box on page 543.

In summary, the general desorption analytical scheme was as follows:

Ether reservoir rinse (2 × 20 mL of distilled ether): Freeze out water at −20 °C. Measure the volume of the decanted ether. Remove the following aliquots:

1. 5 mL for 5-chlorouracil analysis (high-performance liquid chromatography [HPLC]).
2. 5 mL for quinaldic and trimesic acids (HPLC).
3. 5 mL for quinaldic, trimesic, and stearic acids and 5-chlorouracil (capillary gas chromatography–flame ionization detection [GC–FID]).
4. 5 mL for humics analyses.
5. 1 mL for chloroform (GC–electron capture detection [GC–ECD]).

Concentrate the remainder to 1.0 mL and spike with 5 μg of *n*-pentadecane internal standard. Analyze for all Group I compounds by capillary GC–FID.

Ether eluent (6 × 25 mL): Freeze out water. Measure the volume and remove 1 mL for analysis of chloroform. Concentrate the remainder to 1 mL and spike with 10 μg of *n*-pentadecane internal standard. Analyze for all Group I compounds by capillary GC–FID.

Analytical Methods for Model Compounds

Group I	*Description of Methods*
Methyl isobutyl ketone	GC–FID capillary
Furfural	HPLC and GC–FID capillary
Isophorone	HPLC and GC–FID capillary
2,4-Dichlorophenol[a]	GC–FID capillary
Quinoline	GC–FID capillary
Biphenyl	HPLC and GC–FID capillary
1-Chlorododecane	GC–FID capillary and GC–Hall
2,6-Di-*tert*-butyl-4-methylphenol	HPLC and GC–FID capillary
2,4′-Dichlorobiphenyl	GC–FID capillary, GC–ECD, GC–Hall
Phenanthrene	HPLC and GC–FID capillary
Caffeine	HPLC and GC–FID capillary
2,2′,-5,5′-Tetrachlorobiphenyl	GC–FID capillary, GC–ECD, GC–Hall
Anthraquinone	HPLC and GC–FID capillary
Bis(2-ethylhexyl) phthalate	HPLC and GC–FID capillary
Group II	
5-Chlorouracil	HPLC
Stearic acid	Diazomethane derivatization, GC–FID capillary
Quinaldic acid	Diazomethane derivatization, HPLC
	Diazomethane derivatization, GC–FID capillary
	Direct aqueous injection, HPLC (paired ion)
Trimesic acid	Diazomethane derivatization, HPLC
	Diazomethane derivatization, GC–FID capillary
	Direct aqueous injection, HPLC (paired ion)
2,4-Dichlorophenol[a]	GC–FID capillary
Humic substances	UV–visible spectroscopy
Group III	
Glucose	Derivatization, GC–FID or spectrophotometric
Glycine	TAB derivatization, GC–FID, fluorometry
Group IV	
Chloroform	GC–ECD

NOTE: Hall denotes Hall detector.
[a] 2,4-Dichlorophenol appears in both Groups I and II.

Analytical Procedures

Group I	*Amount (mg/L)*
Methyl isobutyl ketone	50
Furfural	50
Isophorone	50
2,4-Dichlorophenol[a]	50
Quinoline	50
Biphenyl	50
1-Chlorododecane	5
2,6-Di-*tert*-butyl-4-methylphenol	50
2,4′-Dichlorobiphenyl	50
Phenanthrene	1
Caffeine	50
2,2′,-5,5′-Tetrachlorobiphenyl	5
Anthraquinone	50
Bis(2-ethylhexyl) phthalate	50
Group II	
5-Chlorouracil	50
Stearic acid	50
Quinaldic acid	50
Trimesic acid	50
2,4-Dichlorophenol[a]	50
Humic substances	2000
Group III	
Glucose	50
Glycine	50
Group IV	
Chloroform	50

[a] 2,4-Dichlorophenol appears in both Groups I and II.

Methanol eluent (2 × 25 mL): Measure the volume and remove 5 mL for humics analyses. Concentrate the remainder to dryness and solvent exchange to methylene chloride to a final volume of 1.0 mL; spike with 20 μg of internal standard. Analyze this fraction by capillary GC–FID for all Group I compounds with special emphasis on caffeine.

0.1 N HCl/ether (2 × 25 mL): Measure the volume and analyze as 0.1 N HCl/methanol.

0.1 N HCl/methanol (2 × 25 mL): Measure the volume and remove aliquots for the following analyses:

1. 5-Chlorouracil.
2. Quinaldic and trimesic acids.
3. Humic substances.
4. 2,4-Dichlorophenol. (After dilution with water, 2,4-dichlorophenol in acid form was extracted with 3 × 25 mL of CH_2Cl_2 and then concentrated to 1.0 mL; 20 μg of *n*-undecane internal standard was added prior to GC determination.)

Concentrate the remaining volume to near dryness by nitrogen blowdown at room temperature. Add several milliliters of distilled ether, concentrate to near dryness in a water bath, and repeat four or five more times to remove HCl. Esterify with diazomethane (2) and analyze by capillary GC–FID for stearic acid and 2,4-dichlorophenol (5-chlorouracil also gave a derivative with diazomethane). Retain all concentrate after GC–FID analysis for possible HPLC analysis.

Saturated HCl/methanol: Measure the volume of eluant and analyze spectrophotometrically for humic substances.

Effluent waters: Measure the volume of water and remove 1 mL for GC–ECD analysis of chloroform. Adjust the pH of the remaining water to >11 and extract with 3 × 100 mL of CH_2Cl_2; adjust the pH to <2 and extract with 3 × 100 mL of CH_2Cl_2. Concentrate the CH_2Cl_2 extracts separately to a final volume of 1.0 mL; spike with 5 μg of *n*-pentadecane and *n*-undecane internal standard. Analyze the basic extract for all Group I compounds by capillary GC–FID. Analyze the acidic fraction for stearic acid after methylation and 2,4-dichlorophenol by capillary GC–FID.

Results and Discussion

Resin Blank Artifacts: Effect of 2 ppm of Residual Chlorine. Three basic types of blank experiments were performed. One was performed to identify any artifacts caused by the presence of 2 ppm of chlorine in blank water used in the separation–concentration procedure, and another was performed to identify any artifacts in the general resin procedure. In addition, a reagent blank was also concentrated and analyzed. The reagent blank was performed in a manner identical to the pure water blanks, except that no resin was included in the procedure. This reagent blank gave an indication of contaminants arising from sources other than the resin (i.e., glassware, water, solvents).

Blank Experiment with Chlorine. This experiment was designed to provide information regarding the effect of chlorine on the production of new chlorinated compounds. Duplicate blank experiments for both pure water and pure water spiked with 2 ppm of chlorine residual were performed by the normal resin separation-concentration procedure. This testing was done for comparative purposes because only compounds that are identified in the extract of the chlorine-spiked water but are not present in the normal extract from pure blank water can be considered as true artifacts caused by the chlorine.

This experiment was performed on XAD-4 quaternary resin in the OH^- form, and desorption was by ethyl ether only (i.e., HCl saturated ether not used). Calcium hypochlorite [$Ca(OCl)_2$] was used to provide the required 2-ppm chlorine concentration. Millipore Super-Q water was salted according to the general procedure and passed over a 10-mL bed volume of resin (approximate dry weight = 6 g at 150 bed volumes/h. The resins were blown with nitrogen (3 lb/in^2) for 10 s to remove residual water and eluted with 3 × 50-mL portions of ethyl ether. Peroxide formation was suppressed by the addition of 2% (v/v) ethanol.

The ether was concentrated by Kuderna–Danish evaporation and then solvent exchanged to hexane to give a final volume of 0.3–0.4 mL. To quantify the compounds present, the samples were spiked with a known amount of anthracene-d_{10} as an internal standard just prior to injection. The concentrated extracts were analyzed by capillary GC–mass spectrometry (GC–MS).

The results of these analyses are summarized in Table I. The estimated amounts of each compound are given in terms of micrograms per gram of dry resin. In a normal resin separation–concentration procedure, these amounts would be the same, independent of the amount of water passed through the resin. Some peaks could not be identified.

Table I. Effect of 2 ppm of Residual Chlorine

Compound	*Reagent Blank (μg/g)*	*With 2 ppm of Cl_2 (μg/g)*	*Without Cl_2 (μg/g)*
Acetic acid	0.7	0.5	0.5
Ethyl acetate	55	30	28
Methyl palmitate	1.8	—	—
Methyl oleate	5.8	—	—
Bis(2-ethylhexyl) phthalate	0.7	0.2	0.2
Farnesol	1.0	1.0	1.5
Ethyl benzoate	—	0.1	0.1
Diisobutyl phthalate	—	0.15	0.1
n-Butyl phthalate	—	0.3	0.2

Clean Water Blank Experiment. This experiment was performed for the purpose of identifying any artifacts that might arise from the resin in the course of normal resin experiments. Four replicate resin blanks were run by the normal resin separation–concentration procedure. Data from this experiment yielded the following conclusions:

1. Ethyl acetate was present in large amounts in each type of resin blank.
2. Trace amounts of other organic compounds were found in each type of resin blank.
3. Compounds identified in the chlorine blanks were also found in resin and reagent blanks.
4. No new chlorine-containing compounds were identified as having originated from the addition of chlorine to the blank water.
5. Many of the compounds identified appeared to have as their origin trace components of the solvents used.
6. None of the alkyl-substituted aromatics identified as artifact contaminants from XAD-2 resin by USEPA–Battelle were identified in this experiment.

Recovery of Lead. This study was conducted by adding lead nitrate to water. The pH was adjusted to 2.5 to avoid chemical precipitation of lead as the hydroxide. Salts were not added in this experiment because the presence of sulfate in the salts might have caused the lead to precipitate as lead sulfate.

The experiment consisted of triplicate runs involving the passage of 1 L of water containing 25 ppb of lead through the 10 cm^3 of resin at a flow rate of 48 bed volumes/h. Analyses of the effluent waters were performed by atomic absorption (furnace). The concentrations of lead (μg/L) in the effluent waters were as follows: resin A, 22.5; resin B, 24.4; resin C, 22.8; method blank 1, 0.18; method blank 2, 0.43; resin blank, 1.03; 15-μg/L check standard, 15.4; 20-μg/L check standard, 20.7; 30-μg/L check standard, 30.6. The average value for resins a, b, and c was 23.2 μg/L, and the 95% confidence interval for these resins was 23.2 $\pm$ 2.53 μg/L.

These results show that the average amount of lead present in the effluents from the three resins was only slightly less than that present in the influent water. The 95% confidence interval on these results includes 25; therefore, at this level, there is no basis to conclude that any of the lead was removed by the resin.

In accordance with conversations with the USEPA, it was necessary to repeat the lead study at pH 6–7 to evaluate the resin under conditions similar to those used for resin adsorption (pH 7). Previous experiments were conducted at pH 2.5 to reduce the likelihood of pH being a significant variable.

Conditions of this experiment were similar to those reported in previous lead studies. Three replicates and a blank were included in the present study. Humic substances were not added to better focus on the effect of pH. Salts also were not added because of the possibility of lead precipitating as lead sulfate.

The concentrations of lead in the effluents were <1 μg/L for replicate numbers 1, 2, and 3 and the resin blank.

Check standards 1, 2, 3, and 4 at a nominal concentration of 25 μg/L also were analyzed and had lead concentrations of 33, 31, 33, and 32 μg/L, respectively.

Although the results for check standards were somewhat high, the extremely low results for the three replicates of lead in the effluent waters from the resins clearly indicate that lead is being removed from the water upon passage through the resin.

To determine whether the mechanism for removal might be due to the precipitation of lead as lead hydroxide and subsequent removal by the resin acting as a filter, samples of water with a pH of 6–7 and a lead concentration of 25 μg/L were prepared and filtered through a 0.45-μm Metricel membrane filter. The results of this experiment were as follows: lead concentration before passage, 26.1 μg/L; lead concentration after passage, 5.01 μg/L.

These results indicate that lead is removed by the membrane filter, and the pH value of 6–7 may be causing lead to precipitate as the hydroxide.

It is uncertain just how relevant the results for these resin studies are to actual drinking waters because lead in drinking water is likely to be in solution and to have had time to reach equilibrium. Tests on actual drinking waters with background levels of lead will provide more definite answers to these questions.

Subsequent conversations with the USEPA indicated the need to determine if the presence of humic substances had any effect on the behavior of lead because humic substances can act as chelating agents for heavy metals. Therefore, an experiment was planned to evaluate the recovery of lead in the presence of humic substances.

The experimental design consisted of setting up duplicate resin columns for the concentration of 1 L of aqueous solution containing 25 μg/L of lead as lead nitrate. Two other resin columns were used to test the 25 μg/L of lead plus 2000 μg/L of humic substances. One resin blank, which contained no lead or humic substances, was used. The pH of these solutions was adjusted to 6–7. No salts were added because of the possibility of precipitating lead as one of the salts, particularly as lead sulfate, which is known to have a very low solubility in water. One major change in these experiments was that each column was rinsed with 2 × 25 mL solution of 0.1 N HCl. The purpose of this rinse solution (prior to the ether rinse) was to remove any lead that may have been physically removed by the resin.

The results of the analyses of the effluent waters and 0.1 N HCl rinses are given in Table II.

These data indicate recoveries of 82–90%, but lead is removed from water because of the initial precipitation of this element. This phenomenon might not occur in real field sampling because all lead compounds should be in dissolved forms.

Concentration of Model Compounds in the Absence of Humic Substances at 120–150 Bed Volumes/h. Resin studies using XAD-4 quaternary resin in the hydroxide form were performed several times at 150 bed volumes/h at various concentrations of the model compounds. The primary objective of these studies was to make a preliminary evaluation of the separation and concentration capacity of XAD-4 quaternary (OH^-) resin for the model compounds. Emphasis was thus placed on the measurement of each compound in the final eluant concentrate. However, whenever convenient, analyses were also made of the aqueous effluent and reservoir to provide for more complete mass balances.

Table III lists the final concentration at which each model compound was tested.

Breakthrough–Flow Rate Studies. Because several of the acidic model compounds showed mixed results, it was decided to study the effect of flow rate and the presence of salts by establishing breakthrough curves. Primary emphasis was given to the evaluation of quinaldic acid, both with and without the presence of salts. The concentration of quinaldic acid was chosen high enough so that each eluant in the breakthrough study could be analyzed by direct injection HPLC.

Table II. Results of Analyses of Effluent Waters and 0.1 N HCl Rinses

	Lead Concentration (ppb)	
Sample	*Effluent Water*	*0.1 N HCl Rinse*
Resin A (without humics)	<1.0	22.3
Resin B (without humics)	<1.0	20.4
Resin C (2000 ppb of humics)	3.8	17.9
Resin D (2000 ppb of humics)	4.0	18.4
Resin blank	<1.0	5.4
Reagent blank	1.1	2.0

Table III. Summary of Percent Recovery and Percent Total Mass Balance of All Model Compounds in Bench-Scale Studies

Model Compound	*Conc. (μg/L)*	*Percent Recovery*[a]	*Total Percent Mass Balance*[b]
Anthraquinone	500	101	101
Biphenyl	596	71.7	71.9
Bis(2-ethylhexyl) phthalate	668	55.2	77.5
Caffeine	50.7	31.6	46.4
Chloroform	48.7	69.9	83.6
1-Chlorododecane	502	42.6	60.0
5-Chlorouracil	5000	91.1[c,d]	91.1[c,d]
2,6-Di-*tert*-butyl-4-methyl-phenol	696	62.2	63.7
2,4′-Dichlorobiphenyl	404	66.7	68.8
2,4-Dichlorophenol	500, 5000	28.8, 66.1[c]	28.8[c]
Furfural	52.4	56.4	75.5
Glycine	19,700	0	87.7[d]
Glucose	19,800	0	88.9[d]
Humic acid[e]	2000	79.9	79.9
Isophorone	500	77.2	77.4
Methyl isobutyl ketone	50.5	102	102
Phenanthrene	1.0	112	112
Quinaldic acid	500, 5000	96.2, 117[c]	96.2, 117[c]
Quinoline	48.4	86.7[d]	86.7[d]
Stearic acid	50	60.1	81.5
2,5,2′,5′-Tetrachlorobiphenyl	172	51.9	67.7
Trimesic acid	500, 5000	33.6, 53.0[c]	34.5, 53.0[c]

NOTE: Studies were performed with salts added, without humic acid, and at a flow rate of 150 bed volumes/h, unless otherwise indicated.

[a] Percent recovery represents the percent of model compound recovered in solvent eluants. Values represent the average of three determinations.

[b] Total percent mass balance represents the total percent recovered from solvent eluants + column effluent + reservoir rinse.

[c] Flow rate was 48 bed volumes/h.

[d] Value is the average of two determinations.

[e] Study was performed in the presence of salts but the absence of other model compounds.

It appeared that changes in flow rate and direct analysis of the acidic compounds were very promising in the concentration–isolation of such compounds on the XAD–4 quaternary resin. Therefore, several experiments were designed to study the effect of flow rate by establishing breakthrough curves.

A gel permeation chromatography apparatus was adapted for use in pumping and fractionating samples in order to study the adsorption characteristics of the XAD–4 quaternary resin in the OH^- form.

A 6-cm × 0.46-cm i.d. column was selected (bed volume = 1 cm^3). Three flow rates of 405, 213, and 103 bed volumes/h were studied. The respective linear velocities of 41, 22, and 10.3 cm/s are shown in Figure 3.

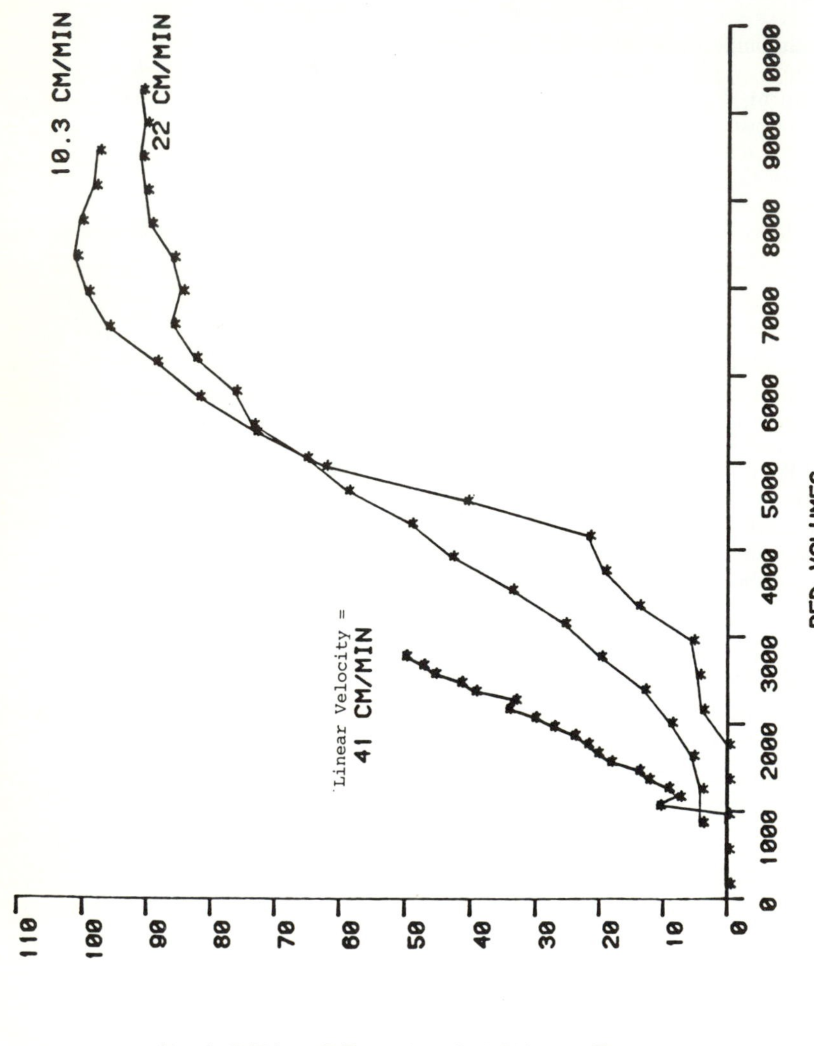

Figure 3. Summary of quinaldic acid (salted H_2O) breakthrough curves (bed volume = 1 cm^3).

At a linear velocity of 41 cm/s, the breakthrough occurred at approximately 800–1000 bed volumes. At 22 cm/s, the breakthrough was at about 1800 bed volumes, and at 10.3 cm/s, the breakthrough was at about 2400 bed volumes.

Figure 4 shows the predicted breakthrough curves for 500 L per 24-h volume using the same linear velocities as those employed in the three experiments for quinaldic acid.

Figure 4 was based on calculations of experimental data, assuming constant linear velocities, residence time, and flow rate. These results are considered to be worst case predictions because of the higher concentrations used and the possibility of channeling having occurred in the small 1-cm^3 experimental column setup.

The results of the breakthrough studies indicated that the optimum flow rate would be approximately 50–100 bed volumes/h. These results were used to scale-up the bench-scale columns for pilot plant studies.

Pilot Plant Design, Experiments, and Results. Design of the pilot plant scale-up was based upon the earlier study involving quinaldic acid breakthrough. Because this compound may represent a worst case, other factors must be considered.

The inside diameter and length of the resin bed were determined by considering the following two factors: (1) residence time and (2) bed volume per unit time or throughput.

It was decided to keep these two factors as close as possible during scale-up from bench-scale studies to the pilot plant studies. The first factor controls the rate of adsorption, and the throughput controls the capacity of the XAD-4 quaternary resin.

The breakthrough studies on quinaldic acid can be summarized as follows:

Flow (mL/min)	*Velocity (cm/min)*	*Residence Time (s)*	*Bed Volume per Hour*
6.75	40.6	9	405
3.66	22.0	16.4	220
1.72	10.3	35.0	103

Figure 3 shows that at 10% breakthrough at 10.3 cm/min, the bed volume is approximately 3200 mL. The project requirement for sampling was 500 L of water to be concentrated during a 24-h period. Therefore, the following calculation would result in the amount of resin needed: flow = 1.72 mL/min; 1.72 mL/min × 60 min/h = 103 bed volumes/h; 103 bed volumes/h × 24 h = 2472 bed volumes; 500 L/2472 bed volumes = ~2 L of resin.

This calculation shows that the bed volume of about 2500 mL found

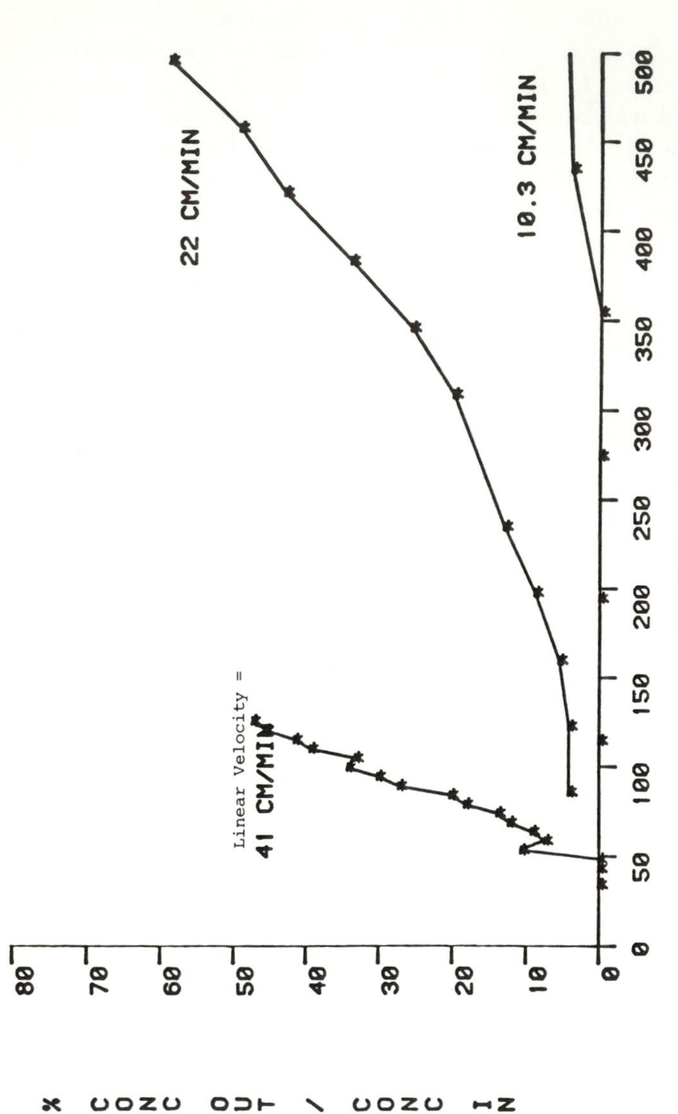

Figure 4. Predicted quinaldic acid (salted H_2O) breakthrough curves for 500-L volumes (bed volume = 1 cm^3).

is close to the bed volume of 3200 mL in the breakthrough study of quinaldic acid.

Because the volume of the resin required has been established (to handle 103 bed volumes/h), the inside diameter (D) and the length (L) of the bed can be determined in order to be comparable to the residence time of the quinaldic breakthrough study. Two approaches can be followed in length and diameter determinations. One involves duplicating the breakthrough or the bench-scale studies. The second approach involves the practical aspects such as L/D and the pressure drop phenomenon. As a general rule, L/D should be more than or equal to 10. Therefore, a 16-in. × 1.0-in. i.d. column was constructed with the following characteristics: surface area = 4.9 cm^2; residence time = 0.57 min; bed volume/h = 103; flow rate = 347 mL/min; linear velocity = 70.8 cm/min; L/D = 16.

A total of four pilot plant studies was performed:

Sample	*Resin Blank*	*Resin 1*	*Resin 2*	*Resin 3*
Salts	Present	Present	Absent	Present
Model compounds	Absent	Present	Present	Present
Humic substances	Absent	Absent	Absent	Present

Many of the major peaks in the acidic eluant samples were polymethyl polysiloxanes (PMPS), which are believed to be artifacts from the column or septa. When a nonacidic solvent such as ether was analyzed, very few siloxane compounds were found.

At least two of the samples contained a compound identified by the library as methyl sulfate. This finding might suggest the presence of organic salts when the acidic eluants are concentrated rather than the presence of inorganic salts. In general, when a compound was found in both the reagent and resin blanks, the concentration of the compound was higher in the resin blank. The source of the PMPS compounds is not known, but one possible explanation could be the stripping of the column liquid phase by the acidic solvents.

Pilot Plant Studies. The experimental work involved in pilot plant studies 1, 2, and 3 was basically the same as the resin blank pilot study. Twenty-one of the 22 model compounds were present in all three studies, but the inorganic salts were present in two studies, and the humic substances were included only in the last experiment. This order was chosen on the basis of the possibility that the humic substances might not be desorbed 100% and therefore alter the effectiveness of the resin in the adsorption of other model compounds.

The desorption steps were performed in the columns three times with 150 mL of solvents as in the resin blank pilot plant study. The resin was shaken each time and then allowed to stand for 20–30 min before it was drained. Volumes were adjusted when necessary to take aliquots for different analyses. Internal standards were added immediately prior to analysis to eliminate the need for exact measurement of the volumes and compensation in sample-size injection.

Table IV provides results of the three runs and the concentration of each model compound in the 500-L pilot plant study.

Conclusions

The procedure used in this study for the isolation–concentration of organic compounds in drinking water demonstrated that the XAD-4 quaternary resin (OH^-) is effective for most neutral, acidic, and semivolatile compounds. Because of the quaternary function of this resin, it was possible to concentrate several acidic compounds without any change in the pH of the sampling water. This feature is an advantage over normal XAD resins for on-site compositing or grab sampling because the problem of continuous acid addition and subsequent pH monitoring is avoided.

Many classes of organic compounds adsorbed by this resin can be desorbed by solvents such as ether (or acidic methanol and ether). The acidic solvents can be concentrated to remove inorganic acid, but some residual inorganic acid always remained in the concentrated eluants. Residual acid, or possible trace of water in the concentrated eluants, caused analytical variances when methylation was attempted. Therefore, quinaldic acid, trimesic acid, and 5-chlorouracil were analyzed by HPLC rather than by GC–FID.

Humic substances were concentrated more than 50-fold on the XAD-4 quaternary resin, but a saturated HCl/methanol solution was required for the desorption. This eluant was not concentrated further because the concentration of humic substances could be measured directly with a spectrophotometer. Total recovery of humic substances was higher in the bench-scale experiments than in the pilot plant studies. On the basis of the pilot plant results (*see* summary of experiment 3), it appears that the adsorption of humic substances was affected by the higher velocity or the loading capacity because 45% was recovered in the effluent water. The higher velocity in the pilot plant studies did not have a similar effect on other compounds such as quinaldic acid, for example, which was recovered at almost 100%. It is believed that caffeine, which was concentrated during bench-scale studies, was also affected by the higher velocity in the pilot plant studies.

The effect of inorganic salts on the concentration of the model compounds appeared to be inconsistent throughout the bench-scale

Table IV. Summary of Mass Balance and Percent Recoveries of All Model Compounds in All Pilot Studies

Model Compound	Amount Spiked (μg)	Resin Blank (μg)	Pilot Plant Study No. 1 (μg)	Pilot Plant Study No. 1 (%)	Pilot Plant Study No. 2 (μg)	Pilot Plant Study No. 2 (%)	Pilot Plant Study No. 3 (μg)	Pilot Plant Study No. 3 (%)
Methyl isobutyl ketone	25,000	—	8,620	34.5	7,030	28.1	6,390	25.6
Furfural	25,000	—	13,400	53.6	14,400	57.6	12,700	50.8
Isophorone	25,000	—	19,300	77.2	16,100	64.4	16,900	67.6
2,4-Dichlorophenol[a]	25,000	49	14,700	59.0	13,400	53.4	16,600	66.5
Quinoline	25,000	—	7,750	31.0	6,850	21.4	5,420	21.7
Biphenyl	25,000	<1	17,100	68.4	15,100	60.4	13,100	52.4
1-Chlorododecane	2,500	4	859	34.4	2,180	87.2	1,050	42.0
2,6-Di-*tert*-butyl-4-methylphenol	25,000	<1	16,200	64.8	12,000	48.0	11,300	45.2
2,4′-Dichlorobiphenyl	25,000	—	15,100	60.4	10,100	40.4	11,800	47.2
Phenanthrene	500	—	263	52.6	283	56.6	223	44.6
Caffeine	25,000	—	6,270	25.1	5,930	23.7	7,160	28.7
2,2′,5,5′-Tetrachlorobiphenyl	2,500	11	1,220	48.8	1,120	44.8	718	28.7
Anthraquinone	25,000	8	13,300	53.2	11,100	44.4	11,400	45.6
Bis(2-ethylhexyl) phthalate	25,000	5	8,690	34.8	9,940	39.8	5,720	22.9
Chloroform	25,000	1170	14,900	60.0	22,900	92.3	19,900	80.2
Stearic acid	25,000	2	29	<1	22	<1	8,120	32.5
Glycine	25,000	NA	20,200	80.7	NA	NA	NA	NA
Humics	1,000,000	—	NA	NA	NA	NA	857,000	85.7
5-Chlorouracil	25,000	618	12,000	48.0	20,800	83.2	16,400	65.6
Quinaldic acid (HPLC)	25,000	—	27,400	109.6	30,200	120.8	24,100	96.4
Trimesic acid	25,000	—	18,900	75.6	23,300	93.1	5,580	22.3

NOTE: NA denotes not analyzed. The totals are sums of micrograms found after adsorption–desorption.

[a] These totals are sums of micrograms found in neutrals analysis and acids analysis (all capillary GC–FID results).

experiments, and several of the model compounds were concentrated less in pilot plant study number 2.

Results of lead studies indicate that dissolved lead at 25 ppb is not isolated by the XAD-4 quaternary resin. However, if lead is precipitated in the drinking water sample, the resin acts as a filter, and the precipitated lead compounds collected on the resin can be dissolved by the acid/methanol eluant. Experiments with the presence of humic substances and lead were conducted because humic substances can act as chelating agents for heavy metals. Approximately 85% of the lead was recovered, but lead was removed from water because of the initial precipitation of this element. This phenomenon may not occur in actual field sampling because all lead compounds should be in a dissolved form.

The XAD-4 quaternary resin was cleaned by Soxhlet extraction and kept wet under water to minimize artifacts from the resin. An oily, yellow residue observed during concentration of the acidic eluants required an additional cleanup step. On the basis of the results of the resin blank experiments, it was necessary to clean the XAD-4 quaternary resin by batch process with saturated HCl/methanol prior to Soxhlet cleaning with solvents. Artifacts such as benzoic acid were found in the resin blank experiments, but, in general, the mass of each artifact (per dry weight of resin) was negligible.

The XAD-4 quaternary resin is recommended for use in the concentration (50-fold and higher) of trace organic compounds in water except carbohydrates, aldehydes, and acidic compounds such as stearic acid and glycine. The breakthrough experiments should be considered for specific compounds to evaluate the effect of mass transfer due to velocity changes. Breakthrough may not occur for large volumes of water samples with trace organic contaminants because of the high capacity of polymeric resins, but high concentrations of specific compounds may break through.

Literature Cited

1. Junk, G. A. *J. Chromatogr.* **1974,** *99,* 745–762.
2. *Fed Regist.* **1985,** *50(193),* 40732–40737.

RECEIVED for review August 14, 1985. ACCEPTED April 2, 1986.

26

High-Performance Concentration System for the Isolation of Organic Residues from Water Supplies

R. B. Baird, C. A. Jacks, and L. B. Neisess

County Sanitation Districts of Los Angeles County, San Jose Creek Water Quality Laboratory, Whittier, CA 90601

Recovery of nonpolar hydrophobic model compounds from a four-resin concentrator system was in the 70% range, whereas hydrophilic organic compounds were not recovered well. The concentrator system consisted of a series of 45–75-μm macroreticular resin columns (MP-1 [anionic], MP-50 [cationic], XAD-2 [nonionic, nonpolar], and XAD-7 [nonionic, moderate polarity]) through which 500-L water samples were pumped with a high-pressure Teflon diaphragm pump. Columns were eluted with acetonitrile; ionic resin columns were also eluted with saturated NaCl, and the salt solutions were extracted with dichloromethane at neutral, acidic, and basic pH. Most model compounds were recovered from MP-1. The hydrophobic materials breaking through this column were usually found on MP-50 and XAD-2. The lower amounts of hydrophilic organics recovered were retained by XAD-2 and XAD-7.

CHEMICAL AND BIOLOGICAL ANALYSES of trace organic mixtures in aqueous environmental samples typically require that some type of isolation–concentration method be used prior to testing these residues; the inclusion of bioassay in a testing scheme often dictates that large sample volumes (20–500 L) be processed. Discrete chemical analysis only requires demonstration that the isolation technique yields the desired compounds with known precision. However, chemical and/or toxicological characterization of the chemical continuum of molecular properties represented by the unknown mixtures of organics in environmental samples adds an extra dimension of the ideal isolation technique:

0065–2393/87/0214/0557$06.00/0

the residues yielded should be representative of the entire mixture of chemicals whose properties are to be tested. Isolation techniques used for concentrating unknown organic mixtures from water samples include solvent extraction (*1, 2*), low-temperature vacuum distillation (*3*), reverse osmosis (*4–6*), adsorption (*7–11*), and purge and trap (*12*). Efforts to concentrate sufficient amounts of organic residues for biological testing have usually relied on reverse osmosis (*13, 14*) or resin adsorption (*15, 16*) to process the large volumes of water needed. Each of these methods offers some advantages over other methods, yet both present some operational drawbacks in that certain classes of organic compounds may be discriminated against. Efficiency evaluations for concentration techniques have relied on either individual compound analysis by gas chromatography–mass spectrometry (GC–MS) and high-performance liquid chromatography (HPLC) or grosser measurements such as dissolved organic carbon (DOC) and mutagenicity recovery. Most of the compounds identified by chromatographic methods have been volatile lipophiles representing less than 20% of the DOC (*17, 18*); few hydrophiles have been identified, usually amino acids and carbohydrates. The bulk of the unidentified residue typically gets categorized as humic, fulvic, or proteinaceous material. It is reasonable to assume that much of the DOC that has proven difficult to isolate and identify is polar or ionic at ambient pH, and the toxicological importance of a large portion of this unknown residue is arguable because of molecular size (*19*).

This chapter presents the results of a study designed to evaluate the ability of a resin-based concentrator to recover a broad spectrum of model compounds specified by the U.S. Environmental Protection Agency (USEPA) (*20*) from 500-L volumes of fortified distilled water. Such an evaluation may indicate the degree to which residues recovered from real samples represent an unknown mixture of organics covering a wide range of properties. The development of the concentrator (*21*) was based on the chromatographic properties of microparticulate, macroporous, ion-exchange, and nonionic resins and their ability to act as efficient liquid chromatographic (LC) adsorbents (*22–27*). Operational parameters of the four resins (anionic MP–1, cationic MP–50, nonpolar XAD–2, and polar XAD–7) in practical application to reclaimed, surface, and ground waters (*28*) were within DOC capacities and optimum ratios of sample volume to resin mass described for various resins (*22–27, 29, 30*); these and other pertinent references were previously reviewed (*21*). The results of the experiments reported herein will be compareu with other measurements of isolation efficiency during system development and application (*21, 28, 31*).

Experimental

Reagents. NaCl (Spectrum Chemical Co.) was fired 4 h at 550 °C in a muffle furnace to remove trace organic contaminants. Acetonitrile (HPLC grade; J. T. Baker), dichloromethane (Burdick and Jackson), 5-dimethylamino-1-naphthalenesulfonyl chloride (dansyl chloride; Aldrich), *O*-*p*-nitrobenzyl-*N*,*N*′-diisopropylisourea (PNBDI; Regis), and *N*-succinimidyl-*p*-nitrophenyl acetate (SNPA; Regis) were used as purchased.

Recovery experiments were conducted with the following standards, which were used as received without further purification: 5-chlorouracil (Calbiochem), furfural (Aldrich), crotonaldehyde (Aldrich), caffeine (Aldrich), isophorone (Aldrich), 2,4-dichlorophenol (Aldrich), anthraquinone (Aldrich), biphenyl (Ultra Scientific), 2,4′-dichlorobiphenyl (Ultra Scientific), 2,6-bis(1,1-dimethylethyl)-4-methylphenol (Aldrich), 2,2′,5,5′-tetrachlorobiphenyl (Ultra Scientific), benzo[*e*]pyrene (Aldrich), bis(2-ethylhexyl) phthalate (Scientific Polymer Products), 4-methyl-2-pentanone (Aldrich), quinoline (Kodak), 1-chlorododecane (Eastman), stearic acid (Kodak), quinaldic acid (Aldrich), trimesic acid (Aldrich), glucose (Aldrich), glycine (Aldrich), and chloroform (Burdick and Jackson).

Concentration System Materials. The concentration system consisted of a Milton-Roy model FR141-144 Teflon diaphragm pump and four 4.9- × 60-cm stainless steel columns connected in series with 3.2-mm stainless steel tubing (Figure 1). Each column was sealed with a 10-μm stainless steel frit held in place by stainless steel washers and a screw-on cap at each end (Figure 2). The first and second columns contained, respectively, MP-1 anion-exchange and MP-50 cation-exchange resins (BioRad). The third and fourth columns contained nonionic, nonpolar XAD-2 and nonionic, moderately polar XAD-7 resins, respectively (Rohm and Haas). The particle size range for all resins was 45-75 μm. The ion-exchange resins (200-400 mesh as purchased) were washed on a 45-μm screen with a stream of deionized water to remove particles less than 45 μm. One-kilogram batches of the XAD resin beads were ground as methanol slurries in a 5-L ball mill by using 4.5-5 kg of burundum cylinders; XAD-2 was ground for about 4 h, but 30 min was sufficient for XAD-7. XAD-2 was wet-sieved in methanol in a continuous flow system previously described (*21*). XAD-7 slurry was wet-sieved with tap water through a 75-μm screen onto a 45-μm screen. Each resin was slurry-packed into its respective column by using suction. After filling, water was pumped through each column, and then additional resin was added to top off the columns.

Resin purification was done on-column while monitoring column effluent at 254 nm to ensure complete elution of contaminants. MP-1 was purified by pumping 1 N HCl, 1 N NaOH, and 1 N HCl followed by distilled water, methanol, acetonitrile, ethyl ether, and methanol. MP-50 was purified by pumping 1 N NaOH, 1 N HCl, and 1 N NaOH followed by water and organic solvents as for MP-1. XAD-2 was purified by pumping 1 N NaOH, 1 N HCl, distilled water, and organic solvents as for MP-1. XAD-7 was purified by pumping methanol, acetonitrile, and distilled water. After resin purification, column blanks were obtained by using the proper elution solvents. Purity criteria included constant chromatographic profiles using GC-flame ionization detection (GC-FID) and HPLC-UV and a negative response in the Ames test (*21*).

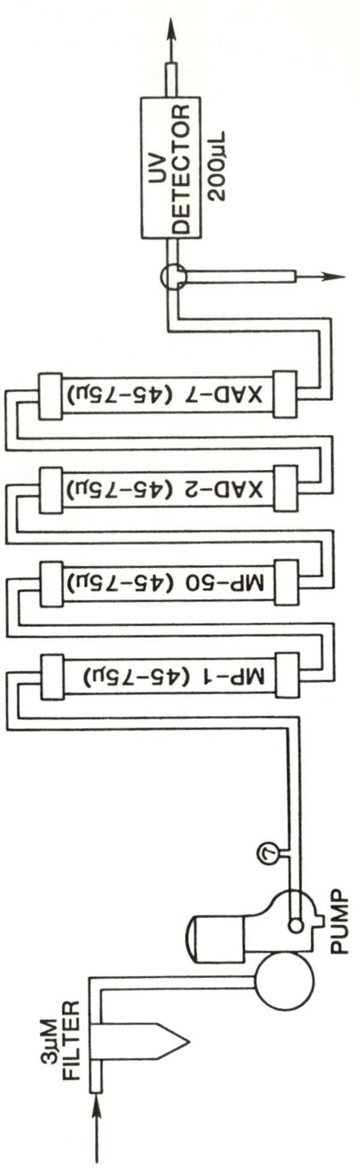

Figure 1. Schematic of resin concentrator.

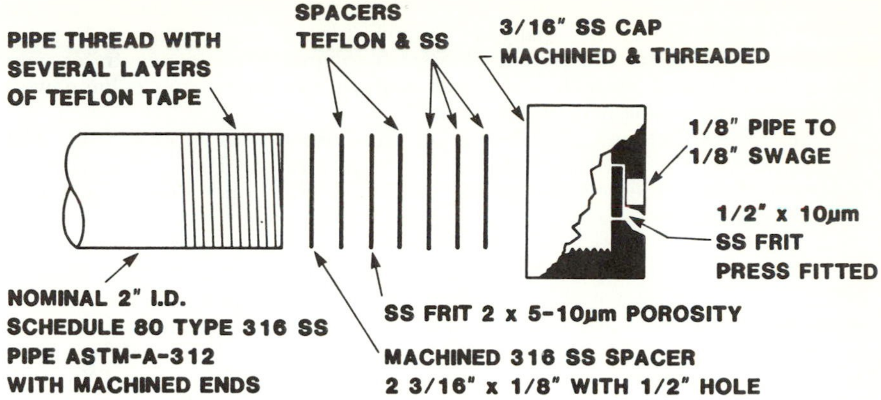

Figure 2. Schematic of stainless steel columns and end fittings.

Sample Processing. Solutions of 70 mg/L of $NaHCO_3$, 120 mg/L of $CaSO_4$, and 47 mg/L of $CaCl_2 \cdot 2H_2O$ were prepared by mixing the reagent chemicals with distilled water in 200-L stainless steel drums. After each drum was filled, the organic standards were added with a 2.5-mL syringe while the sample was stirred. The organic stock solution concentrations were either 5 mg/mL, 0.5 mg/mL, or 50 µg/mL (in acetonitrile, acetonitrile/water, or methanol), depending on the desired concentration in the sample.

Concentration of 500-L samples was usually performed at a flow rate of 150 mL/min and 500 lb/in.^{2}. A piston-type pulse dampener (Hydrodyne) was used to minimize breakdown of resin particles. Flow rates did not exceed 250 mL/min (0.2-cm/s linear flow velocity).

A final 4-L aliquot of concentrator effluent was collected and extracted sequentially with dichloromethane at neutral, acidic (pH 2), and alkaline (pH 11) pH by using two extractions at each pH. The dichloromethane extracts were pooled and further concentrated by using Kuderna-Danish evaporators. Analytical results from this residue were used in mass balance determinations under the conservative assumption that the concentrations of compounds present in the column effluent were relatively constant and that there was good partitioning of each into dichloromethane.

Samples were eluted in the reverse direction by using the Milton-Roy pump with the pulse dampener removed. The eluant flow (50–75 mL/min at 200–300 lb/in.^{2}) was monitored at 254 nm by using an Altex 153 detector with a biochemical flow cell. Elution with each solvent was continued until the detector response returned to base line. All columns were eluted with acetonitrile; this solvent was preceded by 4.5 M NaCl/0.04 M HCl and 0.04 M HCl elutions on the MP-1 column and by 4.5 M NaCl and distilled water elutions on the MP-50 column. The aqueous column effluents were adjusted to pH 2 (MP-1) or pH 11 (MP-50) and then extracted three times with dichloromethane. The acetonitrile column effluents were saturated with NaCl to separate the water, which was extracted twice more with acetonitrile. Fifty percent aliquots of the processed organic solvents from each respective column were concentrated in Kuderna-Danish evaporators to a final volume of about 10 mL (any remaining water was removed as the low-boiling azeotrope in the process) to give 25,000:1

concentrates. The remaining unevaporated aliquots (500:1 concentrates) were saved for possible HPLC analysis. Column blanks were obtained between samples by repeating the elution with fresh solvent; blank concentrates were saved for chemical evaluation. Solvent concentrates from each column were analyzed to determine percent recoveries.

The drums that contained the sample were rinsed with approximately 1000 mL of acetonitrile after the sample was concentrated. The acetonitrile was separated from excess water as just described and concentrated to a final volume of about 10 mL.

Chemical Analysis. Analytical LC was performed on a Hewlett-Packard 1084B HPLC with a variable wavelength UV detector and a Shimadzu dual monochromator spectrofluorometer. Gradient elutions were performed by using water and acetonitrile as the solvents at a flow rate of 1.5 mL/min. The reverse-phase column for all analyses was a 4.6- × 25-mm Whatman PXS 10/25 ODS-2 used at a column oven temperature of 45 °C. Sample injections were made with a Hamilton gas-tight syringe into the 100-μL loop of a Rheodyne 7125 injector. The majority of standards were analyzed directly by HPLC–UV at 215 nm [crotonaldehyde, isophorone, 2,4-dichlorophenol, anthraquinone, biphenyl, 2,4′-dichlorobiphenyl, 2,6-bis(1,1-dimethylethyl)-4-methylphenol, 2,2′,5,5′-tetrachlorobiphenyl, bis(2-ethylhexyl) phthalate, and quinoline] or 271 nm (furfural and caffeine). Five- or ten-microliter injections of the 25,000:1 concentrates were chromatographed by using a 19-min gradient, 8–95% acetonitrile, a 2-min initial hold time, and a 9-min final hold time.

Standards with a poor UV chromophore (4-methyl-2-pentanone, 1-chlorododecane, and stearic acid) were analyzed on a Perkin–Elmer 3920 gas chromatograph with a flame ionization detector and a 30-m SE–54 wall-coated open tubular (WCOT) fused-silica capillary column (J & W Scientific). The injector temperature was 200 °C; the detector interface temperature was 280 °C. The carrier gas was He at 16.5 lb/in.^{2}, and the makeup gas was nitrogen at a flow of 40 cm^3/min. Splitless 1-μL injections of the 25,000:1 concentrates were made by starting with an oven temperature of 45 °C and the oven door open; after 2.75 min, the oven door was closed and the temperature was programmed at 4 °C/min to 280 °C, which was held for 8 min. The syringe was kept in the injection port for 15 s after injection.

Chloroform was determined on the 500:1 concentrate by using a Tracor 550 gas chromatograph with a 310 Hall electroconductivity detector in the halogen mode, connected to a Tekmar LSC–2 purge and trap system. The column was an 8-ft 1% SP–1000 on 60–80-mesh Carbopack B. An oven temperature gradient of 80–220 °C at 8 °C/min after an initial hold of 2 min was used. Fifty microliters of the acetonitrile concentrate was diluted to 5 mL with organic-free water in the purging vessel for the analysis.

Glucose was analyzed by using the Folin–Wu procedure (*32*) modified as follows: 25-mL glass-stoppered tubes were used, the final volume was 15 or 20 mL, and colorimetric analysis was performed at 650 nm. The aqueous and acetonitrile column effluents were analyzed prior to further concentration; 200-μL aliquots were used in the Folin–Wu test. Samples were prepared for Folin–Wu analysis by extracting them with dichloromethane to remove other species and evaporating the remaining aqueous phase to about 5 mL.

Quinaldic, trimesic, and stearic acids were derivatized with PNBDI to facilitate their chromatographic separation and UV detection (*33*). The procedure was as follows: 125 μL of PNBDI (2.6 mg) in dichloromethane was mixed with 200 μL of concentrated column effluent in a 5-mL Reactivial and diluted to 1 mL with acetonitrile; the vial was sealed and heated 2 h at 75 °C; the derivatized samples were analyzed by HPLC-UV at 254 nm by using a 10-min 20–95% acetonitrile gradient with a 9-min final hold.

Glycine was too polar to be resolved on the ODS-2 column and was also a very poor UV chromophore. Fluorescence derivatization using dansyl chloride (*34*) was employed. Derivatives were made by mixing 500 μL of the unextracted MP-1 or MP-50 NaCl effluent with 1000 μL of acetonitrile, mixing for 15 s, and evaporating 200 μL of the organic phase to dryness in a 1-mL Reactivial. Next, 20 μL of pH 10.5 $NaHCO_3$ buffer (0.2 M) and 50 μL of dansyl chloride solution (1.25 μg/μL) were added. The vial was capped, agitated vigorously for 15 s, and heated at 100 °C for 2 min in a block heater. Ten microliters of the cooled contents was analyzed by using a 20-min gradient, 20–40% mobile phase B, and a 5-min final hold. Mobile phase A was 10 mM sodium acetate to which 0.1-mL/L acetic acid was added; the pH was adjusted to 3.0 with phosphoric acid. Mobile phase B was acetonitrile to which 0.1-mL/L acetic acid and 0.77-mL/L phosphoric acid were added. The fluorometer monochromators were set at 298 nm (excitation) and 545 nm (emission).

Benzo[*e*]pyrene was analyzed by using HPLC with fluorescence detection; for the Shimadzu, the optimum wavelengths were 280 nm (excitation) and 394 nm (emission).

A lead blank (500 L of distilled water containing 25 μg/L of $Pb(NO_3)_2$) was pumped onto the ion-exchange columns only. The aqueous eluents were combined with the acetonitrile eluant for each column; NaCl was added to separate the aqueous and organic phases. After separation, the aqueous phase was extracted twice with acetonitrile. The extracts from each column were pooled and reduced in Kuderna–Danish evaporators to a final acetonitrile volume of 4–6 mL and then diluted to a known volume with distilled water for atomic absorption spectroscopic (AAS) determination of lead. AAS analysis was on an IL 951 with an air–acetylene flame; the 217.0-nm lead line was used.

A 500-L solution containing 2 mg/L of free chlorine residual in distilled water was pumped onto the four-column system; the columns were eluted and the eluants were processed as described earlier. This chlorine blank and resin eluant blanks were analyzed by GC–MS by using a Finnigan 4023 with the INCOS data system and a 31,000-compound National Bureau of Standards library. Electron impact spectra were obtained by using an electron energy of 70 eV and a scan time of 1 s for the mass range 33–550 amu. A 30-m WCOT SE-54 fused-silica capillary column (J & W Scientific) was used for separations. Injections were made with the oven at 40 °C and the door open, the injector at 220 °C, and the interface at 270 °C. Two minutes after injection, the door was closed and the temperature was raised ballistically to 60 °C, ramped at 4 °C/min to 280 °C, and held there for 4 min. The split and septum purge valves were closed for injection and opened after 1 min.

Results and Discussion

Average recoveries for model compounds from five repetitive 500-L experiments are shown in Tables I and II. Table I lists data for the low-to-medium polarity organics (broadly classified by retention through more than half of the reverse-phase HPLC gradient) of low aqueous solubility, whereas Table II summarizes data for the organic acids and hydrophilic neutrals of higher polarity. Although these are not

Table I. Distribution of Resin Recoveries of Low-Polarity Model Compounds

Compound (Concentration)	*Average Recovery (%)*[a]			*Total Average Recovery (%)*
	MP-1	*MP-50*	*XAD-2*	
Isophorone (50 μg/L)	nd	3.7 (0.8)	74.1 (14.2)	77.0
Anthraquinone (50 μg/L)	63.8 (20.1)	10.8 (10.6)	nd	74.6
Quinoline (50 μg/L)	nd	68.1 (19.6)	29.4 (2.9)	79.9
2,4-Dichlorophenol (50 μg/L)	73.2 (9.5)	4.6 (0.0)	5.3 (4.2)	76.2
2,4′-Dichlorobiphenyl (50 μg/L)	64.2 (7.3)	9.9 (12.1)	1.7 (1.5)	72.4
2,2′,5,5′-Tetrachlorobiphenyl (5 μg/L)	58.2 (8.7)	7.7 (7.5)	10.8 (10.1)	75.1
1,1′-Biphenyl (50 μg/L)	67.0 (5.6)	14.0 (12.8)	1.6 (0.4)	81.6
Benzo[*e*]pyrene (10 ng/L)	53.3 (24.0)	3.0 (0.0)	nd	53.9
Bis(2-ethylhexyl) phthalate (50 μg/L)	62.5 (37.2)	2.7 (0.0)	6.4 (6.3)	65.7
1-Chlorododecane (50 μg/L)	60.2 (31.1)	14.6 (3.6)	12.6 (13.4)	76.5
5-Chlorouracil (50 μg/L)	77.5 (32.3)	nd	nd	77.5
2,6-Bis(1,1-dimethylethyl)-4-methylphenol (50 μg/L)	33.5	3.5	17.2	54.1[b]
Chloroform (50 μg/L)	----------- composite -----------			67.9 (22.8)

NOTE: Classification as low polarity is based upon elution in the last half of the reverse-phase HPLC gradient described in the Experimental section.

[a] $N = 5$ for these results; standard deviations are given in parentheses; nd indicates not detected. For XAD-7, none of the compounds were detected except for bis(2-ethylhexyl) phthalate, which gave an average percent recovery of 1.0 ± 0.0.

[b] Standard used for spiking was found to be deteriorated after three experiments; $N = 3$ for this result.

rigorous distinctions, there is an obvious difference in the average recoveries of the two groups of compounds. Most of the compounds in Table I, which would be considered lipophilic, showed recoveries greater than 70%. The hydrophiles (Table II), all being water soluble in the grams-per-liter range, were less than 45%, and half were less than 7% recovery.

Results of mass balance and recovery calculations for the components of the experimental system (sample drum, resin columns, and column effluent) are shown in Table III. Resin breakthrough was detected (and at high levels) for only three compounds: furfural,

Table II. Distribution of Resin Recoveries of Organic Acids, High-Polarity Compounds, and Hydrophilic Neutrals

	Average Recovery (%)[b]				
Compound[a]	*MP-1*	*MP-50*	*XAD-2*	*XAD-7*	*Total (%)*
Stearic acid	33.6 (16.7)	nd	10.5 (0.0)	nd	36.3
Quinaldic acid	31.8 (9.0)	nd	na	na	31.8
Trimesic acid	nd	nd	na	na	nd
Glycine	1.9 (2.3)	1.7 (0.9)	nd	nd	3.3
Glucose	nd	nd	nd	nd	nd
Furfural	nd	0.5 (0.2)	4.9 (1.5)	1.3 (0.4)	6.6
Crotonaldehyde	nd	nd	1.0 (0.4)	0.2 (0.0)	0.8
Caffeine	nd	1.5 (0.2)	34.3 (4.0)	6.3 (1.9)	41.8
4-Methyl-2-pentanone	nd	nd	26.8 (15.7)	nd	26.8[c]

NOTE: nd indicates not detected; na indicates not analyzed.
[a] All concentrations were 50 μg/L.
[b] $N = 5$ for these results; standard deviations are given in parentheses.
[c] Compound formed azeotrope with water, and variable losses occurred in solvent reduction.

crotonaldehyde, and caffeine. Breakthrough may have also occurred for glucose, glycine, trimesic acid, and quinaldic acid, but it was not determined: glucose and glycine would not have partitioned into dichloromethane and could not be analyzed directly in the effluents, and analytical problems were encountered with trimesic and quinaldic acids. Sample drum residues of only five of the model compounds were detected sporadically and at low levels (<10% of the amount spiked).

The data summarized in Tables I and II also show the distribution of each compound recovered from the individual resins. The ion-exchange resins, which have a styrene–divinylbenzene cross-link structure similar to XAD-2 (but with much larger pore size), showed significant retention of several of the hydrophobic neutral compounds. However, for several compounds, breakthrough to the macroreticular XAD-2 was observed. XAD-7 yielded only four compounds: phthalate, furfural, crotonaldehyde, and caffeine. For furfural, crotonaldehyde, and caffeine, significant system breakthrough also occurred as just described. The resins used in this system were a priori selected to accommodate a wide range of chemical or physical properties that might be present in a given sample and to provide a resin mass and surface area consistent with processing very large volumes of water

Table III. Mass Balance of Model Compounds Recovered from Concentrator System Components

Compound	*Average Recovery (%)* Resin Columns	Sample Drum	Column Effluent	Total (%)
Isophorone	77.0	nd	nd	77.0
Anthraquinone	74.6	8.8	nd	83.3
Quinoline	79.9	nd	nd	79.9
2,4-Dichlorophenol	76.2	nd	nd	76.2
2,4′-Dichlorobiphenyl	72.4	0.05	nd	72.9
2,2′,5,5′-Tetrachlorobiphenyl	75.1	4.0	nd	79.1
1,1′-Biphenyl	81.6	nd	nd	81.6
Benzo[*e*]pyrene	53.9	2.4	nd	56.5
Bis(2-ethylhexyl) phthalate	65.7	2.6	nd	68.3
1-Chlorododecane	76.5	nd	nd	76.5
5-Chlorouracil	77.5	nd	nd	77.5
2,6-Bis(1,1-dimethylethyl)-4-methylphenol	54.1	nd	nd	54.1
Chloroform	67.9	na	nd	67.9
Stearic acid	36.3	nd	nd	36.3
Quinaldic acid	31.8	na	na	31.8
Trimesic acid	nd	na	na	nd
Glycine	3.3	na	na	3.3
Glucose	nd	nd	nd	nd
Furfural	6.6	nd	57.6	64.7
Crotonaldehyde	0.8	nd	28.4	28.9
Caffeine	41.8	nd	64.7	104.5
4-Methyl-2-pentanone	26.8	nd	nd	26.8

NOTE: nd indicates not detected; na indicates not analyzed.

through the columns. The choice of particle size was intended to increase efficiency of adsorption and elution by minimizing band spreading and increasing effective capacity at a given flow rate.

The results of model compound recovery experiments, in part, support these selection criteria. For example, the anionic resin (MP-1) yielded the best recoveries of the anionic organics (little or no adsorption was observed on subsequent resins); glycine was equally distributed (but poorly recovered) on MP-1 and MP-50 (no adsorption was observed on the nonionic resins). Although some selective adsorption occurred on the lower surface area ionic resins, the nonpolar macroporous XAD-2 showed its retentive power for low-polarity compounds as none were seen to break through to the more polar methacrylate polymer, XAD-7. XAD-7 was included in the system for use with reclaimed and surface waters (*21*) because literature reports indicated that the methacrylate XAD resins had significantly better retention of humics, fulvics, and smaller phenolics (*9, 29, 30*). However, in the model compound evaluations reported here, where the DOC of the solutions was quite low and organic breakthrough to XAD-7 was not induced, the use of XAD-7 was not well-demonstrated. For the four compounds recovered from XAD-7, the percentages retained were not impressive.

The resin columns used in this evaluation had been used previously in a water reuse project for isolating organics from more than 70 samples of 100–200 L volumes of surface, reclaimed, and ground waters (*28*). During that study, the columns were unpacked and the resins were reprocessed one time. The protracted and repetitive use of adsorbents for isolating residues for chemical and biological analyses automatically raises questions about long-term efficiency and cross contamination or carry-over. The approximately 70% recoveries of hydrophobic model compounds observed in this evaluation were consistent with the 70% recoveries of a 27-compound mixture from a series of tap water and reclaimed water samples reported previously during development of this system (*31*). These data are also consistent with the average total organic carbon (TOC) removal of 76% observed during the water reuse study (*21, 28*). The extent of model compound cross contamination was estimated by pumping a 500-L distilled water blank through the system, followed by normal column elution. Analysis revealed a minimal percentage carry-over of a few spiked standards: 0.01% biphenyl; 0.02% 1-chlorododecane; <0.01% dichlorobiphenyl; 0.03% tetrachlorobiphenyl; 0.96% phthalate. This carry-over experiment was also consistent with previous results that showed that carry-over of mutagens between samples did not occur, nor did mutagens break through the system (*28, 31*).

In spite of these apparent consistencies, Table III demonstrates mass balance problems. Except for caffeine, the total mass of each model compound accounted for in this system was significantly below 100%. Although some of the losses may be attributed to poor partitioning from salt solution eluants into dichloromethane for the hydrophiles, the incomplete accounting for the low-polarity compounds is not satisfactorily explained. Independent HPLC analyses of eluants prior to Kuderna–Danish evaporation showed that the losses observed were not due to the solvent evaporation step. The column-to-column variation in recoveries and incomplete mass balance are therefore difficult to explain unless some portion of each compound is irreversibly adsorbed to the resins. These observations may be consistent with those of Aiken (*30*), who reported that as much as 50% of the DOC adsorbed by XAD-8 from high-organic-content natural waters could not be removed. The concept of irreversible adsorption is important in the repetitive use of polymeric adsorbents because it implies a gradual change in the adsorptive nature of the polymers, which could eventually lead to degraded performance. The possibility of declining efficiency supports routine use of evaluation checks such as monitoring UV absorbance of column effluents, chemical and bioassay of column blanks between samples, use of radio-labeled internal standards, and TOC removal calculations. In addition, use of a small guard column filled with MP-1 is recommended as an expendable part of the system when it becomes visibly and irreversibly discolored throughout (*21*).

Inorganic cations, although probably isolated by ion exchange, should not be soluble in the dichloromethane extract of the aqueous eluents and should probably remain therein. The experiment with lead(II) nitrate, which yielded <0.2% of the spiked Pb ion, supported this expectation. Therefore, heavy metal toxicity to bioassay systems should not be a problem for testing organic residues. Conversely, when inclusion of inorganics in a test residue is desirable, other recovery techniques should be considered.

Use of the resins with samples containing free chlorine residual is not recommended. Cheh (*35*) suggested that chlorine may produce mutagenic artifacts on XAD-4. Our experiment with 2-mg/L chlorine residual appeared to promote the release of irreversibly adsorbed spiked standards: Six model compounds were recovered at levels several times higher than those observed in normal blank runs. In addition, many resin artifacts were eluted after exposure to this chlorine level, primarily aromatic and aliphatic acids, aldehydes, and ketones. Stoichiometric dechlorination (ferrous ion) is therefore recommended in order to avoid cross contamination between samples and inclusion of undesirable resin artifacts in the residue to be bioassayed.

Acknowledgments

The authors are grateful to John Gute for the GC–MS characterization. This work was funded under USEPA Contract No. R-806399010.

Literature Cited

1. Wu, C.; Suffet, I. H.; *Anal. Chem.* **1977,** *49,* 231.
2. Yohe, T. L.; Suffet, I. H.; Grochowski, R. J. *Measurement of Organic Pollutants in Water and Wastewater;* ASTM: Philadelphia, 1979; p 47.
3. Midwood, R. B.; Felbeck, G. T., Jr. *J. Am. Water Works Assoc.* **1968,** *60,* 357.
4. Deinzer, M.; Melton, R.; Mitchell, D. *Water Res.* **1975,** *9,* 799.
5. Klein, E.; Eichelberger, J.; Eyer, C.; Smith, J. *Water Res.* **1975,** *9,* 807.
6. Kopfler, F. C.; Coleman, W. E.; Melton, R. G.; Tardiff, R. G.; Lynch, S. C.; Smith, J. K. *Ann. N.Y. Acad. Sci.* **1977,** *298,* 20.
7. Chriswell, C. D.; Ericson, R. L.; Junk, G. A.; Lee, K. W.; Fritz, J. S.; Svec, H. J. *J. Am. Water Works Assoc.* **1977,** *69,* 669.
8. Junk, G. A.; Richard, J. J.; Grieser, M. D.; Witiak, D.; Witiak, J. L.; Arguello, M. D.; Vick, R.; Svec, H. J.; Fritz, J. S.; Calder, G. V. *J. Chromatogr.* **1974,** *99,* 745.

9. Stepan, S. F.; Smith, J. F. *Water Res.* **1977,** *11,* 339.
10. Van Rossum, P.; Webb, R. G. *J. Chromatogr.* **1978,** *150,* 381.
11. Leenheer, J. A.; Huffman, E. W. D., Jr. *J. Res. U.S. Geol. Surv.* **1976,** *4,* 737.
12. Bellar, T. A.; Lichtenburg, J. J. *J. Am. Water Works Assoc.* **1974,** *66,* 739.
13. Loper, J. C.; Lang, D. R.; Schoeny, R. S.; Richmond, B. R.; Gallagher, P. M.; Smith, R. C. *J Toxicol. Environ. Health* **1978,** *4,* 919.
14. *Evaluation of Toxic Effects of Organic Contaminants in Recycled Water;* EPA. U.S. Government Printing Office: Washington, DC, 1978.
15. Rappaport, S. M.; Richard, M. G.; Hollstein, M. C.; Talcott, R. E. *Environ. Sci. Technol.* **1979,** *13,* 957.
16. Baird, R. B.; Gute, J. P.; Jacks, C. A.; Jenkins, R. L.; Neisess, L.; Scheybeler, B.; Van Sluis, R.; Yanko, W. A. *Water Chlorination: Environmental Impact and Health Effects;* Ann Arbor Science: Ann Arbor, MI, 1980.
17. Donaldson, W. T. *Environ. Sci. Technol.* **1977,** *11,* 348.
18. *Drinking Water and Health;* Natural Academy of Sciences: Washington, DC, 1977.
19. Reinard, M. *Environ. Sci. Technol.* **1984,** *18,* 410.
20. Kopfler, F. C.; Ringhand, H. P.; Bull, R. J. *Water Reuse Symposium II—Proceedings;* American Water Works Association: Washington, DC, 1981; p 2282.
21. Baird, R. B.; Jacks, C. A.; Jenkins, R. L.; Gute, J. P.; Neisess, L.; Scheybeler, B. *Chemistry in Water Reuse;* Ann Arbor Science: Ann Arbor, MI, 1981; p 149.
22. Grieser, M. D.; Pietrzyk, D. J. *Anal. Chem.* **1973,** *45,* 1348.
23. Pietrzyk, D. J.; Chu, C. *Anal. Chem.* **1977,** *49,* 757.
24. Pietrzyk, D. J.; Chu, C. *Anal. Chem.* **1977,** *49,* 860.
25. Pietrzyk, D. J.; Kroeff, E. P.; Rotsch, T. D. *Anal. Chem.* **1978,** *50,* 497.
26. Kroeff, E. P.; Pietrzyk, D. J. *Anal. Chem.* **1978,** *50,* 502.
27. Baum, R. G.; Cantwell, F. F. *Anal. Chem.* **1978,** *50,* 280.
28. *Health Effects Study—Final Report;* L.A. County Sanitation Districts: Los Angeles, 1984; NTIS No. PB84191568.
29. Thurman, E. M.; Malcolm, R. L.; Aiken, G. R. *Anal. Chem.* **1978,** *50,* 775.
30. Aiken, G. R.; Thurman, E. M.; Malcolm, R. L.; Walton, H. F. *Anal. Chem.* **1979,** *51,* 1799.
31. Jenkins, R. L.; Jacks, C. A.; Baird, R. B.; Scheybeler, B. J.; Neisess, L. B.; Gute, J. P.; Van Sluis, R. J.; Yanko, W. A. *Water Res.* **1983,** *17,* 1569.
32. *Methods of Analyses, AOAC;* Association of Official Analytical Chemists: Washington, DC, 1965; p 499.
33. Knapp, D. R.; *Handbook of Analytical Derivatization Reactions;* Wiley-Interscience: New York, 1979; p 190.
34. Olson, D. C.; Schmidt, G. J.; Slavin, W. *Chromatogr. Newsl.* **1979,** *7,* 22.
35. Cheh, A. M.; Skochdopole, J.; Cole, L. *Science* **1980,** *207,* 90.

RECEIVED for review August 14, 1985. ACCEPTED February 6, 1986.

27

Continuous Liquid–Liquid Extractor for the Isolation and Concentration of Nonpolar Organic Compounds

for Biological Testing in the Presence of Humic Materials

R. J. Baker and I. H. Suffet

Environmental Studies Institute, Drexel University, Philadelphia, PA 19104

A continuous liquid–liquid extraction sampling device (CLLE) was evaluated for its ability to concentrate nonpolar organics from water into methylene chloride. CLLE recoveries were determined at pH 3 and 7 for mixtures of up to 13 compounds and compared to recoveries of separatory funnel batch extraction (batch LLE). The comparison at pH 3 was made with and without a humic substance present. Linear regression and analysis of variance showed that CLLE and batch LLE recoveries are statistically equivalent for most compounds under the conditions tested and that the presence of the humic material has similar effects on CLLE and batch LLE. CLLE sampling was found to be reproducible and suitable for extracting nonpolar organics from water samples for use in biological testing.

BIOLOGICAL TESTING undertaken to assess the health significance of potable water supplies usually requires the exposure of test species to chemicals at levels higher than those present in the water being tested. One example is the Ames test for mutagenic activity. Direct biological testing of drinking water is not now possible because the total concentration of organic matter in water supplies is usually much less than 10 mg/L as dissolved organic carbon (DOC). It is estimated that 90% of the DOC in drinking water supplies is high molecular weight natural humic material, which is apparently nontoxic (*1*). The remaining 10% of

0065–2393/87/0214/0571$06.00/0

the DOC is characterized as low molecular weight organics. When natural humics are chlorinated, low and high molecular weight materials can be formed. It is not known what percentage of the products are of health significance. Many identifications have been made of the volatile organic fraction of the low molecular weight organics by gas chromatography–mass spectroscopy (GC–MS), whereas only 5–10% of the nonvolatile organics composing the remaining 90% of the DOC has been identified (*2*).

A problem often overlooked in potable water evaluation is that the complement of organic compounds, especially from a surface water source, changes over time. The compounds present and their concentrations are constantly changing (*3*). Thus, it is almost impossible to formulate a typical mixture of organic compounds for a biological testing program. Therefore, a concentrated composite sample representing the water over a period of time should be used for any biological testing program of potable water. This requirement is especially important for trace organic chemicals for which the chronic effect is the primary concern. Thus, large-volume composite samples taken over long periods of time are more representative of the chronic exposures presented to the public.

A recent National Academy of Sciences (NAS) review (*4*) of the available methods for the preparation of water concentrates concluded that no single concentration method is adequate for isolating all the organic constituents from an aqueous sample. Each method suffers from differences in selectivity that are fundamental to that method. Artifacts added during processing are also a major concern. The U.S. Environmental Protection Agency (USEPA) recently evaluated several methods for preparing such concentrates from water for biological testing (*5, 6*).

The isolation method of solvent extraction has been suggested as a potentially feasible process to concentrate trace organic compounds from finished drinking water (*4*). One positive attribute of the solvent extraction method is that its performance for any given compound is theoretically predictable from a partition coefficient of a compound between the water sample and an organic solvent. The partition coefficient can be experimentally determined for any solute in any two-phase solvent system (*7, 8*). Variables of the extraction procedure such as solvent-to-water ratio and the choice of solvents can be adjusted to achieve optimum recovery.

Batch solvent extraction methods (e.g., by separatory funnel) for the preparation of extracts for biological analysis have a number of major problems. A large volume of solvent must be evaporated after the extraction step in order to obtain a sample sufficiently concentrated to be useful for biological testing. Artifacts can occur from solvent impurities, and reactions can occur during the evaporation process.

Continuous liquid–liquid extractors (CLLEs) using a small amount of solvent that is continuously recycled would be more effective for this type of sampling. A CLLE system would be less affected by solvent contaminants, and less solvent would be needed.

An on-line continuous liquid–liquid extraction system has been developed in our laboratories and has been used to isolate nanogram-per-liter concentrations of organophosphate pesticides from natural waters (*9, 10*) and to isolate a broad spectrum of gas chromatographable trace organics present in drinking water (*11–13*). An upgrade of the extractor has been recently developed (*14*). Features of the extractor include solvent recycle and adjustable water and solvent flow rates. The mixing of water and solvent is completed in a Teflon coil. The solvent is separated from the aqueous phase by gravity and is recovered by distillation. All wetted parts are constructed of glass and Teflon. The sampler can operate at 2 L/h as presently designed. A 100-L sample concentrated to 1 mL would be needed for Ames testing and would require 50 h of extraction followed by Kuderna–Danish evaporation to reduce the sample extract from 100 mL to 1 mL. This approach is now being studied to collect samples for the Department of Environmental Protection of New Jersey at different water treatment plants in New Jersey (*15*).

Another scheme proposed for collecting enough concentrate for animal studies is shown in Figure 1. In this scheme for concentrating 135 gal of water to a 5-gal methylene chloride concentrate, reverse osmosis is used to reduce the initial 2000-gal volume of water to 135 gal, and the CLLE is used to further concentrate the retentate to 5 gal for biological testing. The LLE portion of this scheme was simulated in a set of CLLE evaluation experiments. The concentration factor from 135 to 5 gal is 27:1, which was easily attainable with the CLLE units. The water-to-solvent ratio in the extractors was 10:1, and the additional volume reduction was achieved by solvent distillation in the distillation chambers of the CLLE units.

Experiments were designed to evaluate the performance of the CLLE units under conditions similar to those that may be encountered in field sampling while at the same time maintaining control over ionic strength, pH, salt matrix, trace organics present, and humic content. Therefore, Milli-Q water spiked with salts, organic solutes, and humics was used rather than natural water samples. Although no biological testing was done, the primary intended application of samples collected by the CLLE method investigated here is collecting extracts for biological testing. The focus of this work was to evaluate the ability of the CLLE method to collect a representative sample of the organic fraction of interest: low molecular weight nonpolar compounds. Although a commercially prepared humic material does not necessarily simulate

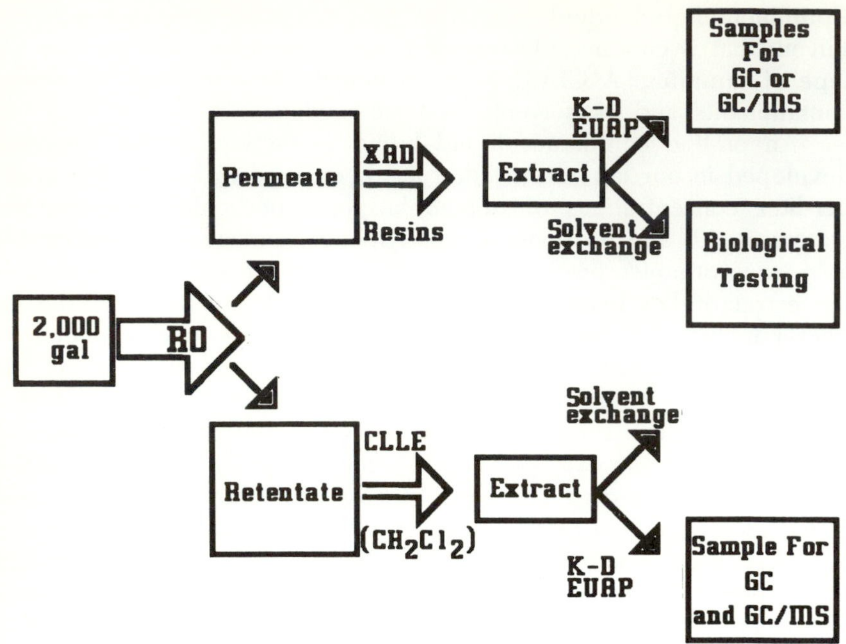

Figure 1. Isolation scheme to collect extracts for animal studies.

natural conditions or variability of humics in the environment, it is used as a model to give some indication of how humiclike materials might affect recoveries and operation of the CLLE equipment.

The objectives of this work were (1) to evaluate the reliability and performance of this CLLE design in a large-scale scheme for the preparation of samples for biological testing, (2) to compare CLLE recovery to batch LLE recovery for a set of organic probes, (3) to evaluate variability in CLLE performance as a function of sample size, and, above all, (4) to evaluate the effects of background humic materials on the extraction process for both CLLE and batch LLE.

Experimental

CLLE Apparatus. Figure 2 is a schematic diagram of the extractor used in this work. Water and the extraction solvent (methylene chloride) are pumped together through a T-joint, and the mixture flows through a mixing coil where extraction takes place. This coil consists of Teflon tubing (1/8 in. o.d., 1/16 in. i.d.), 32 ft long, wrapped around a 1-in. steel pipe for support.

The mixture exits the coil(s) and flows into the separation chamber where the solvent and water mixture is separated by gravity. The lighter water phase rises and flows through the upper outlet. The solvent phase is heavier and flows down through the main body of the extractor. A portion of it fills the float chamber, and the rest flows into the distillation chamber. Because they are

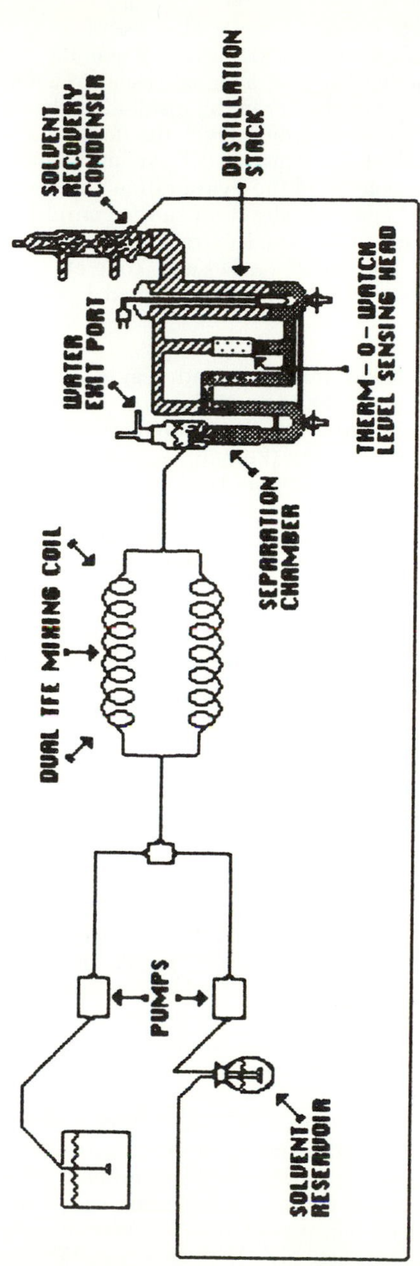

Figure 2. Schematic diagram of the continuous liquid–liquid extractor. (Reproduced with permission from reference 14.)

hydraulically connected, the liquid level of the float and the liquid level of the distillation chamber remain equal while the extractor is in operation. A liquid level sensor (Therm-O-Watch Model L6-1000, I^{2}R Corp.) clipped to the float chamber electronically detects the presence or absence of a glass float, which rises and falls with the methylene chloride level. When the float rises high enough to be detected by the level sensor, a signal is sent to a Therm-O-Watch switching device, which turns on the heater in the distillation chamber. When the solvent is distilled down to a preset minimum level, the float falls below the level sensor and the heater is switched off. This on–off cycling continues during the entire extraction run, and the volume in the evaporation chamber remains relatively constant (70–80 mL). When the extraction run is completed, contents of the distillation chamber constitute the final extract. Initial versions of the extractor were designed by Wu and Suffet (*9–11*) and Yohe et. al (*11*). The most recent modification of the extractor has been described in detail by Baker et al. (*14*).

Materials and Reagents. The materials and reagents for this work are described elsewhere in detail (*14*). The 14 nonpolar organic compounds and salt solutions used in the CLLE evaluation are the following (the functional groups are given in parentheses): acetophenone (ketone), 2,4-dichlorophenol (phenol), biphenyl (aromatic hydrocarbon), diacetone-L-sorbose (sugar), 2-methylnaphthalene (polyaromatic hydrocarbon), caffeine (xanthene), bis(2-ethylhexyl) phthalate (phthalate), isophorone (ketone), quinoline (amine), ethyl cinnamate (ester), 1-chlorodecane (chloroalkane), 2,4-dichlorobiphenyl (polychlorinated biphenyl, anthracene (polyaromatic hydrocarbon), and 2,2′,5,5′-tetrachlorobiphenyl (polychlorinated biphenyl). The pH 7 salt solution contained calcium sulfate (376 mg/L), sodium bicarbonate (214 mg/L), and calcium chloride dihydrate (144 mg/L); the ionic strength was 0.012. The pH 3 salt solution contained calcium sulfate (1800 mg/L), sodium bicarbonate (1050 mg/L), and calcium chloride dihydrate (705 mg/L); the ionic strength was 0.734.

Sample Concentration Experiments. A CLLE quality assurance blank was run by extracting 90 L of Milli-Q water with three CLLE samplers in a parallel configuration and concentrating the composited extract to 4 mL by Kuderna–Danish evaporation. The 22,500-fold concentrate was analyzed by GC–flame ionization detection (GC–FID) and GC–MS. Thirty-two peaks were observed by using GC–FID analysis, but because of their low concentrations, only four contaminants were identified by GC–MS: cyclohexene, 2-cyclohexen-1-one, *n*-butyl phthalate, and bis(2-ethylhexyl) phthalate. Cyclohexene is a solvent preservative that has been identified in commercial high-purity methylene chloride (*16*), and 2-cyclohexen-1-one is its air oxidation product. The phthalates are ubiquitous laboratory contaminants and have also been identified in commercial methylene chloride (*17*).

Three large-scale water samples were prepared in 55-gal stainless steel drums and used for evaluation of the CLLE units. Each solution was greater than 100 L and consisted of an aqueous mixture of inorganic salts and low molecular weight organic solutes. In the first experiment, the sample was adjusted to pH 7, whereas in the two other experiments, the water was adjusted to pH 3. During the first experiment, two batch and two continuous LLEs were completed on 1.35 and 2.3 L, respectively. The remaining 178 L of sample was extracted by the three CLLE samplers used in a parallel configuration. The concentrates were continuously drained from each extractor through a needle valve and collected as a composite sample. The needle valves were set so that

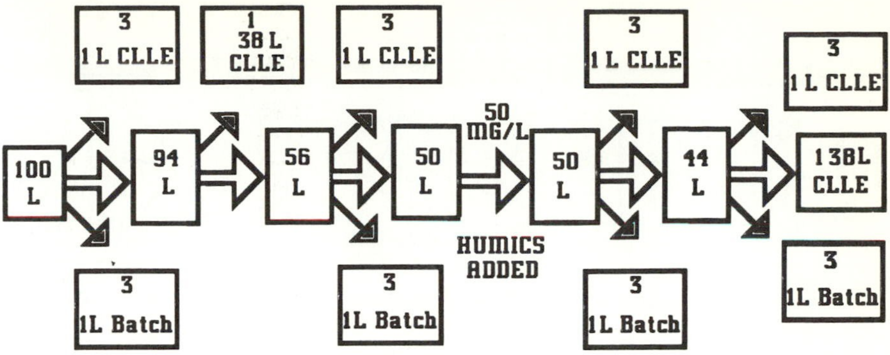

Figure 3. Second large-scale CLLE extraction experiment of 100 L at pH 3.

the final methylene chloride extract volume was 1/27 of the volume of aqueous solution extracted. This procedure served to assess the reliability of the units and the feasibility of using them in large-scale concentration schemes as described in Figure 1. Sampling schemes for the second and third extraction experiment are shown in Figures 3 and 4. Several 12.5-L CLLE extractions were run in the third experiment, as shown in Figure 4. These allowed comparisons to be made of large- versus small-sample CLLE recoveries.

Humic materials (50 mg/L of Fluka humic acids) were added to the drum contents before the second half of each pH 3 extraction, Experiments 2 and 3 (Figures 3 and 4). This step allowed comparisons to be made between extractions with and without humics present.

Results and Discussion

First Experiment, pH 7. The extractors performed without major interruptions during the 178-L, pH 7 extraction. The minor problems encountered were the following:

1. During the extraction, which was conducted over several days, significant growth of filamentous algae and bacteria grew in the sample. The biomass accumulated in the phase separator and caused emulsification of the two phases. This occurrence resulted in some loss of sample concentrate, which would decrease recovery values for the trace organics. Lowering the pH to 3 during the subsequent two extraction experiments decreased, but did not entirely eliminate, the problem.
2. The glass floats, which control the liquid levels in the distillation chambers, leaked and sank and thus caused problems with liquid level control in the distillation chambers. New floats were made, but minor leakage persisted. The float has since been eliminated, and a more sensitive level detecting device is now used.

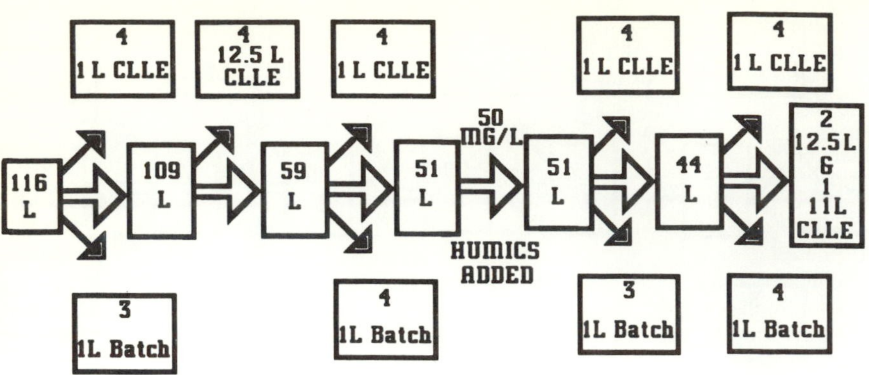

Figure 4. Third large-scale CLLE extraction experiment of 100 L at pH 3.

3. Flow rate control of the solvent pump was difficult because 0.2 L/h is near the low end of the solvent pump's capacity. This problem was resolved before the second experiment by changing the reduction gear ratio of the pumps.
4. Air bubbles in the system carried small amounts of solvent extract out the phase separator with the aqueous phase. This loss was eliminated by moving the pumps to a position lower than the other parts of the extractor, as recommended by the manufacturer, and ensuring that all Teflon connections were airtight.

These problems were overcome by the end of the third extraction experiment, and the extractors can now be operated smoothly for long periods of time with no interruptions.

The recovery data for the CLLE and batch-extracted portions at pH 7 are shown in Table I. The limited number of data points (two each for batch and continuous extraction) made statistical analysis of the data impossible. However, batch and continuous recoveries for most compounds are similar. The pH 7 solution used in the 45-gal CLLE had biological growth in the drum, which caused emulsification problems in the phase separator. Baker et al. (*14*) showed that partition coefficients for nearly all of the test compounds are the same at pH 3 and 7, so extraction efficiency was not expected to decrease by lowering the pH in subsequent experiments. Therefore, pH 3 was chosen for the second and third experiments to minimize the effects of biological growth in the sample. This growth is not a problem when continuous, large-volume sampling is done directly from the water source into the extractors.

Second Experiment, pH 3 (Figure 3). The first pH 3 extraction was run continuously without major interruptions or operational

Table I. Comparison of Batch and Continuous LLE Recovery at pH 7

Compound	Batch ± Range[a]	CLLE ± Range[b]	CLLE[c]
	Percent Recovery		
Acetophenone	72 ± 5.0	82 ± 10.2	56
Isophorone	78 ± 6.8	88 ± 11.1	81
2,4-Dichlorophenol	52 ± 5.9	55 ± 3.6	61
Quinoline	69 ± 3.2	84 ± 5.6	87
Biphenyl	60 ± 27.6	46 ± 7.0	28
Caffeine	40 ± 3.6	42 ± 4.3	28
Bis(2-ethylhexyl) phthalate	101 ± 8.8	39 ± —	35
	Micrograms per Liter[d]		
2,2′,5,5′-Tetra-chlorobiphenyl	114 ± 9.6	70 ± 14.0	134
2,4-Dichlorobiphenyl	41 ± 3.7	41 ± 7.6	30

NOTE: The ranges given are for two values. All water-to-solvent ratios were 10:1. No humic material was present.
[a] $N = 2$; sample volume = 1.35 L; concentration factor = 10:1.
[b] $N = 2$; sample volume = 2.32 L; concentration factor = 10:1.
[c] $N = 1$; sample volume = 178 L; concentration factor = 27:1.
[d] Values are given in micrograms per liter recovered from aqueous solution (initial concentration was unknown).

problems. The phase separator had to be cleaned to remove algae and humic deposition four to five times during the 100-L extraction. This problem would not occur during field sampling because a 0.45–0.70-μm on-line glass fiber filter could be installed at the sample intake. Recovery values for batch and continuous extraction of this sample are listed in Table II. This table includes recoveries before the addition of humics (*14*) and recoveries in the presence of humics.

A linear regression procedure (Figure 5) showed that 1-L batch LLE is equivalent to 1-L CLLE with and without humics present. Some compounds showed lower recoveries for 38-L CLLE than for 1-L CLLE, but statistical evaluation of this result was not possible because only one 38-L extraction was run for each condition (with and without humics). The third experiment was designed to address the effect of sample size on solute recovery and further investigate the effect of humic materials on solute recovery.

Third Experiment, pH 3 (Figure 4). Recoveries for the third large-scale extraction are shown in Table III. This experiment determined that recoveries for 1-L batch LLE, 1-L CLLE, and 12.5-L CLLE samples are equivalent for 10 of 13 compounds tested without humics present (*14*), and recovery differences for the other three are not substantial. Linear regression and analysis of variance (ANOV) procedures (*18*) were used to make these comparisons.

Table II. Recovery for CLLE and Batch Extraction at pH 3 with and without Humic Material: Experiment 2

Compound	*1 L w/o Humics*[a] *Batch* ± *S.D.*	*1 L w/Humics*[b] *Batch*	*1 L w/o Humics*[a] *CLLE* ± *S.D.*	*1 L w/Humics*[c] *CLLE* ± *S.D.*	*38 L w/o*[d] *Humics CLLE*	*38 L w/Humics*[d] *CLLE*
			Percent Recovery			
Acetophenone	89 ± 4	79	85 ± 9	78 ± 4	86	79
Isophorone	89 ± 4	88	86 ± 7	81 ± 6	97	98
2,4-Dichlorophenol	73 ± 8	82	70 ± 13	66 ± 17	98	—
1-Chlorodecane	25 ± 3	35	22 ± 4	24 ± 7	5	9
Biphenyl	52 ± 2	23	45 ± 4	32 ± 2	39	22
Caffeine	37 ± 5	50	33 ± 10	41 ± 13	41	19
			Micrograms per Liter[e]			
2,4-Dichlorobiphenyl	1090 ± 180	—	1000 ± 100	—	806	1042
2,2′,5,5′-Tetra-chlorobiphenyl	114 ± 32	179	99 ± 12	80 ± 20	63	82

NOTE: S.D. denotes standard deviation.
[a] Six 1-L batch extractions were analyzed.
[b] One 1-L batch extraction was analyzed.
[c] Three 1-L CLLE samples were analyzed.
[d] One 38-L CLLE sample was analyzed.
[e] Values are given in micrograms per liter recovered from aqueous solution (initial concentration was unknown).

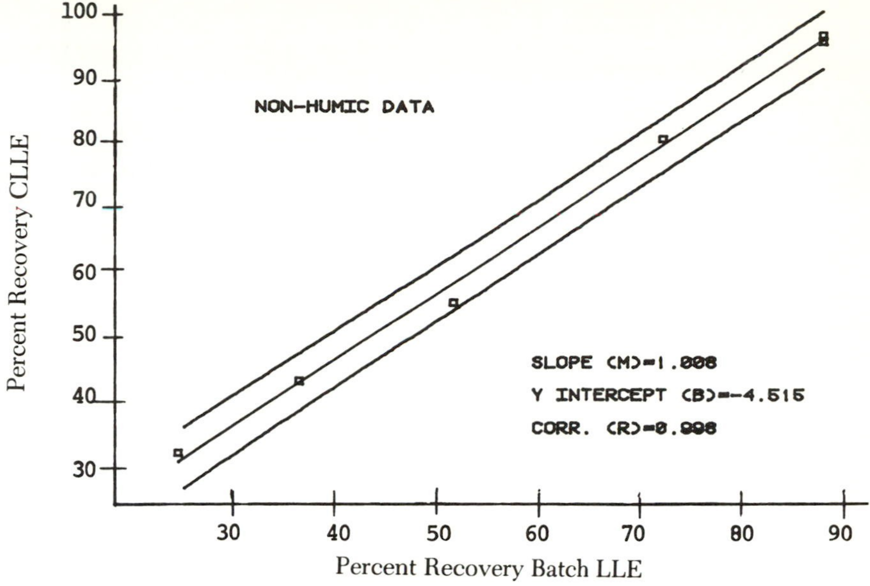

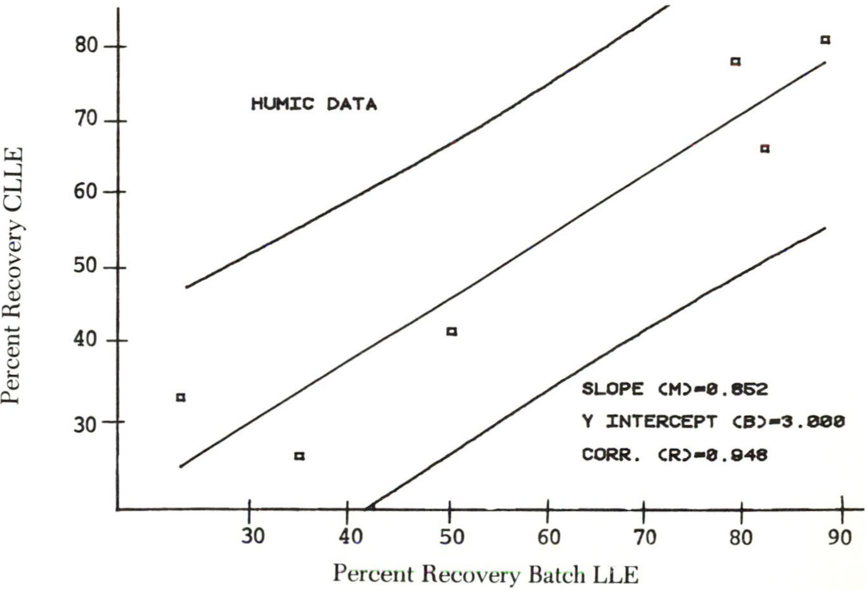

Figure 5. Linear regression analysis of 1-L batch versus 1-L continuous liquid-liquid extraction Experiment 2.

The same statistical procedures were used here to evaluate the effects of humics on batch and continuous LLE. A base extraction procedure (*19*) was required for processing methylene chloride extracts prior to GC injection in order to protect the GC column from contamination by humics. This process led to losses of 2,4-dichlorophenol and the chlorinated biphenyls. Therefore, these compounds were not used in the evaluation of the CLLE in the presence of humics. All other compounds were not affected by the base extraction procedure. The ANOV procedure tested each compound for changes in concentration by comparing early batch extraction recoveries (from freshly prepared solution) to later ones (after the 12.5-L extraction). This process was done separately for Parts 1 and 2. It was therefore possible to test each compound for time-dependent decreasing concentration with and without the presence of humics.

Results of the nested ANOV for Part 1 (without humics) and Part 2 (with humics) of Experiment 3 are listed in Table IV. The following comparisons will be made for Experiment 3 data:

1. 1-L batch LLE versus 1-L CLLE recoveries without humics and with humics.
2. Decrease in concentration of each compound over time in the pH 3 salt solution without humics and with humics.
3. Extraction with humics versus extraction without humics for batch LLE and for CLLE.

1. Batch LLE Versus CLLE. The linear regression procedure described earlier was also applied to the third experiment, and good linear relationships between batch LLE and CLLE recoveries for both Parts 1 and 2 of the experiment without humics and with humics were shown (Figure 6). The ANOV procedure for Part 1 data (*14*) showed that mean recoveries for 10 of 13 compounds are homogeneous when 1-L batch LLE, 1-L CLLE, and 12.5-L CLLE comparisons are made. Slight recovery differences in the other three appear to be caused by minor losses in the CLLE distillation chamber during larger sample extractions because of the longer distillation time. One-liter batch LLE and 1-L CLLE samples were compared in Part 2, and eight of the nine compounds were found to have homogeneous mean recoveries (*see* Figure 6, Part 2). The one exception, quinoline, has low recovery values for both CLLE and batch LLE. This situation made the 3% difference between them (32% and 29%, respectively) proportionally large. This difference, combined with a small analytical error (also 3%) for quinoline, resulted in the difference between means being considered significant by the ANOV procedure. Although this difference is statistically significant, it is not substantial enough to cause problems. Therefore, it has been shown that CLLE and batch recoveries are equivalent,

Table III. Recovery for CLLE and Batch Extraction at pH 3 with and without Humic Material: Experiment 3

Compound	*1 L w/o Humics*[a] *Batch* ± *S.D.*	*1 L w/Humics*[b] *Batch*	*1 L w/o Humics*[c] *CLLE* ± *S.D.*	*1 L w/Humics*[d] *CLLE* ± *S.D.*	*12.5 L w/o Humics*[e] *CLLE* ± *S.D.*	*12.5 L w Humics*[f] *CLLE* ± *Range*
			Percent Recovery			
Acetophenone	94 ± 4	89 ± 6	101 ± 4	94 ± 9	89 ± 3	87 ± 12
Isophorone	81 ± 8	79 ± 5	87 ± 6	83 ± 7	78 ± 3	88 ± 12
2,4-Dichlorophenol	89 ± 10	—	94 ± 13	—	74 ± 8	—
Quinoline	27 ± 4	29 ± 3	27 ± 2	32 ± 3	25 ± 2	25 ± 5
1-Chlorodecane	41 ± 3	38 ± 10	38 ± 9	36 ± 11	33 ± 6	19 ± 3
2-Methylnaphthalene	46 ± 9	23 ± 7	51 ± 8	27 ± 4	46 ± 3	32 ± 7
Biphenyl	64 ± 4	52 ± 9	70 ± 2	54 ± 6	71 ± 3	50 ± 17
Ethyl cinnamate	88 ± 6	82 ± 8	96 ± 3	88 ± 8	90 ± 2	82 ± 14
Diacetone-L-sorbose	50 ± 6	28 ± 9	55 ± 8	30 ± 7	46 ± 5	40 ± 24
Anthracene	91 ± 34	61 ± 16	61 ± 3	65 ± 15	73 ± 16	47 ± 3
			Micrograms per Liter[g]			
2,2′,5,5′-Tetrachloro-biphenyl	37 ± 9	—	22 ± 5	—	15 ± 3	—
2,4-Dichlorobiphenyl	174 ± 11	—	169 ± 15	—	148 ± 11	—

NOTE: S.D. denotes standard deviation.
[a] Six 1-L batch extractions were analyzed.
[b] Seven 1-L batch extractions were analyzed.
[c] Eight 1-L CLLE samples were analyzed (except seven for anthracene).
[d] Seven 1-L CLLE samples were analyzed.
[e] Four 12.5-L CLLE samples were analyzed (except three for anthracene).
[f] Two 12.5-L CLLE samples were analyzed.
[g] Values are given in micrograms per liter recovered from aqueous solution (initial concentration was unknown).

Table IV. ANOV Comparisons of Batch LLE and 1-L CLLE with and without Humics Present: Experiment 3

	Compound Stability		*CLLE vs. Batch LLE*		*Humic vs. Nonhumic*	
Compound	*H_o 1 Nonhumic*	*H_o 1 Humic*	*H_o 2 Nonhumic*	*H_o 2 Humic*	*H_o 3 Batch LLE*	*H_o 3 CLLE*
Acetophenone	Accept	Reject	Reject	Accept	Accept	Accept
Isophorone	Accept	Reject	Accept	Accept	Accept	Accept
2,4-Dichlorophenol	Accept	[a]	Accept	[a]	[a]	[a]
Quinoline	Accept	Accept	Accept	Reject	Accept	Accept
1-Chlorodecane	Accept	Accept	Accept	Accept	Accept	Accept
2-Methylnaphthalene	Reject	Reject	Accept	Accept	Reject	Reject
Biphenyl	Accept	Reject	Accept	Accept	Accept	Reject
Ethyl cinnamate	Accept	Accept	Reject	Accept	Accept	Accept
Diacetone-L-sorbose	Reject	Reject	Accept	Accept	Reject	Reject
2,4-Dichlorobiphenyl	Accept	[a]	Accept	[a]	[a]	[a]
Caffeine[b]	Accept	[c]	Accept	[c]	[c]	[c]
Anthracene	Reject	Accept	Accept	Accept	Accept	Accept
2,2′,5,5′-Tetrachlorobiphenyl	Reject	[a]	Accept	[a]	[a]	[a]

NOTE: Hypotheses were rejected when the probability of H_0 being false was >95. H_0 1: Recovery of the compound does not decline over time (ANOV procedure). H_0 2: Mean values of recovery for 1-L batch LLE = 1 CLLE (ANOV procedure). H_0 3: Mean value of recovery without humics present = mean value of recovery with humics present (ANOV procedure).

[a] The chlorinated phenol and chlorinated biphenyls were not measured in the humic portion of the experiment because of losses from the base extraction procedure.

[b] Caffeine data are from Experiment 2.

[c] Insufficient data were obtained for ANOV procedure.

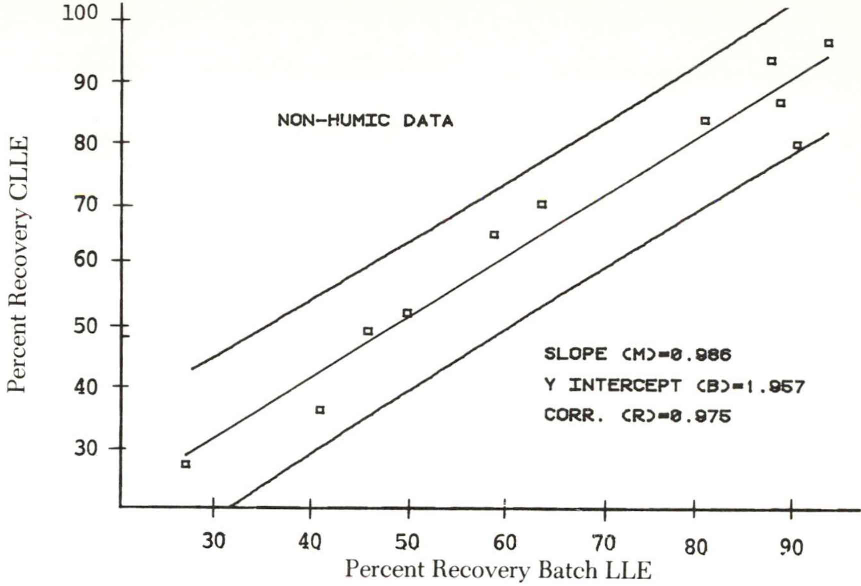

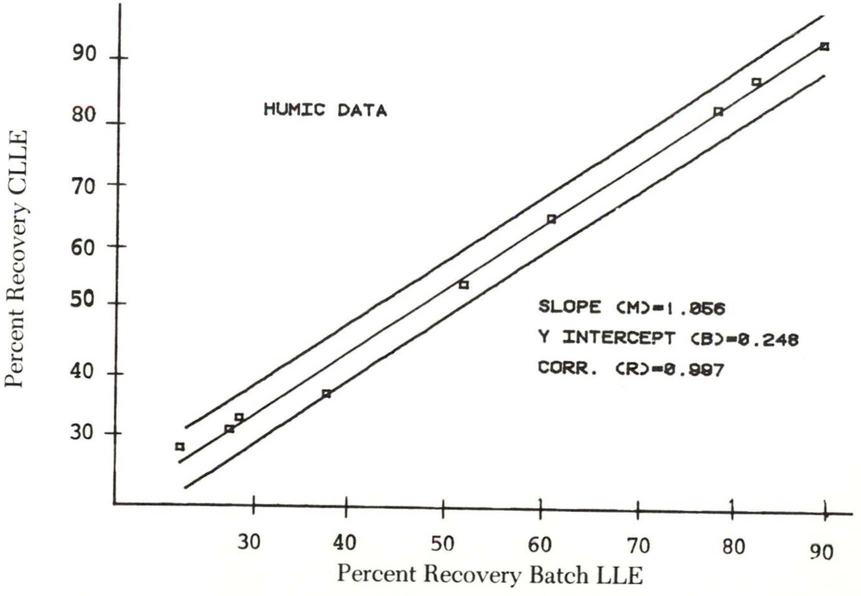

Figure 6. Linear regression analysis of 1-L batch versus 1-L continuous liquid-liquid extraction Experiment 3.

with and without the addition of humics. When volume reduction is required after extraction (for CLLE or batch LLE), some evaporative losses will be experienced for some compounds.

2. Decreases in Concentration over Time. The third experiment was run in two parts. During Part 1, no humic materials were present; at the beginning of Part 2, humics were added. The ANOV procedure tested each compound for changes in concentration by comparing early sample recoveries to later ones. This process was done separately for Parts 1 and 2. It was therefore possible to test each compound for decreasing concentration with and without the presence of humics.

Results of the nested ANOV for Part 1 (*14*) and Part 2 are listed in Table IV. The analysis shows that many of the compounds decrease in recovery over time. Four compounds decreased during Part 1 (without humics), and three others declined after humics were added. Anthracene stopped declining after humics were added; this result suggested a humic–anthracene association that protects anthracene from further losses. Only three of the compounds showed no decline in concentration over the entire duration of the experiment: quinoline, 1-chlorodecane, and ethyl cinnamate. These time-dependent recovery decreases would not occur during on-line sampling of a natural or treated water source because any humic–organic interactions that result in the removal of organics from aqueous solution would already have occurred and humic–organic complexing would be in a state of equilibrium.

3. Extraction with Humics Versus Extraction without Humics. Because the two parts of the experiment were run sequentially and not parallel, any decreases in the concentrations of test compounds over time would lead to lower apparent recoveries in Part 2. This situation had to be taken into account when comparisons were made between Part 1 and Part 2 (humic versus nonhumic) recoveries. Therefore, only recovery data for extraction immediately before and immediately after the addition of humics were used for this comparison (H_0 3 in Table IV). Less than 1 h elapsed between taking these sets of samples; thus, differences between the sets are not attributable to time-dependent decreases in concentration. Figure 7 shows linear regressions of humic versus nonhumic recovery data for batch LLE and CLLE. Three different effects were observed for different compounds when humics were added to the sample:

1. Decreases in recoveries over time began after the addition of humics (three compounds).
2. Decreases in recoveries over time ceased after the addition of humics (one compound).

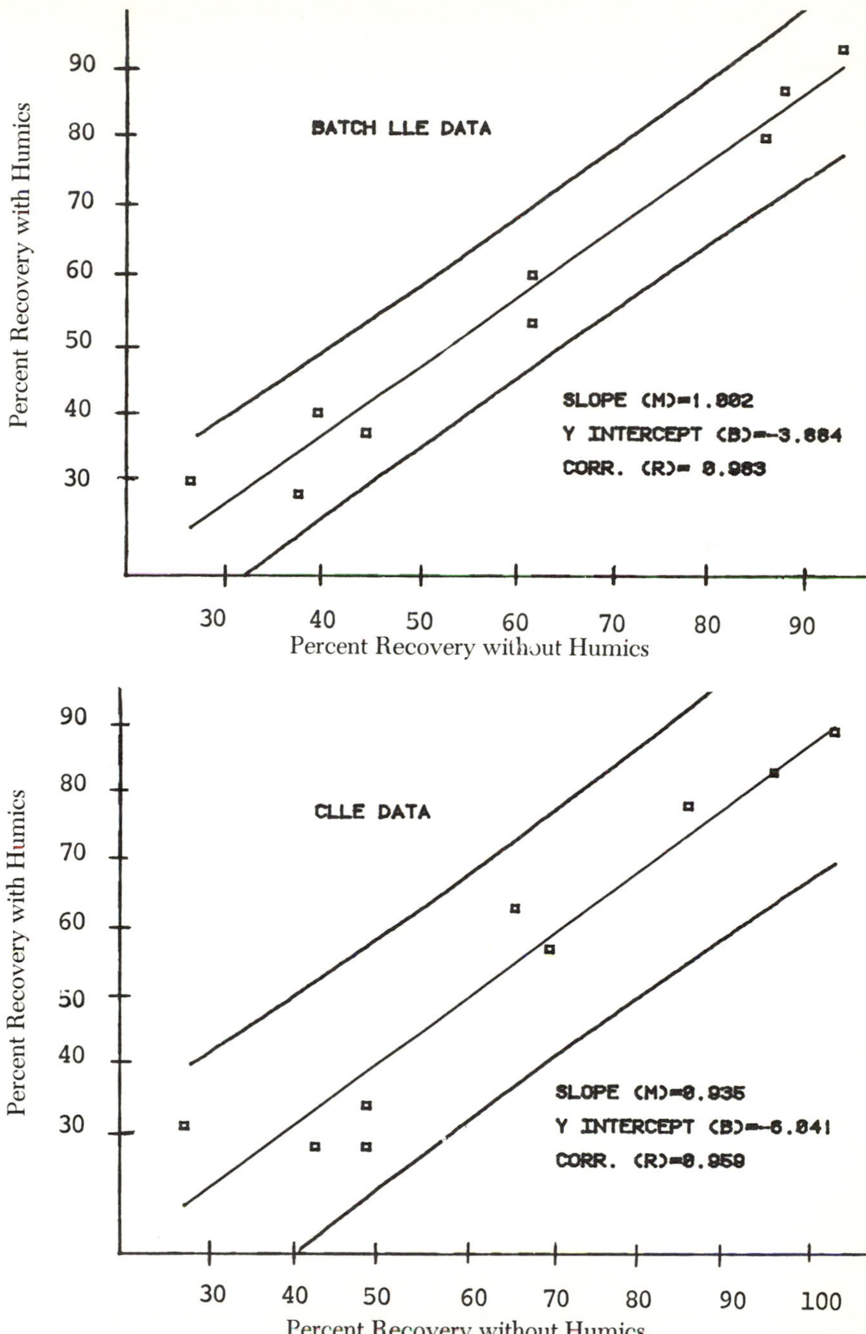

Figure 7. Linear regressions of humic versus nonhumic recovery data for 1-L batch versus 1-L continuous liquid-liquid extractions.

3. Recoveries were different for Part 1 (nonhumic) and Part 2 (humic) for batch LLE, CLLE, or both (three compounds).

The distribution of compounds within these three categories is shown in Table IV. It is not known by what mechanism the humics interfere with recovery of these compounds, but several mechanisms are possible. An association between humics and the organics is one possibility. Carter and Suffet (*20*) found that a high degree of organic–humic complexing is correlated with a high octanol–water partition coefficient (K_{OW}). These compounds all have high K_{OW} values, as evidenced by their high methylene chloride–water partition coefficients. Also, the low pH (pH 3) used in this experiment favors stability of organic–humic complexes because at low pH values, humics are less charged and therefore less polar than at high pH values. The association of humics and organics may be strong enough to decrease the extraction efficiency of some compounds when humics are present. Humics may catalyze degradation of the compounds, either by providing a more suitable environment for bacteria or by directly catalyzing chemical breakdown, for example, hydrolysis. A third possibility is that humic–organic associations may bind to the drum surface; thus, a portion of the humics would be removed from the extraction system and thereby lead to lower apparent extraction recoveries. Evidence by Carlberg and Martinsen (*21*) indicates that the humic–organic equilibrium is very slow for some compounds, requiring from a few days to several months, depending upon the compound. Carter and Suffet (*20*) found that humic–organic associations for some compounds reach equilibrium in less than 1 day. Experimental factors such as mixing could be important in the kinetics of organic–humic associations.

The partitioning of Fluka humic material was studied at pH 3 and pH 7 to ascertain the amount of potential interference with subsequent GC analysis (*19*). Water–methylene chloride partition coefficients were determined by quantification of the humic concentration in pH 3 and pH 7 salt solutions by UV analysis at 254 nm of the humic material before and after methylene chloride extraction. The method followed the procedure of Suffet and Faust (*7*) in which *p*-values (fraction recovered in methylene chloride at 1:1 water:methylene chloride) and *E*-values (fraction recovered in methylene chloride at any specified water-to-solvent ratio) were calculated. Water:methylene chloride (10:1) was used for all *E*-value calculations. The values obtained were as follows: for pH 3, *p*-value = 0.48 and 10:1 *E*-value = 0.07; for pH 7, *p*-value = 0.19 and 10:1 *E*-value = 0.02.

Less humic material is extracted at the high pH because the charge of the humic material increases when the acid groups are dissociated. Humics would adversely affect the GC column and interfere with GC analysis, so base extraction would be needed for samples containing high humics at pH 3 or pH 7 to protect the GC column (*19*). Naturally occurring humic materials may behave differently toward dissolved organics than toward the commercially prepared humics used in this work.

pH 7 Versus pH 3 Extraction. Because all pH 7 extractions were completed without humics present, only nonhumic pH 3 data will be used to compare pH 3 and pH 7 recoveries. Rigorous statistical comparisons between pH 3 and pH 7 extraction recoveries were not possible because of the small number of pH 7 replicate values. However, some semiquantitative observations can be made. As previously reported (*14*), partition coefficients for 12 of the 14 compounds are virtually identical for pH 3 and pH 7. The partition coefficients are the primary controlling factor in extraction recovery for both batch LLE and CLLE. The only operational differences between CLLE operation at pH 3 versus pH 7 were those caused by rapid biological growth at pH 7. This growth may decrease recoveries at pH 7 by metabolic breakdown of compounds and/or by interfering with extraction and separation. Recoveries of four compounds are compared at pH 3 and pH 7 in Table V.

Recovery trends at pH 7 appear to follow those measured at pH 3. Acetophenone shows the most significant pH effect. More pH 7 data would be needed for a statistical comparison. A qualitative review of the data indicates that pH 3 and pH 7 extraction recoveries are not substantially different for batch LLE or for CLLE. The current standard CLLE procedure specifies pH 3 to control biological growth.

Conclusion

A Teflon helix CLLE apparatus was adapted for use in water sampling for biological testing. The CLLE was shown to be sufficiently reliable to be used for long-term composite sampling.

Statistical evaluations showed close agreement between batch LLE and CLLE recoveries for the compounds tested, both with and without humics present. Some decrease in recovery was noted after humics were

Table V. Comparison of Batch and Continuous LLE Percent Recoveries of Compounds at pH 3 and pH 7

Compound	*pH 3 Batch*	*pH 7 Batch*	*pH 3 CLLE*	*pH 7 CLLE*
Acetophenone	93	72	101	81
Isophorone	81	78	82	87
Biphenyl	64	60	70	47
Caffeine	37	40	33	42

added to the aqueous sample for three of the nine compounds for which this comparison was possible.

A pH of 3 was selected as the standard operating pH for the CLLE because of biological growth at pH 7 and salt precipitation at higher pH values. Extraction recoveries at pH 3 appear to be equivalent to those at pH 7 for compounds that do not dissociate in that pH range.

The current version of CLLE apparatus is field ready and suitable for sampling for biological testing, and sampling currently underway will give an indication of the future role of this sampler.

Acknowledgment

This work received financial and technical support from the USEPA, Health Effects Branch, Project Officer Paul Ringhand, under USEPA grant no. CR810484-01-0. Graphics were done by Jeffrey Suffet.

Literature Cited

1. National Academy of Sciences Safe Drinking Water Committee *Drinking Water and Health;* National Academy: Washington, DC, 1977; Vol. 1.
2. National Academy of Sciences Safe Drinking Water Committee *Drinking Water and Health;* National Academy: Washington, DC, 1980; Vol. 3.
3. Suffet, I. H.; Brenner, L; Coyle, J. T.; Cairo, P. R. *Environ. Sci. Technol.* **1978,** *12,* 1315–1322.
4. National Academy of Sciences In *Quality Criteria for Water Reuse;* National Academy: Washington, DC, 1982; Appendix A.
5. Kopfler, F. C. In *Short-Term Bioassays in Analysis of Complex Environmental Mixtures;* Waters, M. D.; Sandhu, S. S.; Huisingh, J. L.; Claxton, L.; Nesnow, S., Eds.; Plenum: New York, 1981; Vol. 2, pp 141–193.
6. Kopfler, F. C.; Ringhand, H. P.; Miller, R. G. In *Organic Pollutants in Water: Sampling, Analysis, and Toxicity Testing;* Suffet, I. H.; Malaiyandi, M., Eds.; Advances in Chemistry 214; American Chemical Society: Washington, DC, 1986; Chapter 20.
7. Suffet, I. H.; Faust, S. D. *J. Agric. Food Chem.* **1972,** *20,* 52–56.
8. Suffet, I. H. *J. Agric. Food Chem.* **1973,** *21,* 591–598.
9. Wu, C.; Suffet, I. H. In *Water Pollution Assessment: Automatic Sampling and Measurement;* American Society for Testing and Materials: Philadelphia, 1975; ASTM STP 582, pp 90–108.
10. Wu, C.; Suffet, I. H. *Anal. Chem.* **1977,** *49,* 231–237.
11. Yohe, T. L.; Suffet, I. H.; Grochowski, R. J. In *Measurement of Organic Pollutants in Water and Wastewater;* Van Hall, C. E., Ed.; American Society for Testing and Materials: Philadelphia, 1979; pp 47–67.
12. Yohe, T. L.; Suffet, I. H.; Coyle, J. T. In *Activated Carbon Adsorption of Organics from the Aqueous Solution;* McGuire, M. J.; Suffet, I. H., Eds.; Ann Arbor Science: Ann Arbor, MI, 1980; Vol. 2, pp 27–69.
13. Yohe, T. L.; Suffet, I. H.; Cairo, P. R. *J. Am. Water Works Assoc.* **1981,** *73,* 402–410.
14. Baker, R. J.; Gibs, J.; Meng, A. K.; Suffet, I. H. *Water Res.,* **1986,** in press.
15. Suffet, I. H., private communication, 1985.

16. Keith, L. H.; Lee, K. W.; Provost, L. P.; Present, D. L. In *Measurement of Organic Pollutants in Water and Wastewater;* Van Hall, C. E., Ed.; American Society for Testing and Materials: Philadelphia, 1979; ASTM STP 686, pp 85–107.
17. Bowers, W. D.; Parsons, M. L.; Clement, R. E.; Eiceman, G. A.; Karasek, F. W. *J. Chromatogr.* **1981,** *206,* 279–288.
18. Snedecor, G. W. *Calculation and Interpretation of Analysis of Variance and Covariance;* Collegiate: Ames, IA, 1934.
19. Gibs, J.; Suffet, I. H. In *Organic Pollutants in Water: Sampling, Analysis, and Toxicity Testing;* Suffet, I. H.; Malaiyandi M., Eds.; Advances in Chemistry 214; American Chemical Society: Washington, DC, 1986; Chapter 19.
20. Carter, C. W.; Suffet, I. H. In *Fate of Chemicals in the Environment;* Swann, R.; Eschenroeder, A., Eds.; ACS Symposium Series 225; American Chemical Society: Washington, DC, 1983; pp 215–229.
21. Carlberg, G. E.; Martinsen, G. E. *Sci. Total Environ.* **1982,** *25,* 245–254.

RECEIVED for review October 17, 1985. ACCEPTED February 5, 1986.

CASE HISTORIES: BIOLOGICAL TESTING OF WATERBORNE ORGANIC COMPOUNDS

28

Biological Testing of Waterborne Organic Compounds

John C. Loper

Department of Microbiology and Molecular Genetics, and Department of Environmental Health, University of Cincinnati, Cincinnati, OH 45267–0524

Increased attention to the possible adverse effects of compounds in environmental waters has been stimulated not only by detection of known toxic chemicals as contaminants, but also by evidence for the presence of multiple unknown genotoxic compounds among waterborne organics. Of numerous genetic tests, bacterial mutagenicity assays have been the most revealing. Examples will be discussed to show that less than 10% of such mutagens have been chemically identified. This situation is true whether the studies involved surface or ground water, industrial wastes, or products of the chlorination of humic acids. Roles of mutagenicity testing will be discussed in relation to evaluating collection procedures, examining origin and fate of mutagens, guiding chemical fractionation of residue mixtures for compound identification, and developing criteria of water quality.

THE CHEMICAL COMPLEXITY of residue organic mixtures in drinking water and the formidable problems in assessing the role of organic compounds in drinking water within the full range of possible toxic effects have been the focus of much research. A priority approach was put forth for compound isolation, identification, and toxicological characterization. The unknown chemicals in these mixtures were ranked for attention on the basis of molecular weight, relative hydrophobicity, and concentration in the water (*1a*).

Our laboratory and those of several others have applied a more direct biological approach. Interest in gaining a general indication of possible adverse effects of drinking water residue organics has led to analyses via various short-term tests of genotoxicity. Most of the progress has occurred through the use of mutagenicity tests, such that

0065–2393/87/0214/0595$06.00/0

over the past few years a far better understanding has been gained of the bioactive properties of waterborne organics as mutagens. This chapter is an abridged overview presenting what biological testing has revealed of the presence of mutagens in water; their nature and origin; and approaches for their prevention, destruction, or removal. Some proposals are presented for future directions.

Background

The perception of waterborne organics has changed during the past 20 years. Chlorination of an urban water supply was introduced in Jersey City, New Jersey, in 1908 (*1b*). Since that time until now and for the foreseeable future, the overriding concern is for a reliably disinfected drinking water free from hazards of waterborne infection. Until recently, the other criteria for water quality were chiefly taste, odor, and color—properties the public could directly recognize. Similarly, there was a strong public rejection of nonbiodegradable detergents when they appeared as foam bubbling out of the faucet. Also, questions were raised by the occasional presence of thousands of fish killed by toxic chemicals released into rivers that were used as drinking water sources.

The unacceptable taste and odor of drinking water from a tributary of the upper Ohio River led Frank Middleton and his collaborators in the U.S. Public Health Service to apply their granular carbon extraction–chloroform elution procedure to obtain residue samples of that water in 1962 (*2*). The first preliminary suggestions that such residues were carcinogenic were presented by Hueper and Payne (*3*) the following year. For whatever reason, those observations appeared to have had little impact.

By the early 1970s, however, the public perception about the origins of cancer had changed dramatically. As early as 1967, Doll (*4*) had documented the geographical patterns of the incidence of different common cancers. On the basis of that documentation and a flood of other data (*5*), it became understood that the large majority of cancer risk is cultural and environmental. In 1974, trihalomethanes (*6, 7*) and other possible carcinogenic substances (*8*) were discovered in water. Epidemiological studies examining cancer incidence in relation to drinking water were widely discussed. (For a recent review of this topic, *see* reference 9.) It became important to determine the presence and to assess the risk to human health of compounds in water that do not have offensive taste, odor, color, or foam and that do not show acute lethal effects on test organisms. It was shown that the great majority of these organic compounds are nonvolatile and require concentration for study.

In the meantime, the analytical chemists continued to extend the range of analysis, and the biologists and biochemists developed an array of biological assays with various toxicological end points. Of the biological assays, the *Salmonella* mutagenesis assay and similar short-term assays proved to be successful detectors of a majority of known chemical mutagens, carcinogens, and procarcinogens. These assays were also rapid and inexpensive relative to other toxicological assays. Although many toxicological tests have been applied to drinking water, the biological data on water residue mixtures and their chemical constituents have accumulated most quickly from these short-term assays of mutagenicity.

Heterogeneity of Mutagens in Water Residues

Many questions remain as to the most appropriate extraction or concentration procedures. Nevertheless, residues from similar water sources have now been obtained by using a variety of techniques. The patterns of mutagenic activity obtained from these residues are sufficiently similar so that it is clear the mutagens are generally not artifacts of concentration. Chemical assays and bioassays of these mixtures typically reveal that among the thousands of largely unknown compounds present at low concentrations, there is a diverse minority of mutagens.

Evidence for this mutagenic diversity comes chiefly from data regarding the following properties: specificity for tester strains, including nitroreductase$^-$ strains; response to microsomal activation systems (S9); additivity of response; class separation; size; chromatography (thin-layer chromatography, reverse-phase high-performance liquid chromatography [HPLC]); UV absorption; and sensitivity to inactivation by heat, alkali, and 4-nitrothiophenol.

Examples are cited in this chapter or appear in other chapters of this book. The tests are those of Ames and co-workers (*10, 11*), who have provided procedures for their use. To conserve sample, often only strains TA98 and TA100 are used in surveys and to guide fractionation. Partially isolated fractions can be characterized further by using strains such as TA1535 and TA1538 and the several nitroreductase-deficient strains developed by Rosenkranz and co-workers (*12, 13*).

Microsomal activation for bacterial assays typically has involved crude supernatant fractions (S9) from livers of rats induced by polychlorobiphenyl mixtures, usually Aroclor 1254. These S9 mixtures contain a spectrum of mixed function oxidases and other enzymes active in biotransformation. In such mixtures, a given compound might be activated to become more mutagenic, may be inactivated, or may remain unaffected. All three types of response have been observed with various water residues (*9, 14*).

Fractionations by all these methods are well-documented by chapters in this book. Our own work has emphasized fractionation by reverse-phase HPLC. By using water-to-acetonitrile elution gradients, mutagenic activity was found predominantly among the mid- to nonpolar fractions (*15*). Among a multitude of peaks absorbing at 254 nm, there were several across a broad range of the elution profile that showed different levels of activity for strains TA98 and TA100. This population included many active fractions that expressed only low levels of mutgenesis (*15*). A feature generally true for these mixtures was that mutagenic subfractions appeared directly additive or antagonistic rather than synergistic (*15, 16*). Such interactive determinations sometimes are complicated by the presence of toxicity in parent mixtures, by technical problems of recovery and reconcentration of subfractions for bioassay, and by low levels of some mutagens in relation to the background mutagenesis inherent in the assay.

Potential Mutagens of Industrial Origin

Several samples of drinking water have yielded residue mixtures with relatively high mutagenic activity. For some of these, mutagenicity has been used as a guide to mutagen isolation. In these cases, the high activity appears to be caused by only one or a few potent mutagens among the myriad of other compounds. Such was the case with the old residue sample from Ohio tributary drinking water, which we obtained from Middleton. By using our *Salmonella* bioassay–HPLC fractionation procedure (*17*), we traced the bulk of the activity in this residue to a single potent promutagen (*18*). This substance was presumptively identified as 3-(2-chloroethoxy)-1,2-dichloropropene (CP), a previously undescribed compound (*19*). The mutagenic potency of CP for TA100, entirely dependent upon the presence of S9, is 75 net revertants/nmol (*19*).

Although CP was the first compound isolated from drinking water residue on the basis of its mutagenic properties, several other laboratories are examining potent mutagens in other residues (*17, 19–22*). Table I lists these studies of mutagen isolation. All of these involve drinking water residues, except for the study by Maruoka and Yamanaka (*22*). Their samples are of urban river water taken from a major tributary that contributes to the drinking water supply of more than 10 million people. Based upon the properties of tester strain specificity and dependence on, or independence of, microsomal activation, this list includes several different mutagens. The study conducted by Heartlein et al. (*20*) implicated spring runoff of agricultural herbicides as the source of an activation-dependent mutagen for strain TA100 (*20*). The pollutant under investigation by Zhou et al. (*21*) is active with strain TA98 in the absence or presence of S9. The one under investigation by Maruoka and Yamanaka (*22*) appears likely to be quite potent. This substance showed its highest mutagenic potency as an S9-dependent mutagen for strain TA1538.

Table I. Isolation of Drinking Water Mutagens

Source	*Activity*	*Compound*	*Ref.*
Industry	S9 dep, TA1535, TA100	3-(2-chloroethoxy)-1,2-dichloropropene	17, 19
Agriculture	S9 dep, TA100	herbicides?	20
City	TA98, + or −S9	?	21
Industry (raw water)	S9 dep, TA1538, TA98	?	22

NOTE: dep denotes dependent.

Although most of these mutagens are in drinking water residues, industrial or agricultural chemical sources are implicated for their origin. It is probable that all of the stable, potent mutagens readily identifiable in drinking water residues have as their origin commercial anthropogenic chemicals or byproducts. Rappaport et al. (*23*) tested organic waste water concentrates from six treatment plants for mutagenicity in the Ames assay. These researchers concluded that all of the definitely positive samples were obtained from plants that treated mixed domestic and industrial wastes, whereas plants that treated wastes strictly from domestic sources were always negative or marginal for mutagenic activity. Similar results involving mutagenic levels that ranged to much higher levels were reported by Hopke and Plewa (*24*) in 1984. A more recent illustration of this pattern is from a study currently directed by Tabor for a neighboring set of facilities: one treating a mixture of domestic and industrial wastes and one treating only domestic wastes. The mutagenic activity in net revertants per liter of waste water influent to the treatment plant for strain TA98 was as follows for domestic-industrial sources: −S9, 90; +S9, 1.2×10^4. For domestic sources only, the following data were obtained (net revertants per liter): −S9, plates showed low activity that was not dose-dependent; +S9, 4×10^2. For TA100, no mutagenic activity was detected. These results were the averages of two assays (*see also* reference 25). Such properties of S9-dependent activity specific for strain TA98 are typically seen among frame shift mutagens of industrial origin. Other data in this study indicate that part of this activity is deposited in the sludge and part is released in the plant stream effluent (*see also* reference 25).

Mutagenic Byproducts of Chlorination

By contrast, numerous studies have shown that raw water sources not affected by industry do not yield microsomal-activation-dependent mutagens active on strain TA98 (*9, 14, 26*). Residues of finished water from these sources typically show direct-acting mutagenesis for TA100 only, or for both TA98 and TA100, which is decreased by the presence of S9 mix. Such mutagenicity arises as a result of chlorination (*9, 14*). The diversity and the chemical nature of this population of mutagens are described by several of the chapters in this book.

Clues as to the identity of some of these compounds may be found in results from mutagenic and chemical studies of other chlorination byproducts. Among these are byproducts of the pulp and paper industry caused during chlorine bleaching processes. These compounds were described in two recent reviews (*27, 28*). A major mutagen produced by the bleaching of softwood kraft pulp has been isolated and identified by Holmbom et al. (*29, 30*) as 3-chloro-4-dichloromethyl-5-hydroxy-2(5*H*)furanone. This newly identified mutagen is extremely potent in assays using strain TA100 −S9; reported values are 2800–10,000 net revertants/nmol. Strains TA1535, TA98, and TA1537 were much less responsive.

Another compound considered to be a major contributor to mutagenicity in such chlorination liquors is 2-chloropropenal. Kringstad et al. (*31*) reported the direct-acting mutagenic activity of this compound for TA1535 to be 320 net revertants/nmol; earlier, Rosen et al. (*32*) reported the activity for TA100 to be about one-third of this value, 113 net revertants/nmol. Other compounds include 1,3-dichloroacetone, 1,1,2,3-tetrachloropropene, and numerous other polychlorinated propenes and propanones identified as *Salmonella* mutagens (*27, 28*), plus resin acids and phenolics detected as mutagenic compounds in a yeast bioassay (*27*).

Similar products have been characterized as mutagenic following the chlorination of humic acids as a possible model for drinking water mutagens (*33*). These include compounds such as 3,3-dichloropropenal, dichloroacetonitrile, and 1,1-dichloro-2-propanone (*34*). To date, however, for humic acids, only a small fraction of the mutagenesis obtained following chlorination has been accounted for by the identified compounds (*33*). Chian et al. (*35*) presented an analysis of the chemical products of humic acid chlorination in relation to those contained in residues of chlorinated water from a surface source.

Mutagen Prevention, Destruction, and Removal

The recognition that many of the numerous and unknown diverse mutagens present in drinking water are byproducts of chlorination has stimulated efforts at their prevention, destruction, or removal. Extensive studies on the chemistry of alternative disinfection procedures have been conducted. These have been directed at the use of other species of chlorine (ClO_2, chloramines) or at combination treatments that require little or no chlorine exposure. Separate studies parallel the observations made of the paper industry mutagens. Early in their work on this subject, Ander et al. (*36*) showed the mutagenicity of bleach process liquors was decreased in the presence of microsomal activation. In a subsequent study, this activity was shown to be alkali sensitive and was reduced when a lime decolorization precipitation step was included or when sulfite bleaching was included (*37*). Mutagenic residues of chlorinated drinking water were then shown to be alkali sensitive and to be susceptible to nucleophiles such as 4-nitrothiophenol. Current applications of these approaches are represented in this book.

Removal of mutagens from drinking water by treatment with granular activated carbon (GAC) also has been examined by several workers. Recent work in our laboratory confirmed observations of others (*38, 39*) that GAC can remove mutagens long after the level of total organic carbon (TOC) in the effluent has risen to a level that parallels that in the influent water (*26*). As well as preexisting mutagens, compounds capable of yielding mutagens upon chlorination also are preferentially removed (*26*). However, apparently these eventually break through during prolonged use of GAC (*38–40*). Chemical and mutagenic analyses were conducted on residues extracted from used GAC taken from the top, middle, and bottom of the GAC column; more polar compounds, including some mutagens, were detected in the residues from the middle and bottom samples (*40*).

Risk Assessment

A feature of this overview has been to highlight the progress of combined biological and chemical analysis in the characterization of waterborne organics. The major social impetus for such work has been unresolved questions on the biological side. What are the risks posed by these organics, primarily to human health and also to the environment?

The *Salmonella* test and similar microbial short-term tests are inadequate for a total assessment of genetic effects, not to mention additional important targets of toxicological or environmental damage. Nevertheless, these mutagenicity tests have served as practical tools both in the characterization of residue mixtures and in the isolation of their constituents. Viewed solely from the perspective of mutagenicity, source waters for drinking purposes can be divided into four categories:

1. Waters contaminated with appreciable levels of stable, potent mutagens that require metabolic activation for their short-term effects. These chemicals appear primarily, if not exclusively, to be of industrial or agricultural origin.
2. Water with relatively insignificant amounts of man-made chemicals but having appreciable TOC levels that yield mutagens as chlorination byproducts. These mutagenic mixtures generally do not require metabolic activation but rather are less active when assayed +S9.
3. Waters that contain appreciable TOC levels and that are also subject to repeated periodic contamination by man-made mutagens. Included in this category are rivers and lakes affected by industrial spills and discharges and by agricultural and urban runoff.
4. Uncontaminated low-TOC water, such as is found in pristine wells.

In setting priorities for the risk assessment of residue mixtures of waterborne organics, it seems prudent to pursue the source of these potent stable mutagens in category 1 waters. Although only a few have been examined to date, the mutagenic data of industrial municipal waste waters suggest major mutagens are being released from point sources. An individual compound isolated as a bacterial mutagen may prove nontoxic in higher organisms, and a compound not detected on the basis of its bacterial mutagenicity may still have significant toxicological effects. Nevertheless, the predictive value of *Salmonella* mutagenesis in the detection of potential carcinogens is high. Stable, microsomal-activation-dependent compounds are among those of greatest potential importance, and it appears they are amenable to isolation from water residue organics for identification by using coupled mutagenic–chemical fractionation.

Considerable information of a general nature is available for uncontaminated water subject to the production of disinfection byproducts. The mutagens produced by drinking water chlorination appear to be numerous, but they exist either at low levels or are of low potency. For both the unresolved mixtures and for the few mutagenic compounds thus far identified, activity is readily reduced or destroyed by treatment with alkali or 4-nitrothiophenol and may be removed by GAC treatment. From water sources subject both to mutagen formation via disinfection and to periodic contamination by toxic chemicals, experimental full-scale GAC treatment systems have provided mutagen-free water.

The other remaining priority is the assessment of ground water. It has been a long time since we could complacently regard ground water as falling under category 4 in this list of water quality. Because of different circumstances, ground water may contain relatively high TOC levels, and in many cases is either contaminated or subject to imminent contamination by toxic, man-made chemicals. Unfortunately, the specific circumstances vary greatly from case to case so that general solutions may not be possible. Local application of chemical and bioassay analyses will be important in the evaluation and preservation of specific ground water resources.

Acknowledgments

Research reported from this laboratory was supported in part by a grant to the University of Cincinnati by the U.S. Environmental Protection Agency (USEPA). This chapter has not been subjected to USEPA review and therefore does not reflect the views of that agency and no official endorsement should be inferred.

Literature Cited

1a. Neal, R. A. *Environ. Sci. Technol.* **1983,** *17,* 113A.
1b. National Research Council *Drinking Water and Health;* National Academy of Sciences: Washington, DC, 1977; pp 4-5.
2. Middleton, F. M.; Pettit, H. H.; Rosen, A. A. *Proc. 17th Ind. Waste Conf. Purdue Univ. Ext. Serv.* **1962,** *122,* 454.
3. Hueper, W. C.; Payne, W. W. *Am. J. Clin. Pathol.* **1963,** *39,* 475.
4. Doll, R. *Proc. R. Soc. Med.* **1972,** *65,* 49.
5. Berg, J. W. In *Origins of Human Cancer;* Hiatt, H. H.; Watson, J. D.; Winsten, J. A., Eds.; Cold Spring Harbor Laboratory: Cold Spring Harbor, NY, 1977; pp 15-19.
6. Rook, J. J. *Water Treat. Exam.* **1974,** *23,* 234.
7. Bellar, T. A.; Lichtenberg, J. J. *J. Am. Water Works Assoc.* **1974,** *66,* 739.
8. U.S. Environmental Protection Agency Report 906/10-74-002; U.S. Government Printing Office: Washington, DC, November 1974.
9. Kool, H. J.; van Kreijl, C. F.; Zoeteman, B. C. J. *CRC Crit. Rev. Environ. Control* **1982,** *12,* 307-359.
10. Ames, B. N.; McCann, J.; Yamasaki, E. *Mutat. Res.* **1975,** *31,* 347.
11. Maron, D. M.; Ames, B. N. *Mutat. Res.* **1983,** *113,* 173.
12. Rosenkranz, H. S.; Speck, W. T. *Biochem. Biophys. Res. Commun.* **1975,** *66,* 520.
13. Rosenkranz, H. S.; McCoy, E. C.; Hermelstein, R.; Speck, W. T. *Mutat. Res.* **1981,** *91,* 103.
14. Loper, J. C. *Mutat. Res.* **1980,** *76,* 241.
15. Loper, J. C.; Tabor, M. W. *Environ. Sci. Res.* **1983,** *27,* 165.
16. Loper, J. C.; Tabor, M. W.; Miles, S. K. In *Water Chlorination: Environmental Impact and Health Effects;* Jolley, R. L.; Brungs, W. A.; Cotruvo, J. A.; Cumming, R. B.; Mattice, J. S.; Jacobs, V. A., Eds.; Ann Arbor Science: Ann Arbor, MI, 1983; Vol. 4, pp 1199-1210.
17. Tabor, M. W.; Loper, J. C. *Int. J. Environ. Anal. Chem.* **1980,** *8,* 197.
18. Loper, J. C.; Tabor, M. W. *Environ. Sci. Res.* **1981,** *22,* 155.
19. Tabor, M. W. *Environ. Sci. Technol.* **1983,** *17,* 324.

20. Heartlein, M. W.; DeMarini, D. M.; Kutz, A. J.; Means, J. C.; Plewa, M. J.; Brockman, H. E. *Environ. Mutagen.* **1981,** *3,* 519.
21. Zhou, S. W.; Xu, F. D.; Liu, J. L., unpublished data.
22. Maruoka, S.; Yamanaka, S. *Environ. Sci. Technol.* **1983,** *17,* 177.
23. Rappaport, S. M.; Richard, M. G.; Hollstein, M. C.; Talcott, R. E. *Environ. Sci. Technol.* **1979,** *13,* 957.
24. Hopke, P. K.; Plewa, M. J.; Stapleton, P. L.; Weaver, D. L. *Environ. Sci. Technol.* **1984,** *18,* 909.
25. Tabor, M. W.; Loper, J. C. *Organic Pollutants in Water: Sampling and Analysis;* Suffet, I. H.; Malaiyandi, M., Eds.; Advances in Chemistry 214; American Chemical Society: Washington, DC, 1986; Chapter 33.
26. Loper, J. C.; Tabor, M. W.; Rosenblum, L.; DeMarco, J. *Environ. Sci. Technol.* **1985,** *19,* 333.
27. Douglas, G. R.; Nestmann, E. R.; McKague, A. B.; Kamra, O. P.; Lee, E. G.-H.; Ellenton, J. A.; Bell, R.; Kowbel, D.; Liu, V.; Pooley, J. *Environ. Sci. Res.* **1983,** *27,* 431.
28. Kringstad, K. P.; Lindstrom, K. *Environ. Sci. Technol.* **1984,** *18,* 236A.
29. Holmbom, B. R.; Voss, R. H.; Mortimer, R. D.; Wong, A. *Tappi* **1981,** *64(3),* 172.
30. Holmbom, B. R.; Voss, R. H.; Mortimer, R. D.; Wong, A. *Environ. Sci. Technol.* **1984,** *18,* 333.
31. Kringstad, K. P.; Ljungquist, P. O.; deSousa, F.; Stromberg, L. M. *Environ. Sci. Technol.* **1981,** *15,* 562.
32. Rosen, I. D.; Segall, Y.; Casida, J. E. *Mutat. Res.* **1980,** *78,* 113.
33. Meier, J. R.; Lingg, R. D.; Bull, R. J. *Mutat. Res.* **1983,** *118,* 25.
34. Coleman, W. E.; Munch, J. W.; Kaylor, W. H.; Streicher, R. P.; Ringhand, H. P.; Meier, J. R. *Environ. Sci. Technol.* **1984,** *18,* 674.
35. Chian, E. S. K.; Giabbai, M. F.; Kim, J. S.; Reuter, J. H.; Kopfler, F. C. In *Organic Pollutants in Water: Sampling and Analysis;* Suffet, I. H.; Malaiyandi, M., Eds.; Advances in Chemistry 214; American Chemical Society: Washington, DC, 1986; Chapter 9.
36. Ander, P.; Eriksson, K.-E.; Kolar, M.-C.; Kringstad, K. *Sven. Papperstidn.* **1977,** *80,* 454.
37. Eriksson, K.-E.; Kolar, M.-C.; Kringstad, K. *Sven. Papperstidn.* **1979,** *82,* 95.
38. Monarca, S.; Meier, J. R.; Bull, R. J. *Water Res.* **1983,** *17,* 1015.
39. Kool, H. J.; van Kreijl, C. F. *Water Res.* **1984,** *18,* 1011.
40. Loper, J. C.; Tabor, M. W.; Rosenblum, L.; DeMarco, J. In *Water Chlorination: Environmental Impact and Health Effects;* Jolley, R. L.; Bull, R. J.; Davis, W. P.; Katz, S.; Roberts, M. H., Jr.; Jacobs, V. A. Eds.; Lewis: Chelsea, MI, 1985; pp 1329–1339.

RECEIVED for review August 14, 1985. ACCEPTED December 17, 1985.

Concentration, Fractionation, and Characterization of Organic Mutagens in Drinking Water

H. J. Kool[1], C. F. van Kreijl, and M. Verlaan-de Vries

National Institute of Public Health and Environmental Hygiene, P.O. Box 150, 2260 AD LEIDSCHENDAM, The Netherlands

A combination of Amberlite XAD-4 and XAD-8 resins is very suitable for concentrating organic mutagens (Ames test positive) in drinking water. Fractionation of these mutagenic organic drinking water concentrates with the aid of Sephadex LH20 revealed that organic mutagens showed a molecular weight in the range of 100–300. The organic mutagens were able to induce chromosomal aberrations in CHO cells. Furthermore, nitro organics, in part, were shown to be responsible for mutagenic activity in organic concentrates prepared from chlorinated drinking water in The Netherlands. Finally, results strongly indicate that nitro organics (halogenated or not) are introduced and/or activated in drinking water after a chlorine treatment.

THE PRESENCE OF ORGANIC CONSTITUENTS IN DRINKING WATER has been known for many years because these substances were found to influence the taste, color, and odor of drinking waters (*1*). The organic constituents consist of compounds of both natural and industrial origin. The natural ones compose the major portion and include mainly undefined fulvic and humic acids (*2*). For the industrial ones, most attention has been paid so far to the volatile nonpolar compounds. In part, this situation is due to analytical (technical) restrictions and to the growing awareness (*3, 4*) that volatile halogenated hydrocarbons are introduced as a result of a chlorine treatment.

To date, hundreds of organic constituents, including several known mutagens and carcinogens, have been identified in drinking water in many countries in the world, but these organics usually are present

[1]Current address: Stichting Waterlaboratorium-Oost, Terborgseweg 138, 7005 BD Doetinchem, The Netherlands

0065-2393/87/0214/0605$06.00/0

below the microgram-per-liter level (*1*, *5*–*7*). In addition, the nonpurgeable fraction, which composes 90–95% of the total organics in the water, has not been identified (*6*, *8*) because this fraction cannot be readily volatilized for its subsequent separation and identification by gas chromatographic–mass spectrometric (GC–MS) analysis. Thus, the majority of the organics in drinking water have not been identified, and this situation is probably also true for the organic mutagens (*9*, *10*). Therefore, advances are being made to combine analytical procedures and biological testing to separate the biological active fraction. This separation will permit the isolation of bioactive subfractions, which we hope will lead to the identification of the bioactive compounds.

Coleman et al. (*11*) investigated what kind of organic compounds could be identified in a mutagenic concentrate of Cincinnati tap water. More than 700 organic compounds could be detected in an Ames test positive concentrate, and 460 of these could be identified. This result shows that a more sophisticated coupled bioassay–chemical fractionation procedure, which will separate the organics not responsible for mutagenic activity from the organic mutagens, is necessary to identify the biological active organics in a complex mixture. Recently Tabor and Loper (*12*) carried out initial partitioning by liquid–liquid extractions, followed by repeated high-performance liquid chromatography (HPLC), for separation into smaller subfractions. Active subfractions (positive in the Ames test) were analyzed by GC–MS and consequently for peak identification. Preliminary results obtained with a carbon–chloroform extract of tap water processed in 1962 from an Ohio River source showed that a polychlorinated aliphatic ether was responsible for the mutagenic activity. More recently, the structure of this mutagenic compound has been identified as 3-(2-chloroethoxy)-1,2-dichloropropene (*13*). The investigation of the organic mutagens in drinking water in The Netherlands is now moving along similar lines, and results of this approach will be presented and discussed in this chapter.

Materials and Methods

XAD Resins. Amberlite XAD-2, -4 and -8 were obtained from Serva GmbH, Heidelberg, Federal Republic of Germany. The resins were purified by repeated Soxhlet extraction for 16 h in (consecutively) methanol, ethyl ether, acetonitrile, and again methanol. A subsample (column packed) of the resin was then eluted with ethyl ether, and the eluate was checked for purity by means of GC analysis (no detectable impurities). The resins were stored in methanol at room temperature.

Bacterial Strains. *Salmonella typhimurium* TA98 and TA100 (*14*, *15*), as well as both nitroreductase-deficient strains TA98NR$^-$ and TA100NR$^-$ (*16*), were used.

XAD Procedure. For the concentration of organic constituents, 150–10,000 L of drinking water was collected per sample. Adsorption of the organic constituents on the XAD resins was carried out as described by Junk et al. (*17*). Glass (25 × 1.5 cm) and stainless steel (60 × 2.5 cm) columns were packed with 20 cm^3 and 250 cm^3 of XAD-2, respectively, or similar amounts of a 1:1 mixture of XAD-4 and XAD-8 (XAD-4/8) as indicated. The columns were washed successively before use with 4 bed volumes of methanol, acetone, methanol, dimethyl sulfoxide (DMSO), and distilled water. For about 7 × 10^3-fold concentration, 150 L of the water was passed over columns containing 20 cm^3 of XAD at a flow rate of maximal 4 bed volumes/min and at a ~21-mL constant temperature of 15 °C. Elution of the adsorbed organic constituents was carried out with a volume of either DMSO or acetone (>1 bed volume, neutral fraction). In some cases, the XAD filtrate was collected after passing the XAD column adjusted to pH 2 with HCl and readsorbed on XAD-4/8. Subsequent elution was carried out with DMSO or acetone (acid fraction). For about 1 × 10^6-fold concentration, four samples of 40,000 L of water were passed over eight columns containing 250 cm^3 of XAD per column at a flow rate of maximal 4 bed volumes/min. Elution of the adsorbed organic constituents was carried out with 350 mL of acetone per column (neutral fraction). The acetone in the drinking water concentrates was removed by rotary evaporation under reduced pressure at 30 °C. The remaining aqueous sample, about 200 mL, was subsequently extracted three times with 200 mL of ethyl ether. The ether extracts of all columns were pooled, the ether was removed by rotary evaporation under reduced pressure at 30 °C, and the dry residue was dissolved in 2–3 mL of isopropyl alcohol.

Freeze-drying. For a 7000-fold concentration, 70 L of drinking water was lyophilized in a Virtus Unitrap II. The dried residue was then divided into equal weights and packed into two columns (25 × 1.5 cm) with a sintered glass filter. The organic material was eluted consecutively with acetone, ether, and DMSO. The ether in the ether eluate was removed by rotary evaporation, and the dried residue was dissolved in DMSO. The DMSO concentrates were sterilized by filtration over a 0.2-μm Teflon filter (Millipore). The acetone and DMSO concentrates were tested in the Ames test.

***Salmonella* Mutagenicity Test (Ames Test).** The methods of bacterial culture, the verification of genetic markers, and the plate incorporation assay were essentially the same as described previously (*14, 15*). Petri dishes (90 mm) containing about 20 mL of 1.2% Noble agar in minimal Vogel Bonner Medium E supplied with excess biotine and

2% bactodextrose (Difco) were used. They were seeded with 3 mL of molten top agar (45 °C) to which the following were added consecutively: 0.1 mL of nutrient broth culture of the bacterial tester strain (containing about 5×10^8 bacteria/mL) and 0.5 mL of S9 mix (as indicated).

Rat liver S9, induced by Aroclor 1254, was obtained from Litton Bionetics. In the S–G mix, 0.075 mL of liver homogenate was added per milliliter of mix. The organic water concentrates were tested in the Ames test in three- to fivefold, and the deviation of the mean in the figures was usually less than 20%. The results were considered significant when a twofold increase above the background and a dose–response effect were observed.

Routine controls were included to check for the presence of histidine and other growth-stimulating substances or possible effects in the sample. First, 0.5 mL of each DMSO concentrate was plated out in the absence of histidine in the top agar and then compared with the normal spontaneous background level. Second, as an internal control, a fixed amount of test mutagen (nitrofurazone) was dissolved in 0.50 mL of each concentrate and tested for possible differences in the mutagenic response.

Finally, as a control for the concentration procedures, similar concentrates (7×10^3 fold) of tap water (The Hague) were assayed for mutagenic activity. This control was always found negative.

Cells, Treatment, and Chromosome Analysis. About 4×10^5 CHO-K1 cells (Flow Laboratories, Scotland) were seeded into Ham's F10 medium (Flow) supplemented with 10% newborn calf serum (Flow). These cells were incubated at 37 °C and 5% CO_2 in 25-cm^3 tissue culture flasks. Twenty-four hours later, the cells were exposed for 1 h at 37 °C to a maximum of 50 μL of drinking water concentrate (neutral fraction) in a total volume of 3 mL of Ham's F10, without serum. As a positive control, 4-methoxyaniline, dissolved in DMSO, was used.

After treatment, the exposure mixture was removed and cells were grown for 18 h at 37 °C in medium plus serum. Then, 1 μg of colcemid (Gibco) was added per tissue culture flask, and cells were incubated for another 2 h at 37 °C. The collection of the cells by trypsinization, their fixation, and the staining of chromosomes with Giemsa were carried out according to standard procedures (*18*). At least 50 metaphases were screened for chromosomal aberrations per treatment group.

Chemical Analyses. Organic halogens of drinking water concentrates were analyzed by microcoulometry (*19*) by direct injection of 100 μL of the organic concentrate.

Fractionation of Drinking Water Concentrates on Sephadex LH20. A glass column (height 80 cm, i.d. 2.5 cm) was packed with Sephadex LH20 in isopropyl alcohol. About 3 mL of an acetone or DMSO–XAD concentrate of drinking water (neutral fraction) was layered on the column, and subsequent gel filtration was performed upside down by using isopropyl alcohol as the solvent. Fractions of 6 mL were collected, and the absorbance at 263 nm was measured. When the first peak was eluted (in general, about 90–100 fractions), the solvent was changed to dioxane/water (7:3 v/v), and A_{263} absorption was measured again. When no more UV-absorbable material was eluted from the column (usually after 150 fractions), the isopropyl alcohol fractions were pooled together and the dioxane/water fractions were pooled in two subfractions, as indicated. Isopropyl alcohol and dioxane were removed by rotary evaporation under reduced pressure at 30–40 °C and 50–60 °C, respectively, and the dry residues were dissolved in 5 mL of acetone and stored at −20 °C prior to mutagenicity testing, if indicated.

Molecular Weight Determination on Sephadex LH20. The glass column (height 40 cm, i.d. 1 cm) was packed with Sephadex LH20 in dioxane/water (7:3) as described previously (*20*). About 1.0 mL of a DMSO–XAD concentrate of drinking water was layered on the column, and subsequent gel filtration was performed by using dioxane/water (7:3) as the solvent. Fractions of 1 mL were collected with an automatic fraction collector. After measuring the absorbance at 263 nm, the fractions were pooled, fivefold diluted in water, reconcentrated on XAD-4/8 (bed volume of 4 mL), and eluted with 5 mL of DMSO. The concentrate was stored at −20 °C prior to mutagenicity testing. Calibration of the column was performed by using two colored markers, namely, vitamin B12 (MW 1355) and nitrofurazone (MW 198).

HPLC and Isolation of Mutagenic Fractions. Analytical and semipreparative reverse-phase HPLC separations were performed by using a water-to-acetonitrile linear gradient (*12*). Separations were carried out on a Hewlett Packard Model 10084 B equipped with an automatic sampling device, a solvent programmer, a variable absorbance detector, and an automatically steered fraction collector. The instrument was fitted with a 3.9-mm × 30-cm prepacked analytical column of 10-μm silica particles bonded with octadecylsilane (Bondapack–C_{18}) for analytical scale. For semipreparative scale separations, the HPLC was fitted with a 7.8-mm × 30-cm prepacked column packed with 10-μm silica particles bonded with octadecylsilane. Samples for HPLC were injected at volumes of 20 μL (flow rate 1 mL/min) and 80 μL (flow rate 4 mL/min), and the absorption was measured at 254 nm. Fractions or subfractions were pooled as indicated, and after reconcentration by ether extraction, they were assayed for mutagenicity in the Ames test.

Results

Effect of Resin Type. The use of the nonpolar XAD resins for the concentration of organic mutagens from drinking water has been known for many years (*9, 10*). As indicated by results of Webb and Rossum (*21, 22*), an equal mixture of XAD-4 and XAD-8 is most effective for concentrating a broad spectrum of organic compounds for subsequent GC–MS analyses. Yamasaki and Ames (*23*), however, found XAD-2 to be superior to XAD-4 during the testing of several model compounds. By using acetone, we have compared the mutagenic activity of organic concentrates of drinking waters obtained by adsorption on either XAD-2 or XAD-4/8. Figure 1 shows a representative example in which XAD-4/8 showed a somewhat higher direct mutagenic activity in the *Salmonella* mutagenicity assay with strain TA98, whereas hardly any difference was observed in promutagenic activity with this strain. Neither direct nor promutagenic activity with strain TA100 was significantly different with both resins (not shown).

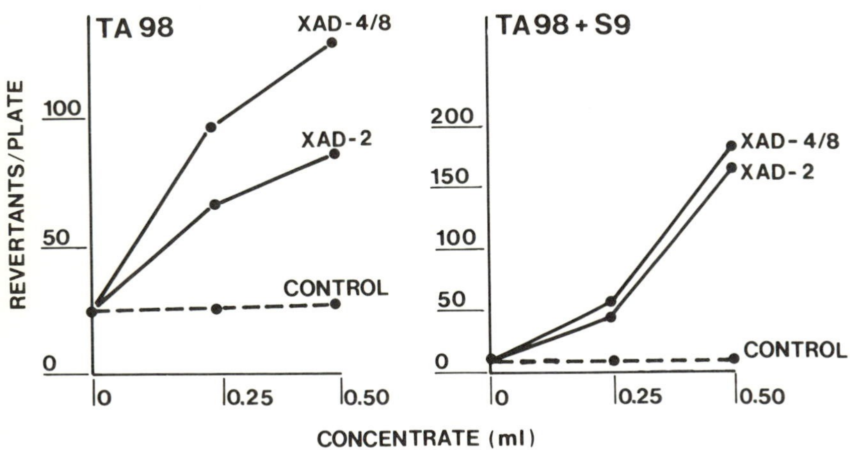

Figure 1. Effect of resin type on the mutagenic activity of drinking water concentrates in the Ames test. The sampling, 7000-fold concentration with either XAD-2 or XAD-4/8, DMSO elution (20 mL, neutral fraction), and subsequent mutagenicity testing were as described in Materials and Methods. Similar concentrates of The Hague tap water were used as controls. Each point represents the average of four plates, and 0.50 mL of concentrate corresponds to 3.5 L of water per plate.

Comparison of the Mutagenic Activity in Lyophilized and XAD-Concentrated Drinking Water. Because of the selective nature of the XAD concentration procedure, only a small fraction of the organic constituents present in drinking water eventually ends up in the organic concentrate, as has been shown with surface water (*24*). Therefore, the mutagenic activity of a lyophilized drinking water concentrate, containing about 80% of the original DOC (value of 5 mg C/L in the dried residue), was compared with the activity of a similar XAD-4/8 concentrate. The results obtained in Figure 2 show that despite the low recovery of the total organics, the XAD method is at least as effective as the freeze-drying technique in concentrating organic mutagens from drinking water. Furthermore, it was shown that acetone was very effective in eluting the organic mutagens. The lyophilized concentrates became toxic for the bacteria above 0.25 mL, an effect that was also observed in the internal control (*see* Material and Methods).

Some Physical–Chemical Properties of Organic Mutagens in Drinking Water. Investigations of many mutagenic drinking waters in The Netherlands have shown previously (*25–28*) the presence of another class of organic mutagens (acid fraction) after readsorption of the acidified XAD filtrate on a second XAD-4/8 column. It was also shown previously (*25*) that ethyl ether elution of XAD-4/8, which is widely used for analytical purposes such as GC–MS, yielded only a minor part of the mutagenic activity. Subsequent elution with acetone, however, eluted the major part of the activity. These results indicate that (1) the major part of the organic mutagens is composed of the somewhat more polar and less volatile organics and (2) the organics already identified by GC–MS in these types of drinking waters with XAD-4/8 ether elution are, in general, not identical to the organic mutagens (*25*).

The organic mutagens concentrated by using the XAD procedure proved not to be gas chromatographable because in a routine preparative GC procedure, less than 10% of the mutagenic activity (Ames test) could be recovered (not shown). To find out whether the heating step in the GC analysis may have caused this poor recovery, an experiment was set up in which a mutagenic drinking water concentrate was heated to 250 °C; this experiment was similar to our routine GC analysis that uses a heated injection system. The results, as depicted in Figure 3, show that after heating a mutagenic drinking water concentrate, mutagenic activity of strain TA98 and TA100 is completely lost. This finding indicates that the organic mutagens in the concentrate showed thermolabile properties under our GC conditions.

Fractionating and Molecular Weight Determination. Another approach to obtain more information on the nature of the responsible

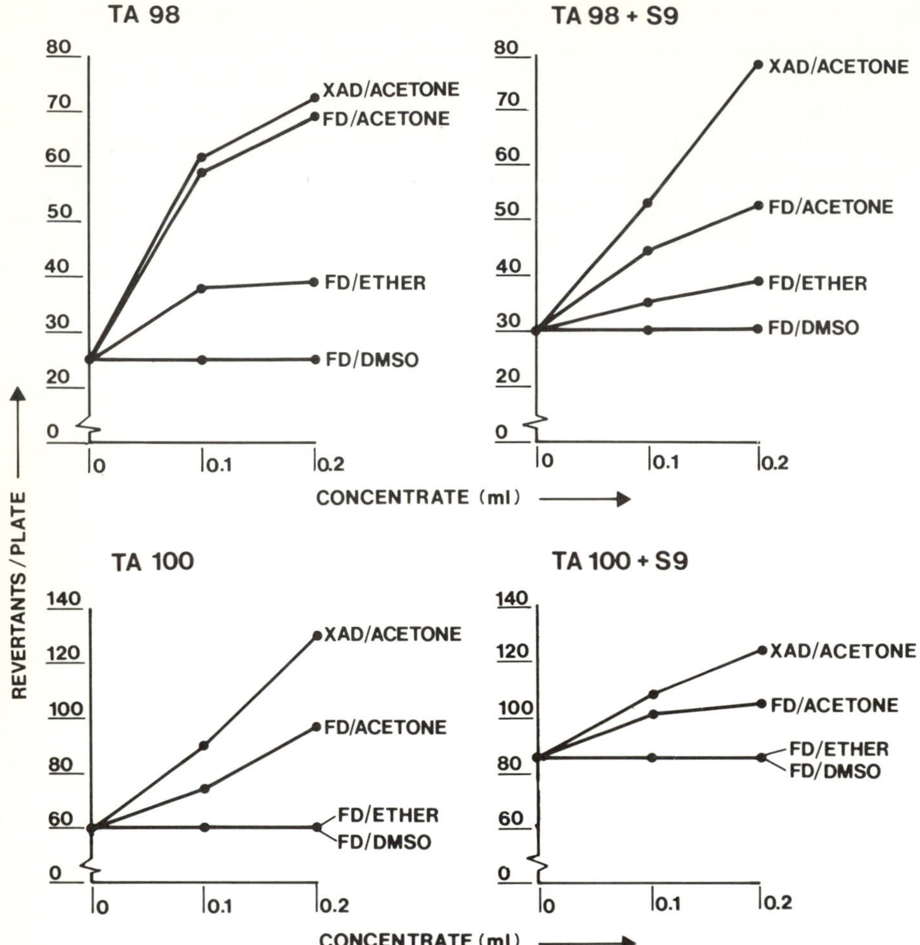

Figure 2. Comparison of mutagenic activity in lyophilized and XAD-concentrated drinking water. The sampling; 7000-fold concentration with either XAD-4/8 (XAD) or freeze-drying (FD); XAD-4/8 elution with acetone (neutral fraction); freeze-drying elution successively with acetone, ether, and DMSO; and subsequent mutagenicity testing with strains TA98 and TA100 were as described in Materials and Methods. Each point represents the average of three plates, and 0.2 mL of concentrate corresponds to 1.4 L of water per plate.

mutagens is to apply thin-layer chromatography (TLC) and HPLC on concentrates of drinking water and look for possible fractionation of the activity. Previous results with TLC showed that the mutagenic activity is found predominantly in one distinct zone (*25*, *26*).

Experiments using gel filtration on Sephadex LH20 according to the methods of Concin et al. (*20*) and designed to obtain information on the molecular weight of the organic mutagens showed for one drinking water concentrate that the organic mutagens present in the neutral fraction had a molecular weight range of 100–300 (*26*). To see whether this result would also hold for XAD concentrates prepared from other chlorinated drinking waters, experiments shown in Figures 4 and 5 were performed. The results obtained with two drinking water concentrates prepared from Meuse River water (Figure 4) and Rhine River water (Figure 5) confirmed our previous results: the bulk of the organic mutagens have a molecular weight in the range of 100–300 because the majority of the activity in drinking water (Figure 4) is detected in the molecular weight range of nitrofurazone (MW 198). Also, after a chlorine treatment (1.5 mg/L of Cl_2; Figure 4), the mutagenic activity strongly increased. Furthermore, the organic mutagens formed after this chlorine treatment are in the same molecular

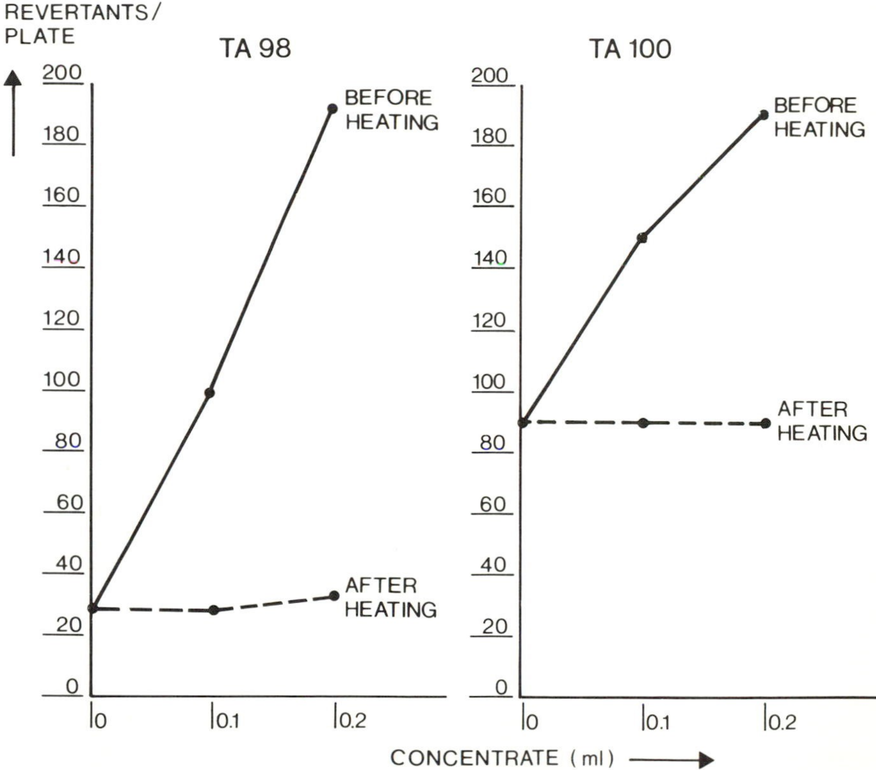

Figure 3. Effect of heating on the activity of a mutagenic drinking water concentrate. A mutagenic XAD-4/8 acetone concentrate (neutral fraction) was heated to 250 °C. The total heating time was 1 h. After this period, the organic residue was dissolved in acetone and retested in the Salmonella *mutagenicity test.*

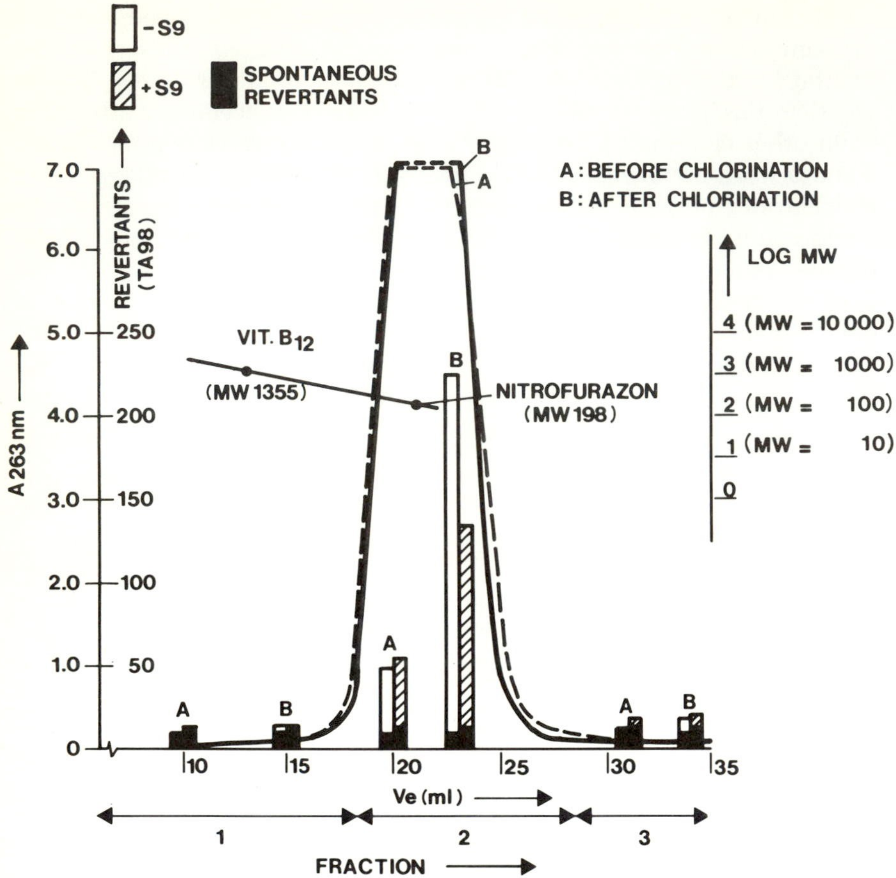

Figure 4. Molecular weight determination of a drinking water concentrate with Sephadex LH20. Sampling, 10^4-fold concentration of drinking water before and after chlorination (1.5 mg/L of Cl_2, Meuse River source) on XAD-4/8, elution with DMSO (neutral fraction), and subsequent gel filtration were as described in Materials and Methods. After measuring the absorbance at 263 mm, the fractions were pooled as indicated. After dilution in water, the fractions were reconcentrated on XAD-4/8, eluted with DMSO, and assayed in the Salmonella *mutagenicity test (strain TA98 ± S9).*

weight range. A similar result was obtained with the mutagenic activity in the neutral and acid fractions in chlorinated drinking water prepared from the Rhine River (Figure 5).

Additional fractionation with Sephadex LH20 using stepwise elution with isopropyl alcohol and dioxane/water was carried out to see whether the majority of the mutagenic activity (neutral fraction) could be separated from the bulk of the organics. The results depicted in

Figure 6 show that this stepwise isopropyl alcohol and dioxane/water (7:3) elution is able to separate most of the strain TA98 mutagenic activity from the TA100 activity and the bulk of organic matter, as detected by 263-mm absorption. Further fractionation of LH20 fractions II and III (Figure 6) was performed by means of repeated linear-gradient HPLC analyses (analytical and semipreparative) in combination with the *Salmonella* mutagenicity test (Figure 7). The results in Figure 7 show that the TA98 ± S9 mutagenic activity is found

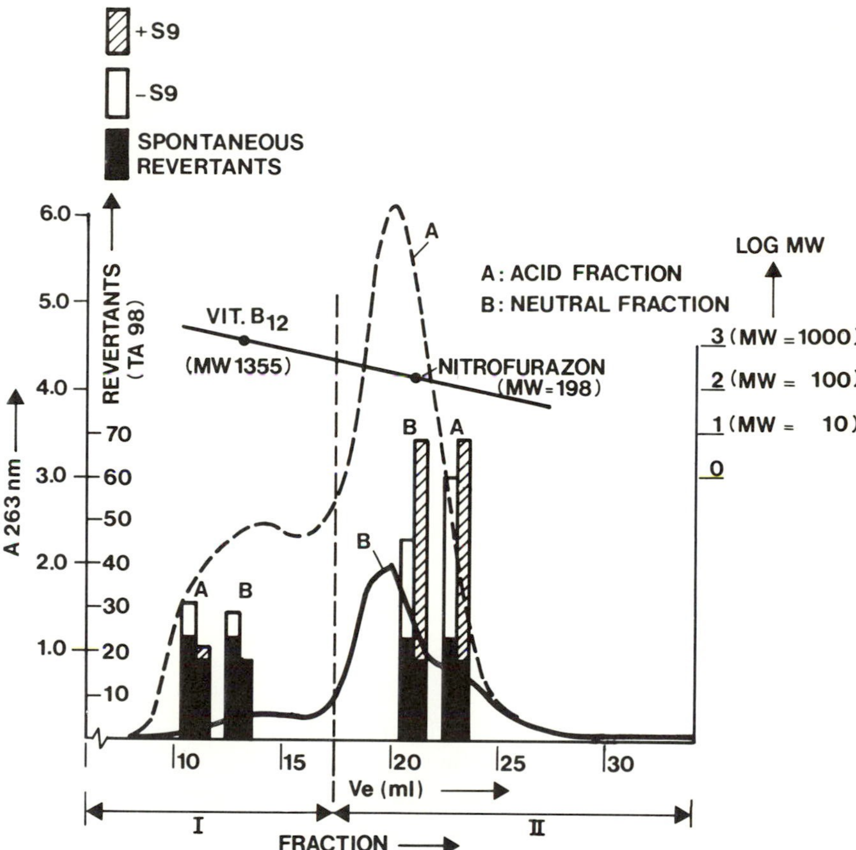

Figure 5. Molecular weight determination of a drinking water concentrate with Sephadex LH20. Sampling, 7 × 10^3-fold concentration of chlorinated drinking water (Rhine River Source), elution with DMSO, and subsequent gel filtration were as described in Materials and Methods. After the absc bance at 263 nm was measured, the fractions were pooled as indicated. After dilution in water, the fractions were reconcentrated on XAD-4/8, eluted with DMSO, and assayed in the Salmonella *mutagenicity test (strain TA98 ± S9).*

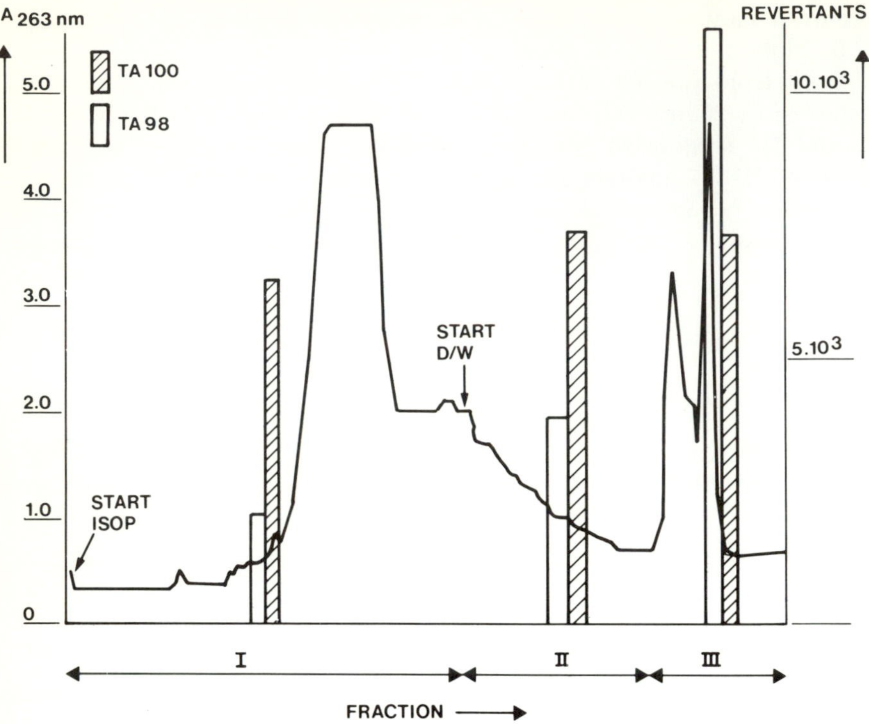

Figure 6. Fractionation of a mutagenic drinking water concentrate with Sephadex LH20. On a Sephadex LH20 column, 1.6 mL of a drinking water concentrate (1×10^6-fold concentrated, neutral fraction) was separated by using stepwise isopropyl alcohol (ISOP) and dioxane/water (D/W) elution as described in Material and Methods. Fractions were pooled as indicated. After reconcentration the fractions were assayed for mutagenic activity in the Salmonella *mutagenicity test and CHO cells (Table I).*

predominantly in two fractions (fractions 3 and 4), and this phenomenon appeared to be reproducible. Further attempts to separate the mutagens in these fractions from the remaining nonmutagenic organics have been unsuccessful.

Introduction of Chromosomal Aberrations by a LH20 Fraction of Drinking Water Concentrate. Because in previous experiments, analysis of XAD-4/8 concentrates of drinking water in in vitro mammalian cell systems was hampered by the toxicity of the concentrates for the cells, an attempt was made to investigate whether the LH20 fractions obtained could be analyzed for the induction of chromosomal aberrations in CHO cells (Table I).

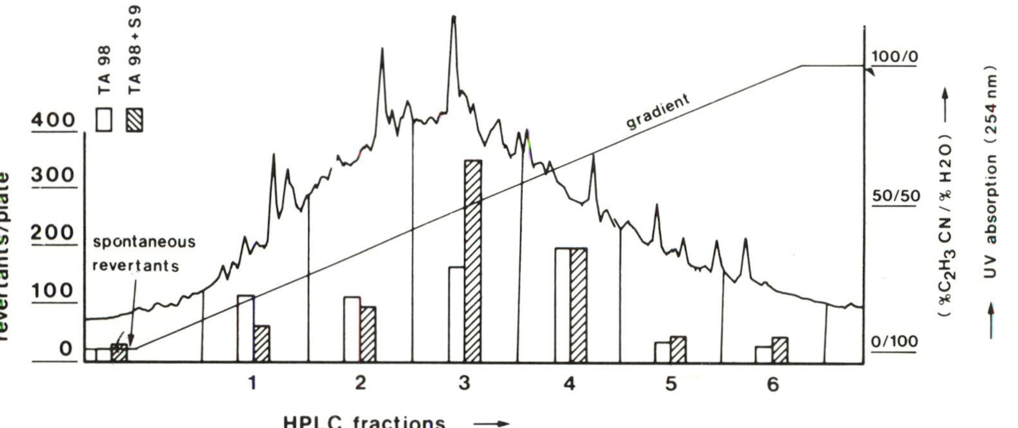

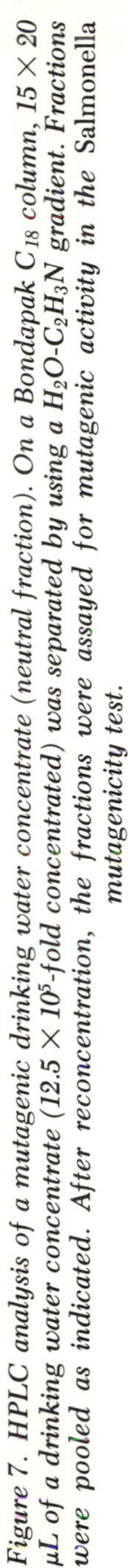

Figure 7. HPLC analysis of a mutagenic drinking water concentrate (neutral fraction). On a Bondapak C_{18} column, 15 × 20 μL of a drinking water concentrate (12.5 × 10^5-fold concentrated) was separated by using a H_2O-C_2H_3N gradient. Fractions were pooled as indicated. After reconcentration, the fractions were assayed for mutagenic activity in the Salmonella *mutagenicity test.*

Table I. Induction of Chromosomal Aberrations in CHO Cells In Vitro by Drinking Water Concentrates

Sample	Equivalent Liters of Water	Salmonella Mutagenicity[a]		Chromosomal Aberrations per 50 Cells					
		−S9	+S9	Gaps	Breaks	Fragments	Exchanges	Terminal Deletions	Total
Experiment 1									
None[b]	—	15	20	4	0	0	0	0	4
Drinking water									
15 μL	3.5	—	—	2	1	0	1	0	4
25 μL	5.8	—	—	5	1	0	0	0	6
50 μL	11.6	150	180	7	1	4	0	4	16
4-Methoxyaniline (5 mg)	—	—	—	2	1	4	4	1	12[c]
Experiment 2[d]									
None	—	28	31	3	0	0	0	0	3
Drinking water (5 μL)	6	260	—	10	22	20	22	21	95
4-Methoxyaniline (5 mg)	—	—	—	4	4	1	12	1	22[c]
Experiment 3[e]									
None	—	24	22	5	0	0	0	0	5
Drinking water									
15 μL	7	140	—	12	11	8	6	7	44
30 μL	14	—	—	7	5	9	17	4	42

NOTE: For chromosomal aberrations, LH20 fractions (a combination of fractions I and II, *see* Figure 6) prepared from chlorinated drinking water were used. 4-Methoxyaniline was included as a positive control. Mutagenicity results refer to revertants induced with strain TA98.

[a] — denotes not tested.

[b] Cells were exposed to 100 μL of DMSO.

[c] Value is chromosomal aberrations per 33 cells.

[d] In Experiment 2, for drinking water samples of 12.5, 25, and 50 μL, no metaphases were observed because of toxicity.

[e] In Experiment 3, for the drinking water sample of 60 μL and the 4-methoxyaniline sample of 5 mg, no metaphases were observed because of toxicity.

Table I shows that a combination of LH20 drinking water fractions II and III (Figure 6) was able to induce chromosomal aberrations in the CHO cells. This experiment was repeated several times with LH20 concentrates throughout the year at the same location; three out of five drinking water concentrates showed increased levels of chromosomal aberrations compared to the control, whereas no toxic effects were observed (Table I). The negative results are not shown. The results indicate that the LH20 fractionation procedure may (in part) be suitable for testing drinking water concentrates for genotoxic effects in mammalian cell systems.

Characterization of Organic Mutagens. The chemical analyses (GC–MS) still leave the identity of the mutagens just discussed untouched. Therefore, further characterization was attempted by using specific enzyme-deficient strains of *Salmonella.* A first indication of the nature of some of these compounds was recently obtained by testing organic drinking water concentrates with the nitroreductase-deficient (NR^-) strains of *Salmonella* that were isolated by Rozenkranz et al. (*16*).

As shown in Figure 8, the direct mutagenic activity with strain TA100 of the three LH20 fractions from Figure 6 is almost completely abolished when the $TA100NR^-$ strain is used. Because (1) hardly any reduction of promutagenic activity with strain TA100 was observed (not shown) and (2) the control experiments demonstrated the capability (~70%) of the two enzyme-deficient strains ($TA98NR^-$, $TA100NR^-$) to detect non-nitroreductase-dependent mutagens, the results with the three fractions are suggestive for the presence of nitro-organics (*16*). For the direct mutagens (strain TA98) in fractions I and II (Figure 8), no reduction or a small reduction was observed; in fraction III, a significant reduction was observed. This result is also suggestive for the presence of nitro-organics (*16*). When fractions obtained by semipreparative HPLC analysis analogous to Figure 7) were analyzed for halogenated hydrocarbons, the results correlated well with the TA98 mutagenic activity (Figure 9).

The presence of mutagenic nitroaromatic compounds in the environment is known because these compounds are byproducts of incomplete combustion processes, and their presence is thus mainly a result of human activities (*16*). In drinking water, however, these kinds of compounds have not been found in sufficient amounts to produce such a mutagenic effect in the Ames test (*24*). For this reason, we investigated whether or not the activities of the nitroreductase-deficient strains were introduced during drinking water treatment, for example, during a chlorine treatment (Figure 10).

Figure 10 shows that stored Meuse River water also contains nitroreductase-dependent mutagens (*see* TA98). This finding indicates

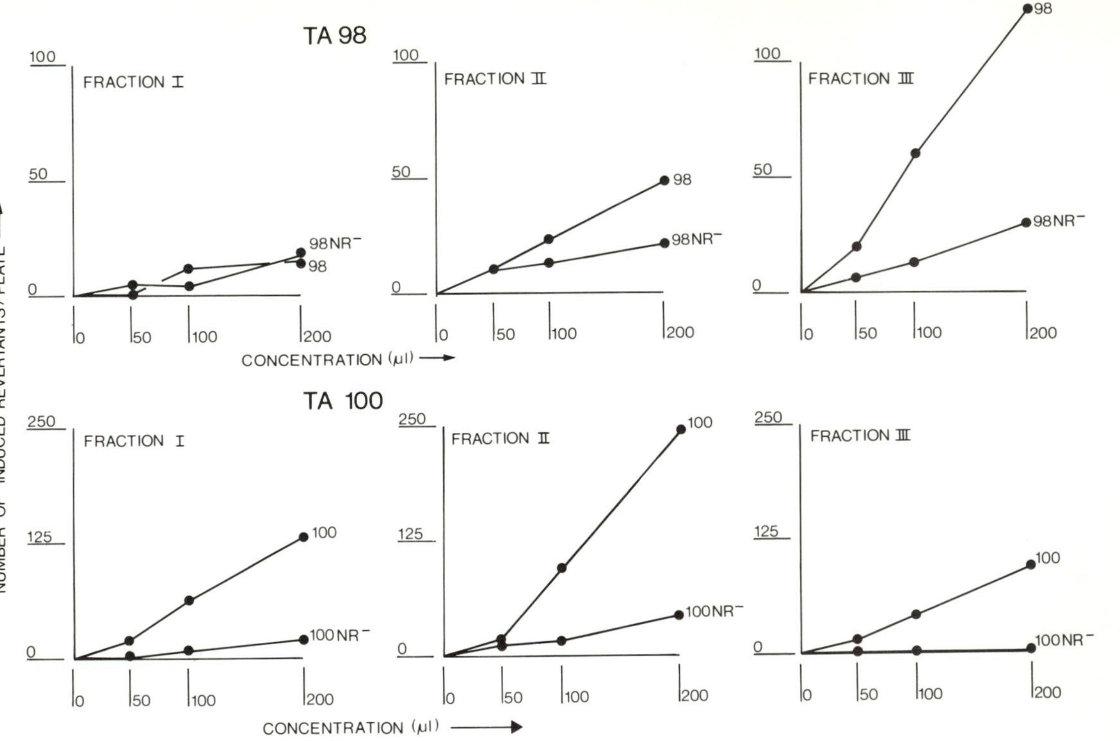

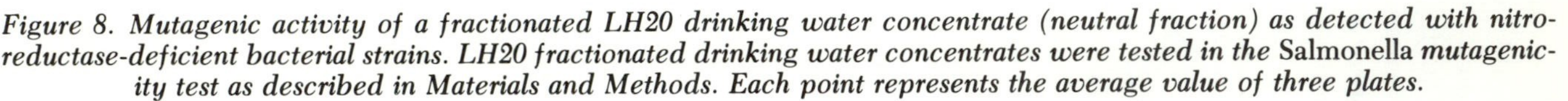

Figure 8. Mutagenic activity of a fractionated LH20 drinking water concentrate (neutral fraction) as detected with nitroreductase-deficient bacterial strains. LH20 fractionated drinking water concentrates were tested in the Salmonella *mutagenicity test as described in Materials and Methods. Each point represents the average value of three plates.*

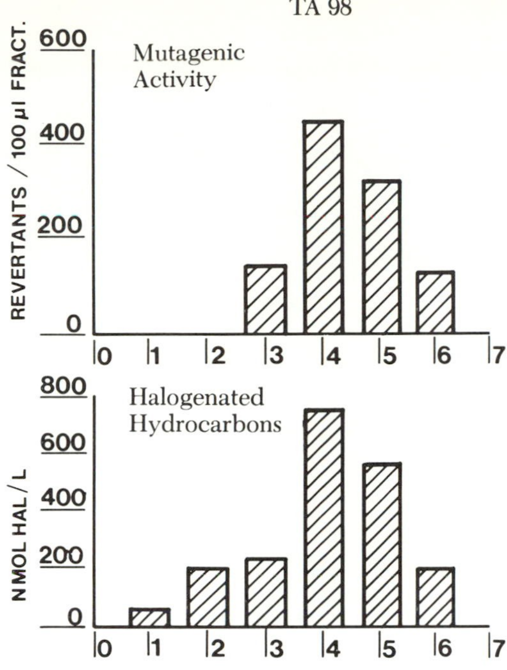

Figure 9. Distribution of halogenated hydrocarbons and mutagenic activity in HPLC fractions of a drinking water concentrate (neutral fraction). HPLC fractions obtained by linear-gradient HPLC analysis (H_2O-C_2H_3N) were tested for mutagenic activity in the Salmonella *mutagenicity test (TA98) and assayed for halogenated hydrocarbon content as described in Materials and Methods.*

that sufficient amounts of mutagenic nitro-organic compounds are present in the Meuse River to show an effect in the Ames test. After a chlorine treatment (1.5 mg/L of Cl_2), new, direct, TA98 mutagens—not nitroreductase-dependent mutagens—are introduced because a significant increase of activity is observed with both TA98 and TA100 strains. The generated, direct, mutagenic activity of strain TA100 after the chlorine treatment, however, is completely abolished with TA100 NR^-. This result indicates that mutagenic nitro-organics are introduced during the chlorine treatment. Further investigations are underway. In regard to the promutagenic activities of all four strains, no differences before and after chlorination were observed.

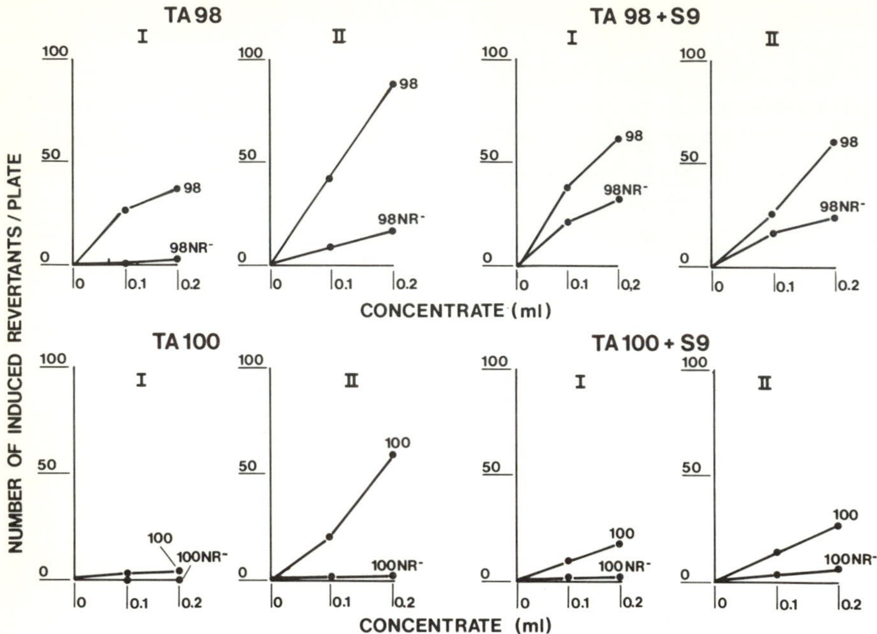

Figure 10. Influence of a chlorine treatment on the mutagenic activity detectable with TA98NR⁻ and TA100NR⁻ strains. Sampling, 7000-fold concentration of water samples before and after a chlorine treatment (1.5 mg/L of Cl_2) on XAD-4/8, elution with DMSO (neutral fraction), and subsequent testing of the DMSO concentrate in the Salmonella *mutagenicity test were as described in Materials and Methods. Each value represents the average of three plates, and 0.2 mL of concentrate corresponds to 1.4 L of water per plate.*

Discussion

In this study, XAD-4/8 resins were shown to be suitable for concentrating organic mutagens in drinking water in comparison to XAD-2 resins and a freeze-drying concentration technique (Figures 1 and 2). Gel filtration on Sephadex LH20 showed that the mutagens in the neutral and acid fractions of different chlorinated drinking waters have molecular weights in the range of 100–300 (Figures 4–6). The results confirmed previous results (*26*). One should, however, be aware that the calibration procedure was performed with two known organic compounds, namely, vitamin B12 and nitrofurazone. Therefore, there is a possibility that the obtained molecular weights of the unknown organic mutagens present in the complex organic mixture may not be reliable because of an adsorption phenomenon. On the other hand, the results are reproducible in time, and, therefore, it is unlikely that adsorption will greatly influence the obtained molecular weight results. Fractionation of XAD-4/8 drinking water concentrates with Sephadex LH20 and stepwise elution with isopropyl alcohol and dioxane/water (70:30 v/v) were able to separate the major part of mutagenic activity of strain TA98 and part of the activity of strain TA100 from the bulk of organic material (Figure 6). However, the amount of organic material measured at 263 nm is an underestimation of the total organic material present in the sample. Purified LH20 fractions obtained in this manner were able to induce chromosomal aberration (Table I) in three out of five drinking water concentrates of one water work. Two concentrates were negative in the assay. This result may be due to changes in the organic composition and water treatment throughout the year.

Additional fractionation of the purified LH20 fractions by means of repeated linear-gradient HPLC analysis according to the methods of Tabor and Loper (*12*) showed that mutagenic activity was found predominantly in two fractions (Figure 7). Further subfractionation with linear-gradient and straight-phase HPLC analysis did not result in a better separation of the organic mutagens from the bulk of the organic material. This result showed that other fractionation procedures have to be applied. In addition, mutagenic concentrates revealed that the organic mutagens possess thermolabile properties (Figure 3), so GC analysis is rather difficult to apply in this way.

A first indication of the nature of a part of the organic mutagens was obtained by testing LH20 fractions with the nitroreductase-deficient strains $TA98NR^-$ and $TA100NR^-$. The results of Figures 8 and 9 show strong indications that nitro-organics, halogenated or not, are, in part, responsible for the mutagenic activity in the organic concentrates prepared from chlorinated drinking water.

Finally, the results obtained with the nitroreductase-deficient strain TA100 indicated that a chlorination step during drinking water treatment is able to produce and/or activate mutagenic nitro-organics. However, much more research is needed to elucidate the precise nature of these nitro (halogenated) aromatics and the mechanisms by which they are formed.

Acknowledgments

We wish to thank the Water Works for their cooperation. Also, the chemical assistance of S. Persad, P. Slingerland, H. den Hollander, and W. Willemse is gratefully acknowledged. W. K. de Raat (MT-TNO Delft, The Netherlands) kindly supplied $TA98NR^-$ and $TA100NR^-$. The research presented in this chapter was supported in part by grant RID 80-1 from The Netherlands Cancer Society (Koningin Wilhelmina Fonds).

Literature Cited

1. Zoeteman, B. C. J. *Sensory Assessment of Water Quality;* Pergamon: Oxford, 1980.
2. Meyers, A. P. Ph.D. Thesis, Technical University Delft, The Netherlands, 1970.
3. Rook, J. J. *J. Water Treat. Exam.* **1974,** *23,* 234.
4. Bellar, T. A.; Lichtenberg, J. J.; Kroner, R. O. *J. Am. Water Works Assoc.* **1974,** *66,* 703.
5. Aichele, D. G.; Contos, D. A.; Foltz., R. L.; Golf, V. R.; Hayes, T. L.; Lin, D. C. K.; Link, P. S.; Lucas, S. V.; Redmont, K. P.; Schweiger, C. A.; Slivon, L. E.; Tabor, J. E.; Thompson, R. M.; Watson, S. C. *GC-MS Analysis of Organics in Drinking Water Concentrates and Advanced Waste Treatment Concentrates. Combined Results on Five Drinking Water Samples from Miami, FL; Philadelphia, PA; New Orleans, LA; Ottumwa, IA; and Seattle, WA;* Preliminary Report. HERL, U.S. EPA, Cincinnati. Bartelle Columbus Laboratories: Columbus OH, 1979; EPA Project No. 68-03-2548.
6. National Academy of Sciences *Drinking Water and Health;* NAS: Washington, DC, 1977; Vols. 1-2.
7. Packham, R. F.; Beresford, S. A.; Fielding, M. *Water Supply and Health;* van Lelyveld, H.; Zoeteman, B. C. J., Eds.; Elsevier Scientific: New York, 1981; pp 167-186.
8. National Academy of Sciences *Drinking Water and Health;* NAS: Washington, DC, 1980, 1981; Vols. 3-4.
9. Kool, H. J.; Van Kreijl; C. F.; Zoeteman, B. C. J. *CRC Crit. Rev. Environ. Control* **1982,** *12,* 307.
10. Loper, J. C. *Mutat. Res.* **1980,** *76,* 241.
11. Coleman, W. E.; Melton, R. G.; Kopfler, F. C.; Barone, K. A.; Aurand, T. A. A.; Jellison, M. G. *Environ. Sci. Technol.* **1980,** *14,* 576.
12. Tabor, M. W.; Loper, J. C. *Int. J. Environ. Anal. Chem.* **1980,** *8,* 197.
13. Tabor, M. W. *Environ. Sci. Technol.* **1983,** *17,* 324.
14. Ames, B. N.; Durston, W. E.; Yamasaki, E.; Lee, F. D. *Proc. Natl. Acad. Sci.* **1973,** *79,* 2281.
15. Ames, B. N.; McCann, J.; Yamasaki, E. *Mutat. Res.* **1975,** *31,* 347.
16. Rosenkranz, H. S.; Mermelstein, R. *Mutat. Res.* **1983,** *114,* 217.
17. Junk, G. A.; Richards, J. J.; Grieser, M. D.; Witiak, D.; Witiak, J. L.; Arguello, M. D.; Vick, R.; Svec, H. J.; Fritz, J. S.; Calder, G. V. *J. Chromatogr.* **1974,** *99,* 745.
18. Evans, H. J.; Riordan, M. C. D. In *Handbook of Mutagenicity Test Procedures;* Killey, B. J.; Legator, M.; Nicolls, W.; Ramel, C., Eds.; Elsevier: Amsterdam, 1977; p 261.
19. Wegman, R. C. C.; Greve, P. A. *Sci. Total Environ.* **1977,** *15,* 137.
20. Concin, R.; Burtscher, E.; Bobleter, O. *J. Chromatogr.* **1980,** *198,* 131.
21. Webb, R. G. *Isolating Organic Water Pollutants, XAD Resins, Urethane Foam, Solvent Extraction;* National Environmental Research Center. Office of Research and Development. U.S. Environmental Protection Agency: Corwallis, OR, 1975; EPA report 660/4-75-003.

22. Van Rossum, P.; Webb, R. G. *J. Chromatogr.* **1978,** *150,* 381.
23. Yamasaki, E.; Ames, B. N. *Proc. Natl. Acad. Sci.* **1977,** *74,* 3555.
24. Kool, H. J.; Van Kreijl, C. F.; Van Kranen, H. J.; De Greef, E. *Chemosphere* **1981,** *10,* 85.
25. Kool, H. J.; Van Kreijl, C. F.; Van Kranen, H. J.; van De Greef, E. *Sci. Total Environ.* **1981,** *18,* 135.
26. Kool, H. J.; Van Kreijl, C. F.; De Greef, E.; van Kranen, H. J. *Environ. Health Perspect.* **1982,** *46,* 207.
27. Kool, H. J.; Van Kreijl, C. F.; Van Oers, H. *Toxicol. Environ. Chem.* **1984,** 7, 111.
28. Van der Gaag, M. A.; Noordsij, A.; van Oranje, J. P. In *Mutagens in Our Environment;* Sorsa, M.; Vaino, H., Eds.; Alan R. Liss: New York, 1982; pp 277-286.

RECEIVED for review August 14, 1985. ACCEPTED December 20, 1985.

Mutagenic Activity of Various Drinking Water Treatment Lines

L. Cognet[1], J. P. Duguet[1], Y. Courtois[2], J. P. Bordet[3], and J. Mallevialle[1]

[1]Lyonnaise des Eaux Central Laboratory, 38 rue du President Wilson 78239 Le Pecq, France
[2]LHVP, 1 Bis rue des Hospitalières ST Gervais 75004 Paris, France
[3]ARLAB, 76–78 rue des Suisses 92007 Nanterre, France

Mutagenic activity of several treatment alternatives in a water treatment pilot plant was studied for 1 year. Water extracts were completed by adsorption of organics on macroreticular resin. Mutagenicity was determinated by the Ames Salmonella *microsome test. Statistical analyses of the data (Wilcoxon signed ranks test and factor analysis of correspondence) were applied to understand the complexity and variations observed in the data. Results showed that ozone decreased or increased mutagenic activity depending on treatment conditions; granular activated carbon (GAC) filtration was less efficient than ozone; and GAC combination and disinfection with chlorine dioxide was less mutagenic than chlorine treatment.*

MACRORETICULAR RESIN SAMPLING TECHNIQUES are used worldwide (*1–3*) for collecting organic constituents from the natural environment. Macroreticular resin (MRR) adsorption has been used for the isolation of trace organics for the identification and determination of mutagenic activity (*4, 5*). Of the several biological tests available, the Ames test is probably the most widely used (*6, 7*) in the water treatment field. Such a technique, when coupled with MRR concentration, is a powerful tool for investigating the effects of various water treatment techniques, for example, ozonation, chlorination, nitrification, and granular activated carbon (GAC) adsorption, because identification of solutes and their mutagenic activity can be determined from the same MRR sample.

Raw water quality varies dramatically with respect to human consumption of micropollutants, both in concentration and composition (*8*).

0065–2393/87/0214/0627$06.00/0

Thus, the collection of a representative sample becomes essential. For this reason, sample collection in this study was done for 2–4 days, and organics were extracted from 150–200 L of water. Although this method dilutes peak concentrations, the real concern with respect to micropollutants is the chronic effect (as opposed to an acute effect). Mutagenicity may result from several of the dissolved organics; thus, the composite sampling procedure allows for the collection of a representative background matrix more closely associated with chronic-type exposure.

The use of the MRR procedure in this study was applied to water samples from a combined ozonation–GAC process in which several ozone doses and ozone contact times were evaluated. The goal was to determine the effects of the combined treatments on the micropollutants and mutagenic activity. In this chapter, data from gas chromatographic (GC) analyses are not reported. Compounds identified by GC–mass spectrometry (GC–MS) and their concentration ranges at the different points of the pilot plant have been published in references 9 and 10. Sampling points allowed for the comparison of the various ozone treatments alone or in combination with GAC and the determination of the effects of postdisinfection.

Mutagenic activity can increase or decrease, depending on ozonation rate (*11*). Data on chlorination usually indicate an increase in mutagenic activity (*12*, *13*, *14*). However, the long-term average increase in mutagenic activity has not been well-defined. For this reason, this experiment was performed for a 1-year period to determine net changes in water quality.

Experimental

Water Treatment. The source of water for this experiment was a pilot plant located on the Seine River upstream from Paris, France (Figure 1). The pilot plant uses an upflow solids contact clarifier (Pulsator, Degremont, Rueil Malmaison, France) followed by rapid sand filtration (RSF). The filtered water is then distributed over four treatment lines to evaluate the efficiency of various ozone–GAC combinations (ozonation rates of 1 or 5 ppm O_3 and 10–30 min of contact time). The GAC used in this study was Calgon F-400 (Calgon Corp.). Disinfection by chlorine or chlorine dioxide completed the process. In this chapter, line 3 treatment was not considered a complete treatment for the water supply. This line was studied to evaluate the efficiency of a high ozonation rate.

Water Samples. Nine monthly MRR samples were collected at 11 sampling points throughout the pilot plant. The sample collector (Concentreur S656, SERES Co., Aix en Provence, France) enabled a composite sample to be collected over several days (*8*) and is based on adsorption of organics on resins (XAD-2 and XAD-8, Rohm and Haas). Factors that influence adsorption capacity in sampling were considered in the design of a two-bed resin column: XAD-8 is more efficient to adsorb humic materials than XAD-2. Thus, an XAD-2 resin bed in series with an XAD-8 resin bed is appropriate to concentrate

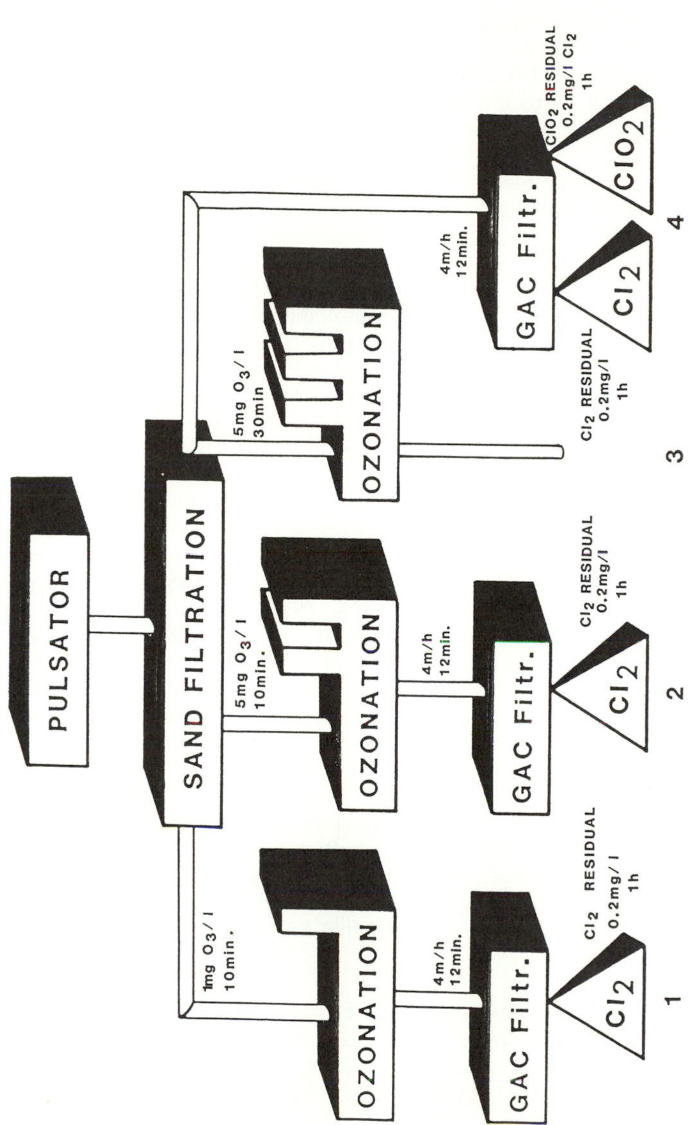

Figure 1. Diagram of Vigneux pilot plant.

micropollutants from water samples for minimum competition of humic materials. All water samples (100–200 L) were contacted with a 100-mL bed of XAD-8 followed by a 100-mL bed of XAD-2. The flow rate was set at 10 bed volumes/h. The pH was adjusted to 2–3 with nitric acid. Sodium thiosulfate (0.1 N) was added to reduce any residual oxidant. Prior to use, the resins were cleaned and never reused for a next sampling. XAD-2 and XAD-8 resins were prepared by Soxhlet extraction with methanol (MeOH) and dichloromethane (DCM) during a 24-h period. Then, three elutions of 1 bed volume of DCM were performed. The third elution was concentrated to a 1-mL extract for GC analysis and Ames testing. No positive mutagenic activity was detected in these blank extracts. When GC profiles showed peaks attributable to resin contaminants, Soxhlet extraction was applied to clean the resins again. After sampling, the adsorbed organics were eluted in the laboratory with DCM and MeOH. The first elution with DCM yielded an extract containing low molecular weight compounds that were gas chromatographable. A subsequent elution of MRR with MeOH was necessary to obtain an extract containing more polar compounds or compounds having higher molecular weights (*11*).

The solvents were then passed over anhydrous sodium sulfate to remove residual water and concentrated to 10 mL by Vigneux distillation evaporation and to 3 mL by evaporation under nitrogen.

The overall concentration factor from the water phase to final concentrate is between 30,000 and 60,000. One milliliter of each concentrate was assayed by the Ames *Salmonella* microsome mutagen. The 1 mL of DCM was evaporated to dryness by evaporation under nitrogen, and the organics were redissolved in DMSO. This last step causes a loss of volatile organics.

Ames Test. *Salmonella tryphimurium* strains TA98 and TA100 were employed according to the Ames test (*15, 16*) to determine the mutagenic activity of the various water samples. Each DCM and MeOH extract was tested with and without S9 (microsomal fraction of activated rat liver with Arochlor 1254) activation. For each assay (16 assays per sampling point), the number of revertants per plate was plotted versus increasing volumes of water extracts injected. Slope values from linear regression of the dose–response curves were calculated and then used in the statistical analysis.

Results were considered positive when (1) a twofold increase of spontaneous revertants occurred for at least one dose and (2) a reproducible dose–response relationship was found. Results were considered uncertain when the first condition was not met and negative when neither condition was met.

Results on the TA100 strain were often uncertain or negative for the first 4 months of the experiment. Therefore, the Ames test on this strain was not employed further (*16*).

Statistical Methods. **WILCOXON TEST.** The test (*17*) is a nonparametric test that enables the resolution of problems without the requirement of probability laws of considered variables. The Wilcoxon test consists of classifying differences between two series with their increasing absolute values; whether these differences are positive or negative is noted. The sum of levels of positive values (W+) is then compared to the sum of negative ones (W−). If no significant differences between the two series are noted, then the statistical distribution of W+ is assimilated to Gauss–Laplace law. A bilateral test of probability is then applied to consider a positive or negative effect of the treatment. The critical probability (CP) then calculated is compared to a threshold of 5%. When CP is greater than 5%, there is no significant difference between the two series (e.g., before and after treatment). In other cases, the two series are significantly different.

Factor Analysis. In this method (*18–20*), mutagenicity values are grouped together in a matrix as follows:

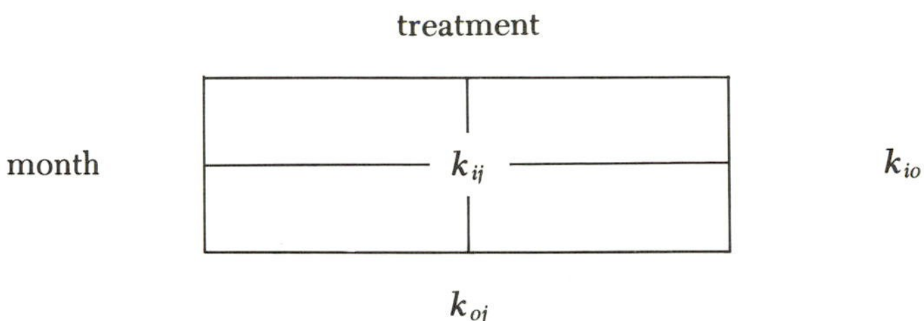

In this representation, k_{ij} is the mutagenicity value (slope of the dose–response curve) for the month i at the treatment point j. A summation of matrices from the various treatments (two types of bacteria, TA98 and TA100; two types of extracts, DCM and MeOH; and two types of microsomal activation, with S9 and without S9) is used in the statistical analysis.

The summation matrix can be considered as two clouds of points in two spaces. One space (treatment space) has a number of points equal to the number of columns. The other space (time space) has the number of points equal to the number of columns, as well as the number of dimensions equal to the number of rows.

The statistical analysis determines the principal axes of inertia of the clouds of data points. In a physical analogy, the axes of inertia of the solid defined by data points determine where the strongest relationships are, that is, which mutagenicity values of a month are more strongly related to a treatment.

When a data set is well-related, the first two or three eigenvalues—the axes of inertia—account for 95% or more of the total inertia. In this study, 80–95% of the inertia is contained in the first three axes. This condition indicates that mutagenic activity can be related to treatment efficiency.

A two-dimensional projection of the clouds of data is made with the two axes of maximum inertia. In this plane, projected treatment points are clustered for each treatment.

Results

Nine sets of data were obtained for the 1-year period. Graphical representations of mutagenicity values were drawn. An example is given for the evolution of mutagenicity at various ozonation rates (Figure 2). Variations of low mutagenicity values were observed during the time of experiment. A decrease of genotoxicity by ozone treatment lines 1 and 2 was observed for the first 2 months and by all ozonation rates for the 9 months. However, these partial conclusions are given from the 36 mutagenicity values of one Ames test assay. The data include 1144 values (slope revertants per liter of water). Simple graphical represen-

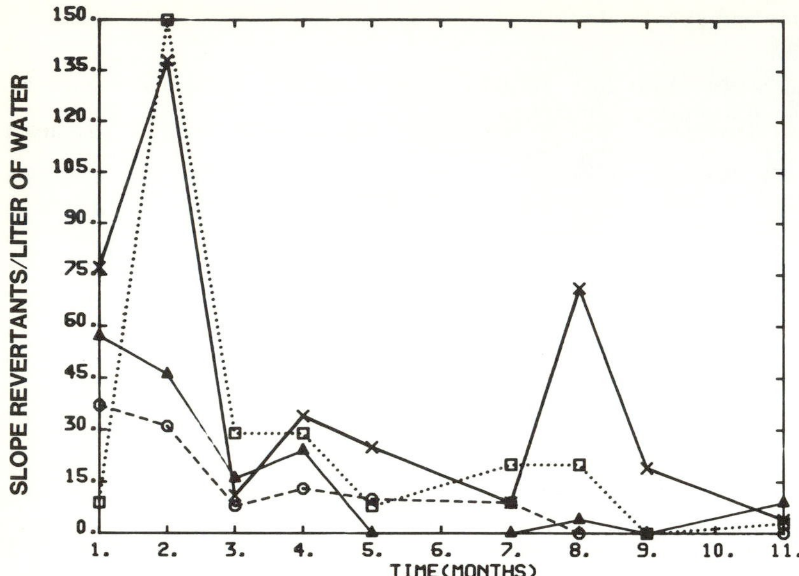

Figure 2. Evolution of mutagenicity with ozonation (data on TA98 strain without S9 mixture on the DCM extract): ×, *RSF;* ▲, O_3 *line 1;* O, O_3 *line 2; and* □, O_3 *line 3.*

tation did not reveal any obvious trends. Therefore, statistical methods were employed to answer the following questions:

1. What is the effect of ozonation on mutagenic activity? Does the activity increase or decrease with ozone dose?
2. What effect did GAC filtration have on mutagenic activity? Did the combination of ozone and GAC result in any significant changes in mutagenic activity?
3. How did final chlorination affect mutagenic activity?
4. Which was the best treatment combination?

Ozonation Effects. The following observations on ozonation were derived by using the Wilcoxon test; mutagenic activity was compared before and after ozonation for 1 year. A significant decrease in mutagenic activity was observed for ozone treatment lines 1 and 2 with the DCM extract (Table I). The highest ozonation rate changed mutagenic activity from that observed in the rapid sand filter. However, as for all the MeOH extracts, no interpretation of the variations and no statistical conclusions for the 1-year period could be drawn.

By using the factor analysis method, the data can be displayed graphically as shown in Figure 3. In this analysis, a trend is observed

Table I. Critical Probabilities (%) for the Evolution of Mutagenicity with Ozonation (Wilcoxon Test Analysis on TA98 Data)

	DCM Extract		*MeOH Extract*	
Ozone Treatment	*W/O S9 Mix*	*With S9 Mix*	*W/O S9 Mix*	*With S9 Mix*
RSF-O_3, line 1	1.2[a]	0.3[a]	84.9	28.9
RSF-O_3, line 2	0.1[a]	0.1[a]	25.8	8.9
RSF-O_3, line 3	69.7	22.2	86.5	40.7

NOTE: No significant decrease of mutagenicity at 5% threshold was found for any of the extracts except as indicated.
[a] A significant decrease of mutagenicity at 5% threshold was found.

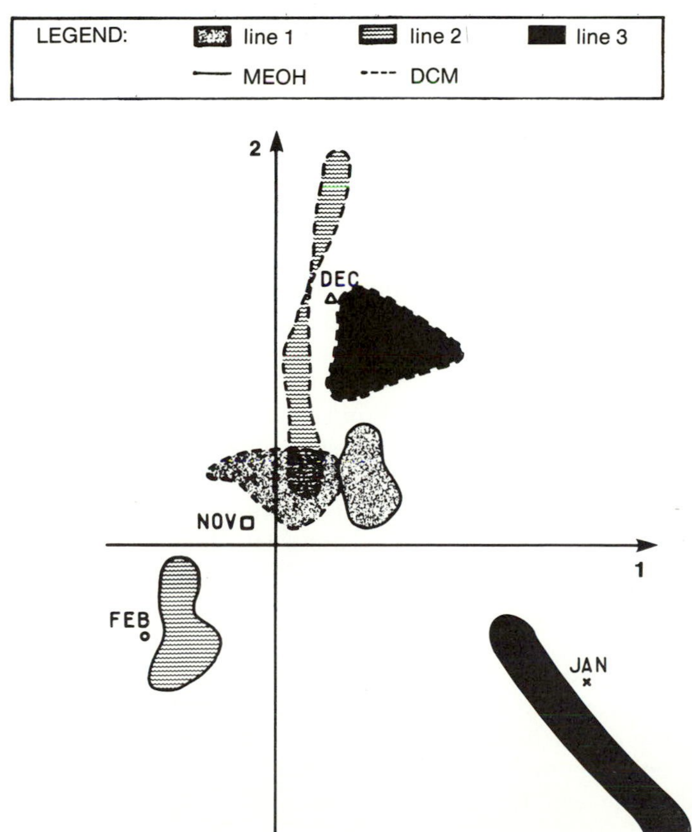

Figure 3. Factor analysis of evolution of mutagenicity after ozonation (data from first 4 months).

for the data clusters of MeOH and DCM extracts. These clusters become further removed from one another as the ozone dose increases. This finding indicates that the difference between mutagenic activity in the two extracts (and thus between the two types of compounds) is a function of the ozonation rate.

A second observation is drawn from comparing the distances between the points representing each month and the clusters of points representing mutagenicity of each ozone treatment. A short distance indicates a higher mutagenic activity for the month at that treatment level. Thus, the best treatment for a month for mutagenicity is the treatment that is the most distant from the month considered.

These distances between months and each treatment level can be weighted to develop a hierarchal monthly classification of treatment efficiency. This classification of treatment efficiency is done for each month by measurement of the distances between months and treatment clusters. A weight of three is attributed to the best treatment (the cluster that is most distant for the month considered), a weight of two is given for the next most distant, and a weight of one is given for the least efficient treatment. By averaging over the year, the weights attributed to each treatment can be summarized, and the most efficient treatment can be defined.

A *relative ideal ozone treatment* for minimizing mutagenic activity can be defined as a treatment for which a weight of three is attributed to each month. The sum of the weights at one treatment level is used to determine a relative ideal ozone treatment. If one process produced the best results on each of the 9 months, it would have a score of $3 \times 9 = 27$ and would be a 100% ideal ozone treatment. The ratios between the sum of the weights and the perfect scores determine the percent of relative ideal ozone treatment. From this type of reasoning, ozone treatment line 1 is a 74% relative ideal ozone treatment and is more efficient than ozone treatments of lines 2 and 3 to remove mutagenicity (Table II). Variations of mutagenicity during ozonation can be attributed to the transformation of organics into byproducts that can be more or less stable and to variations of the organic matrix in the RSF water.

GAC and Ozone–GAC. The Wilcoxon test was applied to compare results before and after GAC filtration. By using data for DCM or MeOH extracts, no interpretation of the variations observed and no statistical conclusions can be drawn. This situation is true for GAC treatment alone and for all combinations of dose, contact time, and GAC (Table III).

Factor analysis was applied, and it integrated data for both DCM and MeOH extracts at the same time. By using the same interpretation

Table II. Weighted Classification of Treatment Lines After Ozonation for Each Month (Factor Analysis of TA98 Data)

Month	*RSF-O_3 Line 1*	*RSF-O_3 Line 2*	*RSF-O_3 Line 3*	*Relative Ideal Treatment*
Sept., May	3	1	2	3
Jan., July, Dec.	3	2	1	3
Feb.	2	1	3	3
Nov., March, June	1	2	3	3
Sum of weights	20	15	17	27
Percent of ideal treatment line	74	55	62	100

Table III. Critical Probabilities (%) for Evolution of Mutagenicity with GAC Filtration (Wilcoxon Test Analysis on TA98 Data)

	DCM Extract		*MeOH Extract*	
GAC Treatment	*W/O S9 Mix*	*With S9 Mix*	*W/O S9 Mix*	*With S9 Mix*
O_3-GAC, line 1	36.8	32.2	43.5	39.5
O_3-GAC, line 2	9.3	40.1	29.8	75.7
RSF-GAC, line 4	45.9	45.9	22.2	96.0

NOTE: No significant decrease of mutagenicity at 5% threshold was found for any of the extracts.

Table IV. Weighted Classification of Treatment Lines After GAC for Each Month (Factor Analysis of TA98 Data)

Month	*RSF-O_3-GAC Line 1*	*RSF-O_3-GAC Line 2*	*RSF-GAC Line 4*	*Ideal Treatment Line*
Sept., Nov., May, March, June, Jan.	3	2	2	3
Feb.	1	2	3	3
July, Dec.	3	2	2	3
Sum of weights	25	18	19	27
Percent of ideal treatment line	89	64	68	100

as that used for the ozone treatment, the following conclusions were made for the 1-year study (Table IV): A difference is found between the ozone-GAC combination line 1 and the GAC filtration of clarified water. The ozone-GAC treatment yields an 89% relative ideal treatment of the clarified water, whereas GAC alone yields 68%. The same difference is observed between the two ozone-GAC combinations (lines 1 and 2).

Disinfection with Chlorine and Chlorine Dioxide. The comparison of the mutagenic activity of the DCM extract before and after disinfection treatment was studied by the Wilcoxon test. No statistical conclusions on disinfection effects can be drawn. However, the MeOH extract showed a significant decrease in mutagenic activity for the line 2 chlorine treatment. Comparison of the two disinfection treatments for the nonozonated GAC filtered water (treatment line 4) shows that chlorine disinfection yields greater mutagenic activity of the DCM extract than chlorine dioxide (Table V).

Complete Treatment Line. The Wilcoxon test was applied to compare mutagenic activity before RSF and after complete treatment lines. Treatment line 4 with chlorine dioxide disinfection significantly decreased the mutagenic activity of the DCM extract. No other statistically significant differences were observed for any other treatment lines from data on DCM or MeOH extracts (Table VI).

Table V. Critical Probabilities (%) for the Evolution of Mutagenicity with Disinfection (Wilcoxon Test Analysis on TA98 Data)

Disinfection Treatment	*DCM Extract*		*MeOH Extract*	
	W/O S9 Mix	*With S9 Mix*	*W/O S9 Mix*	*With S9 Mix*
GAC-Cl_2, line 1	26.7	83.4	55.5	58.9
GAC-Cl_2, line 2	74.1	66.7	0.4[a]	6.1
GAC-Cl_2, line 4	66.0	92.8	0.7[a]	2.1[a]
GAC-ClO_2, line 4	7.2	18.7	27.6	97.6
Cl_2-ClO_2, line 4	0.3[a]	3.3[a]	44.7	17.6

NOTE: No significant decrease of mutagenicity at 5% threshold was found for any of the extracts except as indicated.
[a] A significant decrease of mutagenicity at 5% threshold was found.

Table VI. Critical Probabilities (%) for the Evolution of Mutagenicity with Each Treatment Line (Wilcoxon Text Analysis on TA98 Data)

In-Out Treatment Line	*DCM Extract*		*MeOH Extract*	
	W/O S9 Mix	*With S9 Mix*	*W/O S9 Mix*	*With S9 Mix*
RSF-Cl_2, line 1	10.3	1.4[a]	11.9	1.2[a]
RSF-Cl_2, line 2	63.8	8.9	15.0	23.8
RSF-Cl_2, line 4	31.7	10.7	16.5	0.9[a]
RSF-ClO_2, line 4	3.7[a]	4.1[a]	81.0	63.1

NOTE: No significant decrease of mutagenicity at 5% threshold was found for any of the extracts except as indicated.
[a] A significant decrease of mutagenicity at 5% threshold was found.

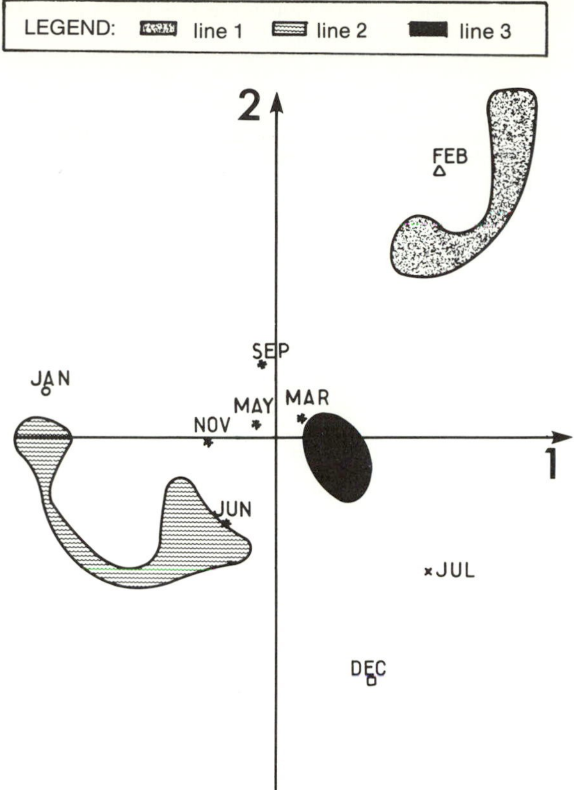

Figure 4. Factor analysis of evolution of mutagenicity on complete treatment lines.

By using the factor analysis method, the mutagenicity data were applied only on water after disinfection. Then, comparisons were made between treatment lines (Figure 4). By using the same interpretation (i.e., assigning weights of 3, 2, or 1) as that used for the ozone and GAC treatment, the conclusions based on ordered classifications of treatment lines for each month and for the year are the following: Treatment line 1 is the best line of the pilot plant and yields a 92% relative ideal complete treatment line, treatment line 2 is 66%, and treatment line 4 is 41% (Table VII).

Conclusions

The evolution of mutagenicity was evaluated for various treatment processes used to study various ozone–GAC treatment combinations. The upstream Seine River contains a low level of organic materials after clar-

Table VII. Weighted Classification of Complete Treatment Lines for Each Month (Factor Analysis of TA98 Data)

Month	*RSF-GAC-* *O_3-Cl_2* *Line 1*	*RSF-GAC-* *O_3-Cl_2* *Line 2*	*RSF-GAC-* *Cl_2* *Line 4*	*Relative* *Ideal* *Treatment*
Nov., Dec., March, May, June, July, Sept.	3	2	1	3
Jan.	3	1	2	3
Feb.	1	3	2	3
Sum of weights	25	18	11	27
Percent of ideal treatment line	92	66	41	100

ification (total organic carbon [TOC]: 1.5–2.0 ppm) and a low level of mutagenic activity. The variations observed here may be specific for these conditions and should be extrapolated with caution.

By using statistical methods for the interpretation, several conclusions were drawn from this complex data set:

1. Low ozonation rate treatments (lines 1 and 2) of this water source were observed to decrease mutagenic activity of low molecular weight and chromatrographable compounds. GAC treatment changed mutagenic activity, but no statistical conclusions could be drawn.
2. The ozone–GAC combination was more effective in decreasing mutagenic activity than the GAC treatment without ozonation.
3. Disinfection with chlorine dioxide treatment of nonozonated GAC-filtered water was less mutagenic than chlorine treatment.
4. No statistically significant effect from chlorination was observed. A classification of treatment line efficiencies on mutagenicity was defined: The best treatment line involves an ozone–GAC combination and a low ozone dose (1 ppm of O_3, 10 min of contact time).

Factor analysis appears to be a powerful method to interpret the data when different parameters (biological, chemical, physical) are included. The next step is to integrate mutagenicity results, chromatographic data, and identified compounds and to use factor analysis on these data for a better understanding of the ozone–GAC combination. Conclusions that can be drawn from this survey are derived from the interpretation of small differences in low mutagenicity values that are produced by low concentrations of organic materials. Relatively few organic compounds were identified in MRR extracts by GC–MS and direct-introduction MS. More disturbing is the presence of supposed MRR contaminants in the extracts. These compounds may be attributed to either the resins or the water sample (acetophenone, alkylbenzenes, etc). Cytotoxicity (measurement of RNA synthesis inhibition on in vitro cultured human cells) and total organic halogens (TOX) versus mutagenicity were analyzed by factorial analysis of correspondence, and no correlation of mutagenicity with these parameters was revealed.

The Vigneux pilot plant survey demonstrates the complexity of mutagenicity behavior after ozonation–GAC treatment in natural waters at very low concentrations of organic materials.

Acknowledgments

The authors thank the Ministère de l'Environnement and the Agence Financière de Bassin Seine Normandie for their financial aid and contribution to this chapter.

Literature Cited

1. Junk, G. A.; Richard, J. J.; Grieser, M. D.; Witiak, D.; Witiak, J. L.; Arguello, M. D.; Vick, P.; Svec, H. J.; Fritz, J. S.; Calder, G. V. *J. Chromatogr.* **1974,** *99,* 745–762.
2. Burnham, A. K.; Calder, G. V.; Fritz, J. S.; Junk, G. A.; Svec, H. J.; Vick, R. *J. Am. Water Works Assoc.* **1973,** *65,* 721–725.
3. Stefan, S. F.; Smith, J. F. *Water Res.* **1977,** *11,* 339–342.
4. Jolley, R. L. *Environ. Sci. Technol.* **1981,** *15,* 874–880.
5. Reinhard, M.; Goudman, N. *Environ. Sci. Technol.* **1982,** *16,* 351–362.
6. Loper, J. C. *Mutat. Res.* **1980,** *76,* 241–268.
7. Van Rossum, P. G.; Willemse, J. M.; Hilner, C.; Alexander, L. *Water Sci. Technol.* **1982,** *14,* 163–173.
8. Suffet, I. H.; Brener, L.; Cairo, P. R. *Water Res.* **1980,** *14,* 853.
9. Bruchet, A.; Tsutsumi, Y.; Duguet, J. P.; Mallevialle, J. Presented before the Division of Environmental Chemistry, American Chemical Society, Philadelphia, August 1984.
10. Mallevialle, J.; Bruchet, A.; Schmitt, E. *Proc. Water Qual. Technol. Conf.* **1983,** *Dec.*
11. Duguet, J. P.; Ellul, A.; Brodard, E.; Mallevialle, J. Presented at the Congrès IOA Bruxelles, September 1983.
12. Meier, J. R.; Ling, R. D.; Bull, R.J. *Mutat. Res.* **1983,** *118,* 25–41.
13. Kool, H. J.; Van Kreis, C. F. *Water Res.* **1984,** *18(8),* 1011–1016.
14. Kool, H. J. *Organic Mutagens and Drinking Water in The Netherlands;* National Institute for Water Supply: The Netherlands, September 1983; Vol. 34.
15. Bruchet, A.; Cognet, L.; Mallevialle, J. Procedings du 31ème Symposium Européen sur l'Analyse des Micropolluants Organiques dans l'Eau, Oslo, September 1983.
16. Ames, B. N.; McCann, J.; Yamasaki, E. *Mutat. Res.* **1975,** *31,* 347–364.
17. Courtois, Y. Personal communication.
18. Lebart, L.; Morineau, A.; Fenelon, J. P. *Traitement des Données Statis-*

tiques. Méthodes et Programme; Ed. Dunod: Paris, 1976; Chapter 2, pp 109–183.
19. Bordet, J. P.; Kokosowski, A. *Analyse Multidimensionnelle et Typologie;* Ed. E.A.P.: Rouen, **1982;** p 447.
20. Hill, M. O. *Appl. Stat.* **1974,** *23(3),* 340.

RECEIVED for review August 14, 1985. ACCEPTED March 26, 1986.

Negative-Ion Chemical Ionization Mass Spectrometry and Ames Mutagenicity Tests of Granular Activated Carbon Treated Waste Water

R. B. Baird, J. P. Gute, C. A. Jacks, L. B. Neisess, M. H. Nellor, J. R. Smyth, and A. S. Walker

County Sanitation Districts of Los Angeles County, San Jose Creek Water Quality Laboratory, Whittier, CA 90601

Granular activated carbon (GAC) removed >70% of the TA98 mutagenicity from secondary effluent during a normal 6-week cycle. Negative-ion chemical ionization (NICI) gas chromatography–mass spectrometry (GC–MS) and silver nitrate derivatization showed that removal of labile organohalides paralleled the mutagen removal, consistent with a causal relationship. Chlorination did not consistently increase or decrease mutagenicity, although NICI showed that some silver-reactive organohalides were reduced and others were formed during chlorination. Silver reaction formation of additional compounds was also demonstrated by NICI; the number was less after either GAC or chlorination. Consistent with the results, it was hypothesized that there is a reservoir of halogenated high molecular weight material that can yield smaller electrophilics and that GAC is effective in removing much of this residue.

THE CHARACTERIZATION OF ORGANIC COMPOUNDS in mutagenic residues from water and waste water is the subject of many literature reports and ongoing studies. Only a very few have successfully identified discrete compounds that appear to be responsible for the mutagenicity (*1–4*). Instead, most studies have identified up to several hundred individual organics, few of which were known mutagens contributing to sample mutagenicity. In a ground water replenishment

0065–2393/87/0214/0641$06.00/0

study, we reported that the known mutagens and major unknowns characterized by gas chromatography/electron impact mass spectrometry (GC/EI-MS) contributed little, if any, to the mutagenicity in resin concentrates of organic residues from ground water, storm water, river water, and reclaimed waste water (*5*). However, the use of 4-nitrothiophenol (NTP) as a selective nucleophile for electrophiles such as organohalides and epoxides (*6*, *7*) and silver nitrate for labile organic halogens showed that chemical derivatization was an effective complement to GC-negative-ion chemical ionization MS (GC/NICI-MS) and Ames mutagen assay for tentatively classifying the chemical nature of some of the mutagens (*5*, *8*) and other reactive electrophiles. Although neither the structure nor sources of the compounds were identified, analyses of mutagenic residues by GC/NICI-MS and Ames assay before and after derivatization were consistent with the contributions of organohalides and epoxides to sample mutagenicity. Several of these compounds were detected in more than one sample type (*5*).

Because granular activated carbon (GAC) effectively removes mutagenic activity from drinking water (*9*) and because these mutagens have yet to be identified, it was desirable to evaluate GAC for removing suspected mutagens from reclaimed water before its use in ground water replenishment. It was also desirable to examine chlorine disinfection as a possible source of these materials because chlorination has been implicated as a cause of toxic organics. This chapter presents data on the removal of mutagens by GAC in a full-scale municipal waste water treatment facility, the effects of chlorination on mutagen formation, and the characterization by chemical derivatization and GC/NICI-MS of reactive electrophiles in these samples.

Experimental

Treatment Plant Operations. The Pomona Water Reclamation Plant is a conventional activated sludge facility treating 10 million gal per day of combined domestic and industrial waste water. Four GAC filters are operated on a staggered 6-week cycle: three units are operated while one is regenerated in a Nichols multiple-hearth furnace. Secondary effluent is split between the three operating filters and gravity fed at a rate providing an empty-bed contact time of 10 min. Each filter is 32 ft long × 16 ft wide × 22 ft high and filled with 80,000 lb of carbon to a depth of 6 feet. During normal operation, the units are periodically backwashed with chlorinated GAC effluent (tertiary effluent) to reduce clogging. For this study, the filters were backwashed with unchlorinated effluent to avoid chlorine contact with organics adsorbed on the carbon.

GAC-filtered effluent containing 10–12 mg/L of NH_3N is routinely disinfected with an average dose of 7 mg/L of chlorine to achieve a typical residual of 3–4 mg/L of chloramine following a 2-h contact time. Effluent turbidity is routinely ≤2 turbidity units and color is ≤10 color units; GAC is normally regenerated at 6-week intervals to maintain this effluent quality.

Sampling and Residue Preparation. Twenty-liter samples were concentrated on a high-performance resin concentrator (*10*). This system consisted of a Teflon diaphragm pump (Milton–Roy), four microparticulate macroporous resin columns (BioRad anionic MP-1 and cationic MP-50, and Rohm and Haas non-ionic XAD-2 and XAD-7), and a UV detector to monitor column effluent (Figure 1). Residue organics were eluted with NaCl solutions and CH_3CN as previously described (*10*); salt solutions were extracted with CH_3CN, and CH_3CN fractions for a given sample were combined and volume reduced by Kuderna-Danish evaporation to achieve a 10^4 concentration factor. Column blanks were obtained by repeating this solvent elution cycle with fresh solvents between samples. Each sample concentrate was fractionated by high-performance liquid chromatography (HPLC) (*5, 11*) by using H_2O/CH_3CN gradients and an analytical reverse-phase column (Whatman ODS-2). The least polar fractions from each sample were isolated at −20 °C until chemical assays and bioassays were performed.

The first set of 14 matched pairs of 2-h composites of GAC influent and effluent was concentrated on the first day of a 6-week GAC contactor operation cycle following regeneration, and the remainder of the sets were sampled throughout the rest of that cycle. No samples were collected during backwash operations. In a subsequent 6-week period, 10 matched pairs of GAC effluent composites were sampled before and after chlorination and were concentrated on the resin system. Chlorinated samples were stoichiometrically dechlorinated with ferrous ammonium sulfate prior to concentration.

Mutagenicity Testing. Residue fractions in CH_3CN were diluted with dimethyl sulfoxide (DMSO) to known volumes and assayed for mutagenicity on *Salmonella typhimurium* strains TA98 and TA100 by using the dose–response plate incorporation procedures recommended by Ames (*12*). Microsomal enzyme (S9 fraction) was not used in these experiments because previous work consistently demonstrated direct-acting mutagenesis (*5*). Net revertants per liter equivalent of sample assayed (rev/L) were calculated from response on duplicate plates of 3–4 doses per assay. A positive (mutagenic) scoring required a net reversion rate of at least twice the spontaneous rate and an observable dose–response. The dose range was 0.075–0.6 L equivalents per plate, based on the original sample volume, concentration factor, and volume of residue applied to the plates.

Derivatization. Residue portions equal to 2% of each total sample volume were treated with equal volumes of 0.1 M $AgNO_3$ in CH_3CN. These mixtures were heated 10 min on a steam bath and cooled. Any precipitates were removed by filtration on Teflon filters; excess silver ion was removed by reverse-phase HPLC.

GC/NICI–MS Analysis. Residue fractions were analyzed by methane GC/NICI–MS before and after silver ion derivatization. A fused-silica SE54 capillary GC column (J & W Scientific, 30 m × 0.25 mm) was used for separations. The injector temperature was 270 °C, and the oven was held at 40 °C for 2 min, ramped ballistically to 60 °C, and programmed for 60–270 °C at 4 °C/min. The final hold was at 270 °C. Helium carrier gas was used at a flow rate of 22 cm/s. A Finnigan 4023-INCOS GC–MS–DS equipped with a PPINICI source was used with the following conditions: emission current, 30 mA; electron multiplier voltage, 1000 V; electron energy, 70 V; manifold, 80 °C; source, 200 °C and 0.45 torr

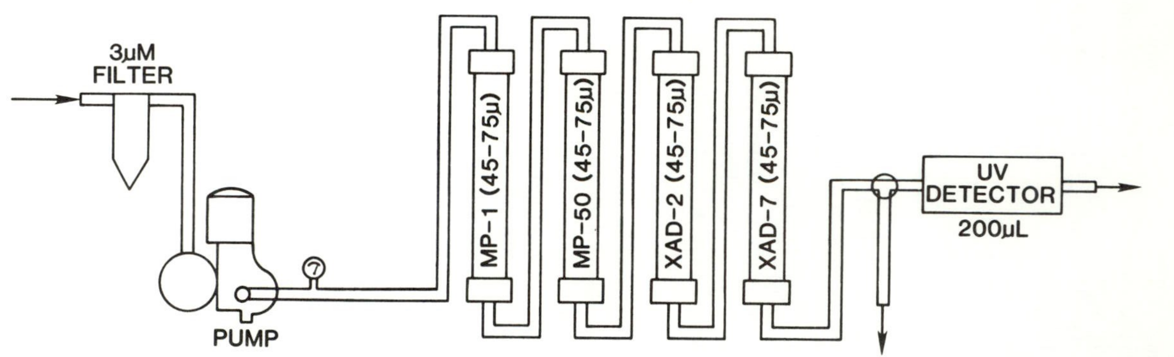

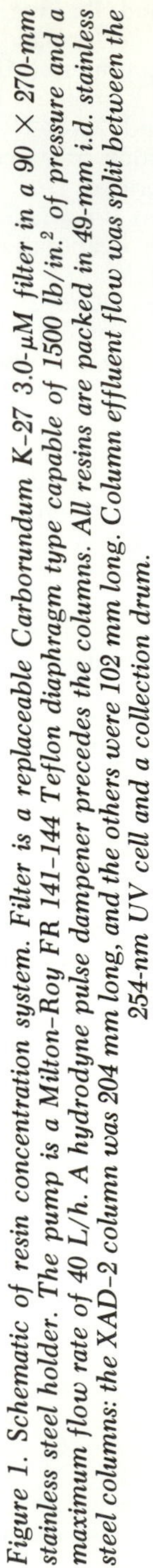

Figure 1. Schematic of resin concentration system. Filter is a replaceable Carborundum K-27 3.0-μM filter in a 90 × 270-mm stainless steel holder. The pump is a Milton-Roy FR 141-144 Teflon diaphragm type capable of 1500 lb/in.^{2} of pressure and a maximum flow rate of 40 L/h. A hydrodyne pulse dampener precedes the columns. All resins are packed in 49-mm i.d. stainless steel columns: the XAD-2 column was 204 mm long, and the others were 102 mm long. Column effluent flow was split between the 254-nm UV cell and a collection drum.

CH_4; and mass range, 33–500 at one scan/s. Decachlorobiphenyl (10 ng/μL) was added as an internal standard prior to derivatization and *p*-chlorobenzophenone was added (20 ng/μL) just prior to injection on the GC–MS to correct for any changes during sample handling and to normalize drift in MS sensitivity. These steps permitted the determination of changes in the concentrations of detected components.

GC–MS runs were stored as files by the data system on discs; FORTRAN routines were written to compare selected parameters in file sets and to reduce the data to summary tables for hard copy output. These routines facilitated the determination of peak areas of components in extracted ion current profiles (EICP) for both total and selected ion chromatograms, calculated the removal of components of interest (e.g., those containing halogen isotopes) by treatment processes (GAC, Cl_2) or derivatization, summarized the occurrence of new components of interest in treatment or derivatization, and calculated the percent of the total ion current represented by a given component. The programs allowed operator discrimination between major and minor components in a file set by preselection of an ion current threshhold for data reduction. For data summarized herein, components were ≥4000 ion counts, which corresponds to a level ≥5% of the internal standard (decachlorobiphenyl) response.

Results and Discussion

Mutagenicity Patterns. Mutagenicity data for the GAC filtration and chlorination unit processes are summarized in Tables I and II, respectively. Removal of mutagenicity by GAC ranged from 71% to 92% for TA98 (avg = 83%) and from 62% to 94% for TA100 (avg = 79%) over the 6-week operational cycle. No temporal trend in mutagen removal

Table I. Summary of GAC Mutagenicity Removal

	GAC Influent (rev/L)		*GAC Effluent (rev/L)*		*Percent Removal*	
Sample	*TA98*	*TA100*	*TA98*	*TA100*	*TA98*	*TA100*
09/27	7970	12100	1890	3020	76	75
09/29	9560	13900	980	860	90	94
10/01	9750	9540	920	1320	91	86
10/05	12000	16000	1770	3350	85	79
10/07	17900	23200	1410	4050	92	83
10/12	7270	13800	1380	2860	81	79
10/18	11700	11400	1590	2840	86	75
10/20	7350	10500	2150	2330	71	78
10/22	8700	5030	740	1130	92	78
10/26	T	T	100	T	—	—
10/28	6610	6010	1020	1160	85	81
11/01	10300	13300	1570	2460	85	82
11/03	7250	7700	2020	2970	72	62
11/05	6120	6340	1170	1780	81	72

NOTE: T denotes toxic.

Table II. Summary of Mutagenicity Results from Chlorination of GAC Effluents

	GAC Influent (rev/L)		*Cl Effluent (rev/L)*		*Percent Change*	
Sample	*TA98*	*TA100*	*TA98*	*TA100*	*TA98*	*TA100*
11/19	1400	2350	1020	2240	−27	−4
11/29	2830	2540	2390	4890	−16	+92
12/01	1390	3540	1250	3390	−11	−4
12/08	2180	3620	980	3660	−55	0
12/13	2740	2330	2592	2670	+4	+4
12/15	420	830	924	1640	+120	+97
12/21	1900	1990	1100	1650	−42	−18
12/27	1490	1290	1500	2170	+1	+68
01/03	670	990	920	1960	+37	+97
01/05	<100	<280	280	800	—	—

was observed, even though historical parameters show that these filters are operationally exhausted and ready for regeneration in this time period. The apparent ability of GAC to continue selectively removing mutagens beyond a practical operational life is consistent with literature reports for drinking water (*9, 13*) and further supports the use of GAC as an effective means of removing unknown organics of possible health significance.

Evaluation of chlorine disinfection (as chloramine residual) as a possible source of mutagenicity in GAC effluents showed that TA98 activity was reduced an average of 30% in 5 of 10 sample pairs, not significantly affected in 2 sample pairs, and increased an average of 70% in 3 sample pairs. For TA100, mutagenicity was reduced during chlorination an average of 17% in 2 pairs, not significantly changed in 3 pairs, and increased an average of more than 90% in 5 pairs. In four of the five experiments showing an increase in TA98 and/or TA100 mutagenicity after chlorination, the mutagenicity prior to chlorination was 50% less than that in matched pairs that showed no increase from chlorination. It may be speculated that chlorine oxidizes (deactivates) some types of mutagens but effectively reacts with available organic matter (precursor) to create more mutagens in a given sample. An alternate hypothesis is that GAC does not consistently remove precursor. Either suggestion is consistent with the observed patterns just described and the literature reports that chlorination increases the mutagenicity of treated waste water, often even after GAC treatment (*4, 5, 14, 15*).

GC/NICI–MS Characterization of GAC Treatment. Six of the GAC influent-effluent matched pairs were analyzed by GC/NICI–MS before and after treatment with silver nitrate. Typical EICPS for GAC influent and effluent before silver derivatization are shown in Figure 2.

Although more than 100 components were detected by NICI-MS in the influents, GAC was effective in removing an average of 40% of these electrophilic components. Figures 3-5 show ion chromatograms demonstrating GAC reduction of compounds of m/z = 35, 37, 79, 81, and 127. Much of the halogen appeared to be labile, as evidenced by reaction with silver nitrate (*16*). Although isotopic ratios for Cl and Br were correct and can be used as additional support for the presence of these functional groups, use of m/z = 127 for I is not as reliable. Figure 5 indicates some anomalous behavior for m/z = 127 components through GAC. Only the peaks labeled with an asterisk were Ag labile and are presumed to contain I. Figures 6 and 7 show the effects of silver on GAC influent and effluent EICP. Figure 8 compares the effects on m/z = 35 and 37, and Table III summarizes categories of major halogen-containing peaks for all sample pairs analyzed. Of the six sample pairs evaluated, up to eight major silver-reactive organohalides per sample were completely removed by GAC, and others were significantly reduced (column 1, Table III). Many of the minor labile halogenated compounds were also removed or reduced by GAC, and

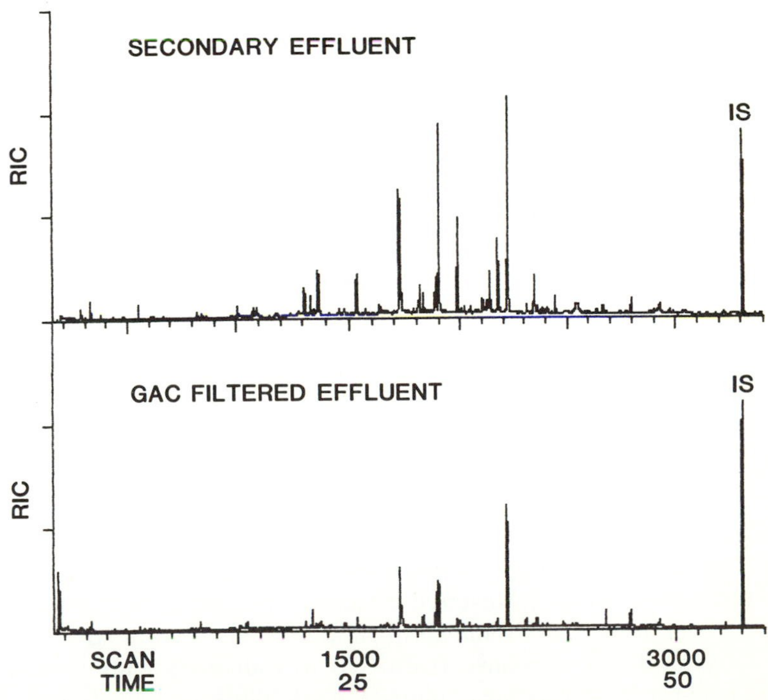

Figure 2. GC/NICI-MS ion current profiles of secondary effluent and GAC-filtered effluent. IS = internal standard decachlorobiphenyl.

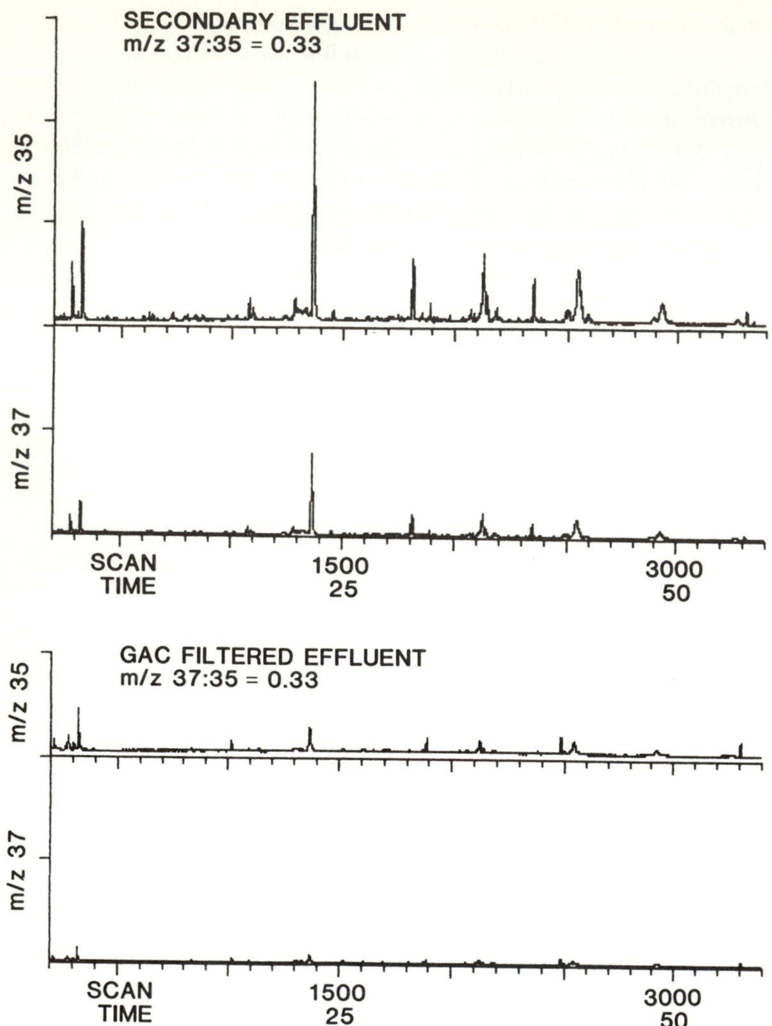

Figure 3. GC/NICI-MS ion current profiles for chlorine isotopes in GC secondary effluent and GAC-filtered effluent. Ion current ratios for ^{35}Cl and ^{37}Cl are indicated for each sample.

a few of these compounds were common to more than one influent. The GAC removal of Ag-reactive organohalides measured by NICI-MS was parallel to the GAC removal of mutagens just described.

GAC effluents contained from two to four early eluting (<10 min on the 30-m SE54 capillary column) Ag-labile components that were not detected in the respective influents. This result suggested either sloughing from or reactions on the carbon. Four major Ag-inert organic halogen components were completely removed by GAC, and several minor inert compounds were reduced by the filters throughout the cycle.

Silver nitrate derivatization revealed the lability of these residues in that many new compounds were created and others were sig-

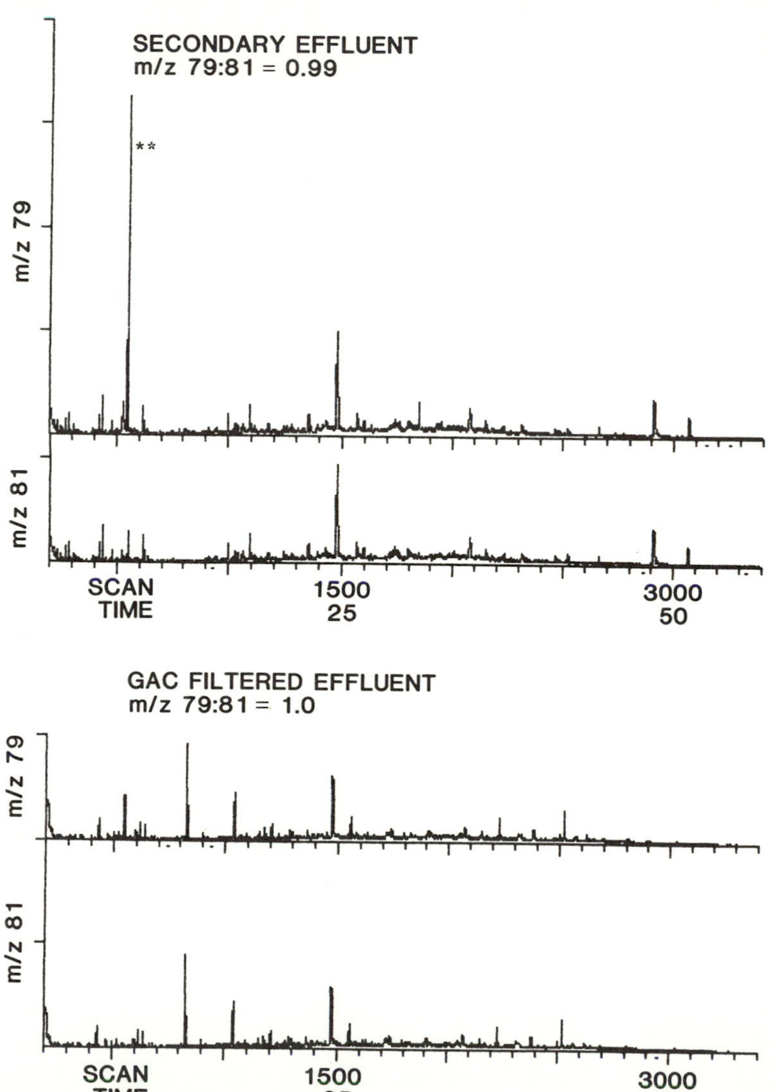

Figure 4. GC/NICI-MS ion current profiles for ^{79}*Br and* ^{81}*Br in secondary effluent and GC-filtered effluent. Ion current ratios were as shown. Note peak labeled ** was not brominated (no* m/z = *81).*

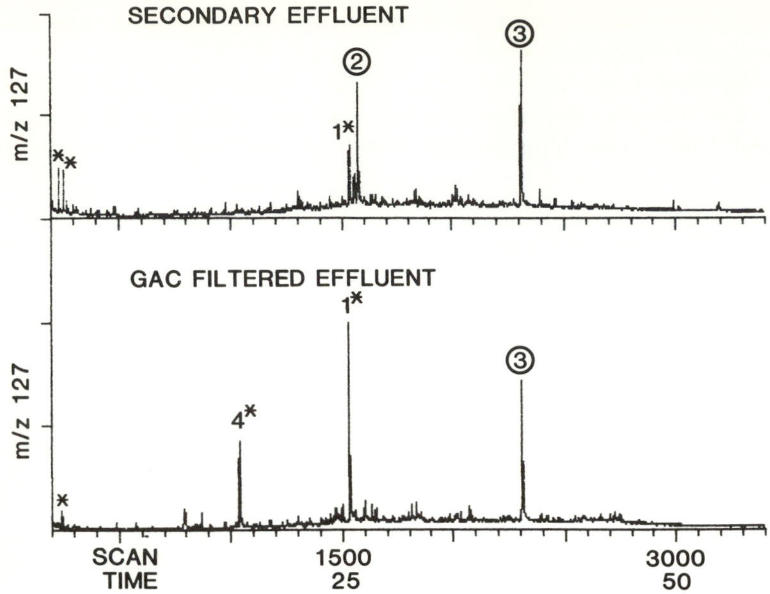

*Figure 5. GC/NICI-MS ion current profiles for ^{127}I in GAC secondary effluent and GAC-filtered effluent. Note peaks 1 and 4 increased after carbon contact. Both were silver labile (labeled *), whereas peaks 2 and 3 were not. The latter two components may not contain I but rather another ion of* m/z = *127. Note early eluting (labeled *) components.*

nificantly increased after silver treatment. These effects are shown clearly in the chromatograms of Figures 6–8. The last two columns of Table III summarize the major components increased by silver but removed by GAC and the number of new components found in influent–effluent pairs, respectively. Except for sample pair B in Table III, the number of new peaks was less after GAC treatment, but the magnitude of individual new components in GAC effluent was not significantly different from those formed by $AgNO_3$ in the influent. Eight of the major new components were common to several of the samples, although sometimes they could be found in the influent for one pair and the effluent for another pair. These results from the $AgNO_3$ experiments lead to an interesting speculation that a reservoir of higher molecular weight organics containing labile halogen exists, and that reaction with silver allows the formation of smaller halogenated organics that are relatively inert to further reaction with silver and are thermally stable enough to be analyzed by GC–MS. Most of the new organohalides formed eluted in the early part of the GC–MS runs. It is not possible to tell from these experiments whether there might be mutagenic byproducts of the hypothesized reactions with $AgNO_3$, but we have previously reported (*5*) an increase in mutagenicity in some ground water and storm water residues after $AgNO_3$ treatment.

The idea that a significant amount of organic halogen is associated with high molecular weight organics is not new. Glaze and Peyton (*17*), McCahill et al. (*18*), and Reinhard (*19*), among others, have described distributions of total organic halogen (TOX) in high molecular weight fractions of water and waste water residues, and Glaze et al. (*20*) have described the uniform molecular weight distribution of TOX formation potential by chlorine disinfection in residues isolated by XAD-2 and activated carbon. It is doubtful that high molecular weight halogenated organic matter is of any direct toxicological significance. However, the fact that a portion of it is reactive with silver nitrate indicates that it is probably labile in other chemical reactions (e.g., hydrolysis, or oxidation by chlorine, ClO_2, ozone, etc.) that could yield smaller halocarbons or other potentially toxic electrophiles.

GC/NICI–MS Characterization of Chlorinated GAC Effluents. Four matched pairs of samples were monitored through the chlori-

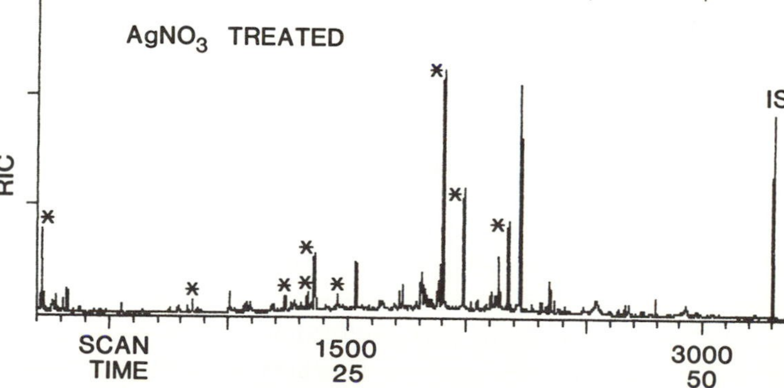

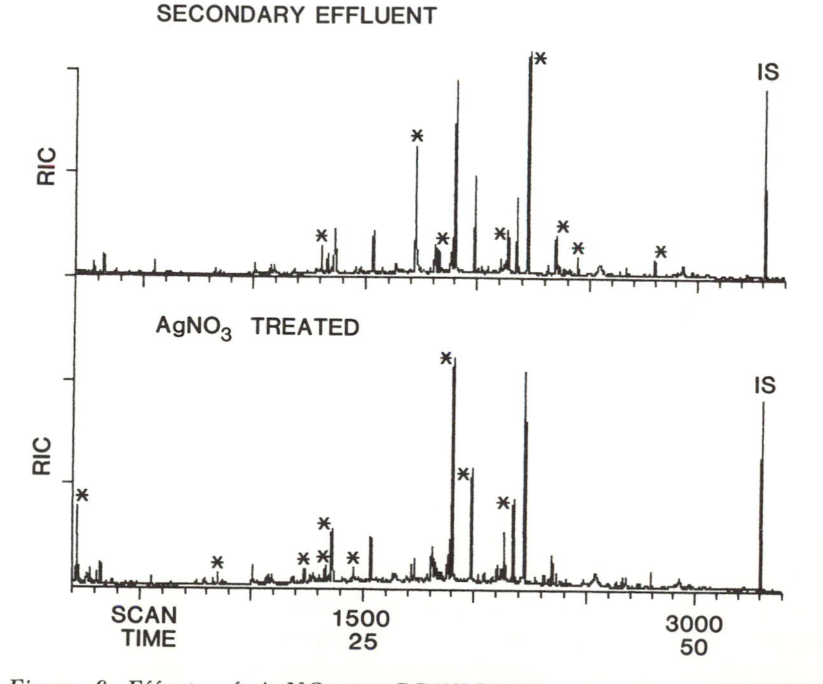

*Figure 6. Effects of $AgNO_3$ on GC/NICI-MS ion current profile of secondary effluent. IS = internal standard. Peaks labeled * in secondary effluent (GAC influent) were reduced or removed by $AgNO_3$. Labeled peaks in bottom trace ($AgNO_3$ treated) were increased or created by $AgNO_3$ treatment.*

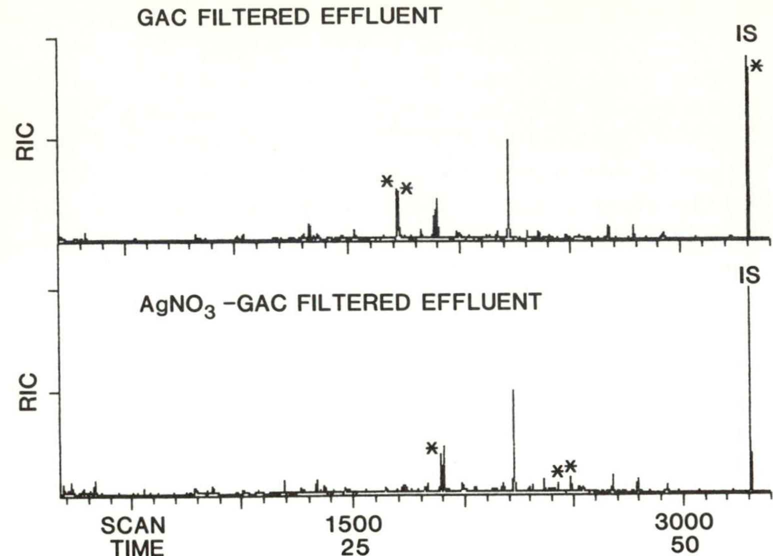

*Figure 7. Effects of $AgNO_3$ treatment on ion current profile of GAC filter effluent. IS = internal standard. Peaks labeled * in untreated sample were removed or reduced by $AgNO_3$. Peaks labeled * in $AgNO_3$-treated sample were created or increased by the treatment.*

nation process by GC/NICI-MS. Figure 9 shows EICPs that demonstrate the effects of chlorination on Ag-labile chlorinated and brominated organics, and Table IV summarizes the categories of the major halogenated compounds detected. Up to nine labile components were removed by chlorination, and several others were reduced (column 1, Table IV). However, as many as nine major halogenated organics per sample were detected only after chlorination (column 2, Table IV), and all of these were Ag labile. Most of these eluted early in the chromatograms. Only one of these sample pairs (C, Table IV) showed an increase in mutagenicity (both TA98 and TA100) after chlorination. However, because Cl_2 is clearly able to destroy some labile organohalides (columns 1 and 3, Table IV) and create others, it is not possible to sort out which, if any, of the peaks detected by GC/NICI-MS may contribute to mutagenicity. If Ag-labile compounds are contributing to sample mutagenicity, then the destruction of some organohalides and the production of others by chlorine may be a reasonable explanation of the results. Careful HPLC fractionation and enrichment techniques followed by mutagen assays and high-resolution GC-MS should be able to clarify this situation.

As in the GAC experiments just described, many new halocarbons were formed by reaction with Ag (column 4, Table IV). Chlorination

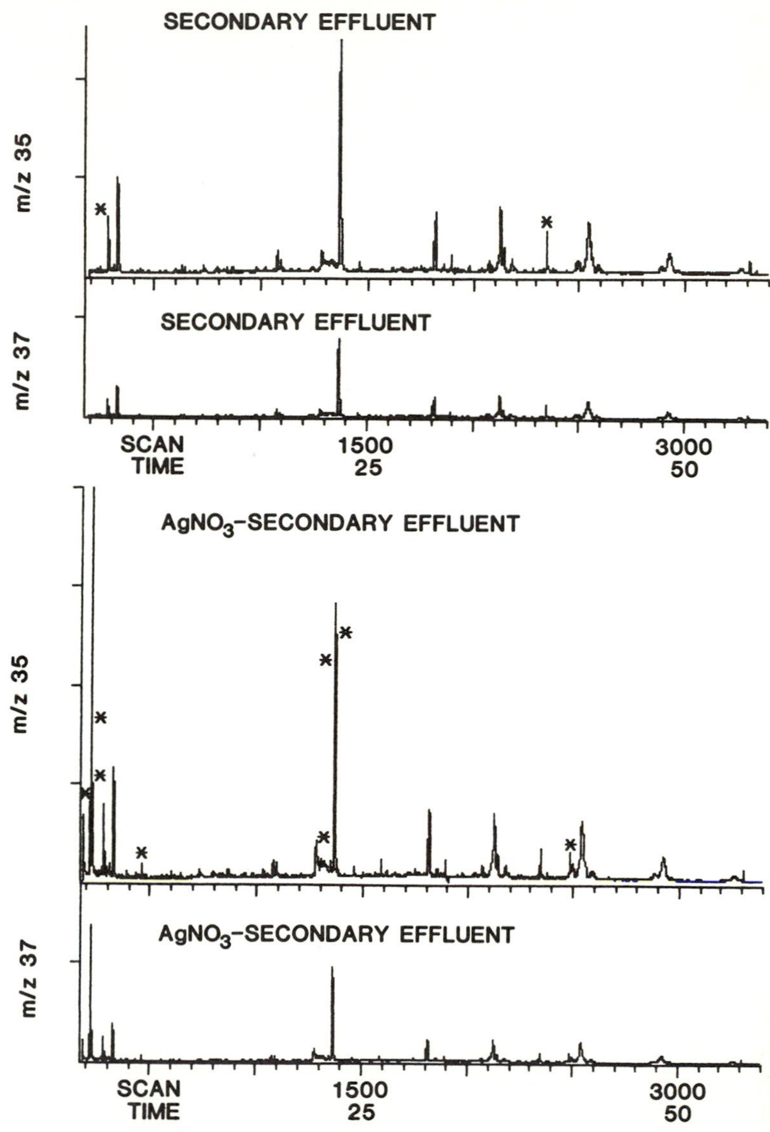

*Figure 8. Effects of $AgNO_3$ on GC–NICI ion current profiles of ^{35}Cl and ^{37}Cl in GAC filter influent (secondary effluent) and effluent (GAC effluent). Peaks labeled * in untreated effluents were reduced by $AgNO_3$. Peaks labeled * in $AgNO_3$-treated effluents were increased by Ag.*
(Continued on next page.)

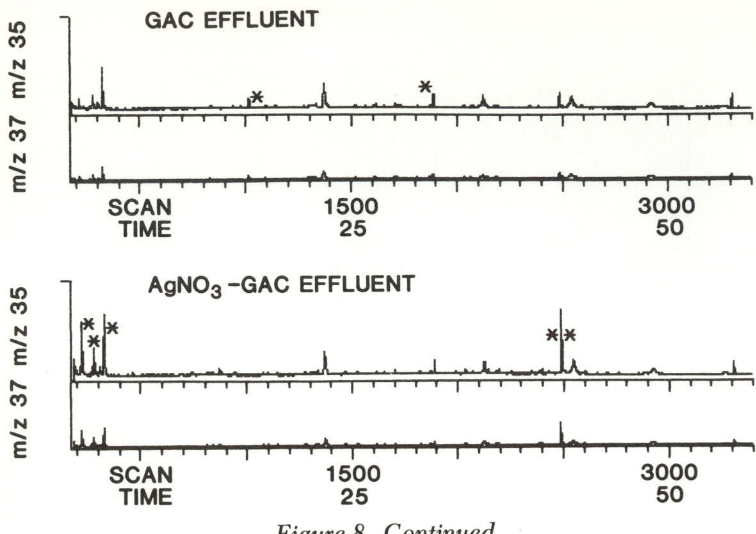

Figure 8. Continued.

Table III. Categories of Major NICI–MS Peaks in GAC Matched-Pair Experiments

Sample Pair	*Reaction Category and Number of Halogenated Peaks*		
	**GAC and *Ag Decrease*[a]	*Ag Increase, GAC Decrease*[b]	*New Peaks *Ag-Inf/*Ag-Eff*[c]
A-09/27	2	2	17/5
B-10/07	4	1	2/15
C-10/18	4	8	8/4
D-10/26	11	4	8/0
E-10/28	8	8	9/3
F-11/05	5	1	9/8

NOTE: Major peaks are ≥4000 ion counts. *GAC means removable by GAC, and *Ag means silver reactive.

[a] Peaks were reduced or removed by GAC and also were decreased by $AgNO_3$ reaction.

[b] Components were decreased by GAC filtration, but sample treatment with $AgNO_3$ caused an increase in the concentration of these.

[c] New peaks were formed by $AgNO_3$ treatment of GAC influent-effluent. Inf denotes influent, and Eff denotes effluent.

did appear to alter the formation potential of new halocarbons by $AgNO_3$ because in two cases (A and B, Table IV), the number of new compounds induced by $AgNO_3$ treatment was smaller after chlorination; in the other two cases, the major $AgNO_3$ reaction products were different after chlorination.

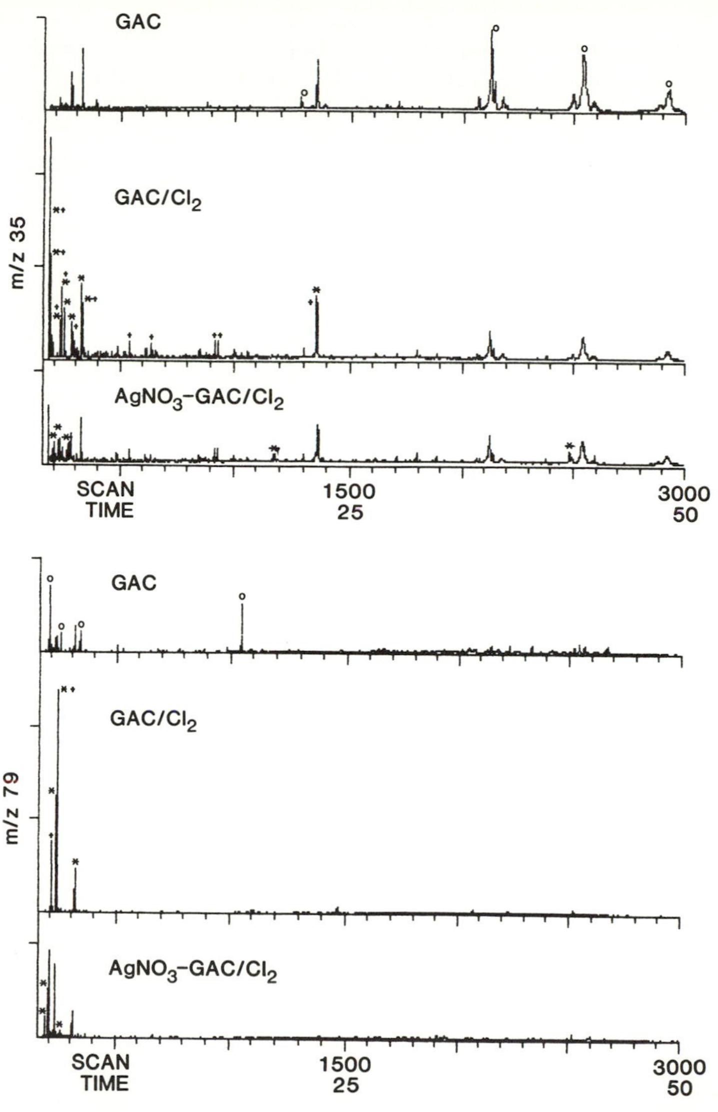

Figure 9. Effects of chlorination and $AgNO_3$ treatment on GAC filter effluent GC/NICI-MS ion current profiles for m/z = 35 *and 79. Peaks labeled* o *in GAC traces were halogenated components decreased or destroyed by chlorination. Peaks labeled + in GAC-Cl_2 traces were produced by chlorination. Peaks labeled * in GAC-Cl_2 traces were reduced by $AgNO_3$. Peaks labeled * in $AgNO_3$ traces were produced by $AgNO_3$.*

Table IV. Categories of Major NICI-MS Peaks in GAC-Cl Experiments

Sample Pair	*Reaction Category and Number of Halogenated Peaks*			
	Ag and Cl_2 Decrease[a]	*New After Cl_2*[b]	*Ag Increase, Cl_2 Decrease*[c]	*New Peaks •Ag-Inf/•Ag-Cl_2*[d]
A	11	7	4	24/6
B	4	2	3	7/5
C	3	3	0	4/15
D	3	9	0	10/10

NOTE: Major peaks are ≥4000 ion counts. •Ag means silver reactive.
[a] Mutagenic on TA98 and TA100 after chlorination.
[b] Chlorination decreased these peaks, as did $AgNO_3$ treatment.
[c] Chlorination decreased these peaks, but $AgNO_3$ treatment increased their concentration.
[d] New peaks were formed by silver treatment of GAC effluent-Cl_2 effluent. Inf denotes influent.

Relationship Between NICI-MS Electrophiles and Mutagenicity. The results of this study and those reported elsewhere (*5, 6, 8*) are consistent with labile organics containing epoxide and halide functional groups contributing to the mutagenicity observed in a wide variety of samples; however, these results do not prove this relationship. That mutagens are electrophiles is generally accepted, and the ability of selective nucleophiles to react with specific classes of mutagenic electrophiles has been demonstrated adequately (*5, 6–8, 15*); silver nitrate is known to react with electrophiles containing labile halogen (I > Br > Cl) by forming insoluble silver halide (*16*). The parallels between loss of specific electrophiles detected by NICI-MS and loss of mutagenicity after GAC filtration or chemical derivatization (reported here and elsewhere) (*5, 8*) are plausible yet only circumstantial evidence for the mutagenicity of these components. The NICI-MS detection of NTP adducts and loss of mutagenicity after derivatization with NTP (*5–7*) does give additional support to this premise but does not eliminate the possibility that silver ion may be participating in reactions with certain mutagens without halide involvement.

Because the estimated concentration of individual halogenated compounds detected in these experiments is in the nanograms-per-liter range or less, one or more of the hundred or so present in a given sample would have to be quite potent in the Ames test to have caused the level of mutagenicity in these sample residues. Mutagenic response to doses ranging from one to several hundred nanograms per assay plate would eliminate most known Ames mutagens from consideration (*21*). However, of the new mutagens that have been identified in water residues (*1, 3, 4, 22*), all were polyhalogenated hydrocarbons (some also were oxygenated) that were potent in the Ames test in this dose range [10^5 revertants/nmol (*3*) to 10^4 revertants/mmol (*1*)]. Furthermore, several studies of similar compound groups (halohydrins, polyhalogenated two- and three-carbon aliphatics, and chlorinated epoxides) demonstrate that many of these are potent Ames mutagens that are active in this dose range (*23–26*). Even though these classes of compounds are known to occur in environmental samples as a result of chlorination or industrial and microbial activity (*5, 21, 26–28*), proof of mutagenicity and possible health effects of the unknowns are incumbent upon structural identification of some of the actual compounds present.

Conclusions

GAC filtration (10-min empty-bed contact time) of municipal secondary effluents can remove approximately 80% of the Ames-active mutagenicity. Consistent with literature reports for drinking water, GAC appears to selectively remove these electrophiles throughout and possibly beyond a normal operational cycle in a waste water treatment plant. There may be a reservoir of labile or reactive halogen-containing high molecular weight molecules that could be a source for production of lower molecular weight toxic molecules. This high molecular weight material also appears to be adsorbed by GAC but may periodically break through the filters. Because this material may be a source of mutagens, especially during chlorination, it may be desirable to monitor GAC filter performances by TOX measurements (*29*).

Consistent with literature reports, chlorination is capable of producing new mutagens; it also appears to destroy other mutagens, even in the presence of ammonia. The extent to which chlorination increases the production of new halogenated compounds is known to depend upon the amount of organic precursors in the water; similarly, the effect on mutagenicity is probably related to the amount and type of precursors available. Hence effective GAC treatment prior to chlorination should minimize the increases in mutagenicity.

On the basis of the results reported here and elsewhere (*5, 8*), much of the mutagenicity detected in residues isolated by a resin concentrator appears to be associated with a plethora of very low level (ca. ng/L) labile halogenated organics (containing Cl, Br, and/or I) and/or epoxides. Although GAC filtration appears to be able to remove mutagenic mixtures containing these components and their precursors, it is still desirable to continue efforts to identify individual mutagens because they may be potent and have human health significance.

Literature Cited

1. Tabor, M. W. *Environ. Sci. Technol.* **1983,** *17,* 324.
2. Loper, J. C.; Tabor, M. W.; Miles, S. K. In *Water Chlorination: Environmental Impact and Health Effects;* Ann Arbor Science: Ann Arbor, MI, 1983; Vol. 4, Chapter 86.

3. Holmbum, B. R.; Voss, R. H.; Mortimer, R. D.; Wong, A. *Tappi* **1981,** *64,* 172.
4. Kringstad, K. P.; Ljungquist, P. O.; de Sousa, F.; Stromberg, L. M. *Environ. Sci. Technol.* **1981,** *15,* 562.
5. *Health Effects Study;* L.A. County Sanitation Districts: Los Angeles, 1984; Final Report; NTIS No. PB84191568.
6. Cheh, A. M.; Carlson, R. E. *Anal. Chem.* **1981,** *53,* 1001.
7. Agarwal, S. C.; Van Duuren, B. L.; Solomon, J. J.; Kline, S. A. *Environ. Sci. Technol.* **1980,** *14,* 1249.
8. Jacks, C. A.; Gute, J. P.; Neisess, L. B.; Van Sluis, R. J.; Baird, R. B. In *Water Chlorination: Environmental Impact and Health Effects;* Ann Arbor Science: Ann Arbor, MI, 1983; Vol. 4, Chapter 89.
9. Monarca, S.; Meier, J. R.; Bull, R. J. *Water Res.* **1983,** *17,* 1015.
10. Baird, R. B.; Jacks, C. A.; Jenkins, R. L.; Gute, J. P.; Neisess, L.; Scheybeler, B. In *Chemistry in Water Reuse;* Ann Arbor Science: Ann Arbor, MI, 1981; Vol. 2, Chapter 8.
11. Baird, R. B.; Gute, J.; Jacks, C.; Neisess, L.; Scheybeler, B.; Van Sluis, R.; Yanko, W. In *Water Chlorination: Environmental Impact and Health Effects;* Ann Arbor Science: Ann Arbor, MI, 1980; Vol. 3, Chapter 80.
12. Ames, B. N.; McCann, J.; Yamasaki, E. *Mutat. Res.* **1974,** *31,* 347.
13. Loper, J. C.; Tabor, M. W. *Environ. Sci. Technol.* in press.
14. Cheh, A. M.; Skochdopole, J.; Koski, P.; Cole, L. *Science* **1980,** *207,* 90.
15. *Advanced Treatment for Wastewater Reclamation at Water Factory 21;* Stanford University: Palo Alto, CA, 1982; Technical Report 267.
16. March, J. *Advanced Organic Chemistry: Reactions, Mechanisms, and Structure;* McGraw Hill: New York, 1968.
17. Glaze, W. H.; Peyton, G. R. In *Water Chlorination: Environmental Impact and Health Effects:* Ann Arbor Science: Ann Arbor, MI, 1978; Vol. 2, p 3.
18. McCahill, M. P.; Conroy, L. E.; Maier, W. J. *Environ. Sci. Technol.* **1980,** *14,* 102.
19. Reinhard, M. *Environ. Sci. Technol.* **1984,** *18,* 410.
20. Glaze, W. H.; Saleh, F. Y.; Kinsteley, W. In *Water Chlorination: Environmental Impact and Health Effects;* Ann Arbor Science: Ann Arbor, MI, 1980; Vol. 3, p 99.
21. Loper, J. C. *Mutat. Res. Rev. Genet. Toxicol.* **1980,** *76,* 241.
22. Kringstad, K. P.; Ljungquist, P. O.; de Sousa, F.; Stromberg, L. M. *Environ. Sci. Technol.* **1983,** *17,* 468.
23. Rosenkranz, S.; Carr, H. S.; Rosenkranz, H. S. *Mutat. Res.* **1974,** *26,* 367.
24. Stolzenberg, S. J.; Hine, C. H. *Environ. Mutagen.* **1980,** *2,* 59.
25. Voogd, C. E.; Van der Stel, J. J.; Jacobs, J. J. J. A. A. *Mutat. Res.* **1981** *89,* 269.
26. Ehrenberg, L.; Hussain, S. *Mutat. Res.* **1981,** *86,* 1.
27. Van Dyke, R. A. *Environ. Health Perspect.* **1977,** *21,* 121.
28. Christman, R. F.; Norwood, D. L.; Millington, D. S.; Johnson, J. D.; Stevens, A. A. *Environ. Sci. Technol.* **1983,** *17,* 625.
29. Quinn, J. E.; Snoeyink, V. L. *J. Am. Water Works Assoc.* **1980,** *72,* 483.

RECEIVED for review August 14, 1985. ACCEPTED December 18, 1985.

32

Techniques for the Fractionation and Identification of Mutagens Produced by Water Treatment Chlorination

Helene Horth, Brian Crathorne, Ray D. Gwilliam, Carole P. Palmer, Jennifer A. Stanley, and Michael J. Thomas

Water Research Centre, Environment, Medmenham Laboratory, Henley Road, Marlow, Bucks, England

Byproducts of the chlorination of treated water, humic acids, and amino acids were shown to be mutagenic to Salmonella typhimurium *strain TA100 in the fluctuation test. A two-stage fractionation procedure, using high-performance liquid chromatography (HPLC) combined with the fluctuation test, was developed. Several mutagenic fractions were obtained from an XAD extract of chlorinated water; this result indicated the presence of several mutagens. The mutagenic activities of extracts of different chlorinated amino acids were compared. The products identified by capillary gas chromatography–mass spectrometry did not account for the mutagenic activity. However, additional products were detected by HPLC with UV detection in extracts of chlorinated tyrosine and phenylalanine.*

CONCENTRATED EXTRACTS OF MANY DRINKING WATERS in the United Kingdom and elsewhere are mutagenic as observed in bacterial assays (*1–3*). Compounds that are widely distributed in the aquatic environment, such as humic substances and amino acids, have been found to react with chlorine to produce mutagenic compounds (*4, 5*). These compounds may account for at least some of the mutagenic activity of extracts of drinking waters.

Because a good correlation is found between mutagenicity (as detected in bacterial assays such as the Ames test) and carcinogenicity (as observed in laboratory animal tests) (*6*), mutagenicity detected in water extracts indicates the possibility of a carcinogenic hazard.

0065–2393/87/0214/0659$06.00/0

However, the risk posed to humans by long-term consumption of low levels of mutagenic compounds in drinking water requires further assessment. This assessment may be achieved by carrying out further biological tests on the extracts, for example, by using mammalian cells and animal tests or by identifying mutagenic compounds in the extracts. Once the chemical identity is known, the significance to health may be evaluated by carrying out further bioassays on the pure compounds and by examination of toxicity data.

A large number of volatile organic compounds have been identified in extracts of drinking waters by gas chromatography–mass spectrometry (GC–MS) (*7*). Although the compounds identified include known mutagens (*8*), these do not appear to be present at sufficiently high concentrations to account for the observed mutagenic activity of the extracts. Because the organic compounds identified in water by GC–MS represent at most 20% of the total organic content, it appears that the activity is due to the remainder, that is, the nonvolatile fraction. An approach to the identification of nonvolatile organic compounds in water involving the use of high-performance liquid chromatography (HPLC) and soft-ionization mass spectrometry has recently been described (*9*).

Two approaches to the identification of mutagenic compounds in drinking water are described in this chapter. The first involves HPLC fractionation of extracts of treated water combined with mutagenicity testing of the fractions. Fractions of a mutagenic extract of chlorinated water were compared with nonmutagenic fractions of the same water sampled before final chlorination to focus on differences between the two, especially substances present only in mutagenic extracts and fractions.

The precursors of mutagenicity produced by chlorination seem to be widespread, naturally occurring substances. If the precursors can be identified, identification of their chlorination products may help to elucidate the nature of the mutagenic activity in treated water. Thus, a complementary approach to that just described involves laboratory chlorination of naturally occurring model compounds at concentrations and conditions that simulate treatment chlorination. The reaction products were first tested for mutagenic activity to identify possible precursors of mutagenicity produced during treatment chlorination. The products were then analyzed by GC–MS and HPLC to try to identify the compounds formed.

Experimental

Extraction Techniques. Extracts of organics from samples of water were prepared by either freeze-drying (12 L) and extraction of the solids with methanol, details of which have been published previously (*10*), or by XAD-2 resin adsorption. Treated water samples (300 L) were passed through columns containing 100 mL of resin, prepared as published previously (*11*), at a flow rate of 500 mL/min. They were then eluted with 250 mL of ethyl ether, and the eluate was concentrated on a Kuderna-Danish apparatus to achieve a concentration factor of 4×10^4. The aqueous samples of chlorinated and unchlorinated model compounds (10–20 L) were passed through a column containing 20 mL of resin at a flow rate of 100 mL/min, eluted with 50 mL of ethyl ether, and concentrated to achieve a concentration factor of 1×10^4.

High-Performance Liquid Chromatography. A Varian 5060 delivery system was used for this work with detection by UV absorption. Either a Varian UV-50 variable wavelength detector or a Hewlett Packard 1040A scanning diode array detector was used. All HPLC columns were packed in our laboratory (*10*) with 5-μm particle size Spherisorb-ODS, Spherisorb-CN (Phase Separations), or 8-μm particle size Zorbax-CN (Dupont Ltd). HPLC columns (20 or 25 cm × 4.6 mm i.d.) were coupled via short lengths of stainless steel capillary tubing (5 cm × 0.25 mm i.d.). Separation conditions were as follows:

FRACTIONATION OF XAD–ETHYL ETHER EXTRACTS OF TREATED WATER. First-stage fractionation was carried out for 25 min on columns containing Zorbax-CN (3 × [20 cm × 4.6 mm i.d.]) with an eluent consisting of a linear gradient from 1% to 50% isopropyl alcohol in *n*-hexane. The injection volume was 50 μL (equivalent to 15 L of water sample), and 10 repeat injections were made to collect enough material for further analysis.

Second stage fractionation was carried out on columns containing Spherisorb-CN (3 × [25 cm × 4.6 mm i.d.]) with an eluent of 3% isopropyl alcohol in *n*-hexane for fraction 2 and 7% isopropyl alcohol in *n*-hexane for fraction 3 (*see* HPLC Fractionation of Extracts of Treated Water and Figures 3 and 4 for details of fractionation). The injection volume was 25 μL (equivalent to 7.5 L of water sample), and 12 repeat injections were made for each fraction.

SEPARATION OF EXTRACTS FROM CHLORINATION OF TYROSINE AND PHENYLALANINE. Separation was by reversed-phase HPLC by using Spherisorb-ODS (25 cm × 4.6 mm i.d.) with an eluent of 35% methanol in water for the chlorinated tyrosine extract and 55% methanol in water for the chlorinated phenylalanine extract.

Bacterial Mutagenicity Assay. Concentrated extracts and fractions were tested for mutagenic activity in two-step bacterial fluctuation assays (*12*) using *Salmonella typhimurium* TA100. Details of the method have been published previously (*1*). Strain TA100 without microsomal activation was used because this strain was found to be the most sensitive to detect mutagenic activity produced as a result of chlorination. Extracts and fractions were tested at 3–4 different dose levels ranging from the equivalent of 0.1–0.8 L of water per milliliter of test medium for extracts of water and aqueous solutions of model compounds, the equivalent of 0.2–3.2 L of water for HPLC fractions, and the equivalent of 0.05–5 μM of amino acids. The methanol extracts of freeze-dried solids were added to the test medium unchanged. XAD-2/ethyl ether extracts and reconcentrated HPLC fractions were transferred to dimethyl sulfoxide (DMSO), the original solvent was removed by evaporation, and the resulting concentrate was added to the test medium. Each assay contained appropriate negative controls, solvent, and positive controls, and each assay was repeated at least once on a different day. The results of the fluctuation assays were analyzed as previously described (*13*). Significance values at or below $p = 0.05$ were taken to indicate significant mutagenicity. The data were plotted as the estimated number of revertants per well versus dose, and the slope values were calculated from the linear part of the curve.

Fractionation of Chlorinated and Unchlorinated Water

Extracts of Treated Water. Water was obtained from a drinking water treatment plant where lowland river water is abstracted. Water treatment consists mainly of slow sand filtration, final chlorination, and partial dechlorination. The mutagenic activities of extracts (freeze-dried–methanol and XAD-2/ethyl ether) of water sampled before and after final chlorination and partial dechlorination were compared. The dose–response graphs of these are shown in Figure 1. Both extracts of chlorinated water were clearly mutagenic at dose levels equivalent to less than 0.2 L of water per milliliter of incubate. In contrast, the extracts of water sampled before final chlorination were not significantly different from the negative controls even when they were tested at levels equivalent to 0.4–0.8 L of water per milliliter of incubate. Procedural blanks from both techniques were satisfactory, that is, they were nontoxic and not significantly mutagenic.

Both freeze-dried/methanol and XAD-2/ethyl ether extracts of water obtained after final chlorination were found to be mutagenic with similar activity (Figure 1). Because XAD adsorption is more suitable for processing large volumes of water, and the extract produced is amenable to GC–MS analysis and HPLC separation, all subsequent work was carried out with this method of extraction.

HPLC Fractionation of Extracts of Treated Water. Capillary GC–MS analysis of the XAD-2/ethyl ether extracts of water sampled before and after final chlorination showed no significant difference that could account for the mutagenic activity observed after chlorination. These results indicate that the mutagenic compounds present in the extracts of drinking water are not readily amenable to analysis by GC–MS. However, the possibility cannot be excluded that the mutagenic compounds were present below the detection limit or that they were masked by other compounds.

Attempts to identify nonvolatile compounds present in water extracts by using soft-ionization mass spectrometry have been reported (*9*). In general, however, the extracts were found to be too complex to be analyzed without using some type of preseparation technique.

Therefore, separation and fractionation of the extracts were carried out by using HPLC in combination with mutagenicity testing to try to isolate fractions containing mutagenic compounds. Identification of the mutagenic compounds in these fractions can then be attempted by using

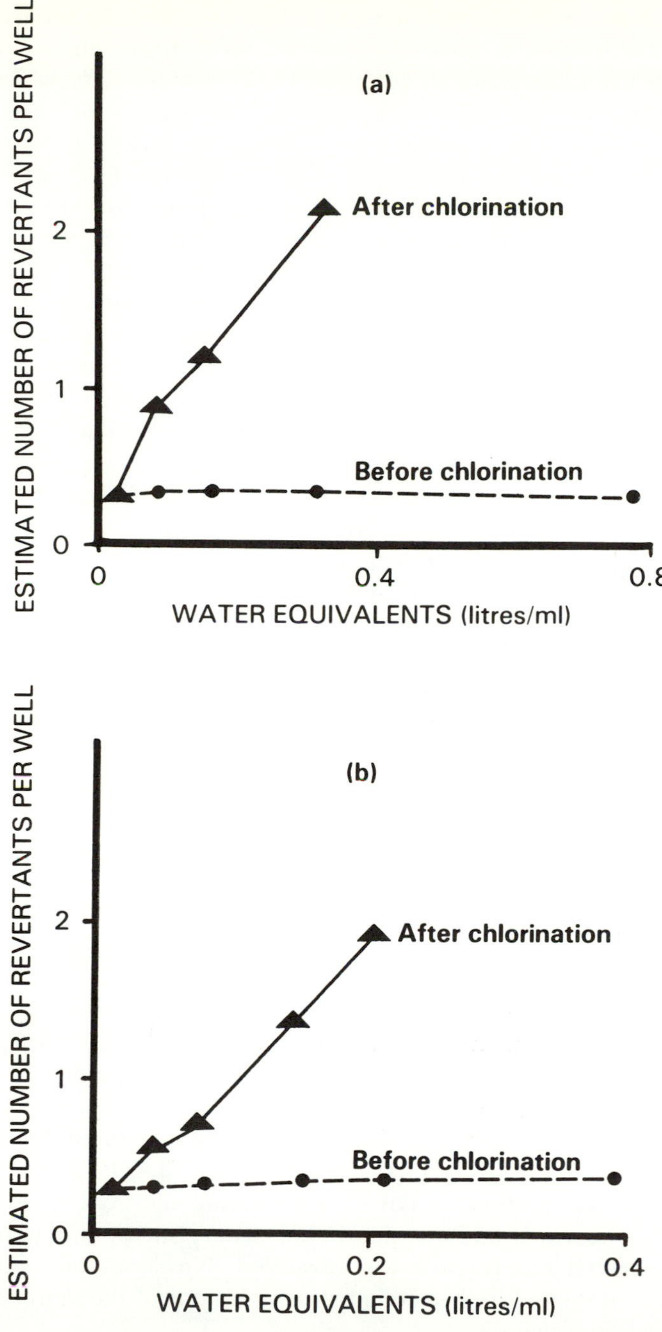

Figure 1. Comparison of TA100 mutagenic activities in treated waters sampled before and after chlorination: (a) XAD-2/ethyl ether extracts and (b) freeze-dried/methanol extracts.

soft-ionization mass spectrometric techniques. HPLC separation of drinking water extracts invariably resulted in complex chromatograms. Even with the most efficient columns, the complex mixtures could not be separated into individual components. One way to increase column efficiency is to couple two or more columns in series. The usefulness of this approach for the separation of water extracts has recently been demonstrated (9). All of the HPLC fractionation described in this section was performed by using a three-column coupled system that provided an efficiency of about 38,000 theoretical plates.

For HPLC fractionation combined with mutagenicity testing, samples were taken at the treatment plant concurrently at points before and after final chlorination. The extract of the water sampled before final chlorination, which was found to be nonmutagenic, was used throughout the fractionation procedure as a blank. The fractionation scheme is shown in Figure 2. The extracts of both water samples were separated, and six fractions of each were collected. Figure 3 shows the chromatograms obtained from HPLC separation of the extracts and indicates where each fraction was collected. The fractions were reconcentrated and tested for mutagenicity. In addition, the parent material and recombined fractions were also tested. This testing was done to check for losses or increases in mutagenicity resulting from sample processing and to check for evidence of synergism and antagonism between separated compounds. As shown in Figure 2, fractions F2, F3, F4, and F5 of chlorinated water were mutagenic, whereas fractions F1 and F6 were not significantly different from the negative controls. Some overall losses in activity were observed as a result of fractionation; these losses can probably be attributed to losses in sample handling during the various concentration steps. No evidence of synergism or antagonism between separated compounds was found, and no mutagenic activity in the fractions of the water sampled before final chlorination was observed.

The chromatograms from the HPLC separation of the extracts of the chlorinated and unchlorinated water, shown in Figure 3, appear to be similar except for the area where fractions F2 and F3 were collected. For this reason, these fractions were selected for further separation as shown in Figure 2. Fractions F2.2 and F2.3 of the second-stage fractionation of F2 were mutagenic, and fractions F3.4, F3.5, and F3.6 of the second-stage fractionation of F3 were mutagenic. Again, slight losses in activity occurred, and no significant synergism or antagonism between separated compounds was observed. No significant mutagenic activity was detected in the second-stage fractions of the water sampled before final chlorination.

Figure 4 shows the chromatograms from HPLC separations of fractions F2 of both chlorinated and unchlorinated samples. These

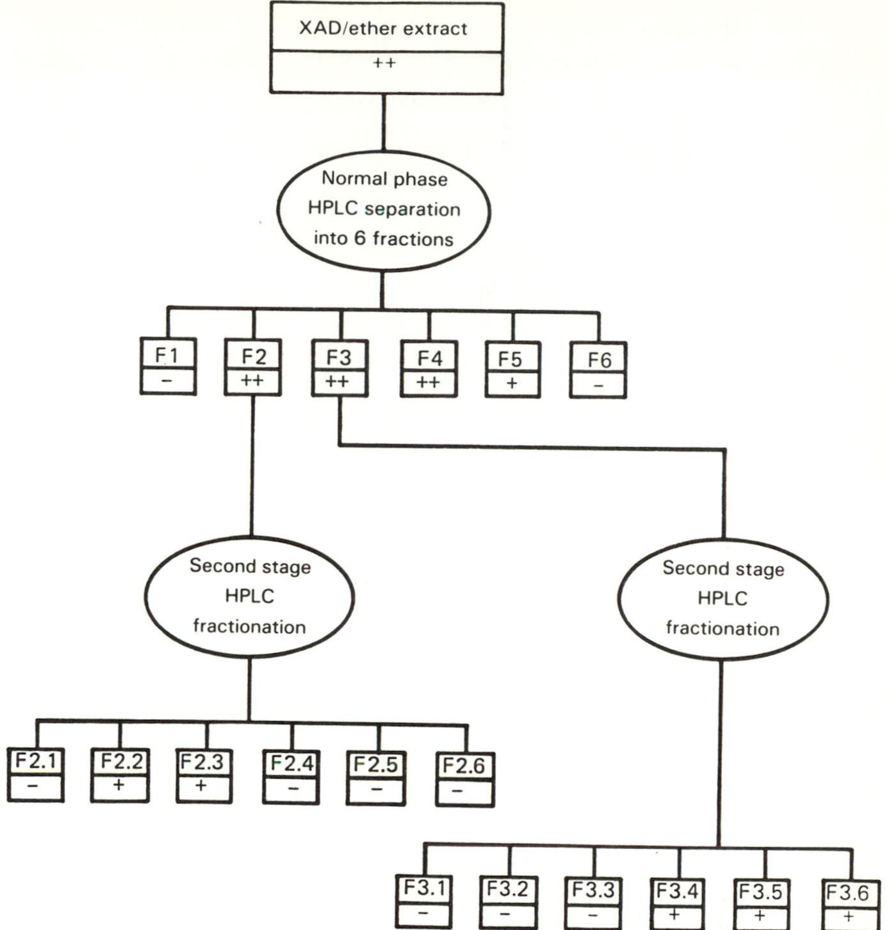

Figure 2. Fractionation scheme for an extract of chlorinated water, indicating TA100 mutagenicity of the resulting fractions. ++, mutagenic ($p = 0.5$–5.0); +, mutagenic ($p < 0.5$); and −, not significantly different from negative controls ($p > 0.05$). The slope values were calculated as revertants per liter equivalent of water per milliliter of incubate. (Reproduced with permission from reference 16. Copyright 1986 Water Research Centre.)

chromatograms show that the fractions are considerably less complex than the original XAD/ethyl ether extracts (Figure 3) with respect to UV-absorbing material. Moreover, a major peak was observed in the mutagenic fraction (F2.3) of the chlorinated water; this peak was not present in the corresponding fraction of the unchlorinated water.

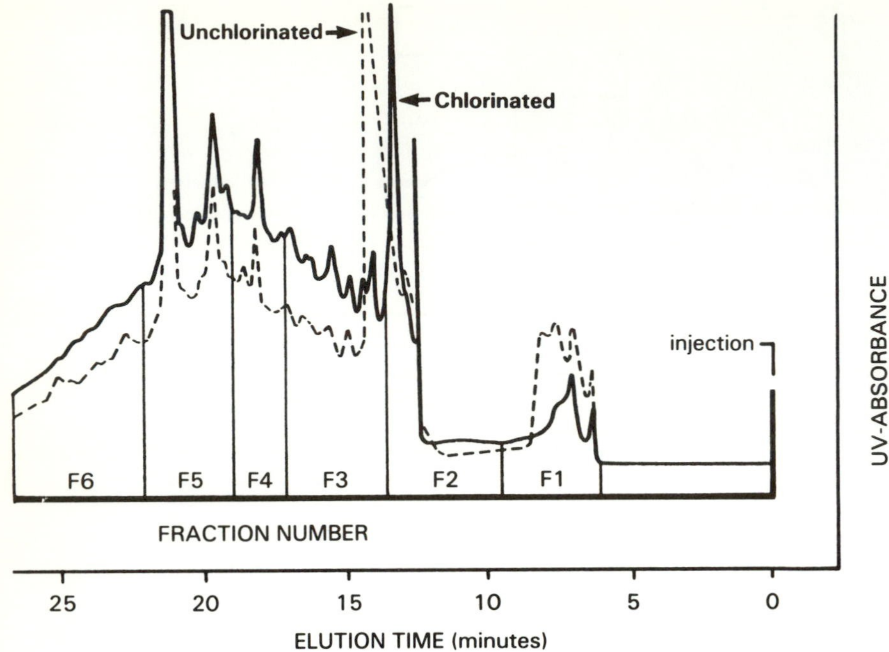

Figure 3. Normal-phase HPLC chromatograms from XAD-2/ethyl ether extracts of chlorinated and unchlorinated water. A 25-min linear gradient was used from 1% to 50% isopropyl alcohol in n*-hexane. (Reproduced with permission from reference 16. Copyright 1986 Water Research Centre.)*

These results demonstrate that a two-stage HPLC fractionation of a highly complex mutagenic extract of drinking water produced less complex mutagenic and nonmutagenic fractions. As indicated by the number of mutagenic fractions, several mutagenic compounds appear to be present. GC–MS analysis of the HPLC fractions indicated no significant differences in the fractions from extracts taken before and after final chlorination. It appears, therefore, that the compounds responsible for the mutagenic activity are not readily amenable to this technique. Other techniques, such as soft-ionization mass spectrometry (using field desorption and fast atom bombardment), are currently being applied to try to identify the mutagenic compounds in the HPLC fractions.

Chlorination of Model Compounds

Humic acids and amino acids have been reported to produce mutagenic activity as a result of chlorination (*4, 5*). However, this work was performed by using relatively high substrate and chlorine concentrations under conditions that are not typical for water treatment chlorination. One aim of this work was to investigate whether or not significant mutagenic activity would result from the chlorination of such compounds at concentrations and conditions that were similar to water treatment chlorination. Furthermore, the investigation was limited to mutagenic compounds that could be concentrated by XAD-2 resin adsorption because this technique had been successfully applied to the concentration of mutagenic compounds from drinking water.

Chlorination of Mixtures of Model Compounds. A laboratory chlorination technique that was compatible with the XAD concentration technique and the mutagenicity assay was developed. Thus, a relatively long chlorine contact time in comparison to sampling time was used to complete the reaction before sampling and to achieve a low residual

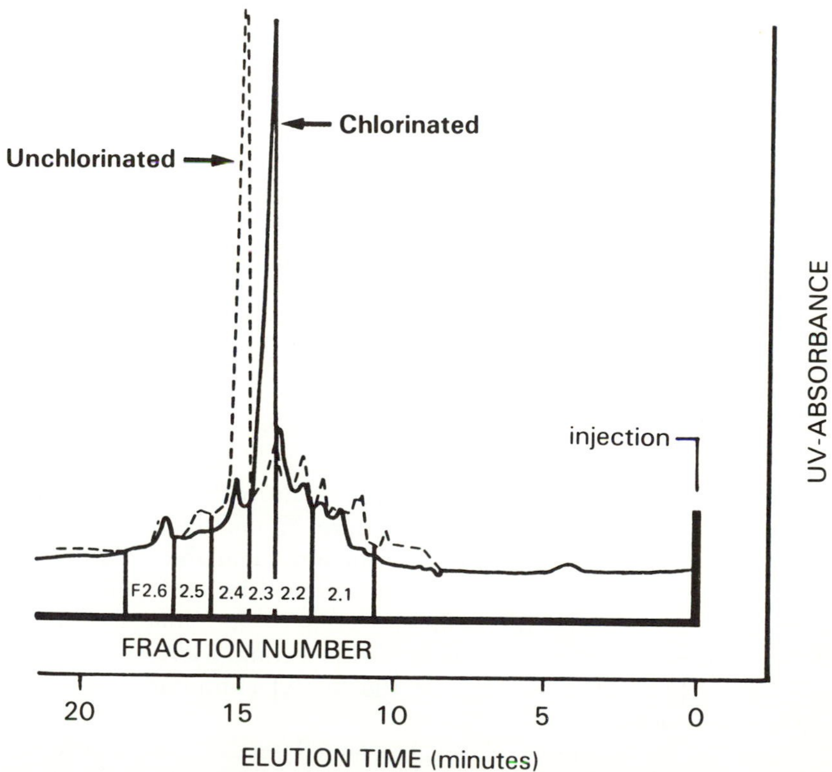

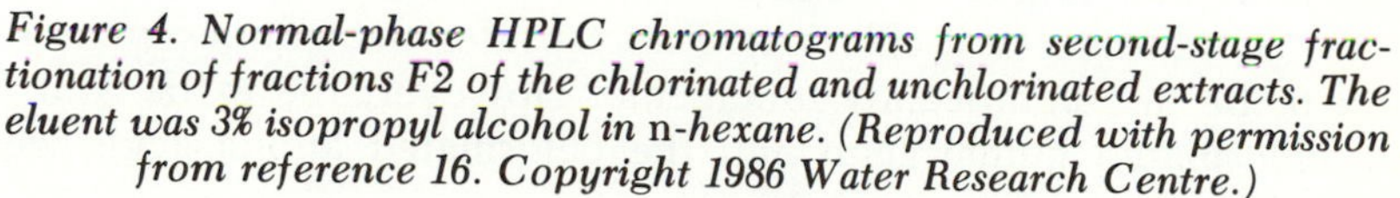

Figure 4. Normal-phase HPLC chromatograms from second-stage fractionation of fractions F2 of the chlorinated and unchlorinated extracts. The eluent was 3% isopropyl alcohol in n*-hexane. (Reproduced with permission from reference 16. Copyright 1986 Water Research Centre.)*

chlorine content without the use of dechlorination [dechlorination agents have been shown to destroy some mutagenic compounds (*13*)]. The selected technique was tested initially on treated water sampled before final chlorination. This water was chlorinated in the laboratory and compared with the water chlorinated at the treatment plant with respect to TA100 activity detected in the XAD-2/ethyl ether extracts. The results showed that the mutagenicity produced by laboratory chlorination was comparable to that produced at the treatment plant. A procedural blank prepared by chlorinating deionized water (which had been passed through activated carbon), followed by XAD-2 adsorption/ethyl ether elution, was neither mutagenic nor toxic.

The following groups of compounds were selected for investigation:

1. Humic acids (Fluka AG).
2. Humic acids (aquatic).
3. Humic acids (peat bogs).
4. L-Amino acids, $2.5 \times 10^{+6}$ molar solution of alanine, arginine, asparagine, aspartic acid, cysteine, cystine, glutamine, glutamic acid, glycine, histidine, hydroxyproline, leucine, lysine, methionine, phenylalanine, proline, serine, threonine, tryptophan, tyrosine, valine.
5. Purines and pyrimidines, 7.5×10^{-6} molar solution of adenine, guanine, xanthine, uric acid, uracil, cytosine, thymine.
6. Nucleosides and nucleotides, 2×10^{-6} molar solution of adenosine, guanosine, cytidine, uridine, thymidine, inosine, adenosine monophosphate, adenosine triphosphate, guanosine monophosphate, uridine monophosphate, thymidine monophosphate, inosine monophosphate.

The concentrations of the solutions were selected to give a realistic level of total organic carbon (i.e., approximately 3 mg/L). The solutions were adjusted to pH 6.2 with phosphate buffer. They were then chlorinated for 24 h at room temperature in the dark with sodium hypochlorite to a residual of <1 mg/L of total available chlorine. The chlorine demand of the solutions was determined in preliminary experiments prior to chlorination of larger samples for concentration by XAD-2 resin adsorption and mutagenicity testing. Corresponding extracts of unchlorinated solutions of the model compounds were also prepared and tested.

None of the extracts from unchlorinated model compounds were significantly different from the negative controls. Extracts of chlorinated purines-pyrimidines and chlorinated nucleosides-nucleotides were also found to be not significantly mutagenic. However, the extracts of all three chlorinated humic acids and the chlorinated amino acids were mutagenic. (Bacterial growth checks were included in the mutagenicity assays, and these indicated that the positive responses were not artifacts caused by enhanced bacterial growth). The dose-response graphs of the mutagenic extracts are shown in Figure 5 together with the dose-response graph of an extract of treated water for comparison.

The mutagenic activities of the chlorinated humic acid and amino acid solutions are of the same order of magnitude as that observed in treated water. (The low numbers of revertants at the higher dose levels indicate toxicity). Therefore, these compounds may be precursors of the mutagenicity that is observed in XAD-2/ethyl ether extracts of treated water and that is the direct result of chlorination in drinking water treatment.

Chlorination of Individual Amino Acids. HPLC analysis of an extract of chlorinated humic acids indicated that the chlorination products compose a highly complex mixture of organic material. Thus, the task of identification of mutagenic products of chlorination would not be simplified by the use of the humic acid model. In contrast, the amino acid model of production of mutagenic compounds can be readily simplified by the use of individual compounds as precursors.

To establish which amino acids were significant precursors of mutagenic activity, the 21 amino acids were chlorinated either in groups or individually. Methionine, tyrosine, phenylalanine, cysteine, and glycine were chlorinated individually because these had been reported to produce mutagenic products of chlorination (5). Cystine was also chlorinated individually because it was the only remaining sulfur-containing amino acid. The rest were divided into two groups according to structure, that is, aliphatic and heterocyclic.

The concentrations of amino acids in the buffered aqueous solutions (pH 6.2) were selected to give a total organic carbon content of 1-3 mg/L, and chlorination conditions were as described for the chlorination of mixtures of model compounds. The chlorinated solutions were concentrated by XAD-2 adsorption/ethyl ether elution, tested for mutagenicity, and analyzed by GC-MS.

Table I shows the results of the mutagenicity assays and some products of chlorination identified in the extracts by GC-MS. With the exception of glycine, mutagenic activity was observed in all the extracts of chlorinated amino acids.

The most potent precursors of mutagenicity were methionine, tyrosine, phenylalanine, and the group of heterocyclic amino acids. Tryptophan, histidine, proline, and hydroxyproline were subsequently chlorinated individually. Of these four, tryptophan and proline were the most significant precursors of mutagenic activity, although some activity was also observed in the extract of chlorinated histidine.

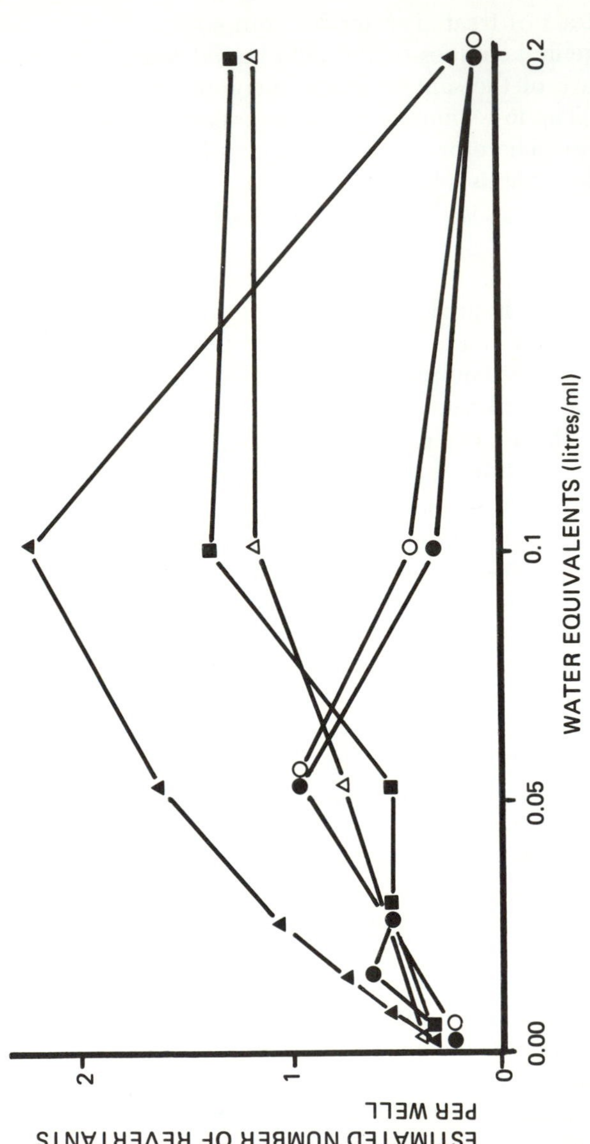

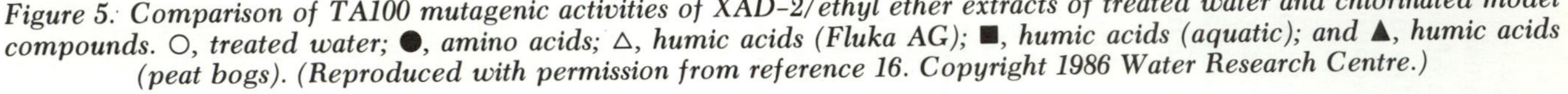

Figure 5: Comparison of TA100 mutagenic activities of XAD-2/ethyl ether extracts of treated water and chlorinated model compounds. ○, *treated water;* ●, *amino acids;* △, *humic acids (Fluka AG);* ■, *humic acids (aquatic); and* ▲, *humic acids (peat bogs). (Reproduced with permission from reference 16. Copyright 1986 Water Research Centre.)*

Table I. Mutagenic Activities of XAD-2/Ethyl Ether Extracts of Chlorinated Amino Acids and Compounds Identified in the Extracts by GC–MS

Amino Acid	*TA100 Activity*[a]	*Chlorination Products Identified*
Glycine	0.010[b]	—
Cystine	0.085	—
Cysteine	0.130	—
Methionine	1.895	dichloroacetonitrile
Tyrosine	2.155	hydroxybenzyl cyanide, chlorohydroxybenzyl cyanide
Phenylalanine	1.920	benzaldehyde, benzyl cyanide, phenylacetaldehyde
Aliphatic group	0.100	—
Heterocyclic group	0.980	indolylacetonitrile

[a] Mutagenic activities are expressed as slope values, revertants per micromole of amino acid per milliliter of incubate.
[b] Value is not significantly different from negative controls.

With the exception of indolylacetonitrile and dichloroacetonitrile, the compounds identified in the XAD-2/ethyl ether extracts of the chlorinated amino acids have also been identified by Glaze et al. (*14*), and Trehy and Bieber (*15*) have shown that dichloroacetonitrile can be produced by chlorination of certain amino acids. Of the compounds identified by GC–MS, only dichloroacetonitrile is a known mutagen (*8*). However, as in extracts of drinking water, dichloroacetonitrile is unlikely to account for a significant proportion of the activity observed in the extract of chlorinated methionine at the level detected. With the exception of chlorohydroxybenzyl cyanide, for which no authentic standard is available, all the compounds identified were tested for mutagenic activity, and no mutagenic response was obtained.

Figure 6 shows chromatograms from reverse-phase HPLC separations of the extracts of chlorinated tyrosine and chlorinated phenylalanine. Authentic standards were used to locate the compounds that had previously been identified by GC–MS. The chromatograms show the presence of several compounds not observed by GC–MS. This result indicates the presence of products of chlorination, which are nonvolatile. Other techniques are currently being employed to identify such compounds, for example, HPLC fractionation techniques combined with mutagenicity testing of the fractions and soft-ionization mass spectrometric analyses of the mutagenic fractions.

Conclusions

A comparison of the mutagenic activity of extracts of water taken from a treatment works before and after final chlorination showed that only

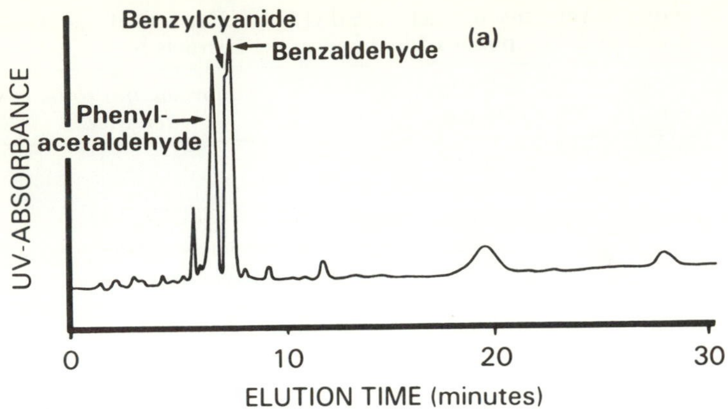

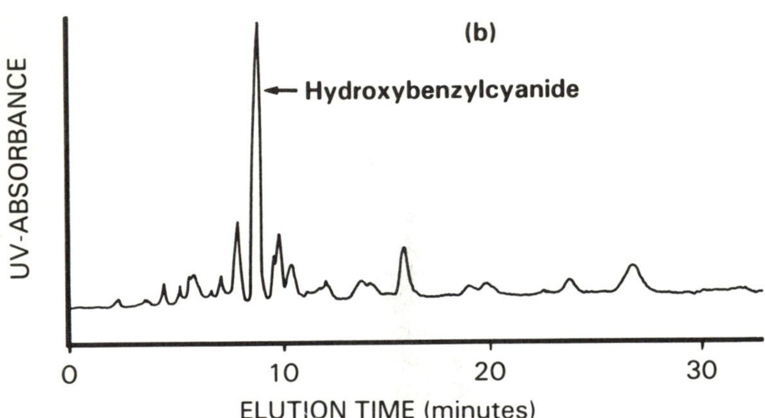

Figure 6. Reverse-phase HPLC chromatograms from XAD-2/ethyl ether extracts of (a) chlorinated phenylalanine and (b) chlorinated tyrosine. An eluent of 55% methanol in water was used in (a), and an eluent of 35% methanol in water was used in (b).

the sample after chlorination was mutagenic to bacterial strain TA100. A two-stage HPLC fractionation procedure was developed and produced a number of mutagenic fractions from the XAD-2/ethyl ether extract. This result indicated the presence of several mutagenic compounds. These fractions were shown to be less complex than the original extract. The mutagenic compounds separated did not appear to be amenable to analysis by GC–MS.

Chlorination of humic acids and amino acids at concentrations and conditions that simulated water treatment chlorination produced mutagenic XAD-2/ethyl ether extracts. Compounds in these extracts may, therefore, account for some of the activity observed in drinking water.

Chlorination of individual amino acids showed that methionine, tyrosine, phenylalanine, tryptophan, and proline produce highly mutagenic extracts. GC-MS analysis of these extracts indicated the presence of only one known mutagen, dichloroacetonitrile, but this mutagen was not present at a level that would be likely to account for the mutagenic activity.

HPLC separation of the extracts of chlorinated tyrosine and phenylalanine indicated the presence of compounds that were not amenable to analysis by GC-MS.

Future work will involve the use of soft-ionization mass spectrometric techniques such as field desorption and fast atom bombardment to try to identify these compounds.

Acknowledgments

This work was carried out under contract to the Department of the Environment. Permission to publish this work is gratefully acknowledged. Thanks are given to H. A. James and T. M. Gibson for GC-MS analyses and to P. Wilcox for advice on mutagenicity testing. We would like to thank W. Kuhn, Engler-Bunte Institute, University of Karlsruhe, for the generous gift of humic acid samples.

Literature Cited

1. Forster, R; Wilson, I. *J. Inst. Water Eng. Sci.* **1981,** *35,* 259-274.
2. Nestman, E. R.; Lebel, G. L.; Williams, D. T.; Kowbel, D. J. *Environ. Mutagens* **1979,** *1,* 337-345.
3. Loper, J. C. *Mutat. Res.* **1980,** *76,* 241-268.
4. Meier, J. R.; Lingg, R. D.; Bull, R. J. *Mutat. Res.* **1983,** *118,* 25-41.
5. Süssmuth, R. *Mutat. Res.* **1982,** *105,* 23-28.
6. McCann, J.; Choi, E.; Yamasaki, E.; Ames, B. N. *Proc. Natl. Acad. Sci.* **1975,** *72,* 5135-5139.
7. Fielding, M.; Gibson, T. M.; James, H. A.; McLoughlin, K.; Steel, C. P. *Technical Report TR159;* Water Research Centre: Marlow, England, 1981.
8. Simmon, V. K.; Tardiff, R. G. In *Water Chlorination: Environmental Impact and Health Effects;* Jolley, R. L.; Gorchev, H.; Hamilton, D. H., Eds.; Ann Arbor Science: Ann Arbor, MI, 1978; Vol. 2, pp 417-431.
9. Crathorne, B.; Fielding, M.; Steel, C. P.; Watts, C. D. *Environ. Sci. Technol.* **1984,** *18,* 797-802.
10. Crathorne, B.; Watts, C. D.; Fielding, M. *J. Chromatogr.* **1979,** *185,* 671-690.
11. James, H. A.; Steel, C. P.; Wilson, I. *J. Chromatogr.* **1981,** *208,* 89-95.
12. Green, M. H. L.; Murial, W. J.; Bridges, B. A. *Mutat. Res.* **1979,** *38,* 33-42.
13. Wilcox, P.; Denny, S. In *Water Chlorination: Chemistry, Environmental Impact and Health Effects;* Jolley, R. L.; Bull, R. J.; Davis, W. P.; Katz, S.;

Roberts, M. H., Jr.; Jacobs, V. A., Eds.; Lewis: Chelsea, MI, 1985; Vol. 5, pp 1341-1353.
14. Glaze, W. H.; Burleson, J. L.; Henderson, J. E., IV; Jones, P. C.; Kinstley, W.; Peyton, G. R.; Rawley, R.; Saleh, F. Y.; Smith, G. *Report No. EPA-600/4-82-072;* U.S. Environmental Protection Agency: Athens, GA; 1982.
15. Trehy, M. L.; Bieber, T. I. In *Advances in the Identification and Analysis of Organic Pollutants in Water;* Keith, L. H., Ed.; Ann Arbor Science: Ann Arbor, MI, 1981; Vol. 2, pp 941-975.
16. Fawell, J. K.; Fielding, M.; Horth, H.; James, H.; Lacey, R. F.; Ridgway, J. W.; Wilcox, P.; Wilson, I. *Health Aspects of Organics in Drinking Water;* Technical Report TR231; Water Research Centre: Marlow, England, 1986.

RECEIVED for review August 14, 1985. ACCEPTED January 27, 1986.

33

New Methods for the Isolation of Mutagenic Components of Organic Residuals in Sludges

M. Wilson Tabor[1] and John C. Loper[1,2]

Department of Environmental Health[1] and Department of Microbiology and Molecular Genetics[2], University of Cincinnati Medical Center, Cincinnati, OH 45267

A general procedure has been developed for the isolation of residue organics from sewage treatment plant sludges for mutagenic assessment. This procedure features milling the sludge with anhydrous sodium sulfate to a homogeneous powder, sequential extractions of the powder with a solvent series from nonpolar to polar, and bioassay of the extracted organics via the Salmonella *microsomal mutagenicity assay. This new method was compared to published U.S. Environmental Protection Agency procedures and solvent extraction procedures for isolating organics from sludges. Results of this study suggested that both published procedures either destroyed labile mutagens in sludges or caused the formation of mutagenic artifacts during the isolations. The new procedure features gentle isolation conditions and produces a homogeneous sample that is easily manipulated for further studies to isolate mutagens for chemical and biological characterization.*

SLUDGES, RESULTING FROM THE TREATMENT of municipal sewage, are complex mixtures of organic and inorganic chemicals, some of which may have biological origins but many of which come from anthropogenic or industrial sources. Many of the residue organics isolated from these sludges have been characterized as toxic and/or mutagenic in bacterial, animal, and plant tester systems (*1–5*). Although some of the chemical constituents of sludges have been identified and partially quantified (*2–4, 6–9*), not much is known about the chemical

0065–2393/87/0214/0675$06.00/0

identity and source of the vast majority of the mutagenic compounds (*5*). Such information is needed to assess the importance of these compounds as potential human health hazards and to develop methods to limit their concentrations in sludges and their release to the environment. Knowledge of mutagenic constituents in sludges is important for human health because of the scale of raw sludge production: 7 million dry tons annually in the United States (*10, 11*).

Progress toward the identification of the mutagens has been slow partly because of the lack of reliable isolation methods for these compounds. The objectives of this study were to isolate residue organic mutagens from municipal sewage treatment plant sludges for chemical and biological characterization and to determine their origin. This chapter describes our first step in that effort: the investigation of methods to extract organic mutagens from sludges.

Experimental

Chemicals. American Society for Testing Materials Type I water and Type IV water (*12*) were generated in our laboratory by using a Continental-Millipore Water Conditioning System, as described previously (*13*). Organic solvents used for extraction were nanograde acetone, benzene, hexane, methanol, methylene chloride, and isopropyl alcohol obtained from Mallinckrodt and high-performance liquid chromatographic (HPLC) grade 1,1,2-trichloro-1,2,2-trifluoroethane (Freon 113) obtained from Fisher Scientific. Sodium sulfate (reagent grade, low in nitrogen and suitable for Kjeldahl analyses, Fisher Scientific) was muffled at 500 °C for 6 h prior to use. All other chemicals were reagent grade or better and were used as supplied.

Samples. Sludge samples were obtained from a municipal sewage treatment plant of the Metropolitan Sewer District in Cincinnati, Ohio. This facility, the U.S. Environmental Protection Agency/Municipal Environmental Research Laboratory—Cincinnati (USEPA/MERL-CIN) Technical and Evaluation Facility at the Mill Creek Sewage Treatment Plant, receives an influent sewage composed primarily (>70%) of industrial discharges. A general outline of the treatment process stream for this facility is shown in Figure 1. Not all effluent discharges from this facility are disinfected via chlorination, as indicated in Figure 1. Primary sludge and secondary sludge samples were taken for the purposes of this study. Primary sludge is the solids and particulate matter settled from treatment plant influent waste water by clarification. The aqueous portion from this treatment process is called primary effluent waste water. Secondary sludge is the solids and particulate matter settled after aerobic activated sludge treatment of the primary effluent waste water.

Samples were collected in 4-kg amber glass jars fitted with Teflon-lined caps and were frozen at −20 °C until processing. At the time of workup, the samples were thawed and allowed to settle overnight at 4 °C, and the supernatant fluids, about 75% of the volume of the parent samples, were decanted. For each sample, the remaining slurry was centrifuged for 20 min at 8000 × *g* at 6 °C. This supernatant fluid was decanted and combined with the previous supernatant fluid and then stored for processing as an aqueous sample (*14, 15*). The

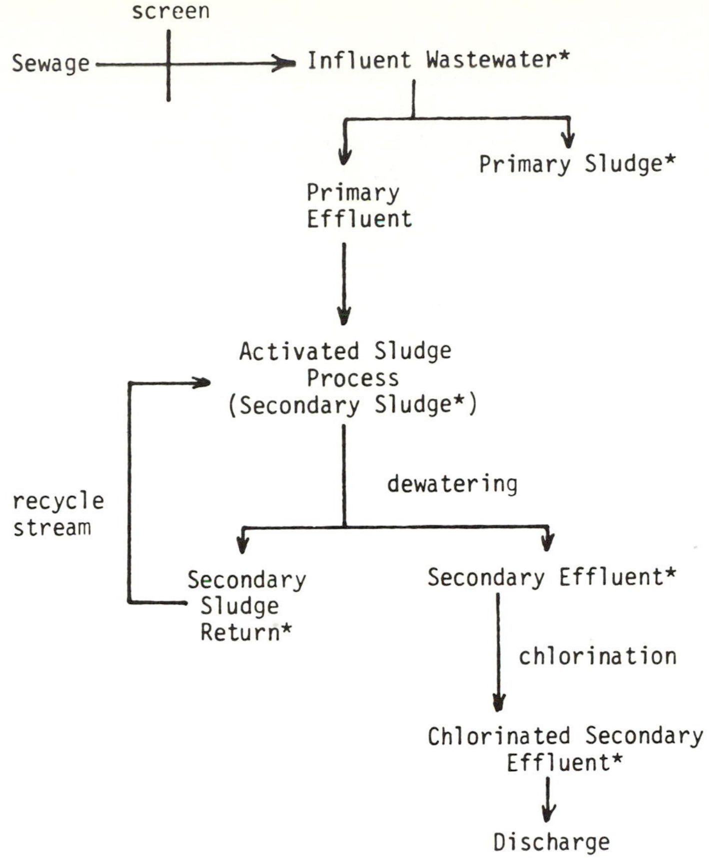

Figure 1. Schematic of municipal sewage treatment of the EPA/MERL–CIN Technical and Evaluation Facility at the Mill Creek Sewage Treatment Plant, Cincinnati, Ohio.

remaining moist pellet was processed as outlined in Figure 2 for the preparation of residue organics for bioassay and further separations. A weighed aliquot of each sludge pellet was oven dried at 125 °C for 24 h at a negative pressure of 30 mmHg and then reweighed for the calculation of dry weight.

Biological Analysis. Tester strains TA90 and TA100 for the *Salmonella* microsomal mutagenicity tests were provided by B. Ames. Mutagenesis assays were conducted as described previously (*14–16*). Characteristic properties of the bacterial tester strains were verified for each fresh stock, and their mutagenicity properties were verified again by using positive and negative controls as part of each experiment as recommended (*16*, *17*). Tests requiring metabolic activation used polychlorinated biphenyl mixture (Aroclor 1254) induced rat liver 9000 × *g* supernatant fraction, S9, from Litton Bionetics. Duplicate bioassays were made for each dose. The bioassay of each extract was repeated within 1 week of the original bioassay. Additionally, each extraction

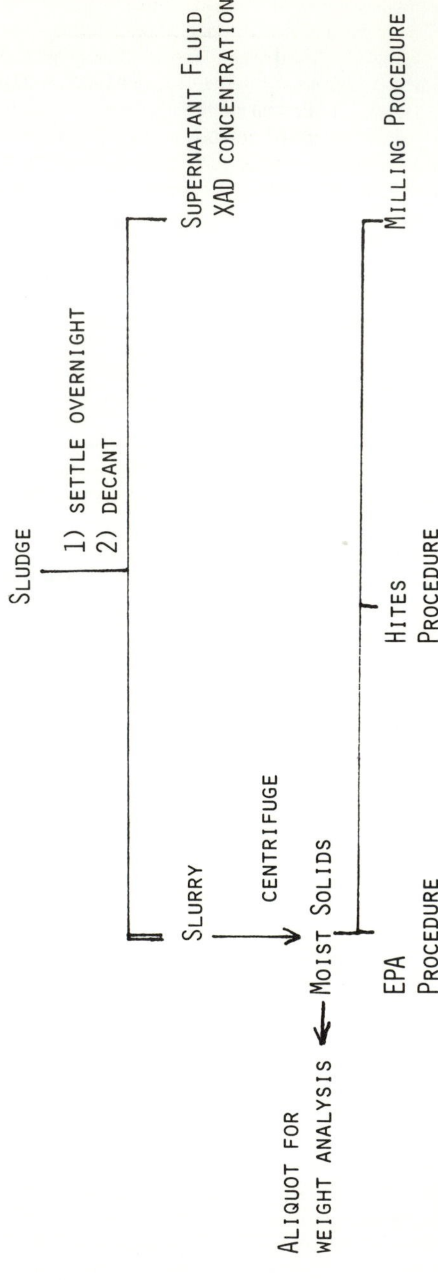

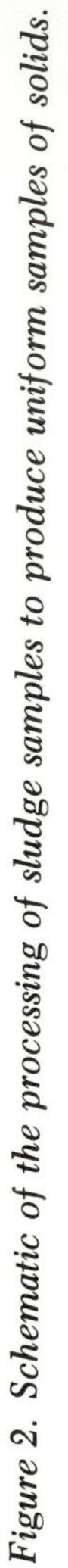

Figure 2. Schematic of the processing of sludge samples to produce uniform samples of solids.

experiment was repeated. The mutagenesis results are averages of these four bioassays, each in duplicate, for each sample extract. The detection of mutagenic activity in experimental samples was based upon a dose-dependent response exceeding the zero-dose spontaneous control value by at least twofold; that is, the ratio of total revertant colonies per plate to spontaneous colonies per plate was ≥2. All recoveries of bioactivity from concentrated residue organic samples were based upon an expression of mutagenesis in terms of net revertant colonies per gram equivalent dry weight of sludge, representative of the original sludge sample. Typical mean revertant colony counts ± standard error, obtained from each group of spontaneous plates and positive control plates from our laboratory for the time period of the experiments described herein, were recently reported (*14*).

Preparation of Residue Organics from Sludges. Three isolation methods were investigated to prepare residue organic samples from the sludges for bioassays and fractionations. These methods are outlined in Figures 3–5.

SOXHLET EXTRACTION. The procedure was based upon that described by Hites and co-workers (*18, 19*). A 50-g aliquot of a moist pellet of a sludge sample was added to a borosilicate glass thimble, which was then placed into a standard Soxhlet extraction apparatus. The sample was extracted for 7 h with 300 mL of isopropyl alcohol, followed by a 7-h extraction with 300 mL of benzene. These extracts were dried separately via passage through a 1- × 25-cm column of anhydrous sodium sulfate. Then each drying column was washed with 100 mL of the respective solvent followed by 50 mL of a hexane:acetone (85:15 by volume) solvent system. These eluates were combined with the respective extracts, and the two sample extracts were concentrated 100-fold via rotary evaporation at 50 °C while a pressure of 30 mmHg was maintained. The samples were concentrated to 10 mL via evaporation at 50 °C under a stream of dry nitrogen. Immediately prior to bioassay, an aliquot of the concentrate was returned to a nitrogen stream and evaporated to dryness; the residue was dissolved in dimethyl sulfoxide.

METHYLENE CHLORIDE EXTRACTION. This procedure was a modified version of the USEPA/Environmental Monitoring Support Laboratory—Cincinnati (USEPA/EMSL-CIN) Method 624S/625S (*20*). By using a Sorvall Omnimixer (Du Pont) at the high-speed setting, a 50-g aliquot of a moist pellet of a sludge sample was homogenized for 1 min in 300 mL of methylene chloride. During all homogenization operations, the sample container was immersed in an ice bath to avoid overheating. The pH of the homogenate was adjusted to ≥11 by the addition of 1.0 N sodium hydroxide, the homogenization was repeated, and the phases were separated by centrifugation at 1500 × g for 10 min. The organic layer was removed, and the residual basic aqueous layer was extracted two additional times with methylene chloride. The combined methylene chloride base-neutral extracts were dried via passage through a column of anhydrous sodium sulfate and then concentrated by using the procedures described in the previous section except that temperatures for evaporation were maintained at 30 °C rather than 50 °C.

The second series of methylene chloride extractions was accomplished by the addition of 300 mL of solvent to the aqueous layer, homogenization for 1 min, then pH adjustment to ≤2 via the addition of 6.0 N hydrochloric acid. Following an additional homogenization for 1 min, the suspension was centrifuged and processed as for the base-neutral extractions. The extrac-

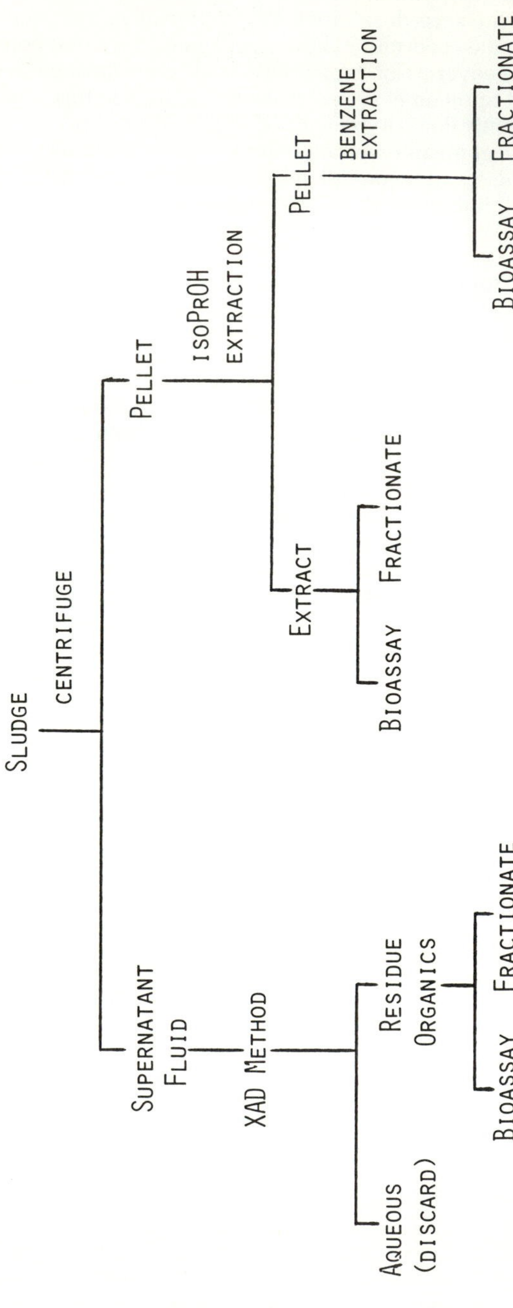

Figure 3. Schematic of the modified EMSL method for the preparation of residue organics from moist pellet sludge samples.

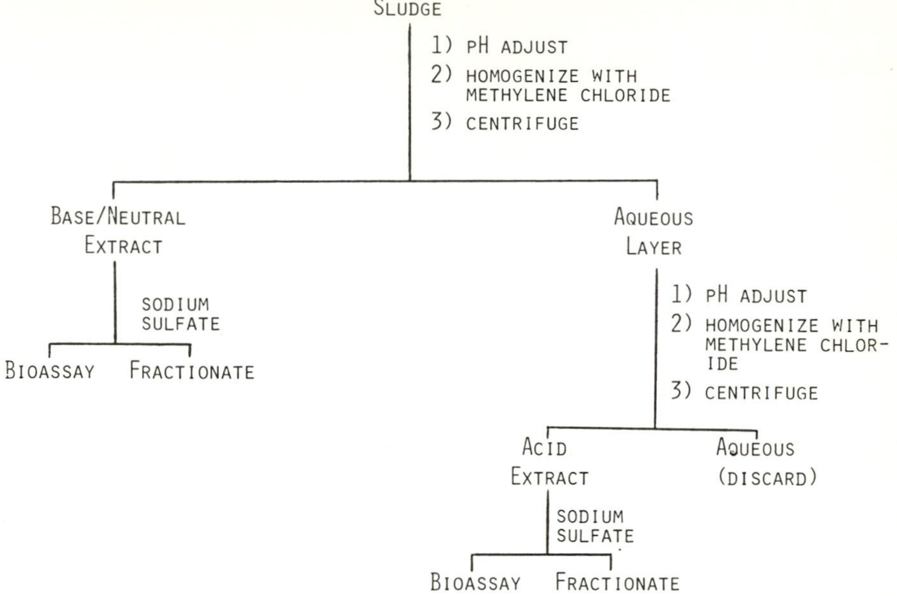

Figure 4. Schematic of the modified Hites method for the preparation of residue organics from moist pellet sludge samples.

tion–centrifugation process was repeated two more times, and the combined methylene chloride acid extracts were dried and concentrated as before.

MILLING-EXTRACTION. A 100-g aliquot of a moist pellet of a sludge sample was mixed with 600 g of anhydrous sodium sulfate. The mixture was added to a 1-L ball mill (U.S. Stoneware) along with 8 1-in. and 60 ½-in. stainless steel bearings. The sample was milled for 5–9 h until the sample had a consistency of flour. Milled samples were stored at −20 °C until further processing. One gram of moist sludge yielded 7 g of milled sample.

Seventy-gram aliquots of the milled sample were added to a glass extraction thimble for Soxhlet extraction. The milled sample was extracted sequentially for 2 h with 300-mL portions of four different solvents: Freon 113, methylene chloride, acetone, and methanol. Following extraction, each solvent extract was dried and then concentrated as described earlier by using a 30 °C evaporation temperature for concentrating the Freon 113 and methylene chloride extracts and a 50 °C evaporation temperature for concentrating the other solvents.

Results and Discussion

The purpose of this chapter is to present a general approach for the isolation of residue organics from sludges for further biological and chemical characterization of the mutagenic constituents of these residues. Three criteria were kept in mind during development and

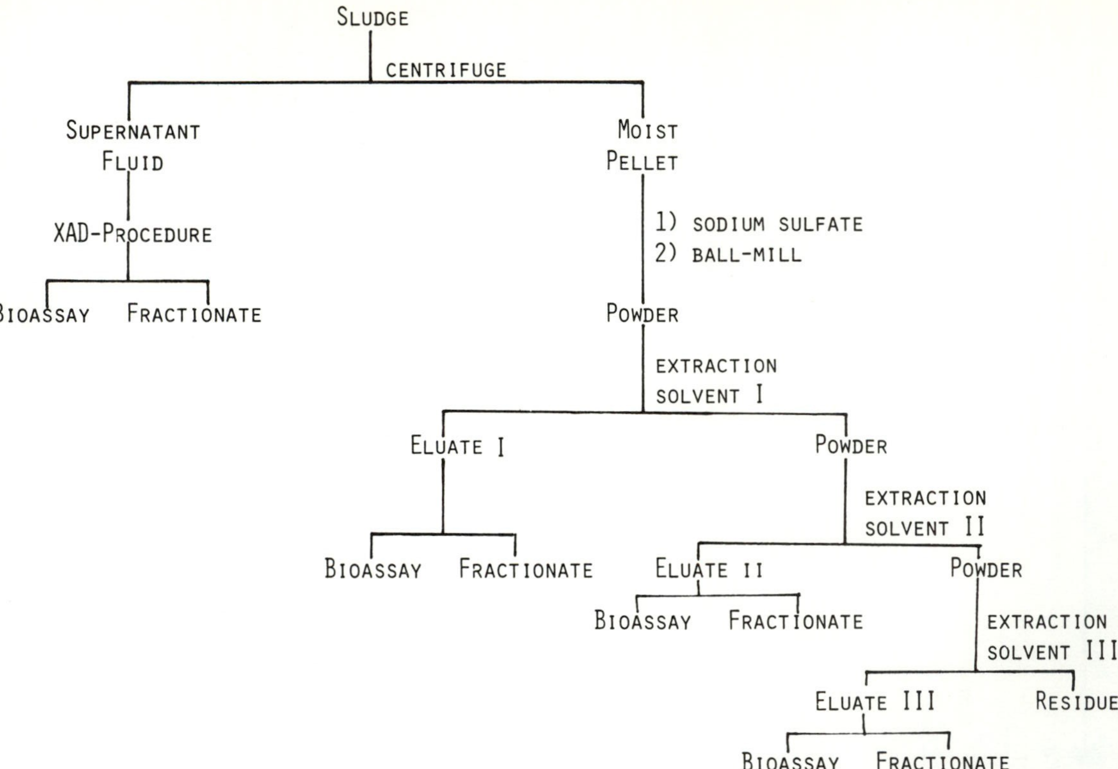

Figure 5. Schematic of the milling procedure for the organics from moist pellet sludge samples.

evaluation: (1) The method should be reasonably quantitative and reproducible. (2) The method should yield residue organics free of artifacts of isolation. (3) The method should be relatively easy and straightforward to apply to a wide variety of sample types. The first two criteria were evaluated in this study in terms of the recovery of mutagenic activity in the residue organics isolated from sludges. The third was evaluated according to the skill required and time involved in sample preparation. By using these three criteria, the method developed for the evaluation of mutagenicity in this study was compared to previously described methods for the isolation of residue organics from sludges and other similar medialike sediments.

Two fundamental approaches have been used by previous investigators for this isolation problem. The first involved the use of one or more solvents for the extraction of the residue organics from the unmodified sludge sample (*18, 19*). The second approach used one or more solvents for the isolation of the residue organics by multiple extraction of the sludge sample that had been pH adjusted prior to extraction (*20*). Originally, these two approaches were developed to screen sludges and/or sediments for organic chemical contamination, for example, priority pollutants, and subsequently were used to isolate residue organics for mutagenic assessment of these mixtures. However, we found that these two previous approaches were inadequate in terms of the three criteria. Therefore, the more general approach described herein was developed.

Preparation of Residue Organics from Sludges. Sludge samples, as obtained from sewage treatment plants, are aqueous slurries that vary as to the amount of suspended solids. Because the objective of our study was to examine the mutagenicity present in the solids portion, a procedure was developed to separate the solids and aqueous phases of the sample. This procedure, outlined in Figure 2, yielded preparations of sludge solids that were uniform for a given sample type in terms of percent moisture content (data not shown). The resulting moist pellets were used for further study.

The three methods investigated in this study to prepare residue organics from sludges for mutagenicity testing and mutagen isolation are outlined in Figures 3–5. Each method was applied to both primary and secondary sludge samples. The residue organics isolated via each procedure were assayed for TA98 and TA100 mutagenic activity in the absence and presence of metabolic activation, S9. Results of the mutagenic assays are summarized in Table I for residue organics isolated by the various methods. No mutagenic activity for tester strain TA100 was observed for any of the sample extracts; however, TA98 mutagenicity varied widely for the residue organics isolated from the

Table I. Residue Organics TA98 Mutagenesis for Primary and Secondary Sludge Extracts Prepared by Various Procedures

Procedure	*Metabolic Activation*	*Primary Sludge Extracts*	*Secondary Sludge Extracts*
Modified Hites	−S9	0 (IA), 0 (B)	88.6 (IA), 8.1 (B)
	+S9	21.3 (IA), 12.4 (B)	100.0 (IA), 6.6 (B)
Modified EMSL	−S9	218.0 (BN), 0 (A)	0 (BN), 23.5 (A)
	+S9	251.0 (BN), 0 (A)	212 (BN), 15.0 (A)
Milling	−S9	70 (F), 41 (MC), 0 (AC), 0 (M)	60 (F), 64 (MC), 0 (AC), 0 (M)
	+S9	14 (F), 32 (MC), 12 (AC), 3 (M)	28 (F), 19 (MC), 20 (AC), 10 (M)

NOTE: Values represent the mutagenesis of net revertants in thousands of colonies and are reported as net revertants per gram of dry weight. The solvents are given in parentheses and are defined as follows: IA, isopropyl alcohol; B, benzene; BN, base-neutral; A, acid; F, Freon; MC, methylene chloride; AC, acetone; and M, methanol.

primary and secondary sludges, depending upon the method employed for isolation. In general, the modified EMSL method yielded the highest level of TA98 mutagenicity for both sample types, whereas the modified Hites method yielded the lowest mutagenic level for the primary extract and approximately the same level as did the milling procedure for the secondary sludge. The discrepancies in these data suggested three possibilities for a given isolation method: (1) destruction of mutagens in a sample, (2) production of mutagenic artifacts during isolation, or (3) differences in extraction efficiency from method to method.

These possibilities could be due to a parameter inherent in a given method. Among these three isolation methods, the differences in parameters include pH, temperature, polarity of extraction solvents, moisture, and particle size. Any single parameter or combination of parameters could be affecting the yield in mutagen levels recovered from the sludge samples. Therefore, a series of experiments was conducted to examine possible effects of these parameters.

Effects of pH on the Recovery of Mutagenic Activity. The effect of pH, characteristic of the modified EMSL method, was examined in the following experiment. Residue organics extracted from a sample of primary sludge first were isolated via the milling procedure and then recombined in proportions representative of the original sample. An aliquot (ALIQ-I) of this reconstituted sample was bioassayed. A second aliquot (ALIQ-II) was spiked into aqueous base of pH 11. Then, the modified EMSL procedure was employed, that is, methylene chloride extraction followed by pH adjustment to 2.0 and reextraction with methylene chloride. The base-neutral and acidic extracts were bioassayed. For the ALIQ-I reconstituted sample, the TA98 net revertants per gram dry weight for −S9 and +S9 were 46.5 and 27, respectively. For the ALIQ-II base-neutral extract, the TA98 net revertants per gram dry weight for −S9 and +S9 were 56 and a positive response, respectively; for the ALIQ-II acidic extract, the values for −S9 and +S9 were 75.5 and 18, respectively. Thus, for ALIQ-II, the total TA98 net revertants per gram dry weight for −S9 and +S9 were 131.5 and 18, respectively. These values represent the mutagenesis of net revertants in thousands of colonies. These results show that the amount of direct-acting (−S9) TA98 mutagenesis was more than doubled, total ALIQ-II versus ALIQ-I, as a result of the organics in the sample being exposed to the extremes of pH of the modified EMSL procedure. This increase, ALIQ-II total, appears to be due to mutagenic artifacts produced from base- and/or acid-catalyzed reactions of nonmutagenic constituents in the original mixture, ALIQ-I. Therefore, these data suggest the chemical environment of this method can significantly alter the sample, probably by contributing to the formation of mutagenic artifacts. Furthermore, these results suggest that an isolation procedure involving extremes of pH is not suitable for the isolation of labile compounds such as mutagens from environmental samples.

Effects of Other Parameters on the Recovery of Mutagenic Activity. The effects of moisture content, particle size, temperature, and solvent polarity on the recovery of residue organics from sludges for mutagenesis testing were examined in the following experiment. A sample of moist pellet, isolated from secondary sludge via the procedure outlined in Figure 2, was divided into five equal aliquots, A–E.

Aliquot A was processed via the milling procedure, Figure 5. The resulting dry powder was extracted as described in the experimental section by using a solvent sequence from nonpolar to polar solvents, that is, Freon 113 to alcohol. The extracted residue organics were bioassayed. As mentioned earlier, no TA100 mutagenesis was observed for any extracts of the secondary sludge samples. The total measured TA98 mutagenic activity ±S9 associated with these extracts is summarized in Table II.

Aliquot B was neither dried nor milled but was extracted directly by the procedure just described for Aliquot A. The extracted residue organics were bioassayed, and the results are summarized in Table II.

Aliquot C was dried for 24 h at ambient temperature in a desiccator under reduced pressure, 50 mmHg, by using previously dried molecular sieves as a desiccant. The dried sample was pulverized in a beaker with

Table II. Effects of Method of Moist Pellet Processing on the Isolation of Mutagenic Residue Organics from Secondary Sludge

Aliquot	*Pellet Processing*[a]	*TA98 Net Revertants per Gram of Dry Weight*[b] −S9	+S9
A	milled with Na_2SO_4	67.0	131.7
B	none, i.e., moist	45.9	44.2
C	desiccated, pulverized	50.1	36.7
D	oven dried, pulverized	23.7	67.7
E	milled with Na_2SO_4	36.1	55.2
—	milled Na_2SO_4	ND[c]	ND[c]

[a] Aliquots A–C and E were processed at ambient temperature; Aliquot D was dried at 110 °C; Aliquots A–D were extracted with the solvent sequence from nonpolar to polar; Aliquot E was extracted with the solvent sequence from polar to nonpolar.
[b] Values represent the mutagenesis of net revertants in thousands of colonies.
[c] No mutagenic activity was detected.

a glass rod before extracting the sample via the same procedure employed for Aliquots A and B. The extracted residue organics were bioassayed, and the results are summarized in Table II.

Aliquot D was dried overnight at 110 °C, pulverized, and then extracted via the same method used for Aliquots A–C. The results of the bioassay of the extracted residue organics are summarized in Table II.

Aliquot E was processed via the same procedure used for Aliquot A, except that the sample was extracted by using the solvent sequence in reverse, that is, polar to nonpolar. The results of the bioassay of this aliquot are summarized in Table II. As one control for this experiment, a sample of sodium sulfate was milled and then extracted via the same solvent sequence used for Aliquots A–D. Bioassay of the extracts of the sodium sulfate showed no detectable net mutagenic activity (Table II).

The results of this experiment show that several important parameters, in addition to pH as shown in the previous experiment, influence the isolation of residue organics from sludge samples when these isolates are to be examined for mutagenic activity. Recovery of mutagenic activity appears to be related to the removal of moisture from the sample. Comparison of the various methods to desiccate the sample—Aliquots A, C, and D—showed that milling with anhydrous sodium sulfate results in the highest recovery of both direct-acting and S9-dependent mutagenic activity. Exposure of the samples to high temperature, Aliquot C versus D, resulted not only in the loss of direct-acting mutagens but also in the recovery of 50% more S9-dependent mutagens. This finding suggests that the S9-dependent mutagens have vapor pressures such that the reduced pressure used for the desiccator resulted in their loss to the desiccant, Aliquot C, but was not sufficient for loss due to the elevated temperature, Aliquot D. An alternative explanation for the differences in recovery of S9-dependent mutagenic activity is that the high temperature of the drying oven could be removing or destroying antagonists in the sample and thereby unmasking additional mutagenic activity. However, it is more likely that both procedures associated with drying, desiccation, or heating are less appropriate than the sodium sulfate procedure because the recovery of mutagenic activity by the procedures applied to both Aliquots C and D was much less than the recovery via the sodium sulfate milling procedure, Aliquot A.

The mutagenic activity of residue organics isolated via the milling–extraction procedure probably is most representative of the bioactivity associated with the residue organics from the sample. This procedure avoids both the high temperatures and the reduced pressures associated with the other two desiccation procedures and thereby circumvents possible losses of mutagens via heat destruction or volatilization. Because extracts of milled sodium sulfate showed no mutagenic activity, the desiccating salt and the solvents used for extraction were eliminated as sources of mutagenic artifacts.

The examination of alternative solvent extraction sequences in this experiment, Aliquot A versus E, showed the sequence from nonpolar to polar solvent to be more efficient in extracting mutagens from the sample (Table II). This result could explain the lower recoveries of mutagenic activity via the modified Hites procedure compared with the recoveries found with the milling procedure (Table I). The published Hites method (*18, 19*) was used in this study, that is, extraction with isopropyl alcohol followed by benzene. Therefore, the results from our experiment suggest that a nonpolar to polar solvent sequence gives better recoveries of mutagenic components from a sample.

Conclusions

The overall objective of this investigation was the isolation of organic mutagens from municipal sewage treatment plant sludges to characterize the mutagens both chemically and biologically. A biological approach to the chemical fractionation of residue organics is the method of choice for the isolation of hazardous compounds from such complex mixtures. Elsewhere in this volume (*15*) and in previous publications (*13–16, 21, 22*), we have described a general analytical fractionation and bioassay procedure for the isolation and identification of mutagens from residue organics isolated from environmental samples. However, a methodological problem was evident when we began our studies of sludges and examined two methods representative of those previously applied to sludges-sediments. As described here, these methods were found inadequate for the preparation of residue organics from sludges for mutagenic component isolation and for compound identification. Our investigation of these procedures suggested the production of mutagenic artifacts of isolation or the destruction of labile mutagens during isolation. However, we have adapted a milling procedure for the purpose that not only features gentle isolation conditions but also produces a homogeneous sample that is easily manipulated. This procedure can be applied to the study of sludges from a variety of municipal sewage treatment plants to assess the potential biohazards associated with these byproducts. Furthermore, the residue organics so isolated can be fractionated via our HPLC method (*15*) to isolate the mutagenic components for identification. Currently, we are applying this procedure to primary and secondary sludges to isolate and identify the mutagenic components.

By providing more information on the identity of mutagenic compounds, their frequency of occurrence, their possible source, and their levels in sewage treatment plant sludges, our research will assist in the definition of waste management practices.

Acknowledgments

Technical assistance by B. Myers, L. Rosenblum, R. Hutchenson, and M. Niemi is gratefully acknowledged. Further, our appreciation is extended to N. Knapp and M. J. Frost for skillful and unstinting labors. This research was supported by USEPA CR810792. This chapter has not been subjected to USEPA review and therefore does not reflect the views of that agency and no official endorsement should be inferred.

Literature Cited

1. Babish, J. G.; Johnson, B. E.; Lisk, D. J. *Environ. Sci. Technol.* **1983,** *17,* 272–277.
2. Hopke, P. K.; Plewa, M. J.; Johnson, J. B.; Weaver, D.; Wood, S. G.; Larson, R. A.; Hinesly, T. *Environ. Sci. Technol.* **1982,** *16,* 140–147.
3. Hopke, P. K., Plewa, M. J.; Stapleton, P. L.; Weaver, D. L. *Environ. Sci. Technol.* **1984,** *18,* 909-916.
4. Mumma, R. D.; Raupach, D. C.; Waldman, J. P.; Tong, S. S. C.; Jacobs, M. L.; Babish, J. G.; Hotchkiss, J. H.; Wszolek, P. C.; Gutenman, W. H.; Bache, C. A.; Lisk, D. L. *Arch. Environ. Contam. Toxicol.* **1984,** *13,* 75–83.
5. Nellor, M. H.; Baird, R. B.; Smyth, J. R. *Health Effects Study;* County Sanitation Districts of Los Angeles County: Whittier, CA, 1984; Final Report.
6. Bedding, N. D.; McIntyre, A. E.; Perry, R.; Lester, J. N. *Sci. Total Environ.* **1982,** *25,* 143–167.
7. Bedding, N. D.; McIntyre, A. E.; Perry, R.; Lester, J. N. *Sci. Total Environ.* **1983,** *26,* 255–312.

8. Strachan, S. D.; Nelson, D. W.; Sommers, L. E. *J. Environ. Qual.* **1983,** *12,* 69-74.
9. Clevenger, T. E.; Hemphill, D. D.; Roberts, K.; Mullins, W. A. *J. Water Pollut. Control Fed.* **1983,** *55,* 1470-1475.
10. Bastian, R. K. In *Municipal Wastewater Sludge Health Effects Research Planning Workshop;* U.S. Environmental Protection Agency: Cincinnati, OH, 1984; pp 2-5, 2-13.
11. *Research Outlook;* Office of Research and Development. U.S. Environmental Protection Agency. U.S. Government Printing Office: Washington, DC, 1983; USEPA-600/9-83-002.
12. *Handbook for Analytical Quality Control in Water and Wastewater Laboratories;* Office of Research and Development. U.S. Environmental Protection Agency: Cincinnati, OH, 1979; USEPA-600/4-79-019, p 2-2.
13. Tabor, M. W.; Loper, J. C. *Int. J. Environ. Anal. Chem.* **1980,** *8,* 197-215.
14. Tabor, M. W.; Loper, J. C. *Int. J. Environ. Anal. Chem.* **1985,** *19,* 281-318.
15. Tabor, M. W.; Loper, J. C. *Organic Pollutants in Water: Sampling and Analysis;* Suffet, I. H.; Malaiyandi, M., Eds.; Advances in Chemistry 214; American Chemical Society: Washington, DC, 1986.
16. Loper, J. C.; Lang, D. R.; Schoney, R. S.; Richmond, B. B.; Gallagher, P. M.; Smith, C. C. *J. Toxicol. Environ. Health* **1978,** *4,* 919-938.
17. Ames, B. N.; McCann, J.; Yamasaki, E. *Mutat. Res.* **1975,** *31,* 347-364.
18. Elder, U. A.; Proctor, B. L.; Hites, R. A. *Biomed. Mass Spectrom.* **1981,** *8,* 409-415.
19. Jungclaus, G. A.; Lopez-Avila, V.; Hites, R. A. *Environ. Sci. Technol.* **1978,** *12,* 88-96.
20. Billets, S.; Lichtenburg, J. J. *Interim Methods for the Measurement of Organic Priority Pollutants in Sludges;* Physical and Chemical Methods Branch. Environmental Monitoring and Support Laboratory. U.S. Environmental Protection Agency: Cincinnati, OH, 1983; pp 1-70.
21. Tabor, M. W. *Environ. Sci. Technol.* **1983,** *17,* 324-328.
22. Loper, J. C.; Tabor, M. W.; Miles, S. K. In *Water Chlorination: Environmental Impact and Health Effects;* Jolley, R. L.; Brungs, W. A.; Cotruvo, J. A.; Cummings, R. B.; Mattice, J. S.; Jacobs, V. A., Eds.; Ann Arbor Science: Ann Arbor, MI, 1983; Vol. 4, pp 1199-1210.

RECEIVED for review August 14, 1985. ACCEPTED December 26, 1985.

TOXICOLOGICAL TESTING-ANALYSIS INTERFACE

34

Risk Assessment and Control Decisions for Protecting Drinking Water Quality

Joseph A. Cotruvo

Criteria and Standards Division, Office of Drinking Water, U.S. Environmental Protection Agency, Washington, DC 20460

This chapter describes risk evaluation processes as they have been applied to drinking water standards and guidelines. Traditional risk assessments and standards are based upon single chemical toxicology. They typically assume that no significant interactions occur at the low levels at which chemicals are commonly found in the environment. Newer evaluation techniques might permit development of standards based upon indications of hazard from exposure to the actual environmental mixtures. New concentration techniques and biological indicator measurements will be the keys to this possible innovation in water regulation. If improvements are expected in the ability to assess risks from the consumption of drinking water, concentrates from several sources of varying quality (including reuse systems) should be tested by these techniques to determine relative qualities of these waters and to compare the results with single toxicology predictions.

SAFETY IS THE PRACTICAL CERTAINTY that injury will not result from a substance when used in the quantity and in the manner proposed for its use (*1*). The goal of all public health and water authorities is to assure that public drinking water supplies are safe, pure, and wholesome in the broadest sense, that is, free from contamination by substances of possible health concern as well as free from adulterants that would detract from the water quality and reduce acceptance by consumers. Not only should public water supplies be safe, but they should be perceived to be safe.

In a sense, drinking water in developed countries can be a closed and totally controllable system. It consists of source, treatment, and transport functions, each of which is subject to contamination stresses, but virtually all of which are controllable by feasible social, political, or technological means, given the willingness to expend the required resources. Thus, in effect, society has almost total control over the quality or composition of the water that is delivered to the consumer's tap. In simple terms, sources of public water systems can be protected from contamination, water can be treated for the removal of undesirable components, and the quality of the treated water can be protected during its transport through distribution mains to consumers (2). In the extreme case, water from any source can be synthesized to any desired composition and packaged for total protection.

Three elements drive public policy decisions to protect health and welfare from perceived or actual risks regardless of origin: (1) identification of the existence or the perception of the existence of the risk; (2) assessment of the likelihood of the risk's existence, the quantification of the magnitude of the risk, and the health significance of the risk; and (3) the feasibility, cost, and effectiveness of the means of abating or managing the risk.

Drinking water quality concerns range from (1) infectious disease risks that are large, obvious, and quantifiable; (2) acute or chronic chemical hazards such as those from arsenic or lead that are infrequent but potentially identifiable in cause and effect when they occur; (3) postulated carcinogenic risks from radionuclides or certain organic chemicals that are largely undetectable and empirically unquantifiable and usually small in magnitude relative to overall cancer incidence rates.

The public perception of the existence of any risk associated with the consumption of essential drinking water can have profound consequences both locally and nationally, including the loss of public confidence in political institutions. It can cause some consumers to shift from public water supplies to private sources, self-provided water treatment, or bottled water.

To a considerable degree, the disorientation can be traced to the nonthreshold hypothesis, that is, the theoretical existence of finite, albeit small, risks at any nonzero exposure and the inability of scientists to unequivocally attest to the absolute safety of the product, even when only a trace of a potential carcinogen has been detected.

Fundamental questions exist on (1) the criteria to be used to identify those substances that have the capacity to increase the risk of human cancer, (2) their mechanisms of action, and (3) the magnitude of the risk posed by episodic or chronic regular exposure. The answers to these questions lead to major public policy determinations based on the validity and significance of the real or postulated effects, the feasibility of the possible risk reduction measures, and the economic and social costs of those measures. In a society with finite resources at its disposal, ultimately the benefits accrued versus the cost burden must be weighed, as well as the trade-off between the value of the chosen course versus other social benefits that have been forgone.

Philosophical Basis for the Determination of the Control Level (Standards)

Each society will have a different legal and procedural framework for making control decisions and determining whether these are advisory or mandatory, depending upon operative laws and tradition. The philosophical bases may be different depending upon the type of contaminant, the mechanism of action, and the significance of the adverse effects. However, the philosophical basis for determination of the control level should be articulated; the public has the right to know the meaning and consequences of any control value.

Among the many bases for establishing control levels are the following: zero or no deliberate addition; no detection by specified analytical methods; natural background level; safe or wholesome level; no unreasonable risk level; no known adverse effect level with a margin or safety; level consistent with a specified risk or probability of harm; technologically and economically feasible level; level achievable by using the best available technology; marginal benefits are greater than marginal control costs; and costs of achieving the level are low and socially acceptable.

The decision-making procedure may include detailed quantitative studies subject to legally specified development and scientific and public reviews, judgmental decisions by experts in council, or legislative determinations.

When public health is at stake, the ideal goal should be to assure against the occurrence or the potential occurrence of any of the adverse effects, with a large margin of safety. On the other hand, all decisions ultimately must reflect economic and technological feasibility; thus, it is probable that selected control levels will differ from ideal goals. However, the risk assessment—the process for determining the extent of the risk and the goal—should be separate and distinct from the risk management—the mechanism for evaluating the feasibility and costs of the controls. In both cases, the assumptions and uncertainties should be clearly stated along with the bases for the conclusions.

Drinking Water as a Source of Risk

Risk factors from drinking water include infectious disease, acute or chronic chemical toxicity, and carcinogenicity. In 1981, the North Atlantic Treaty Organization Committee on Challenges to Modern

Society, in its Report on the Health Aspects of Drinking Water Contaminants (chemicals), concluded that, in general, no adverse health effects have been observed from the consumption of drinking water that has been generated in a controlled public supply and that has met drinking water standards (*3*). Nevertheless, the committee stated that known contamination of drinking water by chemicals from disinfection practices, industrial discharges, hazardous waste disposal, and corrosion of piping is a potential hazard. Adverse health risks have been associated with failure to protect the source, to provide adequate treatment, and to ensure the integrity of the distribution system.

Source. Source waters, including rivers, lakes, and ground waters, can often be selected so as to be free from significant biological contaminants or protected from potentially harmful anthropogenic contaminants. Source waters can be contaminated by a variety of synthetic organic chemicals, usually in trace amounts. Ground waters in the vicinity of improperly designed waste disposal sites have sometimes been found to be heavily contaminated by migrating chemicals, most frequently chlorinated solvents such as trichloroethylene, tetrachloroethylene, 1,1,1-trichloroethane, or carbon tetrachloride and fuel products such as benzene and aliphatic hydrocarbons (*4*). However, the most universally found organic contaminants in surface waters and some ground waters are natural products including humic and fulvic acids, terpenes, tannins, amino acids, peptides, and other cellular debris.

Inorganic contaminants such as common salts or trace toxic substances such as arsenic or cadmium can be present. Nitrates are common in agricultural areas. Among the inorganic natural contaminants of potential significance are localized deposits of arsenic or selenium and widespread sources of radionuclides such as radium and especially radon gas from ground sources. The presence or absence of inorganic ions such as calcium may play a role in reducing the postulated risks of cardiovascular diseases associated with the degree of hardness of drinking water.

Treatment Processes. Technology and operating procedures are available to prevent the introduction of these contaminants, and technology is available to remove all of these contaminants from drinking water; however, consumer costs can be substantial when economies of scale are absent (e.g., small communities).

Many chemicals are added to water to remove contaminants such as organic matter, suspended or dissolved solids, and microbial pathogens. Among those added are alum, iron salts, polymeric coagulant aids, chlorine, and other oxidizing agents, all of which may leave residues or byproducts in the finished water. Chlorine gas often contains chloroform, carbon tetrachloride, or other residues, and it reacts with organic matter in the water to produce trihalomethanes, chloramines, haloacetonitriles, haloacetic acids, halophenols, and a host of other byproducts. Normally, the major source of the synthetic chemicals in treated drinking water is the interaction of chlorine or another oxidizing agent with the natural products already there. To further complicate the risk assessment and control decisions, a preliminary unverified report has indicated that the consumption of water containing these oxidizing agents may have increased cholesterol levels in test animals on high fat and low calcium diets. Thyroid hormone levels may have also been affected (*5*).

Distribution Systems. A substantial amount of contamination of drinking water can occur while the water is in transit to the consumer after treatment. Pipes are made of copper, galvanized iron, asbestos–cement, lead, or plastic, and often polymeric or coal tar coatings are used. All of these are capable of contributing contaminants to the water, especially if the water is corrosive. Lead, copper, cadmium, and polynuclear aromatic hydrocarbons in finished water are primarily problems of water distribution and not source water contamination. Physical deterioration of the distribution system can also permit biological contamination to occur during transit.

Microbial Risks from Drinking Water

The principal risk factors in drinking water (except possibly natural radioactivity) are biological in origin as indicated by the reported and projected evidence of waterborne disease. In the 12-year period from 1971 to 1982 in the United States, there were 392 outbreaks involving almost 86,000 reported cases (Table I) (*6*). Many outbreaks, probably the great majority, go unreported because of the difficulty of detecting the event and identifying the etiology of the occurrences. In one pilot study, only about one-third to one-fifth of the actual outbreaks were being recognized and reported. The diseases include uncharacterized acute gastroenteritis, giardiasis, shigellosis, hepatitis-A, typhoid, and salmonellosis. Fortunately, massive numbers of deaths associated with waterborne cholera and typhoid no longer occur in developed countries as they did in the last century prior to the widespread introduction of filtration and disinfection.

Retrospective identification of risk from waterborne infectious disease is a relatively simpler task compared with carcinogenic risks. Many acute effects can be identified with proper population surveillance, related to probable origin, and quantified. Assessments of microbial risks from theoretical projections would be extremely complex. They

Table I. U.S. Outbreaks and Cases: 1971–82

Year	Giardia	*Bac*	*Virus*	*Chem*	*Unknown*	*Total*
			Outbreaks			
1971–80						
Community PWS	22	17	7	22	55	123
Noncommunity PWS	12	16	15	7	110	160
Individual	5	7	4	9	12	37
1981	9	3	1	5	14	32
1982	12	3	7	2	16	40
Total 1971–82	60	46	34	45	207	392
			Cases			
1971–80						
Community PWS	17090	8663	1820	2886	28928	59387
Noncommunity PWS	2390	2683	1762	645	10783	18263
Individual	72	42	28	63	134	339
1981	297	351	1761	1893	128	4430
1982	561	188	853	18	1836	3456
Total 1971–82	20410	11927	6224	3740	43574	85875

NOTE: PWS denotes public water supply, Bac means bacterial, and Chem means chemical.
SOURCE: Adapted from reference 6.

would have to be based upon individual identifications of all of the many pathogenic microorganisms potentially in water; determination of infective doses to each segment of the population; considerations of complex interactions including age, physical state, and other health stresses; immunity states; determinations of exposed populations at various levels; variability of diagnoses; variability of sources of microorganisms; secondary infection rates; and differentiation between water and food and other sources of similar infections.

All of these intellectually challenging and intriguing theoretical exercises have been obviated, although not without conflicts, by the introduction of two operationally simple and practical treatment techniques: disinfection and filtration.

Numerous studies have shown that these conventional processes can typically remove or inactivate six to eight orders of magnitude of virus (*7*). By using the larger figure (10^8), it has been calculated that treatment of a water containing 300 viral units per 380 L would result in a finished water that contained one infectious virus unit per 120 million L. By application of the same treatment conditions uniformly applied to a source water containing as much as 95,000 viral units per 380 L (a grossly contaminated source), one infectious unit would be present in about 400,000 L of finished drinking water (*8*). Only about 1% of the treated water is ultimately ingested; thus, the probability of consumption of that infectious unit by an individual could be commensurately smaller.

Figure 1 illustrates several additional examples of calculated expected infections per year from a community water supply that contains one polio or Echo 12 virus per 1000 L (*9*). For example, one Echo 12 virus in 1000 L is projected to produce about four infections per year per 1000 persons.

The technical objective would be to assure that a consumer would not be exposed to an infectious dose of a pathogen from the drinking water. Such a dose can range from a few or perhaps only one virulent organism (PFU) for polio virus or certain protozoa such as *Giardia lamblia,* to hundreds of *Shigella* or perhaps millions of opportunistic

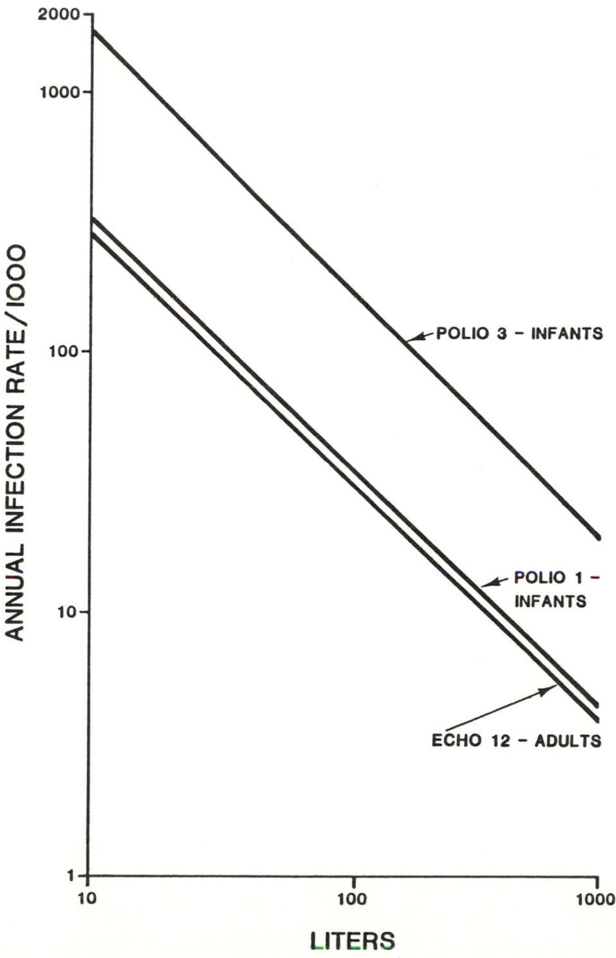

Figure 1. Examples of calculated expected infections per year from a community water supply that contains one polio or Echo 12 virus per 1000 L.

pathogens such as *Pseudomonas* (*10*). Water production systems using available technologies can be sited, built, and operated to reduce the probability of consumer exposure to an infective dose to an extremely low and insignificant risk level. Thus, in the case of biological contaminants of drinking water, the goal of virtually no risk for all practical purposes is within reach in the case of a well-designed and properly operated water production and distribution facility.

In the case of biological contamination, the identification of risk became obvious by experience, the risk assessment was made unambiguous by epidemiology, and the immediate and obvious effectiveness of the risk management decisions demonstrated their wisdom in the absence of elegant quantitative risk extrapolation models and projections of costs per case averted. Costs of water treatment and distribution became trivial relative to almost all other essential commodities, and in the public expectation the biological safety of drinking water became axiomatic.

The fundamental and unresolvable element concerning adverse health effects and trace chemical contamination of drinking water is that in all but a few exceptional cases, three elements—risk identification, risk assessment by epidemiological data, and demonstrable risk management results—may never be available.

Safety and Risk Determination for Chemical Agents

Paracelsus (*11*) observed that "All things are poisons, for there is nothing without poisonous qualities. It is only the dose that makes a thing a poison."

Toxicity has been defined as the intrinsic quality of a chemical to produce an adverse effect (*1*). The toxicology of chemical substances found in drinking water is commonly divided into two broad classes: (1) acute or chronic toxicity and (2) carcinogenicity. However, teratogenic and mutagenic risks could also be considered. The same substance may be capable of causing classic toxic effects and imparting risks of carcinogenicity. The distinguishing characteristic between these categories of effects lies (1) in the probably unverifiable assumption that dose thresholds exist for chronic toxicity effects and (2) in the also unverifiable assumption that dose thresholds do not exist (or have not been demonstrated) for carcinogenic effects. In the case that dose thresholds exist for chronic toxicity effects, the nominal basis for standard setting is to achieve a total daily dose of the substance that is with practical certainty below the level at which any injury would result to any individual in the population. For toxicants assumed to be acting by nonthreshold mechanisms, it follows that some finite risk may exist at any nonzero dose level. Thus, standard setting objectives range from zero, which is not quantifiable and often not practically achievable, to a daily dose level that contributes only a negligible theoretical incremental increase in the lifetime risk of the effect to individuals and/or the population exposed.

The determination of a permissible exposure to a toxic substance requires evaluation of qualitative and quantitative factors including the identification and health significance of the adverse effect; the sensitive members of and the size of the exposed population, biological absorption, distribution, metabolism, and excretion; and the possible additivity, synergism, or antagonism with coexposed substances.

The U.S. National Academy of Sciences (NAS) in the series *Drinking Water and Health* and other writings has described the theory and practice of toxicology and risk assessment and related them to drinking water. These will be used liberally in the following discussions.

Noncarcinogenic Effects: Safety Factors

Numerous substances detected in drinking waters are known to induce toxicity but usually at dose levels much higher than those found in water. Nitrates or nitrites can cause infant methemoglobinemia, lead can affect the hematopoetic or nervous system, cadmium can cause renal damage, and some organohalogens may cause liver toxicity (*12*).

When appropriate data are available from human epidemiology or animal studies, the use of the acceptable daily intake (ADI) concept is a well-accepted procedure for determining concentration levels for standard setting. The ADI of a chemical is defined as the dose that is anticipated to be without lifetime risk to humans when taken daily. The ADI does not guarantee absolute safety, however, and it is not an estimate of risk. The assumption of one threshold for each individual in a large population is simplistic; the population is genetically heterogeneous with a varied history of exposure, prior disease states, nutritional status, and stresses. Thus, it is likely that each individual has a unique threshold, and certain individuals in the population will be at inordinately high risk, whereas others may be at very low risk. The ADI concept is probably not applicable to heavy metals and lipophilic substances, which tend to bioaccumulate (*13*).

The ADI is usually derived from a detailed analysis of the toxicology of the chemical being examined. The no observed adverse effect level (NOAEL) is determined for the most sensitive adverse effect in the test system (usually animals but occasionally humans), and a safety or uncertainty factor is applied to the NOAEL dose to derive the safe level for the general human population.

The ADI is computed by multiplying the experimental NOAEL (in milligrams per kilogram per day) by the weight of a typical adult (70 kg) and dividing by the safety (uncertainty) factor.

$$\text{ADI (mg/person/day)} = \text{NOAEL (mg/kg/day)} \times 70 \text{ (kg/person)}/ \text{safety (uncertainty) factor}$$

Because an ADI is intended to account for total daily intake of the toxicant from all sources, inhalation and food intake as well as water should be accounted for when attempting to arrive at the maximum drinking water level or the adjusted ADI for drinking water at the maximum drinking water level considering only health factors. Thus, in the optimum case when such information is available, the daily uptake from air and the daily intake from food (if 100% uptake is assumed) should be subtracted from the ADI. Finally, for the determination of the acceptable drinking water concentration value, the assumption in the United States is that adults consume 2 L of water per person per day; thus, the final value should be divided by a factor of 2.

$$\text{drinking water target (mg/L)} = \text{ADI (mg/day)} - \text{inhalation (mg/day)} - \text{food (mg/day)}/(2 \text{ L/day})$$

The foregoing calculation is commonly used for determining acceptable lifetime exposures from drinking water to chronic toxicants. In some cases in which the concern is for shorter exposures to young children who may be at higher risk because of a higher water consumption to body weight ratio, the U.S. Environmental Protection Agency (USEPA) has used the 10-kg child and an assumed consumption of 1 L of water per day as the standard exposed individual for calculation purposes. In effect, such calculation introduces an additional safety factor of 3.5.

Some procedures use a dose conversion method involving body surface area (milligrams per square meter) as opposed to weight (milligrams per kilogram). Body surface area is approximately proportional to the two-thirds power of body weight. This relationship may be particularly appropriate for extrapolations from small animals (rats and mice) to humans rather than for data obtained from dogs or monkeys because on this basis, chemicals would be relatively more toxic to larger animals than to smaller ones (*13*).

Figure 2 is a general illustration of a process for the calculation of an ADI for a particular substance (*14*). The solid line to point A is the dose–response curve determined by the multiple dosing experiment. Point A is the highest no observed effect level in milligrams per kilogram per day for the most sensitive adverse end point that was determined from the animal multiple-dose chronic study. Points B, D,

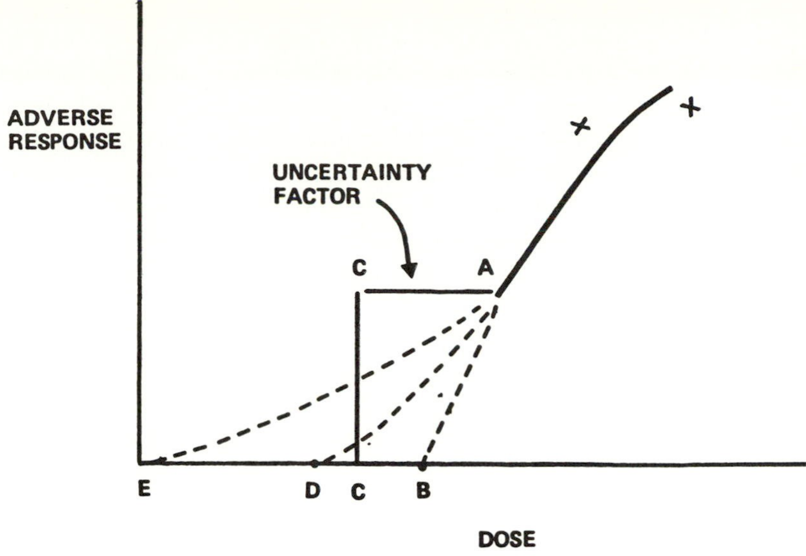

Figure 2. General illustration of a process for the calculation of an ADI for a particular substance.

and E are the presumed thresholds for the adverse effect in the human population if the extrapolated dose–response curves AB, AD, or AE are correct. Point C would be the ADI concentration determined by application of the selected safety (uncertainty) factor to the dose at point A. Because lines AB, AD, and AE are extrapolations, the true dose–response curve in the range of concern is unknown; thus, any of the curves could be correct in a given case. The intention of the standard setter is that AB would be the true curve because the no (actual) response dose would be greater than the calculated ADI value C; thus, the safety (uncertainty) factor was chosen appropriately. However, if AD or AE were the true dose–response curve, then the calculated ADI was too large; thus, the safety (uncertainty) factor was too small, and some members in the human population might suffer the adverse effect. AE indicates a nonthreshold dose–response. The size of the gap between C and B is also of interest because if it were excessively large, overregulation could result in excessive control expenditures without any benefit.

The value of an ADI is entirely dependent on the quality of the experimental data and the judicious selection of the safety (uncertainty) factor, which is entirely judgmental. Among the factors influencing the quality of the experimental data, beyond the mechanics, are the selection of the appropriate animal model as the human surrogate, the number of animals at each dose and the number and range of the doses for acceptable statistical significance of results and shape of the experimental curve, the actual detection of the most sensitive adverse effect (which could be only biochemical change or frank organ damage), the length of the study (lifetime studies versus shorter term studies), and the appropriate route of exposure (inhalation, gavage, ingestion in food or water, etc.). The quality of the experimental evidence determines the magnitude of the safety (uncertainty) factor to be applied.

Safety (Uncertainty) Factors

The safety factor is a number that reflects the degree or amount of uncertainty that must be considered when experimental data are extrapolated to the human population. When the quality and quantity of dose-response data are high, the uncertainty factor is low; when the data are inadequate or equivocal, the uncertainty factor must be larger (*15*).

The original use of the safety factor approach in regulation was by Lehman and Fitzhugh (*13*), who considered that animals may be more resistant to the toxic effects of some chemicals than humans are. They proposed the use of a factor of 10 when extrapolating from animals to humans and the use of another factor of 10 to account for differential sensitivities within the human population (*13*). These are not, however, rigid rules, and they should be applied with a strong infusion of scientific judgment.

The following general guidelines (*15*) have been adopted by the NAS Safe Drinking Water Committee, and they are also used by the USEPA in the development of drinking water standards and guidelines and health advisories.

1. 10 Factor: Valid experimental results from studies on prolonged human ingestion with no indication of carcinogenicity.
2. 100 Factor: Experimental results of studies of human ingestion not available or scanty. Valid results from long-term feeding studies on experimental animals or, in the absence of human studies, on one or more species. No indication of carcinogenicity.
3. 1000 Factor: No long-term or acute human data. Scanty results on experimental animals. No indication of carcinogenicity.

The NAS also examined the application of quantitative models such as log probit and log logistic for human risk assessments for noncarcinogenic substances but found these to be of limited value for contaminants in drinking water (*16*). These models could be used to estimate the risk of a toxic effect, but they require data from lifetime feeding studies with sufficient numbers of animals and with a demonstrated dose–response. Data of this type are seldom available; thus, the NAS concluded that the ADI approach is most useful at this time. However, in those cases in which such data can be obtained, a risk estimate approach can be employed.

An example of an ADI calculation for *p*-dichlorobenzene is provided (*17*). Animal studies at various doses have observed liver and kidney damage, porphyria, pulmonary edema, and spleenic weight changes. Human exposure at high concentrations has been reported to result in pulmonary damage and hemolytic anemia. A 1-year gavage study in the rabbit contained five animals per group dosed between 0 and 1000 mg/kg/day and resulted in weight loss, tremors, and liver pathology. The highest NOAEL was 357 mg/kg/day. A subchronic study indicated a NOAEL of 150 mg/kg in the rat exposed by gavage. Animals received doses of 37.5, 75, 150, 300 or 600 mg/kg/day in corn oil 5 days per week for 13 weeks. No significant differences were observed in food consumption or body weight gain compared with controls for either sex at any dose. At the two highest doses, there was a microscopically detected increase in the incidence and severity of renal cortical degeneration.

By using the experiment just described as the basis for calculations with an additional factor reflecting that the exposure in the experiment occurred for 5 out of 7 days each week, the provisional ADI could be computed as follows:

$$\text{ADI (mg/day)} = (150 \text{ mg/kg/day} \times 70 \text{ kg/person} \times 5/7)/(100 \times 10) = 7.5 \text{ mg/person/day}$$

where 100 is the uncertainty factor appropriate for use with a NOAEL from animal studies with comparable human data and 10 is an additional uncertainty factor because the exposure duration in the experiment was significantly less than lifetime. The assumed daily water intake per person was 2 L/day.

Minimal data were available on food and air contributions to exposure, so an arbitrary designation of 20% was chosen as the maximum allocation from drinking water. Other factors could have been selected.

$$\text{drinking water target} = (\text{ADI} \times \text{water allocations})/(2 \text{ L/day}) = 7.5 \text{ mg/day} \times 20\%/2\text{L/day} = 0.75 \text{ mg/L}$$

More recent data indicate that *p*-dichlorobenzene is carcinogenic in test animals.

Nonthreshold Toxicants

Two fundamentally distinct processes are especially involved in the regulatory decision maker's role in prescribing controls for substances that may cause public health risks without a known safe exposure threshold: (1) assessment of the risk and (2) management of the risk. Risk assessment is the use of a base of scientific research to define the probability of some harm coming to an individual or population as a result of an exposure. Risk management, however, is the public process of deciding what actions to take when risk has been determined to exist. It includes integration of the risk assessment with consideration of engineering feasibility and determination of how to apply the public health official's imperatives to reduce risk in light of legal, social, economic, and political factors (*18*). These two functions should be formally separated within regulatory agencies.

William Ruckelshaus, a former administrator of USEPA, recently described the dilemma in light of all of the uncertainties as follows (*18*):

> When the action . . . has dire economic or social consequences, the person who must make the decision may be sorely tempted to ask for a "reinterpretation" of the data . . . . Risk assessment can be like a captured spy: If you torture it long enough, it will tell you anything you want to know. So it is good public policy to so structure an agency that such temptation is avoided.

The assessment of human cancer risk associated with a substance is a complicated scientific endeavor requiring careful review of all pertinent information by professionals. Such assessment involves primarily the evaluation of clinical, epidemiological, and animal studies as well as short-term tests, structure activity, comparative metabolism pharmacokinetics, and mechanism of action when possible.

The U.S. Office of Science and Technology Policy (OSTP) recently prepared a very comprehensive document reviewing the science and associated principles concerning chemical carcinogens (*19*). It was intended to be of use to regulatory agencies in the United States as a framework for assessing cancer risks. After extensive discussions of basic principles, mechanisms, short-term tests, long-term bioasays, epidemiology, and exposure assessment, it describes a process for using scientific data in the assessment of cancer risk.

The OSTP also described four steps in the risk assessment process:

1. Hazard identification: qualitative evaluations of the agent's ability to produce carcinogenic effects and the relevance to humans.
2. Exposure assessment: the number of individuals likely to be exposed with the types, magnitudes, and durations of the exposure.

3. Hazard or dose–response assessment: the attempt to assemble the hazard and exposure information along with mathematical models to estimate an upper bound on the carcinogenic risk at a given dose.
4. Characterization of the risk associated with human exposure.

Risks from Potential Nonthreshold Toxicants

In 1977, the NAS Safe Drinking Water Committee outlined four principles that it said should be useful in dealing with the assessment of hazards that involve chronic irreversible toxicity or the effects of long-term exposure (*20*). These principles (paraphrased as follows) were intended to apply primarily to cancer risks from substances whose mechanisms involve somatic mutations and may also be applicable to mutagenesis and teratogenesis:

1. Effects in animals, properly qualified, are applicable to man.

   This premise underlies all of experimental biology, but it is often questioned with regard to human cancer. Virtually every form of human cancer has an experimental counterpart, and every form of multicellular organism is subject to cancer. There are differences in susceptibility between animal species, between different strains of the same species, and between individuals of the same strain. However, large bodies of data indicate that exposures that are carcinogenic to animals are likely to be carcinogenic to humans, and vice versa.
2. Methods do not now exist to establish a threshold for long-term effects of toxic agents.

   Thresholds in carcinogenesis that would be applicable to a total population cannot be established experimentally. There is no scientific basis for estimations of safe doses using classic ADI techniques for carcinogens. Experimental bioassays with even large numbers of animals are likely to detect only strong carcinogens. Even negative results in such bioassays do not assure that the agent is unequivocally safe for humans. Therefore, possibly fallible measures of estimating hazard to humans must be used.
3. The exposure of experimental animals to toxic agents in high doses is a necessary and valid method of discovering possible carcinogenic hazards in humans.

   Only dosages that are high in relation to expected human exposures must be given to animals under the experimental conditions that are used. There is no choice but to use numbers of animals that are small relative to exposed human populations, and then to use biologically reasonable models in extrapolating the results to estimate risk at low doses. An incidence as low as 0.01% would represent a risk to 20,000 people in a population of 200 million, whereas the lower limit of reproducibility in common animal studies would be an incidence of 10%. The committee concluded that the best method available today is to assume no threshold and a direct proportionality between dose and tumor incidence. The actual human risk may be greater than predicted by the small animal study because of the longer human lifetime and exposure period.

4. Material should be assessed in terms of human risk rather than as safe or unsafe.

Extrapolation techniques may permit the estimation of upper limits of risk to human populations. To do so, data are needed to estimate population exposure; valid, accurate, precise, and reproducible animal assay procedures are required; and appropriate statistical methods are necessary.

Decisions cannot involve merely risk; benefit evaluations should include the nature, extent, and recipient of the benefits. It is often necessary to accept risks when the benefits warrant the risk, but risks imposed on persons who gain no benefits are generally not acceptable.

The committee concluded the following:

> Mankind is already exposed to many carcinogens whose presence in the environment cannot be easily controlled. In view of the nature of cancer, the long latent period of its development, and the irreversibility of chemical carcinogenesis, it would be highly improper to expose the general population to an increased risk if the benefits were small, questionable, or restricted to limited segments of the population. Such benefit-risk considerations not only must be based on scientific facts but also must be ethical, with as broad a population base as possible used in the decision-making process.

Identification of Compounds Likely To Be Carcinogenic to Humans

The fundamental question of risk assessment for potential human carcinogens requires definition of substances that exceed an evidentiary threshold. Once the scientific evidence establishes a substantial basis for conclusion of known or potential human cancer, it is then in order to determine a procedure for risk quantification. Quantitative risk assessments must always be read with the qualitative evidence of the likelihood of carcinogenicity.

The International Agency for Research on Cancer (IARC) has provided guidelines (*21*) for assessing the epidemiological and animal toxicological data base leading to a conclusion of the strength of the evidence of carcinogenicity of numerous substances. USEPA has recently proposed a similar approach with some added refinement. Three classifications are defined as follows:

1. Carcinogenic to humans. This category includes substances for which there was sufficient evidence from epidemiological studies to support a causal association between exposure and cancer.
2. Probably carcinogenic to humans. This category includes substances for which the evidence ranged from almost sufficient to inadequate at the other extreme. The category was subdivided as follows: (A) at least limited evidence of carcinogenicity to humans and (B) sufficient evidence in animals and inadequate data in humans.
3. Cannot be classified as to its carcinogenicity to humans.

The IARC working group considered that the known chemical properties of a compound and the results from short-term tests could allow its transfer to a higher ranking group.

The definitions of the key terms—sufficient, limited, and inadequate—are provided for both human and animal data.

Data from Humans

1. Sufficient: causal relationship between the agent and human cancer.
2. Limited: a causal relationship is credible, but alternative explanations such as chance, bias, or confounding could not be adequately excluded.
3. Inadequate: one of three conditions including (A) few pertinent data were available; (B) available studies did not exclude chance, bias, or confounding; and (C) some studies did not show evidence of carcinogenicity.

Data from Animals

1. Sufficient: increased incidence of malignant tumors (A) in multiple species or strains; (B) in multiple experiments (preferably by different routes and different doses); or (C) unusual incidence, site, tumor type, or age at onset. Dose-response, short-term tests, and chemical structure may also be factored.
2. Limited: suggestive of carcinogenicity but limited because (A) single species, strain, or experiment; (B) inadequate dosage levels, duration of exposure, follow-up period, poor survival, too few animals, or inadequate reporting; or (C) neoplasms often occurring spontaneously and difficult to classify as malignant by histological criteria alone (e.g., lung and liver tumors in mice).

3. Inadequate: studies cannot be interpreted or the chemical was not carcinogenic within the limits of the test.

I suggest an initial determination of at least sufficient human evidence, limited human evidence supported by animal or short-term tests, or sufficient animal evidence for a conclusion of probable human carcinogenicity as warranting the most conservative regulatory control philosophy. Limited animal evidence without substantial support from short-term tests or mechanistic data indicative of potential human risk would warrant less heroic controls. Yet, these controls would be more protective than those developed from an ADI determination and sufficient to preclude any significant risk in the event that further studies raised the classification.

Risk Extrapolation

Numerous mathematical models have been developed in attempts to estimate potential risks to humans from low-dose exposures to carcinogens. Each model incorporates numerous unverifiable assumptions. Low-dose calculations are highly model dependent, widely differing results are commonly obtained, and none of the models can be firmly justified on either statistical or biological grounds (*22*). Thus, the decision to use this approach and the choice of how to do the calculations are matters of judgment. Among the choices that the decision makers must consider are which model(s) to employ, which assumptions to incorporate, and which acceptable risk to allow.

Numerous bodies, including the NAS Safe Drinking Water Committee, Food Safety Council Scientific Committee, and U.S. Office of Science and Technology Policy, have examined risk extrapolation science and methodology in great detail. The following brief discussion from the NAS Safe Drinking Water Committee report is intended as an introduction to the description of cases in which risk extrapolation procedures have been evaluated in several regulatory decisions involving water.

Risk Calculation Models

All of the mathematical models that relate dose to response rate are either dichotomous response models or time-to-response models (*23*). Dichotomous response models are concerned with whether or not a particular response (tumor) is present by a particular time (e.g., the animal's normal lifetime). In time-to-response models, the relationship between initiation of exposure and the actual occurrence of the response is determined for each animal.

Dichotomous Response Models. It is often assumed that the carcinogenetic process consists of one or more stages at the cellular level beginning with a single-cell somatic mutation at which point the cancer is initiated. The Armitage and Doll multievent theory leads to a model that relates the probability of response, $P(d)$, to the daily dose, d, by $P(d) = 1 - exp[-(\lambda_0 + \lambda_1 d + \lambda_2 d^2 + \ldots \lambda_k d^k)]$ where k represents the number of transitional events in the carcinogenic process, and $\lambda_0, \lambda_1, \ldots \lambda_k$ are unknown nonnegative parameters. For very small values of d, this dose-response rate will be approximately equal to $\lambda_1 d$, assuming λ_0 is the background rate. This model was suggested for use by the Safe Drinking Water Committee, and it is commonly used by USEPA in its analyses of most water exposure risks.

Linear, No-Threshold Model. This simplest model is based on the assumption that risk is directly proportional to the dose: $P(d) = \alpha d$. When it is assumed that the true dose-response curve is convex, linear extrapolation in the low-dose region may overestimate the true risk. However, it is not known if the experimental dose is in the convex region of the curve.

Tolerance Distribution Model. This model assumes that each member of the population at risk has an individual tolerance below which no response will be produced and that these tolerances vary in the members of the population according to some probability distribution (F). The probability distribution is also assumed to involve parameters of location (α) and scale ($\beta \geq 0$) and can be generally denoted by $F(\alpha + \beta \log z)$, where z is the tolerance level to a particular toxic agent. The probability, $P(d)$, that a random individual will suffer a response from a dose, d, is $P(d) = F(\alpha + \beta \log d) = \int_{-\infty}^{\alpha + \beta \log d} dF(x)$.

Therefore, the proportion of the population expected to respond to a specific dose is indicated by the proportion of individuals having tolerances less than this dose level.

Logistic Models. This model is based upon the assumption of a logistic distribution of the logarithms of the individual tolerances.

$$P(d) = F(\alpha + \beta \log d) = [1 + \exp(\alpha + \beta \log d]^{-1}$$

where $\beta \leq 0$. The Committee concluded that tolerance distribution models have little theoretical justification for carcinogenic response.

Hitness Models. Models for radiation-induced carcinogenesis have been proposed on the basis of a target theory that assumes that the site of action has some number of particles ($N \geq 1$) that are hit by k or more radiation particles. The probability of a hit is assumed to be proportional to the dose. The most commonly used versions are the single hit ($N = 1$, $k = 1$), the two hit ($N = 1$, $k = 2$), and the two target ($N = 2$, $k = 1$). Some have suggested a multielement theory of radiation-induced carcinogenesis that involves both a linear and a quadratic dependence upon dose. The possibility of cell killing at high doses can also be included in this model.

Time-to-Tumor Models. Experiments that produce nearly 100% incidence provide little dose-response information. However, examination of times to response may show a monotonic relationship between means or medians and dose levels. Also, early appearing tumors may be more biologically significant and a greater hazard than later tumors.

The Weibull model is a generalization of the one-hit model (*22*):

$P(d) = 1 - \exp(-\beta d^M)$ where M and β are parameters. $\lim_{d \to 0} \frac{P(d)}{d^M} =$ constant. Thus, in the low-dose region, this model becomes linear for $M = 1$, concave for $M < 1$, and convex for $M > 1$.

Case: Radium-226 and Radium-228. The concept of risk projections from experimental dose-response curves has been highly developed in the case of estimating risks to the population from low doses of radiation. Such methods were later extended to estimate risks from other carcinogens in drinking water and other media. Radioactivity can contribute risks from teratogenic, genetic, and somatic (carcinogenic) effects.

Figure 3 illustrates rough dose-response model fits with human data for ionizing radiation and leukemia incidence from atomic bomb survivors (*24*). Data exist down to about the 10^{-5} lifetime risk per person exposed. This value is close to the region of regulatory interest, and relatively small risk differences are predicted by the three illustrated models in the dose range up to two orders of magnitude below the last observed dose value (ca. 5 rad).

Carcinogenic risks are considered to be stochastic effects—those for which the probability of an effect occuring, rather than its severity, is regarded as a function of the dose without threshold.

The basic assumption of the International Commission on Radiological Protection (ICRP) is that for stochastic effects, a linear relationship without threshold is found between dose and the probability of an effect within the range of exposure conditions usually encountered in radiation work. However, ICRP cautions that if the dose is highly sigmoid, the risk from low doses could be overestimated by linear extrapolation from data obtained at high doses. Furthermore, ICRP

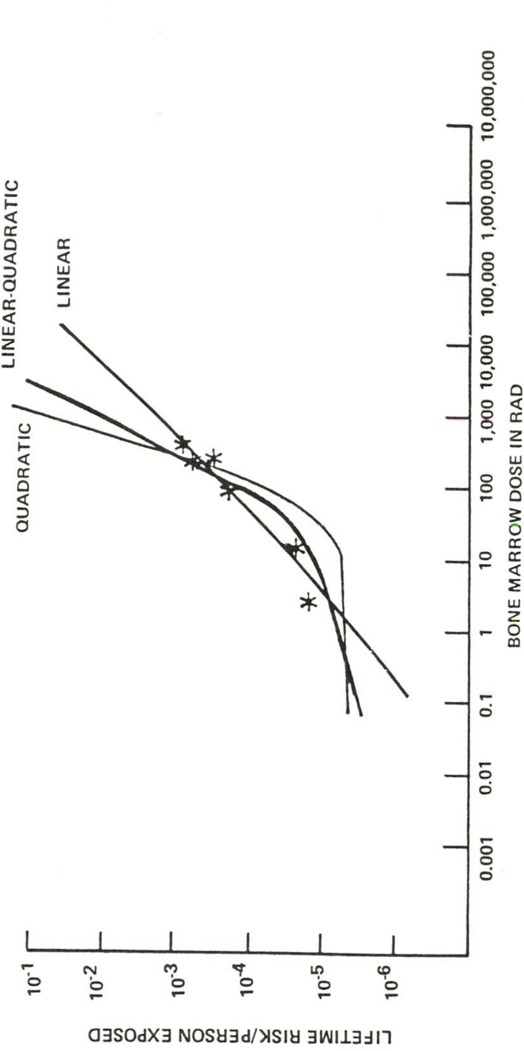

Figure 3. Rough dose–response model fits with human data for ionizing radiation and leukemia incidence from atomic bomb survivors.

states that linear extrapolations from high-dose effects may suffice to assess an upper limit of risk with which the benefits of some activity or the hazards of an alternative activity may be compared. But, in the choice of alternative practices, radiation risk estimates should be used with great caution and with recognition of the possibility that the actual risk at low doses may be lower than that implied by a deliberately cautious assumption of proportionality (*25*).

In the United States, the average background dose of radiation from natural sources is about 100 mrem/year, but it can be approximately double that amount in some localities. It has been estimated that this background may contribute 4.5 to 45 cancers per million people per year. Minute amounts of radioactivity are found in all drinking waters. Some of this natural radioactivity (tritium) comes from cosmic ray bombardment.

In 1976, the USEPA promulgated national drinking water regulations for radionuclides. These regulations included limits for natural radium-226 and -228, gross α-particle activity, and man-made β and photon activity. The background analysis supporting the decision included the following: estimation of the projected cancer risk per dose for an array of possible standard levels, estimation of the total population exposed at a range of dose levels and the number of cancer cases projected to occur because of drinking water, projection of the number of communities affected at various potential standard levels, costs of compliance per community, water rate increases, aggregate national costs, projected number of cancer cases averted, and projected number of lives saved per year by each control strategy.

This analysis is summarized in Tables II and III (*26*). The U.S. drinking water standard for the sum of radium-226 and -228 is 5 pCi/L (*26*). Radium deposits both in bone and to a lesser degree in soft tissue. Ingestion of 2 L/day of drinking water containing radium-226 at 5 pCi/L would result in a skeletal burden of 500 pCi (*27*). The ICRP dose model used in the "Biological Effects of Ionizing Radiation" (BEIR) report predicts an average dose to the bone of about 30 rem/year from a burden of 100,000 pCi; proportionally, a burden of 500 pCi would cause an average dose of 150 mrem/year.

The absolute risk of bone cancer estimated from the BEIR report would be 3 per million man-rem per year. The relative risk of bone cancers expected on the basis of the percent increase in the exposed population was also derived from the BEIR report as 17 per million man-rem per year. This estimate was based on an assumption that bone cancers contributed 4% of the relative risk from total body exposure (excluding leukemia).

About 15% of the absorbed radium is deposited in soft tissues in addition to that deposited in the skeleton (*27*). Again by using the BEIR

Table II. Annual National Cost and Health Savings for Achieving Radium Control Limits

Control Limit (pCi/L)	*Estimated Number of Systems*	*Average Size of Systems*	*Average Cost per System*[a]	*National Cost To Achieve Limit*[b]	*Estimated Total Number of Lives Saved per Year*
9	240	4,200	6.0	1.4	0.6
8	300	5,400	8.0	2.4	1.1
7	370	5,000	9.2	3.4	1.6
6	450	7,450	12.4	5.6	2.5
5[c]	500	8,800	17.5	8.8	3.7
4	670	9,500	21.3	14.0	5.5
3	800	12,000	30.4	24.0	8.2
2	860	12,100	41.6	36.0	11
1	980	18,400	70.2	70.0	15
0.5	1100	20,800	90.2	100.0	20

NOTE: Table includes systems currently exceeding 10 pCi/L.
[a] Values are reported as thousands of dollars per year.
[b] Values are reported as millions of dollars per year.
[c] Interim maximum contaminant level for radium.

Table III. Marginal Cost-Effectiveness of Radium Removal by Zeolite Ion Exchange

Initial Radium Concentration (pCi/L)	*Volume of Water Treated per Person Year (1000 gallons)*	*Annual Cost per Person To Remove 1 pCi/L (dollars)*	*Marginal Cost To Prevent One Cancer (millions of dollars)*
10	3.8	0.57	1.88
9	4.2	0.63	2.09
8	4.7	0.71	2.35
7	5.2	0.78	2.61
6	6.3	0.94	3.14
5	7.5	1.13	3.77
4	9.4	1.41	4.71
3	12.6	1.88	6.28
2	18.9	2.82	9.41
1	36.5	5.48	18.83

report and weighing the risk estimates by organ doses, the absolute risk of bone and soft tissue cancers was determined to be 60% greater than the risk of bone cancer alone; thus, the annual absolute rate would be 1.6×3 per million man-rem per year or 4.8. The relative risk was determined to be 16% greater, or 1.16×17 per million man-rem per year or 20.

Thus, the range of annual risk rates per man-rem is from 4.8 to 20 per million. Because 10 pCi ($2\ L \times 5$ pCi/L) per day would be an ICRP dose to the bone of 0.15 rem (150 mrem), the range of health effects projected at the standard level of 5 pCi/L would be from 0.7 to 3 cancers per year per million exposed persons. By assuming a 70-year lifetime, the lifetime risk would range from approximately 49 to 210 cases (almost all deaths) per million.

The analyst cautioned that the risk estimates are uncertain by a factor of 4 or more, and later model iteration by ICRP and BEIR and adjustments to various parameters have modified those results somewhat (perhaps one-half of the prior risk) (*28*). Nevertheless, when one examines the huge uncertainties incumbent in risk calculations for organic chemicals, it is remarkable in this case that a true risk could be suggested to be within one order of magnitude of the calculated risk.

Case: Chloroform and Trihalomethanes. Chloroform and other trihalomethanes (THMs) are among the principal organic chemical byproducts of the chlorination of drinking water. The four most common THMs are chloroform, bromoform, bromodichloromethane, and chlorodibromomethane. Concentrations approaching 1000 μg/L have been detected in some public water supplies that used surface water sources that contained high concentrations of natural total organic carbon (5–10 mg/L). More typical concentrations in chlorinated finished drinking waters in the United States range from trace to about 540 μg/L.

Although chlorinated drinking water is normally the principal source of human exposure and uptake of THMs, food and inhaled air can also contribute substantial amounts of chloroform. In one study (*29*), mean concentrations (milligrams per year) of animal human uptake of chloroform from drinking water, food, and air were the following (ranges are given in parentheses): 64 (0.73–343), 9 (2–16), and 20 (0.4–204), respectively. The mean concentration (milligrams per year) of animal human uptake of THMs from drinking water was 85; the range was 0.73–572.

Trihalomethanes were regulated in the United States in 1979 (*30*), and Canada (*31*) has also issued guidelines. Both countries examined exposure and risk data, including epidemiology studies, and both used mathematical extrapolation to evaluate potential risks to populations exposed to THMs from drinking water. However, each country used a different logical train to arrive at different but not inconsistent conclusions. The United States established a legally enforceable control requirement for public water systems at 0.10 mg/L to be computed from a running annual average of the sum of the four most common THMs. This value was determined from a balancing of public health considerations and the feasibility of achievement in public water systems in the United States. It reflected the existing and generally available technology for water treatment, which relies heavily on the proven use of chlorine to produce biologically safe water. Canada established a guideline maximum acceptable concentration of 0.35 mg/L for chloroform and concluded that the health hazard posed by substances in drinking water below this level was negligible. Canada also established an objective (goal) concentration for THMs not to exceed 0.0005 mg/L.

The decision by the USEPA to regulate THMs (*30, 32*) was based on several factors:

1. Potential human health risks are present.
2. Drinking water is the major source of human exposure to THMs.
3. THMs are the most ubiquitous synthetic organic chemicals identified in drinking water in the United States.
4. THMs are generally found at the highest concentration of any synthetic organic chemical in drinking water.
5. THMs are inadvertantly introduced during water treatment, and they are readily controlled.
6. Monitoring is feasible.
7. THMs are indicative of the presence of a host of other halogenated and oxidized potentially harmful byproducts of the chlorination process. These byproducts are concurrently produced in greater amounts but cannot be readily characterized or monitored.

The last factor, THMs as indicators of concurrent additional byproduct formation, will probably be shown to be the most significant basis for the regulation; some of these additional byproducts are known to be haloacetonitriles, halogenated phenols, halogenated ethanoic alcohols, aldehydes, and acids (e.g., trichloroacetic acid). None of these were included in the risk calculations that were performed because at the time (1979) they were either unidentified, inadequately tested for their toxicology and carcinogenicity, or unquantified as to population exposure. In addition, all risk evaluations were based on the potential carcinogenicity of chloroform alone because the other three THMs had not been adequately tested for their carcinogenic potential. One of the major assumptions required to include those other THMs in the regulation on an equal weight basis with chloroform was founded on structure–activity relationships and on the greater mutagenic activity for the three brominated compounds in the Ames *Salmonella* test system (*33*).

Another concurrent factor not considered in the risk analysis was the toxicology of residual amounts of the disinfectant species including hypochlorous acid and chloramines related to chlorine that would normally be present as residuals in chlorinated water. The in vivo toxicology of hypochlorite now indicates the formation of haloforms and halonitriles and thus additional risks (*34*).

Carcinogenicity of Chloroform

Chloroform carcinogenicity has been detected in A-strain mice orally dosed at 0.1–1.6 mL/kg every 4 days for 30 doses. In another study, $B_6C_3F_1$-strain mice received oral doses of 138–477 mg/kg/day by corn oil gavage 5 days/week for 78 weeks. Hepatocellular carcinoma was reported in all groups. Osborne–Mendel rats dosed at 90–200 mg/kg/day by corn oil gavage 5 days/week for 78 weeks exhibited kidney epithelial tumors among males. Benign thyroid tumors were seen in females (*35*). Roe (*36*) also reported liver tumors at 17 mg/kg/day and renal tumors at 60 mg/kg/day in mice treated for 96 weeks. On the basis of these data, the NAS concluded that chloroform was an animal carcinogen, and the IARC listed chloroform in Category II—sufficient evidence for carcinogenicity in animals.

More recent studies (*37*) using the Osborne–Mendel rat exposed to chloroform in drinking water at concentrations up to 130 mg/kg/day reported kidney tumors. However, liver tumors were not reported in $B_6C_3F_1$ mice exposed to chloroform in drinking water at concentrations up to 400 mg/kg/day. The differences between the two studies involving $B_6C_3F_1$ mice—corn oil gavage versus drinking water ad libitum and without corn oil—need further examination to explain the effects of the test protocol on the apparent carcinogenicity. It would appear that the latter study (drinking water) would be most relevant as the basis for calculations of potential human risks, and it should lead to much lower estimates than were calculated from the National Cancer Institute (NCI) corn oil gavage study.

Epidemiology

In 1980, the NAS (*38*) reported that a review of 10 epidemiological studies failed to support or refute the results of the positive animal bioassays. This report suggested that chloroform may cause cancer in humans. The NAS stated that any association between THMs and bladder cancer was small and had a large margin of error both because of statistical variance and the nature of the studies that had been conducted. The NAS reached the following conclusion (*38*):

> The methodological complexities inherent in epidemiological studies of human population exposed to multiple contaminants at low concentrations in drinking water make it virtually impossible to establish a causal link between THMs and an increased in cancer of the bladder or any other site. Small differences in cigarette consumption between two population groups could account for the observed associations.

Risk Assessments

At least two attempts have been made to calculate ADI values for chloroform. These attempts assumed the existence of a threshold for the carcinogenicity of chloroform. Roe (*39*) suggested an ADI of 300 μg/L in drinking water. He used the no observed effect level of 17 mg/kg/day in the mouse and a margin of safety of 2000 and assumed consumption of 2 L per person per day.

Health and Welfare Canada calculated a value of 0.02 mg/kg/day or 700 μg/L by using 90 mg/kg/day as the minimum effect dose from the National Cancer Institute bioassay in the rat and applying a 5000-fold safety factor (*40*).

By using a modified linear multistage model, the NAS (1977) concluded that the nominal lifetime incremental risk of cancer falls between 1.5 and 3×10^{-7} per microgram per liter per day. The range of values obtained is the result of calculations made from several data sets.

In 1980, the Ambient Water Quality Criteria USEPA Carcinogen Assessment Group (CAG) by using a further modified linear multistage model as well as a one-hit model, arrived at a value of 0.2 μg/L/day as the upper 95% confidence estimate of the dose in drinking water contributing an excess lifetime risk of 1 in 1 million.

In its chloroform guideline published in 1980, Health and Welfare Canada reported the following risk calculations based upon data from the Osborne–Mendel rat studies (the maximum dose for all models was 0.35 mg/L/day):

Model	*Estimated Maximum Risk*
Probit-log (slope = 1)	$1.6\text{–}4 \times 10^{-8}$/year
Probit-log (actual slope)	1×10^{-9}/lifetime
Linear (one hit)	0.42×10^{-6}/year
Two step	0.267×10^{-6}/year

Comparison of the results of the various linear calculations at 0.10 mg/L (the USEPA drinking water standard) yielded the following results for lifetime (70 year) risk (a consumption of 2 L/day was assumed): NAS 1977, 7.4×10^{-5} to 1.5×10^{-4}; USEPA CAG 1980, 10^{-3}; and Health and Welfare Canada 1979, 0.8×10^{-5}.

These three calculations based upon similar models yielded maximum lifetime risks that differed by about a factor of 100 (10^{-3} to 0.8×10^{-5}). (The lower limit of the risk is zero.) These values should be compared to the approximately 10^{-9} lifetime risk computed by the Probit-log model. Because there is virtually no firm biological basis from which to select among the models (although the one-hit and multistage models seem more appropriate), it is apparent that the predictive value of this type of quantitative procedure in evaluating actual risks was extremely limited in this case. The assessment is seriously weakened further because, of necessity, it was limited to chloroform itself and did not include any other THMs or any of the multitude of additional byproducts of chlorination that would be concurrently present in solution. In this case, the best use of the quantitative assessment is to provide further support for the conclusion that the risk, if any, is small under normal drinking water quality conditions.

This situation provided the opportunity for the decision maker to consider many other factors when the control decision was made. The predominant factors driving the control decision were judgmental and unquantifiable. They included the essentiality of the disinfection process in the control of waterborne disease (which is a large risk), the desire to optimize drinking water quality so as to avoid unnecessary risks, and the consideration of the unique compliance problems of small underfinanced and marginally operated water systems. Ultimately, the decision was driven by the feasibility and costs of treatment process improvements.

USEPA performed post hoc cost–benefit calculations and sensitivity analyses for several regulatory options (*41*). Population exposure calculations were prepared for pre- and postregulation scenarios, and risks were calculated by using the CAG multistage model. A cost of $200,000 per cancer case avoided was assumed on the basis of both earnings and social value.

The analysis was published with the following statement: "EPA feels that benefit–cost analysis using this methodology is more sophisticated than the available data." Figure 4 shows the benefits associated with three maximum contaminant level (MCL) alternatives (0.15, 0.10, and 0.05 mg/L), associating a value of $200,000 per cancer case by using the best estimate of 322 cases avoided (*42*). The costs of regulation alternatives are also shown. The largest vertical distance between the benefit and cost curves represents the maximum net benefit. At this point, the greatest economic efficiency is achieved. This point corresponds to an MCL of 0.105 mg/L (105 μg/L).

Case: Volatile Synthetic Organic Chemicals. The USEPA recently published recommended maximum contaminant levels (RMCLs) for eight volatile synthetic organic chemicals (*43*). Seven of these were treated as potential carcinogens in this proposal. Under the U.S. Safe Drinking Water Act (SDWA), the first step in the regulatory process for

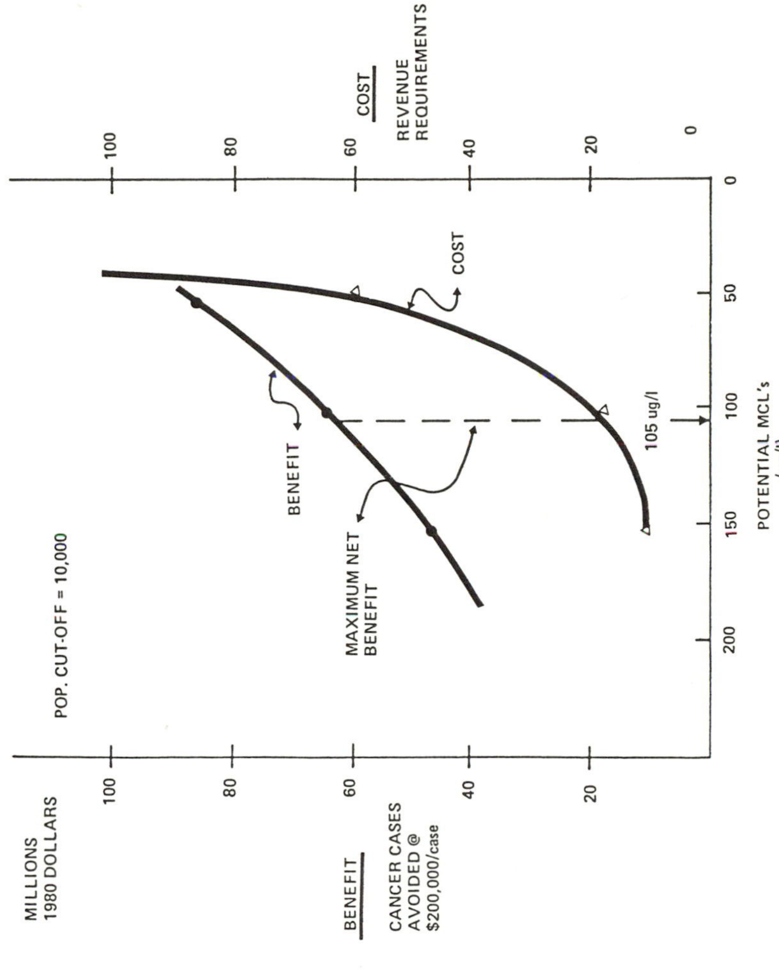

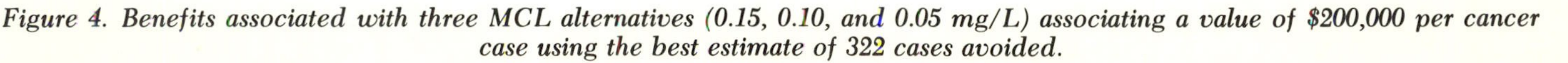

Figure 4. Benefits associated with three MCL alternatives (0.15, 0.10, and 0.05 mg/L) associating a value of $200,000 per cancer case using the best estimate of 322 cases avoided.

establishment of Revised National Primary Drinking Water Regulations requires determination of the level in drinking water that would result in no known or anticipated adverse effects on health. RMCLs are unenforceable health goals. These RMCLs are to be followed by MCLs, which would take costs into consideration and which would become the enforceable drinking water standards. MCLs are to be set as close to the RMCLs as is feasible.

RMCLs for noncarcinogens can be derived from classical ADI calculations using safety factors applied to NOAELs. The only legislative history guidance was provided in a report issued in conjunction with the passage of the Safe Drinking Water Act, which stated the following (*44*): "It [RMCL] must include an adequate margin of safety unless there is no safe threshold. In such a case the RMCL should be set at zero level."

The USEPA considered the following approaches for setting RMCLs for carcinogens:

1. Set the RMCLs at zero.
2. Set the RMCLs at the analytical detection limit.
3. Set the RMCLs at a nonzero level based upon a calculated negligible contribution to lifetime risk.

ALTERNATIVE 1: SET RMCLS AT ZERO. One approach would be to establish RMCLs at zero for substances considered to be nonthreshold toxicants. The existence of a threshold for the action of genotoxic carcinogens cannot be demonstrated by current analytic procedures; thus, it could be conservatively assumed that no threshold exists, absent evidence to the contrary. Because distinctions between mechanisms of action of most carcinogens also cannot be conclusively made at this time, virtually all substances determined to be probable human carcinogens would be assumed to be nonthreshold. Variations of this approach would be to limit the selection of RMCLs at zero only for those substances known to function by genotoxic processes, or perhaps only those determined to be human carcinogens, or only those for which sufficient rather than limited evidence of mammalian carcinogenicity exists.

Setting RMCLs for carcinogens at zero would follow the guidance provided in the legislative history and would express a general philosophy that, as a goal, carcinogens should not be present in drinking water. The USEPA believes that the RMCLs, as a goal, should express the ideal concept that drinking water should be free from avoidable contamination and risk and that quality degradation should not be permitted.

ALTERNATIVE 2: SET RMCLS AT THE ANALYTICAL DETECTION LIMIT. Because of limitations in analytical techniques, it will always be impossible to say with certainty that the substance is not present. In theory, RMCLs at zero will always be unachievable (or at least not demonstrable). Although zero could be the theoretical goal for carcinogens in drinking water, in practice, a goal of achieving the analytical detection limits for specific carcinogens would have to be followed.

The verifiable detection limits (i.e., the RMCLs) would probably fall in the vicinity of 5 μg/L, depending upon the specific chemical. The USEPA suggested that approach was justifiable in that zero is analytically undefinable and the detection limit may be the functional equivalent of zero. Analytical detection limits are moving targets as the state of the art of analytical chemistry progresses, but at least they do provide a measurable target.

ALTERNATIVE 3: SET RMCLS AT A NONZERO LEVEL BASED UPON A CALCULATED NEGLIGIBLE CONTRIBUTION TO LIFETIME RISK. This approach would establish a nonzero level as the RMCL. A level could be selected that would present a negligible risk. In practical terms, such a low nominal risk would effectively preclude any discernible adverse effect on the health of the population. Also, because of the conservative nature of the risk calculation process, such risk may not result in any actual adverse effects on any individual. Just as with analytical detection limits (Alternative 2), a calculated risk target would also be a moving target because calculation methods change and the subjective determination of what is a negligible risk might change.

One possible variation of Alternative 3 would be to set RMCLs as a range of finite risk levels. This alternative would recognize the lack of accuracy and precision of risk calculations and the inherent difficulties in selecting one finite level as the only appropriate health goal in view of the numerous scientific uncertainties of risk estimates.

If a nonzero level is determined as appropriate for the RMCLs, two questions must be considered: (1) What level should be used as representing the no-effect level? (2) How can an adequate margin of safety be incorporated into the finite risk level?

The NAS principles (*Drinking Water and Health, Volume 1*) state that human exposure to carcinogens should be addressed in terms of risk rather than as safe or nonsafe. Because zero is not definable in an analytical sense, rather than speaking in terms of zero concentrations for carcinogens, RMCLs for carcinogens could be set at levels at which the risks are so small that they are considered virtually nonexistent.

The commonly used risk models are generally conservative in their estimation of human risk of exposure to a contaminant. Selection of a target risk based upon a conservative risk model, such as the linearized multistage model, is arguably in accord with the SDWA, which requires the RMCL to be set at a no-effect level with an adequate margin of safety. The decision is judgmental; it is not strictly based upon science but upon a social judgment on what constitutes a negligible risk.

U.S. regulations for environmental contaminants have generally fallen in the 10^{-4} to 10^{-6} lifetime risk range, as calculated from a relatively worst-case linear multistage model. Most of those decisions incorporated consideration of costs and feasibility.

The negligible risk concept considered here is based strictly on individual risk rates and exposure. It does not include other economic or technical considerations that are part of setting the enforceable standards (i.e, the MCLs). The levels for the MCLs (not RMCLs) would thus be considered to be the upper limits of risk that are considered to be acceptable on the basis of current evaluations of the feasibility and costs of controls.

If RMCLs were to be set at a nonzero level, use of the linearized multistage model would often appear to be more appropriate than others to meet the congressional intent. The conservative nature of the model could actually mean that the real risk of exposure was probably lower (e.g., 10^{-7} or 10^{-8}) if any risk actually exists (assuming a nonthreshold mechanism was operative) because the model was structured to be conservative and because of the nature of many of the assumptions in the model.

As an example of what 10^{-6} would mean in terms of the U.S. population, a total of 20 cases of cancer would result if 10% of the population was exposed at a dose level equivalent to a 10^{-6} risk for 70 years. That would be one-third of a cancer case per year as an upper limit in the U.S. population compared with the approximately 500,000 annual cancer deaths that occur. The actual number of cases attributable to that particular substance would probably be less, and perhaps none at all would occur unless some additive or synergistic interaction with other substances resulted in enhanced toxicity.

Table IV lists the results of risk calculations provided in the preliminary proposal for the substances that were proposed as potential carcinogens in the regulatory context at that time (*44*). 1,1-Dichloroethylene was later converted to a listing of equivocal evidence of carcinogenicity. The table includes calculations made by the USEPA CAG and the NAS Safe Drinking Water Committee. These calculations attempt to project concentrations of each chemical in drinking water that, if consumed for a lifetime (70 years) at the rate of 2 L of water per day would contribute an excess lifetime cancer risk of up to 1 in 100,000 and up to 1 in 1,000,000. The quality of evidence of carcinogenicity ranging from sufficient in humans to limited in animals is also included for each chemical. Provisional ADI values calculated from chronic toxicity data only are included for the sake of comparison.

The three sets of risk values listed were calculated by two versions of the linear multistage model generally from the same data, except for vinyl chloride for which CAG 1984 calculation used a different animal

Table IV. Cancer Risk Estimates and Adjusted ADI Calculations for Volatile Organic Compounds

Compound	Adjusted ADI[a] (μg/L)	Projected Upper Limit Excess Lifetime Cancer Risk	Concentration in Drinking Water (μg/L)			Quality of Evidence
			CAG	CAG[a]	NAS	
Trichloroethylene	260	10^{-5}	28	18	45	limited
		10^{-6}	2.8	1.8	4.5	(animal)
Tetrachloroethylene	85	10^{-5}	10	—	35	limited
		10^{-6}	1	—	3.5	(animal)
Carbon tetrachloride	25	10^{-5}	4	2.7	45	sufficient
		10^{-6}	0.4	0.27	4.5	(animal)
1,2-Dichloroethane	260	10^{-5}	9.5	5.0	7.0	sufficient
		10^{-6}	0.95	0.5	0.7	(animal)
Vinyl chloride	60	10^{-5}	20	0.15	10	sufficient
		10^{-6}	2	0.015	1	(human)
Benzene	25	10^{-5}	6.7	—	—	sufficient
		10^{-6}	0.67	—	—	(human)

[a] Values do not consider carcinogenicity.

experiment than CAG 1980 or NAS 1977. Except for vinyl chloride, the several values obtained for each chemical are mathematically and toxicologically insignificantly different, given all of the assumptions and uncertainties. Yet, in the regulatory and economic sense, a factor of 2 to 10 times in the range of interest will have profound effects.

Cothern (*24*) recently performed a series of analyses describing the range of uncertainty inherent in some risk calculations. These range from the inability to precisely estimate the exposure of the cross-sectional population to the substance to interpretation of the scientific data base and to the uncertainties in the models themselves. The authors suggested that the uncertainties in the toxicological data interpretation may be in the range of one to three orders of magnitude, whereas the uncertainty due to the choice of calculation models may exceed six orders of magnitude.

Figure 5 is an illustration of the variability between models for trichloroethylene. At a drinking water concentration of about 50 μg/L, projected incremental lifetime risks range from 10^{-9} (Probit) to 10^{-3} (Weibull).

Case: Ambient Water Quality Criteria. Section 304(a)(1) of the U.S. Clean Water Act of 1977 required the Administrator of the USEPA to publish criteria for water quality (not drinking water) reflecting the latest scientific knowledge on the kind and extent of all identifiable

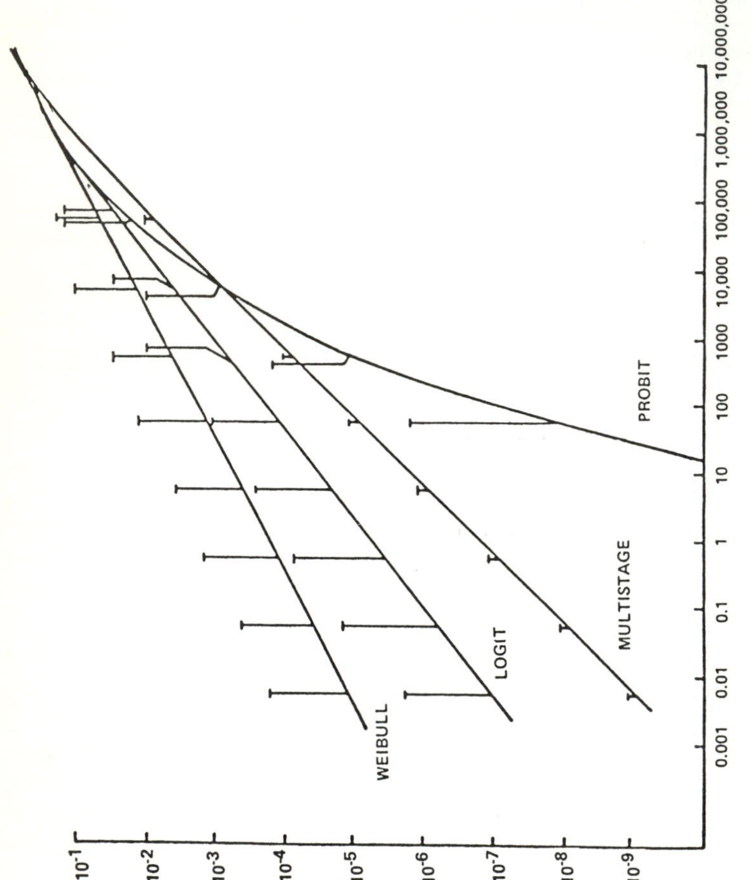

Figure 5. Illustration of variability between models for trichloroethylene.

effects on health and welfare that may be expected from the presence of pollutants in any body of water. Water quality criteria were proposed for 65 pollutants. In section 304, these are nonregulatory scientific assessments of ecological effects. However, when adopted as state water quality standards, they could become enforceable MCLs of a pollutant in ambient waters. In many cases, criteria were determined for the protection of aquatic life as well as for human risks from consumption of fish and from consumption of the untreated water. No provisions are provided in this section of the statute for consideration or incorporation of social and economic factors in setting the water quality criteria.

ADI-type calculations were performed for noncarcinogens. For potential carcinogenic risk calculations, a linearized multistage model or a one-hit model was employed and water concentrations equivalent to calculated risks of 10^{-5}, 10^{-6}, and 10^{-7} were reported. No selection was made as the specific criterion; however, 10^{-5} was suggested as a reasonable value.

Concentration values in micrograms per liter are provided in the box on page 728 for 40 substances at the calculated 10^{-5} risk level at the upper bound. Both a linearized multistage model and the one-hit model (in parentheses) were used (*45*). Many of these values are now being updated.

In the proposed criteria, the linear model was used to calculate the concentrations associated with incremental lifetime risks of 10^{-5}. However, in response to public comment, the USEPA ultimately decided to adopt the linearized multistage model to make full use of all available data. Comparison of the values reported in the box indicates that, for most cases, the concentrations calculated by either model for a given nominal risk are very close.

When applying these values, the USEPA cautioned that the weight of biomedical evidence varied enormously for those chemicals and advised that this finding should not be ignored when applying those target concentrations to local situations.

WHO Guidelines

In 1984, the World Health Organization (WHO) published Guidelines for Drinking Water Quality (*46*) with the goal of the protection of public health by the reduction or elimination of constituents in drinking water that are known to be hazardous to health and well being.

According to WHO, a guideline represents the level of a constituent that ensures an aesthetically pleasing water and does not result in any significant risk to the health of the consumer. The guidelines were derived to protect health, assuming lifelong consumption. WHO also

Chemical	*Water Concentrations (µg/L) Corresponding to a Risk Level of 10^{-5} at the Upper Bound*
Acrylonitrile	0.6 (0.08)
Aldrin	7.4×10^{-4} (5×10^{-5})
Arsenic	0.02
Asbestos	300,000 (fibers/L) (0.05)
Benzene	7
Benzidine	1×10^{-3}
Beryllium	0.1 (0.1)
Carbon tetrachloride	4 (3)
Chloroform	2 (2)
Chlordane	5×10^{-3} (1×10^{-3})
Chloroalkyl ethers	
Bischloromethyl ether	4×10^{-5} (2×10^{-5})
Bischloroethyl ether	0.3 (0.4)
Chlorinated benzenes	
HCB	7×10^{-3} (1×10^{-3})
Chlorinated ethanes	
1,2-Dichloroethane	9 (7)
1,1,2-Trichloroethane	6 (3)
1,1,2,2-Tetrachloroethane	2 (2)
Hexachloroethane	19 (6)
Dichlorobenzene	0.1 (0.02)
DDT	2×10^{-4} (5×10^{-4})
Dichloroethylenes	
1,1-Dichloroethylene	0.3 (1)
Dieldrin	7×10^{-4} (4×10^{-5})
Dinitrotoluene	1 (0.1)
Dioxins	
2,3,7,8-Tetrachlorodibenzodioxin	2×10^{-9} (5×10^{-7})
Diphenylhydrazine	0.4 (0.4)
Heptachlor	3×10^{-3} (2×10^{-4})
Hexachlorobutadene	5 (1)
Hexachlorocyclohexane	
Technical grade	0.1 (0.02)
α-Isomer	0.02 (0.02)
β-Isomer	0.1 (0.03)
γ-Isomer	0.2 (0.05)
Nitrosamines	
Dimethylnitrosamine	1×10^{-2} (3×10^{-2})
Diethylnitrosamine	8×10^{-3} (9×10^{-3})
Dibutylnitrosamine	0.1 (0.01)
Polyaromatic hydrocarbons	3×10^{-2} (1×10^{-2})
Polychlorinated biphenyls	8×10^{-4} (3×10^{-4})
Tetrachloroethylene	8 (2)
Trichloroethylene	27 (21)
Toxaphene	7×10^{-3} (5×10^{-4})
Vinyl chloride	20

NOTE: Many of these values have been superseded by later calculations.
SOURCE: Adapted from reference 45.

cautioned that when a guideline value is exceeded, the public health authority should be consulted for suitable action. WHO concluded that in developing national standards based upon the guidelines, it would be necessary to take into consideration local geographical socioeconomics and dietary and industrial conditions. These considerations could lead to national standards that differ appreciably from the guideline.

Finally, WHO stated that the judgment about safety, or what is an acceptable risk level, is a matter in which society as a whole has a role to play. The final judgment as to whether the benefit of adopting any of the proposed guidelines does or does not justify the risk is for each country to decide.

Guidelines were provided for biological quality, aesthetic quality, radioactivity, inorganic chemicals, and organic chemicals. Among the 20 organic substances considered, 10 of the chemicals were treated as potential carcinogens. These 10 chemicals are the following (guideline values in micrograms per liter are given in parentheses): benzene (10), benzo[*a*]pyrene (0.01), carbon tetrachloride (3), chloroform (30), 1,2-dichloroethane (10), 1,1-dichloroethene (0.3), hexachlorobenzene (0.01), tetrachloroethylene (10), trichloroethylene (30), and 2,4,6-trichlorophenol (10).

The WHO committee decided to set guidelines for carcinogens when reliable data were available from two animal species, preferably when supporting data such as mutagenicity data or population studies were available.

Tentative guideline values were recommended in three cases: carbon tetrachloride, tetrachloroethylene, and trichloroethylene. For these cases, the committee felt that the carcinogenicity data did not justify a full guidance. The committee stated that these tentative values display a greater degree of uncertainty than those derived for the other chemicals.

It was assumed that thresholds for carcinogenicity were either nonexistent or nonmeasurable. A multistage model was used in many cases with a data base similar to the Water Quality Criteria, and often calculated values were rounded off to the nearest digit. The guidelines state "The model is designed to estimate the highest possible upper limit of incremental risk from a lifetime of exposure to a particular daily amount of substance. . . ." An acceptable risk of 1 in 100,000 per lifetime was arbitrarily selected as the criterion.

Finally, the committee cautioned that because the guideline values for the carcinogens were computed from a conservative, hypothetical, mathematical model that could not be experimentally verified, different interpretations should be applied to those values. It stated that the uncertainties involved are considerable, and a variation of about two orders of magnitude (i.e., from 0.1 to 10 times the number) could exist.

In retrospect, it is likely that for many substances the uncertainties could be considerably greater and probably toward the lower risk side, given the design and intent of the conservative calculation model. On the other hand, lack of definitive information on potential synergistic or additive interactions could lead to a reason to take this more conservative approach.

Conclusion

Drinking water can be contaminated from numerous sources by biological, physical, and chemical contaminants at any point from its origin through treatment and distribution to consumers. All of these contaminants are controllable in developed countries. Biological contamination causing waterborne infectious disease is still the most obvious health risk to consumers of public drinking water supplies. Except for relatively isolated cases, in general, risks from natural and anthropogenic chemical contamination do not appear to be great relative to many other sources of risk, although additional contamination problems are being detected as analytical techniques improve and numerous cases of ground water contaminations are being detected. Nevertheless, because control means are available often at low unit costs and countervailing benefits are nonexistent, acceptance of even small risks could be contrary to expectations and good public policy.

Processes for decision making regarding assessment and management of risks should be formally separated to ensure the integrity of the process. Risk assessments are often fraught with many considerable and unresolvable uncertainties. Management decisions must be made in the light of those uncertainties as well as social demands and economic and technological realities.

Numerous assessment methodologies are available to decision makers. ADI calculations, although not quantitatively rigorous, are widely accepted means for determining safe levels for classical threshold toxicants.

Regardless of all of their weaknesses, quantitative extrapolation techniques are at present the only feasible means of attempting to project the consequences (in terms of probabilities) of environmental exposures to potentially nonthreshold toxicants (carcinogens). Numerous calculation models have been suggested, and none of them have a strong biological basis, and none are verifiable in the range of environmental exposure, which can be six orders of magnitude below the animal test dose.

Risk assessment being by nature imprecise must consist to a substantial degree of scientific judgment, and the weight of evidence must constantly be the guide. All of the assumptions and uncertainties in these assessments must be made clear to the decision maker and the public.

Model selection will have a profound effect on low-dose risks, and indications of this variability should be provided. Biological variability must also be described. It is essential to differentiate all of these assumptions and judgments from scientific fact. Indeed, clear distinctions should be drawn between chemicals for which qualitative evidence is overwhelming and those for which the evidence is marginal for additional insight on the quantitative conclusions.

The one-hit or linear multistage models should always be among the models employed because they are usually among the more conservative procedures. Reasonable worst-case assumptions must often be made to err on the side of safety.

Ultimately, decision makers must have access to the best possible scientific analyses with full knowledge of all of the incumbent uncertainties to make the best possible public health judgments in the context of all of the legal, social, political, technological, and economic imperatives that must be weighed.

The decision maker's problem is to operate within all of the uncertainties inherent in the determination of the adequacy of the qualitative evidence of carcinogenicity to humans and in the quantitative projection of risks to exposed populations and arrive at a reasonable public policy conclusion. One very experienced operative in the field drew the following conclusion regarding the goal of determining the risks to public health from potential carcinogens (*47*):

> . . . there is enough substance to the concept of the nonthreshold character of carcinogenic responses to warrant reasonably stringent controls of carcinogens. However, there are enough uncertainties in this concept to regard the actual estimates as being useful mainly as indices of the adequacy of control rather than as accurate estimates of the true number of excess cases that would be produced by a given level of exposure. With this view one would adopt a reasonably stringent yet not excessively stringent target level of control in terms of levels of risk: for example, a lifetime excess cancer risk of one chance in a hundred thousand.

Acknowledgments

This chapter was prepared for the International Agency for Research on Cancer Symposium on Evaluation of Human Health Risks from Drinking Water (December 14, 1984). The opinions expressed in this chapter are those of the author. They do not necessarily reflect the position of the USEPA. The author's perspectives are derived from the assessment activities and interactions with many members of the Criteria and Standards Division and others in the USEPA Office of Drinking Water. These include Larry Anderson, Ambika Bathija, Paul Berger, Richard Cothern, William Coniglio, Penelope Fenner-Crisp,

Susan Goldhaber, Kris Khanna, Victor Kimm, Arnold Kuzmack, William Lappenbusch, William Marcus, Edward Ohanian, Yogindra Patel, David Schnare, and Craig Vogt.

Literature Cited

1. *Principles for Evaluating Chemicals in the Environment;* National Academy of Sciences: Washington, DC, 1975.
2. Cotruvo, J. A. In *Water Chlorination: Environmental Impact and Health Effects;* Jolley, R. L., Ed.; Ann Arbor Science: Ann Arbor, MI, 1980; Vol. 2, pp 1093–1102.
3. Borzelleca, J. In *Drinking Water Pilot Study Summary,* CCMS 130; Bellack, E.; Cotruvo, J. A., Eds.; U.S. Environmental Protection Agency. U.S. Government Printing Office: Washington, DC; EPA 570/9-82-007; p VI 18; *Water Supply and Health;* Van Lelyveld, H.; Zoeteman, B. C. J., Eds.; Elsevier: New York, 1981; pp 205–217.
4. *Fed. Regist.* **1984,** *49(114),* 24333-24335.
5. Bull, R., Director, U.S. Environmental Protection Agency Drinking Water Health Effects Research Laboratory, Personal Communication.
6. Berger, P., Tabulated reports from U.S. Center for Disease Control.
7. *Human Viruses in the Aquatic Environment;* U.S. Environmental Protection Agency. U.S. Government Printing Office: Washington, DC; EPA 570/9-78-006, p 10.
8. Sproul, O. J. In *Viruses in Water;* Berg, G. et al, Eds.; American Public Health Association: Washington, DC, 1976.
9. Calculations by Charles Haas In *Strategies for Control of Viruses in Drinking Water;* by Gerba, C. Office of Drinking Water. Environmental Protection Agency. U.S. Government Printing Office: Washington, DC, 1984.
10. *Drinking Water and Health;* National Academy of Sciences: Washington, DC, 1977; Vol. 1, pp 66–69.
11. Holmstedt, B.; Liljestrand, G. In *Readings in Pharmacology;* Pergamon: New York, 1963.
12. *Fed. Regist.* **1983,** *194,* 45502-45521.
13. Lehman, A. J.; Fitzhugh, O. G. *Assoc. Food Drug Off. Q. Bull.* **1954,** *18,* 33–35.
14. *Fed. Regist. 49(114),* 24337; *Drinking Water and Health;* National Academy of Sciences: Washington, DC, 1977; Vol. 1, p 803.
15. *Drinking Water and Health;* National Academy of Sciences: Washington, DC, 1977; Vol. 1, p 804.
16. *Drinking Water and Health;* National Academy of Sciences; Washington, DC, 1980; Vol. 3, p 35.
17. Draft Criteria Document for Dichlorobenzenes; Office of Drinking Water. Criteria and Standards Division. Environmental Protection Agency. U.S. Government Printing Office: Washington, DC, February 1984.
18. Ruckelshaus, W. D. *EPA J.* **1984,** *Apr.,* 12.
19. *Fed. Regist.* **1984,** *49(100);* 21594-21661.
20. *Drinking Water and Health;* National Academy of Sciences: Washington, DC, 1977; Vol. 1, pp 52–57.
21. *IARC Monogr. Suppl.* **1982,** *4,* 11-14.
22. *Food Cosmet. Toxicol.* **1980,** *18,* 711-734.
23. *Drinking Water and Health;* National Academy of Sciences: Washington, DC, 1977; Vol. 1, pp 47-48, and 1980; Vol. 3, pp 37-45.

24. Land, C. *Science* **1980,** *209,* 1197; Cothern, C. R.; Coniglio, W. A.; Marcus, W. L. In *Assessment of Carcinogenic Risk to the U.S. Population Due to Exposure from Selected Volatile Organic Compounds from Drinking Water via the Ingestion, Inhalation, and Dermal Routes;* Office of Drinking Water. Criteria and Standards Division. U.S. Environmental Protection Agency. U.S. Government Printing Office: Washington, DC, 1984; p 132.
25. *ICRP Publ.* **1977,** *3,* 7.
26. *National Interim Primary Drinking Water Regulations, Radionuclides,* Appendix B; Office of Drinking Water. Environmental Protection Agency. U.S. Government Printing Office: Washington, DC; EPA 570/9-76-003; pp 129–159.
27. Ibid., p 146; and *The Effects on Populations of Exposure to Low Levels of Ionizing Radiation;* Division of Medical Sciences. NRC. National Academy of Sciences: Washington, DC, November 1972.
28. Ibid., p 147.
29. *Statement of Basis and Purpose for an Amendment to the National Interim Primary Drinking Water Regulations on Trihalomethanes;* Office of Drinking Water. Environmental Protection Agency. U.S. Government Printing Office: Washington, DC, August 1979.
30. *Fed. Regist.* **1979,** *44(231),* 68624–68705.
31. *Guidelines for Canadian Drinking Water Quality;* Canadian Government Publishing Centre: Hull, Quebec, 1978; H48–10/1978, p 52.
32. Cotruvo, J. A. *Environ. Sci. Technol.* **1981,** *15(3),* 269.
33. Tardiff, R. G. Presented at the 96th Annual Conference of the American Water Works Association, New Orleans, LA, 1976.
34. Mink, F. L.; Coleman, W. E.; Munch, J. W.; Kaylor, W. H.; Ringhand, H. P. *Bull. Environ. Contam. Toxicol.* **1983,** *30,* 394–399.
35. *Report on the Carcinogenesis Bioassay of Chloroform;* U.S. National Cancer Institute, 1976.
36. Roe, F. J. C. *Preliminary Report of Long-Term Tests of Chloroform in Rats, Mice, and Dogs;* Hazelton Laboratories: Vienna, VA, 1978.
37. Jorgenson, T. A.; Meierhenry, E. F.; Rushbrook, C. A.; Bull, R. J.; Robinson, M.; Whitmice, C. E. *Carcinogenicity of Chloroform in Drinking Water to Male Osborne-Mendel Rats and Female B6C3F1 Mice;* Draft report, SRI International: Menlo Park, CA, 1984.
38. *Drinking Water and Health;* National Academy of Sciences: Washington, DC, 1980; Vol. 3, pp 5–24.
39. *Fed. Regist.* **1979,** *44(231),* 68662.
40. *Guidelines for Canadian Drinking Water Quality,* Supporting Documentation; Minister of Supply and Services, 1980; H48–10/1978-1E, p 668.
41. Kimm, V. J.; Kuzmack, A. M.; Schnare, D. W. In *The Scientific Basis of Health and Safety Regulation;* Crandall, R. W.; Lave, L., Eds.; pp 229–249.
42. *Fed. Regist.* **1980,** *45(49),* 15546.
43. *Fed. Regist.* **1984,** *49(114),* 24330–24355.
44. *Fed. Regist.* **1984,** *49(114),* 24338, 24340.
45. Anderson, E. L. *Risk Analysis;* Vol. 3, No. 4, 1983.
46. *Guidelines for Drinking Water Quality;* World Health Organization: Geneva, 1984; Vol. 1.
47. Albert, R. In *Perceptions of Risk: Proceedings of the 15th Annual Meeting of the National Council on Radiation Protection* March 14–15, 1979; p 12.

RECEIVED for review August 14, 1984. ACCEPTED March 17, 1986.

Guidelines for Canadian Drinking Water Quality

Peter Toft, Murugan Malaiyandi, and J. R. Hickman

Bureau of Chemical Hazards, Environmental Health Directorate, Health Protection Branch, Health and Welfare Canada, Tunney's Pasture, Ottawa, Ontario, Canada, K1A 0L2

The development of Guidelines for Canadian Drinking Water Quality is described. These guidelines are compared with guidelines published by the World Health Organization in 1984. Information is included on drinking water quality in Canada and drinking water consumption habits of Canadians.

AN ADEQUATE SUPPLY OF CLEAN, POTABLE WATER is one of the primary requirements for good health. Traditionally, health hazards associated with water have been the classic waterborne diseases, namely, typhoid, cholera, and hepatitis. The advent, advancement, and practice of the science of bacteriology after the late 18th century led to the recognition of the causes and sources of these diseases, which resulted in the development of disinfection processes and in the recognition of the necessity to prevent public potable water sources from pollution from sewage and postdisinfection contamination.

Following the introduction of disinfection to deal with the problem of bacterial contamination, little was done for many years to investigate other possible effects of disinfection and the link between health effects and drinking water quality. Perhaps the remarkable success in the minimization of the outbreaks of the epidemics just mentioned led to complacency and to a disregard of other potential health problems associated with disinfection processes. Considerable efforts, however, were directed toward reducing levels of specific substances such as iron and manganese and reduction of parameters such as color, turbidity, and hardness. Such efforts were more concerned with reducing the problems of corrosion and scale formation and making the potable water more aesthetically appealing than with health problems.

0065-2393/87/0214/0735$06.00/0

In recent years, advances in analytical methods such as gas chromatography–mass spectrometry and their application to the analysis of drinking water have uncovered a vast number of chemical compounds that had previously gone undetected, many of which are known to manifest toxic properties. Ironically, the discovery and recognition of certain chlorinated organic substances produced by the disinfection process itself gave the stimulus for much of the current research on identification and minimization of organic contaminants in drinking water.

A review committee set up under the auspices of the North Atlantic Treaty Organization in 1979 listed 744 chemical contaminants that had been identified in the drinking water of 14 countries (*1*). Since then, more have been found. However, it has been estimated that more than 80% of the total organic carbon in the drinking water still remains uncharacterized. This growing number of organic chemicals identified in drinking water supplies led to public concern and debate about the potential risks to health.

Jurisdictional Aspects of Drinking Water Supply

According to the British North America Act of 1867, the Canadian legislative base for drinking water derives from the constitutional distribution of powers between the federal and provincial governments. Although this act does not deal specifically with water resources and therefore drinking water supply, judicial interpretation over the years has resulted in a situation in which drinking water is a shared federal–provincial jurisdiction.

Under the British North America Act, the ownership of natural resources, including water, is vested in the provinces that have exclusive jurisdiction over municipal institutions, local water works, and undertakings within the provinces. The municipal or equivalent acts in each of the provinces empower the municipalities to construct, operate, and maintain water systems including potable water supplies. It is apparent, then, that the legal framework for regulating drinking water in Canada differs from that in the United States. There is no Canadian legislation that corresponds to the United States Safe Drinking Water Act.

The Department of National Health and Welfare Act gives reponsibility to this department to investigate and conduct programs related to public health. The Minister of National Health and Welfare has the authority to prescribe, by regulation, standards for foods (and, consequently, water). Although standards have been prescribed for bottled (mineral and spring) waters, standards have not been prescribed for tap water because of the major role traditionally assumed by the provinces.

It would appear, therefore, that the federal and provincial jurisdictions overlap to a considerable degree in regard to drinking water. However, in practice, any action taken has often involved both levels of government, and efforts are complementary rather than overlapping. Generally, the provinces play a lead role in providing an adequate and safe supply of drinking water, whereas the Federal Government of Canada provides leadership in conducting research and in developing standards and guidelines for drinking water quality to protect human health.

There are a few specific cases in which the Federal Government of Canada is solely responsible for drinking water. These include administering the potable water regulations for common carriers (transportation crossing Canadian interprovincial and international borders) and on Canadian coastal and shipping vessels, as well as ensuring an adequate and safe drinking water supply in the Territories, Indian reservations, and military bases.

Drinking Water as a Vehicle for Exposure to Contaminants

The box on page 738 shows the various sources of chemicals to which we may be exposed from our drinking water (*2*). Drinking water is a major route whereby humans are exposed to many of these contaminants.

Table I shows average values for the levels of these contaminants obtained from various national surveys undertaken by the Department of National Health and Welfare. Some water supplies showed significantly higher concentrations of these substances than those in Table I, and in these cases the contribution via water to the total daily intake would be proportionally greater.

Levels for most metals and organics are well below the maximum acceptable concentrations (MAC) specified in Guidelines for the Canadian Drinking Water Quality (1978). However, the highest level of lead found in some localities is higher (79.7 μg/L) than the MAC (50 μg/L).

To evaluate the actual contribution to the total exposure from contaminants via tap water, two general factors have to be considered: (1) the quantity of water consumed per person per day and (2) water-use habits. Under ordinary circumstances, an average adult consumes about 2–5 L of water per day, and this requirement is partly filled by ingesting liquids such as beer, soft drinks, etc. Furthermore, age, activity, and climatic conditions influence the amount of tap water ingested.

As mentioned earlier, information on water quality at the treatment plant is inadequate to assess human exposure to various contaminants. The corrosion of plumbing fixtures and the type of construction materials also contribute to the levels of contaminants in treated water. Although cooking processes at high temperatures accelerate corrosion

Chemicals Commonly Occurring in Drinking Water and Their Sources

Substances influencing the source quality (raw water)

1. Naturally occurring substances: compounds leached from the earth's crust (calcium, heavy metals) and leachates from soils and sediments (humic and fulvic acids).
2. Pollutants from point sources: domestic sewage (detergents), industrial effluents (synthetic organics, metal cyanides, metals, caustic chemicals), landfill waste disposal (metallic ions, chloride, nitrate, nitrite, sulfate, and synthetic organics).
3. Pollutants derived from nonpoint sources: run-off from agricultural lands (fertilizers, pesticides, humic materials), run-off from urban areas (salt, polyaromatic hydrocarbons [PAHs], asbestos), atmospheric fall-out (particulates containing sulfate, nitrate, heavy metals, PAHs, and chlorinated organics).

Substances added intentionally

1. Disinfectants: chlorine, ozone, chloramine, chlorine dioxide, and their byproducts.
2. Coagulants and coagulant aids: iron and aluminum sulfates, activated silicates, alginates, synthetic polyelectrolytes and their impurities, silica, and aluminum electrolyte monomers.
3. pH adjusters: technical grade chemicals and their impurities.
4. Fluoridation agents: fluorides.
5. Corrosion inhibitors: filming amines and their impurities (dicyclohexylamine, *N*-nitroso compounds).
6. Softeners: Calgon products, suds, etc.

Contaminant byproducts

1. Impurities in water treatment chemicals: halogenated hydrocarbons in chlorine, oxides of nitrogen from ozonators, chlorates and chlorites from chlorine dioxide, acrylamide monomers.
2. Disinfection byproducts: trihalomethanes from chlorination, epoxides from ozonation.
3. Substances released from distribution systems and synthetic coatings.
4. Asbestos, metals, and vinyl chloride monomers from certain types of poly(vinyl chloride) piping and PAHs from coal-tar coatings.
5. Substances arising from plumbing fixtures and water treatment devices: metals such as chromium, cadmium, copper, and antimony from plumbing fixtures and bacteria from carbon filters used as water treatment devices.

of the utensils and lead to elevated levels of metallic contaminants, reduced levels of volatile organic contaminants such as trihalomethanes (THMs) will be achieved. Water-softening devices could contribute to increased levels of sodium ions, a factor that may have some deleterious effects on individuals on salt-restricted diets. Hardness is considered to be both an aesthetic and a health parameter. Some epidemiological evidence suggests an inverse correlation between water hardness and cardiovascular disease; the incidence of cardiovascular disease is higher in soft-water communities than in hard-water areas (*3*).

To obtain relevant Canadian data on tap water consumption and use patterns, a survey was undertaken in 1977–78 (*4*). The average daily consumption was found to be 1.34 L/person/day, and this figure agrees well with similar studies conducted in The Netherlands (*5*) and in Great Britain (*6*). Table II shows the volume of tap water consumed by Canadians according to age and sex group. Only 32.3% of the people surveyed ingested amounts in the range of 1–1.5 L/day. Twelve percent consumed quantitites of tap water exceeding 2 L/day, and about 2% consumed more than 3.9 L/day. Practically no difference was observed in tap water consumption between the sexes.

Of particular interest is the consumption of tap water by the young and the elderly, who are likely to be more susceptible to the potential health effects of contaminants (*4*). Table II shows that children consumed less water than adults did. However, when body weight is taken into account, the amount of water (and hence the quantity of contaminants) consumed is significantly higher in children than in adults. This finding is important not only because the dose of contaminants

Table I. Concentration of Contaminants in Canadian Drinking Water Supplies

Contaminant	*Concentration*		*MAC*[a]
	Median	*Range*	
Inorganics			
Cadmium	≤0.02	≤0.02–0.07	5
Cobalt	≤2.0	≤2.0–6.0	—
Chromium	≤2.0	≤2.0–4.1	50
Copper	≤10.0	≤10–900	1000
Nickel	≤2.0	≤2.0–69	—
Lead	≤1.0	≤1.0–79.7	50
Zinc	≤10	≤10–750	5000
Organics			
$CHCl_3$	25	0.00–122	350[b]
Nitrilotriacetic acid	2.82	0.20–20.4	50
Polycyclic aromatic hydrocarbon	3.8	0.05–14.0	—
	2.4	0.05–8.1	—
O-Polycyclic aromatic hydrocarbon	0.91	0.10–1.8	—
	1.0	0.20–2.4	—
DDT	3.0	0.20–8.0	30

NOTE: All concentrations are given in micrograms per liter, except those for polyaromatic hydrocarbons, *O*-polyaromatic hydrocarbons, and DDT, which are given in nanograms per liter.

[a] MACs are specified in Guidelines for Canadian Drinking Water Quality (1978) (now under review).

[b] Value is for THMs.

Table II. Tap Water Consumption in Canada

Age (years)	Females		Males	
	Average	Range[a]	Average	Range[a]
<3	0.69	0.14–1.50	0.50	0.14–1.07
3–5	0.85	0.43–1.47	0.90	0.37–1.57
6–17	1.00	0.33–1.76	1.27	0.50–2.61
18–34	1.33	0.54–2.47	1.43	0.57–2.54
35–54	1.63	0.83–2.62	1.47	0.76–2.57
>54	1.56	0.90–2.28	1.59	0.86–2.29

NOTE: Values are the volume of tap water consumed in Canada per person per day, in liters.
[a] The range is defined by the 10th and 90th percentiles for each age group.

is higher, but also because infants are more susceptible to certain contaminants, for example, lead and nitrate, for which drinking water can be the main vehicle of exposure. In the age group less than 3 years, about 23% of water consumed is in the form of hot beverages in comparison to only 10% in the 3–5-year age group, 17% in the 6–17-year age group, 47% in the young adults (18–34 years), 57% in older adults (35–54 years), and 61% in those over 55 years (*4*).

Many other interesting facts emerged from the survey. Remarkably little difference was found in overall consumption in summer and winter and the form in which water was consumed. A slightly increased consumption in winter was due to increased intake of tea, coffee, and soup. On the other hand, ingestion of homemade beer and wine in summer was twice the amount consumed in winter. Significant differences in water consumption were observed between the various regions of the country. The highest consumption was noticed in Québec, and the Maritime provinces were only slightly lower. Residents of the Prairie provinces drank only slightly more than the Ontarians, who appear to ingest the least quantity of water.

Our greater knowledge about potable water consumption habits will be invaluable in the future to assess the significance of chemical contaminants in water. Not only will the information permit proper guidelines to be established, but the survey also indicated certain areas where public education has to be increased to encourage consumers to modify their water-use habits.

Ground water is the source of water supplied by municipal treatment plants to more than 2 million Canadians (*7*). In addition, there are more than 500,000 drilled wells in Canada that mainly serve rural populations not served by municipal supplies (*8*). Ground water has several advantages over surface waters, for example, constant temperature, consistent quality, relatively dependable constant supply, and on-site availability. Although ground water seems to contain more minerals than surface water, it tends to be free of disease-causing organisms. However, a great deal of concern has been expressed about the occurrence of potentially toxic organic contaminants in ground water in several countries, including the United States (*9*) and The Netherlands (*10*). Several factors that led to the concern are the following:

1. Once a substance has found its way into a ground water aquifer, it is usually not affected by the anoxic, stagnant conditions that prevail underground.
2. Once a substance has reached ground water, it may be transported slowly at a rate determined by adsorption–desorption by the undergound bed materials. Contaminants can travel considerable distances underground and thus make the task of tracing the source of contamination of drinking water wells difficult.
3. Contaminated aquifers may remain contaminated to an unacceptable level for long periods of time.
4. In many cases, ground waters are distributed with minimum or no treatment, and percolation through soil does not remove many chemicals.

The health significance for consumers of contaminated drinking water can be far reaching. In the United States, a number of surveys have been done in recent years. Significant ground water contamination has been reported in 34 states, and at least 33 toxic organic chemicals have been found in drinking water wells, frequently in significant concentrations (*9*). In The Netherlands, a study of all 250 ground water wells showed the presence of trichloromethane, tribromoethene, trichloroethene, and tetrachloroethene at levels sometimes considerably higher than 1.0 μg/L; chlorinated benzenes and 1,1-dichloroethane were also found in a few cases (*10*).

There is a scarcity of Canadian data on ground water quality. In 1979, a survey was undertaken by the Department of National Health and Welfare. Water samples were taken at 30 water treatment plants in 29 municipalities serving 5.5 million consumers, including all major population centers (*11*). Only three of these were ground water sources where minimal or no treatment was practiced. The total organic carbon content of these ground waters was less than half of that found in water supplies taken from rivers and lakes, and the trihalomethane content was also significantly low.

Health authorities around the world have recognized the need for some standard by which to judge the acceptability of water contaminated with traces of organic pollutants. The World Health Organization (WHO), in developing the WHO Guidelines for Drinking Water Quality, has included a number of those organic substances commonly detected in drinking water supplies.

In developing standards to assure that drinking water is safe and wholesome, a number of steps are followed. The essential steps are research→criteria→guidelines→standards. The Department of National Health and Welfare is involved primarily in research and the development of quality standards to protect human health. To meet this challenge, the department maintains an active research program to investigate the toxicological properties of contaminants in water and to assess the potential exposure of people to contaminants from all sources, including water, to assess the risk to human health. The department has, therefore, invested resources in developing analytical methodologies, conducting survey programs, and developing technology to minimize the levels of these contaminants. Research on alternative or improved treatment processes to remove contaminants of concern has also been sponsored. Criteria are developed from dose-response and risk assessment data. These are formal documents summarizing the relevant, valid, and acceptable basis for assessing risks.

The first standards for drinking water quality issued by the Canadian government were promulgated in 1923 and were concerned with the bacteriological quality of drinking water carried on ships in the Great Lakes and other inland waters. Until 1968, Canadian authorities, federal and provincial, relied primarily on standards published by the United States government and by WHO for guidance on other parameters. In 1968, the Department of National Health and Welfare published the first comprehensive Canadian Drinking Water Standards and Objectives, developed by a federal-provincial committee in collaboration with the Canadian Public Health Association (*12*). These standards drew much of their inspiration from those issued by the U.S. Public Health Service in 1962.

By 1974, the need to revise the 1968 standards became apparent because of stimulus by events such as the discovery of chloroform and other THMs in chlorinated drinking water. The detection of various organic constituents at low concentrations made possible by improvements in analytical methodology provided additional impetus.

The current guidelines were published in 1978; they were developed by a Federal-Provincial Working Group set up in 1974 (*13*). Altogether, 61 parameters were reviewed, and guideline values were derived for 51 of them (6 physical, 2 microbiological, 1 radiological, and 42 chemical) (*see* the list on page 743). Two types of limits were defined in the 1978 guidelines. First, an MAC was defined as the upper limit for each parameter. Water containing contaminants above the MAC limit may cause disease or may be aesthetically unpleasant. Conversely, water

Parameters Evaluated During the Revision of the Canadian Drinking Water Standards

Aldrin and dieldrin	Mirex
Ammonia	Nitrate
Antimony	Nitrilotriacetate
Arsenic	Nitrite
Asbestos	Odor
Barium	Parathion
Boron	Pesticides (total)
Cadmium	pH
Calcium	Phenols
Carbaryl	Phthalates
Chlordane	Polychlorinated biphenyls
Chloride	Polyaromatic hydrocarbons
Chromium	Selenium
Color	Silver
Copper	Sodium
Cyanide	Sulfate
DDT	Sulfide
Diazinon	Tannins
Endrin	Taste
Fluoride	Temperature
Hardness	Total dissolved solids
Heptachlor and heptachlor epoxide	Total organic carbon
Iron	Toxaphene
Lindane	THMs
Magnesium	Turbidity
Manganese	2,4-Dichlorophenoxyacetic acid
Methoxychlor	Uranium
Methyl parathion	Zinc

that contains substances at concentrations below their MAC values is considered acceptable for lifelong consumption. Second, an *objective level* was defined as the ultimate quality goal for both health and aesthetic considerations. It was felt that this level would provide a yardstick by which to judge water quality and would serve to discourage pollution.

For those substances for which ingestion may cause adverse health effects, the objective limit was generally set at the detection limit obtainable by a laboratory of good standing by using conventional analytical techniques. For aesthetic parameters, the objective concentration was rather vaguely described as being less than the MAC.

An exception was made for fluoride for which the beneficial effects on dental health warranted recommending an objective concentration of 0.1 mg/L. This concentration was achieved in many cases by adding fluoride ion at the water works.

The Canadian guidelines take account of prevailing socioeconomic factors and are tempered by practical considerations, such as the availability of technological means, to produce water of the desired quality. The guidelines are used by federal, provincial, and territorial authorities; they are not legally enforceable with respect to public water supplies unless promulgated by the appropriate provincial or territorial agency. At present, two provinces have legally enforceable regulations based on the Canadian Drinking Water Guidelines.

Although the current edition of the guidelines is 8 years old, a number of recent developments have indicated a need to revise them before too long. The 1978 guidelines are thought to be adequate in the case of physical, microbiological, and radiological parameters, and no reevaluation is required. Problems have been identified with untreated or private water systems; the current guidelines do not provide advice on these kinds of water supplies (especially in relation to sampling frequency and microbiological safety).

The Federal–Provincial Working Group was reconvened in June 1983 to begin revision of the 1978 guidelines. Some of the commonly occurring inorganic parameters to be reevaluated are calcium, magnesium, sodium, potassium, sulfate, copper, cyanide ion, hardness, and total dissolved solids. The guidelines regarding lead, arsenic, uranyl ion, asbestos, boron, and selenium will also be reviewed and updated.

Activities at the international level relating to drinking water guidelines will now be discussed to put the Canadian effort into perspective.

Previously, two sets of international standards were used: WHO European Standards for Drinking Water (last revised in 1970) and WHO International Standards for Drinking Water (last revised in 1971). Revision of these began in December 1978 and was completed in 1982. The revision formed a part of the International Drinking Water and Sanitation Decade, which has the aim of providing a supply of safe drinking water for all by the year 1990.

Task Groups of Experts were convened by WHO to revise the international drinking water standards. Water technologists, engineers, microbiologists, toxicologists, and chemists from all over the world, representing both developed and lesser developed countries, contributed to the task. The resulting guidelines were published in 1984.

Differences are found between the maximum values in the Canadian guidelines for sodium (not specified), sulfate (500 mg/L), and total dissolved solids (500 mg/L) and those specified by WHO guidelines for sodium (200 mg/L), sulfate (400 mg/L), and total dissolved solids (1000 mg/L). Asbestos, lead, and arsenic were identified for reconsideration because of high public concern about these substances. The WHO guidelines for lead and arsenic are identical to those found in the current

Canadian guidelines, and no limit is set for asbestos in both the WHO standards and the Canadian guidelines.

The 1978 Canadian guidelines specify an MAC of 0.02 mg/L for uranium. This value is about 250 times lower than the previous level (5.0 mg/L) specified in the 1968 guidelines and is based primarily upon the results of chronic studies in rats (although the data were considered to be of poor quality). Studies have been initiated in the Environmental Health Directorate to acquire a more accurate data base, and the need to reevaluate uranium once these results are available has been recognized. The preliminary results suggest that the currently specified levels may be unnecessarily stringent.

A new parameter, aluminum, which is not currently in the Canadian guidelines, has been included in the new WHO guidelines. Aluminum compounds are used extensively in water treatment, although levels are generally less than 0.1 mg/L in distributed water. Above this level, discoloration of water can occur in the presence of iron salts. The WHO limit for aluminum is 0.2 mg/L and is based on aesthetic considerations (*3*).

Organic Contaminants

With few exceptions, drinking water standards for organic substances did not exist until recently. The WHO European and International Drinking Water Standards (1970 and 1971) included pesticides and polycyclic aromatic hydrocarbons (PAHs), and total extractable organics were included in the European edition.

The 1978 Guidelines for Canadian Drinking Water Quality included phenols (for organoleptic reasons), biocides, and THMs. Nitrilotriacetic acid (NTA) was included because of its use as a constituent of laundry detergents, most of which are disposed into surface waters. Studies with rodents have shown that very large doses of NTA can result in an increased incidence of urinary tract tumors. THMs were included because of their production during the process of chlorine disinfection.

The current revision of 1978 guidelines includes the reevaluation of data on phenols, pesticides, and THMs. The Working Group is also considering the need for guidelines for a variety of other organic substances such as benzene, ethylbenzene, xylenes, toluene, chlorobenzenes, trichloroethylene, 1,1-dichloroethylene, tetrachloroethylene, methylene chloride, 1,2-dichloroethane, 1,1,1-trichloroethane, carbon tetrachloride, chlorophenols, dioxins, and benzo[*a*]pyrene, many of which are included in the WHO guidelines. At levels commonly found in drinking water, however, most organic chemicals are not detectable by taste or odor.

In an attempt to ascertain the potential implications of the WHO guidelines in Canada, a comparison is presented in Table III of the limits set by WHO for certain organic compounds and their occurrence in some Canadian potable waters. The data are from a recent survey by the Department of National Health and Welfare (*11*). Water samples were collected from 30 water treatment facilities in 29 different cities across Canada. All the samples were analyzed for 43 compounds, but only 27 were detected at least once in these samples. Ten compounds were detected relatively frequently.

1,1-Dichloroethylene has been shown to cause mammary tumors in both rats and mice and kidney adenocarcinomas in mice (*3*). This compound was found only once during the survey, but it was found at 20 μg/L in treated water, a level significantly higher than the WHO action limit. Further investigation is necessary to determine the significance of this observation.

1,2-Dichloroethane occurred several times in both raw and treated water at levels >10 μg/L, particularly during the summer.

Chloroform concentrations were usually >10 μg/L in treated water and again were generally highest in summer. The values for total THMs were always less than the Canadian MAC of 350 μg/L, but these results suggest there may be difficulties in meeting the recommended WHO limit. (The WHO limit relates to chloroform, whereas the existing Canadian guideline is for total THMs).

Benzene and toluene occurred quite frequently at levels >1 μg/L; occasionally levels >10 μg/L were detected in the corresponding raw and treated water samples from particular treatment plants. Benzene is of concern because of its ability to cause leukemia in humans. A level

Table III. Comparison of Proposed WHO Limits for Certain Organics and Their Observed Levels in Canadian Drinking Waters

Contaminant	*Proposed WHO Limits (μg/L)*	*Incidence in 30 City Surveys*		*Treated Water Observed Concentration (μg/L)*	
		Raw Water (60 samples)	*Treated Water (90 samples)*	*Mean*	*Maximum*
1,1-Dichloroethylene	0.3	0	1	<1	~20
1,2-Dichloroethane	10	10	25	4	30
Chloroform	30	28	87	31	110
Benzene	10	32	55	2.1	47
Monochlorobenzene	10	3	16	<1	5
1,2-Dichlorobenzene	1	nd	nd	<1	1
1,4-Dichlorobenzene	1	5	6	<1	<1
Trichloroethylene	30	48	51	<1	9
Tetrachloroethylene	10	24	39	<1	4

NOTE: nd means none detected.

of 10 μg/L is predicted to carry an additional risk of one leukemia per 100,000 population per lifetime.

Other parameters measured in this survey do not appear to pose widespread potential problems in Canada at the present time.

Monochlorobenzene is widely used as a solvent and industrial intermediate; it can also be formed upon chlorination of benzene-contaminated water, a possible explanation of the higher incidence in treated water than in raw water (*3*). Levels detected in this limited survey, however, were all <5 μg/L.

The dichlorobenzenes are industrial chemical intermediates; 1,4-dichlorobenzene is used as a moth repellent and also in toilet blocks, whereas 1,2-dichlorobenzene is used as a solvent in the chemical industry. Dichlorobenzenes accumulate in tissues, especially fat, and are moderately toxic (*3*). However, levels found in Canadian drinking water supplies were consistently <1 μg/L.

Trichloroethylene was seldom found at levels >5 μg/L, and tetrachloroethylene was never found at levels >4 μg/L, even though both occurred frequently in raw and treated water. Carbon tetrachloride was not detected.

Inorganic Chemical and Aesthetic Characteristics

The recommended Canadian limits for the physical characteristics of drinking water are given in Table IV. They are quite similar to those set by WHO.

The recommended MACs and objective concentrations of certain inorganic substances related to health are shown in Table V. In most cases, the MAC is well above the average daily intake of Canadians through drinking water. The objective concentrations for inorganic

Table IV. Recommended Limits for the Physical Characteristics of Drinking Water

Parameter	*Maximum Acceptable Level*	*Objective Limit*
Color (TCU)	15	<15
Odor	—	inoffensive
pH	6.5–8.5	—
Taste	—	inoffensive
Temperature (°C)	15	<15
Turbidity (NTU)	5	<1

NOTE: TCU = true color unit; NTU = nephelometric turbidity unit.

Table V. Canadian Recommended Limits for Some Inorganic Substances Related to Health

Substance	*MAC*	*Objective Concentration*
Antimony	—	≤0.0002
Arsenic	0.05	≤0.005
Barium	1.00	≤0.100
Boron	5.00	≤0.01
Cadmium	0.005	≤0.001
Chromium	0.05	≤0.0002
Cyanide (free)	0.20	≤0.002
Lead	0.05	≤0.001
Mercury	0.001	≤0.0002
Nitrate (as N)	10.00	≤0.001
Nitrite (as N)	1.00	≤0.001
Selenium	0.01	≤0.002
Silver	0.05	≤0.005
Sulfate	500.00	≤150.00
Uranium	0.02	≤0.001

NOTE: Values are given in milligrams per liter.

Table VI. Canadian Recommended Limits Related to Aesthetic and Other Considerations

Substance	*MAC*	*Objective Concentration*
Chloride	250	250
Copper	1.0	1.0
Iron	0.3	0.05
Manganese	0.05	0.01
Phenols	0.002	0.002
Sulfide (as H_2S)	0.05	0.05
Zinc	5.0	5.0

NOTE: Values are given in milligrams per liter.

parameters are set close to a detection limit achievable by a competent laboratory.

In the case of substances related to aesthetic and other considerations, the Canadian Drinking Water Guidelines (1978) recommend the MACs and objective concentrations given in Tables VI and VII.

Microbiological Considerations

The guidelines provide the following recommendations with respect to microbiological contamination of drinking water:

1. No sample should contain more than 10 total coliform organisms per 100 mL.

Table VII. Recommended Limits for Pesticides

Pesticides[a]	*MAC*	*Objective Concentration*
Aldrin and dieldrin	7×10^{-4}	5×10^{-8}
Carbaryl	7×10^{-2}	5×10^{-4}
Chlordane (total isomers)	7×10^{-3}	5×10^{-8}
DDT (total isomers)	3×10^{-2}	5×10^{-8}
Diazinon	1.4×10^{-2}	1×10^{-6}
Endrin	2×10^{-4}	5×10^{-8}
Heptachlor and heptachlor epoxide	3×10^{-3}	5×10^{-8}
Lindane	4×10^{-3}	1×10^{-7}
Methoxychlor	1×10^{-1}	5×10^{-8}
Methyl parathion	7×10^{-3}	1×10^{-6}
Parathion	3.5×10^{-2}	1×10^{-6}
Toxaphene	5×10^{-3}	5×10^{-8}
2,4-Dichlorophenoxyacetic acid	1×10^{-1}	1×10^{-3}
2,4,5-Trichlorophenoxypropionic acid	1×10^{-2}	1×10^{-3}
Total pesticides	1×10^{-1}	1×10^{-3}

NOTE: Values are given in milligrams per liter.
[a] The limits for each pesticide refer to the sum of all forms present.

2. Not more than 10% of the samples taken in a 30-day period should show the presence of coliform organisms.
3. Not more than two consecutive samples from the same site should show the presence of coliform organisms.
4. None of the coliform detected should be fecal coliform. The objective limit is that no organisms should be detected per 100 mL of the sample.

Conclusion

The existence of a guideline or standard achieves nothing unless it is enforced. This situation implies a need for monitoring and surveillance activities. It is of prime importance that every effort be made to ensure that water quality guidelines are met and that water quality is maintained throughout the distribution system.

Literature Cited

1. *Drinking Water Pilot Study, Committee on the Challenges of Modern Society (NATO/CCMS);* Bellack E.; Cotruvo, J. A., Eds.; Environmental Protection Agency. U.S. Government Printing Office: Washington, DC, 1982.
2. Hickman, J. R.; McBain, D. C.; Armstrong, V. C. *Environ. Monit. Assess.* **1982,** 2, 71–83.
3. *Guidelines for Drinking Water Quality; Health Criteria and Other Supporting Information;* World Health Organization: Geneva, 1984; Vol. 2.
4. *Tap Water Consumption in Canada;* Department of National Health and Welfare: Ottawa, 1981; Report No. 82-EHD-80.

5. Haring, B. J. A. *Tribune du Cebedeau* **1978,** *31*, 349.
6. Hopkin, S. M.; Ellis, J. C. *Drinking Water Consumption in Great Britain;* Medmenham Laboratory, Water Research Centre: U.K., 1980; Technical Report-TR-137.
7. *National Inventory of Municipal Waterworks and Wastewater Systems in Canada;* Environment Canada: Ottawa, 1975.
8. *Canada Water Year Book 1977-78;* Environment Canada: Ottawa, 1979.
9. Council on Environmental Quality *Contamination of Groundwater by Toxic Organic Chemicals;* U.S. Government Printing Office: Washington, DC, 1981.
10. Zoeteman, B. C. J.; Harmsen, K.; Linders, J. B. H. J.; Morra, C. F. H.; Sloof, W. *Chemosphere* **1980,** *9*, 231.
11. Otson, R.; Williams, D. T.; Bothwell, P. D. *J. Assoc. Off. Anal. Chem.* **1982,** *65*, 1370-1374.
12. Health and Welfare Canada *Canadian Drinking Water Standards and Objectives (1968);* Queen's Printer: Ottawa, 1969.
13. Health and Welfare Canada *Guidelines for Drinking Water Quality (1978);* Canadian Government Publishing Centre: Hull, 1979.

RECEIVED for review August 14, 1985. ACCEPTED December 13, 1985.

36

Investigating the Toxicology of Complex Mixtures in Drinking Waters

Richard J. Bull[1]

Toxicology and Microbiology Division, U.S. Environmental Protection Agency, Cincinnati, OH 45268

Assessment of health hazards associated with drinking water obtained from any source that is potentially contaminated requires knowledge of (1) the chemicals that are contaminants of that source, (2) the relative concentrations of those contaminants, (3) some objective indication of a potential for producing adverse health effects, and (4) the relationship between the dose of these chemicals and the adverse effects. In general, the more complete this information is, the more confident one can feel about the assessment of risks. The approach of testing a prepared concentrate of a water sample was evaluated versus identifying all the constitutents and preparing synthetic mixtures of them. Both approaches have substantial drawbacks. However, critical examination of these methods suggests that they might actually be considered as complementary approaches rather than as alternatives.

TOXICOLOGICAL PROBLEMS THAT PEOPLE ENCOUNTER in their environments rarely involve exposure to pure chemicals. Drinking water is a particularly complex situation in this regard. Concerns about hazards to health that might result from drinking water have been historically viewed in the context of individual contaminants. The presence of background contaminants was generally disregarded as long as the water came from a good source because it generally involved low concentrations of individual components that in total usually did not exceed a few milligrams per liter. Analytical techniques with suffi-

[1] Current address: College of Pharmacy, Washington State University, Pullman, WA 99164-6510

cient sensitivity to dependably detect concentrations in this range were primarily confined to pesticides. The broad-spectrum sensitive techniques such as linked gas chromatography and mass spectrometry had yet to be applied to water problems. As these techniques became more generally available, public health officials began to worry about microgram- and even nanogram-per-liter concentrations of individual contaminants in drinking water. In this context, the few milligrams per liter of organic material began to assume more importance. As will be illustrated later, the vast majority of the chemicals that make up the organic fraction of drinking water remain uncharacterized. As a consequence, the concern over the potential toxicity carries with it an amorphous fear that only the toxicology of those compounds that have been specifically identified in drinking waters will be studied, rather than the most important chemicals.

The concern over the uncharacterized organic material in drinking water becomes exacerbated when communities propose potable reuse of municipal waste water. Although this issue is often merely one of convenience to those who would oppose direct potable reuse, it is actually the area of greatest ignorance in water supply. In many respects, direct reuse of waste water for potable purposes represents the same problem that is encountered when water is removed from certain unprotected sources of drinking water. If a community surveys and quantifies the chemical input into its muncipal waste water, designs its treatment plant accordingly, and operates an adequate monitoring program, then the community should be able to produce a drinking water that is at least as good as that produced from other vulnerable sources because it knows what to expect from its source. However, one is still faced with the problem of how to demonstrate the safety of such product water as unequivocally as possible.

Toxicological testing in modern times is very much directed toward individual components. A variety of strong reasons for this approach will be dealt with to some extent later. Historically, this focus on individual components comes from other contexts in which the identification of the individual chemical components that were responsible for the observed effects could be eliminated or exploited more effectively (e.g., pharmacologically). In the case of drinking water, the intent is to generate the data needed to assess hazards to health in the most cost-effective way as possible. In the assessment of risks to human populations, it is as important to be able to estimate the degree of risk as it is to know the type of risk. The question is "How can problems of this kind be approached in a legitimate scientific way?" A second problem is realizing the limitations that must be placed upon the interpretation of data obtained in various ways. This second problem is one to which particular attention must be paid. The alternative is to run the risk of spending large sums of money without being any surer of the answers.

This chapter examines a few of the principles that must guide toxicological testing as well as the extent to which alternative experimental approaches to the complex mixture area come near this ideal or fall short of meeting these principles.

Principles of Toxicological Testing

The basic premise behind toxicological testing of drinking water or its components is that the testing applied can be related to recognized hazards to human health. Three distinct assumptions made in extrapolating from the results of a toxicological test must always be kept in mind. These are the following:

1. A link has been established between the effect observed in the test system and the development of pathology in humans (the response measured in the test system must be recognized as being involved in the development of the disease).
2. The effect observed in the test system should occur at equivalent doses in the test system and in humans after adjustment for differences in metabolism, body weight, surface area, etc. (across-species extrapolation).
3. A reasonable basis exists for extrapolating from doses that produce frank effects to doses that have a very low probability of producing any harm (low-dose extrapolation).

The validity of these assumptions has distinct impacts on estimating the hazard to which humans are exposed. These impacts must not be confused. For example, the consensus among scientists in the area of carcinogenesis is that the initial step in the development of cancer induced by most chemical carcinogens involves a mutagenic event. On this basis, experimental evidence that a chemical is a mutagen in a bacterial test system indicates that the chemical possesses a property often associated with chemical carcinogens. However, such data does not provide any basis for estimating the potency of the chemical as a carcinogen. Therefore, the information provides no basis for estimating the potential carcinogenic risk the chemical might have for humans.

Animals that differ significantly from humans in their physiology and biochemistry may provide a distorted view of the relative hazard that a chemical poses for humans. On the other hand, there are a variety of examples when even mammalian species most closely related to humans respond quite differently to particular chemicals. Nevertheless, some mammalian test systems have been accepted as providing reasonable models of the human, and the points of disagreement are becoming better defined with time. The use of lower species for purposes of quantitatively estimating health hazards to humans should await the day when a similar background of comparison exists. For the present, these species must occupy a level at which a demonstrated effect (e.g., carcinogenesis in fish) may well involve common mechanisms. However, their use must be compared critically for cost-effectiveness with other systems that would occupy a similar position in a testing scheme (e.g., fish carcinogenesis vs. transformation of mammalian cells in vitro would provide presumptive evidence for carcinogenic activity but would not provide information that would be useful to estimate the relative degree of risk).

Discussion of the attributes of various mathematical models that are used to extrapolate from doses producing a frank effect to doses at which the frequency of effect is so low as to be negligible is quite beyond the scope of this discussion. However, there is a trend toward developing extrapolation models that take more specific consideration of biological processes (*1*). Currently, there is a considerable amount of discussion about whether or not to distinguish between genotoxic and nongenotoxic carcinogens. Consequently, it is likely that the mechanism by which a chemical produces its effects will receive increasing emphasis in the future.

Several practical results of the principles just outlined are obtained when one attempts to approach the complex mixture problem experimentally. Probably the most difficult to deal with in the drinking water area is the necessity for testing a contaminant at some dose above which it is likely to be encountered. This method is the only way in which to provide an estimate of the margin of safety. The practical difficulty that this situation raises is the necessity for concentrating the contaminants of concern before they are administered to experimental animals. In the case of health effects that are thought to result from irreversible effects (e.g., the initiation of cancer) there is a straightforward means of estimating the degree of concentration required. In this case, risk is commonly assumed to be a straight-line function of dose at low-response rates. If one considers that 100 animals per dose is the maximum that can be reasonably handled in an experiment and that this amount provides the ability to estimate an additional risk of 1 in 10, then a 10,000-fold concentration would be necessary to reliably estimate one additional cancer death in a population of 100,000 in a lifetime. The approach with health effects that are reversible is generally handled with safety factors that simply specify an acceptable margin of safety. Factors between 100 and 1000 are commonly used with animal data in this case.

Alternative Approaches to the Testing of Complex Mixtures

The critical factors for deciding between approaches have little or nothing to do with the types of toxicological test systems that should be applied. The only way a test procedure would affect these decisions is if it places such demands on the quantity of sample required that the approach becomes cost-ineffective. Consequently, the assumption is made that the testing done will be approximately the same regardless of the approach taken.

Essentially, there are two distinct practical alternatives to the testing of complex mixtures. These are (1) the testing of material concentrated from drinking water and (2) the testing of a synthetic mixture of chemicals that have been identified in drinking water.

In addition to simply describing these approaches, the extent to which interactions between components in the complex mixture are addressed in either approach must be acknowledged. Although many assume the testing of mixtures automatically takes care of toxicological interactions, this situation is not the case. In both of the situations just described, the impact of deleting or adding one more component cannot be clearly predicted. Therefore, one has incomplete information as to interactions with either approach. There may be a way of addressing this issue on a larger scale, but it has yet to be systematized.

Concentrate Approach. The advantages and disadvantages of using concentrates to estimate health hazards due to organic chemicals in drinking water include the following:

Advantages

1. Approach can provide information on potential health hazards that are associated with unidentified water constituents.
2. Approach has the potential for identifying those compounds that represent the most significant hazards in the product water.
3. Approach can be made more cost-effective with experience (and time).
4. Comparison of concentrates from the product water with concentrates from an acceptable source will allow a reasonable estimate of the relative risks associated with the two sources.
5. Short-term tests applied to small samples can add useful information about the variability of the source and product water that could effectively supplement analytical data.

Disadvantages

1. Sample is unique in time and location. Approach limits the ability to generalize from the data to other times and locations.
2. Approach is unable to produce truly representative samples.
3. Approach makes it difficult to assess the performance of the concentration technique.
4. Characterization of the stability of chemicals during preparation and storage of the samples is indirect.
5. Approach uses inadequate methods to characterize contamination of the sample during concentration.
6. Expense is involved in the preparation of sufficient amounts of material to perform long-term studies in rodents.

Concentrate studies provide information on health hazards of chemicals found in the water that are not easily amenable to routine chemical analysis. If positive, these studies will provide impetus to the characterization of components responsible for the effects both chemically and toxicologically. In addition, dealing with each subsequent site with full knowledge of what has been done in prior evaluations can set the stage for more cost-effective approaches in the future. Unfortunately, in the recent past, we have seen a considerable amount of resources devoted to applying cheaper methods to the problem without actually verifying their effectiveness in estimating the degree of risk. Nevertheless, some of the short-term methodologies applied to concentrates can provide some useful information about the variability of both source and finished water. Although not necessarily useful in risk assessment, this information can supplement analytical chemical information concerning variability in the source water in particular. Finally, testing of concentrates of drinking water prepared from municipal waste water and concentrates of the usual acceptable source of drinking water provides one with as direct a means as possible to compare the relative risks of the two drinking waters.

On the other hand, definite disadvantages are associated with the testing of concentrates. First, and perhaps most important, is that a concentrate with an uncharacterized chemical composition may exist as such at only one point in time at a single location. The implications of this shortcoming are the following: (1) The ability to generalize from the data obtained may be quite limited. (2) The results are essentially unverifiable except to the extent that sufficient sample is collected to allow repetitions of the testing. (3) Repetition of testing with the same sample may lead to gratifying results, but the question of how representative the sample tested would be of a sample taken the next day, week, month, or year raises the question of how cost-effective repetitions would really be.

Other shortcomings of the concentrate approach are all simply contributory to this central issue. They all contribute to the uncertainties of how representative the sample is of the water from which it was taken and how representative the water quality was during the sampling period. Because the chemical nature of concentrates cannot be directly defined, much of the quality control information will have to be indirect. For example, the performance of the concentration technique may have to be judged on the basis of spiking the concentrate with a group of innocuous tracer substances to be certain that the character of individual waters is not drastically altering the recovery of chemicals with certain physical and chemical properties. There seems to be no direct way of determining the stability of unidentified chemicals during concentration and storage of samples. Attempts have been made to use the Ames test to determine if chemicals that are reactive in this system are being lost (i.e., judging the stability of the sample by following mutagenic activity during the preparation and storage of the concentrate). However, this methodology is indirect. Similar problems arise from contamination of the sample during concentration and storage. In this case, one can only depend upon appropriate control runs and hope that the contaminants present are not exerting some kind of interactive effect that could alter the results. Finally, the costs associated with the production of concentrates in sufficient quantities to conduct long-term in vivo studies can be prohibitive. For studies to be affordable, it is almost certain that one will have to settle for something that is less than completely representative of the original water. For this reason, concentration techniques employed by different groups must be defined as explicity as possible in terms of recovery of total organic carbon as well as how the technique performs with the matrix being sampled (perhaps again by using chemicals with a defined range of physical and chemical properties as tracers). This determination would assist in making subsequent work more systematic and directed.

Synthetic Mixture Approach. Robert Neal has been the chief proponent of the synthetic mixture approach (2). The advantages and disadvantages of using synthetic mixtures of chemicals as surrogates for mixtures found in actual waters are the following:

Advantages

1. The results can be confirmed by repeating experiments.
2. Approach provides a better base for subsequent development and testing of hypotheses concerning interactions between individual components in the mixture (e.g., synergisms, antagonisms).
3. Approach avoids the large expenditures necessary for the preparation of concentrates.

4. Questions of stability and contamination can be directly assessed analytically.

Disadvantages

1. Ability to generalize results to other waters is limited.
2. Approach does not address the contribution of unidentified compounds to the toxicological hazards associated with the original mixture.
3. Disinfectant byproducts generally predominate over the identified organic chemicals found in finished drinking water. These chemicals may completely overshadow the effects of chemicals that are dependent upon the source of water.
4. Approach makes it necessary to assign an arbitrary limit to the consideration of minor components.

Because this approach deals only with identified components and they are added to the mixture at known concentrations, the results are much more easily verified. To the extent that the identified chemicals are commercially available, this approach avoids the large expenditures necessary for the preparation of concentrates. Problems of stability of the mixture can be directly addressed because of its defined chemical nature. In summary, to the extent that the synthetic mixture can be made representative of the problems associated with a particular water, this approach provides a much better defined and theoretically more controlled situation in which to conduct the toxicological testing.

Again, there are several distinct disadvantages to the synthetic mixture approach. The ability to generalize from the results obtained is no greater than that from the concentrate experiments. Depending upon the criteria one develops to include various chemicals in the mixture, it is likely that such mixtures would contain at least 20 and perhaps up to several hundred individual components. Unless one has preexisting information on the components of the mixture, it will not be possible to determine if the results are due to a summation of the individual effects of the components or if the results are peculiar to the mixture. In other words, any positive effects may represent some unique interactive toxicity that would only be observed at appropriate ratios (and doses) of the individual components. Thus, the addition or deletion of a single component from the mixture has the potential (perhaps not large, but basically unknown) to significantly alter the outcome. In this context, it is not clear what advantages, other than potential cost savings from the reduced testing, that a synthetic mixture approach has over using existing data for identified compounds and summating those results with the results for compounds for which no background toxicological data exists (essentially the recommended approach in proposed U.S. Environmental Protection Agency [USEPA] guidelines).

A major disadvantage that the synthetic mixture has is that it ignores those chemicals that have yet to be identified in drinking water. Summing the concentrations of compounds that have been identified in the most thorough analyses of drinking water indicates that the fraction of the organic material in water that has yet to be identified is between 85% and 95% of the total organic carbon, depending somewhat on the individual water supply. Also, the proportion of those identified is quite often predominated by chemicals that result from the disinfection of water, such as the trihalomethanes, and is not necessarily an indication of the differences in drinking water sources (e.g., a municipal waste water vs. a river).

Discussion

These considerations indicate that there is no perfect solution to the toxicological evaluation of the organic contamination of various water sources or the drinking water that might be prepared from them. A number of pieces of critical information can help decide which of the approaches might be the most cost-effective:

1. The proportion of organic material in the product water that can be identified and quantified.
2. Qualitative and quantitative performance of concentration techniques used singly or in combination.
3. Consideration of which approach would provide a base for characterizing health hazards associated with drinking waters in the long term.
4. Long-term national commitment to the development of a systematic approach to the evaluation of hazards associated with potable reuse of waste water.

Some of the information simply has not been gathered; however, there is also a chance that it exists and simply has not been put together in a recognizable form. The most critical scientific issues are physical–chemical in nature. They involve the technology of concentration and fractionation of organic chemicals from drinking water and the ability to provide a more complete definition of the composition of the organic carbon found in drinking waters from various sources. The other issues are more social and political, but no less critical. One is a question of whether or not there exists a long-term national commitment to systematically develop a means of assessing different drinking water sources and the effectiveness of treatment processes to make these sources acceptable as drinking waters. Such assessment will require considerably more effort in the design of methods to prepare concentrates of contaminants that are representative of the water from which they are derived. The second issue involves the development of a group of toxicological testing methods that can be used to provide the data needed to estimate relative degrees of different types of risks. If that support exists, the concentrate approach is going to be more cost-effective in the long run because what is learned at one site can be used to make evaluations at subsequent sites more efficient. On the other hand, if that commitment does not exist, the synthetic mixture tailored to individual circumstances may be the only affordable approach possible at a local level. In this case, little information gathered from one location would be applicable to another. Consequently, we would probably be better served by evaluating the hazards of individual compounds and summing those hazards as recommended in the proposed USEPA guidelines for assessing risks from complex mixtures. This approach does not necessarily provide an accurate estimation of risk, but data on individual contaminants can be applied to virtually every circumstance.

Despite their expense and the limited ability to generalize from the data that results, concentrate studies are difficult to abandon completely. A literature has developed that illustrates in certain instances, at least, fairly clear-cut indications of adverse health effects (*3*, *4*). There are even more instances in which data have been suggestive of carcinogenic and mutagenic hazards (*5–7*). However, efforts to identify the responsible chemicals have been less than completely successful. For example, in an artificial mixture of byproducts produced from the chlorination of humic acids, less than 7% of the mutagenic activity could be accounted for by extensive chemical analyses (*8*).

Conclusions

The scientific issues relative to the practicality of these two approaches to assessing the overall toxicological risk from water produced from waste water are ones for chemists to address. However, there are rarely simple answers to such complex problems. I suspect that the solution lies somewhere between these approaches. In general, our toxicological data base is probably better for those volatile chemicals that have been identified in drinking water than for the polar nonvolatile materials that are difficult to handle analytically. These polar nonvolatile compounds are apparently concentrated fairly efficiently in some systems that have been examined. An interim solution might be to continue to depend upon an individual chemical approach (as is currently done by USEPA) for those chemicals we can identify and to use the concentrate approach to deal with the unknowns. In the long run, we must understand as much about the chemistry and toxicology of water contaminants as possible. As that research effort proceeds, we should also work to develop the basic mechanistic information that is essential for the prediction of toxicological interactions in a way that is useful in quantitative assessments of risk.

Literature Cited

1. Munro, I. C.; Krewski, D. R. *Food Cosmet. Toxicol.* **1981,** *19,* 549-560.
2. Neal, R. A. *Environ. Sci. Technol.* **1983,** *117,* 113A.
3. Graillot, Cl.; Gak, J. C.; Lancret, C.; Truhaut, R. *Water Res.* **1979,** *13,* 699-710.
4. Truhaut, R.; Gak, J. C.; Graillot, Cl. *Water Res.* **1979,** *13,* 689-697.
5. Loper, J. C.; Lang, D. R.; Schoeny, R. S.; Richmond, B. B.; Gallagher, P. M.; Smith, C. C. *J. Toxicol. Environ. Health* **1978,** *4,* 919-938.
6. Lang, D. R.; Kurzepa, H.; Cole, M. S.; Loper, J. C. *J. Environ. Pathol. Toxicol.* **1980,** *4,* 41-54.
7. Bull, R. J.; Robinson, M.; Meier, J. R. *Environ. Health Perspect.* **1982,** *46,* 215-227.
8. Meier, J. R.; Ringhand, H. P.; Coleman, W. E.; Munch, J. W.; Streicher, R. P.; Kaylor, W. H.; Schenck, K. M. *Mutat. Res.* **1985,** *157,* 111-122.

RECEIVED for review August 14, 1985. ACCEPTED December 16, 1985.

Panel Discussion: Analytical Chemistry–Toxicity Testing Interface

Edited By I. H. (Mel) Suffet (Chairman) and the Panelists

SUFFET: We have the itinerant experts on this podium who have brought to us the problem that the analytical chemists are going to face: What are the best schemes to adapt for concentration of trace contaminants in water and waste water systems? Is it a broad spectrum approach to determine everything that is present, based upon many different isolation methods? Is it an approach to determine everything in a sample as the Master Analytical Scheme proposes? Is it an approach to select specific chemicals for quantitative analysis as priority pollutants? Can it be that some combination of these is the best? The basic question is, what is the best use of analytical chemistry in making judgments about the safety and the quality goals for drinking water?

Phase transfer processes describe the different methodologies that we have available for isolating samples for each of these methods. Which methods are best? A focus of attention is the priority pollutant concept that the USEPA [U.S. Environmental Protection Agency] uses. Does the European community agree with the priority pollutant approach?

I have selected a compound, X, to consider. This compound is present in water and it is not a priority pollutant. This is my favorite compound because it is the compound that is in the water, but we do not know anything about its toxicology yet. If we are going to consider broad spectrum analysis for the general isolation–concentration method, how do we handle it?

Now that the background for the discussion has been developed, I would like to begin by asking the panel, do we analyze only for specific compounds or do we develop a multiple compound protocol? This could be addressed from the analytical and toxicological viewpoints.

GURKA: Well, the specific compound approach obviously wastes most of the information available in the extract. I mean it is bad enough now that the USEPA only considers about 126 priority pollutant com-

0065–2393/87/0214/0763$06.00/0

pounds of the GC [gas chromatographic] volatile fraction, and yet the GC volatile fraction is only 5% or 10% of the total solvent extractables. That leaves open the question of how we do it, but most of the information about the extract is being wasted. That's perfectly obvious.

PIET: To approach this, we probably have to do it in both ways: First, set out the specific compounds that are of great consequence to the environment in its entirety including air, water, and sediment. These might be called real priority pollutants that endanger not only humans but also the environment and the ecosystem itself. So let's find these specific compounds. We have often seen that specific effects are caused by some specific compounds.

SUFFET: You are referring to things such as vinyl chloride, a known carcinogen to humans?

PIET: Yes.

SUFFET: The dioxin isomers, for example?

PIET: Yes, dioxins are compounds that we have not yet identified to cause a specific problem, but we are quite sure that we will find them responsible for problems.

On the other hand, there is the broad spectrum component approach. The component approach is best used to compare environmental systems with each other and to see the extent of pollution or how it may be decreased. With the broad spectrum approach, we want to interface the presence of effect-causing compounds, which might not be a substantial part of the extract, with the presence of other industrial compounds in the extract.

BULL: Well, it seems silly to me to consider neglecting what you can easily analyze chemically. You easily analyze for certain chemicals, and these tend to be the volatile and relatively nonpolar chemicals for which there is a lot of available toxicological information. This does not mean that because you are worried about the whole mixture, you discard what you know about the individual compounds.

SUFFET: But those individual compounds, like the volatiles, which are not very difficult to isolate and collect, do you put them back into a reverse-osmosis extract of a broad spectrum of compounds before you give them to the animals?

BULL: No, I would not. I definitely would not. I would test those separately, because there is no point in cluttering up an experiment that is already cluttered.

What I would really like to see as a bottom line is to deal with those volatile compounds that are usually poorly concentrated by most concentration methods on the basis of what you know about them. That's where the bulk of the toxicological information is available already. You can do a combined experiment using those, if you like, if you can't deal with them on the basis of what you know about them individually. However, compounds such as trichloroethylene, tetrachloroethylene, chlorinated benzene, etc. have received a lot of study individually.

What really is the issue, I think, is the material that has yet to be dealt with analytically. I tend to think these are the polar materials. It doesn't make sense, for example, to run an Ames test on benzene. You already know benzene is a carcinogen. You do not need to run an Ames test on benzene to deal with benzene in drinking water.

The information that you don't have, which bioassays of the concentrate will partially give you, is the information on that fraction of the material that is in the water that you can't deal with analytically by more traditional means.

SUFFET: Well, something like electrophilic materials, the ones I think of are the epoxides: How would you concentrate them? They are not very stable in the laboratory by known concentration methods. Do we know if they're destroyed during sampling? Has anybody tested this? Has anybody taken a series of or should we take a series of these very difficult to analyze electrophiles and put them through the broad spectrum isolation procedures for testing?

TABOR: At the USEPA Workshop at Palo Alto, California, in July 1984 [*see* chapter 2 of this book], it was suggested that every one of the six isolation protocols recommended would be tested with surrogate compounds to validate the procedures, and Dave Brusick of Litton Bionetics and some of the other biologists in the group were talking about lists of compounds on the priority pollutant list and others to reflect compounds with and without known mutagenic activity.

BULL: You're really referring to direct-acting carcinogenic compounds for the most part, the unstable or reactive electrophilic compounds.

SUFFET: I think this audience would appreciate it if you explain the difference between the two types of compounds that are carcinogenic.

BULL: Okay, certain compounds require metabolic activation to electrophilic intermediates, and the others are electrophiles themselves. Both types of compounds can be carcinogenic. Presumably, the latter group [electrophiles] would be the ones that would be destroyed by the concentrating methods. The worry would be that nucleophilic material in the concentrate could react with the electrophiles and destroy them as you concentrate the samples. This would be proportional to the amount of compounds present and the degree and method of concentration. Ultimately, you can expect them to disappear.

Most of our evidence comes from model studies on chlorination byproducts. This is the major source of electrophilic chemicals in most drinking waters. The situation, however, may be different in industrial or other types of waters. Those compounds seem to survive the concentration procedure, at least to the extent that we have been able to deal with that. I think Jack Loper's data on the mutagenic activity of extracts collected by reverse osmosis at the cities in the five-city study support the idea that at least those compounds that you recover are fairly stable yet pretty active mutagenic compounds.

AUDIENCE: What do you do about compounds that are not direct-acting mutagens? I am talking about compounds that are probably in drinking water and that might be tumor promoters yet would not be active in the Ames test.

BULL: This is one of the reasons I was presenting the argument that you have to do whole animal studies not only to judge the degree of hazard but to deal with the many factors that can modify the impact of a mutagenic chemical in the whole animal. Those toxicologists who tend to be generalists are reluctant to use the Ames tests and the shorter term tests for carcinogenesis because there are many toxicological hazards that they are not capable of dealing with at all. I did not get a chance to develop that argument as well as I would have liked. I see the short-term carcinogenic test as not particularly useful in judging hazards associated with a particular water. I see it very useful in telling me something about the variability of that water.

I think at this stage of the game I do not see any other alternative but to go to real animal studies to try to judge the relative risk of this source of water versus a source of water that is already deemed to be acceptable. Not only do you have to worry about the tumor promoters, as you pointed out, but you also have to worry about renal effects, hepatotoxic effects, etc. These kinds of effects have been associated with those chemicals that have been identified in water just as much or even more so perhaps than carcinogenesis.

LOPER: I have a question for Cotruvo. I think that one thing strictly from the bioassay point of view (and that comes down to being 98% Ames test data) is that there is a logical approach from the practical point of view of using Ames test data.

I agree, as Bull just said, the assay does not see a lot of toxicological end points, but for better or for worse, I think we cannot ignore the fact that we are still on the dilemma of the relative importance of threshold versus nonthreshold effects; we are living with the recognized concern about compounds that might have a 5–30-year time in terms of latency for cancer. So you are dealing with a toxicology problem for compounds that have genotoxic effects with long-term gestation periods versus compounds that we approach classically as short-term toxicants, where we can define the functional nonthreshold effect.

We are not going to solve this problem today, but I do not think we can get away from the issue by wishing it away. The Ames test remains highly predictive for a large percentage of compounds that are known to be carcinogenic, and the reality is that the data will be there and you will have to find a way to deal with it. When we started our work, the state-of-the-art collection method of reverse osmosis was used to screen a number of toxicological parameters; it is the nature of the biology we deal with that the Ames test was found to develop reproducible data at a low cost and in a time frame that was less than most of the other toxicological tests. So, if we set up a comparison of short-term screening-test studies, it will generate the same kind of hard numbers for mutagens that somehow you will have to deal with.

I agree with what Bull said with respect to his first two major points; namely, if you focus on water reuse as in the Denver program, you want to consider the source of the water and then, secondly, the actual levels of the known toxicants. If Neal were here I think he would say, "Let's study those compounds that are at high level—identify them and see if they're toxicants in any of the likely biological end points, and set your standards that way."

Finally, when looking at water of different types, we have a problem of deciding, what is the appropriate sample? I think one can divide water into four areas from the mutagenicity data. One is the water source that is clearly industrially contaminated with compounds that we should find out more about. Second, there is an area that involves surface water where waters can be affected by industrial problems of specific chemicals, which we should know more about, but we do not really know which sample to take and how often the compounds might be there because of variability of discharge.

Then, there is a third category, which is water that is not frequently affected by chemical toxicants, but it has a high TOC [total organic carbon] level and THMFP [trihalomethane formation potential]. This type of water introduces all of the problems that we have been dealing with under the THM regulations. The Ames test would assist in determining which disinfection process to use. What do we do about the water that shows mutagenic activity as a result of chlorination? I think that the presentation of Baird [*see* chapter 31 of this book] speaks about what those compounds might be, not by name, but by property.

Finally, my problem, Cotruvo, is I think really we are going to find that we are going to be looking more and more at ground water, partly because of social reasons and partly because of questions that we do not have answers for.

I think ground water is going to be such a heterogeneous, locally determined problem that I do not know what your ideal sample is. I think we will have to focus on a priority procedure that goes through the major chemicals. From the biological side, of less priority are the TA100 mutagens that are not active in the presence of S9 or the compounds that turn over TA100 in the absence of S9 and are decreased in the presence of S9. Of more concern are the S9-dependent compounds active for either 98 or 100. Those are implied to be industrial chemicals.

I think those would be the logical mutagens to pursue on a name-by-name basis. Then go to Ames testing for toxicological assessment, setting limits, whether they are industrial outfalls or whatever; then, having removed them from the problem, you can look at the other category that I described, for example, the question of chlorination byproducts in the absence of those industrial compounds.

But, to pick a representative sample and then try to deal with ground water contaminants of different types across different parts of the country, I think that is a question we have not addressed. That is an area where a lot of the analytical chemistry will have to go on for, unfortunately, much too long.

SUFFET: All right. Thank you, Loper. I appreciate you taking over because you summarized this problem very nicely.

COTRUVO: I think you did summarize the problems very well, Loper, and I think I agree with all of the points you raised. I guess the point I was making in my paper was that we have a small number of choices available to us in terms of what to do in the event that an unacceptable contamination is found. We have a small number of engineering and economic choices. So it really then becomes a question of what are the circumstances that lead you to make the choice of Option A, Option B, or Option C in terms of your public water supply. I think the precise definition of those circumstances is never going to be achievable, but we perhaps could arrive at some set of qualitative principles that would be used to allow one to make a choice in a given case, for example, this water, because of its source and its history, needs a certain kind of treatment applied to it so that there will be the minimum possibility of undesirable chemicals being present in the finished water. Now, we would not know exactly what each one of those chemicals might be, but our bottom line conclusion would be less is better, and the object is to perhaps take the Sontheimerian approach. His point is very simple: to treat the water to try to approach the quality.

SUFFET: Of ground water.

COTRUVO: Of what you would acknowledge to be good quality ground water.

SUFFET: Right.

COTRUVO: The ground water is set as being ideal, one that is uncontaminated by industrial activities. So, in the case of ground water, as an example, I would answer your last question by picking out a representative contaminated ground water. I would try to pick a representative clean ground water as the primary standard.

Then I would deal with it on a chemical-by-chemical basis because really it is a much simpler problem. I mean, traditionally, when one finds a contaminated ground water, one finds 1 or 2 or 5 or 12 or maybe 50, but not 500 or 700 or 900 chemicals. So really that ground water has already undergone some level of treatment by the fact that it is ground water. It has undergone certain biological and chemical processes. Therefore, one can deal with a smaller problem. One can try to pick those substances that are most likely to be the toxic ones and establish limits for those, just as we do now by proposing MCLs [maximum contaminant levels]. But, I guess my bottom line is there is a great amount of methodology that has been developed both for concentration and for assessments of hazard. All of those methods of in vitro and in vivo assessment procedures are available.

There should be some way of establishing a multiple-tier testing system that would allow one to go through those procedures to apply to these concentrates, ultimately arriving at the whole animal bioassay. By these means and by using representative sampling of water types around the country, one could then arrive at the matrix that would lead one to relate source type to quality goal and appropriate treatment in between to achieve the quality goal. Then one can apply that matrix in the other cases that have not been tested by the complete testing routine using the rapid techniques for analyzing what one would anticipate in the nonsophisticated test procedure that was discussed.

There has been a lot of development work done on all of these techniques, both in concentration techniques and the toxicological methodology. A lot of it has been done by you and by our people in the water laboratories and in the toxicology laboratories in Cincinnati. So the proposal is, let's put this all together. Let's start applying it. Let's make some decisions while we are doing more detailed mechanism work. I think we are ready to start putting it together and making some decisions to improve water quality where it is obviously needed.

SUFFET: I would like, while we are on this line of thought, to say there is an alternate proposal which is related to reuse of water. I and some people in the audience such as Kopfler and Bull of USEPA are associated with the Denver Water Department reuse work. Denver is going to treat its waste water for drinking water reuse, and 10% of the water supply is intended to be reused waste water.

They have come up with an idea or a protocol for a decision on the basic question: Is this reuse water safe to drink? And their idea is very simple. On the one hand, Denver has its normal drinking water; they will subject it to all the testing possible. On the other hand, Denver has its reuse water; if they subject it to the same tests, and the reuse water is no worse than the drinking water that the people normally get, they will consider it to be acceptable for drinking. What is the comment from the audience or the panel about that type of proposal?

COTRUVO: Well, there is one basic premise, and that is that your normal water is a good quality water. I mean you are obviously not trying to compare the water unfairly.

SUFFET: That is their premise.

AUDIENCE: It is true some years ago we had recycled water in Dallas, and it was of highest quality or higher quality than the drinking water for the city of Dallas. However, the discussion was related to the propriety of tests to be run on the reuse water. The design and type of test used for drinking water are different than tests that have yet to be developed for recycled water. And the area that we directed our attention to at the time was the area of biology. We detected several viruses in the recycled water that were not detected in the drinking water.

SUFFET: So, on the basis of comparative chemical tests, the reused water was judged to be of higher quality. But, this was not the case for viruses.

AUDIENCE: Chemically yes, because the argument was that the tests used to evaluate the quality of water from a chemical point of view are the same, but not from a bacteriological point of view.

SUFFET: In the Denver project, they are also looking at the bacteriology and virology from this same viewpoint. In virology, you do not know all the organisms, and you have not identified them. We had the unfortunate situation in Philadelphia many years ago with the Legionnaires' disease where we did not know about the organism *Legionella*. There are a lot of organisms that are present that we do not know about, and in this respect there is a definite similarity between virology and the trace chemicals that we do not know about.

PIET: There is some information from Europe, for instance, on the Rhine River in Germany. I know that there are many big cities in Germany and also in our country where water is purified by bank infiltration followed by some ozone treatment; that is a good system and leads to an excellent quality in terms of bacteriology.

What we always try to avoid is chemical treatment of water. In fact, we use chlorination only as a safety measure to have some hygienic control in the distribution system. Yet, when the organic content of the water is even as low as possible, preferably 1 mg of total organic carbon per liter (not more), and you use chlorination, you increase the mutagenicity. That is well-known.

You also increase the adsorbable organic chlorine (AOCl), which is difficult to remove, even by ozone and by bank filtration. Thus, quite a bit of unidentified polar compounds is introduced into drinking water, which you can measure with AOCl and which could have some health effect, and that is the reason why we avoid chlorination even before bank filtration.

According to the theory of Zoeteman, the taste perception of water is a very good method of local control. We had taste tests carried out in Rotterdam, the Hague, and other big cities by the so-called consumer panels—not trained people, just consumers—and we let them rate the taste of the water. Well, we worked it out statistically and showed that the method was excellent. Now we polish the water treatment methods on the basis of taste perception. It is not expensive. It works quite well.

SUFFET: You are bringing up two interesting points: one that Loper referred to, that is, using treatment techniques as a method to control the contaminants and then testing before and after each treatment to see differences. Maybe the best application of Ames testing and other toxicity testing would be on a difference basis, before and after a process. Is there a difference? Is it less? Is that an approach that can be used, especially in treatment situations, where industrial wastes are potentially present in drinking water?

Would anybody on the panel like to make a comment about that?

BULL: I would subscribe to that to some extent, but you are not going to be able to conduct a 2-year bioassay at each plant and each and every treatment process at every water treatment location in the country. So I think that this is where the problem arises, and you require a reasonable time resolution in your analysis. To deal with a day-to-day problem, the short-term tests are appropriate.

As the state of the art in toxicology exists today, I think the only practical application of the long-term or lifetime studies for cancer as well as other end points is a comparison with an acceptable source of drinking water as Cotruvo suggested, or as is being proposed for the Denver project. The most you can hope for is a demonstration that that water is as safe or safer than an acceptable source.

SUFFET: Considered acceptable, but with no toxicological data to say it is acceptable.

BULL: That is what I am saying. That is where you confine your long-term testing. You cannot do long-term testing on every water sample you might like to collect. However, at present and for the foreseeable future, these studies are the only thing that is going to allow you to deal with relative risk. You can use the shorter term test to do the kind of thing you are referring to because they have the time resolution, but you need to deal with them at that level and recognize their limitations.

COTRUVO: That is the control.

JUNK: I am a chemist. I am not a toxicologist. I am going to discuss something out of my field and with complete ignorance. I want to draw an analogy. Thinking in terms of the tremendous difficulties associated with the identification of pathogenic organisms, what did people in the drinking water supply industry do several years ago? They finally came up—and I think maybe I am agreeing with Cotruvo and maybe I am not—with a simple test of the number of the fecal coliform, and they are still using it today.

We do not test for Legionnaires' disease, and every so often something unfortunate comes up like that. But by and large, the fecal coliform test protects us from pathogenic diseases.

The question is, is there something like a mutagenicity test, direct acting or indirect acting, whereby we can test and determine the risk and set an acceptable level for our drinking water supply? I would consider an acceptable level toward humans and an acceptable ecological level. Maybe there is a single test or maybe a battery of simple tests.

I think Cotruvo was saying, let's put all of our knowledge together now and see if we can't come up with those simple tests for a number of different water supplies where we can decide whether we have an acceptable risk level without trying to decide whether this water is better than that water, because as soon as you try to decide that, what are your criteria?

Now, at the same time, I think we should not ignore the fact that we need whole animal tests and we need tests of specific chemicals.

In the same sense, and I will draw the analogy to the bacteriologists. They don't ignore everything else and just say fecal coliform is it. We still need to continue to determine the risk associated with specific chemicals and then the whole combination of those specific chemicals.

Finally, synergism is not always positive. Sometimes it is negative, and we are never drinking pure water. We are always drinking, even out of the ground, a mixture of chemicals. Sometimes that mixture of chemicals is good for us, and sometimes that mixture of chemicals is bad for us.

BULL: I only have one problem with what you said, Junk, and that was when you attempt to judge the level of risk using the Ames test results. You cannot compare this water with that water. I do not think there is any way that we are going to deal with risk with the Ames test or use it in any kind of absolute terms now or in the future. I think the only thing you are going to be able to do is use it in a relative sense to judge the effectiveness of a treatment process within a plant. It will not serve as a standard that can be applied nationally in any meaningful way.

SUFFET: All right. A relative sense.

JUNK: Agreed, agreed.

SUFFET: I want to get this clarified. You are saying that the Ames test can be used to measure differences before and after a process in a relative sense, and you are saying that you cannot judge risk by the Ames test?

BULL: That is right. I do not think you can.

SUFFET: But, by the fecal coliform method, can you judge risk?

BULL: No. To quibble a little bit with the fecal coliform situation, you have, I think, a different situation. You have a fecal coliform test, which is really telling you how effective your treatment has been in terms of killing bacteria and inactivating viruses. You know that if the treatment has been effective, there should be no fecal coliform. It is a fairly universal thing. No single chemical will represent the ability of a treatment process to deal with all chemicals. Therefore, the Ames test does not even have the simple utility of the fecal coliform test. However, very few people would argue that drinking water should contain as few mutagens as possible. In addition, I do not think you have the luxury that chemical problems are necessarily going to be dealt with effectively by only a single type of treatment process that is analogous to disinfection with fecal coliforms.

The other problem is that probably the major source of mutagens in most drinking water comes from one of the treatment processes.

JUNK: But, was that not Cotruvo's point, whether we should treat the water further to make it acceptable or not? The comments I made were within the realm of deciding whether you should treat the water to get it down to an acceptable level or not. In the same sense, with fecal coliform, they finally decided on chlorination. Yet, there would be the European community, in particular, that would say chlorination really is not the way to go; ozone is. Those arguments will still be there.

SUFFET: The European community uses ozone as an alternative to chlorine, and I personally have a problem with the thought process that ozone does not produce other chemicals that might be as toxic or carcinogenic as those produced by chlorination. At present, we do not have good analytical methodologies to identify those ozonation products. So that is my basic difference with the European thesis in terms of using ozone and not too much chlorine.

I want to make one other appropriate comment, and that is the Ames test could be considered as a nonspecific analytical methodology. Really, that is what we are talking about.

Now, Piet mentioned something like organic chlorine as a nonspecific chemical parameter. So the question is, should we go after something like organic chlorine, oxygenated compounds, total epoxides, total something or other, to try to get some correlations here with the specific compounds and develop this correlation?

You also mentioned taste and odor, a subject near and dear to my heart because I have been working in that field. I think the human species is very intriguing in that we do not drink water that does not taste good. There is some background to the history of how humans developed as a race and how they were able to protect themselves by using their noses. Maybe that would be some type of correlation that could be developed. But this is something that has to be looked at.

What about the use of these nonspecific parameters? I know Cotruvo and I have had some go-arounds on organic chlorine as a methodology in the past. What does the panel think about that?

PIET: The nonspecific parameters are used in local and regional control applications. They are probably not general control parameters but more associated with local and regional problems. I think we definitely can use these parameters, and they are very fast.

Another thing is what you just mentioned, that many contaminants are not analyzed according to standardized analytical procedures. That is one of the biggest problems, of course. For water treatment methods, we do not prefer the chemical treatment methods but the physical treatment methods. Thus, we are trying to remove things and not to introduce things. This is important. That is our philosophy.

There is another approach that is quite opposite to this one. Water has to be, according to the WHO [World Health Organization] recommendation, wholesome and agreeable. We are always dealing with bad things, but we do not know so much about good things. Why does water taste poorly, why is it agreeable? When we make water with only minerals, it is not real tasty water; it is not agreeable.

There are some agreeable organic compounds present in ground water. We have tried to analyze the compounds with a nice flavor. Water treatment plants, however, do not apply flavoring to the water.

SUFFET: You can also add vitamin C, I have heard.

PIET: Well, adding things is a matter of controversy in our society. When good things, however, are not in a system, then there might be something wrong in that system. When good things disappear from a system—that is also true with the ecosystem—there is something out of balance in the system. And that is also in water. When the tasty compounds disappear from water, there is something out of balance.

And that is the same, for instance, in the soil and its ecosystem. When soil is polluted, you see some changes in the ecosystem. You can try to measure this, for instance, by determining how microorganisms are affected. I think that is a nice approach; see which good things are in systems, and in the case that these good things disappear, find out why they disappear.

SUFFET: Cotruvo, do you have any comment about the nonspecific idea?

COTRUVO: Well, I think their primary application is probably in the area of process control. About taste and odor, I do not know what one can do on the positive side, but obviously there is concern on the negative side. If one develops adverse tastes, they should be removed. But for things such as TOX [total organic halogen] or other group, a parameter would be useful in evaluating the process that is being used to improve the quality of water and to demonstrate that, in fact, components are being removed down to some desirable level. There may not be any direct correlation between toxicology and that level, but that level may be chosen strictly on the basis of engineering principles and common sense, and that is a positive step.

SUFFET: Common sense is the most positive thing, if it is used.

COTRUVO: Occasionally, it is.

BULL: Cotruvo made my comment for me to a large extent, but one of the things I would like to challenge this group of chemists with is the fact that many of the compounds we are dealing with in the mutagenicity area are electrophiles. Electrophilic compounds seem to be what we are measuring after disinfection, that is, the production of direct-acting mutagens.

A very useful parameter that would tend to take the results a little out of the emotional realm would be a measure for electrophilic compounds, that is, total electrophilic compounds. That should be easy enough to do with these presumably direct-acting materials. There could be a standard method using nucleophilic trapping agents.

SUFFET: The problem is the concentration of the materials and the test at that low concentration. That is the major problem.

BULL: That is the major problem, but from an analytical standpoint, I do not think that is an impossible problem.

SUFFET: Well, the analytical chemists will tell you nothing is impossible. The water treatment people say we can treat anything. So I think it is a challenge for the analytical chemists to come up with a test like that. Albert Cheh came up with some methods to look at change in color using the epoxide as an idea, but it has never been followed through and it has never been done in contaminated systems.

BULL: The other thing I would like to do is reinforce what Cotruvo said about the use of surrogate methods in process control. I think they are probably very useful in a situation where you know what is going on. But when you are trying to take a TOX measurement, which might include anything from dioxin in one sample to trichloroethylene in another, the meaning of that TOX result is entirely different in the two circumstances.

SUFFET: This enthusiasm is great! I would like to thank the panelists. I think they did a very nice job in presenting different viewpoints about the subject, and I thank the panelists and the audience for putting the presentations into a framework.

RECEIVED February 3, 1986.

INDEXES

AUTHOR INDEX

SUBJECT INDEX

A

D

E

H

I

N

Q

R

S

Copy editing and indexing by Karen McCeney
Production by Meg Marshall and Karen McCeney
Jacket design by Pamela Lewis
Managing Editor: Janet S. Dodd

Typeset by McFarland Company, Dillsburg, PA,
and Hot Type Ltd., Washington, DC
Printed and bound by Maple Press Company, York, PA

Titles of Related Interest

Evaluation of Pesticides in Ground Water
Edited by Willa Y. Garner, Richard C. Honeycutt,
and Herbert N. Nigg
ACS Symposium Series 315; 574 pages; ISBN 0-8412-0979-0

Organic Marine Geochemistry
Edited by Mary L. Sohn
ACS Symposium Series 305; 430 pages; ISBN 0-8412-0965-0

Reverse Osmosis and Ultrafiltration
Edited by S. Sourirajan and Takeshi Matsuura
ACS Symposium Series 281; 508 pages; ISBN 0-8412-0921-9

Environmental Sampling for Hazardous Wastes
Edited by Glenn E. Schweitzer and John A. Santolucito
ACS Symposium Series 267; 134 pages; ISBN 0-8412-0884-0

Recent Titles

Pharmacokinetics: Processes and Mathematics
Peter G. Welling
ACS Monograph Series 185; 290 pages; ISBN 0-8412-0967-7

Personal Computers for Scientists: A Byte at a Time
Glenn I. Ouchi
288 pages; ISBN 0-8412-1000-4 (paper), 0-8412-1001-2 (cloth)

For further information contact:
American Chemical Society, Sales Office
1155 16th Street, NW, Washington, DC 20036
Telephone 800-424-6747